Y0-AAY-404

Ermanno Bonucci

Biological Calcification

Ermanno Bonucci

Biological Calcification

Normal and Pathological Processes in the Early Stages

With 48 Figures

Professor Dr. Ermanno Bonucci
Dipartimento di Medicina Sperimentale e Patologia
Università La Sapienza – Policlinico Umberto I
Viale Regina Elena, 324
00161 Roma
Italia
e-mail: ermanno.bonucci@uniroma1.it

Library of Congress Control Number: 2006928438

ISBN-10 3-540-36012-3 Springer-Verlag Berlin Heidelberg New York
ISBN-13 978-3-540-36012-4 Springer-Verlag Berlin Heidelberg New York

Springer is a part of Springer Science+Business Media
springer.com

Editor: Dr. Sabine Schreck, Heidelberg, Germany
Desk Editor: Dr. Jutta Lindenborn, Heidelberg, Germany
Cover design: WMXDesign GmhH, Heidelberg, Germany
Typesetting and production: LE-TEX Jelonek, Schmidt & Vöckler GbR, Leipzig, Germany
39/3100 YL 5 4 3 2 1 0 - Printed on acid-free paper

Wenn Gott in seiner Rechten alle Wahrheit und in seiner Linken den einzigen immer regen Trieb nach Wahrheit, obschon mit dem Zusatze, mich immer und ewig zu irren, verschlossen hielte und spräche zu mir: wähle! Ich fiele ihm mit Demut in seine Linke und sagte: Vater gib! Die reine Wahrheit ist ja doch nur für dich allein.

(Gotthold Ephraim Lessing, Eine Duplik, VIII, 26-27, 1778)

If God clasped all truths in his right hand, and in his left held only the ever-living quest for truth – though I add that I will always, eternally, err – and if he said to me: "Chose", then I would humbly grasp his left hand and answer: "Father, give me this! The pure truth is reserved for you alone".

(translated by Anthony Johnson,
Department of English Studies, University of Pisa)

Preface

This book follows a precursor volume devoted to biological calcification, issued by the CRC Press, Boca Raton (Florida) in 1992. Several basic aspects of the calcification process were analyzed in it by outstanding authors who had unquestioned competence in their respective research areas. Its main aim was that of giving readers access to a series of papers which, even though they discussed divergent aspects of biological calcifications drawn from the study of systems as different as vertebrate skeletons and mollusks, in vitro cultures and unicellular organisms, ectopic calcification and urinary stones, provided elements permitting a coherent approach to a comprehensive view of the calcification process in biological tissues.

Now, almost 15 years after the publication of that book, a great variety of new data from a wide spectrum of biological organisms and systems has enriched our knowledge of the normal and pathological mechanisms which can lead to calcification. Even so, this whole process is still problematic: the new knowledge, concepts and ideas have often suggested that a definitive solution was close at hand, but the local mechanism through which the inorganic substance is laid down in organic matrices continues to be an elusive, largely enigmatic topic.

This tantalizing situation is certainly grounded in the intrinsic difficulties of the problem, but persisting doubts appear to derive from at least two extrinsic issues: first, the topics falling under the general heading 'Biological Calcifications' are exceedingly numerous and various; second, they are necessarily faced by investigators who possess distinct kinds of specialization and competence, and apply different, often highly sophisticated methods to very dissimilar calcified tissues. This potentially productive situation is partly undermined by the fact that the results produced by so many studies are scattered through a daunting number of publications which are quite often repetitive, or express conflicting standpoints, may only be available to a restricted number of those who are interested, are often written in technical language that is extremely specialized, and less than transparent to all readers, and tend to set up palisades round specific scientific domains.

This volume aims to offer a critical revision of the old and recent literature on the various aspects of the calcification process, with special reference to its earliest phases as seen in different tissues and organisms. Its scope extends beyond an updating of previously available data; its chapters have the overall purpose of collecting results from as many different sources as possible and attempting, on the basis of personal experience, to achieve integration in definitive conclusions that in all cases aim to make the whole range of results cohere into a comprehensive theory of calcification mechanisms in biological systems. A great deal of effort has been put into the task of selecting – while avoiding arbitrary or subjective exclusions – the results that have withstood critical attitudes and the test of experience, that are common to different tissues and/or organisms, and/or that appear to be typical of one or many of them, and to report them as simply as possible so that they become fully intelligible to all readers.

To ensure a unified treatment of the various topics as far as this is attainable, I have chosen to be the only author of the book, in spite of the risk of appearing presumptuous and the likelihood of incurring blame for inadvertent omissions and mistakes. That choice has the justification, brought by personal editorial experience, that an unduly large number of authors would otherwise have been needed to cover all the topics to be included, and that the final result would probably have been an oversized volume (if not several volumes) comprising a multiplicity of complete, exhaustive chapters but inevitably lacking a single focus or conclusion, which are, conversely, the overriding aim of this book.

As this writer has been involved for over 40 years in studies on the biological calcification process, mostly carried out using morphological methods, a certain prevalence of morphological data could have found its way into discussions on the various aspects of the problem. Moreover, such a long application to the same topic has meant that some ideas have already taken shape, especially those about the nature of the earliest inorganic particles, the organic-inorganic relationships that underlie crystal development, and the possibility that biological calcifications occur through the same basic mechanism(s) in all organisms. These personal ideas have, however, not been allowed undue prominence; the theories, proposals, opinions, suggestions and conclusions currently available in the literature have been viewed as impersonally as possible. For this reason, a large number of references appear at the end of each chapter. To some readers they may appear redundant, but they do offer those interested – especially young investigators – the chance to personally verify the points made in this book and build up an independent view. In this context, old references have not been omitted. Besides their historical value, they allow the hard-won progress of scientific knowledge to be surveyed, and forestall the risk – less rare than is usually thought – of replicating research that has already been carried out. In line with these concepts, the concluding remarks at the end of chapters have the function of focusing the reader's attention on a set of main topics and problems, not of advancing preconstituted solutions or personal opinions.

In spite of his wish to be inclusive, the author is well aware that omissions and mistakes are a constant hazard; in this sense, I apologize in advance. This book has been written in the conviction that all results – even those that currently appear to be obsolete, incongruous or inconsistent – may play a part in the search for truth, provided they are rooted in serious studies and open minds. Perfection is hard, if not impossible, to attain and – at least in the biological field – what appears today to be an incontrovertible truth may turn out tomorrow to be a mistake. Readers who share this concept will perhaps be kind enough to allow for occasional imperfections.

This book could not have been produced without the help of a number of people, too many to be listed one by one. Yet the author wishes to recall with gratitude the memory of Antonio Ascenzi and Vittorio Marinozzi, the two scientists who introduced him to the study of normal and pathological calcified tissues. In addition, he is so indebted to his direct co-workers for their continuous scientific advice and support and for their technical assistance that he cannot avoid mentioning them and expressing his warmest thanks to Paola Ballanti, Silvia Berni, Carlo Della Rocca, Renato Di Grezia, Martina Leopizi, Patrizia Mocetti, Giuliana Silvestrini and Lucio Virgilii. The invaluable assistance of Anthony Johnson, Department of English Studies, University of Pisa, in improving language style and correcting grammar mistakes, and the friendly suggestions of Dr Mariannina Failla, are gratefully acknowledged. The electronic facilities of the library of the "La Sapienza" University, Rome, have greatly facilitated the writing of this book. The more personal studies have been supported by grants from the "La Sapienza" University, Rome, the Italian National Research Council and the Italian Ministry of Education. The constant support of the author's wife, Anna de Matteis, and the encouragement of his children Francesca and Alessandro, have been indispensable in overcoming several difficult moments along the road.

Rome, Spring 2006 *Ermanno Bonucci*

Contents

Acronyms Used in the Text

ABC	Avidin-biotin complex
AFM	Atomic force microscopy
ADP	Adenosine diphosphate
AHSG	α_2-Heremans-Schmid glycoprotein (or α_2-glycoprotein, or fetuin)
AMP	Adenosine monophosphate
Ank	Ankylosis protein
ATP	Adenosine triphosphate
BAG-75	Acidic glycoprotein-75
BGP	Bone Gla-protein (or osteocalcin, or OC)
BMP	Bone morphogenetic protein
BM-40	Basement membrane-40 (or osteonectin, or ON, or SPARC)
BOI	Bovine osteogenesis imperfecta
BSA	Bovine serum albumin
BSP	Bone sialoprotein
CAP	Cementum attachment protein
CAP-1	Cementum associated peptide-1 or Calcification associated peptide-1
Ca-PL-P	Calcium-acidic phospholipid-phosphate complexes
Cbfa1	Core binding factor α 1 (or Runx2)
CCN	Crustocalcin
CGF	Cementum derived growth factor
CM (cellulose)	Carboxymethyl (cellulose)
Con A	Concanavalin A (lectin)
CRA	Calcium reserve assembly
CVC	Calcifying vascular cell
DBS	Decalcification by block soaking
DCN	Decorin
DEAE (cellulose)	Diethylaminoethyl (cellulose)
DEJ	Dentin-enamel junction
DHT	Dihydrotachysterol
DMP1	Dentin matrix protein 1
DNA	Deoxyribonucleic acid
DPP	Dentin phosphoprotein

DSP	Dentin sialoprotein
DSPP	Dentin sialophosphoprotein
EDTA	Ethylene-diamine-tetraacetic acid
EDX	Energy dispersive X-ray elemental analysis
EELS	Electron energy-loss spectroscopy
EFEM	Energy filtering electron microscopy
EGTA	Ethylene-glycol-bis-(β-aminoethyl ether)N, N'-tetraacetic acid
ELISA	Enzyme-linked immunosorbent assay
ELLA	Enzyme-linked lectin assay
EMSP1	Enamel matrix serine proteinase 1 (or KLK4)
END	Electron nanodiffraction
EP(A)	Extrapallial A (calcium-binding glycoprotein A)
ESD	Electron spectroscopic diffraction
ESI	Electron spectroscopic imaging
ESR	Electron spin resonance
FACIT	Fibril-associated collagen with interrupted triple helices
FGF	Fibroblast growth factor
FRP4	Frizzled-related protein 4
FTIRI	Fourier transform infrared imaging
FTIRM	Fourier transform infrared microspectroscopy
GAMP	Gastrolith associated matrix protein
GLA	Gamma-carboxyglutamic acid
HID-TCH-SP	High iron diamine-thiocarbohydrazide-silver proteinate
HMW-DSP	High molecular weight-DSP
HPA	*Helix pomatia* agglutinin
HPL	*Helix pomatia* lectin
IGFBP	Insulin-like growth factor binding protein
IRS	Infrared spectroscopy
KLK4	Kallikrein 4 (or EMSP1)
LRAP	Leucine-rich amelogenin peptide
MEPE	Matrix extracellular phosphoglycoprotein (or osteoregulin, or OF45)
MGP	Matrix Gla protein
MMP	Metalloproteinase
MPA	Maclura pomifera (lectin)
MSP-1	Mollusk shell protein-1
MV	Matrix vesicle
NMR	Nuclear magnetic resonance
NPP1	Nucleotide pyrophosphatase phosphodiesterase 1 (or PC-1)
OC	Osteocalcin (or bone Gla protein, or BGP)
OC-17	Ovocleidin-17

OC-23	Ovocleidin-23
OC-90	Otoconin-90
OC-116	Ovocleidin-116
OF45	Osteoblast/osteocyte factor 45 (or MEPE)
OI	Osteogenesis imperfecta
oim	Osteogenesis imperfecta mouse
OMP-1	Otolith matrix protein-1
ON	Osteonectin (or BM-40, or SPARC)
OPG	Osteoprotegerin
OPN	Osteopontin
PAP	Peroxidase-antiperoxidase method
PAS	Periodic acid-Shiff
PBS	Phosphate buffered saline
PC-1	Plasma cell membrane glycoprotein-1 (or NPP1)
PCR	Polymerase chain reaction
PEDS	Post-embedding decalcification and staining
PTA	Phosphotungstic acid
PVDF	Polyvinylidene fluoride membrane
RANKL	Receptor activator of NF-kappaB ligand
RMS[Rq]	Root-mean-square roughness
RNA	Ribonucleic acid
RT-PCR	Reverse transcriptase linked - polymerase chain reaction
Runx-2	See Cbfa1
SAED	Selected-area electron diffraction
SAXS	Small-angle X-ray scattering
SCA-1, -2	Struthiocalcin-1, -2
SDS	Sodium dodecyl sulfate
SDS-PAGE	Sodium dodecyl sulfate – polyacrylamide-gel electrophoresis
SEM	Scanning electron microscopy
SIBLING	Small integrin-binding ligand, N-linked glycoprotein
SPARC	Secreted protein, acid and rich in cysteine (or osteonectin, or ON, or BM-40)
SRCT	Synchrotron radiation computed tomography
STEM	Scanning-transmission electron microscopy
TEM	Transmission electron microscopy
TGF-β	Transforming growth factor beta
TIP	Tuftelin interacting protein
TNSALP	Tissue non-specific isoenzyme of alkaline phosphatase
TRAP[1]	Tyrosine-rich amelogenin-peptide
WGA	Wheat germ agglutinin (lectin)
ZIO	Zinc-iodide-osmium

[1] Not to be confused with TRAP: Tartrate resistant acid phosphatase, the enzyme which characterizes the osteoclast.

1 Introduction

The deposition of calcium salts in the organic matrix of specialized tissues, which continues to be called 'calcification' by some and 'mineralization' by others, is a process that is indispensable to life (Mann 1988). The clearest example is bone, which not only supplies a skeleton that is able to provide an organism with a stable shape capable of resisting gravity, but also gives an anchorage for muscles, leverage for bodily movements, protection for soft tissues and organs, a housing for some of these (mainly bone marrow), a source of metabolically crucial ions (especially calcium ions) and a temporary storage space for dangerous substances in circulation that are drained from blood, immobilized and temporarily neutralized (these include radioactive isotopes, lead, aluminum and tetracyclines; Bonucci 2000). Some of these functions are shared by cartilage, which is solely responsible for the skeleton of certain fish, and whose main role in vertebrates is probably that of being an initial model to be replaced by bone during the development of many higher organisms (Engfeldt and Reinholt 1992). Tooth dentin and enamel are other examples of calcified tissues which enhance for aggression and defense and certainly allow many animals to eat (Slavkin and Diekwisch 1996), while statoliths, statoconia, otoliths and otoconia make it possible to keep a balance and perceive changes in speed (Fermin and Igarashi 1986). The calcification process is fundamental for a number of lower organisms (Lowenstam 1981), including unicellular organisms (Pautard 1970; Boyan et al. 1992), and for the formation of vital structures such as eggshells (Arias et al. 1993; Panhéleux et al. 1999), the exoskeletons of insects and crustaceans (Travis 1963; Wilt et al. 2003), the shells of many mollusks (Wheeler 1992; Wilt et al. 2003), and the scales and spicules of certain aquatic organisms (Benson and Wilt 1992; Wilt 2002). Besides all this, calcification characterizes and complicates a number of pathological conditions (Seifert 1970; Daculsi et al. 1992), including the formation of some stones (Kim 1982; Khan 1992) and, above all, the development of a number of vascular lesions (Seifert and Dreesbach 1966).

Clearly, biological calcifications, or biomineralizations, are widely distributed processes that help living organisms in a number of ways. Regrettably, this process, however indispensable, is still poorly understood; in particular, its earliest phases still require a thorough explanation. Knowledge of those phases could be thought of as the foundation from which the whole structure must rise. One result is that research on bone, calcified cartilage, dentin, enamel and other hard tissues is currently carried out as if the ways in which inorganic substances are laid down in them was fully known, whereas a lack of knowledge of some of the earliest steps in the process may lead to misinterpretations of important facets of normal and pathological calcification. It must be made clear that studies are often directed to one hard tissue but then their results are applied to another, as if the presence of the mineral substance was by itself sufficient to guarantee that all calcified tissues share the same calcifying mechanism, in spite of the radical differences that have been found in their structure and composition. On the other hand, comparative studies on different hard tissues are few (Travis et al. 1967; Travis 1968, 1970; Travis and Gonsalves 1969) and incomplete. This makes it objectively difficult to answer the basic question: are there as many calcification processes as calcified tissues, or just one basic, molecular mechanism that is valid for all of them (see Bonucci 1987) – a hypothesis that, on the basis of personal experience, does not seem unfounded.

The many studies and results on biological calcifications that are available in the literature have relied on different methods; they refer to different hard tissues and have come from investigators possessing differing kinds of specialization and competence. Those results have been reported and discussed in a great many books and monographs (Neuman and Neuman 1958; McLean and Urist 1968; Schraer 1970; Bourne 1971; Hancox 1972; Serafini-Fracassini and Smith 1974; Vaughan 1975; Watabe and Wilbur 1976; Simmons and Kunin 1979; Lowenstam 1981; Volcani and Simpson 1982; Hall 1983, 1993; Peck 1983; Linde 1984; Simkiss and Wilbur 1989; Bonucci and Motta 1990; Bonucci 1992; Mann 2001; Veis 2003), the problem being that in most cases they are scattered through a daunting number of publications. The chapters that follows have been written to offer a critical revision of the early and more recent literature on the whole range of biological calcification as seen in different tissues and organisms, with special reference to its earliest phases. On the basis of the available data and of personal experience, a two-pronged effort will be made: first, to select from the vast mass of the results those that appear most pertinent to the early phases of the calcification process, that have withstood critical attack and the test of experience, and are applicable to a range of tissues and/or organisms, and those that appear to be typical of one or more of them; second, to verify the feasibility of integrating the whole range of results into a unified theory on the mechanism of mineralization in biological systems.

References

Arias JL, Fink DJ, Xiao S-Q, Heuer AH, Caplan AI (1993) Biomineralization and eggshells: cell-mediated acellular compartments of mineralized extracellular matrix. Int Rev Cytol 145:217–250

Benson SC, Wilt FH (1992) Calcification of spicules in the sea urchin embryo. In: Bonucci E (ed) Calcification in biological systems. CRC Press, Boca Raton, pp 157–178

Bonucci E (1987) Is there a calcification factor common to all calcifying matrices? Scanning Electron Microsc 1:1089–1102

Bonucci E (1992) Calcification in biological systems. CRC Press, Boca Raton
Bonucci E (2000) Basic composition and structure of bone. In: An YH, Draughn RA (eds) Mechanical testing of bone and the bone-implant interface. CRC Press, Boca Raton, pp 3–21
Bonucci E, Motta PM (1990) Ultrastructure of skeletal tissues. Bone and cartilage in health and disease. Kluwer Academic Publishers, Boston
Bourne GH (1971) The biochemistry and physiology of bone, 2nd edn. Academic Press, New York
Boyan BD, Swain LD, Everett MM, Schwartz Z (1992) Mechanisms of microbial mineralization. In: Bonucci E (ed) Calcification in biological systems. CRC Press, Boca Raton, pp 129–156
Daculsi G, Pouëzat J, Péru L, Maugars Y, LeGeros RZ (1992) Ectopic calcifications. In: Bonucci E (ed) Calcification in biological systems. CRC Press, Boca Raton, pp 365–397
Engfeldt B, Reinholt FP (1992) Structure and calcification of epiphyseal growth cartilage. In: Bonucci E (ed) Calcification in biological systems. CRC Press, Boca Raton, pp 217–241
Fermin CD, Igarashi M (1986) Review of statoconia formation in birds and original research in chick (*Gallus domesticus*). Scanning Electron Microsc 4:1649–1665
Hall BK (1983) Cartilage. Academic Press, New York
Hall BK (1993) Bone. CRC Press, Boca Raton
Hancox NM (1972) Biology of bone. Cambridge University Press, Cambridge
Khan SR (1992) Structure and development of calcific urinary stones. In: Bonucci E (ed) Calcification in biological systems. CRC Press, Boca Raton, pp 345–363
Kim KM (1982) The stones. Scanning Electron Microsc 4:1635–1660
Linde A (1984) Dentin and dentinogenesis. CRC Press, Boca Raton
Lowenstam HA (1981) Minerals formed by organisms. Science 211:1126–1131
Mann S (1988) Molecular recognition in biomineralization. Nature 332:119–124
Mann S (2001) Biomineralization. Principles and concepts in bioinorganic materials chemistry. Oxford University Press, Oxford
McLean FC, Urist MR (1968) Bone. The University of Chicago Press, Chicago
Neuman WF, Neuman MW (1958) The chemical dynamics of bone mineral. University of Chicago Press, Chicago
Panhéleux M, Bain M, Fernandez MS, Morales I, Gautron J, Arias JL, Solomon SE, Hincke M, Nys Y (1999) Organic matrix composition and ultrastructure of eggshell: a comparative study. Br Poult Sci 40:240–252
Pautard FGE (1970) Calcification in unicellular organisms. In: Schraer H (ed) Biological calcification: cellular and molecular aspects. Appleton-Century-Crofts, New York, pp 105–201
Peck WA (1983) Bone and mineral research. Elsevier, Amsterdam
Schraer H (1970) Biological calcification: cellular and molecular aspects. Appleton-Century-Crofts, New York
Seifert G (1970) Morphologic and biochemical aspects of experimental extraosseous tissue calcification. Clin Orthop Relat Res 69:146–158
Seifert G, Dreesbach HA (1966) Die calciphylaktische Artheriopatie. Frankf Z Pathol 75: 342–361
Serafini-Fracassini A, Smith JW (1974) The structure and biochemistry of cartilage. Churchill Livingstone, Edinburgh
Simkiss K, Wilbur S (1989) Biomineralization: cell biology and mineral deposition. Academic Press, San Diego, CA
Simmons DJ, Kunin AS (1979) Skeletal research. An experimental approach. Academic Press, New York

Slavkin HC, Diekwisch T (1996) Evolution in tooth developmental biology: of morphology and molecules. Anat Rec 245:131–150
Travis DF (1963) Structural features of mineralization from tissue to macromolecular levels of organization in the decapod crustacea. Ann N Y Acad Sci 109:117–245
Travis DF (1968) Comparative ultrastructure and organization of inorganic crystals and organic matrices of mineralized tissues. Biology of the mouth. American Association for the Advancement of Sciences, Washington, pp 237–297
Travis DF (1970) The comparative ultrastructure and organization of five calcified tissues. In: Schraer H (ed) Biological calcification: cellular and molecular aspects. Appleton-Century-Crofts, New York, pp 203–311
Travis DF, Gonsalves M (1969) Comparative ultrastructure and organization of the prismatic region of two bivalves and its possible relation to the chemical mechanism of boring. Am Zool 9:635–661
Travis DF, François CJ, Bonar LC, Glimcher MJ (1967) Comparative studies of the organic matrices of invertebrate mineralized tissues. J Ultrastruct Res 18:519–550
Vaughan JM (1975) The physiology of bone, 2nd edn. Clarendon Press, Oxford
Veis A (2003) Mineralization in organic matrix frameworks. Rev Miner.Geochem 54:249–289
Volcani BE, Simpson TL (1982) Silicon and siliceous structures in biological systems. Springer, Berlin Heidelberg New York
Watabe N, Wilbur KM (1976) The mechanisms of mineralization in the invertebrates and plants. The University of South Carolina Press, Columbia, SC
Wheeler AP (1992) Mechanisms of molluscan shell formation. In: Bonucci E (ed) Calcification in biological systems. CRC Press, Boca Raton, pp 179–216
Wilt FH (2002) Biomineralization of the spicules of sea urchin embryos. Zool Sci 19:253–261
Wilt FH, Killian CE, Livingston BT (2003) Development of calcareous skeletal elements in invertebrates. Differentiation 71:237–250

2 Historical Notes

2.1 Introduction

Studies on calcified tissues go back over four centuries. As early as 1543, Andrea Vesalio (the italianized name of André Vésale) described the anatomy of the human skeleton in the opening volume of his *De Humani Corporis Fabrica* (Joannes Oporinus, Basel). Galileo Galilei was probably the first to investigate the links between the volume, structure and dynamics of living bodies and their bones in his 1638 book *Discorsi e dimostrazioni matematiche intorno a due nuove scienze* (Carugo A. and Geymonat L., eds. Boringhieri, Turin 1958), reporting there that bird bones are lighter and, therefore, better structured for flying than those of mammals. Gomez (2002) credits Crisóstomo Martínez with having drawn the first pictures of spongy bone trabeculae (Fig. 2.1), bone marrow nerves and vessels in 1689 after using a single, double-convex optical lens, and Clopton Havers and Leeuwenhoeck with descriptions made in 1691 of cortical bone and Haversian canals, respectively. Ascenzi MG and Lomovtsev (2006) credit Leuwenhoek (1693) (Antoni van Leeuwenhoek, Delft, Nederland, 1632–1723) with the first description of lamellar structures in osteons. In 1743 Nicolas Andry published what was probably the first book on orthopaedics (*L'orthopédie ou l'art de prévenir et de corriger dans les enfants les difformités du corps.* Pierre de Hondt, Paris 1743), and Giovanbattista Morgagni first described pathological changes in bone tissue in his *De sedibus et causis morborum per anatomen indagatis* published in 1761. Further basic observations on bone as a tissue became public towards the end of the seventeenth century, with the perfection of the microscope. References to the old literature from that time to the early years of the twentieth century can be found in de Ricqlès et al. (1991) for bone structure, Rönnholm (1962) for enamel, Towe and Cifelli (1967) for calcareous foraminifera, Grégoire (1957) for mother-of-pearl, and Pautard (1970) for calcification in unicellular organisms (coccolithophorids, foraminifera and other protozoa). The most incisive studies on biological calcification carried out during the first half of the twentieth century were surveyed by Urist (1966), who also published photographs of some of the eminent scientists active in that period. He also included a few references to studies carried out during the nineteenth century or earlier.

These few examples give only a faint impression of the wealth of data accumulated in the literature since the earliest publications on the morphology,

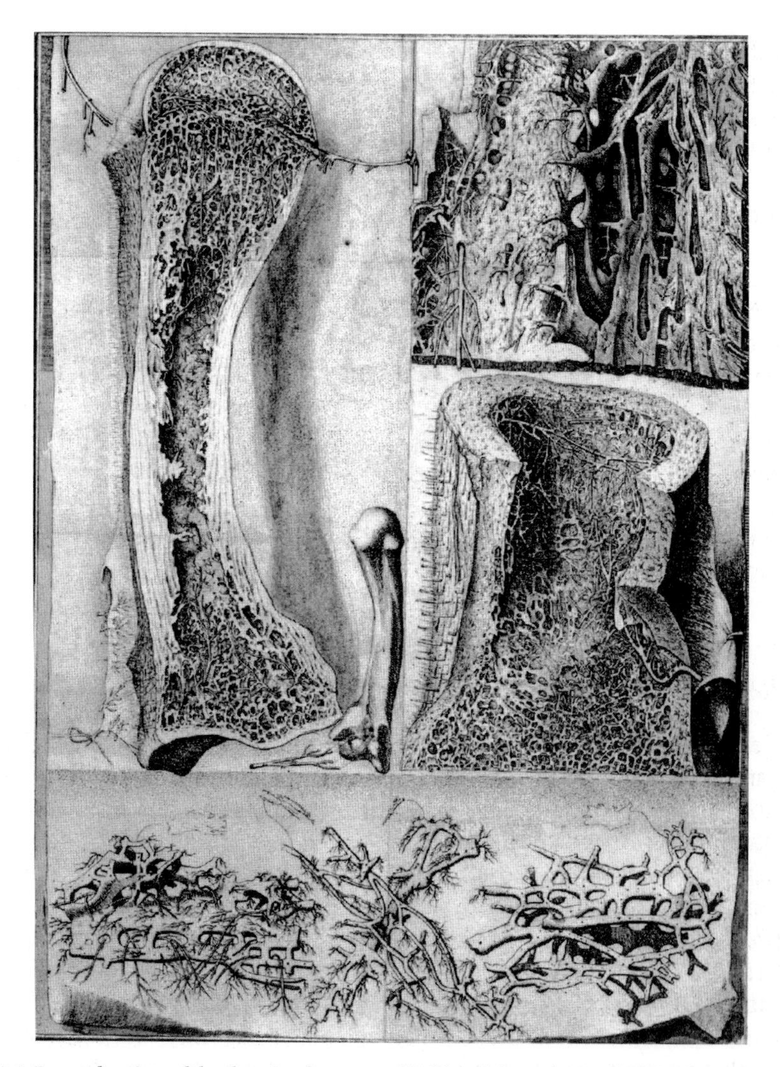

Fig. 2.1. Reproduction of the drawing known as "Table X" from the book "El Atlas Anatomico de Crisóstomo Martínez", by José María López Piñero (Ajuntament de Valencia, Acción Cultural, Delegation de Cultura, 3rd edn, 2001); this is the earliest known representation of spongy bone trabeculae (courtesy of Amando Peydró Olaya and Carmen Carda, University of Valencia)

biochemistry and biophysics of bone and other calcified tissues, and how hard it would be to review them all. If, however, a bibliographic review is limited to publications belonging to the earliest phases of the calcification process, the material in question becomes more manageable, and a review can be made with some confidence that important omissions can be kept to a minimum. Even so, plenty of data are available in the literature on biological calcifications and their basic mechanisms, as may be inferred from the fact that as early as 1923 Freudenberg and György reviewed several independent theories on these topics.

Studies on the early phases of the calcification process can be roughly split into two groups: those carried out up to about 1960 on the basis of histological, histochemical, biochemical and biophysical methods, mostly on topics related to the nature and composition of the inorganic component of hard tissues and the structure of their organic matrix (reviewed by Dallemagne 1951; Neuman and Neuman 1953; Amprino 1955; Carlström 1955; Glimcher 1959; Fitton Jackson 1960; Posner 1960; Urist 1966; Elliott 1973; Wadkins et al. 1974), and those carried out in and after the 1960s by applying more refined methods (reviewed by Taves 1965; Bonucci 1971; Glimcher and Krane 1968; Posner 1969, 1987; Glimcher 1976, 1990, 1992; Posner et al. 1978; Höhling et al. 1990; Bianco 1992; Bonucci 1992; Robey 1996; Daculsi et al. 1997; Boskey 1998; Wilt et al. 2003). Many of these studies should be included in a historical overview because of their year of publication, but most of their results are still valid and pertinent.

2.2 Before 1960

The studies which, strictly speaking, dealt with the early stages of the calcification mechanism were mostly carried out on bone and cartilage, which are the most representative tissues in any vertebrate skeleton. On the basis of the knowledge that the bone mineral is calcium phosphate and that alkaline phosphatase is found in areas of calcification, Robison (1923) suggested that the enzyme could hydrolyze phosphate esters, so increasing the local concentration of inorganic phosphates and promoting calcium phosphate formation. Several findings have, indeed, supported a role for alkaline phosphatase in raising phosphate concentrations in areas of calcification, and it has been shown recently that about 99% of the phosphate produced by the phosphatase activity is laid down in inorganic form (Beertsen and van den Bos 1989). The theory was not, however, thought to be tenable in the simple form given it by Robison. Two of the major objections made were that alkaline phosphatase occurs in tissues which do not calcify and, above all, that the calcifying tissues do not contain amounts of phosphate esters large enough to raise, through their hydrolysis, the local inorganic phosphate concentration above the solubility product of calcium phosphate. It was also added that the supersaturation theory fails to explain all aspects of biological calcifications (Sendroy and Hastings 1927). These criticisms led to the hypothesis that sufficient quantities of phosphate esters may be generated by the process of glycogenolysis (Harris 1932; Gutman and Yu 1950). This proposal stimulated a number of studies on the presence and fate of glycogen in hard tissues (Picard and Cartier 1956a, b; Hirschman and Dziewiatkowski 1966). Schajowicz and Cabrini (1958) found that glycogen is present in all the cells of the hyaline cartilage, that its concentrations rise in hypertrophic chondrocytes of epiphyseal cartilage but fall abruptly before calcification in rapidly growing cartilage, that it is invariably present in cells during membranous ossification but barely present in the osteoblasts of the areas of active osteogenesis, while it is moderately to abundantly present in adjacent mesenchymal cells considered to be pre-osteoblasts. So the theory that phosphoric esters may derive from the glycogen metabolism appeared to be strengthened by the observation that its concentration falls before calcification both in cartilage and bone. Moreover, calcification is blocked by enzymes which inhibit the glycolytic cycle (Gutman and Yu 1950). Sobel et al. (1957), however, did not find any correlation between glycolytic enzymes and cartilage calcification in vitro, and Waldman (1948) had reported that cartilage calcification was intrinsically non-enzymatic, and that phosphorylase and phosphatase systems are not indispensable for the localized deposition of bone mineral to take place.

The observation that too little substrate was available to the enzyme then led to the suggestion that phosphate esters may be an outcome of glycogenolysis (Harris 1932; Gutman and Yu 1950), or that the phosphate donor was ATP (Cartier and Picard 1955a, b, c). Alternatively, it was proposed that the role of alkaline phosphatase could be that of somehow contributing to the synthesis of the organic matrix comprising both collagen fibrils and mucopolysaccharides, rather than that of increasing local phosphate concentrations (Majno and Rouiller 1951; Cabrini 1961; Weidmann 1963).

In the same year in which Robison put forward his alkaline phosphatase theory, two other important findings were reported. First, deJong (1926) observed that bone powder and mineral apatite give similar X-ray diffraction patterns. An image of the hydroxyapatite atomic structure was then supplied by X-ray (Posner et al. 1958) and neutron diffraction (Kay et al. 1964) analysis. Second, Freudenberg and György (1923) observed that the cartilage matrix calcifies in vitro if first treated with calcium chloride and then with a phosphate mixture, but not vice versa; they therefore made the suggestion that the process first requires the binding of calcium to a tissue "colloid", and then phosphate binding, with the possible formation of calcium-phosphoprotein complexes.

Plenty of investigations followed deJong's observation, as reported in detail in the following chapters. The results showed that hydroxyapatite is the mineral most frequently found in upper and lower vertebrate calcified tissues (reviewed by Posner 1960, 1969; Zipkin 1970); that calcium phosphate or other calcium salts occur in unicellular organisms (reviewed by Ennever and Creamer 1967; Pautard 1970; Boyan et al. 1992); and that calcium carbonate, in the form of calcite and aragonite, is the mineral salt most often found in the calcified structures of invertebrates (reviewed by Travis 1970; Lowenstam 1981).

The observation by Freudenberg and György (1923) that the organic matrix of cartilage contains a calcium-binding "colloid" prompted various studies. Robison and Rosenheim (1934), while admitting that the amount of organic phosphates in calcifying matrix is too low to allow them to act as a substrate for alkaline phosphatase, and that the process occurs in extracellular sites, had already hypothesized that an enzymatic "second mechanism" may be involved. More specifically, they suggested that calcification may be triggered by the presence of a factor, called "local factor", which could promote the linkage of calcium and/or phosphate ions, even when their local concentration is too low to allow spontaneous precipitation to occur (reviewed by Sobel 1955). This factor was thought to be a specific component, or complex of components, or even a complex of substances and/or functions, of the organic matrix. The histochemical observation that, at least in cartilage and bone, organic matrix which actually calcifies is PAS-positive (Leblond et al. 1959) and metachromatic (Pritchard 1952) – i.e., contains glycoproteins and acid proteoglycans – led Sobel (1955) to propose that the local factor may be chondroitin sulfate and that, at least in epiphyseal cartilage, a complex of chondroitin sulfate and collagen may account for the initiation of the calcification process. This viewpoint was supported by the observation that cartilage calcification in vitro is inhibited by substances which react with chondroitin sulfate, such as toluidine blue or protamine (Sobel and Burger 1954). The role of chondroitin sulfate as a local factor was, however, questioned on the grounds that it is present in non-calcifying tissues, and is a component of proteoglycans, which appear to inhibit calcification (Chen et al. 1984). Especially because of this second objection, chondroitin sulfate was set aside; only recently has it been reconsidered as having a possible role in calcification (see Sect. 18.3).

In line with these concepts, DiStefano et al. (1953) formulated the working hypothesis that calcification is a process of catalyzed crystallization in which a specific surface or template acts as a center or 'seed' for the induction of crystal formation. At the same time, Neuman and Neuman (1953) specified that calcification is the result of the stereotactic properties of parts of the organic matrix, and that a 'seeding' or nucleation process may be initiated by the binding to organic templates of calcium or phosphate ions mirroring the proper space relationships of the apatite lattice, so that the further aggregation of ions to the 'seeds', or nuclei, would result in the growth and formation of definitive crystals whose orientation therefore corresponds to that of the template. These concepts were then worked on by Glimcher (1959), who suggested that specific steric relationships between reactive amino-acid side-chain groups from adjacent macromolecules within collagen fibrils could act as nucleation centers (see Sect. 18.2).

A close relationship between components of the organic matrix and inorganic substance had been hypothesized as early as 1935 by Caglioti (1935), who showed that the diffractograms of the organic and inorganic components of the femur coincide, actually appearing as closely connected as if they were part of a new, combined reticulum ("quasi-combined reticulum"). Dawson (1946) observed that the fuzziness of bone X-ray diffraction patterns was greatly reduced by trypsin digestion. This finding allowed him to infer that there may be an interference between the organic matrix and the inorganic substance, probably due to the mechanical strain of crystals, or to an interaction between collagen polar groups and inorganic substance. In the same year, Dallemagne and Melon (1946) showed that the birefringence of the total bone is the outcome of a combination of the negative birefringence of mineral crystals with the positive birefringence of the stick type due to both organic and mineral fractions, and a few years later, again on the basis of bone anisotropy, Ascenzi (1950) hypothesized the existence of bonds between organic matrix and inorganic substance, a concept later substantiated by correlating the results of electron microscopy with those of X-ray diffraction and optical birefringence (Caglioti et al. 1956) and now supported by more than a single investigation (Glimcher 1992).

The first electron microscope studies on calcified tissues confirmed the close relationship between organic matrix and mineral substance. Although carried out using replica techniques (Bradley 1960), which provide poorly resolved images, these studies showed a close relationship between collagen periodic binding and the inorganic substance both in bone (Kellenberger and Rouiller 1950; Rutishauser et al. 1950; Ascenzi and Chiozzotto 1955) and in dentin (Rouiller et al. 1952). Moreover, one important finding to emerge was that the inorganic substance is at least partly contained in the interfibrillar spaces (Kellenberger and Rouiller 1950), where it encrusts the surface of collagen fibrils (Robinson 1952). Applying the same techniques, a close organic-inorganic relationship was again shown in the mother-of pearl of mollusks, where plate-like aragonite crystals alternate with, and are united by, transverse bridges containing 'conchiolin', which forms extremely thin organic sheets (Grégoire 1957). The close relationship which links collagen fibrils and inorganic crystals in bone was demonstrated by Robinson and Watson (1952): using electron microscopy and ultrathin sections, they were able to display crystals aligned alongside collagen fibrils, often exactly in line with their periodic banding.

The introduction of new microradiographic (Amprino and Engström 1952) and autoradiographic (Leblond et al. 1959; Lacroix 1960) techniques led to important discoveries on calcification and calcium metabolism in bone and other tissues. Among other major results, microradiographic techniques showed that the calcium content of the primary bone is always higher (between 5 and 20%) than that of secondary bone, and that osteon calcification goes through two stages: in the first, a level of about 70% of matrix calcification is quickly reached; in the second, full calcification is slowly achieved (Amprino and Engström 1952). They also showed that the earliest phases of the calcification process are localized along the line of separation (so-called calcification front) between the already calcified matrix and the thin border of uncalcified matrix (the osteoid border), which is in contact with the cells. Autoradiographic techniques added important information on the transport of ions and organic precursors from the cells to the matrix of bone (Carneiro and Leblond

1959; Lacroix 1960; Linquist et al. 1960; Vincent and Haumont 1960), cartilage (Dziewiatkowski et al. 1957), tooth (Reith and Cotty 1962; Frank 1970; Nagai and Frank 1974; Leblond and Warshawsky 1979) and other hard tissues.

2.3 After 1960

Important insights into the process of biological calcifications came in the course of the second half of the twentieth century. They were widely reported in the literature and most of them are discussed in the following chapters. Some of the investigations in that half-century developed and broadened those carried out in the first half; others brought innovation. The following discoveries merit special consideration because of their possible implications for the early phases of the calcification process (these are listed together with references to the papers where they are reviewed): the structure and composition of calcifying matrices from a biochemical stand point (Glimcher 1990; Robinson et al. 1998; Ameye et al. 2001; Bedouet et al. 2001; Levi-Kalisman et al. 2001), the ultrastructure of the organic matrix and the inorganic substance in bone (Cameron 1963; Glimcher 1990; Hoshi et al. 2000), cartilage (Hunziker and Herrmann 1990; Engfeldt and Reinholt 1992), dentin (Linde 1992), enamel (Warshawsky 1989), exo- and endo-skeleton of invertebrates (Bevelander and Nakahara 1969; Saleuddin 1971; Wilt 1999; Ameye et al. 2000), and in pathological calcifications (Tintut and Demer 2001); the role of collagen fibrils and associated phosphoproteins in bone and dentin (Veis et al. 1972; Carmichael and Dodd 1973; Dickson et al. 1975; Cohen-Solal et al. 1979; Prince et al. 1987; Glimcher 1989; Chang et al. 1996); the presence of proteoglycans in calcifying matrices and their possible function (Kobayashi 1971; Lohmander and Hjerpe 1975; Takagi et al. 1982; Hunter 1991); the localization and role of lipids in matrix calcification and osteogenesis (Wuthier 1973; Boskey and Reddi 1983; Boyan et al. 1989, 1992; Dziak 1992; Goldberg et al. 1995); in vitro studies on matrix calcification (Hunter 1992) and the role of cells in this process (Robey 1992); the solubilization and extraction of matrix crystal-bound proteins (Termine et al. 1981; Butler 1984); the expression, characterization and function of organic, non-collagenous macromolecules in calcifying matrices (Weiner and Traub 1984; Uchiyama et al. 1986; Benson and Wilt 1992; Wheeler 1992; Deutsch et al. 1995; Smith and Nanci 1996; Fincham et al. 1999); the immunohistochemical, ultrastructural localization of these non-collagenous proteins (Bianco 1990; Carlson et al. 1993; Riminucci et al. 1995; McKee and Nanci 1995; Nanci 1999); the structure, function and localization of matrix vesicles (Anderson 1976; Bonucci 1984; Sela et al. 1992); the organic-inorganic relationships during the earliest phases of the calcification process (Bonucci 1969, 2002a; Addadi and Weiner 1985; Bonucci et al. 1988); the loss of non-collagenous components during the course of the process (Nusgens et al. 1972; Lohmander 1976; Glimcher et al. 1977; Overall and Limeback 1988); the expression of growth factors and lymphokines and their function (Gowen 1992, 1994); the nature of bone marrow stem cells and their potential applications (Bianco et al. 2001); the analysis of gene expression (Inaoka et al. 1995; Diekwisch et al. 1997; Karg et al. 1997; Kergosien et al. 1998; Ishigaki et al. 2002); the use of transgenic (Cassella et al. 1994; Paschalis et al. 1996; Snead et al. 1998; Dunglas et al. 2002) and knock-out or null (Xu et al. 1998; Rittling and Denhardt 1999; Delany et al. 2000) animal models; the formulation of new concepts about some pathological changes in normally or pathologically calcified tissues (Riminucci et al. 1997; Bianco and Gehron Robey 1999; Jakoby and Semenkovich 2000; Marie 2001; Chen and Moe 2004) and about the development of metastases in bone (Bonucci 2002b).

The inquiries which have led to the innovations summarized above have produced an impressive numbers of results, but have not yet led to definitive conclusions, and many of the problems faced by early investigators in probing the mechanisms of biological calcifications still await a definitive solution. One paradigmatic case is the persisting doubts about the nature of the mineral substance in biologically calcified systems. With specific reference to bone, and on the basis of the data available till 1942, Dallemagne and Brasseur (1942) concluded that three concepts were prominent at that time: that the bone mineral is a carbonato-apatite whose CO_3 may be substituted by other elements; that it is a calcium carbonate mixed with, or adsorbed by, a hydroxylapatite; or that it is a mixture of tricalcium phosphate and calcium carbonate. In 1987, over 40 years later, the question had still not been definitively settled, and Posner (1987) could only conclude that "the bone mineral is a calcium- and hydroxyl-deficient, hydrogen- and carbonate-containing analogue of hydroxyapatite characterized by structural imperfection and a high surface area". Glimcher (1990) stated that the earliest bone mineral is a very poorly crystalline carbonato-apatite, and that with time its Ca/P ratio increases to approximate that of pure apatite. Eanes (1992) reported that "the principal mineral phase in vertebrate skeletal tissue is a basic calcium phosphate best described as a calcium-deficient carbonato-apatite". Years later, Boskey (1998), in her review on biomineralization conflicts, challenges, and opportunities, shared the concept that the vertebrate mineral is apatitic, but listed the nature of the first mineral deposited in bone and teeth among still unsolved questions. Today, the situation seems little changed, and the time-resisting doubts about the calcification process in vertebrates can be expanded to include practically all living organisms. In their review on coccolith ultrastructure and biomineralization, Young et al. (1999) concluded that, despite the fact that coccolithophorids have often been chosen for laboratory investigations because of the expectation that their relatively simple biological systems might make them easier to study than more complex organisms, the process of coccolith calcification has turned out to be remarkably complex and there is much that remains uncertain.

2.4 Calcification or Mineralization?

The doubts reported above are probably responsible for the ambiguity of the words chosen to refer to the tissue deposition of inorganic substance, which hover uncertainly between 'calcification' and 'mineralization' (or 'biomineralization'). Some researchers consider these words to be synonymous, at least in referring to vertebrates (Engfeldt and Reinholt 1992), but for others their meanings diverge. According to Simkiss (1976), whose starting point is the theoretical concept of 'maintained supersaturation', when a phase change in the extracellular fluid follows the formation of mineral, the fluid finds a new equilibrium by releasing protons at the site of the phase change, and the ions lost as mineral are restored in a free form. If the mineral forms at one site at the expense of ions elsewhere, there will be no release of protons, and the fluid ion composition will stay the same. On this basis, Simkiss argues that a distinction should be made between calcification, which supposedly develops through the first process (that is, production of new mineral by ions present in the system, with ion replenishment by displacing protons), and mineralization which supposedly develops through the second process (that is, by a relocation of mineral products formed elsewhere). This means that the two different processes should be referred to in different ways, and that the term 'mineralization' should be used when appropriate, because many of the minerals found in living organisms contain no calcium at all (Weiner and Addadi 2002; see Chap. 4 and Sect. 13.5.1).

Considering that there are intrinsic difficulties in distinguishing between the two processes, and that calcium ions, although not always involved, are preponderant in them, the single term 'calcification' will be used in this book, with reference to a process which leads to the formation of a solid, stable, amorphous or crystalline inorganic phase in the context of intra- and/or extra-cellular organic structures. This choice is in line with the intention of emphasizing the personal opinion that the calcification and mineralization names refer to the same basic mechanism, that this goes forward under cellular control, that it is not based on exclusively physical processes like those which can lead to the formation of minerals, and that the early inorganic particles which characterize the initial phases of both processes are far from being mineral structures in the strictest sense of the term.

References

Addadi L, Weiner S (1985) Interactions between acidic proteins and crystals: stereochemical requirements in biomineralization. Proc Natl Acad Sci USA 82:4110–4114

Ameye L, Hermann R, Dubois P (2000) Ultrastructure of sea urchin calcified tissues after high-pressure freezing and freeze substitution. J Struct Biol 131:116–125

Ameye L, De Becker G, Killian C, Wilt F, Kemps R, Kuypers S, Dubois P (2001) Proteins and saccharides of the sea urchin organic matrix of mineralization: characterization and localization in the spine skeleton. J Struct Biol 134:56–66

Amprino R (1955) Struttura microscopica e rinnovamento delle ossa. Atti Soc Ital Patol 4:11–68

Amprino R, Engström A (1952) Studies on X ray absorption and diffraction of bone tissue. Acta Anat 15:1–22

Anderson HC (1976) Matrix vesicles of cartilage and bone. In: Bourne GH (ed) The biochemistry and physiology of bone. Academic Press, New York, pp 135–157

Ascenzi A (1950) On the existence of bonds between ossein and inorganic bone fraction. Science 112:84–86

Ascenzi A, Chiozzotto A (1955) Electron microscopy of the bone ground substance using the pseudo-replica technique. Experientia 11:140

Ascenzi MG, Lomovtsev A (2006) Collagen orientation patterns in human secondary osteons, quantified in the radial direction by confocal microscopi. J Struct Biol 153:14–30

Bedouet L, Schuller JM, Marin F, Milet C, Lopez E, Giraud M (2001) Soluble proteins of the nacre of the giant oyster *Pinctada maxima* and of the abalone *Haliotis tuberculata*: extraction and partial analysis of nacre proteins. Comp Biochem Physiol B Biochem Mol Biol 128:389–400

Beertsen W, van den Bos T (1989) Calcification of dentinal collagen by cultured rabbit periosteum: the role of alkaline phosphatase. Matrix 9:159–171

Benson SC, Wilt FH (1992) Calcification of spicules in the sea urchin embryo. In: Bonucci E (ed) Calcification in biological systems. CRC Press, Boca Raton, pp 157–178

Bevelander G, Nakahara H (1969) An electron microscope study of the formation of the nacreous layer in the shell of certain bivalve molluscs. Calcif Tissue Res 3:84–92

Bianco P (1990) Ultrastructural immunohistochemistry of noncollagenous proteins in calcified tissues. In: Bonucci E, Motta PM (eds) Ultrastructure of skeletal tissues. Kluwer Academic Publishers, Boston, pp 63–78

Bianco P (1992) Structure and mineralization of bone. In: Bonucci E (ed) Calcification in biological systems. CRC Press, Boca Raton, pp 243–268

Bianco P, Gehron Robey P (1999) Diseases of bone and the stromal cell lineage. J Bone Miner Res 14:336–341

Bianco P, Riminucci M, Gronthos S, Robey PG (2001) Bone marrow stromal stem cells: nature, biology, and potential applications. Stem Cells 19:180–192

Bonucci E (1969) Further investigation on the organic/inorganic relationships in calcifying cartilage. Calcif Tissue Res 3:38–54

Bonucci E (1971) The locus of initial calcification in cartilage and bone. Clin Orthop Relat Res 78:108–139

Bonucci E (1984) Matrix vesicles: their role in calcification. In: Linde A (ed) Dentin and dentinogenesis. CRC Press, Boca Raton, pp 135–154

Bonucci E (1992) Role of collagen fibrils in calcification. In: Bonucci E (ed) Calcification in biological systems. CRC Press, Boca Raton, pp 19–39

Bonucci E (2002a) Crystal ghosts and biological mineralization: fancy spectres in an old castle, or neglected structures worthy of belief? J Bone Miner Metab 20:249–265

Bonucci E (2002b) Physiopathology of cancer metastases in bone and of the changes they induce in bone remodeling. Rend Fis Acc Lincei 13:181–246

Bonucci E, Silvestrini G, Di Grezia R (1988) The ultrastructure of the organic phase associated with the inorganic substance in calcified tissues. Clin Orthop Relat Res 233:243–261

Boskey AL (1998) Biomineralization: conflicts, challenges, and opportunities. J Cell Biochem 30/31:83–91

Boskey AL, Reddi AH (1983) Changes in lipids during matrix-induced endochondral bone formation. Calcif Tissue Int 35:549–554

Boyan BD, Schwartz Z, Swain LD, Khare A (1989) Role of lipids in calcification of cartilage. Anat Rec 224:211–219

Boyan BD, Swain LD, Everett MM, Schwartz Z (1992) Mechanisms of microbial mineralization. In: Bonucci E (ed) Calcification in biological systems. CRC Press, Boca Raton, pp 129–156

Bradley DE (1960) Replica techniques in applied electron microscopy. J R Microsc Soc 79:101–118

Butler WT (1984) Matrix macromolecules of bone and dentin. Collagen Rel Res 4:297–307

Cabrini RL (1961) Histochemistry of ossification. Int Rev Cytol 2:283–306

Caglioti V (1935) Sulla struttura delle ossa. Atti V Congresso Nazionale Chimica Pura Applicata. Rome, Associazione Italiana di Chimica, pp 320–331

Caglioti V, Ascenzi A, Santoro A (1956) Correlation of electron microscopy with X-ray diffraction and optical birefringence in the study of bone. Stockholm, Proceedings Stockholm Conference on Electron Microscopy, pp 234–237

Cameron DA (1963) The fine structure of bone and calcified cartilage. A critical review of the contribution of electron microscopy to the understanding of osteogenesis. Int Rev Cytol 11:283–306

Carlson CS, Tulli HM, Jayo MJ, Loeser RF, Tracy RP, Mann KG, Adams MR (1993) Immunolocalization of noncollagenous bone matrix proteins in lumbar vertebrae from intact and surgically menopausal cynomolgus monkeys. J Bone Miner Res 8:71–81

Carlström D (1955) X-ray crystallographic studies on apatites and calcified structures. Acta Radiol Suppl 121:1–59

Carmichael DJ, Dodd CM (1973) An investigation of the phosphoprotein of the bovine dentin matrix. Biochim Biophys Acta 317:187–192

Carneiro J, Leblond CP (1959) Role of osteoblasts and odontoblasts in secreting the collagen of bone and dentin, as shown by radioautography in mice given tritium-labelled glycine. Exp Cell Res 18:291–300

Cartier P, Picard J (1955a) La minéralisation du cartilage ossifiable. II. – Le système ATPasique du cartilage. Bull Soc Chim Biol 37:661–675

Cartier P, Picard J (1955b) La minéralisation du cartilage ossifiable. III. – Le mècanisme de la réaction ATPasique du cartilage. Bull Soc Chim Biol 37:1159–1168

Cartier P, Picard J (1955c) La minéralisation du cartilage ossifiable: IV. – La signification de la réaction ATPasique. Bull Soc Chim Biol 37:1169–1176

Cassella JP, Pereira R, Khillan JS, Prockop DJ, Garrington N, Ali SY (1994) An ultrastructural, microanalytical, and spectroscopic study of bone from a transgenic mouse with a COL1.A1 pro-alpha-1 mutation. Bone 15:611–619

Chang SR, Chiego D Jr, Clarkson BH (1996) Characterization and identification of a human dentin phosphophoryn. Calcif Tissue Int 59:149–153

Chen NX, Moe SM (2004) Vascular calcification in chronic kidney disease. Semin Nephrol 24:61–68

Chen C-C, Boskey AL, Rosenberg LC (1984) The inhibitory effect of cartilage proteoglycans on hydroxyapatite growth. Calcif Tissue Int 36:285–290

Cohen-Solal L, Lian JB, Kossiva D, Glimcher MJ (1979) Identification of organic phosphorus covalently bound to collagen and non-collagenous proteins of chicken-bone matrix. Biochem J 177:81–98

Daculsi G, Bouler J-M, LeGeros RZ (1997) Adaptive crystal formation in normal and pathological calcifications in synthetic calcium phosphate and related biomaterials. Int Rev Cytol 172:129–191

Dallemagne MJ (1951) L'os et les mécanismes de sa formation. Les phosphates de calcium, la biochimie de l'ossification et la composition de l'os. J Physiol 43:425–515
Dallemagne MJ, Brasseur H (1942) La diffraction des rayons X par la substance minérale osseuse. Bull Soc R Sci Liége 8/9:1–19
Dallemagne MJ, Melon J (1946) Nouvelles recherches relatives aux propriétés optique de l'os: la biréfringence de l'os minéralisé; relations entre les fractions organiques et inorganique de l'os. J Washington Acad Sci 36:181–195
Dawson JM (1946) X-ray diffraction pattern of bone: evidence of reflexions due to the organic constituent. Nature 157:660–661
de Ricqlès A, Meunier FJ, Castanet J, Francillon-Vieillot H (1991) Comparative microstructure of bone. In: Hall BK (ed) Bone, vol. 3: Bone matrix and bone specific products. CRC Press, Boca Raton, pp 1–78
deJong WF (1926) La substance minérale dans les os. Rec Trav Chim 45:445–446
Delany AM, Amling M, Priemel M, Howe C, Baron R, Canalis E (2000) Osteopenia and decreased bone formation in osteonectin-deficient mice. J Clin Invest 105:915–923
Deutsch D, Catalano-Sherman J, Dafni L, David S, Palmon A (1995) Enamel matrix proteins and ameloblast biology. Connect Tissue Res 32:97–107
Dickson IR, Dimuzio MT, Volpin D, Ananthanarayanan S, Veis A (1975) The extraction of phosphoproteins from bovine dentin. Calcif Tissue Res 19:51–61
Diekwisch TGH, Ware J, Fincham AG, Zeichner-David M (1997) Immunohistochemical similarities and differences between amelogenin and tuftelin gene products during tooth development. J Histochem Cytochem 45:859–866
DiStefano V, Neuman WF, Rouser G (1953) The isolation of a phosphate ester from calcifiable cartilage. Arch Biochem Biophys 47:218–220
Dunglas C, Septier D, Paine ML, Zhu DH, Snead ML, Goldberg M (2002) Ultrastructure of forming enamel in mouse bearing a transgene that disrupts the amelogenin self-assembly domains. Calcif Tissue Int 71:155–166
Dziak R (1992) Role of lipids in osteogenesis: cell signaling and matrix calcification. In: Bonucci E (ed) Calcification in biological systems. CRC Press, Boca Raton, pp 59–71
Dziewiatkowski DD, Di Ferrante N, Bronner F, Okinaka G (1957) Turnover of S^{35}-sulfate in epiphyses and diaphyses of suckling rats. Nature of the S^{36}-labelled compounds. J Exp Med 106:509–524
Eanes ED (1992) Dynamics of calcium phosphate precipitation. In: Bonucci E (ed) Calcification in biological systems. CRC Press, Boca Raton, pp 1–17
Elliott JC (1973) The problems of the composition and structure of the mineral components of the hard tissues. Clin Orthop Relat Res 93:313–345
Engfeldt B, Reinholt FP (1992) Structure and calcification of epiphyseal growth cartilage. In: Bonucci E (ed) Calcification in biological systems. CRC Press, Boca Raton, pp 217–241
Ennever J, Creamer H (1967) Microbiological calcification: bone mineral and bacteria. Calcif Tissue Res 1:87–93
Fincham AG, Moradian-Oldak J, Simmer JP (1999) The structural biology of the developing dental enamel matrix. J Struct Biol 126:270–299
Fitton Jackson S (1960) Fibrogenesis and the formation of matrix. In: Rodahl K, Nicholson JT, Brown EM (eds) Bone as a tissue. McGraw-Hill Book Company, New York, pp 165–185
Frank RM (1970) Autoradiographie quantitative de l'amélogenèse en microscopie électronoque a l'aide de la proline tritiée chez le chat. Arch Oral Biol 15:569–581
Freudenberg E, György P (1923) III. Der Verkalkungsvorgang bei der Entwicklung des Knochens. Ergebn inn Med 24:17–28
Glimcher MJ (1959) Molecular biology of mineralized tissues with particular reference to bone. Rev Mod Phys 31:359–393

Glimcher MJ (1976) Composition, structure, and organization of bone and other mineralized tissues and the mechanism of calcification. In: Greep RO, Astwood EB (eds) Handbook of physiology: Endocrinology. American Physiological Society, Washington, pp 25–116
Glimcher MJ (1989) Mechanism of calcification: role of collagen fibrils and collagen-phosphoprotein complexes in vitro and in vivo. Anat Rec 224:139–153
Glimcher MJ (1990) The nature of the mineral component of bone and the mechanism of calcification. In: Avioli LV, Krane SM (eds) Metabolic bone disease and clinically related disorders. W.B. Saunders Company, Philadelphia, pp 42–68
Glimcher MJ (1992) The nature of the mineral component of bone and the mechanism of calcification. In: Coe FL, Favus MJ (eds) Disorders of bone and mineral metabolism. Raven Press, New York, pp 265–286
Glimcher MJ, Krane SM (1968) The organization and structure of bone, and the mechanism of calcification. In: Gould BS (ed) Biology of collagen. Academic Press, London, pp 67–251
Glimcher MJ, Brickley-Parsons D, Levine PT (1977) Studies of enamel proteins during maturation. Calcif Tissue Res 24:259–270
Goldberg M, Septier D, Lécolle S, Chardin H, Quintana MA, Acevedo AC, Gafni G, Dillouya D, Vermelin L, Thonemann B, Schmalz G, Bissila-Mapahou P, Carreau JP (1995) Dental mineralization. Int J Dev Biol 39:93–110
Gomez S (2002) Crisóstomo Martínez, 1638–1694. The discovery of trabecular bone. Endocrine 17:3–4
Gowen M (1992) Cytokines and bone metabolism. CRC Press, Boca Raton
Gowen M (1994) Cytokines and cellular interactions in the control of bone remodeling. In: Heersche JNM, Kanis JA (eds) Bone and mineral research/8. Elsevier, Amsterdam, pp 77–114
Grégoire C (1957) Topography of the organic components in mother-of-pearl. J Biophys Biochem Cytol 3:797–806
Gutman AB, Yu TF (1950) A concept of the role of enzymes in endochondral calcification. In: Reifenstein EC (ed) Metabolic interrelations. Josiah Macy Jr Foundation, New York, pp 167–190
Harris HA (1932) Glycogen in cartilage. Nature (London) 130:996–997
Hirschman A, Dziewiatkowski DD (1966) Protein-polysaccharide loss during endochondral ossification: immunochemical evidence. Science 154:393–395
Höhling HJ, Barckhaus RH, Krefting E-R, Althoff J, Quint P (1990) Collagen mineralization: aspects of the structural relationship between collagen and the apatitic crystallites. In: Bonucci E, Motta PM (eds) Ultrastructure of skeletal tissues. Kluwer Academic Publishers, Boston, pp 41–62
Hoshi K, Ejiri S, Ozawa H (2000) Ultrastructural, cytochemical, and biophysical aspects of mechanisms of bone matrix calcification. Acta Anat Nippon 75:457–465
Hunter GK (1991) Role of proteoglycan in the provisional calcification of cartilage. A review and reinterpretation. Clin Orthop Relat Res 262:256–280
Hunter GK (1992) In vitro studies on matrix-mediated mineralization. In: Hall BK (ed) Bone, volume 4: Bone metabolism and mineralization. CRC Press, Boca Raton, pp 225–247
Hunziker EB, Herrmann W (1990) Ultrastructure of cartilage. In: Bonucci E, Motta PM (eds) Ultrastructure of skeletal tissues. Kluwer Academic Publishers, Boston, pp 79–109
Inaoka T, Lean JM, Bessho T, Chow JWM, Mackay A, Kokubo T, Chambers TJ (1995) Sequential analysis of gene expression after an osteogenic stimulus: *c-fos* expression is induced in osteocytes. Biochem Biophys Res Commun 217:264–270
Ishigaki R, Takagi M, Igarashi M, Ito K (2002) Gene expression and immunohistochemical localization of osteonectin in association with early bone formation in the developing mandible. Histochem J 34:57–66

Jakoby MG IV, Semenkovich CF (2000) The role of osteoprogenitors in vascular calcification. Curr Opin Nephrol Hypertens 9:11–15

Karg HA, Burger EH, Lyaruu DM, Wöltgens JHM, Bronckers ALJJ (1997) Gene expression and immunolocalisation of amelogenins in developing embryonic and neonatal hamster teeth. Cell Tissue Res 288:545–555

Kay MI, Young RA, Posner AS (1964) Crystal structure of hydroxyapatite. Nature 204: 1050–1052

Kellenberger E, Rouiller C (1950) Die Knochenstruktur, untersucht mit dem Elektronenmikroskop. Schweiz Z Allg Pathol Bakteriol 13:783–788

Kergosien N, Sautier J-M, Forest N (1998) Gene and protein expression during differentiation and matrix mineralization in a chondrocyte cell culture system. Calcif Tissue Int 62: 114–121

Kobayashi S (1971) Acid mucopolysaccharides in calcified tissues. Int Rev Cytol 30: 257–371

Lacroix P (1960) Ca^{45} autoradiography in the study of bone tissue. In: Rodahl K, Nicholson JT, Brown EM (eds) Bone as a tissue. McGraw-Hill, New York, pp 262–279

Leblond CP, Warshawsky H (1979) Dynamic of enamel formation in the rat incisor tooth. J Dent Res 58:950–975

Leblond CP, Lacroix P, Ponlot R, Dhem A (1959) Les stades initiaux de l'ostéogenèse. Nouvelles données histochimique et autoradiographiques. Bull Acad R Med Belg 24: 421–443

Levi-Kalisman Y, Falini G, Addadi L, Weiner S (2001) Structure of the nacreous organic matrix of a bivalve mollusk shell examined in the hydrated state using cryo-TEM. J Struct Biol 135:8–17

Linde A (1992) Structure and calcification of dentin. In: Bonucci E (ed) Calcification in biological systems. CRC Press, Boca Raton, pp 269–311

Linquist B, Budy AM, McLean FC, Howard JL (1960) Skeletal metabolism in estrogen-treated rats studied by means of Ca^{45}. Endocrinology 66:100–111

Lohmander S (1976) Proteoglycans of hyaline cartilage. Thesis. Karolinska Institute

Lohmander S, Hjerpe A (1975) Proteoglycans of mineralizing rib and epiphyseal cartilage. Biochim Biophys Acta 404:93–109

Lowenstam HA (1981) Minerals formed by organisms. Science 211:1126–1131

Majno G, Rouiller C (1951) Die alkalische Phosphatase in der Biologie des Knochengewebes. Histochemische Untersuchungen. Virchows Arch 321:1–61

Marie PJ (2001) Cellular and molecular basis of fibrous dysplasia. Histol Histopathol 16: 981–988

McKee MD, Nanci A (1995) Postembedding colloidal-gold immunocytochemistry of noncollagenous extracellular matrix proteins in mineralized tissues. Microsc Res Technol 31:44–62

Nagai N, Frank RM (1974) Electron microscopic autoradiography of Ca^{45} during dentinogenesis. Cell Tissue Res 155:513–523

Nanci A (1999) Content and distribution of noncollagenous matrix proteins in bone and cementum: relationship to speed of formation and collagen packing density. J Struct Biol 126:256–269

Neuman WF, Neuman MW (1953) The nature of the mineral phase of bone. Chem Rev 53:1–45

Nusgens B, Chantraine A, Lapiere CM (1972) The protein in the matrix of bone. Clin Orthop Relat Res 88:252–274

Overall CM, Limeback H (1988) Identification and characterization of enamel proteinases isolated from developing enamel. Amelogeninolytic serine proteinases are associated with enamel maturation in pig. Biochem J 256:965–972

Paschalis EP, Jacenko O, Olsen B, Mendelsohn R, Boskey AL (1996) FT-IR microscopic analysis identified alterations in mineral properties in bones from mice transgenic for type X collagen. Bone 18:151–156
Pautard FGE (1970) Calcification in unicellular organisms. In: Schraer H (ed) Biological calcification: cellular and molecular aspects. Appleton-Century-Crofts, New York, pp 105–201
Picard J, Cartier P (1956a) La minéralisation du cartilage ossifiable. V. – Glycolyse et glycogénolyse du cartilage. Bull Soc Chim Biol 38:697–706
Picard J, Cartier P (1956b) La minéralisation du cartilage ossifiable. VI. – Influence des phosphodérivés du catabolisme glucidique sur le métabolisme et la minéralisation du cartilage. Bull Soc Chim Biol 38:707–715
Posner AS (1960) The nature of the inorganic phase in calcified tissues. In: Sognnaes RF (ed) Calcification in biological systems. Am Assoc Adv Sci, Washington, pp 373–394
Posner AS (1969) Crystal chemistry of bone mineral. Physiol Rev 49:760–792
Posner AS (1987) Bone mineral and the mineralization process. In: Peck WA (ed) Bone and mineral research 5. Elsevier Science Publisher, Amsterdam, pp 65–116
Posner AS, Perloff A, Diorio AF (1958) Refinement of the hydroxyapatite structure. Acta Crystallogr 11:308–309
Posner AS, Betts F, Blumenthal NC (1978) Properties of nucleating systems. Metab Bone Dis Rel Res 1:179–183
Prince CW, Oosawa T, Butler WT, Tomana M, Bhown AS, Bhown M, Schrohenloher RE (1987) Isolation, characterization, and biosynthesis of a phosphorylated glycoprotein from rat bone. J Biol Chem 262:2900–2907
Pritchard JJ (1952) A cytological and histochemical study of bone and cartilage formation in the rat. J Anat 86:259–277
Reith EJ, Cotty VF (1962) Autoradiographic studies on calcification of enamel. Arch Oral Biol 7:365–372
Riminucci M, Silvestrini G, Bonucci E, Fisher LW, Gehron Robey P, Bianco P (1995) The anatomy of bone sialoprotein immunoreactive sites in bone as revealed by combined ultrastructural histochemistry and immunohistochemistry. Calcif Tissue Int 57:277–284
Riminucci M, Fisher LW, Shenker A, Spiegel AM, Bianco P, Gehron Robey P (1997) Fibrous dysplasia of bone in the McCune-Albright syndrome. Abnormalities in bone formation. Am J Pathol 151:1587–1600
Rittling SR, Denhardt DT (1999) Osteopontin function in pathology: lessons from osteopontin-deficient mice. Exp Nephrol 7:103–113
Robey PG (1992) Cell-mediated calcification in vitro. In: Bonucci E (ed) Calcification in biological systems. CRC Press, Boca Raton, pp 107–127
Robey PG (1996) Vertebrate mineralized matrix proteins: structure and function. Connect Tissue Res 35:131–136
Robinson C, Brookes SJ, Shore RC, Kirkham J (1998) The developing enamel matrix: nature and function. Eur J Oral Sci 106:282–291
Robinson RA (1952) An electron-microscopic study of the crystalline inorganic component of bone and its relationship to the organic matrix. J Bone Joint Surg 34A:389–434
Robinson RA, Watson ML (1952) Collagen-crystal relationships in bone as seen in the electron microscope. Anat Rec 114:383–409
Robison R (1923) The possible significance of hexosephosphoric esters in ossification. Biochem J 17:286–293
Robison R, Rosenheim AH (1934) Calcification of hypertrophic cartilage in vitro. Biochem J 28:684–698
Rönnholm E (1962) The amelogenesis of human teeth as revealed by electron microscopy II. The development of enamel crystallites. J Ultrastruct Res 6:249–303

Rouiller C, Huber L, Rutishauser E (1952) La structure de la dentine. Étude comparée de l'os et de l'ivoire au microscope électronique. Acta Anat 16:16–28
Rutishauser E, Huber L, Kellenberger E, Majno G, Rouiller C (1950) Étude de la structure de l'os au microscope électronique. Arch Sci 3:175–180
Saleuddin ASM (1971) Fine structure of normal and regenerated shell of *Helix*. Can J Zool 49:37–41
Schajowicz F, Cabrini RL (1958) Histochemical studies on glycogen in normal ossification and calcification. J Bone Joint Surg 40A:1081–1092
Sela J, Schwartz Z, Swain LD, Boyan BD (1992) The role of matrix vesicles in calcification. In: Bonucci E (ed) Calcification in biological systems. CRC Press, Boca Raton, pp 73–105
Sendroy J Jr, Hastings AB (1927) Studies on the solubility of calcium salts: III. The solubility of calcium carbonate and tertiary calcium phosphate under various conditions. J Biol Chem 71:797–846
Simkiss K (1976) Cellular aspects of calcification. In: Watabe N, Wilbur KM (eds) The mechanisms of mineralization in invertebrates and plants. The University of South Carolina Press, Columbia, SC, pp 1–31
Smith CE, Nanci A (1996) Protein dynamics of amelogenesis. Anat Rec 245:186–207
Snead ML, Paine ML, Luo W, Zhu D-H, Yoshida B, Ley Y-P, Chen L-S, Paine CT, Burstein JM, Jitpukdeebudintra S, White SN, Bringas P Jr (1998) Transgene animal model for protein expression and accumulation into forming enamel. Connect Tissue Res 38:279–286
Sobel AE (1955) Local factors in the mechanism of calcification. Ann N Y Acad Sci 60:713–731
Sobel AE, Burger M (1954) Calcification XIV. Investigation of the role of chondroitin sulfate in the calcifying mechanism. Proc Soc Exper Biol Med 87:7–13
Sobel AE, Burger M, Deane BC, Albaum HG, Cost K (1957) Calcification XVIII. Lack of correlation between calcification in vitro and glycolytic enzymes. Proc Soc Exper Biol Med 96:32–39
Takagi M, Parmley RT, Toda Y, Austin RL (1982) Ultrastructural cytochemistry and immunocytochemistry of sulfated glycosamiglycans in epiphyseal cartilage. J Histochem Cytochem 30:1179–1185
Taves DR (1965) Mechanisms of calcification. Clin Orthop Relat Res 42:207–220
Termine JD, Belcourt AB, Conn KM, Kleinman HK (1981) Mineral and collagen-binding proteins of fetal calf bone. J Biol Chem 256:10403–10408
Tintut Y, Demer LL (2001) Recent advances in multifactorial regulation of vascular calcification. Curr Opin Lipidol 12:555–560
Towe KM, Cifelli R (1967) Wall ultrastructure in the calcareous foraminifera: crystallographic aspects and a model for calcification. J Paleontol 41:742–762
Travis DF (1970) The comparative ultrastructure and organization of five calcified tissues. In: Schraer H (ed) Biological calcification: cellular and molecular aspects. Appleton-Century-Crofts, New York, pp 203–311
Uchiyama A, Suzuki M, Lefteriou B, Glimcher MJ (1986) Isolation and chemical characterization of the phosphoproteins of chicken bone matrix: heterogeneity in molecular weight and composition. Biochemistry 25:7572–7583
Urist MR (1966) Origins of current ideas about calcification. Clin Orthop Relat Res 44:13–39
Veis A, Spector AR, Zamoscianyk H (1972) The isolation of an EDTA-soluble phosphoprotein from mineralizing bovine dentin. Biochim Biophys Acta 257:404–413
Vincent J, Haumont G (1960) Identification autoradiographique des ostéones métaboliques après administration de Ca^{45}. Rev Franç Clin Biol 5:348–358
Wadkins CL, Luben R, Thomas M, Humphreys R (1974) Physical biochemistry of calcification. Clin Orthop Relat Res 99:246–266
Waldman J (1948) Calcification of hypertrophic epiphyseal cartilage in vitro following inactivation of phosphatase and other enzymes. Proc Soc Exp Biol Med 69:262–263

Warshawsky H (1989) Organization of crystals in enamel. Anat Rec 224:242–262
Weidmann SM (1963) Calcification of skeletal tissues. Int Rev Connect Tiss Res 1:339–377
Weiner S, Addadi L (2002) At the cutting edge. Science 298:375–376
Weiner S, Traub W (1984) Macromolecules in mollusc shells and their function in biomineralization. Phil Trans R Soc London 304B:425–434
Wheeler AP (1992) Mechanisms of molluscan shell formation. In: Bonucci E (ed) Calcification in biological systems. CRC Press, Boca Raton, pp 179–216
Wilt FH (1999) Matrix and mineral in the sea urchin larval skeleton. J Struct Biol 126:216–226
Wilt FH, Killian CE, Livingston BT (2003) Development of calcareous skeletal elements in invertebrates. Differentiation 71:237–250
Wuthier RE (1973) The role of phospholipids in biological calcification: distribution of phospholipase activity in calcifying epiphyseal cartilage. Clin Orthop Relat Res 90: 191–200
Xu T, Bianco P, Fisher LW, Longenecker G, Smith E, Goldstein S, Bonadio J, Boskey A, Heegaard A-M, Sommer B, Satomura K, Dominguez P, Zhao C, Kulkarni AB, Gehron Robey P, Young MF (1998) Targeted disruption of the biglycan gene leads to an osteoporosis-like phenotype in mice. Nat Genet 20:78–82
Young JR, Davis SA, Bown PR, Mann S (1999) Coccolith ultrastructure and biomineralisation. J Struct Biol 126:195–215
Zipkin I (1970) The inorganic composition of bones and teeth. In: Schraer H (ed) Biological calcification: cellular and molecular aspects. Appleton-Century-Crofts, New York, pp 69–103

3 Methodology

3.1 Introduction

Calcified tissues are unique in consisting not only of cells and intercellular organic structures, but also of inorganic substance – mostly calcium phosphate or carbonate – which is located in, and closely connected with, the organic matrix and heightens its strength and hardness. This arrangement, which brings major benefits to the life of organisms, is also beneficial to some types of inquiry, in facilitating a number of biophysical research methods. Against this, it raises problems of other types, because the hardness of calcified tissue makes it difficult to process and section, so ruling out or complicating other types of study, especially morphological investigations. Generally speaking, biophysical methods are mainly applied to the study of inorganic material, biochemical methods to that of organic matrix, and morphological methods to both. It may still, of course, be possible to use one or more methods belonging to one of these categories together with one or more methods belonging to another category.

A detailed description of all these methods lies outside the scope of this book, so brief mention will be made only of those which appear to be the most relevant in connection with the early stages of the calcification process.

3.2 Biophysical Methods

Identification of the inorganic component of the calcified tissues as hydroxyapatite in the case of vertebrate skeletons, as calcite or aragonite in the case of invertebrates, and as other types of inorganic salts in other specific cases (discussed below), has chiefly been achieved using biophysical methods. There are many of these, often used simultaneously on the same specimen (reviewed by Arsenault 1990; Arnold et al. 2001). Some of them require complex kinds of equipment that are not available in all laboratories.

3.2.1 X-ray Diffraction

X-ray diffraction is prominent among these methods in the study of calcified tissues both because it has long been used, and because it has made possible the acquisition of fundamental knowledge on the structure and nature of the organic and inorganic components (deJong 1926; Roseberry et al. 1931; Dallemagne and Brasseur 1942, 1947; Henny and Spiegel-Adolf 1945; Carlström and Finean 1954; Carlström 1955; Donnay and Pawson 1969; Eanes et al. 1970; Ascenzi et al. 1978, 1979; Sakae 1988; Bigi et al. 1996). As has long been known, the reticular organization of crystalline materials has the property of diffracting (or scattering) incident X-rays, that is, each reticular plane diffracts X-rays in a specific way. If a very thin collimated beam of monochromatic X-rays goes through a sample and the diffracted (or scattered) rays are directed by a suitable apparatus to a photographic plate, they will produce, besides a central spot arising from non-diffracted rays, a series of spots (or semicircles or circles, according to the size and orientation of crystals) caused by the X-rays which, because they are in phase, have a particularly strong effect (Fig. 3.1). These spots are distributed along arcs which respect the symmetry of the crystal. If the wavelength of the monochromatic X-rays is known, and the angle between the incident and diffracted (or scattered) beam has been measured, it also becomes possible to measure the distance between the reticular planes and calculate the a, b and c parameters of the crystal elementary cell, which is one way to discover how the crystal is structured. This method can be carried out using *wide-angle X-ray diffraction*, which allows the crystallographic orientation of the mineral particles and their degree of crystallinity to be measured (Wenk and Heidelbach 1999), or using *small-angle X-ray scattering*, which, as it is based on the angular region below about 5° 2θ (using copper K_α radiation), allows the average size, shape and orientation of mineral particles to be determined (Fratzl et al. 1991, 1996; Kinney et al. 2001a; Gupta et al. 2003). X-ray diffraction provides exact values of lattice plane distances, and its accuracy can be increased by applying synchrotron radiation (Ascenzi et al. 1985) which, because of its brilliance, allows the analysis of samples with weak scattering power and the reduction of radiation damage (Zanini et al. 1999).

One disadvantage of X-ray diffraction is the poor degree of localization that can be attained (for a comparison of different X-ray diffraction procedures, see Matsushima et al. 1986). A microbeam X-ray diffraction system with a beam spot of 100 μm in diameter has been used to select small areas of different types of bone and study the preferential orientation of apatite crystallites in relation to mechanical function (Nakano et al. 2002), and a 2- or 10-μm synchrotron microfocus X-ray beam has been adopted for quantitative X-ray texture analysis (Wenk and Heidelbach 1999). Scanning by small-angle X-ray scattering (sSAXS) has recently been proposed to gather information at two length scales – micron and nanometer (Fratzl et al. 1992; Gupta et al. 2003). This method has

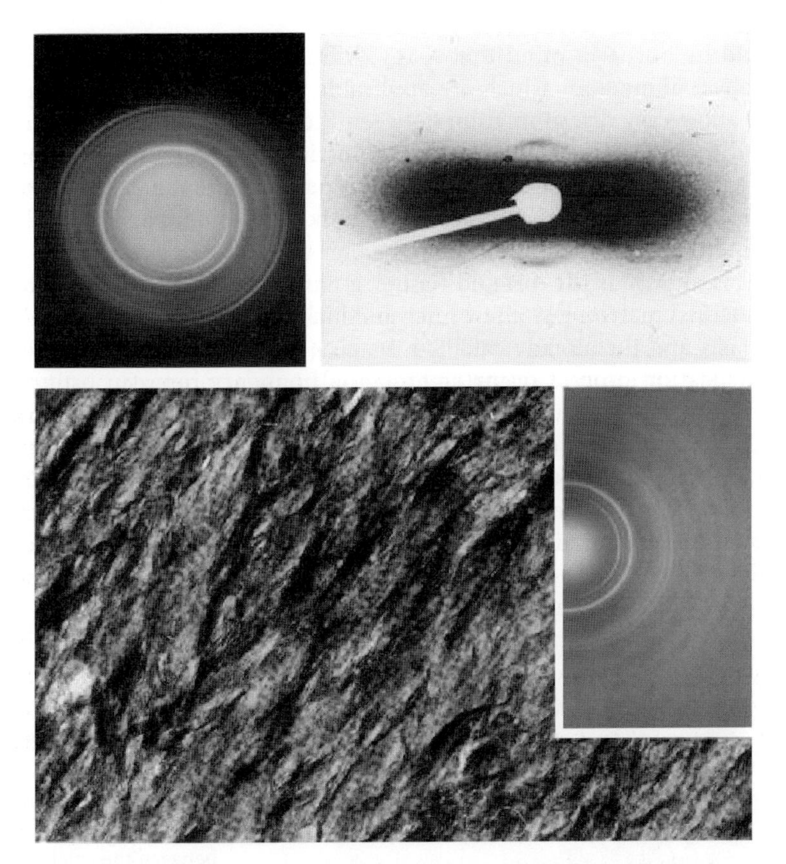

Fig. 3.1. *Above*: X-ray diffraction of compact bone at a high degree of calcification. The wide-angle diffractogram (*left*) shows the 002 reflection of hydroxyapatite; the low-angle diffractogram (*right*) shows a diffraction area that is extensive in a meridional direction. *Below*: compact bone at a high degree of calcification, as seen under the electron microscope: both the inorganic substance in bands, corresponding to the collagen period, and bundles of crystals are recognizable (untreated, ×33,000). *Inset*: electron diffraction of fully calcified bone matrix

been described by Gupta et al. (2003): an X-ray probe beam can be positioned over the sample with microscopic accuracy, and maps of calcified nanostructures can be obtained over an area of 100–200 μm, corresponding to the diameter of the beam. To increase position resolution and to limit the time needed for data acquisition, synchrotron X-ray radiation can be used (Zizak et al. 2003).

The X-ray diffraction method has also been used to obtain information on the organic components of the calcified matrix when they can be crystallized or have an intrinsically crystal-like organization (as may occur, for instance, in the case of collagen fibrils), and on their relationship with the inorganic substance (Dawson 1946; Miller and Wray 1971; Grynpas 1977; White et al. 1977; Weiner and Traub 1980; Lees and Hukins 1992). In this connection, it should be borne in mind that X-ray diffraction and, to some extent, the other physical methods which are applicable to the study of calcified tissues do not allow easy discrimination between the fully calcified and the not yet calcified matrix. This is not due to any intrinsic defect of the method, but to the very different proportion of these types of matrix in calcified tissues. Fully calcified tissue usually makes up the bulk of calcified specimens, whereas uncalcified matrix is found in much smaller amounts. In bone, this consists of what is known as the 'osteoid tissue', a name referring to the thin border of uncalcified matrix, just a few microns thick, which is located between the osteoblasts and the already calcified matrix (Fig. 3.2). The earliest stages of the calcification process occur right at the boundary between osteoid and calcified matrix, where morphological studies reveal 'calcification nodules' or 'calcification islands', i. e., early, spherical or elongated aggregates of crystals which, viewed as a whole, form the 'calcification front' (see Sect. 5.2.1 for further details). Because the osteoid border is very thin, its physical properties and chemical characteristics are easily masked by those of the more plentiful calcified matrix. This is why studies on the composition and structure of the

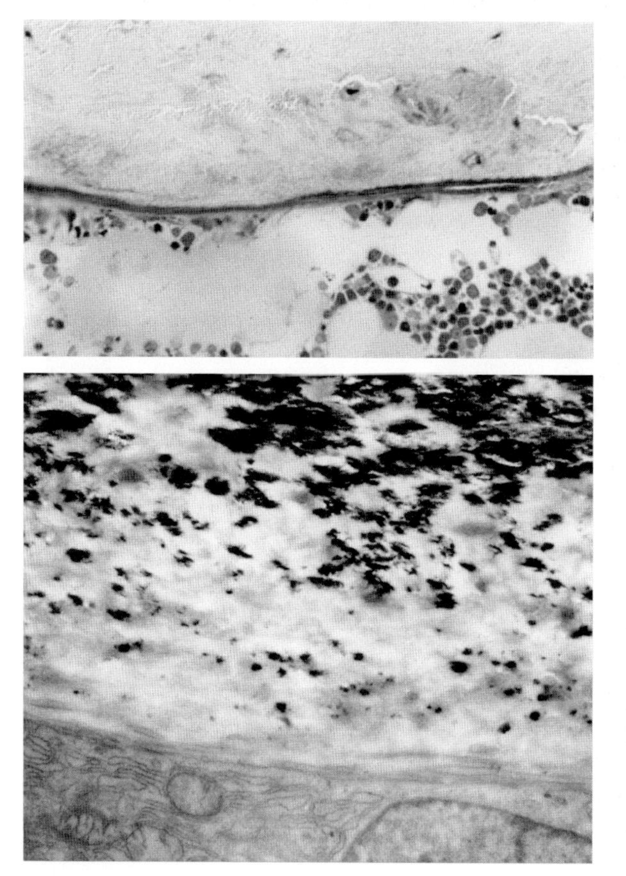

Fig. 3.2. *Above*: histological appearance of quiescent bone surface; a thin osteoid border is interposed between the calcified matrix (*above*) and the bone marrow (*below*); no active osteoblasts are visible. Azure II-methylene blue, ×80. *Below*: ultrastructural appearance of an osteogenic area; a wide osteoid border is interposed between fully calcified matrix (*above*) and an osteoblast (*partly visible below*); the small, irregularly scattered, electron-dense areas correspond to calcification nodules, which, taken as a whole, mark out the calcification front. Note that their size increases with the calcification of the matrix. Unstained, ×12,000

osteoid border, and on how it can be distinguished from the calcified matrix, often require the two structures to be dissected and isolated (Bonucci et al. 1970) prior to separate investigation – at least if they are not examined under the light or electron microscope (Schenk 1974). The same is true of predentin, which is the equivalent of the osteoid tissue in bone, and of growth cartilage, where a calcification front can easily be made out a few microns away from the border of the chondrocyte lacunae, whereas it is less true of other tissues like enamel and invertebrate calcified structures, where calcification seems to occur through synthesis of the organic matrix.

3.2.2 Neutron Diffraction

Neutron diffraction is a recognized method for the study of the structure of crystals and/or biological macromolecules, especially those which give weak reflections. It has the advantage that measurements can be obtained deep within a specimen without its structure being changed by the few neutrons that are absorbed. Neutron diffraction has often been used to study a number of biological problems, such as the degree of hydroxyapatite crystallinity (Girardin et al. 2000), crystal alignment with respect to the collagen fibrils of calcified tendons (White et al. 1977), and the equatorial diffraction spacing of calcified and uncalcified collagen fibrils as a function of water content (Bonar et al. 1985; Lees and Mook 1986). Inelastic neutron scattering has been used to control the presence of OH-ions in isolated bone crystals (Loong et al. 2000). New kinds of neutron diffraction equipment have recently been suggested (see Kurihara et al. 2004).

3.2.3 Selected-area Electron Diffraction (SAED)

SAED is based on the same principle as X-ray diffraction but, as it is used under the electron microscope (Fig. 3.1), it permits the study of very small areas such as early crystals and calcification nodules (Baud and Badonnel 1970; Steve-Bocciarelli 1973; Landis and Glimcher 1978; Arnold et al. 1999). Electron diffraction methods have recently been refined by the introduction of the *electron nanodiffraction* (END) procedure (reviewed by Cowley (2004). END is based on the observation that the strong electromagnetic lenses used in electron microscopes can be demagnified to small bright electron sources (electronic energy of a few hundred thousand eVs) so as to produce electron-probes of sub-nanometer diameter which, when focused on a thin specimen, produce diffraction patterns from areas less than 1 nm in diameter. Even if the usefulness of the END method for the analysis of crystal ultra-high resolution appears quite obvious, it does not seem to have been applied to the study of calcifying tissues, possibly because of the poor availability of suitable instruments (scanning transmission electron microscope, or STEM, optimized for the recording of END patterns).

3.2.4 Energy Dispersive X-ray Elemental Analysis (EDX)

Electron diffraction can be combined with the EDX, or electron probe X-ray analysis, a method in which highly focused incident electrons generate X-ray excitations whose characteristic energies indicate the presence of specific atomic elements. This method has often been used for the quantitative evaluation of elements such as Ca, P, Mg, K and S in calcified tissues (reviewed by Lewinson and Silbermann 1990). Electron probe analysis, in its turn, can be used in conjunction with backscattered electron imaging, so as to select and identify the microareas under study and to correlate better elemental concentrations with morphology (Gomez and Boyde 1994; Gomez et al. 1996): the correlated analysis is conducted in a scanning electron microscope equipped with a backscattered electron image detector and an energy-dispersive X-ray analytical system. The analysis is carried out on the carbon-coated surface of the specimen from which thin sections were taken, and the best spatial X-ray resolution that can be obtained is between 1 and 3 μm.

Together with EDX, or in its place, *electron energy-loss spectroscopy* (EELS) can be used to achieve a high spatial resolution of elemental distribution in the tissue. It is known that incident electrons lose a determinate amount of energy when they interact with the inner-shell electrons of atomic elements in the specimen. Electrons that have lost and those that have not lost energy can be separated if they pass through a magnetic prism: those that have lost more energy are deflected to a greater extent than those that have lost less energy or none at all; this makes it possible to obtain and image a gradient spectrum of energy-loss electrons. The formation of energy spectra by EELS allows constitutive atomic elements to be detected and mapped. The method has a subnanometric spatial resolution and is highly sensitive (a few scores of atoms can be detected). EELS is achieved with a dedicated STEM or a TEM equipped with a spectrometer behind the camera chamber.

3.2.5 Nuclear Magnetic Resonance (NMR) Spectroscopy

NMR spectroscopy can be used to study the structure of small proteins and protein domains, as well as the carbohydrate chains of glycoproteins, independently of their amorphous or crystalline organization (Wuthrich 1989). When a concentrated protein solution is placed in a strong magnetic field, the hydrogen spin becomes aligned with it. The alignment can be altered by applying radiofrequency pulses of electromagnetic radiation, with the result that, when the hydrogen spin returns to its aligned state, radiation is emitted which can be displayed as a spectrum and measured. This radiation depends not only on the misaligned, excited state of one hydrogen nucleus, but also on that of the hydrogen nuclei closest to it, so allowing the measurement of small signal shifts brought about by interacting atoms and the differentiation of signals produced by hydrogen nuclei in different amino acid residues. The shift size corresponds to the distance between any pair of such atoms so that, as long as the amino acid sequence in a protein is know, its three-dimensional reconstruction becomes feasible. A technique combining NMR and deuterium isotope exchange has been used to measure water content in cortical bone (Fernandez-Seara et al. 2004). Moreover, ^{31}P solid state NMR spectroscopy has been used to study the presence and localization of protonated phosphate in bone crystals (Wu et al. 1994).

3.2.6 Proton Nuclear Magnetic Resonance Microscopy

Nuclear magnetic resonance (NMR) microscopy should be discussed in the chapter dedicated to morphological technique, because it is able to show the shape and size of trabeculae and marrow spaces of the spongy bone. It can also, however, be considered a biophysical technique, because it is based on NMR procedures and therefore allows verification of the intrinsic conformation of structures as small as individual calcium-phosphate crystals. As a morphological method, it relies on the observation that bone trabeculae and bone marrow differ in their magnetic permeability and cause a spatial non-uniformity of the magnetic field within the volume being measured, so that the resulting spread in resonance frequency reduces the decay time constant of the time domain proton signal in bone marrow which, in its turn, changes proportionally to the width of bone marrow spaces (Wehrli et al. 1993). NMR microscopy can thus be used to carry out histomorphometric measurements with the advantage that unfixed, unembedded bone samples can be evaluated (Wehrli et al. 1993; Jokisch et al. 1998). It should be pointed out that histomorphometric measurements on spongy bone, even when carried out on fixed, embedded specimens of the iliac crest, give very reliable results (Parfitt et al. 1987; Ballanti et al. 1990; Parisien et al. 1997) and are much simpler than NMR techniques. In addition, light microscope histomorphometry is usually carried out in conjunction with the evaluation of the vertebral bodies or the whole skeleton by densitometry (Hansson and Roos 1986; Duriez et al. 1988). This is a useful clinical technique which has, however, proved to be a poor predictor of abnormal histological features (Laval-Jeantet et al. 1981), especially those which depend on osteomalacia (Hruska et al. 1978).

NMR microscopy has been used to study the development of cartilage in vitro (Potter et al. 1998, 2001), as well as its structure before and after calcification (Potter et al. 2002). Moreover, ^{31}P NMR has been applied to the study of the bone mineral density (Wu et al. 2000a,b) and also to the structure and composition of newly formed crystals of bone, enamel and synthetic hydroxyapatite (Bonar et al. 1991; Rey et al. 1991; Kuhn et al. 2000; Wu et al. 2002).

3.2.7 Electron Spin Resonance (ESR)

The crystalline fraction of tissue mineral can be detected with ESR spectrometry: the irradiation of the sample with ionizing radiation (^{60}Co source) produces stable paramagnetic centers connected with radiation-induced defects in the hydroxyapatite crystal lattice. The total number of defects, divided by the total ash content of the sample, is a good indicator of the mineral's degree of crystallinity (Ostrowski et al. 1973, 1975). This method has been applied to the study of crystallinity in osteonic bone (Ascenzi et al. 1977), enamel (Roufosse et al. 1976), and other normal and pathologically calcified tissues (Ostrowski et al. 1972).

3.2.8 Energy Filtering Electron Microscopy (EFEM)

The presence and distribution of elements in a calcified tissue, and the structure and organization of crystals, can be clarified using an energy filtering electron microscope (EFEM) operating in an *electron spectroscopic diffraction* (ESD) or an *electron spectroscopic imaging* (ESI) mode (for the spatial resolution and detection limits of the EFEM, see Grogger et al. 2003). Several modifications and new techniques have recently been suggested to increase the power, sensitivity and range of applications of the EFEM (Egerton 2003; Leapman 2003; Feng et al. 2004).

As reported by Ottensmeyer and Arsenault (1983), ESD is based on energy discrimination between elastically and inelastically scattered electrons, an elaboration of the dark field image produced by the scanning transmission electron microscope (STEM). By filtering out elastically scattered electrons without energy loss (zero-loss filtering), diffraction patterns can be obtained in which the background produced by inelastic scattered electrons is cut to a minimum, so that even faint Debye-Scherrer diffraction rings can be made out (Reimer et al. 1990; Barckhaus et al. 1991). Electron spectroscopic imaging (ESI) with zero-loss filtering can be used to improve the contrast and resolution of the ultrastructures. This is the outcome of the rise in scattering contrast and the fall in the chromatic aberration attributable to inelastically scattered electrons. Studies carried out with these spectroscopic methods, especially by Arsenault and Ottensmeyer (1983, 1988) and by Höhling and coworkers (Barckhaus et al. 1991; Plate et al. 1992, 1994; Wiesmann et al. 1993; Arnold et al. 1997, 2001), have produced important results on the early stages of the calcification process in bone, dentin and cartilage. *Spectroscopic electron tomography* allows recovery of the signals produced by inelastic interaction between electrons and matter, which are proportional to the local concentration of a single element. This is based on the use of the EFEM in an ESI mode, STEM-EDX mapping and STEM-EELS spectrum mapping (reviewed by Möbus et al. 2003). The tomographic nanoscale reconstruction of 3D chemical maps is a further option.

3.2.9 Infrared Spectroscopy (IRS)

Infrared spectroscopy can provide information on both the inorganic and organic components of calcified matrices. The coupling of a light microscope with an IR spectrometer has allowed spectra to be recorded for distinct microscopic tissue areas whose resolution can be increased by the addition of an array detector. Fourier transform infrared microspectroscopy (FTIRM) and imaging (FTIRI) have both been used to study the degree of crystallinity of bone inorganic substance (Pleshko et al. 1991) and isolated apatite crystals (Kim et al. 1996), the spatial distribution of collagen fibrils in normal and osteoporotic bone (Dziedzic-Goclawska et al. 1982), the maturation of organic and inorganic components in dentin (Magne et al. 2001), changes in mineral and matrix content and composition in osteonic bone (Paschalis et al. 1996), alterations in mineral content and crystallinity in osteopontin knockout mice (Boskey et al. 2002), and several other biological problems. This method has been briefly described by Boskey (1998): sections of bone are placed on a BaF_2 infrared window; an aperture limits the area seen by the IR radiation, and the FTIRM spectra of that area are recorded using a Fourier transform infrared microscope equipped with a computerized *xy* stage, coupled with a spectrometer in transmission mode. Sub-bands of acid phosphate, phosphate in a non-stoichiometric apatite, and phosphate in a stochiometric apatite can be identified by curve-fitting the phosphate region of the spectra (900–1,200 cm^{-1}). On the basis of X-ray diffraction analysis, the percentage areas of the sub-bands can be correlated with apatite crystal size and perfection. The inorganic (phosphate bands) to organic (amide I band) ratio, and carbonate (CO_3) to phosphate (PO_4) ratios can be calculated on the basis of these integrated areas.

3.2.10 Raman Microspectroscopy

This is a nondestructive, vibrational technique which has been applied to the study of mineralized tissues (Carden and Morris 2000). It is based on Raman's effect, which can be summarized as follows. An incident radiation (usually a near-infrared laser excitation) interacts with a molecule that is at its lowest vibrational state so inducing an increase in its energy level. This is unstable and tends to return to its initial vibrational state, in which case a photon is emitted of the same frequency as the incident one. The level to which the radiation returns may be higher than the initial one, in which case a photon is emitted whose energy is equal to that of the incident photon less the difference between the initial and final energy states of the molecule. Alternatively, the incident radiation may interact with a molecule that was already in an excited vibrational state. In this case, a higher energy level is reached and, when the molecule tends to return to its original vibrational state, a photon is emitted whose energy is equal to the incident one plus the difference between the original and final energetic state. In all cases, the spectrum signals the radiations emitted at a microscale degree of resolution. Imaging with chemical composition contrasts has recently been introduced (Tarnowski et al. 2004).

3.2.11 Atomic Force Microscopy (AFM)

The atomic force microscope, otherwise called the scanning force microscope, may be considered to operate at a level intermediate between biophysical and morphological methods. While it has been defined as belonging "to a family of proximal probe microscopy techniques used for probing surface topography and properties on the atomic-molecular scale" (Reich et al. 2001), i.e., a system which permits intermolecular forces to be evaluated both in air and in solution, it can also image protein surfaces, cell membrane components, and macromolecular crystals at subnanometer resolution (Paloczi et al. 1995; Raspanti et al. 1997; Eppell et al. 2001; Tong et al. 2003). Very sketchily, AFM is based on scanning the specimen surface using a sharp, very small tip mounted on a cantilever with a very low spring constant. The system can be moved over the sample by a computer-controlled device; the surface topography produces cantilever deflections that can be measured by a laser-photodiode apparatus. The procedure can be carried out by direct contact of the tip with the sample surface (contact mode) or, to minimize the possibility of damage being done to the tip by surface roughness, by choosing the tapping mode, in which there is no direct interaction between the tip and the sample surface. The use of atomic force microscopy in, and its application to, biological research has recently been reviewed (Siedlecki and Marchant 1998; Reich et al. 2001; Santos and Castanho 2004). Immunohistochemical antigen-antibody reactions have been proposed to identify specific types of molecule in an image, to map their composition, and to detect possible compositional changes (Stroh et al. 2004). Impressive images of individual cartilage aggrecan molecules and their constituent glycosaminoglycan chains have been produced via atomic force microscopy by Ng et al. (2003). One advantage of this method is the possibility to image in aqueous solutions biological structures which are fully hydrated and native (Drake et al. 1989).

3.2.12 Other Biophysical Techniques

Although less frequently used, other biophysical techniques have been applied to the study of the composition and structure of calcified tissues: high-resolution synchrotron radiation computed tomography (SRCT) and small-angle X-ray scattering (SAXS) have been applied together to the study of dentin crystal shape in normal teeth and in dentinogenesis imperfecta (Kinney et al. 2001b); X-ray photoemission has been used to study the characteristics of solid-liquid interfaces as a function of the biomedical application of biomaterials (Sodhi 1996). Other biophysical techniques are available which can be applied to the study of calcified tissues, including spectroscopic electron tomography (Möbus et al. 2003), electron diffraction and high-resolution transmission electron microscopy (Thomas and Midgley 2004), and atomic absorption spectroscopy (Smeyers-Verbeke and Verbeelen 1985).

3.3 Biochemical Methods

So many biochemical methods are now available that they could only be reported in detail in a book exclusively dedicated to that topic. Many of them have been used to study calcified tissue components and the calcification mechanism. They have led to a good knowledge of the nature, composition and structure of the components of the calcified matrix, as well as the interactions between them, and of the metabolic activities of calcified tissue cells (Eastoe and Eastoe 1954; Shapiro 1970a, b; Termine and Eanes 1972; Wuthier 1975; Eanes 1979).

3.3.1 Electrophoresis

Sodium dodecyl sulfate polyacrylamide-gel electrophoresis (SDS-PAGE) is a powerful procedure which has been and still is used in a lot of studies on calcified matrix proteins of vertebrates and invertebrates (e.g., among the most recent, Marin et al. 2001). It is based on treating proteins with SDS, a negatively charged detergent which causes protein unfolding by binding to their hydrophobic regions, and with β-mercaptoethanol which breaks disulfide linkages, so that polypeptides can be analyzed separately. The later electrophoresis of negatively charged molecules treated in this way on polyacrylamide gel allows them to be fractionated on the basis of their size and charge; the largest of them are those with the highest negative charge and highest drag.

To increase the degree of resolution achievable with this method, *two dimensional polyacrylamide-gel electrophoresis* (O'Farrell 1975) has also been used (e.g., Hecker et al. 2003). It is similar to SDS-PAGE, but is carried out in two steps. During the first, proteins are treated with mercaptoethanol and with urea to break disulfide linkages, and with an uncharged detergent, so that polypeptides keep their own charge. They are then separated by isoelectric focusing, a procedure based on the fact that each protein at a specific pH value (isoelectric point) is not charged and therefore does not migrate to an electric field. So, if charged polypeptides migrate through a polyacrylamide-gel in which a pH gradient has been created, they will stop where the pH value is equal to their isoelectric point. After this step, SDS is added and the now negatively charged proteins are submitted to electrophoresis in a direction perpendicular to the former one. The now charged molecules migrate according to their size and charge and can be detected in the gel by using a suitable protein stain.

Proteins fractionated with SDS-PAGE or two-dimensional polyacrylamide-gel electrophoresis can be further analyzed using *western blotting*, a method that, like southern and northern blotting, is often used for the study of nucleic acids. The fractionated peptides (or nucleic acids) are transferred from the electrophoretic gel on to an adhesive sheet of nitrocellulose paper (nylon) by buffer permeation. They are then exposed to a specific antibody which has been coupled with a detectable enzyme or a fluorescent dye or, in some cases, a radioactive isotope, all of which can be easily detected. This method has been used, together with other procedures, to study annexins, metalloproteinases and other components in matrix vesicles (Genge et al. 1992; D'Angelo et al. 2001), to verify the distribution of small keratan sulfate proteoglycans in predentin and dentin (Yamauchi et al. 2000), to detect the expression of enamel proteins in porcine and rat incisors (Nanci et al. 1998; Fukae et al. 2001), and for several other types of study on calcified tissues.

3.3.2 Chromatography

Chromatography is a method used for protein analysis at least as often as electrophoresis. In its simplest form, a drop of the protein solution to be examined is applied to a sheet of absorbent material such as a suitable type of paper (paper chromatography), glass or plastic covered with a thin layer of cellulose or other material (thin-layer chromatography). If a mixture of two or more solvents flows through the sheet, the organic molecules present in the drop move along the sheet with a speed and over a distance that depend on their degree of solubility in the solvents, so accumulating in different areas of the sheet, where they can be demonstrated with a suitable stain.

Use is often made of *column chromatography*, based on the separation of proteins according to the different levels of permeability they have within the porous, permeable material of a column through which they flow. Flow retardation depends on the characteristics of the proteins and the material: small molecules, in fact, penetrate into pores better, and are slowed down more, than large ones. The protein fractions can be collected separately as the solution drips from the column. Several cross-linked, inert polysaccharides are commercially available for this type of chromatography (*gel-filtration chromatography*), with a wide range of pore size which allows the fractionation of proteins with very different molecular weights. Gel-filtration chromatography has often been used, in some cases recently (e.g., Lakshminarayanan et al. 2003), for studies on calcified tissues.

Other types of column chromatography are available, which can also be used in series. Chromatography on hydroxyapatite columns is based on the differing degrees of peptide adsorption on apatite (Tiselius et al. 1956; Bernardi et al. 1972). In what is known as *ion-exchange chromatography* (Lohmander 1972), the solid matrix through which the protein solution permeates consists of positively or negatively charged, small beads which retard molecules with the opposite charge. A mixture of positively charged diethylaminoethylcellulose (DEAE-cellulose) and negatively charged carboxymethylcellulose (CM-cellulose) is generally used in this type of column. The reaction of proteins with the ion-exchange matrix depends on the pH and ionic strength of the flowing solution. Suitable variations in these parameters allow protein separation to be achieved. Ion-exchange chromatography has yielded important results in calcified tissues (e.g., Dimuzio and Veis 1978; Yamauchi et al. 2000).

Another type of chromatography – *affinity chromatography* – is based on the possibility of inducing a covalent link between the insoluble matrix of the column and a specific ligand, which can be the substrate of an enzyme or an antibody. The former will bind the specific enzyme, the latter the specific epitope, if present in the solution flowing through the column. Both types of complex can then be eluted separately. Affinity chromatography is a highly efficient procedure that has often been used in studying calcified tissues (e.g., Hauschka et al. 1986).

High-performance liquid chromatography is based on columns of special materials. These usually consist of very small (< 10 μm in diameter), silica-based spheres which can be compressed so as to reduce the irregular spaces which are often present in other types of material and cause irregular permeation through the column, so reducing resolution. The flow rate through such compressed material is very low. High pressure must be applied to the column to reduce the experimental time, which makes the use of reinforced steel cylinders imperative. An example of the use of this procedure for studying calcified tissues has been reported by Canalis et al. (1988), who adapted it to the identification and purification of growth factors from adult bovine bone.

Electrophoresis or chromatography can be used to obtain what is called *protein fingerprints*. To achieve this, proteins are first treated with substances (enzymes or chemicals) that cleave their molecules between specific amino-acid residues. A small number of peptides are usually produced which can be separated by electrophoresis or chromatography, so giving a 'map' or 'fingerprint', which is typical of each protein. The peptides can be further separated and identified by treatment with a chemical reagent which only forms covalent bonds with the amino group at the peptide amino-terminus. This amino acid is then cleaved from the molecule by treatment with a weak acid and can be identified by chromatography. If the procedure is replicated, all the amino acids in the peptide chain can be separated and identified. The procedure is demanding and time-consuming, but automatic devices are available and many technical improvements have emerged.

3.3.3 Enzyme-linked Assay

Other procedures are available for verifying the presence of specific molecules in tissues, and most of them can be applied to calcified tissues. One such method, which is often used for clinical purposes and has been carried out on calcified tissues to demonstrate the presence in them of a number of molecules and their nature, is the enzyme-linked immunosorbent assay, or *ELISA*. This method is based on the demonstration of an antigenic substance by its reaction with its specific antibody, in the same way as in immunohistochemistry (see below). Among its many applications in calcified tissues, that reported by Foged et al. (1996) is a good example of the method's potential. They used ELISA for the in vitro measurements of the collagenolytic activity of isolated osteoclasts, i. e., the amount of collagen $\alpha 1$(I) fragments these produced and freed in the culture medium as bone was resorbed. They injected into rabbits a synthetic peptide, 8AA, comprising an eight amino acid sequence, which is a specific part of the C-telopeptide of the $\alpha 1$ chain of type I collagen. The raised 8AA antibodies were used for a competitive binding, either to 8AA-molecules, previously immobilized by coupling with thyroglobulin using glutaraldehyde, or to antigenic $\alpha 1$(I) fragments present in the medium. As the osteoclast collagenolytic activity raised the concentration of collagen fragments, and antibodies reacted with them, the amount of antibodies available for reaction with immobilized 8AA-thyroglobulin antigens fell; this could be evaluated and measured by incubations with peroxidase-conjugated anti-rabbit immunoglobulin and a chromogenic substrate.

A method comparable to ELISA is the enzyme-linked lectin assay, or *ELLA*, where lectins are used instead of antibodies. ELLA has been used by Ameye et al. (2001), who also report a list of lectins that bind to carbohydrates, to distinguish the saccharide fractions of the organic matrix in the spines of the echinoid *Paracentrotus lividus*.

3.3.4 Density Gradient Fractionation

Although biochemical studies on calcified tissues have produced an impressive quantity of invaluable results, they do reveal limitations, mostly arising from the fact that the identification and discrimination of sites where the various tissue components are located are difficult – in some cases impossible. Tissue specimens are, in fact, often homogenized and submitted to the extraction of their components, so that structural details vanish.

Fractionation of structures of different weight or size can be obtained by tissue homogenization and *density gradient fractionation*. Tissues and cells may be disrupted and homogenized when using methods as different as ultrasonication, osmotic shock, enzyme digestion or, in the case of calcified matrices, grinding or crushing them in an agate mortar in liquid nitrogen. The homogenate is then submitted to ultracentrifugation in a sucrose solution with various gradients of sucrose, so that homogenate components separate out according to their size and shape, i. e., their sedimentation coefficient. The very sensitive method of *equilibrium sedimentation* can be used to heighten the degree of fractionation: the homogenate is sedimented through a very concentrated sucrose (or cerium chloride) solution characterized by steep gradients of density: the homogenate components stop where their density is equal to that of the solution. Several important scientific insights into calcified tissues have been gained by using these methods. This is evident in the following few examples. Based on the observation that crystal maturation is accompanied by a rise in mineral content and a fall in organic constituents, and that, as a result, bone density reflects mineral maturation over time, the density gradient fractionation of homogenized calcified matrix has been used to evaluate specific components in dentin and bone fractions from uncalcified through fully calcified matrix (Engfeldt and Hjerpe 1974, 1976; Hjerpe and Engfeldt 1976; Bonar et al. 1991; Sodek et al. 2000). A number of studies on calcified matrix components and, specifically, on isolated crystals have been carried out with this method (Robinson 1952; Johansen and Parks 1960; Bocciarelli 1970; Termine et al. 1973; Weiner and Price 1986; Weiner and Traub 1989, 1991; Moradian-Oldak et al. 1991; Rey et al. 1991; Weiner et al. 1991; Wachtel and Weiner 1994; Ziv and Weiner 1994; Sodek et al. 2000; Eppell et al. 2001; Su et al. 2003; Tong et al. 2003), although the homogenization of calcified matrices by mechanical or chemical means rises some suspicions about the full integrity of the isolated crystals (see Sect. 5.1). By enzymatic digestion of the matrix (a non-enzymatic method of isolation is available; see Wuthier et al. 1978, 1985), followed by centrifugation, matrix vesicles have been isolated from normal cartilage (Ali et al. 1970; Ali 1979; Deutsch et al. 1981), human atherosclerotic aorta (Hsu and Camacho 1999), calcified vessels of rabbits fed with a high cholesterol diet (Hsu et al. 2000), experimental aneurysms in sheep (Martin et al. 1992) and developing dentin (Slavkin et al. 1972). The isolation of mitochondrial granules from Ca^{2+}-loaded mitochondria gave a definitive demonstration that they contained hydroxyapatite (Weinbach and von Brand 1965).

To increase the selectivity of biochemical analysis in calcified tissues, this has often been carried out on isolated structures separated from whole samples by manual *dissection* (Ascenzi 1983). Pugliarello et al. (1970), for instance, evaluated the chemical modifications which occur during calcification in dissected, isolated osteons at different degrees of calcification. They found that while collagen content is not altered by ongoing calcification, as shown by the constant amount of hydroxyproline, non-collagenous nitrogen and proteoglycan concentrations fall. The same technique was used to evaluate the density of osteoid tissue with respect to calcified matrix (Bonucci et al. 1970). By dissecting bovine predentin from dentin, it was documented that the former does not contain highly phosphorylated phosphoproteins (Jontell and Linde 1983), and that porcine predentin contains considerable amounts of glycosaminoglycans, presumably in the form of proteoglycans (Linde 1973).

3.3.5
Sequential Dissociative Extraction

One biochemical method which merits special mention for the important repercussions its findings have had is the sequential dissociative extraction procedure. Termine et al. (1980a,b) found that, in intact fetal calf bone and fetal bovine molar dentin, extraction with a 4 M guanidine-HCl solution containing protease inhibitors yields a minor protein fraction (from non-calcified organic components), while extraction with the same solution containing EDTA yields over two thirds of total non-collagenous proteins (from calcified matrix). In the same year, Linde et al. (1980) reported similar results in rat incisor dentin. These were confirmed in bone and other calcified tissues by a number of studies which showed not only that non-collagenous proteins are present in calcified tissues, but, above all, that they are so tightly bound to crystals that they can be extracted only after the removal of the inorganic substance (Smith and Leaver 1981; Termine 1981; Termine et al. 1981; Belcourt et al. 1982; Diamond et al. 1982; Butler 1984; Fisher and Termine 1985). Years before, Shapiro (1970a, b) had reported that acidic phospholipids of bone and of elasmobranch and teleost skeletal tissues can only be extracted after decalcification.

3.3.6
In Vitro Systems

Cell culture systems are an effective tool for studying the origin and function of cells, the biosynthesis of organic macromolecules, the way these can interact with the cells themselves or other molecules, the processes which occur at the cell membrane, and a number of other cellular activities and properties which can lead to the synthesis of an organic matrix and to the precipitation in it of inorganic substance. Cell cultures are mostly studied with biochemical methods, but biophysical and morphological methods may also be suitable for evaluating their constituents. A great number of procedures have been adopted for these types of study and the results obtained from their use as models for normal and pathological calcification are correspondingly great (reviewed by Robey 1992) and defy summary in a few lines.

In vitro organ/tissue cultures can be applied not only to the study of the cell role in calcification, but also of the effects on it of specific organic and/or inorganic molecules (Harell et al. 1976; D'Errico et al. 1995; Cheng et al. 1996) which can be blocked or modified to verify their impact on the calcification process (Handley and Lowther 1976; Ikeda et al. 1986; Neufeld and Boskey 1994; Moradian-Oldak et al. 2002). The Millipore diffusion chamber can be used to control the induction or inhibition of calcification by cells and chemicals (Simmons et al. 1982).

The cell-free, *gel diffusion-precipitation* method appears to be a valuable tool for the characterization of the effects of different proteins on calcification. As reviewed by Silverman and Boskey (2004), it is based on the slow diffusion of inorganic ions (chiefly calcium and phosphate) from adjoining solutions or gels into the zone of the gel which contains the molecule under study, whose nucleating, inhibiting, accelerating, or retarding effects (depending on solution concentration, temperature, pH, etc.) can be evaluated according to the quantity and quality of the accumulated ions.

3.4 Morphological Methods

Plenty of information on calcified tissues has been obtained using morphological methods. There have been so many light microscope descriptions of the morphology and structure of hard tissues that a detailed report is not feasible, and references must be restricted to textbooks of histology, microscopic anatomy and pathology. Besides histological and cytological, light microscope studies, other morphological techniques have been used to advantage, such as polarization microscopy (Dallemagne and Melon 1946; Schmidt 1952; Ascenzi and Bonucci 1964; Saleuddin and Wilbur 1969; Giraud-Guille 1988), polarization microscopy using circularly polarized light (Portigliatti Barbos et al. 1987), transmission electron microscopy (Glimcher and Krane 1968; Reith 1968; Travis 1968; Matthews 1970; Warshawsky 1985; Hunziker and Herrmann 1990; Hoshi et al. 2000), scanning electron microscopy (Boyde and Hobdell 1969a, b; Boyde 1974; Marotti 1990), microradiography (Boivin and Meunier 2002) and histochemistry (Everett and Miller 1974; Doty and Schofield 1976, 1990). During the last few decades, new techniques, or refinements of old ones, have been used to study hard tissues. They comprise high voltage electron microscopy (Lee et al. 1986; Landis et al. 1992, 1993), high resolution electron microscopy (Selvig 1973; Boothroyd 1975; Kerebel et al. 1976; Daculsi and Kerebel 1978; Hayashi 1992, 1993), selected-area dark field electron microscopy (Grove et al. 1972; Arsenault 1988), cryo-immuno electron microscopy (Nanci et al. 1994) and other ultrastructural cryotechniques (Schraer and Gay 1977; Landis and Glimcher 1982; Landis et al. 1977; Hunziker and Schenk 1987; Beniash et al. 2000; Levi-Kalisman et al. 2001), confocal laser scanning microscopy (Laitala and Väänänen 1993; Deutsch et al. 2002), immunocytochemistry and immunohistochemistry (Takagi et al. 1982, 1983, 1990; Bianco 1990; McKee and Nanci 1995; Riminucci et al. 1995; Septier et al. 1998; Nanci et al. 2000), in situ hybridization (Shibata et al. 2002; Silvestrini et al. 2003), polymerase chain reaction, or PCR (Cao et al. 1993; Murayama et al. 2002; Landis et al. 2003), and several other related techniques. The reverse transcriptase linked polymerase chain reaction, or RT-PCR, has recently been modified so as to preserve tissue and cellular structures without generation of significant artifacts and to improve transcript detection (Stamps et al. 2003).

3.4.1 Preparative Procedures

Because morphological methods yield images that may be considered equivalent to the constitutive microscopic and ultramicroscopic structures of tissues, some comment should be made on the possibility that they represent artifactually modified structures. Morphological investigations actually require several preliminary treatments of the tissue specimens to prepare them for microscopic examination; each of these treatments can to some extent modify tissue composition and organization.

Any structural changes produced by preparative procedures are, surprisingly, often ignored as long as the final pictures have a satisfactory microscopic appearance. Even more surprising is the observation that pictures of deeply modified tissues may be prized more highly, and be taken to be more representative of actual structural characteristics, than pictures of tissues in which most of the components have been kept intact. This may happen, for instance, with electron microscope pictures of tissues prepared by routine methods, which imply the loss of many components, compared with pictures of cryo-preserved tissues embedded in water soluble resins, in which most of the components survive unharmed: the former show well-contrasted, sharp structures, while the latter appear poorly resolved and partly masked. Because a high proportion of the results, and the consequent conclusions, reported in this volume have been obtained with morphological techniques, a thorough discussion of preparative and other possible artifacts seems appropriate.

The most frequent tissue changes are those due to fixation, dehydration and embedding. In the case of calcified tissues, a decalcification procedure may be needed which itself gives rise to fresh artifacts. This disadvantage is partly compensated for by the often overlooked fact that the calcified organic structures are largely protected from the dangerous effects of fixation and dehydration by the inorganic substance in which they are embedded.

3.4.2 Fixation

Almost all morphological methods require preliminary fixation to avoid both the damage caused by the autolysis which occurs naturally in any dead tissue, and the changes which would inevitably be caused by the processing of a fresh tissue. Regrettably, the fixation procedure is itself a cause of structural and chemical alterations (Arsenault 1990). Given that the choice of a fixative depends on the nature of the structures that must be fixed, and that there are obvious parameters (pH, temperature, osmolarity, tissue volume, and so on) that cannot be neglected, two aldehydes – formaldehyde and glutaraldehyde – are the fixatives of choice (Reale and Luciano 1970). The first is generally used at a concentration of 4% (10% of the 30–40% commercial solution), the second at a concentration of 2.0–2.5%, both diluted in a buffer at pH 7.2. The penetrating properties of formaldehyde are stronger than those of glutaraldehyde, which explains why the former is mainly used when the specimen is soaked in the fixative, the latter when fixation is carried out by perfusion of a living organism, that is, by circulating the fixative solution in a vascular bed. Fixation by specimen soaking has the advantage of great simplicity, and the disadvantage of slow penetration of the fixative into the tissue, so that the central part of the specimen is fixed later, and often less completely, than the peripheral part (the fixation process can be accelerated by using microwave irradiation; Arana-Chavez and Nanci 2001). Fixation by perfusion has the advantage that all tissue structures are reached by the fixative rapidly and with the cells still living, and the disadvantage that the method can only be applied to experimental animals.

Both formaldehyde and glutaraldehyde react with tissue chemical groups, so introducing a number of cross-links most of which can, however, be eliminated by prolonged washing. They, on the other hand, may preserve the secondary structure of proteins instead of denaturing them (Mason and O'Leary 1991). Some sort of fixation, due to dehydration, is produced by ethyl alcohol or acetone, which may cause structural distortions because of their dehydrating effect. Ethanol is usually adopted for the fixation of glycogen, which is soluble in aqueous solvents (Schajowicz and Cabrini 1958). Several ions and molecular complexes (osmium tetroxide, chromium sulfate or bichromate, ruthenium hexammine trichloride, malachite green) are chosen for special fixation purposes, such as electron microscopy or lipid fixation (Casley-Smith 1963; Pourcho et al. 1978). In any case, a quantitative X-ray microanalysis of the calcium-phosphate granules contained in mitochondria of damaged myocardium showed that, compared with alcoholic fixation, fixation with 1% osmium in phosphate buffer resulted in a tenfold reduction in Ca and P of just formed granules and 30% reduction in the advanced ones (Hagler et al. 1981).

Special attention has rightly been dedicated to methods of *cryofixation*, or *cryofixation-substitution* (Ryan 1992), which can eliminate changes due to the reaction of tissue components with water and chemical substances. They can, however, induce ion dislocations in the tissue (Appleton et al. 1985; Appleton 1987). A carbon sandwich technique has recently been suggested to avoid the electron beam-induced charging of this type of specimen which causes image shift and limits image resolution (Gyobu et al. 2004). Several methods have been proposed for the fixation of acid proteoglycans, which can easily be lost during specimen processing. Engfeldt and Hjertquist (1968) have found a total loss of 70% of ^{35}S from sections when the growth cartilage is fixed in glutaraldehyde, immersed in cacodylate buffer, post-fixed in osmium tetroxide, immersed in Tyrode's solution, and dehydrated in ethanol (followed by treatment with propylene oxide) or acetone, the majority being lost in osmium tetroxide; when glutaraldehyde contained 0.5% cetylpyridinium chloride, a total of only 7% of ^{35}S was lost. Cationic substances like cationic dyes can be mixed with a fixative to improve the preservation of anionic components in the tissue (Szirmai 1963). Anhydrous fixation, for instance by exposing the tissue to acrolein (Landis et al. 1980) or osmium vapors, has the advantage that water contact, and the solubilization of tissue components in it, are avoided (Landis and Glimcher 1978), and the disadvantage that fixation is rather superficial.

Fourier transform infrared microspectroscopic studies of Aparicio et al. (2002) have shown that the fixatives which induce minor changes in apatite crystallinity and the inorganic/organic ratio are those used in non-aqueous solutions. The use of these fixatives would even yield better results than cryosectioning or using non-fixed material.

3.4.3 Embedding

Other artifacts can be caused by tissue embedding. This procedure is adopted because unembedded tissues are too soft, plastic and elastic to be cut with a microtome. With embedding, these physical properties change and the embedded tissue acquires properties of hardness and plasticity that allow sections of appropriate thickness to be cut for microscopic study. Paraffin, the embedding substance most commonly used in histology, does not mix with water and can only penetrate tissue after its interstitial water has been removed (dehydration) by soaking the specimens in substances (ethanol, acetone) which must, in their turn, be removed and substituted with paraffin solvents. Xylene or chloroform are used for this 'clearing' process. During these treatments, tissue components, and, above all, unsaturated lipids, may be solubilized and lost. This phenomenon occurs whenever hydrophobic substances (paraffin, celloidin, methacrylates, epoxy resins) are used for embedding. It can largely be avoided by using water-miscible substances (for instance, gelatin or carbowax for light microscopy, LR-white or Lowicryl for electron microscopy), or by methods of cryofixation and cryosectioning. Tissues treated by these methods are, however, hard to section. The changes introduced by fixation and embedding in tissue components, and their possible loss, are often easy to detect under the electron microscope. This is true, for instance, of cartilage matrix proteoglycans, whose ultrastructural morphology changes (from filament-like to granular) with the type of tissue processing (see Sect. 9.3.1). The embedding media that best preserve crystallinity and the inorganic to organic ratio are the hydrosoluble resins LR White and Spurr, the epoxy resin Araldite, and methyl methacrylate (Aparicio et al. 2002).

3.4.4 Decalcification

Another possible source of artifacts is decalcification, which is often needed when calcified tissues must be sectioned with a microtome. The irregular hardness caused by the presence of inorganic substance in these tissues does, in fact, make sectioning difficult or impossible. Decalcification can be skipped if small calcified specimens are embedded in epoxy resins. This technique, which is common in electron microscopy, yields polymerized blocks which have about the same hardness as that of the calcified matrix and can therefore be sectioned with ultramicrotomes and diamond knives. Obviously, in this case the inorganic substance masks the organic components, and decalcification will have to be used if the latter must be studied.

Several *decalcification methods* are available (reviewed by Callis and Sterchi 1998); the most frequent choice is to soak the calcified specimen in an acidic solution (acetic acid, hydrochloric acid, nitric acid or another acid), or in a buffered solution of chelating agents, those most often used being ethylenediaminetetracetic acid (EDTA, also called Sequestrene or Versene) and ethyleneglycol-bis-(β-aminoethyl ether)N,N'-tetraacetic acid (EGTA). These procedures are very simple, but they can have a major impact on the composition and organization of the decalcified specimen, so much so that decalcification is considered one of the most frustrating, most demanding and least understood procedures (Callis and Sterchi 1998). It can, in fact, create a number of artifacts through the removal of varying amounts of organic molecules together with inorganic ions (Schajowicz and Cabrini 1955; Vigliani and Marotti 1963; Boonstra et al. 1990; Hosoya et al. 2005), as shown by the decrease in chromatin staining and by the loss of matrix proteoglycans (Weatherford and Mann Jr 1973; Ippolito et al. 1981; Kiviranta et al. 1984; Campo and Betz 1987). The artifacts brought about by decalcification obviously cause greater problems when the organic matrix of hard tissues is studied under the electron microscope (Bonucci and Reurink 1978), as is clearly shown by the morphology of the collagen fibrils, which appear dissociated and disaggregated (Fig. 3.3), and by the fact that only the presence of the *lamina limitans* allows recognition of the previously calcified cartilage matrix (Scherft 1972). It is striking that the antigenic reactivity seems to persist with many of the decalcifying agents (Athanasou et al. 1987; Arber et al. 1996).

Several methods have been proposed to improve the preservation of organic material during decalcification. A correct fixation does much to forestall decalcification artifacts (Kiviranta et al. 1984); in this connection, because the compactness of the calcified tissues hinders the penetration of the fixative solution, the perfusion fixation method seems to give better results than

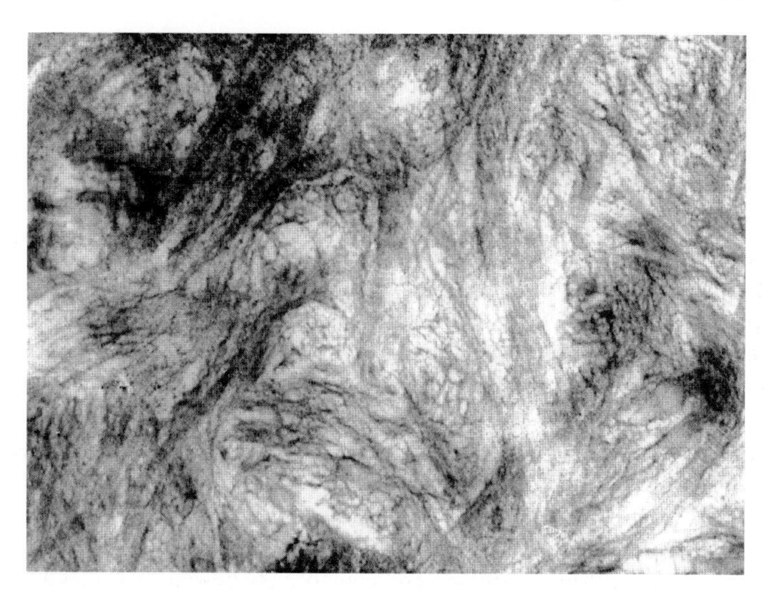

Fig. 3.3. Detail of trabecular bone matrix decalcified before embedding by prolonged immersion in EDTA: the collagen fibrils are highly dissociated. Uranyl acetate and lead citrate, ×54,000

immersion methods, especially when the technique adopted is regional perfusion by circulating the solution in the lower extremities from the abdominal aorta (Trudel et al. 2000). Incidentally, perfusion of a decalcifying solution can be used to achieve decalcification of the whole skeleton (Trudel et al. 2000). Buffered 4% formaldehyde or 2% glutaraldehyde, possibly followed by osmium tetroxide postfixation, are the most frequently used perfusion fixation mixtures (Reale and Luciano 1970; McKee et al. 1991). Methods of cryofixation and cryosubstitution can bring major benefits, as is confirmed by the preservation of proteoglycans in their native, extended state (Hunziker and Schenk 1984). Comparing the decalcifying solutions, a minimum loss of proteoglycans and rapid decalcification have been reported with 5% (Ippolito et al. 1981) or 20% (Weatherford and Mann Jr 1973) formic acid, whereas others have reported good results with alcoholic formalin fixation and aqueous (Jubb and Eggert 1981) or alcoholic (Scott and Kyffin 1978; Dickson and Jande 1980) EDTA decalcification. Baird et al. (1967) have obtained acceptable ultrastructural results by using glutaraldehyde fixation and decalcification in cold 0.1 M solutions of disodium or tetrasodium EDTA containing 4% glutaraldehyde and adjusted to pH 7.2–7.4 with sodium hydroxide or versenic acid, respectively. Satisfactory results have been reported by Warshawsky and Moore (1967), who used fixation by perfusion of 2.5% glutaraldehyde and decalcification by specimen immersion in 4.13% disodium EDTA for 14–21 days, followed by washing in phosphate buffer for 2 days and postfixation in 1% osmium tetroxide.

The loss of anionic molecules during fixation and decalcification can largely be avoided by the cationic-dye stabilization method, which relies on addition to the fixative of cationic substances such as the cationic dyes alcian blue, acridine orange, cupromeronic blue, polyethyleneimine, ruthenium hexammine trichloride, ruthenium red, safranin O, or toluidine blue O (Shepard and Mitchell 1976a, b; Takagi 1990; Shepard 1992; Engfeldt et al. 1994; Király et al. 1996; Ueda et al. 1997, 2001; Leng et al. 1998). The cationic-dye method gives an ultrastructural pattern comparable with that which can be obtained using high pressure freezing, freeze substitution and low temperature embedding (Engfeldt et al. 1994). The same effect has been reported with aldehyde fixation followed by treatment with *N,N*-dimethylformamide (Kagami et al. 1990). Freeze-substitution staining of cartilage with alcian blue has been suggested as a method capable of preserving proteoglycans and suitable for electron microscopy (Maitland and Arsenault 1989).

In spite of these refinements, some distortion of ultrastructures, enzyme inactivation and/or loss of organic material must be expected as unwelcome results of decalcification. These artifacts can largely be avoided with the post-embedding decalcification methods which had initially been proposed for electron microscopy (Bonucci 1967, 1969) but can also be used for light microscopy (Bonucci 1970, 1971; Bonucci and Reurink 1978). Because of their intrinsic interest, these methods will now be considered in some detail.

Post-embedding decalcification and staining (PEDS) can either be carried out by floating sections from fixed, resin-embedded tissues on a decalcified solution, or by immersion in it of the resin blocks containing the fixed tissue. Both methods make it possible to differentiate the organic structures by treating the decalcified ultrathin (about 75.0 nm thick) sections with the common 'stains' used in electron microscopy (uranyl acetate, lead citrate, phosphotungstic acid, silver nitrate, etc.; for methods of staining and other electron microscope methods in biology see Hayat 1989). Semithin (about 1 μm thick) sections can be stained using any of a wide spectrum of histological and histochemical methods (Bianco et al. 1984).

As previously noted, either ultrathin sections for electron microscopy or semithin sections for light microscopy can be floated for a few minutes on a decalcifying solution consisting of 1% hydrochloric acid, 2.2% formic acid, or 10% EDTA. To achieve this, a thin plastic disk with a central hole of the same diameter as the electron microscope grids is floated on the decalcifying solution, as suggested by Marinozzi (1961). By manipulating a platinum wire loop, or by using the disk as if it was a wire loop, sections can be retrieved from the water of the knife trough and transferred to the surface of the liquid meniscus in the hole. Using the same technique, they can be moved from the decalcifying solution to a washing solution, then to a stain solution, and again to a washing solution, to be collected one last time by touching them with the electron microscope grid or with a histological cover slip. In this way, sections can be moved from one solution to another without dispersion, loss or impairment.

Decalcification can also been carried out after sections have been mounted on the grids by floating them side down on a decalcifying solution (Nylen and Omnell 1962; Rönnholm 1962; Travis 1968; Travis and Gonsalves 1969; Frank 1979). This method has the disadvantage that impurities can be formed by reaction of the metal of the grids with decalcifying and staining solutions.

The second post-embedding decalcification and staining method (decalcification by block soaking, or DBS) is similar to the first one but a little more time-consuming (Bonucci and Gherardi 1975). After the fixed specimen has been embedded in an epoxy resin, the block containing it is trimmed so that the tissue surfaces are exposed, and is then soaked for a day or more in a decalcifying solution (usually 2% formic acid). The decalcified block is then washed in distilled water, dehydrated, and again submitted to embedding using the same epoxy resin. In this way the decalcified specimen acquires the same properties as a soft tissue and can be sectioned very easily.

These two methods are comparable as far as the maintenance of organic structures is concerned (Bonucci and Reurink 1978; Goldberg et al. 1980); each has its advantages and disadvantages. The PEDS method has the advantage that comparison between non-decalcified and decalcified, serial sections from the same specimens is possible, and that the same area can be examined before and after decalcification; it has the disadvantage that the presence of inorganic substance in the specimen makes it hard to cut sections, which may show several scratches and an irregular thickness. The DBS method has the advantage that a hard specimen is transformed into a soft one, so that sections of suitable thickness and free from scratches can be easily obtained; it has the disadvantage that comparison with non-decalcified sections is impossible and that specimens as small as possible must be used.

The excellence of the morphological results which can be obtained with either the PEDS or DBS method is shown by the fact that, in either case, sections contrasted with uranyl acetate and lead citrate appear as they do in undecalcified sections and cell ultrastructure is perfectly preserved. This is clearly shown by comparing the effects produced on the bone matrix by decalcification when this is carried out before and after embedding, a procedure that can be carried out on the same specimen (Fig. 3.4): the bone matrix decalcified before embedding is stained by azure II-methylene blue, probably because it is dissociated (as confirmed by electron microscopy; Figs. 3.3 and 3.4) and its reacting groups are available to the stains, whereas the bone matrix decalcified after embedding is not stained, because it is not dissociated (Fig. 3.4) and its groups are not available to the stains (see Chap. 16 for further details). After the PEDS method, the calcification front is still clearly recognizable because the aggregates of crystals which correspond to calcification nodules are replaced by very similar aggregates of crystal ghosts, and these, in their turn, have the same morphology as the untreated crystals (see Fig. 3.5 and Sect. 16.2.1).

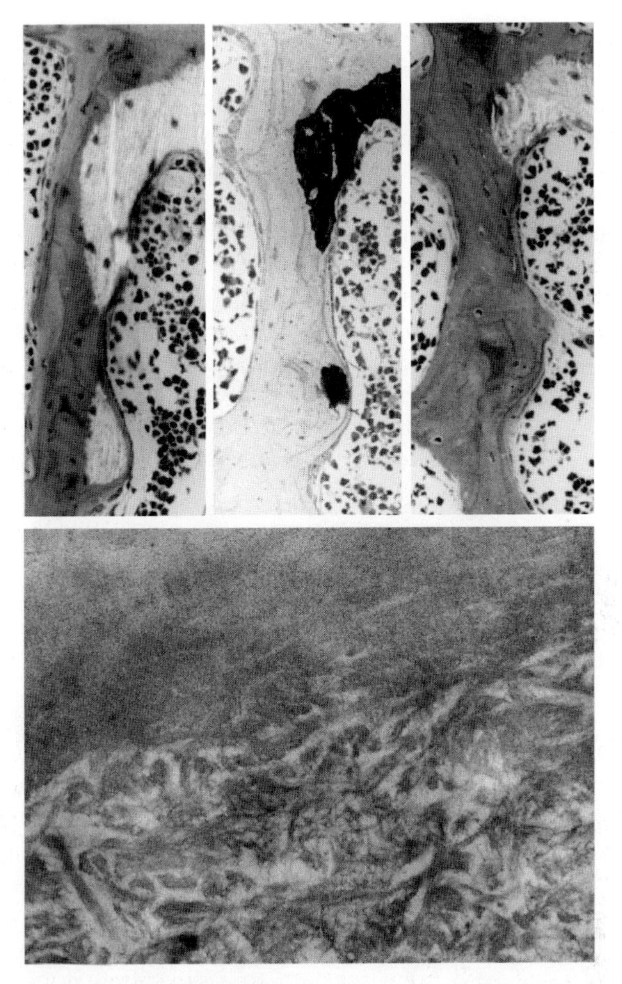

Fig. 3.4. *Above*: serial sections of metaphyseal bone trabeculae. *Left*: specimen partly decalcified by 10 min immersion in EDTA before embedding in Araldite; section stained with azure II-methylene blue: only the decalcified matrix is stained, probably because it is dissociated and its chemical groups have become available to stains. *Center*: same specimen; the von Kossa method confirms that the unstained matrix (now *black*) had not been decalcified. *Right*: the same partly decalcified, Araldite-embedded specimen was soaked again in EDTA till decalcification was complete; section stained by von Kossa method and azure II-methylene blue: the previously von Kossa-positive (now decalcified and von Kossa-negative) matrix (*upper right corner*) is unstained because its structure has been preserved by the resin, so that reacting groups are not available. ×250. *Below*: boundary between the matrix decalcified before embedding, whose collagen fibrils are highly dissociated (*below*), and the matrix decalcified after embedding, whose collagen fibrils are closely aggregated and hard to recognize (*above*). Uranyl acetate and lead citrate, ×60,000

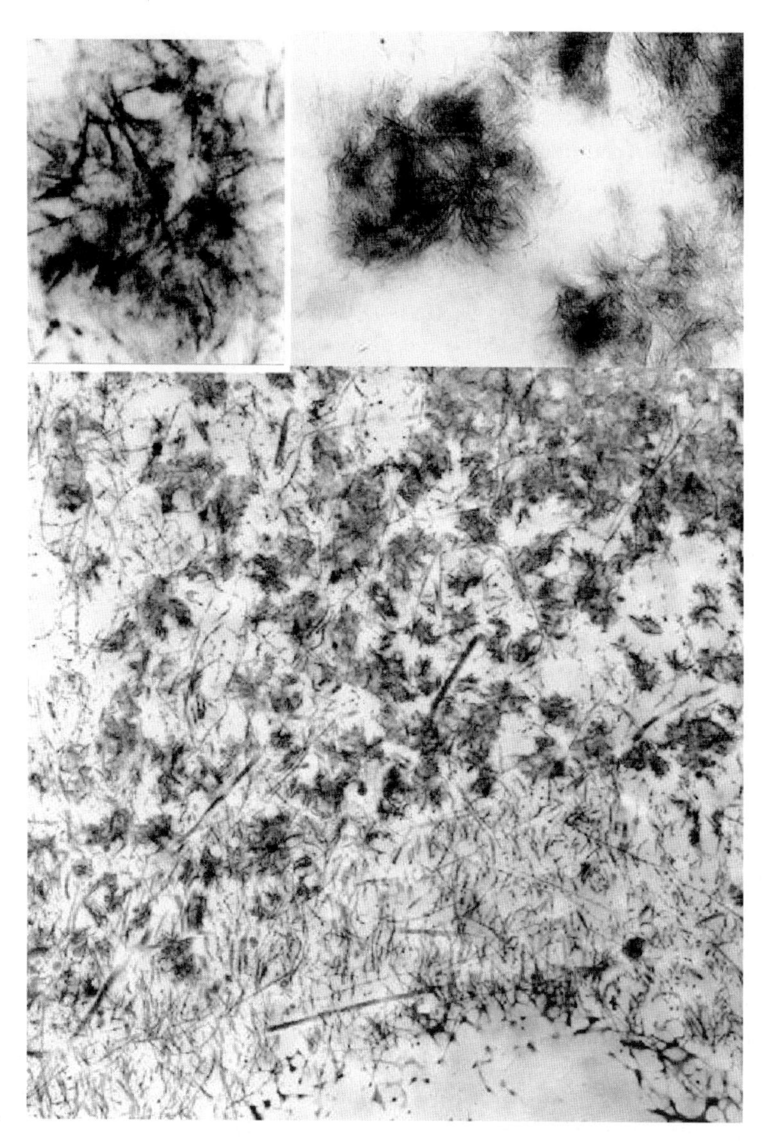

Fig. 3.5. Calcification front in growth cartilage; section decalcified and stained according to the PEDS method. Note that the cartilage seems undecalcified, due to the aggregates of crystal ghosts that have replaced the calcification nodules (see Sects. 16.3 and 18.4 for further details). The border of a chondrocyte lacuna is partly visible below. *Left inset*: detail of a decalcified and stained calcification nodule: it now takes the form of an aggregate of crystal ghosts. Specimen embedded in Araldite, soaked in 1% formic acid for 24 h, and embedded again in Araldite; section stained with uranyl acetate and lead citrate, ×18,000 and, *inset*, ×90,000. *Right inset*: untreated cartilage calcification nodules showing filament-like crystals. ×60,000

Decalcification has the aim of unmasking organic components otherwise covered by the inorganic substance; other methods aim to free crystals of organic material. Treatment with 95% hydrazine has been used to completely deproteinate bone under nearly anhydrous conditions and moderate heating, without inducing changes in crystals' structural properties (Termine et al. 1973). Although this method is considered safe as far as crystal structure is concerned, so that it has been used for sophisticated studies (equilibrium solubility, NMR, FTIRM, FTIRI) on crystal composition, the fact that treated crystals still contain 1–2% of proteins (Kuhn et al. 2000) and, above all, that their links with the organic matrix have been disconnected, raises some doubts on one hand about the efficacy of the method, and on the other, about the statement that crystals – in spite of their extreme smallness and thinness – are not modified by the removal of proteins.

3.4.5
Histochemistry and Immunohistochemistry

Once tissue samples have been processed, specific molecules can be detected in them by treating sections with suitable chemicals. Methods are so numerous that they can only be reported in textbooks of histology, histochemistry and electron microscopy. A few of them, however, are especially useful in the study of calcified tissues and are worthy of brief mention.

The *periodic acid-Schiff method* (PAS) is used to detect the glycoproteins which have free, vicinal hydroxide groups in their molecule. These groups are oxidized to aldehydes by treatment with 1% periodic acid (for no more than 5 min, to avoid further oxidation to carboxy groups) and the aldehydes are then stained red by the sulfate fucsin contained in Schiff's solution (Kasten 1960). The reaction needs to be suitably controlled.

A number of PAS-like methods are available. The periodic acid-silver nitrate methenamine (Marinozzi 1961; Rambourg 1971) and the acidic phosphotungstic acid (Marinozzi 1967) methods are the most interesting of them because they can also be used for electron microscopy.

Acidic molecules can be detected under the microscope by reacting them with basic dyes, which also stabilize them during fixation and decalcification (see Sect. 3.4.4). Alcian blue (Quintarelli et al. 1964; Ippolito et al. 1983), acridine orange and toluidine blue are the cationic dyes most often used for light microscopy, but several other molecules of the same type are available (Takagi 1990). Some of them are metachromatic, that is, when they bind to acidic, polymeric molecules bestow them a colour different from their own (for metachromasia, see Szirmai 1963). Ruthenium red, thorium dioxide, colloidal iron, colloidal gold and other cationic, electron-dense molecules can be used for electron microscopy (Groot 1982; Curran et al. 1965).

The low specificity of all these methods can be increased by pH control (at pH below 2.8 only sulfate and phosphate groups are dissociated enough to react with basic dyes) or by digestion with specific enzymes (Takagi 1990).

Immunohistochemistry appears to be one of the most effective methods for the specific detection of individual proteic molecules. In its simplest form, it is based on the immunological reaction between an antigen and its specific antibody (primary antibody); the immuno-complex is then reacted with an antiglobulin (secondary antibody) which, in its turn, is conjugated to a substance which can be detected under the microscope. In light microscopy (Colvin et al. 1995), this substance can be a peroxidase (peroxidase-antiperoxidase, or PAP, method), a lectin, a fluorochrome, an enzyme, like alkaline phosphatase, or the very sensitive avidin-biotin complex (ABC method). Avidin, an eggwhite glycoprotein which can be conjugated to a secondary antibody, has high affinity for the vitamin biotin, which can be conjugated to one of the substances mentioned above. Some of these methods can be applied to elecron microscopy, but the protein A-colloidal gold method is the most often used (Griffiths 1993; Bendayan 1995): the Fc portion of IgG binds the protein A, which in its turn can be conjugated to an electron-dense molecule such as peroxidase or, preferably, colloidal gold. The immuno-complex is pointed out by the presence of gold nanoparticles, easily recognizable under the electron microscope. The use of gold nanoparticles of different sizes allows the simultaneous localization of two antigens to be achieved.

3.5 Concluding Remarks

Most of the techniques reported above are the same as those currently used for the study of soft tissues. A chapter on the methodologies applied to calcified tissues is justified, therefore, not so much by the fact that special methods are needed, as by the recognition that the study of inorganic material may require rather dedicated techniques, and that decalcification may be needed for the analysis of the organic matrix. The following points should be stressed.

- Biophysical methods, ranging from the long-lasting X-ray diffraction to the present resonance and spectroscopy techniques, have been applied to the study of calcified tissue structures to encompass the level of nanoparticles and atoms. They have produced a number of invaluable results, but the facilities required for their use are not easily available. On the other hand, the achievement of such extremely fine levels of analysis does not necessarily lead to a better understanding of the structure and composition of calcified tissues.

- Biochemical methods have led to the discovery of so many data that any comment would be superfluous. Most of our knowledge on the properties of calcified organic matrices has been attained through biochemical researches.

- Biophysical methods are carried out on either whole tissue specimens or on selected parts of them, and biochemical methods usually comprise dissociation, solubility and extraction of tissue constituents. As a result, both types of analysis have the drawback that a precise structural localization cannot always be attained, and that the results can only yield 'mean' data hardly referable to the individual heterogeneous structures which comprise whole samples.
- To overcome at least partly the problem of the poor selectivity, density gradient fractionation and manual dissection have been used to isolate and separate different components of the calcified tissues, including single crystals. There are some doubts, however, that after fractionation these maintain the characteristics they have in the whole tissue. With the same aim, sequential dissociative extraction has been applied to the study of organic components of calcified matrix.
- Morphological investigations are usually carried out on fixed and embedded specimens. Fixation and embedding can cause extraction artifacts, most of which can be avoided by cryo- or anhydrous fixation and embedding in water soluble resins and, in the case of polyanionic components, like acid proteoglycans, by fixation in the presence of cationic substances (cationic dyes).
- Morphological investigations very often need the unmasking of organic structures by decalcification, a procedure that has the advantage of reducing the hardness of the calcified tissues and avoiding the difficulties connected with their sectioning.
- Several extraction artifacts are produced when decalcification is carried out by soaking tissue specimens in a decalcifying solution. The post-embedding decalcification and staining (PEDS) methods are the most effective in keeping artifacts to a minimum.

References

Ali SY (1979) Isolation of cartilage matrix vesicles. In: Simmons DJ, Kunin AS (eds) Skeletal research. An experimental approach. Academic Press, New York, pp 109–119

Ali SY, Sajdera SW, Anderson HC (1970) Isolation and characterization of calcifying matrix vesicles from epiphyseal cartilage. Proc Natl Acad Sci 67:1513–1520

Ameye L, De Becker G, Killian C, Wilt F, Kemps R, Kuypers S, Dubois P (2001) Proteins and saccharides of the sea urchin organic matrix of mineralization: characterization and localization in the spine skeleton. J Struct Biol 134:56–66

Aparicio S, Doty SB, Camacho NP, Paschalis EP, Spevak L, Mendelsohn R, Boskey AL (2002) Optimal methods for processing mineralized tissues for Fourier transform infrared microspectroscopy. Calcif Tissue Int 70:422–429

Appleton J (1987) X-ray microanalysis of growth cartilage after rapid freezing, low temperature freeze drying and embedding in resin. Scanning Microsc 1:1135–1144

Appleton J, Lyon R, Swindin KJ, Chesters J (1985) Ultrastructure and energy-dispersive X-ray microanalysis of cartilage after rapid freezing, low temperature freeze drying, and embedding in Spurr's resin. J Histochem Cytochem 33:1073–1079

Arana-Chavez VE, Nanci A (2001) High-resolution immunocytochemistry of noncollagenous matrix proteins in rat mandibles processed with microwave irradiation. J Histochem Cytochem 49:1099–1109

Arber JM, Arber DA, Jenkins KA, Battifora H (1996) Effect of decalcification and fixation in paraffin-section immunohistochemistry. Appl Immunohistochem 4:241–248

Arnold S, Plate U, Wiesmann H-P, Kohl H, Höhling H-J (1997) Quantitative electron-spectroscopic diffraction (ESD) and electron-spectroscopic imaging (ESI) analyses of dentine mineralisation in rat incisors. Cell Tissue Res 288:185–190

Arnold S, Plate U, Wiesmann H-P, Stratmann U, Kohl H, Höhling H-J (1999) Quantitative electron spectroscopic diffraction analyses of the crystal formation in dentine. J Microsc 195:58–63

Arnold S, Plate U, Wiesmann H-P, Straatmann U, Kohl H, Höhling H-J (2001) Quantitative analyses of the biomineralization of different hard tissues. J Microsc 202:488–494

Arsenault AL (1988) Crystal-collagen relationships in calcified turkey leg tendons visualized by selected-area dark field electron microscopy. Calcif Tissue Int 43:202–212

Arsenault AL (1990) The ultrastructure of calcified tissues: methods and technical problems. In: Bonucci E, Motta PM (eds) Ultrastructure of skeletal tissues. Kluwer Academic Publishers, Boston, pp 1–18

Arsenault AL, Ottensmeyer FP (1983) Quantitative spatial distributions of calcium, phosphorus, and sulfur in calcifying epiphysis by high resolution electron spectroscopic imaging. Proc Natl Acad Sci USA 80:1322–1326

Arsenault AL, Ottensmeyer FP, Heath IB (1988) An electron microscopic and spectroscopic study of murine epiphyseal cartilage: analysis of fine structure and matrix vesicles preserved by slam freezing and freeze substitution. J Ultrastruct Mol Struct Res 98:32–47

Ascenzi A (1983) Microscopic dissection and isolation of bone constituents. In: Kunin AS, Simmons DJ (eds) Skeletal research. An experimental approach, vol 2. Academic Press, New York, pp 185–236

Ascenzi A, Bonucci E (1964) A quantitative investigation of the birefringence of the osteon. Acta Anat 44:236–262

Ascenzi A, Bonucci E, Ostrowski K, Sliwowski A, Dziedzic-Goclawska A, Stachowicz W, Michalik J (1977) Initial studies on the crystallinity of the mineral fraction and ash content of isolated human and bovine osteons differing in their degree of calcification. Calcif Tissue Res 23:7–11

Ascenzi A, Bonucci E, Ripamonti A, Roveri N (1978) X-ray diffraction and electron microscope study of osteons during calcification. Calcif Tissue Res 25:133–143

Ascenzi A, Bonucci E, Generali P, Ripamonti A, Roveri N (1979) Orientation of apatite in single osteon samples as studied by pole figures. Calcif Tissue Int 29:101–105

Ascenzi A, Bigi A, Koch MH, Ripamonti A, Roveri N (1985) A low-angle X-ray diffraction analysis of osteonic inorganic phase using synchrotron radiation. Calcif Tissue Int 37:659–664

Athanasou NA, Quinn J, Heryet A, Woods CG, McGee JO (1987) Effects of decalcification agents on immunoreactivity of cellular antigens. J Clin Pathol 40:874–878

Baird IL, Winborn WB, Bockman DE (1967) A technique of decalcification suited to electron microscopy of tissues closely associated with bone. Anat Rec 159:281–290

Ballanti P, Bonucci E, Della Rocca C, Milani S, Lo Cascio V, Imbimbo B (1990) Bone histomorphometric reference values in 88 normal Italian subjects. Bone Miner 11:187–197

Barckhaus RH, Höhling HJ, Fromm I, Hirsch P, Reimer L (1991) Electron spectroscopic diffraction and imaging of the early and mature stages of calcium phosphate formation in the epiphyseal growth plate. J Microsc 162:155–169

Baud CA, Badonnel M-C (1970) Electron-microscope and electron-diffraction study of experimental cutaneous calcinosis. Clin Orthop Relat Res 69:55–65

Belcourt AB, Fincham AG, Termine JD (1982) Acid-soluble bovine fetal enamelins. J Dent Res 61:1031–1032

Bendayan M (1995) Colloidal gold post-embedding immunocytochemistry. Prog Histochem Cytochem 29:1–159

Beniash E, Traub W, Veis A, Weiner S (2000) A transmission electron microscope study using vitrified ice sections of predentin: structural changes in the dentin collagenous matrix prior to mineralization. J Struct Biol 132:212–225

Bernardi G, Giro M-G, Gaillard C (1972) Chromatography of polypeptides and proteins on hydroxyapatite columns: some new developments. Biochim Biophys Acta 278:409–420

Bianco P (1990) Ultrastructural immunohistochemistry of noncollagenous proteins in calcified tissues. In: Bonucci E, Motta PM (eds) Ultrastructure of skeletal tissues. Kluwer Academic Publishers, Boston, pp 63–78

Bianco P, Ponzi A, Bonucci E (1984) Basic and "special" stains for plastic sections in bone marrow histopathology, with special reference to May-Grunwald Giemsa and enzyme histochemistry. Basic Appl Histochem 28:265–279

Bigi A, Gandolfi M, Koch MHJ, Roveri N (1996) X-ray diffraction study of in vitro calcification of tendon collagen. Biomaterials 17:1195–1201

Bocciarelli DS (1970) Morphology of crystallites in bone. Calcif Tissue Res 5:261–269

Boivin G, Meunier PJ (2002) The degree of mineralization of bone tissue measured by computerized quantitative contact microradiography. Calcif Tissue Int 70:503–511

Bonar LC, Lees S, Mook HA (1985) Neutron diffraction studies of collagen in fully mineralized bone. J Mol Biol 181:265–270

Bonar LC, Shimizu M, Roberts JE, Griffin RG, Glimcher MJ (1991) Structural and composition studies on the mineral of newly formed dental enamel: a chemical, X-ray diffraction, and ^{31}P and proton nuclear magnetic resonance study. J Bone Miner Res 6:1167–1176

Bonucci E (1967) Fine structure of early cartilage calcification. J Ultrastruct Res 20:33–50

Bonucci E (1969) Further investigation on the organic/inorganic relationships in calcifying cartilage. Calcif Tissue Res 3:38–54

Bonucci E (1970) Fine structure and histochemistry of "calcifying globules" in epiphyseal cartilage. Z Zellforsch 103:192–217

Bonucci E (1971) The locus of initial calcification in cartilage and bone. Clin Orthop Relat Res 78:108–139

Bonucci E, Gherardi G (1975) Histochemical and electron microscope investigations on medullary bone. Cell Tissue Res 163:81–97

Bonucci E, Reurink J (1978) The fine structure of decalcified cartilage and bone: a comparison between decalcification procedures performed before and after embedding. Calcif Tissue Res 25:179–190

Bonucci E, Ascenzi A, Vittur F, Pugliarello MC, de Bernard B (1970) Density of osteoid tissue and osteones at different degree of calcification. Calcif Tissue Res 5:100–107

Boonstra WD, ten Bosch JJ, Arends J (1990) Protein and mineral release during in vitro demineralization of bovine dentine. J Biol Buccale 18:43–48

Boothroyd B (1975) Observations on embryonic chick-bone crystals by high resolution transmission electron microscopy. Clin Orthop Relat Res 106:290–310

Boskey AL (1998) Biomineralization: conflicts, challenges, and opportunities. J Cell Biochem 30/31:83–91

Boskey AL, Spevak L, Paschalis E, Doty SB, McKee MD (2002) Osteopontin deficiency increases mineral content and mineral crystallinity in mouse bone. Calcif Tissue Int 71:145–154

Boyde A (1974) Transmission electron microscopy of ion beam thinned dentine. Cell Tissue Res 152:543–550

Boyde A, Hobdell MH (1969a) Scanning electron microscopy of primary membrane bone. Z Zellforsch 93:98–108

Boyde A, Hobdell MH (1969b) Scanning electron microscopy of lamellar bone. Z Zellforsch 93:213–231

Butler WT (1984) Matrix macromolecules of bone and dentin. Collagen Rel Res 4:297–307

Callis G, Sterchi D (1998) Decalcification of bone: literature review and practical study of various decalcifying agents, methods, and their effects on bone histology. J Histotechnol 21:49–58

Campo RD, Betz RR (1987) Loss of proteoglycans during decalcification of fresh metaphyses with disodium ethylendiaminetetraacetate (EDTA). Calcif Tissue Int 41:52–55

Canalis E, McCarthy T, Centrella M (1988) Isolation of growth factors from adult bovine bone. Calcif Tissue Int 43:346–351

Cao X, Genge BR, Wu LNY, Buzzi WR, Showman RM, Wuthier RE (1993) Characterization, cloning and expression of the 67-kDa annexin from chick growth plate cartilage matrix vesicles. Biochem Biophys Res Comm 197:556–561

Carden A, Morris MD (2000) Application of vibrational spectroscopy to the study of mineralized tissues (review). J Biomed Opt 5:259–268

Carlström D (1955) X-ray crystallographic studies on apatites and calcified structures. Acta Radiol Suppl 121:1–59

Carlström D, Finean JB (1954) X-ray diffraction studies on the ultrastructure of bone. Biochim Biophys Acta 13:183–191

Casley-Smith JR (1963) Some observations on the fixation and staining of lipids. J R Microsc Soc 81:235–238

Cheng S-L, Zhang S-F, Avioli LV (1996) Expression of bone matrix proteins during dexamethasone-induced mineralization of human bone marrow stromal cells. J Cell Biochem 61:182–193

Colvin RB, Bhan AK, McCluskey RT (1995) Diagnostic immunocytochemistry, 2nd edn. Raven Press, New York

Cowley JM (2004) Applications of electron nanodiffraction. Micron 35:345–360

Curran RC, Clark AE, Lovell D (1965) Acid mucopolysaccharides in electron microscopy. The use of the colloidal iron method. J Anat 99:427–434

D'Angelo M, Billings PC, Pacifici M, Leboy PS, Kirsch T (2001) Authentic matrix vesicles contain active metalloproteases (mmp). A role for matrix vesicle-associated mmp-13 in activation of transforming growth factor-beta. J Biol Chem 276:11347–11353

D'Errico JA, MacNeil RL, Strayhorn CL, Piotrowski BT, Somerman MJ (1995) Models for the study of cementogenesis. Connect Tissue Res 33:9–17

Daculsi G, Kerebel B (1978) High-resolution electron microscope study of human enamel crystallites: size, shape, and growth. J Ultrastruct Res 65:163–172

Dallemagne MJ, Brasseur H (1942) La diffraction des rayons X par la substance minérale osseuse. Bull Soc R Sci Liége 8/9:1–19

Dallemagne MJ, Brasseur H (1947) La nature du sel principal de l'os étudiée par la diffraction des rayons X. Experientia 3:469–471

Dallemagne MJ, Melon J (1946) Nouvelles recherches relatives aux propriétés optique de l'os: la biréfringence de l'os minéralisé; relations entre les fractions organiques et inorganique de l'os. J Wash Acad Sci 36:181–195

Dawson JM (1946) X-ray diffraction pattern of bone: evidence of reflexions due to the organic constituent. Nature 157:660–661
deJong WF (1926) La substance minérale dans les os. Rec Trav Chim 45:445–448
Deutsch D, Bab I, Muhlrad A, Sela J (1981) Purification and further characterization of isolated matrix vesicles from rat alveolar bone. Metab Bone Dis Rel Res 3:209–214
Deutsch D, Leiser Y, Shay B, Fermon E, Taylor A, Rosenfeld E, Dafni L, Charuvi K, Cohen Y, Haze A, Fuks A, Mao Z (2002) The human tuftelin gene and the expression of tuftelin in mineralizing and nonmineralizing tissues. Connect Tissue Res 43:425–434
Diamond AG, Triffitt JT, Herring GM (1982) The acid macromolecules in rabbit cortical bone tissue. Arch Oral Biol 27:337–345
Dickson IR, Jande SS (1980) Effects of demineralization in an ethanolic solution of triethylammonium EDTA on solubility of bone matrix components and on ultrastructural preservation. Calcif Tissue Int 32:175–179
Dimuzio MT, Veis A (1978) Phosphophoryns – Major noncollagenous proteins of rat incisor dentin. Calcif Tissue Res 25:169–178
Donnay G, Pawson DL (1969) X-ray diffraction studies of echinoderm plates. Science 166:1147–1150
Doty SB, Schofield BH (1976) Enzyme histochemistry of bone and cartilage cells. Progr Histochem Cytochem 8:1–38
Doty SB, Schofield BH (1990) Histochemistry and enzymology of bone-forming cells. In: Hall BK (ed) Bone. Vol. 1: The osteoblast and osteocyte. The Telford Press, Caldwell, pp 71–102
Drake B, Prater CB, Weisenhorn AL, Gould SAC, Albrecht TR, Quate CE, Cannell DS, Hansma HG, Hansma PK (1989) Imaging crystals, polymers, and processes in water with the atomic force microscope. Science 243:1586–1589
Duriez J, De Guembeker C, Duriez R (1988) Confrontation des densitométries mono- et biphotonique avec l'histomorphométrie. J Médec Nucl Biophys 12:99–106
Dziedzic-Goclawska A, Rozycka M, Czyba J-C, Sawicki W, Moutier R, Lenczowski S, Ostrowski K (1982) Application of the optical Fourier transform for analysis of the spatial distribution of collagen fibers in normal and osteopetrotic bone tissue. Histochemistry 74:123–137
Eanes ED (1979) Enamel apatite: chemistry, structure and proteins. J Dent Res 58:829–834
Eanes ED, Lundy DR, Martin GN (1970) X-ray diffraction study of the mineralization of turkey leg tendon. Calcif Tissue Res 6:239–248
Eastoe JE, Eastoe B (1954) The organic constituents of mammalian compact bone. Biochem J 57:453–459
Egerton RF (2003) New techniques in electron energy-loss spectroscopy and energy-filtered imaging. Micron 34:127–139
Engfeldt B, Hjerpe A (1974) Density gradient fractionation of dentine and bone powder. Calcif Tissue Res 16:261–275
Engfeldt B, Hjerpe A (1976) Glycosaminoglycans and proteoglycans of human bone tissue at different stages of mineralization. Acta Pathol Microbiol Scand Sect A 84:95–106
Engfeldt B, Hjertquist S-O (1968) Studies on the epiphysial growth zone. I. The preservation of acid glycosaminoglycans in tissues in some histochemical procedures for electron microscopy. Virchows Arch B Cell Pathol 1:222–229
Engfeldt B, Hjerpe A, Reinholt FP, Wikström B (1985) Distribution of matrix vesicles in the lower part of the epiphyseal growth plate. In: Butler WT (ed) The chemistry and biology of mineralized tissues. Ebsco Media, Inc., Birmingham, AL, pp 356–359
Engfeldt B, Reinholt FP, Hultenby K, Widholm SM, Müller M (1994) Ultrastructure of hypertrophic cartilage: histochemical procedures compared with high pressure freezing and freeze substitution. Calcif Tissue Int 55:274–280

Eppell SJ, Tong W, Katz JL, Kuhn L, Glimcher MJ (2001) Shape and size of isolated bone mineralites measured using atomic force microscopy. J Orthop Res 19:1027–1034

Everett MM, Miller WA (1974) Histochemical studies on calcified tissues II. Amino acid histochemistry of developing dentine and bone. Calcif Tissue Res 16:73–88

Feng J, Somlyo AV, Somlyo AP (2004) A system for acquiring simultaneous electron energy-loss and X-ray spectrum-images. J Microsc 215:92–99

Fernandez-Seara MA, Wehrli SL, Takahashi M, Wehrli FW (2004) Water content measured by proton-deuteron exchange NMR predicts bone mineral density and mechanical properties. J Bone Miner Res 19:289–296

Fisher LW, Termine JD (1985) Noncollagenous proteins influencing the local mechanisms of calcification. Clin Orthop Relat Res 200:362–385

Foged NT, Delaissé J-M, Hou P, Lou H, Sato T, Winding B, Bonde M (1996) Quantification of the collagenolytic activity of isolated osteoclasts by enzyme-linked immunosorbent assay. J Bone Miner Res 11:226–237

Frank RM (1979) Tooth enamel: current state of the art. J Dent Res 58 (special issue B):684–693

Fratzl P, Fratzl-Zelman N, Klaushofer K, Vogl G, Koller K (1991) Nucleation and growth of mineral crystals in bone studied by small-angle X-ray scattering. Calcif Tissue Int 48:407–413

Fratzl P, Groschner M, Vogl G, Plenk H Jr, Eschberger J, Fratzl-Zelman N, Koller K, Klaushofer K (1992) Mineral crystals in calcified tissues: a comparative study by SAXS. J Bone Miner Res 7:329–334

Fratzl P, Schreiber S, Klaushofer K (1996) Bone mineralization as studied by small-angle X-ray scattering. Connect Tissue Res 34:247–254

Fukae M, Tanabe T, Yamakoshi Y, Yamada M, Ujiie Y, Oida S (2001) Immunoblot detection and expression of enamel proteins at the apical portion of the forming root in porcine permanent incisor tooth germs. J Bone Miner Metab 19:236–243

Genge BR, Cao X, Wu LN, Buzzi WR, Showman RW, Arsenault AL, Ishikawa Y, Wuthier RE (1992) Establishment of the primary structure of the major lipid-dependent Ca2+ binding proteins of chicken growth plate cartilage matrix vesicles: identity with anchorin CII (annexin V) and annexin II. J Bone Miner Res 7:807–819

Girardin E, Millet P, Lodini A (2000) X-ray and neutron diffraction studies of crystallinity in hydroxyapatite coatings. J Biomed Mater Res 49:211–215

Giraud-Guille M-M (1988) Twisted plywood architecture of collagen fibrils in human compact bone osteons. Calcif Tissue Int 42:167–180

Glimcher MJ, Krane SM (1968) The organization and structure of bone, and the mechanism of calcification. In: Gould BS (ed) Biology of collagen. Academic Press, London, pp 67–251

Goldberg M, Noblot MM, Septier D (1980) Effets de deux méthodes de démineralisation sur la préservation des glycoprotéines et des protéoglycanes dans les dentines intercanaliculaires et péricanaliculaires chez le cheval. J Biol Buccale 8:315–330

Gomez S, Boyde A (1994) Correlated alkaline phosphatase histochemistry and quantitative backscattered electron imaging in the study of rat incisor ameloblasts and enamel mineralization. Microsc Res Technol 29:29–36

Gomez S, Lopez-Cepero JM, Silvestrini G, Mocetti P, Bonucci E (1996) Matrix vesicles and focal proteoglycan aggregates are the nucleation sites revealed by the lanthanum incubation method: a correlated study on the hypertrophic zone of the rat epiphyseal cartilage. Calcif Tissue Int 58:273–282

Griffiths G (1993) Fine structure immuno-cytochemistry. Springer, Berlin Heidelberg New York

Grogger W, Schaffer B, Krishnan KM, Hofer F (2003) Energy-filtering TEM at high magnification: spatial resolution and detection limits. Ultramicroscopy 96:481–489

Groot CG (1982) An electron microscopic examination for the presence of acid groups in the organic matrix of mineralization nodules in foetal bone. Metab Bone Dis Rel Res 4:77–84

Grove CA, Judd G, Ansell GS (1972) Determination of hydroxyapatite crystallite size in human dental enamel by dark-field electron microscopy. J Dent Res 51:22–29

Grynpas M (1977) Three-dimensional packing of collagen in bone. Nature 265:381–382

Gupta HS, Roschger P, Zizak I, Fratzl-Zelman N, Nader A, Klaushofer K, Fratzl P (2003) Mineralized microstructure of calcified avian tendons: a scanning small angle X-ray scattering study. Calcif Tissue Int 72:567–576

Gyobu N, Tani K, Hiroaki Y, Kamegawa A, Mitsuoka K, Fujiyoshi Y (2004) Improved specimen preparation for cryo-electron microscopy using a symmetric carbon sandwich technique. J Struct Biol 146:325–333

Hagler HK, Lopez LE, Murphy ME, Greico CA, Buja LM (1981) Quantitative X-ray microanalysis of mitochondrial calcification in damaged myocardium. Lab Invest 45:241–247

Handley CJ, Lowther DA (1976) Inhibition of proteoglycan biosynthesis by hyaluronic acid in chondrocytes in cell culture. Biochim Biophys Acta 444:69–74

Hansson T, Roos B (1986) Age changes in the bone mineral of the lumbar spine in normal women. Calcif Tissue Int 38:249–251

Harell A, Binderman I, Guez M (1976) Tissue culture of bone cells: mineral transport, calcification and hormonal effects. Israel J Med Sci 12:115–123

Hauschka PV, Mavrakos AE, Iafrati MD, Doleman SE, Klagsbrun M (1986) Growth factors in bone matrix. Isolation of multiple types by affinity chromatography on heparin-Sepharose. J Biol Chem 261:12665–12674

Hayashi Y (1992) High resolution electron microscopy in the dentino-enamel junction. J Electron Microsc 41:387–391

Hayashi Y (1993) High resolution electron microscopic study on the human dentine crystal. J Electron Microsc 42:141–146

Hayat MA (1989) Principles and techniques of electron microscopy. Biological applications, 3rd edn. McMillan Press, Houndmills, Basingstoke, Hampshire

Hecker A, Testenière O, Marin F, Luquet G (2003) Phosphorylation of serine residues is fundamental for the calcium-binding ability of Orchestin, a soluble matrix protein from crustacean calcium storage structures. FEBS Lett 535:49–54

Henny GC, Spiegel-Adolf M (1945) X-ray diffraction studies on fish bones. Am J Physiol 144:632–636

Hjerpe A, Engfeldt B (1976) Proteoglycans of dentine and predentine. Calcif Tissue Res 22:173–182

Hoshi K, Ejiri S, Ozawa H (2000) Ultrastructural, cytochemical, and biophysical aspects of mechanisms of bone matrix calcification. Acta Anat Nippon 75:457–465

Hosoya A, Hoshi K, Sahara N, Ninomiya T, Akahane S, Kawamoto T, Ozawa H (2005) Effects of fixation on the immunohistochemical localization of bone matrix proteins in fresh-frozen bone sections. Histochem Cell Biol 123:639–646

Hruska KA, Teitelbaum SL, Kopelman R, Richardson CA, Miller P, Debman J, Martin K, Slatopolsky E (1978) The predictability of the histological features of uremic bone disease by non-invasive techniques. Metab Bone Dis Rel Res 1:39–44

Hsu HH, Camacho NP (1999) Isolation of calcifiable vesicles from human atherosclerotic aortas. Atherosclerosis 143:353–362

Hsu HH, Camacho NP, Sun F, Tawfik O, Aono H (2000) Isolation of calcifiable vesicles from aortas of rabbits fed with high cholesterol diets. Atherosclerosis 153:337–348

Hunziker EB, Herrmann W (1990) Ultrastructure of cartilage. In: Bonucci E, Motta PM (eds) Ultrastructure of skeletal tissues. Kluwer Academic Publishers, Boston, pp 79–109

Hunziker EB, Schenk RK (1984) Cartilage ultrastructure after high pressure freezing, freeze substitution, and low temperature embedding. II. Intercellular matrix ultrastructure – Preservation of proteoglycans in their native state. J Cell Biol 98:277–282

Hunziker EB, Schenk RK (1987) Structural organization of proteoglycans in cartilage. In: Wight TN, Mecham P (eds) Biology of proteoglycans. Academic Press, Orlando (FL), pp 155–185

Ikeda K, Matsumoto T, Morita K, Kurokawa K, Ogata E (1986) Inhibition of in vitro mineralization by aluminum in a clonal osteoblastlike cell line, MC3T3-E1. Calcif Tissue Int 39:319–323

Ippolito E, La Velle S, Pedrini V (1981) The effect of various decalcifying agents on cartilage proteoglycans. Stain Technol 56:367–373

Ippolito E, Pedrini VA, Pedrini-Mille A (1983) Histochemical properties of cartilage proteoglycans. J Histochem Cytochem 31:53–61

Johansen E, Parks HF (1960) Electron microscopic observations on the three dimensional morphology of apatite crystallites of human dentine and bone. J Biophys Biochem Cytol 7:743–746

Jokisch DW, Patton PW, Inglis BA, Bouchet LG, Rajon DA, Rifkin J, Bolch WE (1998) NMR microscopy of trabecular bone and its role in skeletal dosimetry. Health Phys 75:584–596

Jontell M, Linde A (1983) Non-collagenous proteins of predentine from dentinogenically active bovine teeth. Biochem J 214:769–776

Jubb RR, Eggert FM (1981) Staining of demineralised cartilage. II. Quantitation of articular cartilage proteoglycan after fixation and rapid demineralisation. Histochemistry 73:391–396

Kagami A, Takagi M, Hirama M, Sagami Y, Shimada T (1990) Enhanced ultrastructural preservation of cartilage proteoglycans in the extended state. J Histochem Cytochem 38:901–906

Kasten FH (1960) The chemistry of Schiff's reagent. Int Rev Cytol 10:1–100

Kerebel B, Daculsi G, Verbaere A (1976) High-resolution electron microscopy and crystallographic study of some biological apatite. J Ultrastruct Res 57:266–275

Kim H-M, Rey C, Glimcher MJ (1996) X-ray diffraction, electron microscopy, and Fourier transform infrared spectroscopy of apatite crystals isolated from chicken and bovine calcified cartilage. Calcif Tissue Int 59:58–63

Kinney JH, Pople JA, Marshall GW, Marshall SJ (2001a) Collagen orientation and crystallite size in human dentin: a small angle X-ray scattering study. Calcif Tissue Int 69:31–37

Kinney JH, Pople JA, Driessen CH, Breunig TM, Marshall GW, Marshall SJ (2001b) Intrafibrillar mineral may be absent in dentinogenesis imperfecta type II (DI-II). J Dent Res 80:1555–1559

Király K, Lammi M, Arokoski J, Lapveteläinen T, Tammi M, Helminen H, Kiviranta I (1996) Safranin O reduces loss of glycosaminoglycans from bovine articular cartilage during histological specimen preparation. Histochem J 28:99–107

Kiviranta I, Tammi M, Jurvelin J, Säämänen A-M, Helminen HJ (1984) Fixation, decalcification, and tissue processing effects on articular cartilage proteoglycans. Histochemistry 80:569–573

Kuhn LT, Wu Y, Rey C, Gerstenfeld LC, Grynpas MD, Ackerman JL, Kim H-M, Glimcher MJ (2000) Structure, composition, and maturation of newly deposited calcium-phosphate crystals in chicken osteoblast cultures. J Bone Miner Res 15:1301–1309

Kurihara K, Tanaka I, Niimura N, Refai Muslih M, Ostermann A (2004) A new neutron single-crystal diffractometer dedicated for biological macromolecules (BIX-4). J Synchrotron Radiat 11:68–71

Laitala T, Väänänen K (1993) Proton channel part of vacuolar H^+-ATPase and carbonic anhydrase II expression is stimulated in resorbing osteoclasts. J Bone Miner Res 8:119–126
Lakshminarayanan R, Valiyaveettil S, Rao VS, Kini RM (2003) Purification, characterization, and in vitro mineralization studies of a novel goose eggshell matrix protein, ansocalcin. J Biol Chem 278:2928–2936
Landis WJ, Glimcher MJ (1978) Electron diffraction and electron probe microanalysis of the mineral phase of bone tissue prepared by anhydrous techniques. J Ultrastruct Res 63:188–223
Landis WJ, Glimcher MJ (1982) Electron optical and analytical observations of rat growth plate cartilage prepared by ultracryomicrotomy: the failure to detect a mineral phase in matrix vesicles and the identification of heterodispersed particles as the initial solid phase of calcium phosphate deposited in the extracellular matrix. J Ultrastruct Res 78:227–268
Landis WJ, Hauschka BT, Rogerson CA, Glimcher MJ (1977) Electron microscopic observations of bone tissue prepared by ultracryomicrotomy. J Ultrastruct Res 59:185–206
Landis WJ, Paine MC, Glimcher MJ (1980) Use of acrolein vapors for the anhydrous preparation of bone tissue for electron microscopy. J Ultrastruct Res 70:171–180
Landis WJ, Hodgens KJ, McKee MD, Nanci A, Song MJ, Kiyonaga S, Arena J, McEwen B (1992) Extracellular vesicles of calcifying turkey leg tendon characterized by immunocytochemistry and high voltage electron microscopic tomography and 3-D graphic image reconstruction. Bone Mineral 17:237–241
Landis WJ, Song MJ, Leith A, McEwen L, McEwen BF (1993) Mineral and organic matrix interaction in normally calcifying tendon visualized in three dimensions by high-voltage electron microscopic tomography and graphic image reconstruction. J Struct Biol 110:39–54
Landis WJ, Jacquet R, Hillyer J, Zhang J (2003) Analysis of osteopontin in mouse growth plate cartilage by application of laser capture microdissection and RT-PCR. Connect Tissue Res 44:28–32
Laval-Jeantet AM, Chateau JY, Bergot C, Laval-Jeantet M, Kuntz D (1981) Comparison de la radiodensitométrie et de l'histomorphométrie dans l'étude de l'os normal et patologique. Pathol Biol 29:155–161
Leapman RD (2003) Detecting single atoms of calcium and iron in biological structures by electron energy-loss spectrum-imaging. J Microsc 210:5–15
Lee DD, Landis WJ, Glimcher MJ (1986) The solid, calcium-phosphate mineral phases in embryonic chick bone characterized by high-voltage electron diffraction. J Bone Miner Res 1:425–432
Lees S, Hukins DWL (1992) X-ray diffraction by collagen in the fully mineralized cortical bone of cow tibia. Bone Mineral 17:59–63
Lees S, Mook HA (1986) Equatorial diffraction spacing as a function of water content in fully mineralized cow bone determined by neutron diffraction. Calcif Tissue Int 39:291–292
Leng C-G, Yu Y, Ueda H, Terada N, Fujii Y, Ohno S (1998) The ultrastructure of anionic sites in rat articular cartilage as revealed by different preparation methods and polyethyleneimine staining. Histochem J 30:253–261
Levi-Kalisman Y, Falini G, Addadi L, Weiner S (2001) Structure of the nacreous organic matrix of a bivalve mollusk shell examined in the hydrated state using cryo-TEM. J Struct Biol 135:8–17
Lewinson D, Silbermann M (1990) Ultrastructural localization of calcium in normal and pathologic cartilage. In: Bonucci E, Motta PM (eds) Ultrastructure of skeletal tissues. Kluwer Academic Publishers, Boston, pp 129–152

Linde A (1973) Glycosaminoglycans of the odontoblast-predentine layer in dentinogenically active porcine teeth. Calcif Tissue Res 12:281–294

Linde A, Bhown M, Butler WT (1980) Noncollagenous proteins of dentin. A re-examination of proteins from rat incisor dentin utilizing techniques to avoid artifacts. J Biol Chem 255:5931–5942

Lohmander S (1972) Ion exchange chromatography of glucosamine and galactosamine in microgram amounts with quantitative determination and specific radioactivity assay. Biochim Biophys Acta 264:411–417

Loong C-K, Rey C, Kuhn LT, Combes C, Wu Y, Chen SH, Glimcher MJ (2000) Evidence of hydroxyl-ion deficiency in bone apatites: an inelastic neutron-scattering study. Bone 26:599–602

Magne D, Weiss P, Bouler JM, Laboux O, Daculsi G (2001) Study of the maturation of the organic (type I collagen) and mineral (nonstoichiometric apatite) constituents of a calcified tissue (dentin) as a function of location: a Fourier transform infrared microspectroscopic investigation. J Bone Miner Res 16:750–757

Maitland ME, Arsenault AL (1989) Freeze-substitution staining of rat growth plate cartilage with Alcian blue for electron microscopic study of proteoglycans. J Histochem Cytochem 37:383–387

Marin F, Pereira L, Westbroek P (2001) Large-scale fractionation of molluscan shell matrix. Protein Expr Purif 23:175–179

Marinozzi V (1961) Silver impregnation of ultrathin sections for electron microscopy. J Biophys Biochem Cytol 9:121–133

Marinozzi V (1967) Réaction de l'acide phosphotungstique avec la mucine et les glycoprotéines des plasmamembranes. J Microsc 6:68A

Marotti G (1990) The original contribution of the scanning electron microscope to the knowledge of bone structure. In: Bonucci E, Motta PM (eds) Ultrastructure of skeletal tissues. Kluwer Academic Publishers, Boston, pp 19–39

Martin BJ, Thomas S, Greenhill NS, Ryan PA, Davis PF, Stehbens WE (1992) Isolation and purification of extracellular matrix vesicles from blood vessels. Prep Biochem 22:87–103

Mason JT, O'Leary TJ (1991) Effects of formaldehyde fixation on protein secondary structure: a calorimetric and infrared spectrospic investigation. J Histochem Cytochem 39:225–229

Matsushima N, Tokita M, Hikichi K (1986) X-ray determination of the crystallinity in bone mineral. Biochim Biophys Acta 883:574–579

Matthews JL (1970) Ultrastructure of calcifying tissues. Am J Anat 129:451–458

McKee MD, Nanci A (1995) Postembedding colloidal-gold immunocytochemistry of noncollagenous extracellular matrix proteins in mineralized tissues. Microsc Res Technol 31:44–62

McKee MD, Nanci A, Landis WJ, Gotoh Y, Gerstenfeld LC, Glimcher MJ (1991) Effects of fixation and demineralization on the retention of bone phosphoprotein and other matrix components as evaluated by biochemical analyses and quantitative immunocytochemistry. J Bone Miner Res 6:937–945

Miller A, Wray JS (1971) Molecular packing in collagen. Nature 230:437–439

Möbus G, Doole RC, Inkson BJ (2003) Spectroscopic electron microscopy. Ultramicroscopy 96:433–451

Moradian-Oldak J, Weiner S, Addadi L, Landis WJ, Traub W (1991) Electron imaging and diffraction study of individual crystals of bone, mineralized tendon and synthetic carbonate apatite. Connect Tissue Res 25:219–228

Moradian-Oldak J, Gharakhanian N, Jimenez J (2002) Limited proteolysis of amelogenin: toward understanding the proteolytic processes in enamel extracellular matrix. Connect Tissue Res 43:450–455

Murayama E, Takagi Y, Ohira T, Davis JG, Greene MI, Nagasawa H (2002) Fish otolith contains a unique structural protein, otolin-1. Eur J Biochem 269:688–696
Nakano T, Kaibara K, Tabata Y, Nagata N, Enomoto S, Marukawa E, Umakoshi Y (2002) Unique alignment and texture of biological apatite crystallites in typical calcified tissues analyzed by microbeam X-ray diffractometer system. Bone 31:479–487
Nanci A, Kawaguchi H, Kogaya Y (1994) Ultrastructural studies and immunolocalization of enamel proteins in rodent secretory stage ameloblasts processed by various cryofixation methods. Anat Rec 238:425–436
Nanci A, Zalzal S, Lavoie P, Kunikata M, Chen W-Y, Krebsbach PH, Yamada Y, Hammarström L, Simmer JP, Fincham AG, Snead ML, Smith CE (1998) Comparative immunochemical analyses of the developmental expression and distribution of ameloblastin and amelogenin in rat incisors. J Histochem Cytochem 46:911–934
Nanci A, Mocetti P, Sakamoto Y, Kunikata M, Lozupone E, Bonucci E (2000) Morphological and immunocytochemical analyses on the effects of diet-induced hypocalcemia on enamel maturation in the rat incisor. J Histochem Cytochem 48:1043–1057
Neufeld EB, Boskey AL (1994) Strontium alters the complexed acidic phospholipid content of mineralizing tissues. Bone 15:425–430
Ng L, Grodzinsky AJ, Patwari P, Sandy J, Plaas A, Ortiz C (2003) Individual cartilage aggrecan macromolecules and their constituent glycosaminoglycans visualized via atomic force microscopy. J Struct Biol 143:242–257
Nylen MU, Omnell K-Å (1962) The relationship between the apatite crystals and the organic matrix of rat enamel. Fifth International Congress for Electron Microscopy, Academic Press, New York, p QQ-4
O'Farrell PH (1975) High-resolution two-dimensional electrophoresis of proteins. J Biol Chem 250:4007–4021
Ostrowski K, Dziedzic-Goclawska A, Stachowicz W, Michalik J (1972) Sensitivity of the electron spin resonance technique as applied in histochemical research on normal and pathological calcified tissues. Histochemie 32:343–351
Ostrowski K, Dziedzic-Goclawska A, Stachowicz W, Michalik J (1973) Application of electron spin resonance in research on mineralized tissues. Clin Orthop Relat Res 97:213–224
Ostrowski K, Dziedzic-Goclawska A, Sliwowski A, Wojtczak L, Michalik J, Stachowicz W (1975) Analysis of the crystallinity of calcium phosphate deposits in rat liver mitochondria by electron spin resonance spectroscopy. FEBS Lett 60:410–413
Ottensmeyer FP, Arsenault AL (1983) Electron spectroscopic imaging and Z-contrast in tissue sections. Scanning Electron Microsc 4:1867–1875
Paloczi GT, Smith BL, Hansma PK, Walters DA (1995) Rapid imaging of calcite crystal growth using atomic force microscopy with small cantilevers. Appl Phys Lett 73:1658–1660
Parfitt AM, Drezner MK, Glorieux FH, Kanis JA, Malluche H, Meunier PJ, Ott SM, Recker RR (1987) Bone histomorphometry: standardization of nomenclature, symbols, and units. J Bone Miner Res 2:595–610
Parisien M, Cosman F, Morgan D, Schnitzer M, Liang X, Nieves J, Forese L, Luckey M, Meier D, Shen V, Lindsay R, Dempster DW (1997) Histomorphometric assessment of bone mass, structure, and remodeling: a comparison between healthy black and white premenopausal women. J Bone Miner Res 12:948–957
Paschalis EP, DiCarlo E, Betts F, Sherman P, Mendelsohn R, Boskey AL (1996) FTIR microspectroscopic analysis of human osteonal bone. Calcif Tissue Int 59:480–487
Plate U, Höhling HJ, Reimer L, Barckhaus RH, Wienecke R, Wiesmann H-P, Boyde A (1992) Analysis of the calcium distribution in predentine by EELS and of the early crystal formation in dentine by ESI and ESD. J Microsc 166:329–341

Plate U, Arnold S, Reimer L, Höhling H-J, Boyde A (1994) Investigation of the early mineralisation on collagen in dentine of rat incisors by quantitative electron spectroscopic diffraction (ESD). Cell Tissue Res 278:543–547

Pleshko N, Boskey A, Mendelsohn R (1991) Novel infrared spectroscopic method for the determination of crystallinity of hydroxyapatite minerals. Biophys J 60:786–793

Portigliatti Barbos M, Carando S, Ascenzi A, Boyde A (1987) On the structural simmetry of human femurs. Bone 8:165–169

Potter K, Butler JJ, Adams C, Fishbein KW, McFarland EW, Horton WE, Spencer RG (1998) Cartilage formation in a hollow fiber bioreactor studied by proton magnetic resonance microscopy. Matrix Biol 17:513–523

Potter K, Landis WJ, Spencer RGS (2001) Histomorphometry of the embryonic avian growth plate by proton nuclear magnetic resonance microscopy. J Bone Miner Res 16:1092–1100

Potter K, Leapman RD, Basser PJ, Landis WJ (2002) Cartilage calcification studied by proton nuclear magnetic resonance microscopy. J Bone Miner Res 17:652–660

Pourcho RG, Bernstein MH, Gould SF (1978) Malachite green: application in electron microscopy. Stain Technol 53:29–35

Pugliarello MC, Vittur F, de Bernard B, Bonucci E, Ascenzi A (1970) Chemical modifications in osteones during calcification. Calcif Tissue Res 5:108–114

Quintarelli G, Scott JE, Dellovo MC (1964) The chemical and histochemical properties of alcian blue II. Dye binding of tissue polyanions. Histochemie 4:86–98

Rambourg A (1971) Morphological and histochemical aspects of glycoproteins at the surface of animal cells. Int Rev Cytol 31:57–114

Raspanti M, Alessandrini A, Ottani V, Ruggeri A (1997) Direct visualization of collagen-bound proteoglycans by tapping-mode atomic force microscopy. J Struct Biol 119:118–122

Reale E, Luciano L (1970) Fixierung mit Aldehyden. Ihre Eignung für histologische und histochemische Untersuchungen in der Licht- und Elektronenmikroskopie. Histochemie 23:144–170

Reich Z, Kapon R, Nevo R, Pilpel Y, Zmora S, Scolnik Y (2001) Scanning force microscopy in the applied biological sciences. Biotechnol Adv 19:451–485

Reimer L, Fromm I, Naundorf I (1990) Electron spectroscopic diffraction. Ultramicroscopy 32:80–91

Reith EJ (1968) Ultrastructural aspects of dentinogenesis. In: Symons NBB (ed) Dentine and pulp: their structure and reactions. Livingstone, Edinburgh, pp 19–41

Rey C, Beshah K, Griffin R, Glimcher MJ (1991) Structural studies of the mineral phase of calcifying cartilage. J Bone Miner Res 6:515–525

Riminucci M, Silvestrini G, Bonucci E, Fisher LW, Gehron Robey P, Bianco P (1995) The anatomy of bone sialoprotein immunoreactive sites in bone as revealed by combined ultrastructural histochemistry and immunohistochemistry. Calcif Tissue Int 57:277–284

Robey PG (1992) Cell-mediated calcification in vitro. In: Bonucci E (ed) Calcification in biological systems. CRC Press, Boca Raton, pp 107–127

Robinson RA (1952) An electron-microscopic study of the crystalline inorganic component of bone and its relationship to the organic matrix. J Bone Joint Surg 34-A:389–434

Rönnholm E (1962) III. The structure of the organic stroma of human enamel during amelogenesis. J Ultrastruct Res 3:368–389

Roseberry HH, Hastings AB, Morse JK (1931) X-ray analysis of bone and teeth. J Biol Chem 90:335–407

Roufosse A, Richelle LJ, Gilliam OR (1976) Electron spin resonance of organic free radicals in dental enamel and other calcified tissues. Archs Oral Biol 21:227–232

Ryan KP (1992) Cryofixation of tissues for electron microscopy: a review of plunge cooling methods. Scanning Microsc 6:715–743

Sakae T (1988) X-ray diffraction and thermal studies of crystals from the outer and inner layers of human dental enamel. Arch Oral Biol 33:707–713
Saleuddin ASM, Wilbur KM (1969) Shell regeneration in *Helix pomatia*. Canad J Zool 47:51–53
Santos NC, Castanho MARB (2004) An overview of the biophysical applications of atomic force microscopy. Biophys Chem 107:133–149
Schajowicz F, Cabrini RL (1955) The effect of acids (decalcifying solutions) and enzymes on the histochemical behaviour of bone and cartilage. J Histochem Cytochem 3:122–129
Schajowicz F, Cabrini RL (1958) Histochemical studies on glycogen in normal ossification and calcification. J Bone Joint Surg 40-A:1081–1092
Schenk RK (1974) Ultrastruktur des Knochens. Verh Dtsch Ges Pathol 58:72–83
Scherft JP (1972) The lamina limitans of the organic matrix of calcified cartilage and bone. J Ultrastruct Res 38:318–331
Schmidt WJ (1952) Zur Polarisationsoptik des Knorpelkewebes. Z Zellforsch 37:534–546
Schraer H, Gay CV (1977) Matrix vesicles in newly synthesizing bone observed after ultracryotomy and ultramicroincineration. Calcif Tissue Res 23:185–188
Scott JE, Kyffin TW (1978) Demineralization in organic solvents by alkylammonium salts of ethylenediaminetetra-acetic acid. Biochem J 169:697–701
Selvig KA (1973) Electron microscopy of dental enamel: analysis of crystal lattice images. Z Zellforsch 137:271–280
Septier D, Hall RC, Lloyd D, Embery G, Goldberg M (1998) Quantitative immunohistochemical evidence of a functional gradient of chondroitin 4-sulphate/dermatan sulphate, developmentally regulated in the predentine of rat incisor. Histochem J 30:275–284
Shapiro IM (1970a) The phospholipids of mineralized tissues I. Mammalian compact bone. Calcif Tissue Res 5:21–29
Shapiro IM (1970b) The phospholipids of mineralized tissues II. Elasmobranch and teleost skeletal tissues. Calcif Tissue Res 5:30–38
Shepard N (1992) Role of proteoglycans in calcification. In: Bonucci E (ed) Calcification in biological systems. CRC Press, Boca Raton, pp 41–58
Shepard N, Mitchell N (1976a) Simultaneous localization of proteoglycan by light and electron microscopy using toluidine blue O. A study of epiphyseal cartilage. J Histochem Cytochem 24:621–629
Shepard N, Mitchell N (1976b) The localization of proteoglycan by light and electron microscopy using safranin O. A study of epiphyseal cartilage. J Ultrastruct Res 54:451–460
Shibata S, Fukada K, Suzuki S, Ogawa T, Yamashita Y (2002) In situ hybridization and immunohistochemistry of bone sialoprotein and secreted phosphoprotein 1 (osteopontin) in the developing mouse mandibular condylar cartilage compared with limb bud cartilage. J Anat 200:309–320
Siedlecki CA, Marchant RE (1998) Atomic force microscopy for characterization of the biomaterial interface. Biomaterials 19:441–454
Silverman L, Boskey AL (2004) Diffusion systems for evaluation of biomineralization. Calcif Tissue Int E-pub ahead of print
Silvestrini G, Mocetti P, Di Grezia R, Berni S, Bonucci E (2003) Localization of the glucocorticoid receptor mRNA in cartilage and bone cells of the rat. An in situ hybridization study. Eur J Histochem 47:245–252
Simmons DJ, Kent GN, Jilka RL, Scott DM, Fallon M, Cohn DV (1982) Formation of bone by isolated, cultured osteoblasts in millipore diffusion chambers. Calcif Tissue Int 34:291–294

Slavkin HC, Croissant R, Bringas P Jr (1972) Epithelial-mesenchymal interactions during odontogenesis. III. A simple method for the isolation of matrix vesicles. J Cell Biol 53:841–849

Smeyers-Verbeke J, Verbeelen D (1985) Determination of aluminum in bone by atomic absorption spectroscopy. Clin Chem 31:1172–1174

Smith AJ, Leaver AG (1981) Distribution of the EDTA-soluble non-collagenous organic matrix components of rabbit incisor dentine. Arch Oral Biol 26:643–649

Sodek KL, Tupy JH, Sodek J, Grynpas MD (2000) Relationships between bone protein and mineral in developing porcine long bone and calvaria. Bone 26:189–198

Sodhi RNS (1996) Application of surface analytical and modification techniques to biomaterial research. J Electron Spectrosc Relat Phenom 81:269–284

Stamps AC, Terrett JA, Adam PJ (2003) Application of in situ reverse transcriptase-polymerase chain reaction (RT-PCR) to tissue microarrays. J Nanobiotechnol 1:3–7

Steve-Bocciarelli D (1973) Apatite microcrystals in bone and dentine. J Microsc (Paris) 16:21–34

Stroh C, Wang H, Bash R, Ashcroft B, Nelson J, Gruber H, Lohr D, Lindsay SM, Hinterdorfer P (2004) Single-molecule recognition imaging microscopy. Proc Natl Acad Sci USA 101:12503–12507

Su X, Sun K, Cui FZ, Landis WJ (2003) Organization of apatite crystals in human woven bone. Bone 32:150–162

Szirmai JA (1963) Quantitative approaches in the histochemistry of mucopolysaccharides. J Histochem Cytochem 11:24–34

Takagi M (1990) Ultrastructural cytochemistry of cartilage proteoglycans and their relation to the calcification process. In: Bonucci E, Motta PM (eds) Ultrastructure of skeletal tissues. Kluwer Academic Publishers, Boston, pp 111–127

Takagi M, Parmley RT, Toda Y, Austin RL (1982) Ultrastructural cytochemistry and immunocytochemistry of sulfated glycosamiglycans in epiphyseal cartilage. J Histochem Cytochem 30:1179–1185

Takagi M, Parmley RT, Denys FR (1983) Ultrastructural cytochemistry and immunocytochemistry of proteoglycans associated with epiphyseal cartilage calcification. J Histochem Cytochem 31:1089–1100

Takagi M, Hishikawa H, Hosokawa Y, Kagami A, Rahemtulla F (1990) Immunohistochemical localization of glycosaminoglycans and proteoglycans in predentin and dentin of rat incisors. J Histochem Cytochem 38:319–324

Tarnowski CP, Ignelzi MA Jr, Wang W, Taboas JM, Goldstein SA, Morris MD (2004) Earliest mineral and matrix changes in force-induced muskuloskeletal disease as revealed by Raman microspectroscopic imaging. J Bone Miner Res 19:64

Termine JD (1981) Integral matrix proteins of fetal bone. In: Ascenzi A, Bonucci E, de Bernard B (eds) Matrix vesicles. Wichtig, Milan, pp 155–159

Termine JD, Eanes ED (1972) Comparative chemistry of amorphous and apatitic calcium phosphate preparations. Calcif Tissue Res 10:171–197

Termine JD, Eanes ED, Greenfield DJ, Nylen MU (1973) Hydrazine-deproteinated bone mineral. Physical and chemical properties. Calcif Tissue Res 12:73–90

Termine JD, Belcourt AB, Christner PJ, Conn KM, Nylen MU (1980a) Properties of dissociatively extracted fetal tooth matrix proteins. I. Principal molecular species in developing bovine enamel. J Biol Chem 255:9760–9768

Termine JD, Belcourt AB, Miyamoto MS, Conn KM (1980b) Properties of dissociatively extracted fetal tooth matrix proteins. II. Separation and modification of fetal bovine dentin phosphoprotein. J Biol Chem 255:9769–9772

Termine JD, Belcourt AB, Conn KM, Kleinman HK (1981) Mineral and collagen-binding proteins of fetal calf bone. J Biol Chem 256:10403–10408

Thomas JM, Midgley PA (2004) High-resolution transmission electron microscopy: the ultimate nanoanalytical technique. Chem Commun (Camb) June 7:1253–1267
Tiselius A, Hjertén S, Levin Ö (1956) Protein chromatography on calcium phosphate columns. Arch Biochem Biophys 65:132–165
Tong W, Glimcher MJ, Katz JL, Kuhn L, Eppell SJ (2003) Size and shape of mineralites in young bovine bone measured by atomic force microscopy. Calcif Tissue Int 72:592–598
Travis DF (1968) Comparative ultrastructure and organization of inorganic crystals and organic matrices of mineralized tissues. Biology of the mouth. American Association for the Advancement of Sciences, Washington, pp 237–297
Travis DF, Gonsalves M (1969) Comparative ultrastructure and organization of the prismatic region of two bivalves and its possible relation to the chemical mechanism of boring. Am Zool 9:635–661
Trudel G, Seki M, Uhthoff HK (2000) Optimization of perfusion decalcification for bones and joints in rats. Anat Rec 260:222–227
Ueda H, Toriumi H, Leng C-G, Ohno S (1997) A histochemical study of anionic sites in the intermediate layer of rat femoral cartilage using polyethyleneimine at different pH levels. Histochem J 29:617–624
Ueda H, Baba T, Toriumi H, Ohno S (2001) Anionic sites in articular cartilage revealed by polyethyleneimine staining. Micron 32:439–446
Vigliani F, Marotti F (1963) Modificazioni microradiografiche ed istochimiche del tessuto osseo nella decalcificazione con Na_4EDTA. Clin Ortop 15:521–534
Wachtel E, Weiner S (1994) Small-angle X-ray scattering study of dispersed crystals from bone and tendon. J Bone Miner Res 9:1651–1655
Warshawsky H (1985) Ultrastructural studies on amelogenesis. In: Butler HT (ed) The chemistry and biology of mineralized tissues. EBSCO Media, Birmingham, pp 33–45
Warshawsky H, Moore G (1967) A technique for the fixation and decalcification of rat incisors for electron microscopy. J Histochem Cytochem 15:542–549
Weatherford TW, Mann WV Jr (1973) A method of microscopic evaluation of effects of demineralizers on complex carbohydrates of rat tissues. J Microsc 99:91–100
Wehrli FW, Ford JC, Chung HW, Wehrli SL, Williams JL, Grimm MJ, Kugelmass SD, Jara H (1993) Potential role of nuclear magnetic resonance for the evaluation of trabecular bone quality. Calcif Tissue Int 53:S162-S169
Weinbach EC, von Brand T (1965) The isolation and composition of dense granules from Ca^{++}-loaded mitochondria. Biochem Biophys Res Comm 19:133–136
Weiner S, Price PA (1986) Disaggregation of bone into crystals. Calcif Tissue Int 39:365–375
Weiner S, Traub W (1980) X-ray diffraction study of the insoluble organic matrix of mollusk shells. FEBS Lett 111:311–316
Weiner S, Traub W (1989) Crystal size and organization in bone. Connect Tissue Res 21:259–265
Weiner S, Traub W (1991) Organization of crystals in bone. In: Suga S, Nakahara H (eds) Mechanisms and phylogeny of mineralization in biological systems. Springer, Tokyo, pp 247–253
Weiner S, Arad T, Traub W (1991) Crystal organization in rat bone lamellae. Fed Eur Biochem Soc 285:49–54
Wenk H-R, Heidelbach F (1999) Crystal alignment of carbonated apatite in bone and calcified tendon: results from quantitative texture analysis. Bone 24:361–369
White SW, Hulmes DJS, Miller A, Timmins PA (1977) Collagen-mineral axial relationship in calcified turkey leg tendon by X-ray and neutron diffraction. Nature 266:421–425
Wiesmann H-P, Plate U, Höhling H-J, Barckhaus RH, Zierold K (1993) Analysis of early hard tissue formation in dentine by energy dispersive X-ray microanalysis and energy filtering transmission electron microscopy. Scann Microsc 7:711–718

Wu Y, Glimcher MJ, Rey C, Ackerman JL (1994) A unique protonated phosphate group in bone mineral not present in synthetic calcium phosphates. Identification by phosphorus-31 solid state NMR spectroscopy. J Mol Biol 244:423–435

Wu Y, Chesler DA, Glimcher MJ, Garrido L, Wang J, Jiang HJ, Ackerman JL (2000a) Multinuclear solid-state three-dimensional MRI of bone and synthetic calcium phosphates. Proc Natl Acad Sci USA 96:1574–1578

Wu Y, Ackerman JL, Chesler DA, Li J, Neer RM, Wang J, Glimcher MJ (2000b) Evaluation of bone mineral density using three-dimensional solid state phosphorus-31 NMR projection imaging. Calcif Tissue Int 62:512–518

Wu Y, Ackerman JL, Kim H-M, Rey C, Barroug A, Glimcher MJ (2002) Nuclear magnetic resonance spin-spin relaxation of the crystals of bone, dental enamel, and synthetic hydroxyapatites. J Bone Miner Res 17:472–480

Wuthier RE (1975) Lipid composition of isolated epiphyseal cartilage cells, membranes and matrix vesicles. Biochim Biophys Acta 409:128–143

Wuthier RE, Linder RE, Warner GP, Gore ST, Borg TK (1978) Non-enzymatic isolation of matrix vesicles: characterization and initial studies on ^{45}Ca and ^{32}P-orthophosphate metabolism. Metab Bone Dis Rel Res 1:125–136

Wuthier RE, Chin JE, Hale JE, Register TC, Hale LV, Ishikawa Y (1985) Isolation and characterization of calcium-accumulating matrix vesicles from chondrocytes of chicken epiphyseal growth plate cartilage in primary culture. J Biol Chem 260:15972–15979

Wuthrich K (1989) Protein structure determination in solution by nuclear magnetic resonance spectroscopy. Science 243:45–50

Yamauchi M, Uzawa K, Katz EP, Lopes MM, Verdelis K, Cheng H (2000) Distribution of small keratan sulfate proteoglycans in predentin and dentin. In: Goldberg M, Boskey A, Robinson C (eds) Chemistry and biology of mineralized tissues. American Academy of Orthopaedic Surgeons, Rosemont, pp 305–310

Zanini F, Lausi A, Savoia A (1999) The beamlines of ELETTRA and their application to structural biology. Genetica 106:171–180

Ziv V, Weiner S (1994) Bone crystal sizes: a comparison of transmission electron microscopic and X-ray diffraction line width broadening techniques. Connect Tissue Res 30:165–175

Zizak I, Roschger P, Paris O, Misof BM, Berzlanovich A, Bernstorff S, Amenitsch H, Klaushofer K, Fratzl P (2003) Characteristics of mineral particles in the human bone/cartilage interface. J Struct Biol 141:208–217

4 The Nature and Composition of the Inorganic Phase

4.1 Introduction

The nature and composition of the inorganic phase in calcified biological tissues were first studied using X-ray diffraction and other physical techniques. These were mainly applied to the study of compact bone, probably because the sheer compactness of this calcified biological tissue compared with all others makes it easier to prepare specimens for physical examination. The results obtained from compact bone have often been extrapolated to cover other calcified tissues, but it must be stressed that clear-cut differences can be found between different minerals, and even for the same type of mineral in different calcified tissues. This is shown, for example, by comparing calcium phosphate in bone, calcified cartilage and enamel, or calcium carbonate as it occurs in various lower organisms. Boyan et al. (1992) reported that as many as 31 biogenic mineral compounds, most of them including calcium, are distributed across all phyla, and Mann (1988) found that over 40 different minerals can be identified in living organisms. Most of these minerals are calcium salts, but other ions can contribute to the formation of intra- and extracellular inorganic substance. Iron can form magnetite in some bacteria and in the teeth of some mollusks (limpets and chitons), barium can be found as sulfate in the organules of some desmids, strontium contributes to the formation of spines in *acantharians*, and silica is found in unicellular organisms and plants (Mann 2001). It is just the variability of the inorganic substance composition that lead many to prefer the term mineralization instead of calcification; this topic has been discussed in Sect. 2.4.

Of all these mineral compounds, those which form the bulk of, and are found most often in, inorganic substances in the animal kingdom will be considered in the following pages, with special reference to calcium phosphate and calcium carbonate. Silicates present in higher and lower plants and in terrestrial and marine organisms are not discussed; several articles and chapters of books can be consulted on that topic (Arnott and Pautard 1970; Arnott 1976; Leadbeater and Riding 1986; Mann 2001; Perry 2003; Sumerel and Morse 2003; Foo et al. 2004).

4.2 Vertebrates

The normally calcified tissues of vertebrates are those of the skeleton (bone and growth cartilage), of teeth (dentin, cementum, enamel) and, as an aging process, some cartilaginous segments like the rib, trachea and larynx cartilages, and tendons in some animals.

4.2.1 Bone

The pioneering studies of deJong (1926) and Roseberry et al. (1931) were the first to show that the inorganic substance of bone consists of very small particles whose X-ray diffractograms resemble those of polycrystalline samples of natural hydroxyapatite. This is the most likely reason why inorganic particles came to be called "crystals" or "crystallites", even in referring to developing bone salt which, as noted by Arnott and Pautard (1967), is usually said to consist of crystals, without any proof being offered that any portion of the area is specifically crystalline. One of the conclusions of the in vitro studies of Eanes and Meyer (1977) was that the first crystals formed in solution show marked divergences from apatite in their morphology, composition, structure, and solubility, and that only with maturation do they take on apatite-like features. Setting this problem aside for a moment, the term "crystals" will be used from now on to refer to the inorganic particles found in calcifying and calcified areas, independently of their degree of crystallinity and without any strictly crystallographic implications. It must be stated, however, that, as outlined by Eanes and Posner (1970) in their review of the structure and chemistry of inorganic substance in bone, it is generally accepted that in bone the inorganic substance is crystalline – irrespective of its stage of formation – and that it is apatitic in structure and composition, a conclusion supported by a number of X-ray diffraction studies and selected area electron diffraction (Lefèvre et al. 1937; Bale 1940; Carlström 1955; Trautz 1955; Wallgren 1957; Urist and Dowell 1967; Lénárt et al. 1968, 1971, 1979; Landis and Glimcher 1978). The inorganic particles of bone have been considered to be crystals; this is the most likely reason why they have been regarded from a strictly mineralogical standpoint as structures possessing the same atomic configuration as hydroxyapatite, whose unit cell is a rhombic prism with an *a*-axis of 943 nm and a *c*-axis of 688 nm, and with a formula of $Ca_{10}(PO_4)_6(OH)_2$ (often simplified as $Ca_5(PO_4)_3OH$; see Posner 1987).

The inorganic ions which go to form bone crystals are not, however, calcium and phosphate alone, as the formula reported above seems to suggest. Carbonate, which accounts for about 5% of the total weight of bone ash, is found together with magnesium, sodium, potassium, and other ions, even if their exact position within the crystal unit is not known (Eanes and Posner 1970). The Ca/P molar ratio is itself variable; it ranges between 1.57 and 1.71 (Woodard 1962), and therefore only rarely reaches the theoretical value of 1.67 the bone inorganic substance should have, if the formula given above is correct (Mellors 1964; Glimcher 1990). On the other hand, Zipkin (1970), from an analysis of the literature, reported that the Ca/P values, given as weight ratio, range between 2.09 and 2.25 in adult human bone, and that values from 1.82 to 1.98 are specific for fetal human bone. Variations may depend on the fact that the Ca/P molar ratio increases from a value of 1.35 in the earliest inorganic deposits (osteoid calcification nodules) to 1.60 in the heavily calcified regions (Wergedal and Baylink 1974) or, according to Landis and Glimcher (1978), from 1.60–1.70 to 1.81–1.97.

Several theories have been put forward to explain the non-stoichiometry of bone hydroxyapatite, such as lattice substitutions, excess ion adsorption on the crystal surface, deficiency of calcium ions, and the addition of a second phase (Posner 1969, 1987; Eanes and Posner 1970; Elliott 1973). Apart from its non-stoichiometry, bone apatite differs from natural apatites in leading to the formation of pyrophosphate when heated to 200–600 °C, in containing some tightly-bound water of crystallization, and in being associated with considerable amounts of extraneous ions (Glimcher and Krane 1968). Several different constituent ions may, in fact, be substituted in the crystal lattice of bone apatite – to give three examples, hydroxyl or phosphate ions may be replaced by carbonate ions, hydroxyl ions by chloride or fluoride ions, and calcium ions by magnesium ions (McConnell 1952; Trautz 1955; François and Herman 1961; Posner 1969, 1987; Elliott 1973; Young 1974). Strontium, barium, zinc, iron, lead, aluminum, bromine are common trace elements in biological apatites (Posner 1987; Zylberberg et al. 1992). As a result, the values of the *a* and *c* axes within the crystal unit cell, as well as the *a*/*c* ratio, may change with the type of substituting ions (Smith and Smith 1976).

It can be concluded that bone "hydroxyapatite" is by no means identical with natural hydroxyapatite. The various formulas that have been suggested for the inorganic phase of bone, as well as the inorganic composition of adult human bone, dentin and enamel, have been reviewed by Zipkin (1970). According to Arnold et al. (2001), the primary crystals of developing hard tissues (bone, dentin, enamel) are apatitic but their crystal lattice may contain so many distortions that they should be viewed as belonging to a state intermediate between amorphous and crystalline; in other words, they have a paracrystalline character, comparable with biopolymers. These lattice fluctuations appear to decrease with the loss of organic material in the matrix and with crystal maturation. Wheeler and Lewis (1977) too reported on the paracrystalline nature of bone apatite: they calculated that the paracrystalline mean distance fluctuations are 1.5 and 2.9% for the basal and prism planes, respectively, and that the corresponding paracrystalline sizes are 2.2 and 0.7 nm. They also observed that heating above 600 °C increases the degree of crystalline regularity. As discussed below, it is known that at this temperature, or above, the structural characteristics of bone crystals are completely subverted (Fig. 4.1).

The structure and composition of crystals can vary. They become more apatite-like with age and maturation (Smith and Smith 1976; Eanes and Meyer 1977), even if the high degree of crystallinity of natural or synthetic apatites is never approached. X-ray diffraction and infrared spectroscopy concur in indicating that the degree of crystallinity increases with bone age (Posner et al. 1965; Burnell et al. 1980). Bonar et al. (1983) used density fractionation to reduce the heterogeneity of bone and select bone fractions of increasing density, hence of increasing degree of calcification and, presumably, tissue age; they confirmed that the degree of crystallinity increases with tissue age, and therefore with mineral age, as well as animal age – a clear indication that changes in bone inorganic substance occur even after calcification is complete

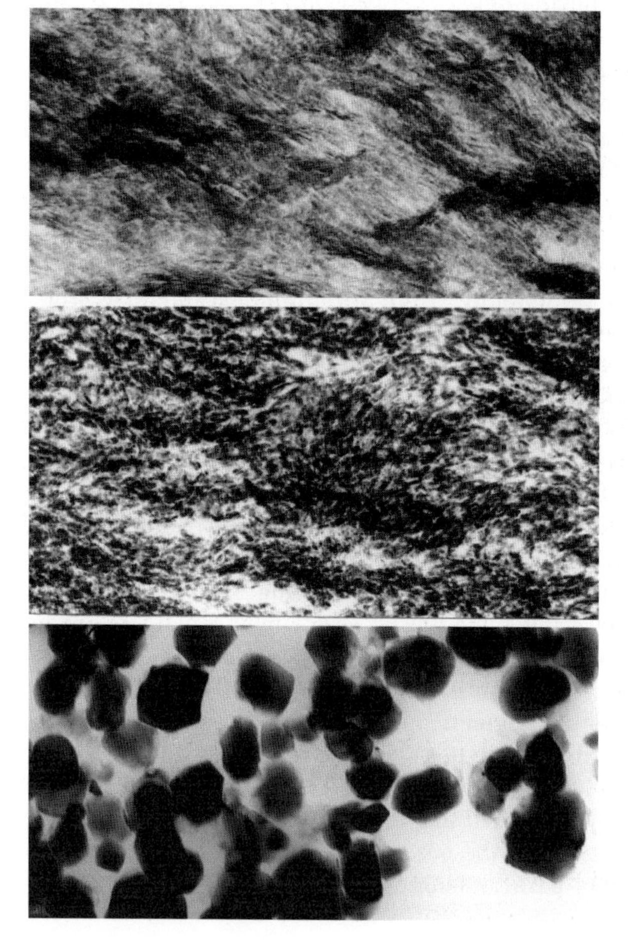

Fig. 4.1. Ultrastructural picture of natural compact ox bone (*above*), and of the same type of bone after heating at 350–550 °C (*middle*) and at over 650 °C (*below*). Untreated, ×60,000

or nearly complete. Wu et al. (2002), by ^{31}P solid state nuclear magnetic resonance spin-spin relaxation studies, found that, in bone, enamel and synthetic hydroxyapatite crystals, a significant fraction of the protonated HPO_4^{2-} has a superficial location, while the unprotonated PO_4^{3-} is concentrated within the apatite lattice. Ascenzi et al. (1977) found that in isolated osteons at different degrees of calcification the crystallinity coefficient – defined as the ratio of the number of radiation-induced paramagnetic defects in the crystal lattice of hydroxyapatite to the total ash content – increases with the degree of matrix calcification, i. e., with mineral age. They reported that human osteons at the initial stage of calcification contained 57% ash when their crystallinity coefficient was 40.6, whereas human fully calcified osteons contained 60% ash when their crystallinity coefficient was 52.1.

For all these reasons, the composition of bone crystals has been the topic of a large number of investigations, which have not, however, led to a definitive solution. It has gradually become clear that although the close similarity between the bone inorganic substance and hydroxyapatite is undeniable, the unit cell dimensions of bone crystals have been given a range of different values (Caglioti 1935; Gruner et al. 1937; Stuehler 1937; Bale 1940; Hendricks and Hill 1942) and their true atomic configuration is still an unresolved issue.

It was realized by several investigators that the main cause of confusion was due to the broad, fuzzy reflections of the X-ray diffraction of bone, caused by the extreme smallness of the crystals and their haphazard orientation (reviewed by Engström 1960). The fact that crystals are mixed with the components of the organic matrix does not seem to account for the low degree of resolution of the diffraction rings, because their appearance is not improved by hydrazine deproteinization (Termine et al. 1973). What can be said is that the lack of order in crystal structure attributable to the presence of carbonate could be partly responsible for lowering the resolution of the X-ray diffraction (Blumenthal et al. 1975). It must be added that X-ray diffractograms of the various isomorphic types of apatite (hydroxyapatite, carbonatoapatite, fluorapatite, oxyapatite, and so on) are so similar (Hirschman et al. 1953) that it is very hard, or impossible, to distinguish between them. Neuman and Neuman (1953) clearly stated that any preparation of basic calcium phosphate with a Ca/P ratio between 1.33 and 2.0 shows the X-ray diffraction of the apatite lattice, and that this should not be considered a compound but a spatial arrangement of atoms common to a number of minerals. The difficulty is increased by the fuzziness of the bone diffraction reflections. It has been reported that sharpness can be increased by heating bone to 500–600 °C, which produces sharp apatite diffractograms (Carlström and Finean 1954), but this procedure turns the inorganic bone substance into a carbonatoapatite (Dallemagne and Brasseur 1942) and induces changes in its molecular arrangement (Fig. 4.1; Bonucci and Graziani 1976; Sakae 1988; Rogers and Daniels 2002). It is striking that the diffuse diffraction lines of developing enamel crystals become sharper as calcification progresses, indicating crystal growth, or even perfection (Nylen et al. 1963).

To understand fully the many difficulties connected with the study of inorganic substance in biological hard tissues, it must also be borne in mind that phase transitions may be induced by the technical procedures used to study the tissue, so that a lack of one substance may simply be due to its transformation into another. As reported above, the composition of the inorganic substance can change, especially with reference to concentrations of carbonate and magnesium (Zipkin 1970), to age and crystal maturation, as well as to species and types of hard tissue. X-ray diffractograms, therefore, cannot differentiate between calcified and uncalcified matrix, or even between the bulk of mature crystals and the immature ones found during the earliest stages of calcification, which make up a very small proportion of a hard tissue.

Further problems arise from the possibility that, before hydroxyapatite forms, there may be a precursor. In this connection, brushite ($CaHPO_4 \cdot 2H_2O$) has been suggested as a substance that could undergo hydrolysis to hydroxyapatite (Neuman and Neuman 1958), but the observation has been considered an outcome of the preparation procedure (Betts et al. 1979). The same type of transformation could be applied to amorphous calcium phosphate, as reviewed several times (Posner 1969; Eanes and Posner 1970; Elliott 1973; Eanes 1975, 1992; Posner et al. 1978). Actually, amorphous calcium phosphate has been found to be present for a short period in in vitro preparations before its autocatalytic (Goldberg and Septier 1985) transformation into crystalline apatite through the formation of octocalcium phosphate (Eanes et al. 1965; Eanes and Posner 1965). It has also been found in vivo (Eanes et al. 1967; Termine and Posner 1967; Termine and Eanes 1972; Termine et al. 1973), and Harper and Posner (1966) have reported that inorganic bone substance from mature cows, humans and rats contains as much as 30–40% of non-crystalline calcium phosphate. Grynpas et al. (1984) could not, however, find any solid phase of amorphous calcium phosphate in embryonic or posthatch chick bone. In spite of the many in vivo and in vitro studies dedicated to this topic (Termine et al. 1970; Boskey and Posner 1974), various doubts persist, especially about the way amorphous calcium phosphate could subsist for long in bone without spontaneously converting into crystalline apatite. Although it might conceivably be stabilized by phosphatidylserine and other acidic phospholipids (Wuthier and Eanes 1975), magnesium (Boskey and Posner 1974), pyrophosphate (Fleisch et al. 1968) and other factors (Blumenthal et al. 1977), not one of these has actually been identified as a physiological regulator (Wuthier et al. 1972). One explanatory hypothesis is that amorphous calcium phosphate is only very transitorily present during the earliest stage of calcification (Posner and Betts 1975); a second is that the X-ray diffractograms interpreted as pointing to the presence of an amorphous phase in bone might have been misunderstood (Posner et al. 1975).

Amorphous or very fine granular material has been noted several times under the electron microscope, especially in areas of initial calcification (Fig. 4.2), but it is not known whether that material can be identified as amorphous cal-

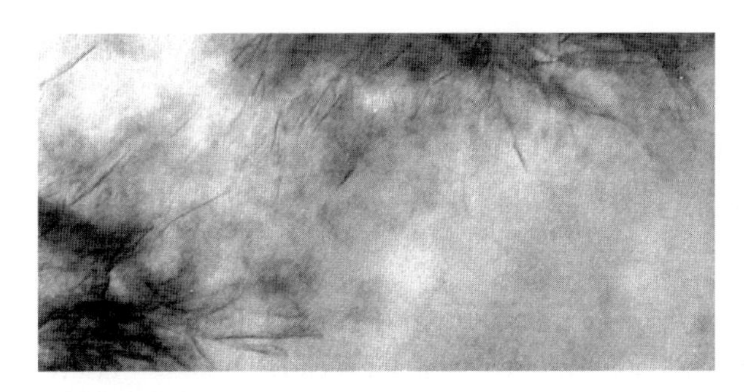

Fig. 4.2. Calcification front in embryonic bone: detail of two calcification nodules and of amorphous, electron-dense substance in between; note early filament-like crystals. Untreated, ×210,000

cium phosphate. One possibility is that at least some of these findings (Molnar 1959; Fitton Jackson and Randall 1956; Fitton Jackson 1957) arose from artefacts, because they have never been replicated. In any case, there have been several reports of an amorphous or finely granular, electron-dense background in areas of initial bone calcification (Ascenzi and Benedetti 1959; Hancox and Boothroyd 1965; Ascenzi and Bonucci 1966; Gay 1977; Bonucci 2002), where Landis and Glimcher (1978) did not observe any electron diffraction pattern. Ascenzi et al. (1965) and Ascenzi and Bonucci (1966) reported that the formation of bone crystals is due to the end-to-end fusion of nanocrystals. According to Höhling and coworkers (Höhling et al. 1971a, b, 1990, 1995; Höhling 1989a), early crystals appear as rows of dots which later fuse into crystals (see Sects. 4.2.2, 4.2.3 and 18.4).

It should be borne in mind that, according to the mineralization principles, the transformation of amorphous calcium phosphate into hydroxyapatite should involve the production of octocalcium phosphate, which is the true precursor phase of hydroxyapatite (Brown and Chow 1976) and should, therefore, necessarily be found in areas of initial calcification. Its presence has, indeed, been reported in bone by Münzenberg and Gebhardt (1973). The topic has been discussed by Elliott (1973), who concluded that the diffractograms do not authorize either the exclusion or the admission of the presence of octocalcium phosphate in bone.

Bigi et al. (1997) make the case that the crystalline phase of both cortical (i. e., compact) and trabecular (i. e., spongy) bone consists of poorly crystalline carbonated apatite, and that this shows a lower Ca/P molar ratio and carbonate content and a greater thermal conversion to β-tricalcium phosphate in trabecular than in cortical bone.

4.2.2
Cartilage, Dentin, Cementum, Tendon

All that is known about the nature and composition of bone crystals can be applied to crystals in cartilage (Urist and Dowell 1967), dentin (Steve-Bocciarelli 1973; Posner and Tannenbaum 2000), cementum (D'Errico et al. 1995) and calcified tendons (Clark 1931). The earliest cartilage crystals, however, have a Ca/P ratio lower than that of lamellar bone (Urist and Dowell 1967), and amorphous, electron-dense areas have been described as a background to cartilage calcification nodules (Bonucci 2002). Structural studies have shown that the early cartilage mineral phase is very poorly crystalline and rich in unstable nonapatitic phosphate ions (Rey et al. 1991). The electron energy-loss spectroscopy studies of Plate et al. (1992) have shown that the calcium content at the predentin-dentin border is higher than in the middle region of predentin, and the electron spectroscopic diffraction studies by the same Authors have pointed to the presence of lattice defects in early crystallites, which appear as chains of dots. This dotted structure of early dentin crystals, very like that described in early bone (Sect. 4.2.1) and enamel (Sect. 4.2.3) crystals, has been reported several times by Höhling et al. (1971a, b, 1995, 1997). Mature dentin crystals have a Ca/P ratio, given as weight ratio, of 2.01–2.07 (Zipkin 1970). Peritubular dentin, i. e., the thin cylinder of calcified matrix which is interposed between the intertubular dentin and the odontoblast processes, shows calcium, phosphate and magnesium ion levels higher than those of intertubular dentin (see review by Jones and Boyde 1984).

4.2.3 Enamel

The composition of enamel crystals resembles that of bone crystals (Nylen et al. 1963; Cuisinier et al. 1992), with Ca/P values of 2.08–2.15 (molar ratio = 1.62; Zipkin 1970). Patel and Brown (cited by Eanes 1979) reported a Ca/P value of 2.08 (Ca/P molar ratio of 1.62) in crystals of human enamel, which contain considerable amounts of other elements, such as Mg, Na, K, CO_2, Cl, F. The molar Ca/P ratios of early apical enamel range between 0.99 and 1.46, whereas those of the first enamel interrod elements rise from 1.24 at the ameloblast-enamel boundaries to about 1.40 at the dentin-enamel junction (Landis et al. 1988). The three most immature preparations obtained by fractionation of the enamel of unerupted porcine teeth showed a Ca/P ratio of 1.41, 1.44 and 1.47 (Bonar et al. 1991).

Enamel crystals are thought to consist of poorly crystalline, non-stoichiometric carbonated hydroxyapatite (Landis et al. 1988; Simmer and Fincham 1995); the degree of crystalline perfection, which is higher than that of mature bone crystals from the onset, increases with maturation (Bonar et al. 1991). According to Nanci and Smith (1992), enamel crystals differ only slightly from bone crystals: young enamel crystals are relatively rich in carbonate, and their CO_2/Ca ratio falls as they age (see also Rönnholm 1962). Mature enamel crystals do, in fact, contain considerably lower concentrations of carbonate – and of magnesium – than bone or dentin crystals (discussed by Daculsi et al. 1997). Diekwisch et al. (1995) reported that early enamel crystals are octocalcium phosphate or tricalcium phosphate; Aoba et al. (1998) supported the view that the first phase of enamel calcification requires the formation of an octocalcium phosphate-like precursor which, however, accounts for less than 2% of fully developed enamel inorganic substance.

The presence of amorphous calcium phosphate in enamel is a controversial issue. An electron microscope study of enamel organs showed a thin layer of amorphous material with small electron-dense granules along the distal surface of secretory ameloblasts and in tubulo-vesicular structures at the distal end of these same cells (Takano et al. 1990). Similar granules were found in the intercellular spaces of the inner enamel epithelium of incisors in calcium-loaded rats; they gradually converted to fine needle-shaped figures in the presecretory ameloblast layer (Takano et al. 1996). The electron diffraction patterns of amorphous calcium phosphate, which is viewed as the precursor of enamel hydroxyapatite, were found in the initial enamel layer of three-day postnatal mouse molars (Diekwisch 1998).

In discussing the modalities of enamel crystal formation, the position of Höhling (1989b) is that the earliest inorganic structures cannot be the suggested platelet-like structures which have a diameter of about 2 nm, a width of more than 10 nm and an even greater length, and that much smaller entities must exist. He reported that, as in bone (Sect. 4.2.1) and dentin (Sect. 4.2.2, the early stages of the calcification process in enamel are distinguished by the presence of linear strands including dot-like substructures which embody the earliest stage of crystal formation. The same views were expressed by R.W. Fearnhead, who suggested that the dots may signal a mixture of organic and inorganic material (see his discussion of Höhling's paper; Höhling 1989b, p. 331).

High resolution electron microscopy allows the determination of the periodic lattice images of the crystals. In enamel, these clearly show the heterogeneity of the crystal structure (Brès et al. 1985) and the presence of crystalline defects such as dislocations (Selvig 1970, 1973, 1975; Kerebel et al. 1976a, 1978, 1979; Voegel and Frank 1977; Daculsi and Kerebel 1978; Lee and LeGeros 1985; Cuisinier et al. 1992; Houllé et al. 1997). These defects, which have been thought to play a role in caries development (Takuma et al. 1987), can be located at the crystal edge or in the crystal core, or may display a shape like a spiral staircase; according to Selvig (1970), they are most often observed near the midline of crystals. Some of these dislocations may be due to mechanical interference from neighbouring growing crystals, while others may be artifacts produced by section cutting or by sublimation under the electron beam. Moreover, as pointed out by R.W. Fearnhead in discussing the work of Kerebel et al. (1979), periodic lattice images do not directly display the crystal lattice, but arise from interference between electrons which may combine or cancel each other out, depending on whether they are in or out of phase. As result, small variations in section thickness may induce different periodic images. Two findings do, however, suggest that the crystal configuration may itself differ in different parts of the crystals.

First, it has been shown that acid dissolution preferentially proceeds along the crystal *c* axis, probably coinciding with dislocations (Little 1959; Nylen 1964;

Johnson 1966; Arends et al. 1975; Jongebloed et al. 1975; Kerebel et al. 1976b; Daculsi and Kerebel 1977; Voegel and Frank 1977; Daculsi et al. 1979; Lee and LeGeros 1985; Hayashi 1995), so suggesting that the chemistry of the central area of crystals differs slightly from that of the rest (Jongebloed et al. 1975). A similar dissolution pattern has been noted in synthetic apatite, where the dimensions of the hole were not consistent (Simmelink and Abrigo 1989), and during caries (Voegel and Frank 1977; Tohda et al. 1987). Second, a number of electron microscope studies on enamel, including fossil enamel from various geological ages (Kakei et al. 2001), have shown the presence of a "dark" or "dense" line (Fig. 4.3) along the crystal axis (Rönnholm 1962; Nylen et al. 1963; Frazier 1968; Daculsi and Kerebel 1978; Marshall and Lawless 1981; Cuisinier et al. 1992; Miake et al. 1993; Bonucci et al. 1994a; Aoba 1996). According to Cuisinier et al. (1992), this dense line seems to be associated with the initial

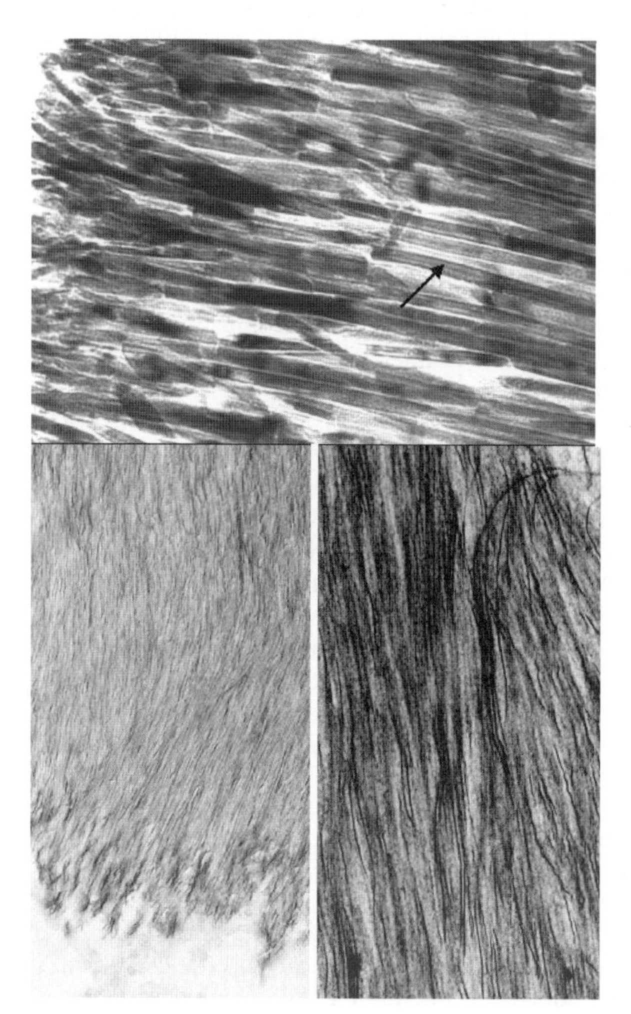

Fig. 4.3. *Above*: immature enamel crystals, some of which show an axial, electron-dense line (*arrow*). Untreated, ×105,000. *Below*: immature enamel; note the long, filament- and ribbon-like crystals. Unstained, ×50,000 and ×95,000

growth process of the crystals, but its exact structure and function have not yet been determined. Brown et al. (1987) and Miake et al. (1993) suggested that it may be the first, octocalcium phosphate-like precursor buried inside the growing crystal. This was also the view taken by Aoba (1996) who, on the basis of high resolution studies of early enamel and synthetic crystals, suggested that isolated apatite units grow along thin, ribbon-like crystals of octocalcium phosphate and later slowly fuse, so giving rise to hexagonal-shaped crystals. Rönnholm (1962) went beyond this, suggesting that the dark line may embody the organic material which acts as a template for crystal formation, or that it may be the earliest crystal formed, which would then remain as an axial core in the final crystal. Selvig (1970), however, rejected this view because, as previously indicated by Nylen and Omnell (1962), the high degree of regularity in the striations rules out the possibility of foreign inclusions, such as organic material, within the crystals (see, however, what is reported in Sects. 16.6 and 16.7 about intracrystalline organic material). According to Nylen et al. (1963), who stress that the dark line disappears at true focus, it may be no more than an interference effect due to a phase discontinuity in the crystal. Daculsi and Kerebel (1978) are of the opinion that the dark line is a focal phenomenon that is no longer visible at high resolution. According to Marshall and Lawless (1981), the dark line is not a dislocation, stacking fault, or lattice twin boundary: it is a planar defect involving a single 100 plane of the hydroxyapatite structure and most likely springs from a substitution in the lattice by carbonate ions or an independent but compatible calcium phosphate phase. A dense line has been described in the hydroxyapatite crystals of fossil enamel (Kakei et al. 2001), alveolar bone (Cuisinier et al. 1987; Kakei et al. 1997), and dental calculi (Kakei et al. 2000).

4.3 Lower Vertebrates

These organisms have an endoskeleton resembling a mammalian skeleton (Lopez 1970; Dickson 1982) and comprising hydroxyapatite, and an exoskeleton, known more correctly as a 'dermal skeleton' (Zylberberg et al. 1992), which is part of ecto-mesodermal structures like scales, scutes, fin rays, spines and osteoderms. The organization of these structures is rather complex, and shows a high level of variability according to the degree of their evolutionary differentiation. Details can be found in an excellent review by Zylberberg et al. (1992).

Apatitic calcium phosphate is the main constituent not only of the endoskeleton, but also of all the calcifiable structures of the dermal skeleton (Cooke 1967; Meinke et al. 1979; Zylberberg and Nicolas 1982; Levrat-Calviac 1986; Ikoma et al. 2003). Landis and Géraudie (1990) found that the inorganic phase of the bony fin rays of the *Oncorhynchus mykiss* trout consists of poorly crystalline hydroxyapatite with a Ca/P molar ratio of 1.0–1.4. As in higher vertebrates, carbonate, fluorine and chlorine can substitute for hydroxyl ions, while trace elements such as strontium, lead, aluminum, barium, cobalt and magnesium may be present too, some of them absorbed from the polluted environment (reviewed by Zylberberg et al. 1992).

Both the endoskeleton and the dermal skeleton of lower vertebrates have much in common with the osseous tissue of higher vertebrates. Some differences have been recorded in the dermal skeleton, mainly due to the occurrence in it of distinctive structures such as enameloid, isopedine and ganoine.

4.3.1 Enameloid

The name enameloid refers to an aprismatic, enamel-like, highly calcified tissue which forms a thin cap covering the outer tooth surface in Osteichthyes (bony fish), Chondrichthyes (rays and sharks) and some Amphibia (frogs, newts and toads) (Fincham et al. 1999). It is thought to have the same ectodermal origin and structure as high vertebrate enamel; this is certainly true as far as its high degree of calcification is concerned. Various controversies do, however, surround the nature of this tissue, as is clearly shown by the persistence of the name 'enameloid', alongside the many other names it has been given (Garant 1970; Kemp and Park 1974). Not all investigators consider it to be an enamel-like tissue; some believe it to be of ectodermal origin, while others favor a mesodermal origin (reviewed by Kemp and Park 1974). Doubts probably derive from the fact that the tissue is not homogeneous, and might even change from species to species if, as seems likely, elasmobranch enameloid is distinct from teleost enameloid (Sasagawa 2002). Fearnhead (1979) has reported that enameloid consists of structures pertaining to both enamel and dentin which, depending on the species, are mixed in different proportions; for this reason, he has supported the hypothesis, previously put forward by Shellis and Miles (1976), that enameloid is comparable to the narrow enamel-dentin junctional region in mammalian teeth. The conical teeth of bony fish contain cup enameloid, but also collar enameloid (Sasagawa and Ishiyama 1988); this appears to be of mesodermal origin (it may be a modified form of dentin), and consists of an outer layer that is nearly as strongly calcified as cup enameloid, and a less calcified, inner enameloid. In the pike, Herold (1974) described a very thin (0.5 μm thick) layer of enameloid, finding it similar to true ectodermal enamel and located directly above a 2–3 μm thick layer where enameloid and dentin structures are mixed. The poorly defined limits between enameloid and dentin probably explain why structures characteristic of the latter have been attributed to the former (Inage 1975; Sasagawa and Ishiyama 1988).

As in enamel itself, the inorganic substance of cup enameloid mainly consists of carbonated hydroxyapatite, which in some species contains a high proportion of carbonated fluorapatite (discussed by Fincham et al. 1999) and in others high concentrations of iron (Suga et al. 1992, 1993) – apparently, regardless of feeding habits. Miake et al. (1991) found high levels of fluoride (2.5 wt % or more) in elasmobranch enameloid, and low levels in teleostean enameloid, in the early stages of calcification. According to Kerebel and Le Cabellec (1980), cup enameloid differs from dentin not only because of its greater electron density, but also in having a higher F and a lower Mg content. It is striking that high resolution electron microscopy has revealed a dark line, very like the dark line found in vertebrate enamel, in the enameloid crystals of the carp (Miake et al. 1991). Lee and LeGeros (1985) have reported that the frequency of crystal defects in enameloid differs little from that found in enamel.

4.3.2 Isopedine

The name isopedine refers to the basal, lamellar layer of collagen fibrils which is organized as a plywood-like structure and forms the basal plate of teleost scales. The distinctive feature of this structure, which resembles lamellar bone, is its variable degree of calcification, which may be incomplete or totally absent (Meunier and François 1980). It is covered by a superficial layer, subdivided into an "external layer", which consists of woven bone tissue with thin collagen fibrils, and an "outer limiting layer", which has no collagen fibrils, or very few, and is rich in mucosubstances (Zylberberg and Nicolas 1982). Structural patterns similar to those of teleost scales can be found in other components of the dermal skeleton, such as the scales and osteoderms of amphibia and reptilia, which display variably calcified basal plates and superficial layers that contain no collagen fibrils but are rich in acid proteoglycans (Levrat-Calviac and Zylberberg 1986). The crystals of isopedine seem to have the same composition as those of bone.

The superficial layer of the scales of primitive Actinopterygii is distinguished by its highly calcified tissue, *ganoine*, which is thought to resemble enamel.

4.4 Invertebrates

The endo- and/or exoskeletal frames of invertebrates are the most widespread calcified structures in the biological domain. In spite of the vast spectrum of representative animals, the composition of the inorganic substance included in their organism shows little variation. A few do comprise special inorganic components such as sulfates, silicates, iron, magnesium and copper salts, oxalate and other elements (Travis 1968; Pautard 1970; Boyan et al. 1992; Weiner and Addadi 2002), but their inorganic substance consists mostly of calcium carbonate. This may be arranged in a poorly crystalline form, or as one of the crystalline polymorphs, calcite or aragonite, or, occasionally, vaterite (Travis

1968, 1970; Lowenstam 1981; Wheeler 1992; Falini et al. 1996; Aizenberg et al. 2002). In bivalve mollusks such as oysters, mussels and clams, whose valves consist of three layers (an outer periostracum, a middle prismatic layer, and an inner nacreous, or mother-of-pearl, layer), calcite is mainly found in the prismatic layer, whereas aragonite is mainly found in the nacreous layer (Travis 1968). Calcium carbonate is also found as calcite in three of the four major layers (the outer epicuticle, a thin sclerotinized, sometimes calcified layer; the thick exocuticle, which is sclerotinized and calcified; the very thick endocuticle, which is calcified; and the inner membranous layer, which is not calcified) of the exoskeleton of Crustacea (Travis 1963). Calcite is also the typical component of sea urchin embryo spicules, whereas aragonite forms most of the skeleton of corals. Calcium phosphate can constitute as much as 90% of the shell inorganic substance in some Brachiopoda (*Lingula unguis*; Kelly et al. 1965).

The presence of an amorphous inorganic phase in the calcified structures of invertebrates raises the same problems already discussed with bone. Amorphous calcium carbonate, later converted to calcite, has been described as the early phase of calcification in sea urchin spicules (Beniash et al. 1997; Wilt 1999; Politi et al. 2004) and in larval shells of marine bivalves (Weiss et al. 2002). Acidic proteins extracted from the aragonite layer of mollusk shell induce the in vitro formation of amorphous calcium carbonate prior to its transformation into aragonite (Gotliv et al. 2003). Electron-dense material resembling inorganic substance, which fails to provide any crystalline pattern when examined using selected-area electron diffraction, but is removed by EDTA decalcification, is present in the smallest (about 0.1 μm wide) spicules of sponges (Travis 1970). Spicules from the calcareus sponge *Clathrina* consist of amorphous calcium carbonate in one layer and calcite in another (Aizenberg et al. 1996, 2002). It has been suggested that specialized proteic macromolecules stabilize the amorphous inorganic material (Aizenberg et al. 1996, 2002), as further discussed below. According to Wilt et al. (2003), the protein-amorphous calcium carbonate complex may be analogous to the paracrystalline form of calcium phosphate, which is intermediate between amorphous and crystalline and is later converted to hydroxyapatite, as described by Arnold et al. (2001) in bone, dentin and enamel (see Sects. 4.2.1, 4.2.2 and 4.2.3). On the other hand, on the basis of the observation that amorphous calcium carbonate is produced by a number of organisms from many taxa, and that it is unstable and relatively soluble, and, therefore, hard to detect, Weiner et al. (2003) have suspected that it is far more widely distributed than is usually supposed, and that, among other functions, it may be a transient precursor phase of crystalline calcium carbonate (see also Weiner et al. 2005).

Calcium carbonate or carbonate-hydroxyapatite are the inorganic components of crystals and calcareous concretions found in many non-skeletal tissues and structures, such as otoliths (Söllner et al. 2003), eggshells (Dennis et al. 1996) and the pineal gland (Mabie and Wallace 1974).

4.5 Unicellular Organisms

Calcification in unicellular organisms differs from calcification in tissues of organisms showing a multicellular grade of organization, mainly because the process occurs within cells or in close apposition to their membrane, so that it depends directly on the activity of the organism itself and to some extent proceeds independently of that of other cells or the intercellular matrix (Pautard 1970). The unicellular organisms most often studied appear to be bacteria (Fig. 4.4) and protozoa.

Certain bacteria can, in fact, develop calcifications (Ennever et al. 1981); these may be either intra- or extracellular (Lie and Selvig 1974), or both (Gonzales and Sognnaes 1960; Zander et al. 1960; Lo Storto et al. 1990). Carbonated apatite (calcite or aragonite) is mainly induced by marine bacteria (Pautard 1970), whereas calcium phosphate is the calcium salt most often found in oral bacteria (Ennever 1960). An analytical electron microscope study of *Bacteri-*

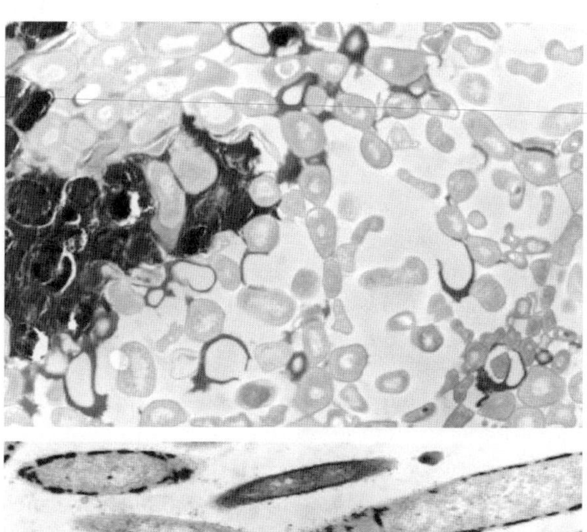

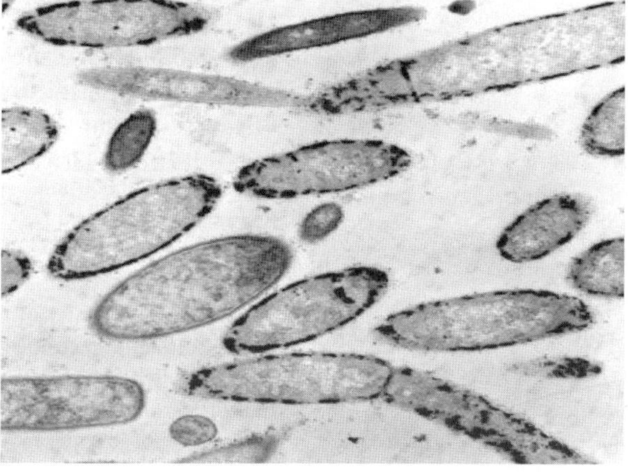

Fig. 4.4. *Above*: dental plaque; the inorganic substance (*black*) completely masks some bacteria (*middle left*) and forms a coat around others. Unstained, ×5,000. *Below*: alkaline phosphatase activity in dental plaque bacteria: most microorganisms show positive enzymatic reaction (*black*) along their membrane. Gomori's method, ×6,000

onema matruchotii has shown amorphous calcium phosphate acting as the earliest intrabacterial mineral phase (Takazoe and Itoyama 1980), whereas the X-ray diffraction of intra-microbial needle-shaped crystals yielded hydroxyapatite diffractograms (Ennever and Creamer 1967). On the other hand, energy-dispersive X-ray microanalysis and Fourier transform IR spectroscopy of the mineral associated with nanobacteria – the smallest bacteria found so far – showed that it corresponds to carbonate apatite (Kajander and Çiftçioglu 1998). For a comprehensive review of microbial calcification and its mechanisms, see Boyan et al. (1992).

One interesting example of biological diversity is given by what are known as 'magnetic' or 'magnetotactic bacteria' (Fig. 4.5), microorganisms which live in salt or fresh water, or in soil (Fassbinder et al. 1990; Schüler and Baeuerlein 1998), and contain intracellular crystals of magnetite (Fe_3O_4) or greigite (Fe_3S_4) (Schüler and Frankel 1999; Arakaki et al. 2002; Matsunaga and Okamura 2003).

The foraminifera are very diffuse protozoa (Towe and Cifelli 1967) whose calcified remnants – together with the coccoliths of the calcifying algae Coccolitophorids (Young et al. 1999) – are the most representative components of

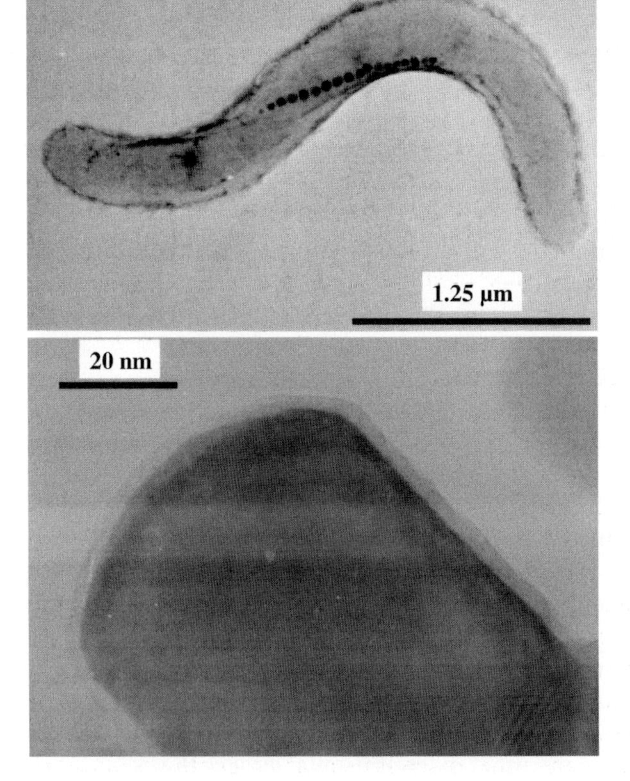

Fig. 4.5. *Above*: *Magnetospirillum magneticum* AMB-1 with the characteristic magnetosome chain; below: part of a magnetosome. ×28,000 and ×115,000 (courtesy of Tadashi Matsunaga, Department of Biotechnology, Tokyo University of Agriculture and Technology; from Trends Microbiol 11:536-541, 2003, with permission of Elsevier)

pelagic sediments. Their calcified structures contain calcium carbonate, mainly in the form of calcite, although aragonite may occur in some of them (Todd and Blackmon 1956). Magnesium, and other trace elements, may be among the components of their inorganic substance (Krinsley 1960). Other protozoa contain inorganic inclusions in their cytoplasm. Of these, *Spirostomum ambiguum* contains characteristic spherical inclusions which can be stained for calcium and phosphorus, bind ^{32}P and ^{45}Ca, and produce X-ray diffractograms whose reflections resemble those of dried bone powder, so pointing to the presence of crystalline apatite (reviewed by Pautard 1970, 1976).

4.6 Pathological Calcifications

The pathological calcification of soft tissues is quite common (Gatter and McCarty 1967). Its etiology and pathogenesis vary (Eisenstein et al. 1960), but its most frequent causes are the hypercalcemic states of hyperparathyroidism (Irnell et al. 1970; Bergdahl and Boquist 1973; Cohen et al. 1997), hypervitaminosis D (Hass et al. 1958; Kent et al. 1958; Eisenstein and Zeruolis 1964), hyperphosphatemic states (Delmez and Slatopolsky 1992; Slatopolsky et al. 2001; Indridason and Quarles 2002), the generalized calciphylaxis (Wilmer and Magro 2002) of renal failure, atherosclerosis (Tintut and Demer 2001), and a number of other diseases, including neoplasias (Ahmed 1975) and topical calciphylaxis and calcergy (Gabbiani and Tuchweber 1970; Johnson and Alkek 1970; Boivin et al. 1987).

Hydroxyapatite has been found to be the most representative inorganic component involved in ectopic calcifications of many tissues (Gatter and McCarty 1967). This is true of breast carcinoma (Ahmed 1975) and other tumors, tumoral calcinosis (Boskey et al. 1983), calcifications of the lung secondary to several pulmonary diseases (Martin et al. 1971), to primary hyperparathyroidism (Cohen et al. 1997), or to long-term dialysis therapy (Bestetti-Bosisio et al. 1984), calcification experimentally induced in the skin of dihydrotachysterol (DHT) sensitized rats (Boivin and Tochon-Danguy 1976), or in their lung (Bonucci et al. 1994b), skeletal muscles (Bonucci and Sadun 1972), the myocardium (Bonucci and Sadun 1973), the aorta (Bonucci and Sadun 1975) and other arteries (Seifert and Dreesbach 1966). Besides hydroxyapatite, octocalcium phosphate and whitlockite have been reported as occurring in pathological calcification of the lung (Martin et al. 1971) and aorta (Höhling et al. 1968), and calcium pyrophosphate is found in soft tissues, especially the periarticular ones, in cases of tophaceous pseudogout, or calcium pyrophosphate dihydrate crystal deposition disease (Keen et al. 1991).

Among various interesting examples of ectopic calcification, there are those that occur in many *tumors*, whether they primarily develop in the skeleton or derive from soft tissues which do not normally calcify. Neither the bone produced in osteosarcomas and other osseous neoplasms, nor the calcification sometimes found in chondrosarcoma and benign cartilage tumors, nor the many forms of calcification that may originate in and from the soft mesenchymal components of the skeleton and teeth, differ substantially from those that are found in normal calcified tissues. Some of the phases of the calcification process may, however, be altered, heightened or attenuated, or may occur in unusual substrates (such as secretion products, membranes and mitochondria), so making possible the study of features that would otherwise be hard to investigate. An example of this opportunity is given by *keratins*: these products of epidermal origin can undergo calcification with the formation of hydroxyapatite (Blakey et al. 1963; Blakey and Lockwood 1968). Keratin calcification has been described in many carcinoma and other tumor-like lesions of epithelial origin (Bonucci et al. 1979).

Calcification can also occur in non-tumoral keratinized structures such as baleens, the horny-like structures in whales which form a complex apparatus for the filtration of planctonic food. Baleens contain appreciable quantities of calcium phosphate, like that occurring in the inorganic phase of bone (Pautard 1965).

4.6.1 Mitochondria

Pathological calcifications often involve mitochondria. It is known that these cellular organelles play an active role in the normal cytoplasmic transport of calcium ions (Patriarca and Carafoli 1968; Carafoli 1969; Lehninger 1970; Selwyn et al. 1970; Carafoli and Lehninger 1971; Borle 1973; Nicholls and Crompton 1980), which may accumulate within them (Greenawalt et al. 1964; Thomas and Greenawalt 1968; Shapiro and Lee 1975a, b; Ramachandran and Bygrave 1978). Low affinity and high affinity Ca^{2+}-binding sites have been described in mitochondria (Reynafarje and Lehninger 1969), even if high affinity binding appears to be the result of energy-linked Ca uptake (Southard and Green 1974). Other divalent cations may be accumulated too (Greenawalt and Carafoli 1966; Somlyo et al. 1974). As a result, inorganic deposits may form inside the mitochondria of calcified or normal soft tissues, or of pathologically altered ones (Fig. 4.6). They consist of calcium phosphate, with traces of magnesium, sulfur and silicon (Hagler et al. 1979). These deposits are easily recognizable under the electron microscope as electron-dense, roundish, granular concretions (Heggtveit et al. 1964; Thomas and Greenawalt 1968; Bonucci et al. 1973; Sayegh and Abousy 1977). They may be associated – occasionally in a single mitochondrion, but mostly in several mitochondria – with clusters of needle-like crystals (Palladini and Carbone 1966; Thomas and Greenawalt 1968; Saladino et al. 1969; Carafoli et al. 1971; Brandt and Bässler 1972; Somlyo et al. 1974). High resolution microincineration (Thomas and Greenawalt 1968; Sayegh et al. 1974) shows that the granules contain inorganic material, and energy dispersive X-ray analysis (Sutfin et al. 1971; Hirashita et al. 1980; Hargest

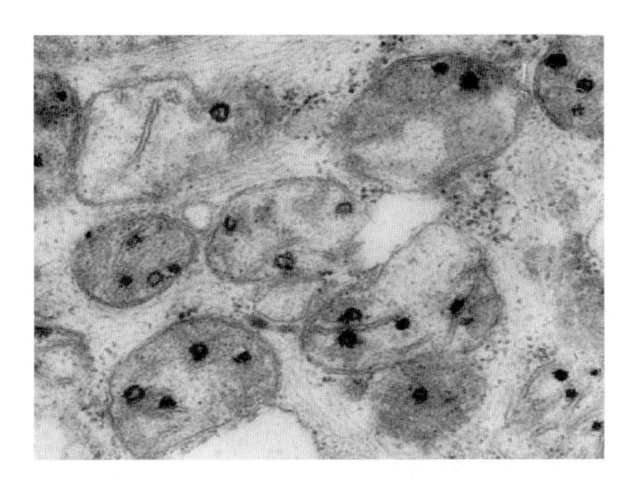

Fig. 4.6. Mitochondria of the myocardium of a DHT-intoxicated rat: most of them contain electron-dense, granular or rosette-like inorganic granules; uranium and lead, ×65,000

et al. 1985) confirms that they include calcium and phosphate with a Ca/P mass ratio of 2.15 (Ali et al. 1978). X-ray diffraction shows that the inorganic substance in the granules is hydroxyapatite (Weinbach and von Brand 1965), and electron diffraction suggests that they consist of β-calcium phosphate or whitlockite (Thomas and Greenawalt 1968). An electron spin resonance spectroscopic study (Ostrowski et al. 1975) did, however, show that the crystalline phase could not exceed 2.33–3.5% of the inorganic phase (assuming a crystal size of 25 nm), and X-ray diffraction and infrared spectroscopy support a non-crystalline pattern (Becker et al. 1976).

Inorganic granules must not be confused with the "dense bodies" that are normally found in mitochondria (Greenawalt and Carafoli 1966; Woodhouse and Burston 1969). These two structures may appear simultaneously in the same mitochondria (Saetersdal et al. 1977). The relationship between them is dubious: X-ray microanalysis has shown that calcium concentrations in dense bodies are themselves variable (Pasquali-Ronchetti et al. 1969; Knowles et al. 1972) and research by Peachey (1964) suggests that they may be transformed into inorganic granules by the addition of calcium ions. Greenawalt and Carafoli (1966) have, however, shown that numbers of inorganic granules clearly exceed those of normal mitochondrial dense bodies, and Woodhouse and Burstone (1969) have reported that, in the metastatic calcification of the myocardium, calcium is initially deposited in the cristae, not in dense bodies.

4.6.2 Vascular Calcification

Vascular calcification is a frequent complication in hypercalcemic and/or hyperphosphatemic states (Tomson 2003) and in diseases like atherosclerosis (Shioi et al. 2000) and calciphylaxis (Seifert and Dreesbach 1966). Vascular calcification, as well as ectopic calcifications in other soft tissues, were once considered to be the outcome of a passive process of calcium and phosphate deposition secondary to degenerative changes of the tissue or to an excessive local concentration of calcium and phosphate ions. Hence the terms 'metastatic' and 'dystrophic' calcification, the first indicating calcification in normal tissues with high local Ca and P concentrations, and the second calcification in injured tissues with normal Ca and P levels (Eisenstein et al. 1960; Reif and Lange 1972). Vascular calcification has been considered to be dystrophic when associated with atherosclerosis and metastatic when associated with hypercalcemic and hyperphosphatemic states. As noted later (Chap. 14), these concepts have been undermined by the discovery that there are analogies between the calcification of arteries and the osteogenetic process, and the understanding that cells and matrix proteins may have a primary role in inducing calcium and phosphate deposition in the vascular walls (Boström and Demer 2000; Shioi et al. 2000; Boström 2001; Demer and Tintut 2003).

The nature and composition of the inorganic deposits found in the aorta and other arteries appear to differ little from those of inorganic substance in normally calcifying skeletal tissues. In human atherosclerotic coronary arteries, energy-dispersive X-ray microanalysis showed that the chemical composition of calcified sites was identical with that of bone hydroxyapatite (Fitzpatrick et al. 1994). Tomazic (2001) has analyzed the results obtained with a range of techniques on isolated and deproteinated calcific deposits from various cardiovascular segments. The inorganic fraction, or what he called "bioapatite" – a defective hydroxyapatite with substantial traces of sodium, magnesium, carbonate and fluoride – was significantly more soluble than hydroxyapatite, and, while its X-ray diffraction suggested an apatitic pattern, its variable degree of crystallinity showed that crystallization was time- and flow dynamics-dependent.

4.7 Congenital or Acquired Diseases; Genetically Engineered Animals

Information of primary importance on the calcification process and its mechanism can be gathered through the study of acquired and, especially, congenital skeletal diseases. There are many of these, both in humans (McKusick 1960; Sillence et al. 1979; Rimoin 1996) and animals (Bargman et al. 1972; Silberberg and Lesker 1975; Kleinman et al. 1977); they may arise spontaneously or after induced genetic mutations (transgenic, knockout and null animals). Acquired or genetic human skeletal diseases have, however, been studied more often from a clinical than a biological viewpoint. A few of them do, in any case, deserve detailed consideration, above all those involving changes in the organic-inorganic relationships which allow the role of the organic components in calcification to be examined. Transgenic and knockout animals, on the other hand, are mostly used to determine the effects induced in calcified tissues through the modification or deprivation of a protein or an enzyme.

It should be borne in mind that the nature and composition of the inorganic phase are not the main targets of this type of study.

Characterized as it is by the extreme fragility of bones (hence the name *fragilitas ossium hereditaria*), *osteogenesis imperfecta* takes a number of different clinical forms (Albright and Grunt 1971; Bauze et al. 1975) resulting from hundreds of mutations (Steinmann et al. 1986) in the collagen type I genes which code for the pro $\alpha 1(I)$ and pro $\alpha 2(I)$ chains. Characterized by collagen fibrils and apatite crystals thinner than normal, the disease can also develop in bovines (Fisher et al. 1987) and lambs (Holmes et al. 1964), and has been induced in transgenic mice (Cassella et al. 1994). Very few studies have been directed to the characteristics of the inorganic substance in this diseases. Bretlau et al. (1970) observed quantitative rather than qualitative changes, and, in some types of osteogenesis imperfecta, Cassella and Ali (1992) and Cassella et al. (1995) observed a Ca/P ratio lower than that of normal, matched bone. They also reported that an abnormal mineral formation was suggested by stromal calcification concomitant with intramitochondrial inclusions containing Ca and P. Electron-probe analysis of bone samples prepared by cryofixation or conventional fixation showed a Ca/P ratio lower than in age- and site-matched controls; the lowest value (Ca/P = 1.49 vs 1.69) was that found in osteogenesis imperfecta type II (Sarathchandra et al. 1999). FT-IRM analysis of the enamel in a case of osteogenesis imperfecta type IV B showed changes in the mineral environment of carbonate and phosphate ions and, apparently, in the size of crystals (Bohic et al. 1998).

A variety of clinical forms have been recognized for inherited tooth diseases such as *dentinogenesis imperfecta*, which is often associated with osteogenesis imperfecta, and *amelogenesis imperfecta*. The former is a dominant, autosomal dentin disease, the latter an X-linked, autosomal, dominant and recessive enamel disease. Both are characterized by an abnormal development of the organic matrix and by its defective calcification; as in osteogenesis imperfecta, both show a quantitatively reduced, though qualitatively normal, inorganic substance (Wright et al. 1993), but in amelogenesis imperfecta (Robinson et al. 2003) as well as in enamel hypoplasia (Batina et al. 2004) the morphology of crystals has been reported to be altered by the impairment of their initiation, fusion and growth.

Several types of *genetically engineered animals* have been produced to examine the impact of proteins on the calcification process. The development of crystals thinner and more irregularly aligned than those found in controls (e. g., Fratzl et al. 1996), and the presence of amorphous calcified material (Cassella et al. 1994), have been found in osteogenesis imperfecta mouse models, but again the changes in the calcification process appear to be independent of alterations in the nature and composition of the inorganic phase.

4.8 Concluding Remarks

Of the results discussed above, a few deserve to be extrapolated and highlighted.

- X-ray diffraction, selected area electron diffraction and other biophysical and biochemical studies alike go to show that the inorganic substance found in many biological systems (bone, dentin, cementum, enamel, calcified tendons, the exo- and endo-skeletons of lower vertebrates and pathological calcifications) closely resembles, but is not identical with, natural polycrystalline apatites.
- Besides calcium and phosphate, other ions are present in the inorganic substance; these include carbonate, chloride, fluoride, magnesium, strontium, barium, zinc, iron, lead and aluminum, whose exact position is not yet known.
- In bone ash, carbonate makes up about 5% of the total weight, so that the inorganic phase can be described as a defectively crystalline, non-stoichiometric carbonated hydroxyapatite; carbonate concentrations are high in invertebrate skeletal tissues, where the inorganic substance is mostly calcium carbonate, whether arranged in a poorly crystalline form, or as one of the crystalline polymorphs, calcite or aragonite.
- Knowledge of the exact structure and composition of the inorganic substance is partly impeded by the poor quality of definition of its X-ray diffractograms, which has been attributed to the smallness of the inorganic particles. Actually, preparations of basic calcium phosphate that utilize a range of different Ca/P ratios can all give rise to the X-ray diffraction pattern of the apatite lattice; moreover, the structure and composition of the inorganic particles may vary and become more apatite-like with age, and degree of maturation. Despite this, the high degree of crystallinity of natural or synthetic apatites is never approached.
- The crystals located in areas of initial calcification appear as chains of small dots; their structure and configuration do not coincide with those found in areas of advanced calcification.
- The electron diffractograms of areas of initial calcification are of amorphous type and the early inorganic particles may well represent a paracrystalline form of calcium phosphate, intermediate between amorphous and crystalline, which takes time to convert to apatite at a later stage.
- Amorphous calcium phosphate, and later octocalcium phosphate, may be present in areas of early calcification, as further suggested by reports of dense, finely granular material under the electron microscope. Amorphous calcium carbonate, intermediate between amorphous and crystalline and later converted to calcite, has been described as the early phase of calcifica-

tion in sea urchin spicules and other invertebrate skeletal tissues. Octocalcium phosphate has been found in pathological soft tissue calcification.

- The results obtained from the bulk of inorganic substance in fully calcified areas do not necessarily apply to the much fewer inorganic particles which are located at the calcification front: their shape and size (see below) and, most clearly, composition, all change with growth and maturation. Only later do they acquire the configuration that makes them apatite-like crystals.
- Due to their resemblance to apatites, all the inorganic particles present in biological tissues have been called "crystals" or "crystallites"; these may be misleading terms, but for ease of reference they have been adopted in this volume in discussing all the inorganic particles found in calcified matrices.
- The crystals can have a number of crystalline defects, such as dislocations. Some crystals, especially those found in enamel, show an axial 'dark' or 'dense' line, whose exact structure and function have not yet been determined; acid dissolution preferentially proceeds along the crystal *c* axis.
- Crystals present in acquired or experimentally induced pathological calcifications do not differ from those found in normally calcified tissues. Mitochondrial calcifications, vascular calcifications, congenital skeletal diseases and calcifications in genetically engineered animals, may all supply useful models for the study of the mechanisms at work in the calcification process.

References

Ahmed A (1975) Calcification in human breast carcinomas: ultrastructural observations. J Pathol 117:247–251

Aizenberg J, Lambert G, Addadi L, Weiner S (1996) Stabilization of amorphous calcium carbonate by specialized macromolecules in biological and synthetic precipitates. Adv Mater 8:222–226

Aizenberg J, Lambert G, Weiner S, Addadi L (2002) Factors involved in the formation of amorphous and crystalline calcium carbonate: a study of an ascidian skeleton. J Am Chem Soc 124:32–39

Albright JA, Grunt JA (1971) Studies of patients with osteogenesis imperfecta. J Bone Joint Surg 53A:1415–1425

Ali SY, Wisby A, Gray JC (1978) Electron probe analysis of cryosections of epiphyseal cartilage. Metab Bone Dis Rel Res 1:97–103

Aoba T (1996) Recent observations on enamel crystal formation during mammalian amelogenesis. Anat Rec 245:208–218

Aoba T, Komatsu H, Shimazu Y, Yagishita H, Taya Y (1998) Enamel mineralization and an initial crystalline phase. Connect Tissue Res 38:129–137

Arakaki A, Webb J, Matsunaga T (2002) A novel protein tightly bound to bacterial magnetic particles in *Magnetospirillum magneticum* strain AMB-1. J Biol Chem 278:8745–8750

Arends J, van den Berg PJ, Jongebloed WL (1975) Dissolution of hydroxyapatite and fluorapatite single crystals. In: Colloques Internationaux C.N.R.S. (ed) Physico-chimie et cristallographie des apatites d'intérêt biologique. Centre National de la Recherche Scientifique, Paris, pp 389–395

Arnold S, Plate U, Wiesmann H-P, Straatmann U, Kohl H, Höhling H-J (2001) Quantitative analyses of the biomineralization of different hard tissues. J Microsc 202:488–494

Arnott HJ (1976) Calcification in higher plants. In: Watabe N, Wilbur KM (eds) The mechanisms of calcification in the invertebrates and plants. University of South Carolina Press, Columbia, pp 55–78

Arnott HJ, Pautard FGE (1967) Osteoblast function and fine structure. Israel J Med Sci 3:657–670

Arnott HJ, Pautard FGE (1970) Calcification in plants. In: Schraer H (ed) Biological calcification: cellular and molecular aspects. Appleton-Century-Crofts, New York, pp 375–446

Ascenzi A, Benedetti EL (1959) An electron microscopic study of the foetal membranous ossification. Acta Anat 37:370–385

Ascenzi A, Bonucci E (1966) The osteon calcification as revealed by the electron microscope. In: Fleisch H, Blackwood HJJ, Owen M (eds) Calcified tissues 1965. Springer, Berlin Heidelberg New York, pp 142–146

Ascenzi A, Bonucci E, Steve Bocciarelli D (1965) An electron microscope study of osteon calcification. J Ultrastruct Res 12:287–303

Ascenzi A, Bonucci E, Ostrowski K, Sliwowski A, Dziedzic-Goclawska A, Stachowicz W, Michalik J (1977) Initial studies on the crystallinity of the mineral fraction and ash content of isolated human and bovine osteons differing in their degree of calcification. Calcif Tissue Res 23:7–11

Bale WF (1940) A comparative roentgen-ray diffraction study of several natural apatites and the apatite-like constituent of bone and tooth substances. Am J Roentgenol 93:735–774

Bargman GJ, Mackler B, Shepard TH (1972) Studies of oxidative energy deficiency I. Achondroplasia in the rabbit. Arch Biochem Biophys 150:137–146

Batina N, Renugopalakrishnan V, Casillas Lavín PN, Guerrero JCH, Morales M, Garduño-Juárez R, Lakka SL (2004) Ultrastructure of dental enamel afflicted with hypoplasia: an atomic force microscopic study. Calcif Tissue Int 74:294–301

Bauze RJ, Smith R, Francis MJO (1975) A new look at osteogenesis imperfecta. A clinical radiological and biochemical study of forty-two patients. J Bone Joint Surg 57B:2–12

Becker GL, Termine JD, Eanes ED (1976) Comparative studies of intra- and extramitochondrial calcium phosphates from the hepatopancreas of the blue crab (*Callinectes sapidus*). Calcif Tiss Res 21:105–113

Beniash E, Aizenberg J, Addadi L, Weiner S (1997) Amorphous calcium carbonate transforms into calcite during sea urchin larval spicule growth. Proc R Soc London B 264:461–465

Bergdahl L, Boquist L (1973) Secondary hypercalcemic hyperparathyrodism. Virchows Arch Abt A Path Anat 358:225–239

Bestetti-Bosisio M, Cotelli F, Schiaffino E, Sorgato G, Schmid C (1984) Lung calcification in long-term dialysed patients: a light and electronmicroscopic study. Histopathology 8:69–79

Betts F, Trotta R, Goldberg MR, Posner AS (1979) Non-apatite mineral in actively calcifying tissue. Orthop Transact 3:201–202

Bigi A, Cojazzi G, Panzavolta S, Ripamonti A, Roveri N, Romanello M, Noris Suarez K, Moro L (1997) Chemical and structural characterization of the mineral phase from cortical and trabecular bone. J Inorg Biochem 68:45–51

Blakey PR, Lockwood P (1968) The environment of calcified components in keratins. Calcif Tissue Res 2:361–369

Blakey PR, Earland DC, Stell JGP (1963) Calcification of keratin. Nature 198:481

Blumenthal NC, Betts F, Posner AS (1975) Effect of carbonate and biological macromolecules on formation and properties of hydroxyapatite. Calcif Tissue Res 18:81–90

Blumenthal NC, Betts F, Posner AS (1977) Stabilization of amorphous calcium phosphate by Mg and ATP. Calcif Tissue Res 23:245–250

Bohic S, Heymann D, Pouëzat JA, Gauthier O, Daculsi G (1998) Transmission FT-IR microspectroscopy of mineral phases in calcified tissues. C R Acad Sci Paris 321:865–876

Boivin G, Tochon-Danguy HJ (1976) Étude chez le rat d'une calcinose cutanée induite per calciphylaxie locale II. Aspects biophysiques de la substance minérale. Ann Biol Anim Bioch Biophys 16:869–878

Boivin G, Walzer C, Baud CA (1987) Ultrastructural study of the long-term development of two experimental cutaneous calcinoses (topical calciphylaxis and topical calcergy) in the rat. Cell Tissue Res 247:525–532

Bonar LC, Roufosse AH, Sabine WK, Grynpas MD, Glimcher MJ (1983) X-ray diffraction studies of the crystallinity of bone mineral in newly synthesized and density fractionated bone. Calcif Tissue Int 35:202–209

Bonar LC, Shimizu M, Roberts JE, Griffin RG, Glimcher MJ (1991) Structural and composition studies on the mineral of newly formed dental enamel: a chemical, X-ray diffraction, and ^{31}P and proton nuclear magnetic resonance study. J Bone Miner Res 6:1167–1176

Bonucci E (2002) Crystal ghosts and biological mineralization: fancy spectres in an old castle, or neglected structures worthy of belief? J Bone Miner Metab 20:249–265

Bonucci E, Graziani G (1976) Comparative thermogravimetric, X-ray diffraction and electron microscope investigations of burnt bones from recent, ancient and prehistoric age. Rend Fis Acc Lincei 59:517–532

Bonucci E, Sadun R (1972) An electron microscope study on experimental calcification of skeletal muscle. Clin Orthop Relat Res 88:197–217

Bonucci E, Sadun R (1973) Experimental calcification of the myocardium. Ultrastructural and histochemical investigations. Am J Pathol 71:167–192

Bonucci E, Sadun R (1975) Dihydrotachysterol-induced aortic calcification. A histochemical and ultrastructural investigation. Clin Orthop Relat Res 107:283–294

Bonucci E, Derenzini M, Marinozzi V (1973) The organic-inorganic relationship in calcified mitochondria. J Cell Biol 59:185–211

Bonucci E, De Matteis A, Anceschi C (1979) Histochemical and electron microscopical investigations on the calcified keratin in the horn pearls of a glans carcinoma (calcified keratin). Basic Appl Histochem 23:93–102

Bonucci E, Lozupone E, Silvestrini G, Favia A, Mocetti P (1994a) Morphological studies of hypomineralized enamel of rat pups on calcium-deficient diet, and of its changes after return to normal diet. Anat Rec 239:379–395

Bonucci E, Silvestrini G, Ballanti P, Della Rocca C, Mocetti P (1994b) Dihydrotachysterol-induced lung calcification in the rat. It J Miner Electrol Metab 8:12–22

Borle AB (1973) Calcium metabolism at the cellular level. Fed Proc 32:1944–1950

Boskey AL, Posner AS (1974) Magnesium stabilisation of amorphous calcium phosphate: a kinetic study. Mater Res Bull 9:907–916

Boskey AL, Vigorita VJ, Sencer O, Stuchin SA, Lane JM (1983) Chemical, microscopic, and ultrastructural characterization of the mineral deposits in tumoral calcinosis. Clin Orthop Relat Res 178:258–269

Boström K (2001) Insights into the mechanism of vascular calcification. Am J Cardiol 88:20E-22E

Boström K, Demer LL (2000) Regulatory mechanisms in vascular calcification. Crit Rev Eukaryot Gene Expr 10:151–158

Boyan BD, Swain LD, Everett MM, Schwartz Z (1992) Mechanisms of microbial mineralization. In: Bonucci E (ed) Calcification in biological systems. CRC Press, Boca Raton, pp 129–156

Brandt G, Bässler R (1972) Die Wirkung der experimentellen Hypercalcämie durch Dihydrotachysterin auf Drüsenfunktion und Verkalkungsmuster der Mamma. Licht-, elektronenmikroskopische und chemisch-analytische Untersuchungen. Virchows Arch Abt A Path Anat 356:155–172
Brès EF, Barry JC, Hutchison JL (1985) High-resolution electron microscope and computed images of human tooth enamel crystals. J Ultrastruct Res 90:261–274
Bretlau P, Jorgensen MB, Hohansen H (1970) Osteogenesis imperfecta. Light and electron microscopic studies of the stapes. Acta Oto-Laryngol 69:172–184
Brown WE, Chow LC (1976) Chemical properties of bone mineral. Ann Rev Mater Sci 6:213–236
Brown WE, Eidelman N, Tomzaic BB (1987) Octocalcium phosphate as a precursor in biomineral formation. Adv Dent Res 1:306–313
Burnell JM, Teubner EJ, Miller AG (1980) Normal maturation changes in bone matrix, mineral, and crystal size in the rat. Calcif Tissue Int 31:13–19
Caglioti V (1935) Sulla struttura delle ossa. Atti V Congr. Naz. Chimica Pura Applicata, 320–331. Rome, Associazione Italiana di Chimica
Carafoli E (1969) Calcium ion transport in mitochondria. Biochem J 116:2–3
Carafoli E, Lehninger AL (1971) A survey of the interaction of calcium ions with mitochondria from different tissues and species. Biochem J 122:681–690
Carafoli E, Tiozzo R, Pasquali-Ronchetti I, Laschi R (1971) A study of Ca^{2+} metabolism in kidney mitochondria during acute uranium intoxication. Lab Invest 25:516–527
Carlström D (1955) X-ray crystallographic studies on apatites and calcified structures. Acta Radiol Suppl 121:1–59
Carlström D, Finean JB (1954) X-ray diffraction studies on the ultrastructure of bone. Biochim Biophys Acta 13:183–191
Cassella JP, Ali SY (1992) Abnormal collagen and mineral formation in osteogenesis imperfecta. Bone Miner 17:123–128
Cassella JP, Pereira R, Khillan JS, Prockop DJ, Garrington N, Ali SY (1994) An ultrastructural, microanalytical, and spectroscopic study of bone from a transgenic mouse with a COL1.A1 pro-alpha-1 mutation. Bone 15:611–619
Cassella JP, Garrington N, Stamp TCB, Ali SY (1995) An electron probe X-ray microanalytical study of bone mineral in osteogenesis imperfecta. Calcif Tissue Int 56:118–122
Clark JH (1931) A study of tendons, bones, and other forms of connective tissue by means of X-ray diffraction patterns. Am J Physiol 98:328–337
Cohen AM, Maxon HR, Goldsmith RE, Schneider HJ, Wiot JF, Loudon RG, Altemeier WA (1997) Metastatic pulmonary calcification in primary hyperparathyroidism. Arch Intern Med 137:520–522
Cooke PH (1967) Fine structure of the fibrillar plate in the central head scale of the striped killifish *Fundulus majalis*. Trans Am Microsc Soc 86:273–279
Cuisinier F, Bres EF, Hemmerle J, Voegel J-C, Frank RM (1987) Transmission electron microscopy of lattice planes in human alveolar bone apatite crystals. Calcif Tissue Int 40:332–338
Cuisinier FJG, Steuer P, Senger B, Voegel JC, Frank RM (1992) Human amelogenesis I: High resolution electron microscopy study of ribbon-like crystals. Calcif Tissue Int 51:259–268
D'Errico JA, MacNeil RL, Strayhorn CL, Piotrowski BT, Somerman MJ (1995) Models for the study of cementogenesis. Connect Tissue Res 33:9–17
Daculsi G, Kerebel B (1977) Some ultrastructural aspects of biological apatite dissolution and possible role of dislocations . J Biol Buccale 5:203–218
Daculsi G, Kerebel B (1978) High-resolution electron microscope study of human enamel crystallites: size, shape, and growth. J Ultrastruct Res 65:163–172

Daculsi G, Kerebel B, Kerebel LM (1979) Mechanisms of acid dissolution of biological and synthetic apatite crystals at the lattice pattern level. Caries Res 13:277–289

Daculsi G, Bouler J-M, LeGeros RZ (1997) Adaptive crystal formation in normal and pathological calcifications in synthetic calcium phosphate and related biomaterials. Int Rev Cytol 172:129–191

Dallemagne MJ, Brasseur H (1942) La diffraction des rayons X par la substance minérale osseuse. Bull Soc Roy Sci Liége 8/9:1–19

deJong WF (1926) La substance minérale dans les os. Rec Trav Chim 45:445–448

Delmez JA, Slatopolsky E (1992) Hyperphosphatemia: its consequences and treatment in patients with chronic renal disease. Am J Kidney Dis 19:303–317

Demer LL, Tintut Y (2003) Mineral exploration: search for the mechanism of vascular calcification and beyond: the 2003 Jeffrey M. Hoeg Award lecture. Arterioscler Thromb Vasc Biol 23:1739–1743

Dennis JE, Xiao S-Q, Agarwal M, Fink DJ, Heuer AH, Caplan AI (1996) Microstructure of matrix and mineral components of eggshells from white leghorn chickens (*Gallus gallus*). J Morphol 228:287–306

Dickson GR (1982) Ultrastructure of growth cartilage in the proximal femur of the frog, *Rana temporaria*. J Anat 135:549–564

Diekwisch TGH (1998) Subunit compartments of secretory stage enamel matrix. Connect Tissue Res 38:101–111

Diekwisch TGH, Berman BJ, Gentner S, Slavkin HC (1995) Initial enamel crystals are not spatially associated with mineralized dentine. Cell Tissue Res 279:149–167

Eanes ED (1975) Amorphous intermediates in the formation of biological apatites. In: Colloques Internationales C.N.R.S. (ed) Physico-chimie et cristallographie des apatites d'intérêt biologique. Centre National de la Recherche Scientifique, Paris, pp 295–301

Eanes ED (1979) Enamel apatite: chemistry, structure and proteins. J Dent Res 58:829–834

Eanes ED (1992) Dynamics of calcium phosphate precipitation. In: Bonucci E (ed) Calcification in biological systems. CRC Press, Boca Raton, pp 1–17

Eanes ED, Meyer JL (1977) The maturation of crystalline calcium phosphates in aqueous suspensions at physiologic pH. Calcif Tissue Res 23:259–269

Eanes ED, Posner AS (1965) Kinetics and mechanism of conversion of non-crystalline calcium phosphate to crystalline hydroxyapatite. Trans N Y Acad Sci 28:233–241

Eanes ED, Posner AS (1970) Structure and chemistry of bone mineral. In: Schraer H (ed) Biological calcification: cellular and molecular aspect. Appleton-Century-Crofts, New York, pp 1–26

Eanes ED, Gillessen IH, Posner AS (1965) Intermediate states in the precipitation of hydroxyapatite. Nature (London) 208:365–367

Eanes ED, Termine JD, Posner AS (1967) Amorphous calcium phosphate in skeletal tissue. Clin Orthop Relat Res 53:223–235

Eisenstein R, Zeruolis L (1964) Vitamin-D induced aortic calcification. Arch Path 77:27–35

Eisenstein R, Trueheart RE, Hass GM (1960) Pathogenesis of abnormal tissue calcifications. In: Sognnaes RF (ed) Calcification in biological systems. American Association for the Advancement of Sciences, Washington, pp 281–305

Elliott JC (1973) The problems of the composition and structure of the mineral components of the hard tissues. Clin Orthop Relat Res 93:313–345

Engström A (1960) Ultrastructure of bone mineral. In: Rodahl K, Nicholson JT, Brown EM (eds) Bone as a tissue. McGraw-Hill, New York, pp 251–261

Ennever J (1960) Intracellular calcification by oral filamentous microrganisms. J Periodontol 31:304–307

Ennever J, Creamer H (1967) Microbiological calcification: bone mineral and bacteria. Calcif Tissue Res 1:87–93

Ennever J, Streckfuss JL, Goldschmidt MC (1981) Calcifiability comparison among selected microorganisms. J Dent Res 60:1793–1796
Falini G, Albeck S, Weiner S, Addadi L (1996) Control of aragonite or calcite polymorphism by mollusk shell macromolecules. Science 271:67–69
Fassbinder JW, Stanjek H, Vali H (1990) Occurrence of magnetic bacteria in soil. Nature 343:161–163
Fearnhead RW (1979) Matrix-mineral relationships in enamel tissues. J Dent Res 58:909–916
Fincham AG, Moradian-Oldak J, Simmer JP (1999) The structural biology of the developing dental enamel matrix. J Struct Biol 126:270–299
Fisher LW, Eanes ED, Denholm LJ, Heywood BR, Termine JD (1987) Two bovine models of osteogenesis imperfecta exhibit decreased apatite crystal size. Calcif Tissue Int 40:282–285
Fitton Jackson S (1957) The fine structure of developing bone in the embryonic fowl. Proc R Soc B 146:270–280
Fitton Jackson S, Randall JT (1956) Fibrogenesis and the formation of matrix in developing bone. Ciba Found Symp on Bone Structure and Metabolism, pp 47–62
Fitzpatrick LA, Severson A, Edwards WD, Ingram RT (1994) Diffuse calcification in human coronary arteries. Association of osteopontin with atherosclerosis. J Clin Invest 94:1597–1604
Fleisch H, Russell RGG, Bisaz S, Termine JD, Posner AS (1968) Influence of pyrophosphate on the transformation of amorphous to crystalline calcium phosphate. Calcif Tissue Res 2:49–59
Foo CWP, Huang J, Kaplan DL (2004) Lessons from seashells: silica mineralization via protein templating. Trends Biotechnol 22:577–585
François P, Herman H (1961) Le composé minéral fondamental des tissus calcifiés. II. Les sels osseux contiennent un phosphate de calcium différent de l'hydroxylapatite. Bull Soc Chim Biol 43:643–649
Fratzl P, Paris O, Klaushofer K, Landis WJ (1996) Bone mineralization in an osteogenesis imperfecta mouse model studied by small-angle X-ray scattering. J Clin Invest 97:396–402
Frazier PD (1968) Adult human enamel: an electron microscopic study of crystallite size and morphology. J Ultrastruct Res 22:1–11
Gabbiani G, Tuchweber B (1970) Studies on the mechanism of calcergy. Clin Orthop Relat Res 69:66–74
Garant PR (1970) An electron microscopic study of the crystal-matrix relationship in the teeth of the dogfish *Squalus acanthias* L. J Ultrastruct Res 30:441–449
Gatter RA, McCarty DJ (1967) Pathological tissue calcifications in man. Arch Path 84:346–353
Gay CV (1977) The ultrastructure of the extracellular phase of bone as observed in frozen thin sections. Calcif Tissue Res 23:215–223
Glimcher MJ (1990) The nature of the mineral component of bone and the mechanism of calcification. In: Avioli LV, Krane SM (eds) Metabolic bone disease and clinically related disorders. W.B. Saunders, Philadelphia, pp 42–68
Glimcher MJ, Krane SM (1968) The organization and structure of bone, and the mechanism of calcification. In: Gould BS (ed) Biology of collagen. Academic Press, London, pp 67–251
Goldberg M, Septier D (1985) Improved lipid preservation by malachite green-glutaraldehyde fixation in rat incisor predentine and dentine. Arch Oral Biol 30:717–726
Gonzales HA, Sognnaes RF (1960) Electron microscopy of dental calculus. Science 131:156–158

Gotliv BA, Addadi L, Weiner S (2003) Mollusk shell acidic proteins: in search of individual functions. Chem Biochem 4:522–529
Greenawalt JW, Carafoli E (1966) Electron microscope studies on the active accumulation of Sr^{++} by rat-liver mitochondria. J Cell Biol 29:37–61
Greenawalt JW, Rossi CS, Lehninger AL (1964) Effect of active accumulation of calcium and phosphate ions on the structure of rat liver mitochondria. J Cell Biol 23:21–38
Gruner JW, McConnel D, Armstrong WD (1937) The relationship between the crystal structure and chemical composition of enamel and dentine. J Biol Chem 121:771–781
Grynpas MD, Bonar LC, Glimcher MJ (1984) Failure to detect an amorphous calcium-phosphate solid phase in bone mineral: a radial distribution function study. Calcif Tissue Int 36:291–301
Hagler HK, Sherwin L, Buja LM (1979) Effect of different methods of tissue preparation on mitochondrial inclusions of ischemic and infarcted canine myocardium. Transmission and analytic electron microscopic study. Lab Invest 40:529–544
Hancox NM, Boothroyd B (1965) Electron microscopy of the early stages of osteogenesis. Clin Orthop Relat Res 40:153–161
Hargest TE, Gay CV, Schraer H, Wasserman AJ (1985) Vertical distribution of elements in cells and matrix of epiphyseal growth plate cartilage determined by quantitative electron probe analysis. J Histochem Cytochem 33:275–286
Harper RA, Posner AS (1966) Measurement of non-crystalline calcium phosphate in bone mineral. Proc Soc Exp Biol Med 122:137–142
Hass GM, Trueheart RE, Taylor B, Stumpe M (1958) An experimental histologic study of hypervitaminosis D. Am J Pathol 34:395–431
Hayashi Y (1995) High resolution electron microscopy of enamel crystallites demineralized by initial dental caries. Scann Microsc 9:199–206
Heggtveit HA, Herman L, Mishra RK (1964) Cardiac necrosis and calcification in experimental magnesium deficiency. A light and electron microscopic study. Am J Pathol 45:757–782
Hendricks SB, Hill WL (1942) The inorganic constituents of bone. Science 96:255
Herold RCB (1974) Ultrastructure of odontogenesis in the pike (*Esox lucius*). Role of dental epithelium and formation of enameloid layer. J Ultrastruct Res 48:435–454
Hirashita A, Nakamura Y, Okumura E, Kuwabara Y (1980) Microanalysis of mitochondrial granules. Microanalysis of mitochondrial granules in bone cells incident to experimental tooth movement. Acta Histochem Cytochem 13:343–358
Hirschman A, Sobel AE, Fankuchen I (1953) Calcification X. An X-ray diffraction study of calcification in vitro in relation to composition. J Biol Chem 204:13–18
Höhling HJ (1989a) Special aspects of biomineralization of dental tissues. In: Oksche A, Vollrath L (eds) Handbook of microscopic anatomy. Springer, Berlin Heidelberg New York, pp 475–524
Höhling HJ (1989b) Do conformities exist between the earliest crystal formations in enamel and those of the collagen-rich hard tissues? In: Fearnhead RW (ed) Tooth enamel V. Florence Publishers, Tsurumi, pp 322–334
Höhling HJ, Fearnhead RW, Lotter G (1968) The mineral components in aortic "calcification" studied by X-ray and electron diffraction combined with electron microscopy. German Med Monthly 13:135–138
Höhling HJ, Scholz F, Boyde A, Heine HG, Reimer L (1971a) Electron microscopical and laser diffraction studies of the nucleation and growth of crystals in the organic matrix of dentine. Z Zellforsch 117:381–393
Höhling HJ, Kreilos R, Neubauer G, Boyde A (1971b) Electron microscopy and electron microscopical measurements of collagen mineralization in hard tissues. Z Zellforsch 122:36–52

Höhling HJ, Barckhaus RH, Krefting E-R, Althoff J, Quint P (1990) Collagen mineralization: aspects of the structural relationship between collagen and the apatitic crystallites. In: Bonucci E, Motta PM (eds) Ultrastructure of skeletal tissues. Kluwer Academic Publishers, Boston, pp 41–62

Höhling HJ, Arnold S, Barckhaus RH, Plate U, Wiesmann HP (1995) Structural relationship between the primary crystal formation and the matrix macromolecules in different hard tissues. Discussion of a general principle. Connect Tissue Int 33:171–178

Höhling HJ, Arnold S, Plate U, Stratmann U, Wiesmann HP (1997) Analysis of general principle of crystal nucleation, formation in the different hard tissues. Adv Dent Res 11:462–466

Holmes JR, Baker JR, Davies ET (1964) Osteogenesis imperfecta in lambs. Vet Rec 76:980–984

Houllé P, Voegel JC, Schultz P, Steuer P, Cuisinier FJG (1997) High resolution electron microscopy: structure and growth mechanisms of human dentin crystals. J Dent Res 76:895–904

Ikoma T, Kobayashi H, Tanaka J, Walsh D, Mann S (2003) Microstructure, mechanical, and biomimetic properties of fish scales from *Pagrus major*. J Struct Biol 142:327–333

Inage T (1975) Electron microscopic study of early formation of the tooth enameloid of a fish (*Hoplognathus fasciatus*). I. Odontoblasts and matrix fibers. Arch Histol Jap 38:209–227

Indridason OS, Quarles LD (2002) Hyperphosphatemia in end-stage renal disease. Adv Ren Replace Ther 9:184–192

Irnell L, Werner I, Grimelius L (1970) Soft tissue calcification in hyperparathyroidism. Acta Med Scand 187:145–151

Johnson NW (1966) Differences in the shape of human enamel crystallites after partial destruction by caries, EDTA and various acids. Archs Oral Biol 11:1421–1424

Johnson WC, Alkek DS (1970) Histopathology and histochemistry of cutaneous calciphylaxis. Clin Orthop Relat Res 69:75–86

Jones SJ, Boyde A (1984) Ultrastructure of dentin and dentinogenesis. In: Linde A (ed) Dentin and dentinogenesis, vol. I . CRC Press, Boca Raton, pp 81–134

Jongebloed WL, Molenaar I, Arends J (1975) Morphology and size-distribution of sound and acid-treated enamel crystallites. Calcif Tissue Res 19:109–123

Kajander EO, Çiftçioglu N (1998) Nanobacteria: an alternative mechanism for pathogenic intra- and extracellular calcification and stone formation. Proc Natl Acad Sci USA 95:8274–8279

Kakei M, Nakahara H, Tamura N, Itoh H, Kumegawa M (1997) Behavior of carbonate and magnesium ions in the initial crystallites at the early developmental stages of the rat calvaria. Ann Anat 179:311–316

Kakei M, Nakahara H, Kumegawa M, Yoshikawa M, Kunii S (2000) Demonstration of the central dark line in crystals of dental calculus. Biochim Biophys Acta 1524:189–195

Kakei M, Nakahara H, Kumegawa M, Mishima H, Kozawa Y (2001) High-resolution electron microscopy of the crystallites of fossil enamels obtained from various geological ages. J Dent Res 80:1560–1564

Keen CE, Crocker PR, Brady K, Hasan N, Levison DA (1991) Calcium pyrophosphate dihydrate deposition disease: morphological and microanalytical features. Histopathology 19:529–536

Kelly PG, Oliver PTP, Pautard FGE (1965) The shell of *Lingula unguis*. In: Richelle LJ, Dallemagne MJ (eds) Calcified tissues. Université de Liège, Liège, pp 337–345

Kemp NE, Park JH (1974) Ultrastructure of the enamel layer in developing teeth of the shark *Carcharhinus menisorrah*. Arch Oral Biol 19:633–644

Kent SP, Vawter GF, Dowben RM, Benson RE (1958) Hypervitaminosis D in monkeys; a clinical and pathologic study. Am J Pathol 34:37–59

Kerebel L-M, Le Cabellec MT (1980) Enameloid in the teleost fish Lophius. An ultrastructural study. Cell Tissue Res 206:211–223

Kerebel B, Daculsi G, Verbaere A (1976a) High-resolution electron microscopy and crystallographic study of some biological apatite. J Ultrastruct Res 57:266–275

Kerebel B, Daculsi G, Verbaere A (1976b) Ultrastructural and crystallographic study of biological apatites. J Ultrastruct Res 57:263–275

Kerebel B, Daculsi G, Kerebel LM (1978) Apports de la méthode d'amincissement ionique à l'étude en haute résolution des cristaux d'émail dentaire humain. C R Acad Sci Paris 286:1903–1906

Kerebel B, Daculsi G, Kerebel LM (1979) Ultrastructural studies of enamel crystallites. J Dent Res 58:844–850

Kleinman HK, Pennypacker JP, Brown KS (1977) Proteoglycan and collagen of "achondroplastic" (cn/cn) neonatal mouse cartilage. Growth 41:171–177

Knowles JC, Weavers B, Cooper EH (1972) Accumulation of calcium in the intramitochondrial dense bodies in mice. Exp Cell Res 73:230–233

Krinsley D (1960) Trace elements in the tests of planktonic Foraminifera. Micropaleontology 6:297–300

Landis WJ, Géraudie J (1990) Organization and development of the mineral phase during early ontogenesis of the bony fin rays of the trout *Oncorhynchus mykiss.* Anat Rec 228:383–391

Landis WJ, Glimcher MJ (1978) Electron diffraction and electron probe microanalysis of the mineral phase of bone tissue prepared by anhydrous techniques. J Ultrastruct Res 63:188–223

Landis WJ, Burke GY, Neuringer JR, Paine MC, Nanci A, Bai P, Warshawsky H (1988) Earliest enamel deposits of the rat incisor examined by electron microscopy, electron diffraction, and electron probe microanalysis. Anat Rec 220:233–238

Leadbeater BSC, Riding R (1986) Biomineralization in lower plants and animals. Systematics association, vol 30. Oxford University Press, Oxford

Lee DD, LeGeros RZ (1985) Microbeam electron diffraction and lattice fringe studies of defect structures in enamel apatites. Calcif Tissue Int 37:651–658

Lefèvre ML, Bale WF, Hodge H (1937) The chemical nature of the inorganic portion of fetal tooth substance. J Dent Res 16:85–101

Lehninger AL (1970) Mitochondria and calcium ion transport. Biochem J 119:129–138

Lénárt G, Bidló G, Pintér J (1968) Use of X-ray diffraction method in investigations of mineral substances of bone and callus. Acta Biochim Biophys Acad Sci Hung 3:305–316

Lénárt G, Bidló G, Pintér J (1971) X-ray diffraction investigation on the growing zone of long bones. Acta Biochim Biophys Acad Sci Hung 6:307–309

Lénárt G, Pflüger G, Bidló G, Pintér J, Fischerleitner F (1979) Kristallographische Untersuchung der Verlängerungskallus. Arch Orthop Traumat Surg 93:303–305

Levrat-Calviac V (1986) Etude comparée des ostéodermes de *Tarentola mauritanica* et de *T. neglecta* (Gekkonidae, Squamata). Arch Anat Microsc Morphol Exp 75:29–43

Levrat-Calviac V, Zylberberg L (1986) The structure of the osteoderms in the gekko: *Tarentola mauritanica.* Am J Anat 176:437–446

Lie T, Selvig KA (1974) Calcification of oral bacteria: an ultrastructural study of two strains of Bacterionema matruchotii. Scand J Dent Res 82:8–18

Little JJ (1959) Electron microscope studies in human dental enamel. J Microsc Soc 78:58–66

Lo Storto S, Di Grezia R, Silvestrini G, Cattabriga M, Bonucci E (1990) Studio morfologico ultrastrutturale di tartaro sopragengivale. Minerva Stomatol 39:83–89

Lopez E (1970) L'os cellulaire d'un poisson téléostéen "*Anguilla anguilla* L." I. Étude histocytologique et histophysique. Z Zellforsch 109:552–565

Lowenstam HA (1981) Minerals formed by organisms. Science 211:1126–1131

Mabie CP, Wallace BM (1974) Optical, physical and chemical properties of pineal gland calcification. Calcif Tiss Res 16:59–71
Mann S (1988) Molecular recognition in biomineralization. Nature 332:119–124
Mann S (2001) Biomineralization. Principles and concepts in bioinorganic materials chemistry. Oxford University Press, Oxford
Marshall AF, Lawless KR (1981) TEM studies of the central dark line in enamel crystallites. J Dent Res 60:1773–1782
Martin J-C, Le Bouffant L, Durif S, Henoc P, Normand C, Policard A (1971) Identification cristallographique et ultrastructure des calcification pulmonaires pathologiques. Path Biol 19:735–742
Matsunaga T, Okamura Y (2003) Genes and proteins involved in bacterial magnetic particle formation. Trends Microbiol 11:536–541
McConnell D (1952) The crystal chemistry of carbonate apatites and their relationship to the composition of calcified tissues. J Dent Res 31:53–63
McKusick VA (1960) Eritable disorders of connective tissues. The C.V. Mosby Company, St. Louis
Meinke DK, Skinner HCW, Thomson KS (1979) X-ray diffraction of the calcified tissues in *Polypterus*. Calcif Tissue Int 28:37–42
Mellors RC (1964) Electron probe microanalysis I. Calcium and phosphorus in normal human cortical bone. Lab Invest 13:183–195
Meunier FJ, François Y (1980) L'organisation spatiale des fibres collagènes et la minéralisation des écailles des Dipneustes actuels. Bull Soc Zool Fr 105:215–226
Miake Y, Aoba T, Moreno EC, Shimoda S, Prostak K, Suga S (1991) Ultrastructural studies on crystal growth of enameloid minerals in Elasmobranch and Teleost fish. Calcif Tissue Int 48:204–217
Miake Y, Shimoda S, Fukae M, Aoba T (1993) Epitaxial overgrowth of apatite crystals on the thin-ribbon precursor at early stages of porcine enamel mineralization. Calcif Tissue Int 53:249–256
Molnar Z (1959) Development of the parietal bone of young mice 1. Crystals of bone mineral in frozen-dried preparations. J Ultrastruct Res 3:39–45
Münzenberg KJ, Gebhardt M (1973) Brushite octacalcium phosphate, and carbonate-containing apatite in bone. Clin Orthop Relat Res 90:271–273
Nanci A, Smith CE (1992) Development and calcification of enamel. In: Bonucci E (ed) Calcification in biological systems. CRC Press, Boca Raton, pp 313–343
Neuman WF, Neuman MW (1953) The nature of the mineral phase of bone. Chem Rev 53:1–45
Neuman WF, Neuman MW (1958) The chemical dynamics of bone mineral. University of Chicago Press, Chicago
Nicholls DG, Crompton M (1980) Mitochondrial calcium transport. FEBS Lett 111:261–268
Nylen MU (1964) Electron microscope and allied biophysical approaches to the study of enamel remineralization. J Microsc 83:135–141
Nylen MU, Omnell K-Å (1962) The relationship between the apatite crystals and the organic matrix of rat enamel. QQ-4. New York, Fifth International Congress for Electron Microscopy, Academic Press
Nylen MU, Eanes ED, Omnell K-Å (1963) Crystal growth in rat enamel. J Cell Biol 18:109–123
Ostrowski K, Dziedzic-Goclawska A, Sliwowski A, Wojtczak L, Michalik J, Stachowicz W (1975) Analysis of the crystallinity of calcium phosphate deposits in rat liver mitochondria by electron spin resonance spectroscopy. FEBS Lett 60:410–413
Palladini G, Carbone A (1966) Ultrastruttura della calcificazione distrofica renale da sublimato. Experientia 22:585

Pasquali-Ronchetti I, Greenawalt JW, Carafoli E (1969) On the nature of the dense matrix granules of normal mitochondria. J Cell Biol 40:565–568
Patriarca P, Carafoli E (1968) A study of the intracellular transport of calcium in rat heart. J Cell Physiol 72:29–38
Pautard FGE (1965) Calcification of baleen. In: Richelle LJ, Dallemagne MJ (eds) Calcified tissues. Université de Liège, Liège, pp 347–357
Pautard FGE (1970) Calcification in unicellular organisms. In: Schraer H (ed) Biological calcification: cellular and molecular aspects. Appleton-Century-Crofts, New York, pp 105–201
Pautard FGE (1976) Calcification in single cells: with an appraisal of the relationship between *Spirostomum ambiguum* and the osteocyte. In: Watabe N, Wilbur KM (eds) The mechanisms of mineralization in the invertebrates and plants. University of South Carolina Press, Columbia,SC, pp 33–53
Peachey LD (1964) Electron microscopic observations on the accumulation of divalent cations in intramitochondrial granules. J Cell Biol 20:95–109
Perry CC (2003) Silification: the processes by which organisms capture and mineralize silica. Rev Miner Geochem 54:297–327
Plate U, Höhling HJ, Reimer L, Barckhaus RH, Wienecke R, Wiesmann H-P, Boyde A (1992) Analysis of the calcium distribution in predentine by EELS and of the early crystal formation in dentine by ESI and ESD. J Microsc 166:329–341
Politi Y, Arad T, Klein E, Weiner S, Addadi L (2004) Sea urchin spine calcite forms via a transient amorphous calcium carbonate phase. Science 306:1161–1164
Posner AS (1969) Crystal chemistry of bone mineral. Physiol Rev 49:760–792
Posner AS (1987) Bone mineral and the mineralization process. In: Peck WA (ed) Bone and mineral research 5. Elsevier Science Publisher, Amsterdam, pp 65–116
Posner AS, Betts F (1975) Synthetic amorphous calcium phosphate and its relation to bone mineral structure. Acc Chem Res 8:273–281
Posner AS, Harper RA, Muller SA, Menczel J (1965) Age changes in the crystal chemistry of bone apatite. Ann N Y Acad Sci 131:737–742
Posner AS, Blumenthal NC, Boskey AL, Betts F (1975) Synthetic analogue of bone mineral formation. J Dent Res 54:B88-B93
Posner AS, Betts F, Blumenthal NC (1978) Properties of nucleating systems. Metab Bone Dis Rel Res 1:179–183
Posner AS, Tannenbaum PJ (2000) The mineral phase of dentin. In: Linde A (ed) Dentin and dentinogenesis. CRC Press, Boca Raton, pp 17–36
Ramachandran C, Bygrave FL (1978) Calcium ion cycling in rat liver mitochondria. Biochem J 174:613–620
Reif W, Lange HP (1972) Vergleichende morphologische und chemische Untersuchungen zur dystrophischen Verkalkung und mineralisation der Kaninchenniere beim Infarkt und während der Regeneration der postischämischen Nephrose. Beitr Path 145:221–248
Rey C, Beshah K, Griffin R, Glimcher MJ (1991) Structural studies of the mineral phase of calcifying cartilage. J Bone Miner Res 6:515–525
Reynafarje B, Lehninger AL (1969) High affinity and low affinity binding of Ca++ by rat liver mitochondria. J Biol Chem 244:584–593
Rimoin DL (1996) Molecular defects in the chondrodysplasias. Am J Med Genet 63:106–110
Robinson C, Shore RC, Wood SR, Brookes SJ, Smith DA, Wright JT, Connell S, Kirkham J (2003) Subunit structures in hydroxyapatite crystal development in enamel: implications for amelogenesis imperfecta. Connect Tissue Res 44:65–71
Rönnholm E (1962) The amelogenesis of human teeth as revealed by electron microscopy II. The development of enamel crystallites. J Ultrastruct Res 6:249–303

Rogers KD, Daniels P (2002) An X-ray diffraction study of the effects of heat treatment on bone mineral microstructure. Biomaterials 23:2577–2585

Roseberry HH, Hastings AB, Morse JK (1931) X-ray analysis of bone and teeth. J Biol Chem 90:335–407

Saetersdal TS, Myklebust R, Berg Justesen N-P, Engedal H, Olsen WC (1977) Calcium containing particles in mitochondria of heart muscle cells as shown by cryo-ultramicrotomy and X-ray microanalysis. Cell Tissue Res 182:17–31

Sakae T (1988) X-ray diffraction and thermal studies of crystals from the outer and inner layers of human dental enamel. Arch Oral Biol 33:707–713

Saladino AJ, Bentley PJ, Trump BF (1969) Ion movements in cell injury. Effect of amphotericin B on the ultrastructure and function of the epithelial cells of the toad bladder. Am J Pathol 54:421–466

Sarathchandra P, Kayser MV, Ali SY (1999) Abnormal mineral composition of osteogenesis imperfecta bone as determined by electron probe X-ray microanalysis on conventional and cryosections. Calcif Tissue Int 65:11–15

Sasagawa I (2002) Mineralization patterns in elasmobranch fish. Microsc Res Tech 59:396–407

Sasagawa I, Ishiyama M (1988) The structure and development of the collar enameloid in two teleost fishes, *Halichoeres poecilopterus* and *Pagrus major*. Anat Embryol (Berl) 178:499–511

Sayegh FS, Abousy A (1977) Mitochondrial granule distribution in tooth germ cells. Anat Record 189:451–466

Sayegh FS, Davis RW, Solomon GC (1974) Mitochondrial role in cellular mineralization. J Dent Res 53:581–587

Schüler D, Baeuerlein E (1998) Dynamics of iron uptake and Fe_3O_4 biomineralization during aerobic and microaerobic growth of *Magnetospirillum gryphiswaldense*. J Bacteriol 180:159–162

Schüler D, Frankel RB (1999) Bacterial magnetosomes: microbiology, biomineralization and biotechnological applications. Appl Microbiol Biotechnol 52:464–473

Seifert G, Dreesbach HA (1966) Die calciphylaktische Artheriopatie. Frankf Z Path 75:342–361

Selvig KA (1970) Periodic lattice images of hydroxyapatite crystals in human bone and dental hard tissues. Calcif Tissue Res 6:227–238

Selvig KA (1973) Electron microscopy of dental enamel: analysis of crystal lattice images. Z Zellforsch 137:271–280

Selvig KA (1975) Resolution of the hydroxyapatite crystal lattice in bone and dental enamel by electron microscopy. In: Colloques Internationaux C.N.R.S. (ed) Physico-chemie et cristallographie des apatites d'intérêt biologique. Centre National de la Recherche Scientifique, Paris, pp 41–49

Selwyn MJ, Dawson AP, Dunnett SJ (1970) Calcium transport in mitochondria. FEBS Lett 10:1–5

Shapiro IM, Lee NH (1975a) Calcium accumulation by chondrocyte mitochondria. Clin Orthop Relat Res 106:323–329

Shapiro IM, Lee NH (1975b) Effects of Ca^{2+} on the respiratory activity of chondrocyte mitochondria. Arch Biochem Biophys 170:627–633

Shellis RP, Miles AEW (1976) Observations with the electron microscope on enameloid formation in the common eel (*Anguilla anguilla*: Teleostei). Proc R Soc London B 194:253–269

Shioi A, Mori K, Jono S, Wakikawa T, Hiura Y, Koyama H, Okuno Y, Nishizawa Y, Morii H (2000) Mechanism of atherosclerotic calcification. Z Kardiol 89:75–79

Silberberg R, Lesker P (1975) Skeletal growth and development of achondroplastic mice. Growth 39:17–33

Sillence DO, Horton WA, Rimoin DL (1979) Morphologic studies in the skeletal dysplasias. A review. Am J Pathol 96:811–870

Simmelink JW, Abrigo SC (1989) Crystal morphology and decalcification patterns compared in rat and human enamel and synthetic hydroxyapatite. Adv Dent Res 3:241–248

Simmer JP, Fincham AG (1995) Molecular mechanisms of dental enamel formation. Crit Rev Oral Biol Med 6:84–108

Slatopolsky E, Brown A, Dusso A (2001) Role of phosphorus in the pathogenesis of secondary hyperparathyroidism. Am J Kidney Dis 37:S54-S57

Smith CB, Smith DA (1976) An X-ray diffraction investigation of age-related changes in the crystal structure of bone apatite. Calcif Tissue Res 22:219–226

Söllner C, Burghammer M, Busch-Nentwich E, Berger J, Schwarz H, Riekel C, Nicolson T (2003) Control of crystal size and lattice formation by starmaker in otolith biomineralization. Science 302:282–286

Somlyo AP, Somlyo AV, Devine CE, Peters PD, Hall TA (1974) Electron microscopy and electron probe analysis of mitochondrial cation accumulation in smooth muscle. J Cell Biol 61:723–742

Southard JH, Green DE (1974) High affinity binding of Ca^{++} in mitochondria: a reappraisal. Biochem Biophys Res Comm 59:30–37

Steinmann B, Nicholls A, Pope FM (1986) Clinical variability of osteogenesis imperfecta reflecting molecular heterogeneity: cysteine substitutions in the $\alpha 1$(I) collagen chain producing lethal and mild forms. J Biol Chem 261:8958–8964

Steve-Bocciarelli D (1973) Apatite microcrystals in bone and dentine. J Microsc (Paris) 16:21–34

Stuehler R (1937) Über den Feinbau des Knochens. Fortschr Röntgenstr 57:231–264

Suga S, Taki Y, Ogawa M (1992) Iron in enameloid of perciform fish. J Dent Res 71:1316–1325

Suga S, Taki Y, Ogawa M (1993) Fluoride and iron concentrations in the enameloid of lower teleostean fish. J Dent Res 72:912–922

Sumerel JL, Morse DE (2003) Biotechnological advances in biosilification. Progr Mol Subcell Biol 33:225–247

Sutfin LV, Holtrop ME, Ogilvie RE (1971) Microanalysis of individual mitochondrial granules with diameters less than 1000 Angstrom. Science 174:947–949

Takano Y, Yamamoto T, Domon T, Wakita M (1990) Histochemical, ultrastructural, and electron microprobe analytical studies on the localization of calcium in rat incisor ameloblasts at early stage amelogenesis. Anat Record 228:123–131

Takano Y, Hanaizumi Y, Ohshima H (1996) Occurrence of amorphous and crystalline mineral deposits at the epithelial-mesenchymal interface of incisors in the calcium-loaded rat: implication of novel calcium binding domains. Anat Record 245:174–185

Takazoe I, Itoyama T (1980) Analytical electron microscopy of *Bachterionema matruchotii* calcification. J Dent Res 59:1090–1094

Takuma S, Tohda H, Tanaka N, Kobayashi T (1987) Lattice defects in and carious dissolution of human enamel crystals. J Electron Microsc 36:387–391

Termine JD, Eanes ED (1972) Comparative chemistry of amorphous and apatitic calcium phosphate preparations. Calcif Tissue Res 10:171–197

Termine JD, Posner AS (1967) Amorphous/crystalline interrelationships in bone mineral. Calcif Tissue Res 1:8–23

Termine JD, Peckauskas RA, Posner AS (1970) Calcium phosphate formation in vitro II. Effects of environment on amorphous-crystalline transformation. Arch Biochem Biophys 140:318–325

Termine JD, Eanes ED, Greenfield DJ, Nylen MU (1973) Hydrazine-deproteinated bone mineral. Physical and chemical properties. Calcif Tissue Res 12:73–90

Thomas RS, Greenawalt JW (1968) Microincineration, electron microscopy, and electron diffraction of calcium-phosphate-loaded mitochondria. J Cell Biol 39:55–76

Tintut Y, Demer LL (2001) Recent advances in multifactorial regulation of vascular calcification. Curr Opin Lipidol 12:555–560

Todd R, Blackmon P (1956) Calcite and aragonite in Foraminifera. J Paleontol 30:270–290

Tohda H, Takuma S, Tanaka N (1987) Intracrystalline structure of enamel crystals affected by caries. J Dent Res 66:1647–1653

Tomazic BB (2001) Physiochemical principles of cardiovascular calcification. Z Kardiol 90:68–80

Tomson C (2003) Vascular calcification in chronic renal failure. Nephron Clin Pract 93:c124–130

Towe KM, Cifelli R (1967) Wall ultrastructure in the calcareous foraminifera: crystallographic aspects and a model for calcification. J Paleontol 41:742–762

Trautz OR (1955) X-ray diffraction of biological and synthetic apatites. Ann N Y Acad Sci 60:698–713

Travis DF (1963) Structural features of mineralization from tissue to macromolecular levels of organization in the decapod crustacea. Ann N Y Acad Sci 109:117–245

Travis DF (1968) The structure and organization of, and the relationship between, the inorganic crystals and the organic matrix of the prismatic region of *Mytilus edulis*. J Ultrastruct Res 23:183–215

Travis DF (1970) The comparative ultrastructure and organization of five calcified tissues. In: Schraer H (ed) Biological calcification: cellular and molecular aspects. Appleton-Century-Crofts, New York, pp 203–311

Urist MR, Dowell TA (1967) The newly deposited mineral in cartilage and bone matrix. Clin Orthop Relat Res 50:291–308

Voegel JC, Frank RM (1977) Stages in the dissolution of human enamel crystals in dental caries. Calcif Tissue Res 24:19–27

Wallgren W (1957) Biophysical analyses of the formation and structure of human fetal bone. Acta Paed Suppl. 113

Weinbach EC, von Brand T (1965) The isolation and composition of dense granules from Ca^{++}-loaded mitochondria. Biochem Biophys Res Comm 19:133–136

Weiner S, Addadi L (2002) At the cutting edge. Science 298:375–376

Weiner S, Levi-Kalisman Y, Raz S, Addadi L (2003) Biologically formed amorphous calcium carbonate. Connect Tissue Res 44:214–218

Weiner S, Sagi I, Addadi L (2005) Choosing the crystallization path less traveled. Science 309:1027–1028

Weiss IM, Tuross N, Addadi L, Weiner S (2002) Mollusc larval shell formation: amorphous calcium carbonate is a precursor phase for aragonite. J Exp Zool 293:478–491

Wergedal JE, Baylink DJ (1974) Electron microprobe measurements of bone mineralization rate in vivo. Am J Physiol 226:345–352

Wheeler AP (1992) Mechanisms of molluscan shell formation. In: Bonucci E (ed) Calcification in biological systems. CRC Press, Boca Raton, pp 179–216

Wheeler EJ, Lewis D (1977) An X-ray study of the paracrystalline nature of bone apatite. Calcif Tissue Res 24:243–248

Wilmer WA, Magro CM (2002) Calciphylaxis: emerging concepts in prevention, diagnosis, and treatment. Semin Dial 15:172–186

Wilt FH (1999) Matrix and mineral in the sea urchin larval skeleton. J Struct Biol 126: 216–226

Wilt FH, Killian CE, Livingston BT (2003) Development of calcareous skeletal elements in invertebrates. Differentiation 71:237–250

Woodard HQ (1962) The elementary composition of human cortical bone. Health Phys 8:513–517

Woodhouse MA, Burston J (1969) Metastatic calcification of the myocardium. J Path 97:733–736

Wright JT, Duggal MS, Robinson C, Kirkham J, Shore R (1993) The mineral composition and enamel ultrastructure of hypocalcified amelogenesis imperfecta. J Craniofac Genet Dev Biol 13:117–126

Wu Y, Ackerman JL, Kim H-M, Rey C, Barroug A, Glimcher MJ (2002) Nuclear magnetic resonance spin-spin relaxation of the crystals of bone, dental enamel, and synthetic hydroxyapatites. J Bone Miner Res 17:472–480

Wuthier RE, Eanes ED (1975) Effect of phospholipids on the transformation of amorphous calcium phosphate to hydroxyapatite in vitro. Calcif Tissue Res 19:197–210

Wuthier RE, Bisaz S, Russell RGG, Fleisch H (1972) Relationship between pyrophosphate, amorphous calcium phosphate, and other factors in the sequence of calcification in vivo. Calcif Tissue Res 10:198–206

Young RA (1974) Implications of atomic substitution and other structural details in apatites. J Dent Res 53:193–203

Young JR, Davis SA, Bown PR, Mann S (1999) Coccolith ultrastructure and biomineralisation. J Struct Biol 126:195–215

Zander HA, Hazen SP, Scott DB (1960) Mineralization of dental calculus. Proc Soc Exper Biol (NY) 103:257–260

Zipkin I (1970) The inorganic composition of bones and teeth. In: Schraer H (ed) Biological calcification: cellular and molecular aspects. Appleton-Century-Crofts, New York, pp 69–103

Zylberberg L, Nicolas G (1982) Ultrastructure of scales in a teleost (*Carassius auratus* L.) after use of rapid freez-fixation and freeze-substitution. Cell Tissue Res 223:349–367

Zylberberg L, Géraudie J, Meunier F, Sire J-Y (1992) Biomineralization in the integumental skeleton of the living lower vertebrates. In: Hall BK (ed) Bone, volume 4: Bone metabolism and mineralization. CRC Press, Boca Raton, pp 171–224

5 The Shape of Inorganic Particles

5.1 Introduction

Analysis of the shape of the inorganic particles found in calcified tissues demands special attention because it may well provide clues about the mechanism that leads to the formation of one or other type of inorganic structure. It can be stated from the outset that the shape of these particles (from now on called 'crystals') usually evolves with their maturation, or may be changed by an abnormally structured organic matrix or the pathological availability, addition or subtraction of inorganic ions. In all cases, simple descriptive morphological data can be converted into information germane to an understanding of the mechanisms of the calcification process.

5.2 Vertebrates

For obvious reasons, most of the studies on the shape of inorganic crystals have been carried out in vertebrate skeletal structures and, precisely, in human bone. The calcification process, as it occurs in rats and mice, is usually considered to be equivalent.

5.2.1 Bone

The shape of crystals in bone and other calcified tissues has given rise to a crucial query, which still continues to sustain discussion: are bone crystals elongated, needle-like structures or flat platelet?

The earliest indications that the inorganic particles in fully calcified bone are shaped as thin, elongated, rod-like structures was obtained from polarized light studies and then confirmed by X-ray diffraction studies. Optical anisotropy was long known to be a fundamental property of bone. It was shown by Schmidt (1933) not only that bone has an intrinsic birefringence induced by both organic and inorganic components, but also that the latter have a form birefringence which complies with Wiener's law for rod-like composite bodies. This conclusion was confirmed by quantitative studies of Dallemagne and

Mélon (1946); they showed that the birefringence of total bone is the result of the combination of the negative birefringence of the inorganic crystals and the positive birefringence of rod-like bodies comprising both organic and inorganic fractions. Ascenzi (1948a, b, 1949) and Ascenzi and Bonucci (1964) also carried out quantitative evaluations of birefringence in whole, in decalcified and in anorganic bone (bone deprived of its organic constituents), finding that bone is a complex system consisting of at least three phases, that the osteon matrix's degree of birefringence falls as calcification goes forward, and that this depends not only on the deposition of calcium salts, but also on the falling content of the matrix interstitial water. Their conclusions were that the fall in birefringence could only be attributed to a slight degree to the negative intrinsic birefringence of the inorganic substance, that the pre-osseous bone matrix may be considered a composite, rod-like body according to Wiener's law, and that the form birefringence of calcifying bone suggests that its apatite crystals are rod-shaped.

A further indication that bone crystals are shaped like rods or needles, whose longest axis coincides with their crystallographic *c*-axis, was provided by X-ray diffraction carried out on bone (Clark 1931; Finean and Engström 1953; Carlström and Finean 1954; Carlström 1955; Fernández-Morán and Engström 1957; Carlström and Glas 1959; Matsushima et al. 1984), but also on calcified tendons (Clark 1931) and teeth (Carlström 1955). Small angle X-ray scattering was used in some cases (Finean and Engström 1953; Matsushima et al. 1984); in others, line-broadening measurements on oriented specimens were chosen (Carlström and Glas 1959).

A number of electron microscope studies have confirmed that bone crystals may have a rod-, needle-, or filament-like shape (Figs. 5.1, 5.2 and 5.3) (Scott and Pease 1956; Fernández-Morán and Engström 1957; Knese and Knoop 1958; Ascenzi and Benedetti 1959; Ascenzi et al. 1963, 1965, 1967; Warshawsky 1985). On the basis of personal experience, and in agreement with the results of Boothroyd (1975a), these crystals are only rarely straight structures, as the names 'needle' or 'rod' seem to suggest; much more often, they are curved or irregular and the term 'filament' would fit them better. To avoid complications, the terms 'needle', 'rod' and 'filament' will be used from now on as synonyms.

The agreement of X-ray diffraction and electron microscopy in indicating bone crystals as needle-like structures shows that this shape is not due to, or altered by, the chemicals used for processing calcified tissues for electron microscopy (Fig. 5.3). No significant changes in crystal shape seem to be produced by fixation, dehydration and embedding, or even by uranyl acetate and lead citrate staining (Bonucci 1975). Yet the exposure of crystals to aqueous solutions can cause their decalcification (Boothroyd 1964) or produce changes in their size, especially when calcium and inorganic phosphorus are used to limit dissolution or recrystallization (Eppell et al. 2001). On the other hand, needle-shaped crystals have been described in bone prepared anhydrously in organic solvents (Landis et al. 1977a) or by ultracryomicrotomy (Landis et al.

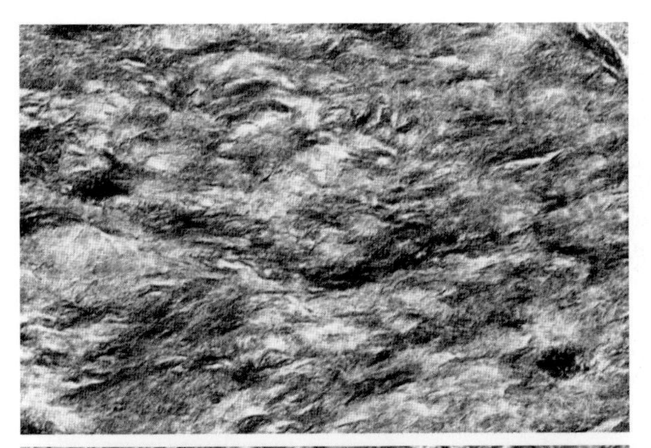

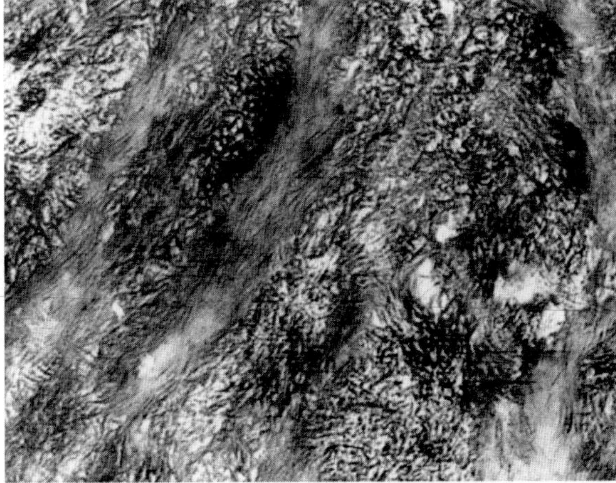

Fig. 5.1. *Above*: fully calcified osteon: the inorganic substance mainly consists of filament-like crystals; the granular inorganic bands can hardly be made out. Untreated, ×35,000. *Below*: spongy bone of the chick embryo: needle-like crystals are located between irregularly arranged, poorly calcified collagen fibrils; no granular inorganic bands are visible. Untreated, ×85,000

1977b), or in fresh, air-dried compact bone sectioned without embedding (Ascenzi et al. 1968). In this case, the electron density of the axial part of the crystals was found to be lower than that of the peripheral part, as if the crystal composition was not homogeneous.

It should be noted that the early mineral deposits in newly synthesized bone did not generate any electron diffraction pattern corresponding to a calcium phosphate solid phase, probably due to the very poor degree of crystallinity of the inorganic deposits (discussed in Sect. 4.2.1), rather than to their low volume (Landis and Glimcher 1978). Nanoparticles, corresponding to foci of crystal inception, have been described in areas of early calcification as rows of dots, which later fuse into needle-shaped crystals (discussed in Sect. 18.4).

Needle-shaped crystals have not been the only kind reported in bone. Studies carried out with small angle X-ray scattering (SAXS) have not only confirmed that needle-shaped crystals occur in bone (Fratzl et al. 1991, 1992), but have also shown that it contains platelet-like crystals (Ziv and Weiner 1994). Electron microscopy, too, has shown that, besides needle-shaped crystals, and often

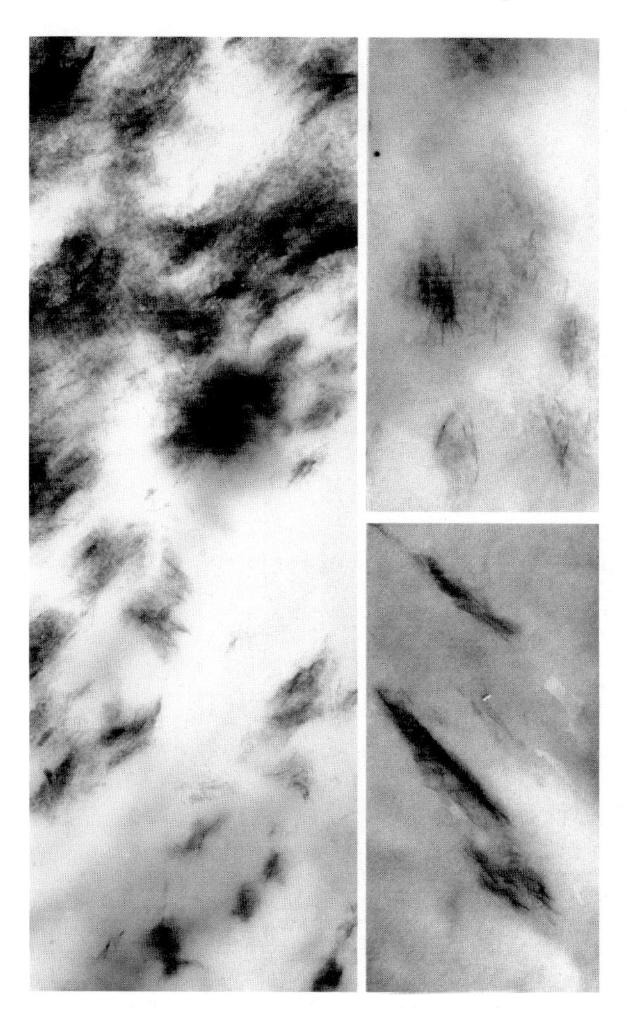

Fig. 5.2. *Left*: calcification front in compact bone; calcification nodules and calcification islands (both shown at high magnification on *upper and lower right*, respectively) are the only recognizable calcified structures; neither show the inorganic substance in bands, which is only recognizable where crystal aggregates have begun to coalesce (*upper half*). Unstained, ×33,000 and, *insets*, ×90,000

together with them, crystals with a platelet-like shape occur in bone (Fig. 5.3). Robinson RA (1952), Robinson RA and Watson (1952a) and Robinson RA and Cameron (1964) were among the first to suggest that the bone crystals are platelet-like, a shape more consistent with their being octocalcium phosphate than hydroxyapatite (Elliott 1973). Jackson et al. (1978) reported that bone crystals have a plate-like shape on the basis of bright and dark field images of unfixed, unembedded rabbit, ox, and human bone, and Arsenault and Grynpas (1988) reported the same findings using the same technique in cartilage and bone. On the basis of the results of TEM analysis of normal and osteoporotic bone, Rubin et al. (2003) identified plate-like crystals, with irregular edges and low electron density, and tablet-like crystals, probably no more than plate-like crystals cut widthwise, with distinct boundaries and a high electron density.

Fig. 5.3. *Above*: calcification front in medullary bone: the calcification nodules consist of needle- and filament-like crystals, whose shape and size have not been changed by staining with uranyl acetate and lead citrate; ×75,000. *Lower left*: detail of incompletely calcified osteon showing the inorganic substance in bands; note that single components of the bands appear as platelet-like crystals. Needle-like crystals are rare. Untreated, ×96,000. *Lower right*: detail of fractured compact bone; structures, probably collagen fibrils encrusted with platelet-like crystals, protrude from the fracture surface. Untreated, ×96,000

The percentages of needle- and platelet-like crystals vary with the bone matrix site, its degree of calcification, and the bone type. Differences in crystal shape emerge clearly when the calcification front – i.e., the zone of early calcification – in the osteoid tissue of compact (osteonic) bone is compared with that of bones displaying a rather loose framework of collagen fibrils and rapid calcification, such as embryonic bone, woven bone and the medullary bone of birds (this is the bone that fills the bone marrow space of long bones in birds under the action of estrogens, and is removed by osteoclasts just before egg deposition to yield calcium for egg shell calcification; see Bloom et al. 1942). Hardly any platelet-like crystals are found in either case, but in bone with a loose collagen texture, the calcification front can be recognized by the prominence of roundish aggregates ('calcification nodules') of almost radially arranged, needle-shaped crystals (Figs. 5.2 and 5.3). By contrast, in bone with a compact collagen texture, the calcification front can be identified through its elongated, electron-dense areas ('calcification islands'; Fig. 5.2) comprising two types of inorganic structure: many needle-shaped crystals, whose long axis is oriented in the same direction as that of collagen fibrils, and a few patches of very small, electron-dense granules (about 1.0 nm across) (Ascenzi et al. 1965) which, as calcification increases, forms electron-dense bands. These bands run transversally through adjacent collagen fibrils and have a close relationship with their period (Figs. 5.2, 5.3, and Sect. 7.2); their width roughly corresponds to that of the hole zones of the collagen fibrils. For the sake of simplicity, and without reference to their being crystalline or not, these distinctive bands will be called 'granular inorganic bands', and will be considered to consist of 'granular inorganic substance', from now on (further discussed below and in Sect. 16.2).

Variations in the percentages of needle- and platelet-like crystals can clearly be observed in osteon compact bone. As mentioned above (Sect. 2.2), the microradiographic method shows that calcium content is 5–20% higher in primary bone than in secondary bone, and that osteon calcification in secondary bone occurs in two stages: an initial stage in which a level of about 70% of matrix calcification is rapidly reached, and a second stage in which calcification is slowly completed, while remaining 5–20% below that of the primary periosteal bone (Amprino and Engström 1952). So if microradiographically selected osteons are examined under the electron microscope, matrix zones with progressively higher degrees of calcification can be compared (Ascenzi et al. 1978), ranging from the calcification front at the osteoid border (Scott and Pease 1956; Hancox and Boothroyd 1965) to incompletely calcified osteons, and from these to fully calcified osteons (Ascenzi et al. 1965) and primary bone (Ascenzi et al. 1967). Using this system, it has been found that the incompletely calcified matrix of young osteons shows two different types of mineral substance: long needle-shaped crystals whose long axis runs roughly parallel to that of the collagen fibrils, and predominant granular inorganic bands (Figs. 5.1, 5.2 and 5.3 and Sect. 7.2). As calcification proceeds, and the osteon is fully calcified, these bands become less and less clearly visible, as if they were masked by the high amounts of needle-shaped crystals (Fig. 5.1).

Speaking from personal experience, and as shown in Fig. 5.3, the granular inorganic bands often seem to merge into platelet-like crystals, as if the latter derived from a maturation of the former. Changes in crystal shape (and size) with maturation are well known (e.g., in enamel; see Sect. 6.2.6) and are considered possible in bone, although they would be too slight to be easily recognized (Posner et al. 1965). It remains to be determined whether granular inorganic bands and platelet-like crystals are different, unrelated structures, or whether the former merge into the latter as calcification proceeds.

The percentage of needle- and platelet-like crystals and granular inorganic bands changes not only with the degree of calcification, but also with the type of bone. Platelet-like crystals have chiefly been described in the matrix of compact lamellar bone, especially when incompletely calcified (Fig. 5.3 and Sect. 7.2), whereas needle-like crystals have mainly been reported in the matrix of embryonic and woven bone, and in the medullary bone of birds (Fig. 5.3). A small-angle X-ray scattering study of mineral crystals in different types of bone and species showed that needle-shaped crystals are mainly present in mouse and rat bone, whereas plate-shaped crystals are mainly found in adult human bone and calcified turkey leg tendons (Fratzl et al. 1996a). In a comparative study by SAXS, Fratzl et al. (1992) found platelet-like crystals in calcified turkey tendons and needle-like crystals in bone from calvaria, femurs and the iliac crest of the mouse, rat and dog. In alveolar bone, Cuisinier et al. (1987) found both platelet- and needle-like crystals; the same findings were reported in bone prepared anhydrously by means of acrolein vapours (Landis et al. 1980) and in human woven bone (Su et al. 2003).

The relationships that may exist between these three types of inorganic structure (needle-like crystals, platelet-like crystals and granular inorganic bands) have not been examined in depth; on the basis of personal experience, needle-shaped crystals and granular inorganic bands are recognizable under the electron microscope much more easily than platelet-like crystals, which are hard to find in optimally preserved specimens and can sometimes be detected at the margin of fractures (Fig. 5.3).

Stereoscopic studies carried out with goniometers that make it possible to examine single crystals from different angles have not only supported the existence of platelet-like crystals in bone, but have gone farther by assuming that needle-shaped crystals are actually no more than platelet-like crystals seen from one side. Bocciarelli (1970), who examined both fixed and embedded, air-dried and unembedded bone and isolated bone crystals by means of ultrastructural stereoscopy, reported that, besides needle-like crystals, platelet-like crystals are easily recognizable. She was one of the first to suggest that the former are a sideways view of the latter. This was also the opinion of Johansen and Parks (1960) and of Boothroyd (1975a), who used the same stereoscopic method and dark field imaging, and of Landis et al. (1977b), who examined the bone of embryonic chicks and young mice processed by ultracryomicrotomy. Fernández-Morán and Engström (1957), who reported that most bone crystals are needle-shaped, found thin flakes with a fine internal structure revealed by regular striations, in some areas. They supposed that these flakes are formed by the lateral aggregation of rod-shaped particles. Höhling et al. (1971a) were of the opinion that the early inorganic deposits appear as rows of dots, which later fuse into needle-shaped crystals, which in turn fuse into platelet-like structures. According to Meyer et al. (1972), who induced crystal formation on in vitro seeds, the process starts with the deposition of amorphous inorganic material; this is followed by the development of needle-like structures which eventually convert into 1–2 μm platelets. On the other hand, Brandenburger and Schinz (1945) were of the opinion that, at least theoretically, the crystals might be either isometric hexagons or plate-like structures, and that the smallest of them might have a needle-like shape.

The existence of platelet-like crystals in bone has chiefly been substantiated by studies of individual crystals (Fig. 5.4) that had been separated and isolated from the calcified matrix using a procedure that can be summarized as follows: grinding or crushing bone in an agate mortar in liquid nitrogen; treatment with sodium hypochlorite to oxidize the organic phase; sonication to disperse the resultant particles; suspension first in water equilibrated with synthetic calcium phosphate and then in 100% ethanol; further sonication; separation of crystal aggregates from individual crystals by gravity setting in ethanol;

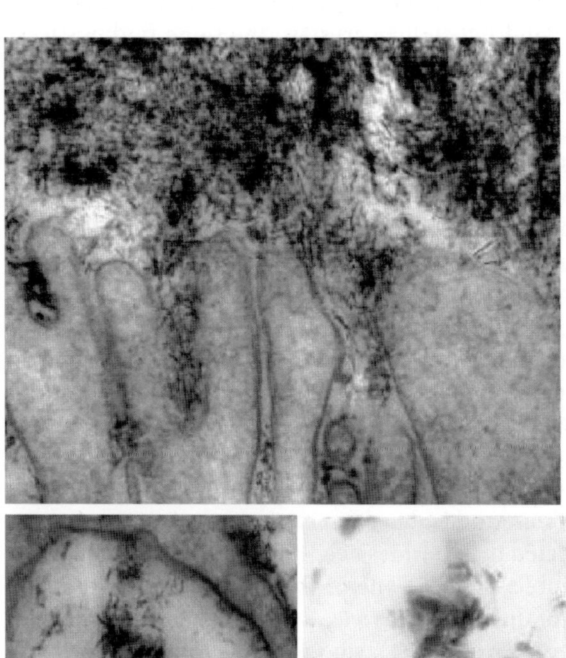

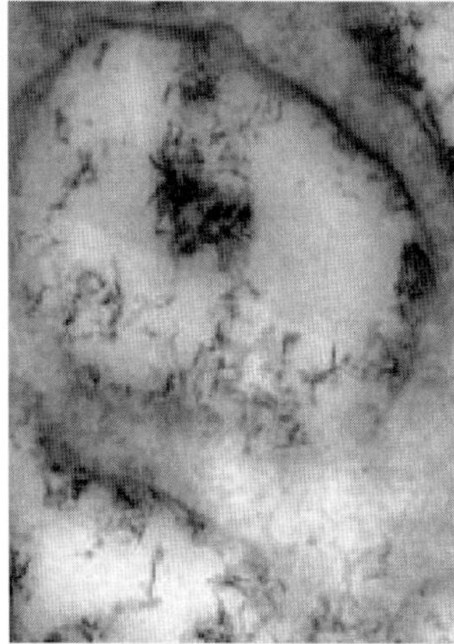

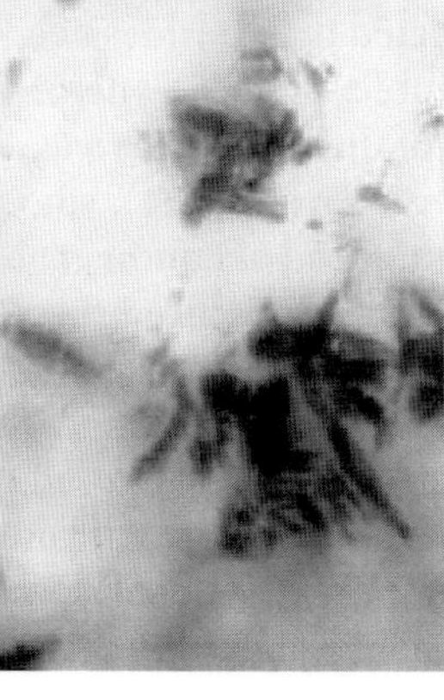

Fig. 5.4. *Above*: area of osteoclastic bone resorption: note the dissociated, loosened, needle-like crystals, partly contained in the channels of the brush border; no platelet-like crystals are detectable. Untreated, ×96,000. *Lower left*: detail of an osteoclast cytoplasmic vacuole, containing needle-like crystals and irregular, poorly defined structures which might be crystals on the way to dissolution; no platelet-like crystals are detectable. Untreated, ×90,000. *Lower right*: crystals isolated from formaldehyde-fixed compact bone; pellet postfixed with osmium tetroxide: some crystals show a polygonal, plate-like shape. Untreated, ×110,000

study by electron microscopy (Weiner and Price 1986; Weiner and Traub 1989, 1991; Weiner et al. 1991), high resolution electron microscopy (Su et al. 2003), small-angle X-ray scattering (Wachtel and Weiner 1994), or X-ray diffraction line width broadening technique (Ziv and Weiner 1994). Crystals isolated with the method of Kim H-M et al. (1995) (based on low temperature, non-aqueous solution and low power plasma ashing for the degradation of the organic matrix) were also reported to be platelets (Eppell et al. 2001;Tong et al. 2003). Termine et al. (1973) found that crystals isolated from hydrazine-deproteinated bone are platelet-like. Moradian-Oldak et al. (1991) reported the same result after applying the same method to compact bone.

The fact that these studies have been carried out on isolated crystals dissociated from bone by mechanical and chemical treatments (Robinson RA 1952; Johansen and Parks 1960; Bocciarelli 1970; Termine et al. 1973; Weiner and Price 1986; Weiner and Traub 1989, 1991; Moradian-Oldak et al. 1991; Weiner et al. 1991; Wachtel and Weiner 1994; Ziv and Weiner 1994; Eppell et al. 2001; Su et al. 2003; Tong et al. 2003) raises doubts that the crystal shape may have been changed and that the platelet-like shape may be the results of the removal of crystals from their biological environment. This doubt is reinforced when it is considered how thin and small these crystals are, and how easy it would be to alter their physical state, an interpretation made more plausible by their low electron-density, fine granularity, poor definition and irregular ultrastructure after isolation (Fig. 5.4). It is interesting that in a recent study on the shape and size of bone crystals isolated using the method of Kim H-M et al. (1995) and examined by atomic force microscopy, the isolated platelet-like structures were called 'mineralites' instead of crystallites (Eppell et al. 2001). The reasons for this were not only that they had a poor stoichiometry and low degree of crystallinity, but also that they turned out to be only one or two unit cells in thickness, a value too small for them to be classified as crystals (S.J. Eppel, personal communication). It might be speculated that platelet-like crystals derive from the granular inorganic bands when these are fragmented by the microtome knife or by extraction methods. This topic is further discussed in Sect. 16.2.

The separation and dissociation of isolated crystals from the bone matrix take place naturally at sites of osteoclastic bone resorption (Fig. 5.4). Several electron microscope studies have, in fact, reported free, individual crystals – described as 'needles' and 'filaments', never as 'platelets' – in the disintegrating layer of the bone matrix but also in the space between this layer – from which they have been dissociated – and the brush border of osteoclasts, as well as between the infoldings of the brush border and within cytoplasmic vacuoles (Scott and Pease 1956; Dudley and Spiro 1961; Gonzales and Karnovsky 1961; Bonucci 1974, 1981; Lucht 1972; Göthlin and Ericsson 1976; Holtrop and King 1977; Pierce 1989; Holtrop 1991). Some of these crystals, especially those contained in cytoplasmic vacuoles, are mixed with a few, irregular, granular, low electron-dense structures (Fig. 5.4). Even if their shape is different, these might be considered to be equivalent to the platelet-like crystals isolated by mechanical or biochemical methods. Alternatively, they might be viewed as needle-shaped crystals whose shape has been modified by still incomplete solubilization and dissolution. Similar results have been reported for the odontoclast resorption of enamel (Sahara et al. 1998).

Before any conclusion about the shape of bone crystals can be drawn from these results, it should be recalled that, just as the ultrastructure of crystals may be changed by the mechanical or chemical procedures used in the laboratory to isolate them, and as they may be fractured by cutting sections with the ultramicrotome, so the ultrastructure of crystals located in areas of bone resorption may have been changed by osteoclast activity and/or the effects of the environment. It is, in fact, accepted that bone resorption occurs in two phases, the first consisting of the decalcification of the bone matrix, the second of the extracellular digestion of its organic components. A wide variety of contributions to the literature (Blair and Ghandur-Mnaymneh 1985; Vaes 1988; Blair et al. 1989; Teti et al. 1989; Bekker and Gay 1990; Marks and Popoff 1990; Sundquist et al. 1990; Väänänen et al. 1990, 1996, 2000; Laitala and Väänänen 1993; Teti 1993; Baron 1995; Hall and Chambers 1996; Katsunuma 1997; Salo et al. 1997; Schlesinger et al. 1997; Rousselle and Heymann 2002) unanimously support the view that osteoclastic bone resorption occurs in an acidic environment and that in the earliest phase an osteoclast proton pump acidifies the compartment encompassed by the bone surface and the osteoclast brush border, so inducing decalcification of the bone matrix, while in the second phase osteoclast-synthesized metalloproteases digest the unmasked collagen fibrils (Vaes 1988; Delaissé et al. 1993; Hill et al. 1995; Holliday et al. 1997). The irregular, low electron-dense, platelet-like structures found in vacuoles of the osteoclast cytoplasm could therefore correspond either to needle-shaped crystals undergoing osteoclastic disaggregation in an acidic environment, or to true but modified platelet-like crystals. This ambiguity strengthens the possibility that the shape of chemically or mechanically isolated crystals may itself have been changed by the disaggregating procedures. As far as we know, the simple disconnection of such thin crystals from the matrix organic framework could potentially induce radical changes in their shape and structure.

A digression seems appropriate at this point to discuss some aspects of the just mentioned process of osteoclastic bone resorption, because they might have a bearing on the question of what the true shape of the crystals is. If bone resorption occurs through a process of acidic decalcification of the bone matrix, as documented by the studies cited above, two discrepancies become apparent: first, the presence of free, isolated, extremely thin crystals at the bone-osteoclast interface, as well as between the infoldings of the brush border and in cytoplasmic vacuoles, is incompatible with an acidic environment which would very quickly solubilize them (see also Hancox and Boothroyd 1961); second, the presence, within the matrix that is undergoing resorption, of collagen fibrils that, even if dissociated, are still masked by crystals (Bonucci 1974) is again in contrast with an acidic environment, because they too should be decalcified and should appear as nude fibrils, which, however, have only been found in exceptional cases (never in personal experience) at the resorption front (Hancox and Boothroyd 1963). At present, these discrepancies are hard to reconcile with the acidic decalcification of the matrix. To do so, it might be proposed that the osteoclast acidic solubilization of the disaggregated apatite crystals increases the calcium and phosphate ion concentration in the resorption compartment to such an extent that crystal reprecipitation may occur during specimen processing for electron microscopy. This hypothesis seems implausible for the following reasons: a) any free calcium and phosphate ions would be rapidly washed away when the specimen is soaked in aqueous fixative solutions, whereas crystals continue to be visible under the electron microscope after fixation with aldehydes and repeated washing in distilled water; b) the probability that three or more individual ions in solution undergo simultaneous collision is low (Engfeldt and Reinholt 1992); c) the reconstituted crystals would probably have different shape and size with respect to those found in whole bone, and the gradient of their concentration – higher near the unresorbed bone matrix than near the brush border – would probably be disrupted; d) if the reprecipitation hypothesis is accepted, then it would be necessary to conclude that all the crystals seen under the electron microscope may have the same artifactual origin. Another possible explanation has been put forward by Schenk et al. (1967): on the basis of their observation that crystals appear between the infoldings of the brush border of the chondroclast if the tissue is fixed with phosphate buffer solutions but not if it is fixed with solutions prepared with cacodylate buffer, they suggested that crystals are produced by the reaction of calcium ions with the phosphate groups of the buffer. According to personal experience, there are no substantial differences in the ultrastructure of apatite crystals in tissues fixed with phosphate or cacodylate solutions. The fine structure of bone matrix, as also of bone matrix undergoing resorption, of dissociated inorganic crystals, and of collagen fibers, was not affected by the type of fixation (simple fixation with OsO4 or double fixation with aldehydes plus OsO4), or by the buffers (phosphate or cacodylate) used to prepare the fixative solutions (Bonucci 1974). Similarly, the early crystals of calcification nodules located in the osteoid tissue of embryonic chick skull showed the same ultrastructure after the specimens were fixed in different ways using formaldehyde, glutaraldehyde and osmium tetroxide in various combinations with sodium cacodylate or sodium phosphate buffers (Boothroyd 1975b). In a study on the odontoclast resorption of the enamel in human deciduous teeth, Sahara et al. (1998) found that the crystals lying in the enamel prisms at the resorption surface were more loosely packed and electron-lucent than those of controls. They hypothesized that the prisms were partially decalcified and that crystals were liberated from them. This hypothesis seems to be in conflict with the observation that in partially decalcified bone a sharp demarcation exists between the calcified and decalcified matrix (Robinson RA and Watson 1952b), i. e., the matrix is either completely decalcified or completely undecalcified.

The inconsistency of the presence of dissociated nanocrystals in an acidic pH led Bonucci (1974) to suggest that bone resorption may occur in two phases. The first could consist of the enzymatic digestion and alteration of the organic components of the calcified matrix; this would lead to the loosening of collagen fibrils, which would become argyrophilic, and to the release of dissociated crystals (most of which, as discussed in Sect. 16.2, are located outside collagen fibrils). The second would be marked by the phagocytosis of the released crystals and by their ultimate dissolution in the acidic environment of the cytoplasmic vacuoles. This hypothesis seems to be in conflict with the idea that enzymes can only spread through the dense calcified matrix of bone after it has been decalcified (Neuman et al. 1960). Collagenase can, however, be stored as a latent proenzyme in the mineralized matrix (Eeckhout and Delaisse 1988). Moreover, the calcified matrices seem more permeable to macromolecules than has been supposed so far. It is known, for instance, that during the maturation of tooth enamel its organic components are enzymatically digested and almost completely removed in spite of its compactness. In addition, changes in the soluble organic matrix prior to inorganic removal have been documented during the resorption of fish acellular bone by osteoclast-like cells (Weiss and Watabe 1978, 1979).

In spite of many persistent doubts, all the data (personal and in the literature) seem to converge toward the conclusion that two types of inorganic structures (needle-, rod- or filament-like on one hand, and platelet-like on the other) do exist in bone. It must still be determined whether the granular inorganic bands are a third form of inorganic substance aggregation or no more than platelet-like crystals at the initial stage of formation, or, alternatively, whether the platelet-like crystals are no more than fragments of disintegrating granular inorganic bands. As further discussed below and reported in Table 5.1, the prevalence of one of these structures depends on the type of bone (spongy or compact), degree of aggregation of collagen fibrils (loose or compact) and stage of calcification (initial or final). The needle-shaped crystals are the only crystals present in calcification nodules in bones with loose collagen texture and many of them constitute the calcification islands and coexist with the granular inorganic bands within compact bone. The platelet-like crystals are mainly found in compact bone, where they coexist with needle-like crystals. The relationships between needle- and platelet-like crystals, on one hand, and platelet-like crystals and granular inorganic bands on the other, and how they evolve with aging, are further discussed in the following chapters.

5.2.2 Tendons

The leg tendons of certain avian species – most specifically the domestic turkey, *Meleagris gallopavo* – calcify as a function of age and provide an opportunity to examine non-calcified, fully calcified and transitionally calcified portions of

Table 5.1. Main distribution of crystal types in bone and calcified tendons

Type of tissue	Aggregation of collagen fibrils	Stage of calcification	Site of crystals	Type of crystals
Woven bone		Initial	Calcification nodules	Needles
Spongy bone	Loose			
Medullary bone		Final	Bone matrix	Mostly needles few platelets
		Initial	Calcification islands	Needles; granular bands
Compact bone	Dense			
		Final	Bone matrix	Mostly platelets; needles
		Initial	Matrix vesicles	Needles; platelets
Tendon	Dense			
		Final	Tendon matrix	Platelets

the same tendon (Abdalla 1979; Yamauchi and Katz 1993; Vanden Berge and Storer 1995; Landis and Silver 2002). So they are considered good models for the study of bone calcification (Bigi et al. 1988; Arsenault 1992). In areas of early calcification, the earliest detectable inorganic deposits have been described as needle-like crystals (Traub et al. 1992) or as fine needles or plates associated with vesicular structures (Landis 1985; Landis and Arsenault 1989; Landis and Song 1991). The crystals found in fully calcified portions are shaped like those in fully calcified bone. Crystals have, in fact, been described as having a needle- (Nylen et al. 1960; Myers and Engström 1965; Krefting et al. 1980; Arsenault 1988; Bigi et al. 1988) or platelet-like (Heywood et al. 1990; Moradian-Oldak et al. 1991; Fratzl et al. 1992; Landis et al. 1993) shape; these two types have been reported as occurring together in the same calcified tendon (Höhling et al. 1976a, 1990; Landis et al. 1991; Lees et al. 1994), the latter having been described more often than the former. As discussed in bone, at least some of the needle-shaped crystals are considered to be platelet-like crystals viewed from one side (Lees et al. 1994).

Platelet-like crystals have been found by applying electron microscopy to single collagen fibrils shredded by sonication from calcified, mortar-crushed, turkey leg tendons (Weiner and Traub 1986; Traub et al. 1989; Landis et al. 1991; Wachtel and Weiner 1994) and their presence has been confirmed by high resolution transmission electron microscopy of isolated crystals (Heywood et al. 1990) and tapping mode atomic force microscopy of collagen fibrils teased with forceps from tendon zones at various degrees of calcification (Siperko and Landis 2001). So, as in the case of bone, descriptions of tendon crystals as platelets have at least partly been based on electron microscope studies of isolated collagen fibrils or crystals which had been mechanically separated from calcified tendons by grinding, scraping, mortar crushing and/or treatment with chemical substances (Weiner and Traub 1986; Traub et al. 1989; Landis et al. 1991; Moradian-Oldak et al. 1991; Wachtel and Weiner 1994). As already discussed, these isolation methods may create artifacts and justify a suspicion that the platelet-like shape is not the true shape of tendon crystals. Their existence is, however, accepted by most investigators. The findings of Landis et al. (1993), who used high-voltage electron microscopic tomography and graphic image reconstruction of embedded calcified tendons, point to the same conclusion.

5.2.3 Dentin

Dentin crystals at an early stage of formation have been described by Höhling and coworkers (Höhling et al. 1970, 1971b, 1995; Plate et al. 1992, 1994; Wiesmann et al. 1993; Arnold et al. 1997, 1999; Stratmann et al. 1997) as chains of minute dots which coalesce into needle-shaped crystals and later into platelet-like ones. According to Kinney et al. (2001), crystals have an approximately needle-like shape near the pulp, but become more plate-like in shape near the dentino-enamel junction. Tubule-like crystals have been reported in ion beam thinned dentin (Boyde 1974), and filament-like ones in cryopreserved dentin (Carter et al. 2000).

Crystals in mature dentin look very like those in compact bone. Johansen and Parks (1960), who studied dentin and bone using stereoscopic methods and dark field imaging, reported the presence in both tissues of platelet-like crystals which appeared as needles or tubules when viewed sideways. Schroeder and Frank (1985) found platelet-like crystals in dentin. Linde and Goldberg (1993), in their review of dentinogenesis, reported that crystals in intertubular dentin are plate-shaped, taking the form of two flat parallelepiped plates. Butler (1998), in his review of dentin matrix proteins, reported plate-like carbonate apatite crystals. Weiner et al. (1999) treated dentin powder with 5.0 wt % NaOCl solution for two weeks to degrade the organic matrix; the mixture was then sonicated and fractionated on the basis of density. They described platelet-like crystals in the densest fraction, which was assumed to correspond to peritubular dentin.

5.2.4 Cementum

Crystals in cementum are thought not to differ from those in bone. Hayashi (1987) reported filament-like crystals at the calcification front of cellular and acellular cementum and needle-like crystals in advanced calcified matrix. Arzate et al. (2000) found filament- and needle-like crystals in the mineral-like tissue deposited in vitro by human cementoma-derived cells.

5.2.5 Cartilage

The crystals found in early calcification sites in the growth cartilage are either located in matrix vesicles or assembled in roundish aggregates (calcification nodules) which, viewed as a whole, make up the calcification front (Fig. 3.5). The former are discussed below, together with the matrix vesicles of other tissues; the latter resemble, but are more plentiful than, those found at the bone calcification front. A number of reports have stressed that the crystals which constitute the cartilage calcification nodules have a needle-, rod-, or filament-like shape (Robinson RA and Cameron 1956; Scott and Pease 1956; Takuma 1960; Aho and Isomäki 1962; Bonucci 1967, 1969, 1971; Thyberg 1974). Landis and Glimcher (1982) and Arsenault and Hunziker (1988) have, however, objected that, as demonstrated in bone, in cartilage too the transition of rod-like crystal images into platelet-like ones can be shown by the serial tilting of the electron microscope stage. Arsenault and Grynpas (1988) reported that the same phenomenon can be seen in synthetic, poorly crystalline hydroxyapatite. Kim H-M et al. (1996) isolated crystals from the calcified zone of chicken and bovine subarticular cartilaginous growth plates and deproteinated them using a low temperature reaction with hydrazine and plasma ashing, on the assumption that this treatment would not change crystal shape. The isolated crystals appeared under the electron microscope as thin, curved, rectangular platelets and, surprisingly, the needle-shaped crystals so often reported in intact calcified cartilage were not found; strangely enough, although the crystals had been deproteinated, 1–2% proteins were still present. Barckhaus and Höhling (1978) described needle-like crystals in the central zone of the calcification nodules and platelet-like crystals at their periphery. Shitama (1979) found two different crystal shapes in an electron microscope study of calcified articular cartilage: slender, twisted, curved crystals, mainly located in the developing epiphysis, and short, needle-like, slightly curved crystals found in the calcified articular cartilage. Höhling et al. (1976b) reported platelet-like crystals in the freeze-dried, unstained cartilage of the guinea pig. Electron microscope investigations of the growth plate cartilage prepared by the quick-freeze, deep etching replica technique and conventional methods, showed that both needle- and platelet-like crystals could be recognized in matrix vesicles (Akisaka et al. 1998).

When sites of early calcification (calcification nodules) in intact, undamaged cartilage (specifically, cartilage as processed for electron microscopy, i. e., fixed in aldehyde alone, dehydrated in ethanol and embedded in an epoxy resin, but not decalcified or stained) are examined under the electron microscope, practically all the crystals appear as thin, elongated structures resembling filaments rather than needles or rods and, in any case, the few platelet-like structures that can be found are much thinner than those seen in bone. A detailed description of the crystals within calcification nodules found in cartilage processed in that way has been given by Bonucci (2002): practically all the crystals, which stand against an electron-dense, amorphous or granular background, are described as needle- or filament-like structures. Just a few of them are branched, have a variable electron density and thickness – the least dense being the thickest, are mixed with low electron-dense, filamentous structures which protrude from the periphery of the nodules and may have a swollen extremity, and in rare cases are arranged in pairs (an arrangement found in dentin calcification nodules by Hayashi 1984) or outline platelet-like polygonal figures. The possible meaning of these figures is discussed by Bonucci (2002) and is considered again in Sect. 18.4. The platelet-like crystals described in bone and tendons have not been found in the calcification nodules of the cartilage.

5.2.6 Enamel

While the crystals in dentin are quite like those in bone, the crystals in dental enamel have a distinctive shape, which is of the greatest interest. Nylen et al. (1963), using electron microscopy and X-ray diffraction, found that early enamel consists of very long (at least 100 μm, according to Daculsi et al. 1984), thin, ribbon- or plate-like crystals (Fig. 4.4), which grow to take on the form of hexagonal rods. These crystals may show a helical or spiral structure (Travis 1968a; Warshawsky 1989), which has been ascribed to dehydration during specimen preparation (Suzuki et al. 1998), with a periodicity of 25–55 nm (Suzuki et al. 1998). Early enamel crystals are intermingled with dentin crystals (Arsenault and Robinson 1989) and no calcification front comprising calcification nodules or islands can be made out.

The ribbon-like shape of early enamel crystals, their growth into polygonal structures resembling flattened hexagons, and their orientation with the *c*-axis parallel to that of the prisms that contain them have been repeatedly confirmed (Rönnholm 1962; Travis and Glimcher 1964; Frazier 1968; Travis 1968a; Selvig 1973; Jongebloed et al. 1975; Daculsi and Kerebel 1978; Kerebel et al. 1979; Cuisinier et al. 1992; Pergolizzi et al. 1995; Aoba 1996a, b). Warshawsky and Nanci (1982) agreed that immature enamel crystals are thin, flat, twisting ribbons, but remarked that images interpreted as hexagons may be due to perspective image distortion. Warshawsky et al. (1987) and Warshawsky (1989) later supported the idea that the hexagonal shape of mature crystals is an optical artefact, because true hexagons should project on the screen or photographic plates as octagonal profiles. This concept was challenged by Kallenbach (1990), who claimed that crystals of hexagonal shape do project as hexagons and only rarely as octagons. So, while there is general agreement that early enamel crystals are ribbon-like, controversy about the true *a*- to *b*-axis shape of mature crystals continues (Nanci and Smith 1992).

It is worth stressing that there is a difference between early, or immature, and late, or mature, enamel crystals. The former are invariably described as thin, flat, long structures that resemble ribbons; the latter are long, too, but their *a* and *b* axes increase with maturation, so that they very rapidly acquire a definitive parallelepiped shape. The atomic force microscopy studies of Kirkham et al. (1998) have shown that the ribbon-like crystals of secretory enamel have greater surface roughness, and are more irregular than, the crystals of maturing enamel, showing spherical sub-structures of 20–30 nm diameter arranged along the *c*-axis. During the maturation process, crystals must conform to the neighbouring ones within the available space, and this may affect their outline. Kerebel et al. (1979) have suggested that adjacent enamel crystals can undergo fusion; however, goniometric tilting of the apparently fused crystals showed that they could be separated into individual crystals (Dong and Warshawsky 1996).

5.3 Lower Vertebrates

The shape of the inorganic structures varies to some extent according to the tissue in which they are located, but on the whole it is similar to that of high vertebrate crystals (Zylberberg et al. 1992). Early X-ray diffraction studies by Henny and Spiegel-Adolf (1945) showed that there are no substantial differences between bones of fishes and those of vertebrates (lamb, rat, cat). This is in agreement with the observation that in early calcification sites in fish scales the inorganic substance consists of needle-shaped crystals (Yamada J and Watabe 1979; Olson and Watabe 1980) and that crystals of exactly this shape are recognizable in the figures published by Glimcher (1959) in his paper on the molecular biology of calcified tissues. In the external layer of the teleost scales – the first layer to be calcified – initial mineral deposits are described as dense rodlets made up of rows of dots (Schönbörner et al. 1979). Needle-shaped crystals have been found in fish dentin both at an early stage of calcification and in fully calcified matrix (Yamada M and Ozawa 1978; Lees and Prostak 1988). Needle- and platelet-like crystals have been reported in bony structures (Landis and Géraudie 1990) and in the superficial layers of the scales (Sire 1988). Needle-shaped and "flaky" crystals have been found in the fibrillary plates of the scales in *Carassius auratus* (Onozato and Watabe 1979). Lee and LeGeros (1985) have reported dislocations in shark enamel resembling those of human enamel.

5.3.1 Enameloid

The crystals found in enameloid are initially needle-shaped (Wakita 1993), but take on a hexagonal cross section in fully calcified areas (Garant 1970). This pattern may, however, vary, probably in response to fluoride content

(LeGeros and Suga 1980). Miake et al. (1991) have reported that elasmobranch enameloid, which contains high levels of fluoride, is distinguished by the initial formation of thin crystals that have a prismatic hexagonal cross-section; these crystals then fuse into thick hexagons, whereas teleost enameloid, which has low levels of fluoride, is marked out by the initial formation of ribbon-like crystals which later grow into flat hexagonal prisms. Slavkin and Diekwisch (1996) found that the initial calcification of shark and ray enameloid shows minute crystalline precipitates at the ameloblast basal lamina. In tooth germs of the fish *Hoplognatus fasciatus*, Inage (1975) found small enameloid crystals that were needle-like and tubular, initially appearing in the circumference of collagen fibers and then in their central portion, and finally becomimg mixed with platelet-like crystals.

5.3.2 Isopedine

Isopedine is characterized by crystals resembling those of bone (Zylberberg et al. 1992), while ganoine, which is hypermineralized, contains filament-like crystals similar to those of vertebrate enamel (Zylberberg et al. 1992), a tissue with which ganoine can be closely identified (Sire et al. 1987; Sire 1994).

Inorganic crystals which provide X-ray diffractograms consistent with hydroxyapatite particles like those found in bone have been described in whale baleen by Pautard (1965). They are described as filaments of different densities which can be aggregated into clusters (Pautard 1975), and may be denser at the periphery than at the axial core, as if they were structured as tubules (Arnott and Pautard 1968).

5.4 Invertebrates

A number of studies have been carried out on the shape taken by the inorganic substance in invertebrates, primarily to understand its nature, structure, arrangement and mechanism of formation, but also because the identification and classification of invertebrates largely depend on the shape of their calcified structures (see reviews by Towe and Cifelli 1967; Pautard 1970; Travis 1970; Benson and Wilt 1992; Wheeler 1992). Polarizing microscope and X-ray diffraction studies have provided a great deal of information on the characteristics of these structures, but have left open the basic problem whether each of them consists of a single crystal (in some cases wider than several microns), or whether each is a polycrystalline structure with crystals so perfectly aligned that they behave as a single crystal (Okazaki and Inoue 1976). This problem has been studied in several different lower organisms. From his crystallographic analysis of coccoliths of the alga *Coccolithus huxleyi* (for a review on coccolithophorids see Young et al. 1999), Watabe (1967) concluded that, in spite of a very complicated microarchitecture, their entire structure was a single crystal of calcite. He did, however, add the comment that the term 'crystal' was being used in a crystallographic sense, and that each single crystal could actually consist of many smaller crystals oriented in parallel. Pautard (1970), in his review on calcification in unicellular organisms, made the similar comment that the appearance of a complex calcified structure such as a single crystal may depend on a perfect alignment of many small crystals in adjacent planes. He concluded that on the basis of the results available at that time, it was not possible to determine whether the larger crystals had atoms arranged in a continuous lattice, or whether they consisted of a mosaic of single, small crystals aligned with such precision as to convey the optical picture of a single crystal. Towe (1967), on the basis of an electron microscope study of replicas of natural and broken surfaces of echinoid skeletal plates, concluded that their inner portion has the morphology of a single calcite crystal, whereas the outer portion is a polycrystalline aggregate with a preferential orientation. Towe and Cifelli (1967) commented that it is difficult to distinguish between a single calcite crystal and a polycrystalline mosaic, adding that the plate-like units attributed to calcified structures in Foraminifera are not individual crystals but parts of a larger crystal mosaic. Towe and Thompson (1972) carried out an electron microscope investigation on *Mytilus* and *Mercenaria*, concluding that the calcite and aragonite prisms and bricks found in some shell carbonates are single crystals rather than aggregates of smaller subunits. No crystalline structures were apparent under SEM examination of intact or fractured spicules of sea urchin larvae, which behave optically as single crystals of calcite although they have greater flexural strength than single calcite crystals and they fracture as if they were made up of many microcrystals (Okazaki and Inoue 1976). The coral skeletal units, or fibres, described for long as single crystal of aragonite, have been shown to be composite structures built by superposition of micron-thick layers which, in addition, include organic components at a submicrometre scale (Cuif and Dauphin 2005).

By now it seems clear that the viewpoint adopted mainly depends on the method of study: a skeletal element that behaves as a single crystal in polarized light may show a polycrystalline arrangement when examined by X-ray diffraction or electron microscopy.

In this connection, the inorganic components of mollusc shells, sea urchin spicules, and crustacean exoskeletons are typical. Early ultrastructural studies, carried out with the replica technique, had shown that the calcified structures appeared as single crystals, although at least some of them appeared to be polycrystalline. Studies carried out on the mother-of-pearl of several species (Cephalopoda, Gastropoda, Pelecypoda) showed that it consists of successive nacreous lamellae within which polygonal structures corresponding to tabular aragonite crystals are visible (Grégoire 1957; Watabe 1965). The use of ultrathin sections confirmed this observation and showed the degree of complexity crystal arrangements may reach in invertebrate skeletal structures.

Wong and Saleuddin (1972) studied the shell of the fresh water snail, *Helisoma duryi duryi.* This consists of three calcareous layers: the outer prismatic, about 30 μm thick, the middle cross-lamellar, about 100 μm thick, and the myostracum, about 10 μm thick (different numbers of layers have been reported in other shells; see Saleuddin 1971). In the prismatic layer the crystals are elongated in columns and lie perpendicular to the shell surface; in the cross-lamellar layer the crystals are arranged parallel to each other and form sheets (second-order lamellae) which come together to form larger units (first-order lamellae); in the myostracum the crystals are arranged in columns. In bivalve molluscs, in which three major layers are found (an outer sclerotinized periostracum, an intermediate mineralized prismatic layer, and an inner mineralized nacreous layer), Bevelander and Nakahara (1969a) found that the nacreous layer consists of tabular crystalline material arranged in parallel lamellae. In *Mytilus edulis* (Travis 1968a, b) and in *Crassostrea virginica* (Travis and Gonsalves 1969), the inorganic substance of the prismatic layer is said to consist of closely packed calcite crystals structurally organized into roughly rectangular prisms which are divided from each other by 30 nm wide uncalcified regions. The crystals, which are separated by very thin (2.2 nm thick) electron-lucent spaces, are parallel to one another with their long axis parallel to the long axis of the prism, and the form normally taken by them in the young shell is that of platelets, although they may appear as needles in thick sections (Travis 1968a). In comparative studies of skeletal structures of various invertebrate species, Travis (1970) described several forms of inorganic organization: in sponge spicules, both amorphous and well-ordered, oriented, rectangular, polycrystalline calcite crystals are found; in crustaceans, spherulitic-like clusters of calcite crystals looking like the crystals in bone mark out the early stage of calcification, whereas thin, long plate-like crystals are found in the heavily calcified skeleton; in the echinoderm endoskeleton, polycrystalline aggregates form well-ordered, sheet-like layers whose crystals, which may be either plate-like, or else needle- or rod-like, lie in highly ordered, closely packed, parallel rows; in mollusks, well organized plate-like crystals are recognizable in the shell prismatic layer. In the natural regenerating shell of *Helix pomatia*, Saleuddin (1971) reported similar results and described the crystals of the regenerated shell as large and stacked like bricks. Long, needle-shaped, single aragonite crystals which are enclosed in an envelope have been reported in the ligament of *Mytilus edulis* and *Pinctada radiata* by Bevelander and Nakahara (1969b). In *Lingula unguis*, whose mineral mainly consists of calcium phosphate, Kelly et al. (1965) found small round particles at the onset of calcification, then needle-shaped particles, and later needle-shaped crystals which may ultimately become packed into wider, rectangular crystals. An atomic force microscope study carried out on the shell prismatic layer of the bivalves *Pinna* and *Pinctada*, whose calcitic prisms have been considered simple prisms on the basis of some morphological and mineralogical characteristics, showed that they are composed of elongated crystallites, in their turn subdivided into smaller rounded units (Dauphin, 2003). In the coral *Galaxea fascicularis*, four types of crystals have been found: nano, acicular, lamellar, and fusiform crystals (Clode and Marshall 2003).

Plate-like, polygonal, calcite or aragonite crystals are found in extraskeletal calcified tissues, such as otoliths, eggshells, the pineal gland, kidney and pancreatic stones, several pathological conditions.

5.5 Unicellular Organisms

Among unicellular organisms, a number of *bacteria* can induce calcium phosphate precipitation with the formation of needle-like hydroxyapatite crystals (Theilade et al. 1976; Ennever et al. 1981; Lo Storto et al. 1990; Boyan et al. 1992). This is especially evident with oral bacteria, the most studied of which is *Bacterionema matruchotii* (alias, *Corynebacterium matruchotii*) (Ennever and Creamer 1967), and of nanobacteria (reviewed by Kajander and Çiftçioglu 1998; Cisar et al. 2000). In vitro calcification of the strain 14266 of *Bacterionema matruchotii* was distinguished by the presence of crystals appearing as straight rods or needles, or as low electron-dense broad profiles; in some cases they were slightly curved needles or plates (Lie and Selvig 1974). Needle-shaped crystals are described close to bacteria in the early calcification sites of salivary stones (Höhling and Schöpfer 1968; Theilade et al. 1976), where platelet-like crystals can also be found (Höhling et al. 1969). Aggregates of filamentous crystals, some of which were denser at the periphery than at the axial core, as if they were structured as tubules, have been described by Pautard (1970) in *Spirostomum ambiguum*.

The shape of crystals associated with Magnetotactic bacteria appears to be species-specific (Fig. 4.5): magnetite crystals appear as cubo-octahedra, truncated hexahedral or octahedral prisms, or have a tooth-like shape; greigite crystals include cubo-octahedra and rectangular prisms (Bazylinski et al. 1994). Tooth-, arrowhead-, bullet-shaped magnetosomes and truncated hexa-octahedral magnetosomes have been described (Mann et al. 1987, 1990; Buseck et al. 2001). Cubo-octahedral, elongated cubo-octahedral and hexagonal magnetosomes have been illustrated by Mann (2001).

The calcification of Foraminifera tests can lead to different types of inorganic organization. According to Pautard (1970), besides uncalcified organic tests, the following organization can be found: agglutinated tests of collected particles cemented with an organic or a calcareous cement; porcelaneous calcification with many small random or agglutinated aragonite crystals; microgranular calcification with calcite arranged in ordered, unrelated domains; small crystals of calcite arranged in domains which are either oriented radially or with their *c*-axis perpendicular to the test; relatively large spicules of calcite; tests appearing as single crystals of calcite. Towe and Cifelli (1967) studied specimens of Foraminifera calcareous wall by electron microscopy (replica method), finding finely layered arrays of plate-like crystals; they also described needle- or lath-like calcite crystals in the porcelaneous wall.

The coccolithophorids form coccoliths, i. e., microscopic, rather complicated skeletal structures (see Sect. 13.5.3) whose optical properties are those of a single crystal of calcite. They have very different microarchitectural organizations, which have been discussed and illustrated by Young et al. (1999) and Marsh (1999).

Crystals with an axial reduced density, resembling those found in bone and enamel, have been described by Arnott and Pautard (1968) and Pautard (1970) in calcified structures of *Spirostomum ambiguum* and of whale baleens.

5.6 Pathological Calcifications

Needle-shaped crystals like those found in bone, dentin and cartilage have been described in pathological, ectopic calcifications of soft tissues. This is true of the crystals found in calcified areas of the human breast carcinoma (Ahmed 1975) and in tumoral calcinosis (Boskey et al. 1983). In animals sensitized by dihydrotachysterol (DHT) administration, needle-shaped crystals have been found in various pathologically calcified tissues, such as skeletal muscles (Bonucci and Sadun 1972), the myocardium (Bonucci and Sadun 1973), the skin (Boivin 1975; Boivin and Tochon-Danguy 1976), the lung (Eggermann and Kapanci 1971), the vagal nerve (Robinson MJ et al. 1968) and the cornea (Valouch et al. 1974). Needle-shaped crystals were also found in calcified areas of the brain in a case of Sturge-Weber disease (Guseo 1975) and in calcified liver mitochondria of carbon tetrachloride-intoxicated rats (Bonucci et al. 1973), where spherical, rosetta-like calcified concretions were found too. Needle-shaped crystals have been found, together with cuboid crystals, in arthritic cartilage (Ali 1985), where highly polymorphic and polymetric, calcium pyrophosphate dihydrate crystals were recognized (Boivin and Lagier 1983). These have a feathery or brush-like pattern in cases of calcium pyrophosphate dihydrate deposition disease according to Keen et al. (1991). Characteristically long, thin, filament-like crystals have been described in calcifications of the lung (Bonucci et al. 1994) induced in rats by DHT administration. Strands of dot-like structures, similar to those described by Höhling and coworkers (Höhling et al. 1971a, 1995, 1997; Höhling 1989) in normally calcified bone, dentin and enamel, have been reported by Boivin (1975) in DHT-induced calcification of the skin. Amorphous inorganic substance has been reported in skin calcifications induced by scleroderma (Daculsi et al. 1983) or by DHT administration (Boivin and Tochon-Danguy 1976).

5.6.1 Calcified Mitochondria

The inorganic deposits which accumulate in mitochondria are easily recognizable under the electron microscope as roundish aggregates of electron-dense granules (Fig. 4.6), sometimes showing a cockade-like arrangement (Greenawalt et al. 1964; Heggtveit et al. 1964; Greenawalt and Carafoli 1966; Cameron et al. 1967; Thomas and Greenawalt 1968; Vasington and Greenawalt 1968; Martin and Matthews 1969; Saladino et al. 1969; D'Agostino and Chiga 1970; Matthews 1970; Carafoli et al. 1971; Duffy et al. 1971; Thyberg and Friberg 1971; Bonucci and Sadun 1972, 1973; Holtrop 1972; Kutsuna 1972; Bonucci et al. 1973; Boivin 1975; Hagler et al. 1981; Landis and Glimcher 1982; Lewinson and Silbermann 1990). These granular aggregates may be associated – occasionally in a single mitochondrion, more often in several mitochondria belonging to one cell or several adjacent cells – with clusters of needle-like crystals; no platelet-like crystals have been reported (Palladini and Carbone 1966; Gritzka and Trump 1968; Woodhouse and Burston 1969; Bonucci and Sadun 1972, 1973; Lin 1972; Oksanen and Poukka 1972; Bonucci et al. 1973; Cavallero et al. 1974; Bonucci 1975; Hagler et al. 1979). The relationship between granular and crystalline aggregates is still not understood, but the presence of single crystals or very small aggregates of crystals in mitochondria containing no granular aggregates (Bonucci et al. 1973) suggests that the two types of intramitochondrial inorganic deposits come to be formed independently. It is of interest that rat liver mitochondria accumulate Sr^{2+} with formation of granular aggregates or needle-like crystalline structures (Greenawalt and Carafoli 1966) very like those deriving from calcium phosphate accumulation.

The mitochondria in pathologically altered cells may display varying degrees of calcification, ranging from a few granules to plenty of crystals (Palladini and Carbone 1966; Carafoli and Tiozzo 1968; Gritzka and Trump 1968; Legato et al. 1968; Ghidoni et al. 1969; Saladino et al. 1969; Woodhouse and Burston 1969; Berry 1970; Carafoli et al. 1971; Bonucci and Sadun 1972, 1973, 1975; Brandt and Bässler 1972; Oksanen and Poukka 1972; Ruigrok and Elbers 1972; Bonucci et al. 1973; Cavallero et al. 1974; Marinozzi et al. 1977; Oberc and Engel 1977; Susheela and Kharb 1990; Kim KM 1994).

5.6.2 Vascular Calcification

Vascular calcifications can display various types of crystals. Needle-like crystals resembling those in cartilage and bone have been reported in the aorta of vitamin D-treated rabbits (Eisenstein and Zeruolis 1964) and in arteriosclerotic aortas (Daoud et al. 1985; Tanimura et al. 1986), where crystals formed roundish aggregates shaped like the calcification nodules of bone and cartilage (Paegle 1969). In this case, platelet-like crystals were present too (Paegle 1969).

Needle-shaped crystals have been found in rabbit aorta cross-linked with 2% glutaraldehyde and implanted subcutaneously in rats for one month (Paule et al. 1992). Very long, thin crystals resembling filaments rather than needles have been found in the calcified aorta of DHT-sensitized rats (Bonucci and Sadun 1975). Needle-shaped crystals, together with amorphous and laminar inorganic material, have been reported in calcified human aortic valves (Kim KM and Huang 1971), while granules with a concentric laminar structure have been recognized in the lung calcifications of dialyzed patients (Bestetti-Bosisio et al. 1984).

5.7
Congenital or Acquired Diseases; Genetically Engineered Animals

Very few investigations have been carried out on the shape of crystal in congenital or acquired diseases and in genetically modified animals. Furuya et al. (2000) studied the ultrastructure of the growth cartilage in the knee and spine of 'twy' mice, which are autosomal recessive mutant animals developing abnormal calcifications, especially in the cartilage and tendons of limbs and spine. They did not report any difference between these calcified zones and those in controls; both types of animal displayed the feature of pericellular chains of granules containing Ca and P which were only partly solubilized by decalcification. Similar pericellular structures were described in the epiphyseal cartilage of normal rats by Cabrini and Klein-Szanto (1973), who did not rule out the possibility that they might have been artefacts, by Quacci et al. (1990), who considered them to be closely connected with proteoglycans and to represent the morphological equivalent of a Ca-P amorphous phase, and by Bonucci and Silvestrini (1994), who were of the opinion that they could be an outcome of fixation on proteoglycan molecules. Needle-like crystals, similar to but thinner than those in controls, have been observed in the cortical bone of oim/oim mice which produce only $\alpha 1$(I) collagen homotrimers and provide a model for human osteogenesis imperfecta (Fratzl et al. 1996b). Loosely packed, randomly oriented crystals which, however, had the usual ribbon-like shape, have been described in hypoplastic enamel in two cases of amelogenesis imperfecta-nephrocalcinosis syndrome (Phakey et al. 1995). Short crystal segments and abnormally large crystals have been described in hypoplastic amelogenesis imperfecta and in hypomaturation amelogenesis imperfecta, respectively (Robinson C et al. 2003). Comparative atomic force microscopy investigations of normal enamel and hypoplastic enamel – enamel hypoplasia is a hereditary defect in which enamel is deficient in quality or in quantity – showed significant differences at the enamel surface: large, compact apatite granules were found in normal enamel, whereas loosely packed, very small grains characterized the hypoplastic enamel and the difference was reflected in the value of the RMS (Rq) – root-mean-square roughness – estimated to be 116.50 in the first case and 103.80 nm in the second (Batina et al. 2004).

5.8 Concluding Remarks

The following results deserve close attention:

- The earliest inorganic deposits fail to generate any electron diffraction pattern signalling a calcium phosphate solid phase, either because of their poor degree of crystallinity or because they are amorphous.
- The shape of apatite crystals varies with the stage of calcification and type of calcified tissue. Two shapes are dominant: needle-, rod- or filament-like, and platelet-like.
- Needle-, rod- or filament-like almost radially arranged crystals are the main components of the calcification nodules which constitute the calcification front in bones with loosely aggregated collagen fibrils (embryonal, woven and medullary bone), as well as in dentin and cartilage; they are also found in unicellular organisms and several pathologically calcified tissues.
- The early stages of calcification in compact bone are marked by the presence in the osteoid tissue of calcification islands, comprising needle-like crystals running parallel to the collagen fibrils.
- In more advanced stages of calcification, granular, intrinsically electron-dense bands (granular inorganic bands) run transversally across, and in line with, the period of adjacent collagen fibrils in bone, tendon and dentin. They are partially obscured by needle-shaped crystals as the degree of calcification is completed.
- Platelet-like crystals have mainly been reported in fully calcified, lamellar compact bone and in calcified tendons, i. e., tissues containing closely aggregated collagen fibrils and granular inorganic bands.
- Most of the investigations that attest the platelet-like shape of bone and tendon crystals have been carried out on crystals isolated from the calcified matrix by mechanical or chemical methods; these may have altered their shape.
- The differences between needle- and platelet-like crystals, and the relationship – if any – of the latter with granular inorganic bands, have not been studied in depth and remain undefined.
- Early enamel crystals have a ribbon-like shape which turns into a polygonal (hexagonal) shape with maturation. Bone crystals can also change their shape during maturation, even if to a much lower degree than enamel crystals.
- Invertebrate crystals appear as large polygonal, monocrystalline structures which, however, prove to be polycrystalline under the electron microscope; the single units have a rod- or needle-like shape.

- The needle-like crystals mainly consist of phosphate apatite, whereas the platelet-like and polygonal ones mainly contain carbonate apatite.
- The shape of the crystals does not depend on their chemical composition alone, but also on their age and degree of maturation. Crystals that initially show the shape of needles or filaments may take on a polygonal shape with maturation.
- The microarchitecture and state of aggregation of the organic matrix where crystals are formed can mold their shape.
- In pathological calcifications, crystals may display specific morphological characteristics and mitochondria may contain roundish aggregates of inorganic granules.
- Crystal shape is not significantly changed by treatment with electron microscope ‘stains’ such as uranyl acetate and lead citrate, or by fixation, dehydration and embedding, although they may be easily decalcified.
- The choice made in defining crystals as needle-, rod-, ribbon-, filament- or platelet-like is not a semantic question alone, but comprises the question of how crystals are formed, as further discussed below. The adoption of terms like “platelet”, “rod” and “needle” seem to be derived less from their shape than from the preconceived mineralogical idea that, if they are crystals, they must be rigid structures.

References

Abdalla O (1979) Ossification and mineralization in the tendons of the chicken (*Gallus domesticus*). J Anat 129:351–359

Ahmed A (1975) Calcification in human breast carcinomas: ultrastructural observations. J Path 117:247–251

Aho AJ, Isomäki AM (1962) Electron microscopic observations on experimental callus formation in rats. Acta Path Microbiol Scand103–105

Akisaka T, Nakayama M, Yoshida H, Inoue M (1998) Ultrastructural modifications of the extracellular matrix upon calcification of growth plate cartilage as revealed by quick-freeze deep etching technique. Calcif Tissue Int 63:47–56

Ali SY (1985) Apatite-type crystal deposition in arthritic cartilage. Scanning Electron Microsc 4:1555–1566

Amprino R, Engström A (1952) Studies on X ray absorption and diffraction of bone tissue. Acta Anat 15:1–22

Aoba T (1996a) Recent observation on enamel crystal formation during mammalian amelogenesis. Anat Rec 245:208–218

Aoba T (1996b) Recent observations on enamel crystal formation during mammalian amelogenesis. Anat Rec 245:208–218

Arnold S, Plate U, Wiesmann H-P, Kohl H, Höhling H-J (1997) Quantitative electron-spectroscopic diffraction (ESD) and electron-spectroscopic imaging (ESI) analyses of dentine mineralisation in rat incisors. Cell Tissue Res 288:185–190

Arnold S, Plate U, Wiesmann H-P, Stratmann U, Kohl H, Höhling H-J (1999) Quantitative electron spectroscopic diffraction analyses of the crystal formation in dentine. J Microsc 195:58–63

Arnott HJ, Pautard FGE (1968) The inorganic phase of bone: a re-appraisal (abstr.). Calcif Tissue Res 2:2

Arsenault AL (1988) Crystal-collagen relationships in calcified turkey leg tendons visualized by selected-area dark field electron microscopy. Calcif Tissue Int 43:202–212

Arsenault AL (1992) Structural and chemical analyses of mineralization using the turkey leg tendon as a model tissue. Bone Mineral 17:253–256

Arsenault AL, Grynpas MD (1988) Crystals in calcified epiphyseal cartilage and cortical bone of the rat. Calcif Tissue Int 43:219–225

Arsenault AL, Hunziker EB (1988) Electron microscopic analysis of mineral deposits in the calcifying epiphyseal growth plate. Calcif Tissue Int 42:119–126

Arsenault AL, Robinson BW (1989) The dentino-enamel junction: a structural and microanalytical study of early mineralization. Calcif Tissue Int 45:111–121

Arzate H, Alvarez-Perez MA, Alvarez-Fregoso O, Wusterhaus-Chavez A, Reyes-Gasga J, Ximenez-Fyvie LA (2000) Electron microscopy, micro-analysis, and X-ray diffraction characterization of the mineral-like tissue deposited by human cementum tumor-derived cells. J Dent Res 79:28–34

Ascenzi A (1948a) Contributo allo studio delle proprietà ottiche dell'osso umano normale. II. Sulla birifrangenza totale. R C Accad Naz Lincei (Classe Sci Fis Mat Nat) 5:100–107

Ascenzi A (1948b) Contributo allo studio delle proprietà ottiche dell'osso umano normale. III. Sulla birifrangenza di forma e la birifrangenza propria. R C Accad Naz Lincei (Classe Sci Fis Mat Nat) 5:171–180

Ascenzi A (1949) Quantitative researches on the optical properties of human bone. Nature (London) 163:604

Ascenzi A, Benedetti EL (1959) An electron microscopic study of the foetal membranous ossification. Acta Anat 37:370–385

Ascenzi A, Bonucci E (1964) A quantitative investigation of the birefringence of the osteon. Acta Anat 44:236–262

Ascenzi A, François C, Steve Bocciarelli D (1963) On the bone induced by estrogen in birds. J Ultrastruct Res 8:491–505

Ascenzi A, Bonucci E, Steve Bocciarelli D (1965) An electron microscope study of osteon calcification. J Ultrastruct Res 12:287–303

Ascenzi A, Bonucci E, Steve Bocciarelli D (1967) An electron microscope study on primary periosteal bone. J Ultrastruct Res 18:605–618

Ascenzi A, Bonucci E, Steve Bocciarelli D (1968) Fine structure of the bone mineral in different experimental conditions. In: Steve Bocciarelli D (ed) Fourth Regional Conference on Electron Microscopy, vol 2. Tipografia Poliglotta Vaticana. Rome, pp 431–432

Ascenzi A, Bonucci E, Ripamonti A, Roveri N (1978) X-ray diffraction and electron microscope study of osteons during calcification. Calcif Tissue Res 25:133–143

Barckhaus RH, Höhling HJ (1978) Electron microscopical microprobe analysis of freeze dried and unstained mineralized epiphyseal cartilage. Cell Tissue Res 186:541–549

Baron R (1995) Molecular mechanisms of bone resorption. An update. Acta Orthop Scand 66:66–70

Batina N, Renugopalakrishnan V, Casillas Lavín PN, Guerrero JCH, Morales M, Garduño-Juárez R, Lakka SL (2004) Ultrastructure of dental enamel afflicted with hypoplasia: an atomic force microscopic study. Calcif Tissue Int 74:294–301

Bazylinski DA, Garratt-Reed AJ, Frankel RB (1994) Electron microscopic studies of magnetosomes in magnetotactic bacteria. Microsc Res Tech 27:389–401

Bekker PJ, Gay CV (1990) Biochemical characterization of an electrogenic vacuolar proton pump in purified chicken osteoclast plasma membrane vesicles. J Bone Miner Res 5:569–579
Benson SC, Wilt FH (1992) Calcification of spicules in the sea urchin embryo. In: Bonucci E (ed) Calcification in biological systems. CRC Press, Boca Raton, pp 157–178
Berry JP (1970) Néphrocalcinose éxperimentale par injection de parathormone. Etude au microanalyseur à sonde électronique. Nephron 7:97–116
Bestetti-Bosisio M, Cotelli F, Schiaffino E, Sorgato G, Schmid C (1984) Lung calcification in long-term dialysed patients: a light and electronmicroscopic study. Histopathology 8:69–79
Bevelander G, Nakahara H (1969a) An electron microscope study of the formation of the nacreous layer in the shell of certain bivalve molluscs. Calcif Tissue Res 3:84–92
Bevelander G, Nakahara H (1969b) An electron microscope study of the formation of the ligament of *Mytilus edulis* and *Pinctada radiata*. Calcif Tissue Res 4:101–112
Bigi A, Ripamonti A, Koch MHJ, Roveri N (1988) Calcified turkey leg tendon as structural model for bone mineralization. Int J Biol Macromol 10:282–286
Blair HC, Ghandur-Mnaymneh L (1985) Macrophage-mediated bone resorption occurs in an acidic environment. Calcif Tissue Int 37:547–550
Blair HC, Teitelbaum SL, Ghiselli R, Gluck S (1989) Osteoclastic bone resorption by a polarized vacuolar proton pump. Science 245:855–857
Bloom W, Bloom MA, McLean FC (1942) Calcification and ossification. Medullary bone changes in the reproductive cycle of female pigeons. Anat Rec 83:443–466
Bocciarelli DS (1970) Morphology of crystallites in bone. Calcif Tissue Res 5:261–269
Boivin G (1975) Étude chez le rat d'une calcinose cutanée induite par calciphylaxie locale I. – Aspects ultrastructuraux. Arch Anat Microsc Morphol Expér 64:183–205
Boivin G, Lagier R (1983) An ultrastructural study of articular chondrocalcinosis in cases of knee osteoarthritis. Virchows Arch [Pathol Anat] 400:13–29
Boivin G, Tochon-Danguy HJ (1976) Étude chez le rat d'une calcinose cutanée induite per calciphylaxie locale II. Aspects biophysiques de la substance minérale. Ann Biol Anim Bioch Biophys 16:869–878
Bonucci E (1967) Fine structure of early cartilage calcification. J Ultrastruct Res 20:33–50
Bonucci E (1969) Further investigation on the organic/inorganic relationships in calcifying cartilage. Calcif Tissue Res 3:38–54
Bonucci E (1971) The locus of initial calcification in cartilage and bone. Clin Orthop Relat Res 78:108–139
Bonucci E (1974) The organic-inorganic relationships in bone matrix undergoing osteoclastic resorption. Calcif Tissue Res 16:13–36
Bonucci E (1975) The organic-inorganic relationships in calcified organic matrices. Physico-chimie et cristallographie des apatites d'intérêt biologique. Centre National de la Recherche Scientifique, Paris, pp 231–246
Bonucci E (1981) New knowledge on the origin, function and fate of osteoclasts. Clin Orthop Relat Res 158:252–269
Bonucci E (2002) Crystal ghosts and biological mineralization: fancy spectres in an old castle, or neglected structures worthy of belief? J Bone Miner Metab 20:249–265
Bonucci E, Sadun R (1972) An electron microscope study on experimental calcification of skeletal muscle. Clin Orthop Relat Res 88:197–217
Bonucci E, Sadun R (1973) Experimental calcification of the myocardium. Ultrastructural and histochemical investigations. Am J Pathol 71:167–192
Bonucci E, Sadun R (1975) Dihydrotachysterol-induced aortic calcification. A histochemical and ultrastructural investigation. Clin Orthop Relat Res 107:283–294

Bonucci E, Silvestrini G (1994) Morphological investigation of epiphyseal cartilage after glutaraldehyde-malachite green fixation. Bone 15:153–160

Bonucci E, Derenzini M, Marinozzi V (1973) The organic-inorganic relationship in calcified mitochondria. J Cell Biol 59:185–211

Bonucci E, Silvestrini G, Ballanti P, Della Rocca C, Mocetti P (1994) Dihydrotachysterol-induced lung calcification in the rat. It J Miner Electrol Metab 8:12–22

Boothroyd B (1964) The problem of demineralisation in thin sections of fully calcified bone. J Cell Biol 20:165–173

Boothroyd B (1975a) Observations on embryonic chick-bone crystals by high resolution transmission electron microscopy. Clin Orthop Relat Res 106:290–310

Boothroyd B (1975b) Observations on embryonic chick-bone crystals by high resolution transmission electron microscopy. Clin Orthop Relat Res 106:290–310

Boskey AL, Vigorita VJ, Sencer O, Stuchin SA, Lane JM (1983) Chemical, microscopic, and ultrastructural characterization of the mineral deposits in tumoral calcinosis. Clin Orthop Relat Res 178:258–269

Boyan BD, Swain LD, Everett MM, Schwartz Z (1992) Mechanisms of microbial mineralization. In: Bonucci E (ed) Calcification in biological systems. CRC Press, Boca Raton, pp 129–156

Boyde A (1974) Transmission electron microscopy of ion beam thinned dentine. Cell Tissue Res 152:543–550

Brandenburger E, Schinz HR (1945) Über die Natur der Verkarkungen bei Mensch und Tier und das Verhalten der anorganischen Knochensubstanz im Falle der hauptsächlichen menschlichen Knochenkrankheiten. Helv Med Acta 12:1–63

Brandt G, Bässler R (1972) Die Wirkung der experimentellen Hypercalcämie durch Dihydrotachysterin auf Drüsenfunktion und Verkalkungsmuster der Mamma. Licht-, elektronenmikroskopische und chemisch-analytische Untersuchungen. Virchows Arch Abt A Path Anat 356:155–172

Buseck PR, Dunin-Borkowski RE, Devouard B, Frankel RB, McCartney MR, Midgley PA, Pósfai M, Weyland M (2001) Magnetic morphology and life on Mars. Proc Natl Acad Sci USA 98:13490–13495

Butler WT (1998) Dentin matrix protein. Eur J Oral Sci 106(suppl 1):204–210

Cabrini RL, Klein-Szanto AJP (1973) The presence of perichondrocytic rods in the tibial epiphyseal plate of the rat. J Microsc (Paris) 17:219–222

Cameron DA, Paschall HA, Robinson RA (1967) Changes in the fine structure of bone cells after the administration of parathyroid extract. J Cell Biol 33:1–14

Carafoli E, Tiozzo R (1968) A study of energy-linked calcium transport in liver mitochondria during CCl_4 intoxication. Exp Molec Pathol 9:131–140

Carafoli E, Tiozzo R, Pasquali-Ronchetti I, Laschi R (1971) A study of Ca^{2+} metabolism in kidney mitochondria during acute uranium intoxication. Lab Invest 25:516–527

Carlström D (1955) X-ray crystallographic studies on apatites and calcified structures. Acta Radiol Suppl 121:1–59

Carlström D, Finean JB (1954) X-ray diffraction studies on the ultrastructure of bone. Biochim Biophys Acta 13:183–191

Carlström D, Glas J-E (1959) The size and shape of the apatite crystallites in bone as determined from line-broadening measurements on oriented specimens. Biochim Biophys Acta 35:46–53

Carter DH, Scully AJ, Hatton PV, Davies RM, Aaron JE (2000) Cryopreservation and image enhancement of juvenile and adult dentine mineral. Histochem J 32:253–261

Cavallero C, Spagnoli LG, Di Tondo U (1974) Early mitochondrial calcification in the rabbit aorta after adrenaline. Virchows Arch A Path Anat Histol 362:23–39

Cisar JO, Xu D-Q, Thompson J, Swaim W, Hu L, Kopecko DJ (2000) An alternative interpretation of nanobacteria-induced biomineralization. Proc Natl Acad Sci USA 97:11511–11515

Clark JH (1931) A study of tendons, bones, and other forms of connective tissue by means of X-ray diffraction patterns. Am J Physiol 98:328–337

Clode PL, Marshall AT (2003) Skeletal microstructure of *Galaxea fascicularis* exsert septa: a high-resolution SEM study. Biol Bull 204:146–154

Cuif I-P, Dauphin T (2005) The two-step mode of growth in the scleractinian coral skeletons from the micrometer to the overall scale. J Struct Biol 150:319–331

Cuisinier F, Bres EF, Hemmerle J, Voegel J-C, Frank RM (1987) Transmission electron microscopy of lattice planes in human alveolar bone apatite crystals. Calcif Tissue Int 40:332–338

Cuisinier FJG, Steuer P, Senger B, Voegel JC, Frank RM (1992) Human amelogenesis I: High resolution electron microscopy study of ribbon-like crystals. Calcif Tissue Int 51:259–268

D'Agostino AN, Chiga M (1970) Mitochondrial mineralization in human myocardium. Am J Clin Pathol 53:820–824

Daculsi G, Kerebel B (1978) High-resolution electron microscope study of human enamel crystallites: size, shape, and growth. J Ultrastruct Res 65:163–172

Daculsi G, Faure G, Kerebel B (1983) Electron microscopy and microanalysis of a subcutaneous heterotopic calcification. Calcif Tissue Int 35:723–727

Daculsi G, Menanteau J, Kerebel LM, Mitre D (1984) Length and shape of enamel crystals. Calcif Tissue Int 36:550–555

Dallemagne MJ, Melon J (1946) Nouvelles recherches relatives aux propriétés optique de l'os: la biréfringence de l'os minéralisé; relations entre les fractions organiques et inorganique de l'os. J Washington Acad Sci 36:181–195

Daoud AS, Frank AS, Jarmolych J, Franco WT, Fritz KE (1985) Ultrastructural and elemental analysis of calcification of advanced swine aortic atherosclerosis. Exp Mol Pathol 43:337–347

Dauphin Y (2003) Soluble organic matrices of the calcitic prismatic shell layers of two pteriomorphid bivalves. J Biol Chem 278:15168–15177

Delaissé JM, Eeckhout Y, Neff L, François-Gillet C, Henriet P, Su Y, Vaes G, Baron R (1993) (Pro)collagenase (matrix metalloproteinase-1) is present in rodent osteoclasts and in the underlying bone-resorbing compartment. J Cell Sci 106:1071–1082

Dong W, Warshawsky H (1996) Lattice fringe continuity in the absence of crystal continuity in enamel. Adv Dent Res 10:232–237

Dudley HR, Spiro D (1961) The fine structure of bone cells. J Biophys Biochem Cytol 11:627–649

Duffy JL, Meadow E, Suzuki Y, Churg J (1971) Acute calcium nephropathy. Early proximal tubular changes in the rat kidney. Arch Path 91:340–350

Eeckhout Y, Delaisse JM (1988) The role of collagenase in bone resorption. An overview. Path Biol 36:1139–1146

Eggermann J, Kapanci Y (1971) Experimental pulmonary calcinosis in the rat. Ultrastructural and morphometric study. Lab Invest 24:469–482

Eisenstein R, Zeruolis L (1964) Vitamin-D induced aortic calcification. Arch Path 77:27–35

Elliott JC (1973) The problems of the composition and structure of the mineral components of the hard tissues. Clin Orthop Relat Res 93:313–345

Engfeldt B, Reinholt FP (1992) Structure and calcification of epiphyseal growth cartilage. In: Bonucci E (ed) Calcification in biological systems. CRC Press, Boca Raton, pp 217–241

Ennever J, Creamer H (1967) Microbiological calcification: bone mineral and bacteria. Calcif Tissue Res 1:87–93

Ennever J, Streckfuss JL, Goldschmidt MC (1981) Calcifiability comparison among selected microorganisms. J Dent Res 60:1793–1796

Eppell SJ, Tong W, Katz JL, Kuhn L, Glimcher MJ (2001) Shape and size of isolated bone mineralites measured using atomic force microscopy. J Orthop Res 19:1027–1034

Fernández-Morán H, Engström A (1957) Electron microscopy and X-ray diffraction of bone. Biochim Biophys Acta 23:260–264

Finean JB, Engström A (1953) The low-angle scatter of X-rays from bone tissue. Biochim Biophys Acta 11:178–189

Fratzl P, Fratzl-Zelman N, Klaushofer K, Vogl G, Koller K (1991) Nucleation and growth of mineral crystals in bone studied by small-angle X-ray scattering. Calcif Tissue Int 48:407–413

Fratzl P, Groschner M, Vogl G, Plenk H Jr, Eschberger J, Fratzl-Zelman N, Koller K, Klaushofer K (1992) Mineral crystals in calcified tissues: a comparative study by SAXS. J Bone Miner Res 7:329–334

Fratzl P, Schreiber S, Klaushofer K (1996a) Bone mineralization as studied by small-angle X-ray scattering. Connect Tissue Res 34:247–254

Fratzl P, Paris O, Klaushofer K, Landis WJ (1996b) Bone mineralization in an osteogenesis imperfecta mouse model studied by small-angle X-ray scattering. J Clin Invest 97:396–402

Frazier PD (1968) Adult human enamel: an electron microscopic study of crystallite size and morphology. J Ultrastruct Res 22:1–11

Furuya S, Ohtsuki T, Yabe Y, Hosoda Y (2000) Ultrastructural study on calcification of cartilage: comparing ICR and twy mice. J Bone Miner Metab 18:140–147

Garant PR (1970) An electron microscopic study of the crystal-matrix relationship in the teeth of the dogfish *Squalus acanthias* L. J Ultrastruct Res 30:441–449

Ghidoni JJ, Liotta D, Thomas H (1969) Massive subendocardial damage accompanying prolonged ventricular fibrillation. Am J Pathol 56:15–29

Glimcher MJ (1959) Molecular biology of mineralized tissues with particular reference to bone. Rev Mod Phys 31:359–393

Gonzales F, Karnovsky MJ (1961) Electron microscopy of osteoclasts in healing fractures of rat bone. J Biophys Biochem Cytol 9:299–316

Göthlin G, Ericsson JLE (1976) The osteoclast. Review of ultrastructure, origin, and structure-function relationship. Clin Orthop Relat Res 120:201–231

Greenawalt JW, Carafoli E (1966) Electron microscope studies on the active accumulation of Sr^{++} by rat-liver mitochondria. J Cell Biol 29:37–61

Greenawalt JW, Rossi CS, Lehninger AL (1964) Effect of active accumulation of calcium and phosphate ions on the structure of rat liver mitochondria. J Cell Biol 23:21–38

Grégoire C (1957) Topography of the organic components in mother-of-pearl. J Biophys Biochem Cytol 3:797–806

Gritzka TL, Trump BF (1968) Renal tubular lesions caused by mercuric chloride. Electron microscopic observations: degeneration of the pars recta. Am J Pathol 52:1225–1277

Guseo A (1975) Ultrastructure of calcification in Sturge-Weber disease. Virchows Arch A Path Anat Histol 366:353–356

Hagler HK, Sherwin L, Buja LM (1979) Effect of different methods of tissue preparation on mitochondrial inclusions of ischemic and infarcted canine myocardium. Transmission and analytic electron microscopic study. Lab Invest 40:529–544

Hagler HK, Lopez LE, Murphy ME, Greico CA, Buja LM (1981) Quantitative X-ray microanalysis of mitochondrial calcification in damaged myocardium. Lab Invest 45:241–247

Hall TJ, Chambers TJ (1996) Molecular aspects of osteoclast function. Inflamm Res 45:1–9

Hancox NM, Boothroyd B (1961) Motion picture and electron microscope studies on the embryonic avian osteoclast. J Biophys Biochem Cytol 11:651–661

Hancox NM, Boothroyd B (1963) Structure-function relationships in the osteoclast. In: Sognnaes RF (ed) Mechanisms of hard tissue destruction. American Association for the Advancement of Science, Washington, pp 497–514
Hancox NM, Boothroyd B (1965) Electron microscopy of the early stages of osteogenesis. Clin Orthop Relat Res 40:153–161
Hayashi Y (1984) Crystal growth in calcifying fronts during dentinogenesis. Acta Anat 118:13–17
Hayashi Y (1987) Ultrastructure of cementum formation on partially formed teeth in dogs. Acta Anat 129:279–288
Heggtveit HA, Herman L, Mishra RK (1964) Cardiac necrosis and calcification in experimental magnesium deficiency. A light and electron microscopic study. Am J Pathol 45:757–782
Henny GC, Spiegel-Adolf M (1945) X-ray diffraction studies on fish bones. Am J Physiol 144:632–636
Heywood BR, Sparks NH, Shellis RP, Weiner S, Mann S (1990) Ultrastructure, morphology and crystal growth of biogenic and synthetic apatites. Connect Tissue Res 25:103–119
Hill PA, Docherty AJP, Bottomley KMK, O'Connell JP, Morphy JR, Reynolds JJ, Meikle MC (1995) Inhibition of bone resorption *in vitro* by selective inhibitors of gelatinase and collagenase. Biochem J 308:167–175
Höhling HJ (1989) Special aspects of biomineralization of dental tissues. In: Oksche A, Vollrath L (eds) Handbook of microscopic anatomy. Springer, Berlin Heidelberg New York, pp 475–524
Höhling HJ, Schöpfer H (1968) Morphological investigations of apatitic nucleation in hard tissue and salivary stone formation. Naturwissensch 55:545
Höhling HJ, Pfefferkorn G, Radicke J, Vahl J (1969) Elektronenmikroskopische Untersuchungen zur organischen Matrix und Kristalbildung in meschlichen Speichelsteinen. Deut Zahnärztl Z 24:663–670
Höhling HJ, Schöpfer H, Neubauer G (1970) Elektronenmikroskopie und Laserbeugungs-Untersuchungen zur Charakterisierung der organischen Matrix im Speichelstein und Hartgewebe. Z Zellforsch 108:415–430
Höhling HJ, Kreilos R, Neubauer G, Boyde A (1971a) Electron microscopy and electron microscopical measurements of collagen mineralization in hard tissues. Z Zellforsch 122:36–52
Höhling HJ, Scholz F, Boyde A, Heine HG, Reimer L (1971b) Electron microscopical and laser diffraction studies of the nucleation and growth of crystals in the organic matrix of dentine. Z Zellforsch 117:381–393
Höhling HJ, Barckhaus RH, Krefting ER, Schreiber J (1976a) Electron microscopic microprobe analysis of mineralized collagen fibrils and extracollagenous regions in turkey leg tendon. Cell Tissue Res 175:345–350
Höhling HJ, Steffens H, Stamm G (1976b) Transmission microscopy of freeze dried, unstained epiphyseal cartilage of the guinea pig. Cell Tissue Res 167:243–263
Höhling HJ, Barckhaus RH, Krefting E-R, Althoff J, Quint P (1990) Collagen mineralization: aspects of the structural relationship between collagen and the apatitic crystallites. In: Bonucci E, Motta PM (eds) Ultrastructure of skeletal tissues. Kluwer Academic Publishers, Boston, pp 41–62
Höhling HJ, Arnold S, Barckhaus RH, Plate U, Wiesmann HP (1995) Structural relationship between the primary crystal formation and the matrix macromolecules in different hard tissues. Discussion of a general principle. Connect Tissue Int 33:171–178
Höhling HJ, Arnold S, Plate U, Stratmann U, Wiesmann HP (1997) Analysis of general principle of crystal nucleation, formation in the different hard tissues. Adv Dent Res 11:462–466

Holliday LS, Welgus HG, Fliszar CJ, Veith GM, Jeffrey JJ, Gluck SL (1997) Initiation of osteoclast bone resorption by interstitial collagenase. J Biol Chem 272:22053–22058
Holtrop ME (1972) The ultrastructure of the epiphyseal plate. II. The hypertrophic chondrocyte. Calcif Tissue Res 9:140–151
Holtrop ME (1991) Light and electronmicroscopic structure of osteoclasts. In: Hall BK (ed) Bone. CRC Press, Boca Raton, pp 1–29
Holtrop ME, King GJ (1977) The ultrastructure of the osteoclast and its functional implications. Clin Orthop Relat Res 123:177–196
Inage T (1975) Electron microscopic study of early formation of the tooth enameloid of a fish (*Hoplognathus fasciatus*). I. Odontoblasts and matrix fibers. Arch Histol Jpn 38:209–227
Jackson SA, Cartwright AG, Lewis D (1978) The morphology of bone mineral crystals. Calcif Tissue Res 25:217–222
Johansen E, Parks HF (1960) Electron microscopic observations on the three dimensional morphology of apatite crystallites of human dentine and bone. J Biophys Biochem Cytol 7:743–746
Jongebloed WL, Molenaar I, Arends J (1975) Morphology and size-distribution of sound and acid-treated enamel crystallites. Calcif Tissue Res 19:109–123
Kajander EO, Çiftçioglu N (1998) Nanobacteria: an alternative mechanism for pathogenic intra- and extracellular calcification and stone formation. Proc Natl Acad Sci USA 95:8274–8279
Kallenbach E (1990) Evidence that apatite crystals of rat incisor enamel have hexagonal cross sections. Anat Rec 228:132–136
Katsunuma N (1997) Molecular mechanisms of bone collagen degradation in bone resorption. J Bone Miner Metab 15:1–8
Keen CE, Crocker PR, Brady K, Hasan N, Levison DA (1991) Calcium pyrophosphate dihydrate deposition disease: morphological and microanalytical features. Histopathology 19:529–536
Kelly PG, Oliver PTP, Pautard FGE (1965) The shell of *Lingula unguis*. In: Richelle LJ, Dallemagne MJ (eds) Calcified tissues. Université de Liège, Liège, pp 337–345
Kerebel B, Daculsi G, Kerebel LM (1979) Ultrastructural studies of enamel crystallites. J Dent Res 58:844–850
Kim H-M, Rey C, Glimcher MJ (1995) Isolation of calcium-phosphate crystals of bone by non-aqueous methods at low temperature. J Bone Miner Res 10:1589–1601
Kim H-M, Rey C, Glimcher MJ (1996) X-ray diffraction, electron microscopy, and Fourier transform infrared spectroscopy of apatite crystals isolated from chicken and bovine calcified cartilage. Calcif Tissue Int 59:58–63
Kim KM (1994) Cell death and calcification of canine fibroblasts *in vitro*. Cells Mater 4:247–261
Kim KM, Huang S (1971) Ultrastructural study of calcification of human aortic valve. Lab Invest 25:357–366
Kinney JH, Pople JA, Marshall GW, Marshall SJ (2001) Collagen orientation and crystallite size in human dentin: a small angle X-ray scattering study. Calcif Tissue Int 69:31–37
Kirkham J, Brookes SJ, Shore RC, Bonass WA, Smith DA, Wallwork ML, Robinson C (1998) Atomic force microscopy studies of crystal surface topology during enamel development. Connect Tissue Res 38:91–100
Knese K-H, Knoop A-M (1958) Elektronenoptische Untersuchungen über die periostale Osteogenese. Z Zellforsch 48:455–478
Krefting E-R, Barckhaus RH, Höhling HJ, Bond P, Hosemann R (1980) Analysis of the crystal arrangement in collagen fibrils of mineralizing turkey tibia tendon. Cell Tissue Res 205:485–492

Kutsuna F (1972) Electron microscopic studies on isoproterenol-induced myocardial lesion in rats. Jpn Heart J 13:168–175

Laitala T, Väänänen K (1993) Proton channel part of vacuolar H^+-ATPase and carbonic anhydrase II expression is stimulated in resorbing osteoclasts. J Bone Miner Res 8:119–126

Landis WJ (1985) Temporal sequence of mineralization in calcifying turkey leg tendon. In: Butler WT (ed) The chemistry and biology of mineralized tissues. Ebsco Media, Inc., Birmingham, AL, pp 360–363

Landis WJ, Arsenault AL (1989) Vesicle- and collagen-mediated calcification in the turkey leg tendon. Connect Tissue Res 22:35–42

Landis WJ, Géraudie J (1990) Organization and development of the mineral phase during early ontogenesis of the bony fin rays of the trout *Oncorhynchus mykiss*. Anat Rec 228:383–391

Landis WJ, Glimcher MJ (1978) Electron diffraction and electron probe microanalysis of the mineral phase of bone tissue prepared by anhydrous techniques. J Ultrastruct Res 63:188–223

Landis WJ, Glimcher MJ (1982) Electron optical and analytical observations of rat growth plate cartilage prepared by ultracryomicrotomy: the failure to detect a mineral phase in matrix vesicles and the identification of heterodispersed particles as the initial solid phase of calcium phosphate deposited in the extracellular matrix. J Ultrastruct Res 78:227–268

Landis WJ, Silver FH (2002) The structure and function of normally mineralizing avian tendons. Comp Biochem Physiol A Mol Integr Physiol 133:1135–1157

Landis WJ, Song MJ (1991) Early mineral deposition in calcifying tendons characterized by high voltage electron microscopy and three-dimensional graphic imaging. J Struct Biol 107:116–127

Landis WJ, Paine MC, Glimcher MJ (1977a) Electron microscopic observations of bone tissue prepared anhydrously in organic solvents. J Ultrastruct Res 59:1–30

Landis WJ, Hauschka BT, Rogerson CA, Glimcher MJ (1977b) Electron microscopic observations of bone tissue prepared by ultracryomicrotomy. J Ultrastruct Res 59:185–206

Landis WJ, Paine MC, Glimcher MJ (1980) Use of acrolein vapors for the anhydrous preparation of bone tissue for electron microscopy. J Ultrastruct Res 70:171–180

Landis WJ, Moradian-Oldak J, Weiner S (1991) Topographic imaging of mineral and collagen in the calcifying turkey tendon. Connect Tissue Res 25:181–196

Landis WJ, Song MJ, Leith A, McEwen L, McEwen BF (1993) Mineral and organic matrix interaction in normally calcifying tendon visualized in three dimensions by high-voltage electron microscopic tomography and graphic image reconstruction. J Struct Biol 110:39–54

Lee DD, LeGeros RZ (1985) Microbeam electron diffraction and lattice fringe studies of defect structures in enamel apatites. Calcif Tissue Int 37:651–658

Lees S, Prostak K (1988) The locus of mineral crystallites in bone. Connect Tissue Res 18:41–54

Lees S, Prostak KS, Ingle VK, Kjoller K (1994) The loci of mineral in turkey leg tendon as seen by atomic force microscope and electron microscopy. Calcif Tissue Int 55:180–189

Legato MJ, Spiro D, Langer GA (1968) Ultrastructural alterations produced in mammalian myocardium by variation in perfusate ionic composition. J Cell Biol 37:1–12

LeGeros RZ, Suga S (1980) Crystallographic nature of fluoride in enameloids of fish. Calcif Tissue Int 32:169–174

Lewinson D, Silbermann M (1990) Ultrastructural localization of calcium in normal and pathologic cartilage. In: Bonucci E, Motta PM (eds) Ultrastructure of skeletal tissues. Kluwer Academic Publishers, Boston, pp 129–152

Lie T, Selvig KA (1974) Calcification of oral bacteria: an ultrastructural study of two strains of Bacterionema matruchotii. Scand J Dent Res 82:8–18

Lin JJ (1972) Intramitochondrial calcification in infant myocardium. Occurrence in a case of coarctation of aorta. Arch Pathol 94:366–369

Linde A, Goldberg M (1993) Dentinogenesis. Crit Rev Oral Biol Med 4:679–728

Lo Storto S, Di Grezia R, Silvestrini G, Cattabriga M, Bonucci E (1990) Studio morfologico ultrastrutturale di tartaro sopragengivale. Min Stomat 39:83–89

Lucht U (1972) Osteoclasts and their relationship to bone as studied by electron microscopy. Z Zellforsch 135:211–228

Mann S (2001) Biomineralization. Principles and concepts in bioinorganic materials chemistry. Oxford University Press, Oxford

Mann S, Sparks NHC, Blakemore RP (1987) Structure, morphology and crystal growth of anisotropic magnetite crystals in magnetic bacteria. Proc R Soc London B 231:477–487

Mann S, Sparks NHC, Board RG (1990) Magnetotactic bacteria: microbiology, biomineralization, palaeomagnetism and biotecnology. Adv Microbiol Physiol 31:125–181

Marinozzi V, Derenzini M, Nardi F, Gallo P (1977) Mitochondrial inclusions in human cancer of the gastrointestinal tract. Cancer Res 37:1556–1563

Marks SC Jr, Popoff SN (1990) Ultrastructural biology and pathology of the osteoclast. In: Bonucci E, Motta PM (eds) Ultrastructure of skeletal tissues. Kluwer Academic Publishers, Boston, pp 239–252

Marsh ME (1999) Coccolith crystals of *Pleurochrysis carterae*: crystallographic faces, organization, and development. Protoplasma 207:54–66

Martin JH, Matthews JL (1969) Mitochondrial granules in chondrocytes. Calcif Tissue Res 3:184–193

Matsushima N, Akiyama M, Terayama Y, Izumi Y, Miyake Y (1984) The morphology of bone mineral as revealed by small-angle X-ray scattering. Biochim Biophys Acta 801:298–305

Matthews JL (1970) Ultrastructure of calcifying tissues. Am J Anat 129:451–458

Meyer JL, Eick JD, Nancollas GH, Johnson LN (1972) A scanning electron microscopic study of the growth of hydroxyapatite crystals. Calcif Tissue Res 10:91–102

Miake Y, Aoba T, Moreno EC, Shimoda S, Prostak K, Suga S (1991) Ultrastructural studies on crystal growth of enameloid minerals in Elasmobranch and Teleost fish. Calcif Tissue Int 48:204–217

Moradian-Oldak J, Weiner S, Addadi L, Landis WJ, Traub W (1991) Electron imaging and diffraction study of individual crystals of bone, mineralized tendon and synthetic carbonate apatite. Connect Tissue Res 25:219–228

Myers HM, Engström A (1965) A note on the organization of hydroxyapatite in calcified tendons. Exp Cell Res 40:182–185

Nanci A, Smith CE (1992) Development and calcification of enamel. In: Bonucci E (ed) Calcification in biological systems. CRC Press, Boca Raton, pp 313–343

Neuman WF, Mulryan BJ, Martin GR (1960) A chemical study of osteoclasis based on studies with yttrium. Clin Orthop Relat Res 17:124–134

Nylen MU, Scott DB, Mosley VM (1960) Mineralization of turkey leg tendon. II. Collagen-mineral relations revealed by electron and X-ray microscopy. In: Sognnaes RF (ed) Calcification in biological systems. American Association for the Advancement of Sciences, Washington, pp 129–142

Nylen MU, Eanes ED, Omnell K-Å (1963) Crystal growth in rat enamel. J Cell Biol 18:109–123

Oberc MA, Engel WK (1977) Ultrastructural localization of calcium in normal and abnormal skeletal muscle. Lab Invest 36:566–577

Okazaki K, Inoue S (1976) Crystal property of the larval sea urchin spicule. Dev Growth Differ 18:413–434

Oksanen A, Poukka R (1972) An electron microscopical study of nutritional muscular degeneration (NMD) of myocardium and skeletal muscle in calves. Acta Pathol Microbiol Scand Sect A 80:440–448

Olson OP, Watabe N (1980) Studies on formation and resorption of fish scales. IV: Ultrastructure of developing scales in newly hatched fry of the sheepshead minnow, *Cyprinodon variegatus* (Atheriniformes: Cyprinodontidae). Cell Tissue Res 211:303–316

Onozato H, Watabe N (1979) Studies on fish scale formation and resorption III. Fine structure and calcification of the fibrillary plates of the scales in *Carassius auratus* (Cypriniformes: cyprinidae). Cell Tissue Res 201:409–422

Paegle RD (1969) Ultrastructure of calcium deposits in arteriosclerotic human aortas. J Ultrastruct Res 26:412–423

Palladini G, Carbone A (1966) Ultrastruttura della calcificazione distrofica renale da sublimato. Experientia 22:585

Paule WJ, Bernick S, Strates B, Nimni ME (1992) Calcification of implanted vascular tissues associated with elastin in an experimental animal model. J Biomed Mater Res 26:1169–1177

Pautard FGE (1965) Calcification of baleen. In: Richelle LJ, Dallemagne MJ (eds) Calcified tissues. Université de Liège, Liège, pp 347–357

Pautard FGE (1970) Calcification in unicellular organisms. In: Schraer H (ed) Biological calcification: cellular and molecular aspects. Appleton-Century-Crofts, New York, pp 105–201

Pautard FGE (1975) The structure and genesis of calcium phosphates in vertebrates and invertebrates. In: Colloque Int. C.N.R.S.N° 230 (ed) Physico-chimie et cristallographie des apatites d'intérêt biologique. Centre National de la Recherche Scientifique, Paris, pp 93–100

Pergolizzi S, Anastasi G, Santoro G, Trimarchi F (1995) The shape of enamel crystals as seen with high resolution scanning electron microscope. It J Anat Embryol 100:203–209

Phakey P, Palamara J, Hall RK, McCredie DA (1995) Ultrastructural study of tooth enamel with amelogenesis imperfecta an AI-nephrocalcinosis syndrome. Connect Tissue Res 32:253–259

Pierce AM (1989) Attachment to and phagocytosis of mineral by alveolar bone osteoclasts. J Submicrosc Cytol Pathol 21:63–71

Plate U, Höhling HJ, Reimer L, Barckhaus RH, Wienecke R, Wiesmann H-P, Boyde A (1992) Analysis of the calcium distribution in predentine by EELS and of the early crystal formation in dentine by ESI and ESD. J Microsc 166:329–341

Plate U, Arnold S, Reimer L, Höhling H-J, Boyde A (1994) Investigation of the early mineralisation on collagen in dentine of rat incisors by quantitative electron spectroscopic diffraction (ESD). Cell Tissue Res 278:543–547

Posner AS, Harper RA, Muller SA, Menczel J (1965) Age changes in the crystal chemistry of bone apatite. Ann N Y Acad Sci 131:737–742

Quacci D, Dell'Orbo C, Pazzaglia UE (1990) Morphological aspects of rat metaphyseal cartilage pericellular matrix. J Anat 171:193–205

Robinson C, Shore RC, Wood SR, Brookes SJ, Smith DA, Wright JT, Connell S, Kirkham J (2003) Subunit structures in hydroxyapatite crystal development in enamel: implications for amelogenesis imperfecta. Connect Tissue Res 44:65–71

Robinson MJ, Strebel RF, Wagner BM (1968) Experimental tissue calcification IV. Ultrastructural observations in vagal calciphylaxis. Arch Path 85:503–515

Robinson RA (1952) An electron-microscopic study of the crystalline inorganic component of bone and its relationship to the organic matrix. J Bone Joint Surg 34-A:389–434

Robinson RA, Cameron DA (1956) Electron microscopy of cartilage and bone matrix at the distal epiphyseal line of the femur in the newborn infant. J Biophys Biochem Cytol 2:253–263

Robinson RA, Cameron DA (1964) Bone. In: Kurtz SM (ed) Electron microscopic anatomy. Academic Press, New York, pp 315–340

Robinson RA, Watson ML (1952a) Collagen-crystal relationships in bone as seen in the electron microscope. Anat Rec 114:383–409

Robinson RA, Watson ML (1952b) Collagen-crystal relationships in bone as seen in the electron microscope. Anat Rec 114:383–409

Rönnholm E (1962) The amelogenesis of human teeth as revealed by electron microscopy II. The development of enamel crystallites. J Ultrastruct Res 6:249–303

Rousselle AV, Heymann D (2002) Osteoclastic acidification pathways during bone resorption. Bone 30:533–540

Rubin MA, Jasiuk I, Taylor J, Rubin J, Ganey T, Apkarian RP (2003) TEM analysis of the nanostructure of normal and osteoporotic human trabecular bone. Bone 33:270–282

Ruigrok TJC, Elbers PF (1972) The effects of calcium acetate on mitochondria in the perfused rat liver I: Accumulation of Ca^{++} and concomitant swelling. Cytobiologie 5:51–64

Sahara N, Ashizawa Y, Nakamura K, Deguchi T, Suzuki K (1998) Ultrastructural features of odontoclasts that resorb enamel in human deciduous teeth prior to shedding. Anat Rec 252:215–228

Saladino AJ, Bentley PJ, Trump BF (1969) Ion movements in cell injury. Effect of amphotericin B on the ultrastructure and function of the epithelial cells of the toad bladder. Am J Pathol 54:421–466

Saleuddin ASM (1971) Fine structure of normal and regenerated shell of *Helix*. Can J Zool 49:37–41

Salo J, Lehenkari P, Mulari M, Metsikkö K, Väänänen HK (1997) Removal of osteoclast bone resorption products by transcytosis. Science 276:270–273

Schenk RK, Spiro D, Wiener J (1967) Cartilage resorption in the tibial epiphyseal plate of growing rats. J Cell Biol 34:275–291

Schlesinger PH, Blair HC, Teitelbaum SL, Edwards JC (1997) Characterization of the osteoclast ruffled border chloride channel and its role in bone resorption. J Biol Chem 272:18636–18643

Schmidt WJ (1933) Der Feinbau der anorganischen Grundmasse der Knochengewebes. Ber Oberhess Ges Natur u Heilk, Naturwiss Abt 15:219–247

Schönbörner AA, Boivin G, Baud CA (1979) The mineralization processes in teleost fish scales. Cell Tissue Res 202:203–212

Schroeder L, Frank RM (1985) High-resolution transmission electron microscopy of adult human peritubular dentine. Cell Tissue Res 242:449–451

Scott BL, Pease DC (1956) Electron microscopy of the epiphyseal apparatus. Anat Rec 126:465–495

Selvig KA (1973) Electron microscopy of dental enamel: analysis of crystal lattice images. Z Zellforsch 137:271–280

Shitama K (1979) Calcification of aging articular cartilage in man. Acta Orthop Scand 50:613–619

Siperko LM, Landis WJ (2001) Aspects of mineral structure in normally calcifying avian tendon. J Struct Biol 135:313–320

Sire J-Y (1988) Evidence that mineralized spherules are involved in the formation of the superficial layer of the elasmoid scale in cichlids *Cichlasoma octofasciatum* and *Hemichromis bimaculatus* (Pisces, Teleostei): an epidermal active participation? Cell Tissue Res 253:165–172

Sire J-Y (1994) Light and TEM study of nonregenerated and experimentally regenerated scales of *Lepisosteus oculatus* (Holostei) with particular attention to ganoine formation. Anat Rec 240:189–207

Sire J-Y, Géraudie J, Meunier FJ, Zylberberg L (1987) On the origin of ganoine: histological and ultrastructural data on the experimental regeneration of the scales of *Calamoichthys calabaricus* (Osteichthyes, Brachyopterygii, Polypteridae). Am J Anat 180:391–402

Slavkin HC, Diekwisch T (1996) Evolution in tooth developmental biology: of morphology and molecules. Anat Rec 245:131–150

Stratmann U, Schaarschmidt K, Wiesmann HP, Plate U, Höhling HJ, Szuwart T (1997) The mineralization of mantle dentine and of circumpulpal dentine in the rat: an ultrastructural and element-amalytical study. Anat Embryol 195:289–297

Su X, Sun K, Cui FZ, Landis WJ (2003) Organization of apatite crystals in human woven bone. Bone 32:150–162

Sundquist K, Lakkakorpi P, Wallmark B, Väänänen K (1990) Inhibition of osteoclast proton transport by bafilomycin A_1 abolishes bone resorption. Biochem Biophys Res Comm 168:309–313

Susheela AK, Kharb P (1990) Aortic calcification in chronic fluoride poisoning: biochemical and electronmicroscopic evidence. Exp Mol Pathol 53:72–80

Suzuki K, Sakae T, Kozawa Y (1998) Helix structure of ribbon-like crystals in bovine enamel. Connect Tissue Res 38:113–117

Takuma S (1960) Electron microscopy of the developing cartilaginous epiphysis. Arch Oral Biol 2:111–119

Tanimura A, McGregor DH, Anderson HC (1986) Calcification in atherosclerosis. I. Human studies. J Exp Pathol 2:261–273

Termine JD, Eanes ED, Greenfield DJ, Nylen MU (1973) Hydrazine-deproteinated bone mineral. Physical and chemical properties. Calcif Tissue Res 12:73–90

Teti A (1993) Biology of osteoclast and molecular mechanisms of bone resorption. It J Miner Electrol Metab 7:123–133

Teti A, Blair HC, Schlesinger P, Grano M, Zambonin-Zallone A, Kahn AJ, Teitelbaum SL, Hruska KA (1989) Extracellular protons acidify osteoclasts, reduce cytosolic calcium, and promote expression of cell-matrix attachment structures. J Clin Invest 84:773–780

Theilade J, Fejerskov O, Horsted M (1976) A transmission electron microscopic study of 7-day old bacterial plaque in human tooth fissures. Arch Oral Biol 21:587–598

Thomas RS, Greenawalt JW (1968) Microincineration, electron microscopy, and electron diffraction of calcium-phosphate-loaded mitochondria. J Cell Biol 39:55–76

Thyberg J (1974) Electron microscopic studies on the initial phases of calcification in guinea pig epiphyseal cartilage. J Ultrastruct Res 46:206–218

Thyberg J, Friberg U (1971) Ultrastructure of the epiphyseal plate of the normal guinea pig. Z Zellforsch 122:254–272

Tong W, Glimcher MJ, Katz JL, Kuhn L, Eppell SJ (2003) Size and shape of mineralites in young bovine bone measured by atomic force microscopy. Calcif Tissue Int 72:592–598

Towe KM (1967) Echinoderm calcite: single crystal or polycrystalline aggregate. Science 157:1048–1050

Towe KM, Cifelli R (1967) Wall ultrastructure in the calcareous foraminifera: crystallographic aspects and a model for calcification. J Paleontol 41:742–762

Towe KM, Thompson GR (1972) The structure of some bivalve shell carbonates prepared by ion-beam thinning. A comparison study. Calcif Tissue Res 10:38–48

Traub W, Arad T, Weiner S (1989) Three-dimensional ordered distribution of crystals in turkey tendon collagen fibers. Proc Natl Acad Sci USA 86:9822–9826

Traub W, Arad T, Weiner S (1992) Origin of mineral crystal growth in collagen fibrils. Matrix 12:251–255

Travis DF (1968a) Comparative ultrastructure and organization of inorganic crystals and organic matrices of mineralized tissues. Biology of the mouth. American Association for the Advancement of Sciences, Washington, pp 237–297

Travis DF (1968b) The structure and organization of, and the relationship between, the inorganic crystals and the organic matrix of the prismatic region of *Mytilus edulis*. J Ultrastruct Res 23:183–215
Travis DF (1970) The comparative ultrastructure and organization of five calcified tissues. In: Schraer H (ed) Biological calcification: cellular and molecular aspects. Appleton-Century-Crofts, New York, pp 203–311
Travis DF, Glimcher MJ (1964) The structure and organization of, and the relationship between the organic matrix and the inorganic crystals of embryonic bovine enamel. J Cell Biol 23:447–497
Travis DF, Gonsalves M (1969) Comparative ultrastructure and organization of the prismatic region of two bivalves and its possible relation to the chemical mechanism of boring. Am Zool 9:635–661
Väänänen HK, Karhukorpi E-K, Sundquist K, Wallmark B, Roininen I, Hentunen T, Tuukkanen J, Lakkakorpi P (1990) Evidence for the presence of a proton pump of the vacuolar H^+-ATPase type in the ruffled borders of osteoclasts. J Cell Biol 111:1305–1311
Väänänen HK, Salo J, Lehenkari P (1996) Mechanism of osteoclast-mediated bone resorption. J Bone Miner Metab 14:187–192
Väänänen HK, Zhao H, Mulari M, Halleen JM (2000) The cell biology of osteoclast function. J Cell Sci 113:377–381
Vaes G (1988) Cellular biology and biochemical mechanism of bone resorption. A review of recent developments on the formation, activation, and mode of action of osteoclasts. Clin Orthop Relat Res 231:239–271
Valouch P, Obenberger J, Vrabec F (1974) Experimental corneal calcification. An ultrastructural study. Ophthal Res 6:6–14
Vanden Berge JC, Storer RW (1995) Intratendinous ossification in birds: a review. J Morphol 226:47–77
Vasington FD, Greenawalt JW (1968) Osmotically lysed rat liver mitochondria. Biochemical and ultrastructural properties in relation to massive ion accumulation. J Cell Biol 39:661–675
Wachtel E, Weiner S (1994) Small-angle X-ray scattering study of dispersed crystals from bone and tendon. J Bone Miner Res 9:1651–1655
Wakita M (1993) Current studies on tooth enamel development in lower vertebrates. Kaibogaku Zasshi 68:399–409
Warshawsky H (1985) Ultrastructural studies on amelogenesis. In: Butler HT (ed) The chemistry and biology of mineralized tissues. EBSCO Media, Birmingham, pp 33–45
Warshawsky H (1989) Organization of crystals in enamel. Anat Rec 224:242–262
Warshawsky H, Nanci A (1982) Stereo electron microscopy of enamel crystallites. J Dent Res 61:1504–1514
Warshawsky H, Bai P, Nanci A (1987) Analysis of crystalline shape in rat incisor enamel. Anat Rec 218:380–390
Watabe N (1965) Studies on shell formation. XI. Crystal-matrix relationships in the inner layer of mollusk shells. J Ultrastruct Res 12:351–370
Watabe N (1967) Crystallographic analysis of coccolith of *Coccolithus huxleyi*. Calcif Tissue Res 1:114–121
Weiner S, Price PA (1986) Disaggregation of bone into crystals. Calcif Tissue Int 39:365–375
Weiner S, Traub W (1986) Organization of hydroxyapatite crystals within collagen fibrils. Fed Eur Biochem Soc 206:262–266
Weiner S, Traub W (1989) Crystal size and organization in bone. Connect Tissue Res 21:259–265
Weiner S, Traub W (1991) Organization of crystals in bone. In: Suga S, Nakahara H (eds) Mechanisms and phylogeny of mineralization in biological systems. Springer, Tokyo, pp 247–253

Weiner S, Arad T, Traub W (1991) Crystal organization in rat bone lamellae. Fed Eur Biochem Soc 285:49–54

Weiner S, Veis A, Beniash E, Arad T, Dillon JW, Sabsay B, Siddiqui F (1999) Peritubular dentin formation: crystal organization and the macromolecular constituents in human teeth. J Struct Biol 126:27–41

Weiss RE, Watabe N (1978) Studies on the biology of fish bone – II. Bone matrix changes during resorption. Comp Biochem Physiol 61A:245–251

Weiss RE, Watabe N (1979) Studies on the biology of fish bone. III. Ultrastructure of osteogenesis and resorption in osteocytic (cellular) and anosteocytic (acellular) bones. Calcif Tissue Int 28:43–56

Wheeler AP (1992) Mechanisms of molluscan shell formation. In: Bonucci E (ed) Calcification in biological systems. CRC Press, Boca Raton, pp 179–216

Wiesmann H-P, Plate U, Höhling H-J, Barckhaus RH, Zierold K (1993) Analysis of early hard tissue formation in dentine by energy dispersive X-ray microanalysis and energy filtering transmission electron microscopy. Scanning Microsc 7:711–718

Wong V, Saleuddin ASM (1972) Fine structure of normal and regenerated shell of *Helisoma duryi duryi*. Can J Zool 50:1563–1568

Woodhouse MA, Burston J (1969) Metastatic calcification of the myocardium. J Pathol 97:733–736

Yamada J, Watabe N (1979) Studies on fish scale formation and resorption I. Fine structure and calcification of the scales in *Fundulus heteroclitus* (Atheriniformes: cyprinodontidae). J Morphol 159:49–66

Yamada M, Ozawa H (1978) Ultrastructural and cytochemical studies on the matrix vesicle calcification in the teeth of the killifish, *Oryzias latipes*. Arch Histol Jpn 41:309–323

Yamauchi M, Katz EP (1993) The post-translational chemistry and molecular packing of mineralizing tendon collagens. Connect Tissue Res 29:81–98

Young JR, Davis SA, Bown PR, Mann S (1999) Coccolith ultrastructure and biomineralisation. J Struct Biol 126:195–215

Ziv V, Weiner S (1994) Bone crystal sizes: a comparison of transmission electron microscopic and X-ray diffraction line width broadening techniques. Connect Tissue Res 30:165–175

Zylberberg L, Géraudie J, Meunier F, Sire J-Y (1992) Biomineralization in the integumental skeleton of the living lower vertebrates. In: Hall BK (ed) Bone, vol 4: Bone metabolism and mineralization. CRC Press, Boca Raton, pp 171–224

6 The Size of Inorganic Particles

6.1 Introduction

Among the parameters that distinguish hard tissue crystals, size is one of the hardest to measure, not only because it is intrinsically variable, especially with crystal maturation, but also because the apparent values for crystals dimensions typically shift with orientation. Many studies have targeted the measurement of crystal size, chiefly by X-ray diffraction or by direct evaluation of crystal dimensions through electron microscope images (discussed by Ziv and Weiner 1994). With X-ray diffraction, the small size of crystals and several technical pitfalls have made for poorly defined diffractograms and may reflect crystal coherence lengths rather than the true dimensions of single crystals (Ziv and Weiner 1994). It is true that the precision of the technique can be improved (e.g., Zanini et al. 1999), but the values for size as calculated so far are highly approximate. With direct evaluation, several problems can be raised by the changes in crystal structure discussed earlier – changes often brought about by tissue processing prior to electron microscopy – and second, by difficulties intrinsic to image evaluation (discussed by Boothroyd 1975; Arsenault 1990). These may derive from the imprecision in securing the true focus or from an adequate degree of photographic contrast, or else from the position of structures relative to the electron beam, i.e., how their images are oriented with respect to the photographic plane. If the structures to be measured under the electron microscope are rod- or needle-like, their thickness can be measured easily and directly, as it corresponds to their transversal dimension and does not change significantly with their orientation. If they are considered to be platelet-like, their thickness can be measured with an acceptable degree of accuracy, as long as the most electron-dense, rod- or needle-like figures are taken to correspond to a platelet projection viewed sideways. By contrast, measurement of their width is a demanding task, because their orientation may be oblique with respect to the photographic plane. Similarly, it is hard, if not impossible, to measure their length, whatever their shape. The value for this dimension may, in fact, vary, not only because of changes in crystal orientation with respect to the electron beam, but also because their length may go beyond the limits of the microscope field; this sometimes happens, for instance, with enamel crystals. It need hardly be pointed out that the three dimensions which define crystal size can rarely, if ever, be measured for the same crystal unit.

As a result, the values for size reported in the literature – especially those for crystal length – must be viewed with caution and be interpreted as no more than approximate mean values.

6.2 Vertebrates

Most of the evaluations of crystal dimensions have been carried out on vertebrate bones, where difficulties in making measurements are increased by the likelihood that needle- and platelet-like crystals occur side by side in the same area.

6.2.1 Bone

Crystal dimensions (Table 6.1) were first evaluated by means of X-ray diffraction and related techniques, which were mainly carried out on compact bone. Although sharp diffraction rings were never obtained from normal, untreated tissues, the conclusion was reached that rod-shaped bone crystals are approximately 2.0–6.5 nm in diameter and 22–70 nm in length (Clark 1931; Stuhler 1938; Finean and Engström 1953; Carlström and Finean 1954; Carlström 1955; Fernández-Morán and Engström 1957; Carlström and Glas 1959; Matsushima et al. 1984; Fratzl et al. 1991, 1992; Ziv and Weiner 1994). Neuman and Neuman (1953), who thought that bone crystals have a platelet-like shape, reported thickness values of 2.5–5.0 nm, approximate length values of 40.0 nm, and width values close to those for length. Bigi et al. (1997) found that crystals in trabecular bone are smaller than those in compact bone. Sodek et al. (2000) fragmented porcine calvaria and long bones into particles of 20 μm and separated them by density gradient sedimentation into fractions of increasing density which were then examined by X-ray diffraction. They found that crystal length rose with density in calvaria (from about 14.5 to about 16.2 nm), while in long bones it increased to a plateau (from about 13.7 to about 15.0 nm), after which it fell back to its initial values, while crystal width fell with density in calvaria (from about 6.1 to about 5.7 nm) and was almost steady in long bones. Fratzl et al. (1996a) reported that the crystal thickness, as determined by small-angle X-ray scattering, rose with age to 3–4 nm, depending on the species in question.

Further information on the size of the bone crystals was obtained with the electron microscope. The earliest ultrastructural studies had low reliability, because of the rather coarse techniques available at that time – either the pseudo-replica method or the blender dissociation of the calcified matrix (Kellenberger and Rouiller 1950; Rutishauser et al. 1950; Rouiller et al. 1952a, b; Robinson 1952; Schwarz and Pahlke 1953; Ascenzi and Chiozzotto 1955). Neither technique was fully capable of showing the fine structural details of the

Table 6.1. Mean dimensions (in nm) of crystals in various types of bone

Type of bone	Source	Thickness	Width	Length	Reference
Cortical	Human	2.5–5.0	35–40	35–40	Robinson and Watson (1952)[a]
Compact	Various	3–4	–	20	Fernández-Morán and Engström (1957)[a,e]
Periosteal	Rat, foetus	–	–	18.5	Knese and Knoop (1958)[a]
Cranial vault	Rat, foetus	3.0	–	22.5	Ascenzi and Benedetti (1959)[a]
Parietal	Mouse	3.0–5.0	–	17.5	Molnar (1959)[a,d]
Alveolar	Human	2.5–4.0	–	100	Johansen and Parks (1960)[a]
Medullary	Pigeon, duck	1.0–4.5	–	> 50	Ascenzi et al. (1963)[a]
Osteon	Ox	4.0–4.5	–	> 60	Ascenzi and Bonucci (1966)[a]
Embryonic	Chick	–	1.0–5.0	40.0	Glimcher and Krane (1968)[a]
Osteon	Ox	2.0–4.0	60–70	Hundreds	Bocciarelli (1970)[a]
Periosteal	Chick, mouse	1.5–3.0	–	40–50	Landis et al. (1977a)[a,c]
Periosteal	Chick, mouse	3.0	–	45	Landis et al. (1977b)[a,b]
Osteon	Ox	4.0–4.5	–	> 60	Ascenzi et al. (1978)[a,e]
Compact	Various	5.0	–	32–36	Jackson et al. (1978)[a,d,e]
Osteon	Ox	4.0–4.5	–	> 60	Ascenzi et al. (1979)[e]
Alveolar	Human	–	–	47	Cuisinier et al. (1987)[f]
Cortical	Rat	–	5	17.0	Arsenault and Grynpas (1988)[d]
Cortical	Rat	–	–	17.0	Arsenault (1989)[d]
Tibia	Rat	2.0–2.5	10.4–18.7[i]	18.0-36.6[i]	Weiner and Traub (1989)[a,h]
Cortical	Rat	–	169–217	298–366	Ziv and Weiner (1994)[a,e,h]
Cortical	Ox (4-8 y. old)	0.61	10.2	10.2	Eppell et al. (2001)[g,h]
Trabecular	Human	7.7	27	57	Rubin et al. (2003)[a]
Woven	Human	–	11.5–19.2	18.1–31.6	Su et al. (2003)[a,h]
Cortical	Ox (3-5 w. old)	2.0	6.0	9.0	Tong et al. (2003)[g,h]

[a] Transmission electron microscopy
[b] Cryomicrotomy
[c] Anhydrous preparation
[d] Selected-area dark field electron microscopy
[e] X-ray diffraction
[f] High resolution transmission electron microscopy
[g] Atomic force microscopy
[h] Dispersed crystals
[i] Increasing with age

crystals; this explains why Robinson (1952) incorrectly reported that bone crystals measure 50 × 25 × 10 nm.

More accurate, but still variable, results were achieved when ultrathin sections became a viable option. Ascenzi and Benedetti (1959) found that bone crystals in the cranial vault of rat foetuses measure about 3.0–4.0 × 22.5 nm. Knese and Knoop (1958) reported that the length of bone crystals is about 18.5 nm, whereas Cuisinier et al. (1987) found a mean value of 47 nm. Fernández-Morán and Engström (1957), Molnar (1959), and Ascenzi and Bonucci (1966) found that crystals in compact bone were 3.0–5.0 nm thick and between 5.0 and over 100 nm long. Ascenzi et al. (1978, 1979) reported values of 40 nm for the length of the inorganic structures that they considered to be located at the level of the collagen main bands, and 4.0–4.5 nm for the thickness of needle-shaped crystals. Glimcher and Krane (1968) reported values of 1.0–5.0 nm (mean value about 3.0 nm) for the thickness of rod-like bone crystals, and 40.0 nm as their maximum length. In anhydrously-prepared bone, needle-shaped crystals were reported to be 1.5–3.0 nm thick and 40–50 nm long (Landis et al. 1977a); almost the same values (3.0 × 45 nm) were found in bone prepared by ultracryomicrotomy (Landis et al. 1977b). Ascenzi et al. (1963) found that crystals in the medullary bone of birds are 1.0–4.5 nm in width and at least 50 nm in length. Robinson and Watson (1952), who considered crystals to be platelets, reported that bone crystals are 2.5–5.0 nm thick and 35–40 nm long and wide. Bocciarelli (1970) reported that platelet-like crystals are 2.0–4.0 nm in cross section, "several hundred" nm in length, and 60–70 nm in width. Fernández-Morán and Engström (1957) found that the flake-like cross sections of bone crystals were 2.5–5.0 nm thick and 30–50 nm in diameter. Jackson et al. (1978) reported values of 5.0 nm for the thickness and between 32 and 36 nm for the *c*-axial length of platelet-like crystals in various types of bone. Rubin et al. (2003) found that plate-like crystals in normal human bone measure 7.7 × 27 × 57 nm. Johansen and Parks (1960), who also considered that bone crystals are platelets, measured their narrowest and longest profile dimensions and reported values of 2.5–4.0 × 100 nm. Su et al. (2003) found that isolated crystals of woven bone are platelets 11.5–19.2 nm wide and 18.1–31.6 nm long; they did not report the dimensions of the needle-shaped crystals that are described as lying close to collagen fibrils and whose length often exceeded the fibril dark bands. Again in isolated crystals, Ziv and Weiner (1994) found that crystal size increases with age and reported a length ranging from 298 (three-week-old rats) to 366 nm (two-year-old rats), and a width ranging from 169 to 217 nm, respectively. Using atomic force microscopy, Eppell et al. (2001) measured the size of crystals (mineralites) isolated using the anhydrous method of Kim et al. (1995) and found a mean thickness of 0.61 ± 0.19 nm, a mean width of 10 ± 2 nm and a mean length of 12 ± 2 nm. Using the same method, Tong et al. (2003) reported values of 2 × 6 × 9 nm for mineralites that had been isolated from young postnatal bovine bone; they found that the isolated mineralites of this type of bone were smaller than those isolated from mature bovine bone. Arsenault (1989) reported that crystals of cortical bone have a mean *c*-axial length of 17.0 ± 5.0 nm, and Arsenault and Grynpas (1988) found values of 5 × 12–17 nm in rat bone. Weiner and Traub (1989) isolated crystals from rat tibiae and found that their thickness varied little, measuring 2.0–2.5 nm, whereas their width ranged with age from 10.4 to 18.7 nm, and their length from 18.0 to 36.6 nm. Arsenault (1988) did, however, report that the *c*-axial length determined by selected-area dark-field imaging falls within a range of 11–17 nm and stated that the higher values obtained by bright-field electron microscopy depend on the aggregation of the 11–17 nm microcrystals.

It is striking that crystal dimensions, especially thickness, show little impact from treatment with the "stains" (uranyl acetate and lead citrate) used to increase ultrastructural contrast under the electron microscope (Bonucci 1975).

6.2.2 Tendon

Crystal dimensions in calcified turkey tendons (Table 6.2) were first measured by Nylen et al. (1960), who found crystals less than 5.0 nm thick and between 20 and 150 nm long. Myers and Engström (1965) calculated a width of 5 nm and a *c*-axial length of 35 nm. Arsenault (1988) reported that the crystals in calcified tendons have a mean *c*-axial length similar to that of bone crystals (between 11 and 17 nm); he later specified a value of 14.2 ± 4.3 nm (Arsenault 1989). Lees et al. (1994) found that needle-shaped crystals have an average thickness of 7 nm and a maximum length of at least 90 nm, while tablet-like crystals have an average length of 58 nm. Using scanning small angle X-ray scattering, a gradient in mineral thickness has been found running through the initial 200 μm from

Table 6.2. Mean dimensions (in nm) of crystals in calcified turkey tendons

Thickness	Width	Length	Reference
5.0	–	20–150	Nylen et al. (1960)
5.0	–	35	Myers and Engström (1965)
25–30	–	30–40	Weiner and Price (1986)[a]
–	–	14.2	Arsenault (1989)
5	20	35	Heywood et al. (1990)[a]
4.6	30–45	40–170	Landis et al. (1993)[c]
7	–	58–90	Lees et al. (1994)
–	30–40	–	Siperko and Landis (2001)[b]
3.3–2.5	–	–	Gupta et al. (2003)

[a] Isolated crystals
[b] Tapping mode atomic force microscopy
[c] High-voltage electron microscopy

the calcification front in tendon longitudinal sections cut parallel to the axis of the collagen fibrils, after which the value for thickness remained almost steady, between 2.3 and 2.5 nm (Gupta et al. 2003). Using tapping mode atomic force microscopy, Siperko and Landis (2001) found mineral plates 30–40 nm wide, which were sometimes aligned to form larger aggregates (475–600 nm long and 75–90 nm wide); Landis et al. (1993), using high-voltage electron microscopic tomography and graphic image reconstruction, found plate-like crystals measuring 4–6 × 30–45 × 40–170 nm. In the disaggregated collagen fibrils of calcified tendons, Weiner and Price (1986) found platelet-like crystals measuring 25–30 × 30–40 nm, and Heywood et al. (1990) found that crystals isolated from calcified tendons measured 5 × 20 × 35 nm.

6.2.3 Dentin

Dentin and dentinogenesis are closely related to bone and osteogenesis, respectively (Linde and Goldberg 1993), so it is not surprising that the size of crystals in dentin (Table 6.3) is basically close to that of crystals in bone. In intertubular dentin, they have been reported to measure 2–3 nm in thickness and 60 nm in length (Linde and Goldberg 1993), but in peritubular dentin they measure 9.7 × 25.5 × 36 nm (Schroeder and Frank 1985). Johansen and Parks (1960), who considered that crystals in human dentin are platelets, measured their narrowest and longest profile dimensions, reporting values of 2.0–3.5 × 100 nm. Hoshi et al. (2001) found that crystals in molar dentin are approximately 10 nm wide and 50–100 nm long, and Arsenault (1989) found that

Table 6.3. Mean dimensions (in nm) of dentin crystals

Type of dentin	Source	Thickness	Width	Length	Reference
Intertubular	Human	2–3	–	60	Linde and Goldberg (1993)[a]
Peritubular	Human	9.76	25.57	36	Schroeder and Frank (1985)[b]
Coronal	Human	22.0–3.5	–	100	Johansen and Parks (1960)[a]
Intertubular	Human	3.2	29	–	Daculsi et al. (1978)[b]
Intertubular	Human	–	10	50–100	Hoshi et al. (2001)[c]
Molar	Rat	–	–	110	Arsenault (1989)[d]
Peritubular	Human	9	25	36	Schroeder and Frank (1985)[b]
Premolar	Dog	–	25–45.8	–	Hayashi (1987)[a]
Intertubular	Human	5	–	–	Kinney et al. (2001)[e]
Canine	Human	–	–	40–60	Abe et al. (1991b)[a,e]

[a] Transmission electron microscopy
[b] High resolution transmission electron microscopy
[c] Focused ion beam and energy-filtering transmission electron microscopy
[d] Selected-area dark field electron microscopy
[e] Small angle X-ray scattering

their c-axial length is 11.0 ± 3.0 nm. Daculsi et al. (1978) found that dentin crystals have an approximate thickness of 3.2 nm and a width of 29 nm. Schroeder and Frank (1985) described dentin crystals as consisting of $9 \times 25 \times 36$ nm platelets. Hayashi (1987) reported that dentin crystals, in the zone of dog roots where the cementum runs parallel to the root surface, are 2.50 nm wide, while the width of mature crystals in the zone where cementum runs perpendicular to the root surface is 4.58 nm. Kinney et al. (2001) found that the crystals which were needle-like near the pulp and progressed to a platelet-like shape near the dentino-enamel junction were about 5.0 nm thick, a value which showed no change with crystal position. Abe et al. (1991) performed a comparative study on normal dentin and globular dentin from patients with familial and sporadic X-linked hypophosphatemic rickets or regional odontodysplasia. They found that in normal dentin crystal size along the c-axis ranged between 40 and 60 nm, much less than that of crystals in globular dentin (125–235 nm).

6.2.4 Cementum

Hayashi (1987) reported that crystals in dog cementum, measured in the root zone, where cementum runs parallel to the root surface, are 1.67 nm wide, whereas mature crystals in the zone where the cementum runs perpendicular to the root surface have a width as great as 2.92 nm.

6.2.5 Epiphyseal Cartilage

In the epiphyseal cartilage of young kittens, Scott and Pease (1956) found crystals about 5.0 nm thick and 40–50 nm long, and, in cartilage in the phalangeal epiphyses of the mouse, Takuma et al. (1960) found crystals 5.0 nm wide and 20.0–80.0 nm long. Arsenault and Grynpas (1988), who thought that the crystals of rat cartilage are platelet-like, reported that they are 5.0 nm thick and 12–17 nm long, the same values they found in bone and in synthetic poorly crystalline hydroxyapatite. The electron microscope and dark field imaging studies of Arsenault and Hunziker (1988) showed that the rod-like crystals are between 30 and 80 nm long, and that they consist of aggregates of smaller crystals. Hunziker and Herrmann (1990) reported that the hydroxyapatite crystal units in cartilage are very small, only about 5 nm in diameter and about 12 nm in length, and that they form plate-like and needle-like aggregates that spread through the cartilage matrix. Bonucci (1967) reported that the needle-like crystals of the cartilage calcification nodules have an average thickness of 1.8 nm and a length ranging between 50 and 160 nm.

6.2.6 Enamel

Rather variable values have been reported for enamel crystal size (Table 6.4). Probably because of the extreme hardness of this tissue, X-ray diffraction studies were initially carried out on enamel powder, which probably accounts for the fact that they were even considered spherical (reviewed by Rönnholm 1962). X-ray diffractograms from oriented dental enamel specimens with a good

Table 6.4. Mean dimensions (in nm) of enamel crystals

Source	Thickness	Width	Length	Reference
Human	33	86–101	–	Frazier (1968)[a]
Human	–	36.6	32.1	Grove et al. (1972)[a,b]
Rat, mature enamel	25	45	–	Selvig and Halse (1972)[a]
	–	37	–	Jongebloed et al. (1975)[a]
Human	26.3	68.3	–	Daculsi and Kerebel (1978)[c]
Human and cat	2.0–3.7	–	–	Weiss et al. (1981)[a]
Human immature	1.5	–	–	Rönnholm (1962)[a]
Human mature	17.0	40.0	160.0	
Human immature	1.0	50–60	–	Nylen et al. (1963)[a]
Human mature	25–30	50–60	–	
Bovine embryo immature	1.9	40	130	Travis and Glimcher (1964)[a]
Bovine embryo mature	16	55	162	
Rat D.E.J. (2 μm)	22.6	62.3	–	Glick and Eisenmann (1973)[a]
Rat inner (10 μm)	15.3	39.8	–	
Rat inner (30 μm)	13.7	38.9	–	
Rat interm. (50 μm)	12.5	36.3	–	
Rat outer (80 μm)	10.5	32.4	–	
Human immature	1.5	15	–	Kerebel et al. (1979)[c]
Human mature	16	61	–	
Rat immature	18	–	–	Bonucci et al. (1994)[a]
Rat mature	33.7	–	–	

[a] Transmission electron microscopy
[b] Selected-area dark field electron microscopy
[c] High resolution electron microscopy

alignment of the crystallites failed to produce better results: Glas and Omnell (1960) reported dimensions of 41 × 160 nm for crystals that they considered to be most probably rod-shaped. The replica technique for electron microscopy was also used, and crystal dimensions reported were 10 × 65 × 400 nm (Hall 1958) (for values obtained before 1960 by X-ray diffraction or electron microscopy, see Glas and Omnell 1960 and Rönnholm 1962).

When enamel sections could be examined under the electron microscope, the crystal size was more accurately determined, but their recorded dimensions still varied considerably (Rönnholm 1962). Crystal length showed the greatest variation: values of 200 nm or more (Frank and Voegel 1975), 500 nm (Simmelink et al. 1974), 600 nm (Frank et al. 1967), or 100 μm (Daculsi et al. 1984) have been reported. This depends on the fact that crystals as long as those of enamel may easily fail to keep a single direction along their whole length, and may extend beyond the section plane and/or the microscopic field, the result being that they may appear to be shorter than they really are (Warshawsky and Nanci 1982). It is also possible that one row of partly overlapping crystals has been considered as one long crystal because sections are excessively thick or because of poor electron microscope resolution (discussed by Rönnholm 1962). This makes it difficult to recognize gaps and/or overlaps in consecutive crystals, which may therefore appear as single, extremely long structures (Travis 1968a). Another suggestion has been that enamel crystals are so long that they run uninterruptedly from the dentin-enamel junction to the surface enamel. The great length of enamel crystals has been convincingly demonstrated by Daculsi et al. (1984) and McKee et al. (1988), who measured isolated crystals: although their isolation method includes a treatment of the enamel sample which may cause ruptures and changes in crystal structure, the view that these crystals are extremely long appears to be well-founded (Warshawsky 1989). The outcome is that crystal thickness and width can be evaluated with a high degree of statistical significance, but no accurate measurement of crystal length is possible (Ziv and Weiner 1994).

In spite of these difficulties, many attempts to measure enamel crystal size have been made. In mature human enamel, Frazier (1968) reported that crystals are about 33 nm thick and 86–101 nm wide. In the same type of enamel, Grove et al. (1972) found that crystals which have a rod-like shape in bright field examination appear to be rectangular in dark field, with a mean length of 32.1 nm and a mean width of 36.6 nm. In the mature enamel of rat incisors, Selvig and Halse (1972) found crystals about 25 nm thick and 45 nm wide. Jongebloed et al. (1975) believed that the thickness and width of enamel crystals cannot easily be differentiated and only reported the "average diameter", specifying a figure of 37 nm for sound enamel. Kerebel et al. (1979) reported that human enamel crystals have a mean thickness of 26.3 nm and a mean width of 68.3 nm. In the enamel of newborn cat and human foetuses, Weiss et al. (1981) found that the developing crystals had a thickness of at least 2.0 nm, with maximum values of 2.5–3.7 nm.

It must be considered that the size of enamel crystals changes significantly with their maturation. In comparing human and rat incisor enamel, Glick (1979) observed that in both cases crystal size increases linearly as a function of distance from ameloblasts. By fractionating enamel crystals of different density, Menanteau et al. (1984) showed an increase in their width directly correlated with protein degradation, in line with the increasing degree of calcification. In human enamel, Rönnholm (1962) found that the first crystals are fairly long plates or ribbons about 1.5 nm thick, whereas the most mature crystals are 17.0 nm thick, 40.0 nm wide, and 160.0 nm long. In young enamel, Nylen et al. (1963) reported that the early crystals are about 1.0 nm thick, while their length is already that of definitive crystals; in adult enamel they found that mature crystals, which are considered to be hexagonal, are 25–30 nm thick and 50–60 nm wide. Travis and Glimcher (1964) reported that the most recently deposited crystals located close to the ameloblasts in embryonic bovine enamel are plates measuring approximately 1.9 × 40 × 130 nm, and that mature crystals have dimensions of about 16 × 55 × 162 nm. Glick and Eisenmann (1973) measured crystal size from the outer enamel to the dentin-enamel junction (DEJ) and found that it rises with crystal age: thickness rose from 10.5 nm in outer enamel (80 μm from DEJ), to 12.5 nm in intermediate enamel (50 μm from DEJ), to 13.7 and 15.3 nm in inner enamel (30 μm and 10 μm from DEJ, respectively), and to 22.6 nm at the DEJ. The corresponding values for width were 32.4, 36.3, 38.9, 39.8 and 62.3 nm. Daculsi and Kerebel (1978) found average thickness values of 26.3 nm and average length values of 68.3 nm; they stressed that at first crystals grow rapidly in length and slowly in thickness, after which their thickness rises until their dimensions at maturity are reached. Using a similar method, Kerebel et al. (1979) measured the dimensions of crystals at various degrees of development (evaluated on the basis of their distance from ameloblasts and the numbers of their lattice planes). They found that young crystals (located near the ameloblasts and having just one lattice plane) were 1.5 nm thick and 15 nm wide, whereas mature crystals (far from ameloblasts and with at least twenty lattice planes) were 16 nm thick and 61 nm wide. Using the same measurement method as Rönnholm (1962), Bonucci et al. (1994) found that the mean crystal thickness in incisors of control rats was about 18.0 nm in the superficial zone, and between 30.7 and 33.7 nm in the intermediate and dentinal zones.

The variability of the results reported above may depend on differences in the source, site and degree of maturation of the enamel examined. They clearly confirm the early data of Rönnholm (1962) that enamel crystals very quickly attain their definitive width and length, and that the further rise in inorganic substance with enamel maturation is mainly due to an increase in crystal thickness.

6.3 Lower Vertebrates

In lower vertebrates, the crystals of the endoskeleton resemble those found in the calcified bone and cartilage of higher vertebrates (Dickson 1982). Crystals in the external layer of the elasmoid scales examined after freeze-fixation and freeze-substitution measure between 30 and 100 nm in length (Zylberberg and Nicolas 1982). The needle-shaped crystals of developing scales in newly hatched fry of *Cyprinodon variegatus* (Atheriniformes: Cyprinodontidae) are 5–8 nm wide and up to 10 nm long (Olson and Watabe 1980). In areas of loosely packed collagen fibrils in fish bone, Glimcher (1959) described rod-shaped crystals that were 1.5–4.0 nm thick and 20.0–40.0 nm wide.

The crystals of *enameloid* resemble those of higher vertebrate enamel and appear to have the same size. Garant (1970) reported that the hexagonal cross section of enameloid crystals in the dogfish *Squalus acanthias* L. is between 20 and 100 nm across and between 100 and 150 nm in length. The crystals located in the non-collagenous tissues of the *ganoine* are 250 nm long and 8 nm thick (Zylberberg et al. 1992). Crystals in baleen are similar to those found in bone (Pautard 1965); they are reported to be 5.0 nm wide (Pautard 1975).

6.4 Invertebrates

There have been relatively few reports on crystal dimensions in invertebrate calcified tissues (Table 6.5). In normal and regenerated shells of *Helisoma duryi duryi*, Wong and Saleuddin (1972) found parallel arranged crystals about 50–70 nm thick and between 0.6 and 0.8 μm long. Watabe (1963) reported that in the nacreous layer of the mollusk shell the individual crystals are 0.3 μm wide and 0.6 μm long. Crystals of the nacreous layer of the shell of *Helix pomatia* are 0.3–1.8 μm long (Saleuddin 1971). Mean values of 15 μm in diameter (*a*- and *b*-directions) and 0.5 μm in height (*c*-direction) are reported by Blank et al. (2003) on the basis of the literature. In the calcareous wall of foraminifera, Towe and Cifelli (1967) reported that, although the boundary of single crystals is poorly defined in the replicas and their thickness is hard to assess, many crystals are in the range of 0.2 μm; they suspect, however, that the plate-like units do not represent individual crystals but are parts of a larger crystal mosaic, a hypothesis supported by later measurements of crystals in various mollusks. Travis and Gonsalves (1969) found that in *Mytilus edulis* the crystals of the prisms in sagittal section are 33 nm wide and about 14 nm thick, and those in cross section are 70 nm long and 14 nm thick. This indicates that the calcite crystals are roughly rectangular (70 × 33 × 14 nm) and plate-like (Travis 1968b). The same authors (Travis and Gonsalves 1969) found that in *Crassostrea virginica* calcite crystals (collected in prisms) are 600 nm long and 100 nm thick when cut in sagittal sections, and that the calcite crystals in the oyster are larger than in *Mytilus*,

Table 6.5. Mean dimensions in nm or in μm (bold characters) of invertebrate crystals

Sourse	Thickness	Width	Length	Reference
Elliptio compl. nacre		**0.3**	**0.6–0.8**	Watabe (1963)
Foraminifera	**0.2**			Towe and Cifelli (1967)
Mytilus edulis prisms	14	33	70	Travis (1968b)
Mytilus edulis prisms	12.2[a]–15.2[b]	22.6[c]	68.0	Travis (1968a)
Mytilus edulis prisms	14[d] 14[e]	33[d] 70[e]		Travis and Gonsalves (1969)
Crassostrea virg. prisms	100	–	600	Travis and Gonsalves (1969)
Oyster	100	300	600	Travis and Gonsalves (1969)
Sponge	11.1–20.4	–	37.0–100.0	Travis (1970)
immature	1.8	17.8	41.7	
Crustacea				Travis (1970)
mature	3.2	25.4	141.6	
platelet	6.7	–	48.2	
Echinoderm				Travis (1970)
needles	14.5	–	196.0	
Helix pomatia nacre			**0.3–1.8**	Saleuddin (1971)
From literature	**15**	**15**	**0.5**	Blank et al. (2003)
Pinna nobilis prisms		150–180		Dauphin (2003)
Pinctada margariti fera prisms		110	250–400	Dauphin (2003)

[a] Thicker dimension in lateral-longitudinal profile
[b] Thicker dimension in cross-sectional profile
[c] Wider dimension in surface-longitudinal profile
[d] Sagittal section
[e] Cross section

but have a similar rectangular shape (600 × 300 × 100 nm). Travis (1970) found crystals in skeletal spicules of sponges that were between 37.0 and 100.0 nm long and between 11.1 and 20.4 nm thick. In an electron microscope study of the crustacea exoskeleton, she found that young, platelet-like crystals average 1.8 nm in thickness, 17.8 nm in width and 41.7 nm in length, whereas mature crystals measure 3.2, 25.4 and 141.6 nm, respectively (Travis 1970); in the echinoderm endoskeleton she reported platelet-like crystals averaging 6.7 nm in thickness and 48.2 nm in length, and needle- or rod-shaped crystals averaging 14.5 and 196.0 nm in length. The shell prisms of the bivalve *Pinna nobilis* consist of oblique and elongated crystallites measuring from 150 to 180 nm in width, and those of the bivalve *Pinctada margaritifera* of elongated crystallites measuring about 110 nm in width and from 250 to 400 nm in length; in both cases the crystallites are subdivided into smaller, rounded subunits (Dauphin, 2003). In *Lingula unguis*, whose shells contain about 50% of inorganic salt, 90% of which is phosphate, Kelly et al. (1965) reported that the round particles which distinguish the initial phase of calcification are 5.0 nm thick, that the subsequent needle-shaped crystals are 5–8 nm wide and 15–30 nm long, and that the mature crystals measure about 20 × 100 nm.

6.5 Unicellular Organisms

The crystals which characterize the calcification of *Spirostomum ambiguum* are about 5.0 nm wide whereas their length is indeterminate (Pautard 1975).

The magnetosomes of *magnetotactic bacteria* are reported to have rather narrow size range, going from approximately 35 to 120 nm (Bazylinski et al. 1994). A diameter of 42–45 nm was reported by Schule and Bauerlein (1998).

6.6 Pathological Calcification

Studies on pathological calcification mostly target the changes that may occur in the components of the organic matrix, whereas calcification is considered in coarsely quantitative terms, i.e., the presence or absence of calcified areas, and the rise or fall in their numbers and size. From the many pictures that illustrate the reports of electron microscopic studies on skeletal disease, it may be inferred that crystals found in pathological calcifications do not differ significantly from normal ones. This idea is misleading, as shown by measurements carried out in rat pups nursed by mothers on a low calcium diet and weaned with that calcium-deficient diet: these animals developed hypocalcified enamel whose crystals were only 10.4 nm thick in the superficial layer, between 16.9 and 25.1 nm in the intermediate zone, and between 16.9 and 20.9 nm in the dentinal zone – values that are all sharply lower than those of the enamel in recovery animals or in the sound enamel of controls (immature crystals 18 nm; mature crystals 33.7 nm) (Bonucci et al. 1994).

The same scarcity of information is found with the mitochondria of normal or pathologically calcified tissues. In the mitochondria of osteoblasts, preosteoblasts and periosteal fibroblasts found in chick and mouse bone prepared anhydrously in organic solvents, Landis et al. (1977a) reported intrinsically electron-dense granules whose diameter ranged between 40 and 100 nm. At higher magnification, some granules appeared to consist of dense, circumferentially located small particles, approximately 5–10 nm in diameter, surrounding a less dense central region. In personal experience, intrinsically electron-dense, intramitochondrial granules have a mean diameter of 65 nm and consist of nanoparticles, about 10 nm across. In frozen thin sections of medullary bone induced in the Japanese quail by estradiol administration, the cells showed mitochondria containing large numbers of 20–80 nm electron-dense granules made up of 5.0–7.5 nm subparticles (Gay and Schraer 1975).

In *vascular calcification*, Paegle (1969) reported that the largest, platelet-like crystals found in arteriosclerotic human aortas measured 8 × 80 × 200 μm.

In *calcification of the skin* induced by calciphylaxis or calcergy, needle-like crystalline structures were found measuring 10–15 nm in width and 100–150 nm in length (Boivin et al. 1987).

6.7
Congenital or Acquired Diseases; Genetically Modified Animals

Crystal size is often altered in congenital or acquired diseases and in genetically modified animals, but exact crystal dimensions are rarely reported. This is particularly true of studies in humans: congenital skeletal diseases are rare in themselves, and studies are limited to a very few cases (sometimes only one); the available material is scarce and often, at least in personal experience, poorly preserved, especially with reference to aborted fetuses. For these reasons, the available results are not above criticism.

A study by Vetter et al. (1991) included a relatively high number of cases: 8 cases of osteogenesis imperfecta (OI) type I (classified after Sillence et al. 1979), 4 cases of OI type II, 11 cases of OI type III and 14 cases of OI type IV. The study, which was carried out by using X-ray diffraction on bone powder, showed that hydroxyapatite crystals were significantly lower in size (15.7 nm calculated along the *c* axis vs 20.2 nm in controls) in fetuses with OI type II, all of which had died at birth, that crystal size was unusually low only in children and adolescents with OI type III (18.3 and 21.1 nm, respectively, vs control values of 22.9 and 25.3 nm) and IV (20.2 and 21.2 nm), whereas crystal size in OI type I was unusually low only in children (20.6 nm), returning to normal (23.0 nm) in adolescents, and that there was a trend toward greater crystal size with age in both OI patients and controls. An electron microscope study of periosteal bone of a newborn died for osteogenesis imperfecta (Fig. 4.11) also showed crystals thinner than in controls (Bonucci and De Matteis, 1968).

A fall in crystal size was also found by the X-ray diffraction of bone from two bovine osteogenesis imperfecta models (BOI-Australia and BOI-Texas), where crystal length fell from 50.0 nm in controls to 37.2 nm in BOI-A and 42.1 nm in BOI-T (Fisher et al. 1987). A study relying on the small-angle X-ray scattering of bone in oim/oim mice (osteogenesis imperfecta mouse, i.e., mouse homozygous for a null mutation in its COL 1A2 gene of type I collagen, which fails to synthesize functional pro $\alpha 2$(I) collagen chains and synthesizes only homotrimers of $\alpha 1$(I) collagen chains; see Phillips et al. 2000) showed crystals that not only had an altered mineral composition (Phillips et al. 2000) but were also thinner and more variably aligned than those in controls (Fratzl et al.

1996b). Again, in oim/oim mice, a fall in the degree of crystallinity, showing no response to the bisphosphonate alendronate treatment, was found in the primary vs the secondary spongiosa (Camacho et al. 2003).

Differences in crystal size have also been reported in amelogenesis imperfecta by Kerebel and Daculsi (1977), who found two crystal size periods of 25 nm and 50.0 nm vs normal value of 37.0 nm. Abnormal enamel crystals with a diameter approximately twice that found in controls and a length of about 1 μm – failing to cover the entire enamel thickness – have been found in transgenic mice overexpressing ameloblastin (Paine et al. 2003).

Mineral maturity (crystal size and perfection), measured with Fourier transform infrared microspectroscopy and infrared imaging, was significantly higher in osteopontin-knockout mice (Boskey et al. 2002) and significantly lower in the secondary ossification centers and cortical bone of TGFβ1 null mice, but not in the calcification areas of the growth plate and metaphyses (Atti et al. 2002).

6.8 Concluding Remarks

A few points should be stressed:

- Crystal size is very hard to measure because of the smallness of single crystals, and especially because of their variable orientation.
- Uncertainty about crystal size is heightened by the fact that measurements have partly been carried out on crystals isolated by mechanical and chemical methods which may have altered their structure.
- In spite of technical refinements, only approximate mean values of crystal size are attainable.
- Values for crystal thickness and width in many calcified tissues, above all those in vertebrates, suggest that both needle- and platelet-like shapes can be found.
- The length of needle-shaped crystals in bone may exceed the collagen period.
- The earliest crystals may appear either as granular, filament-like structures that are hard to measure, or as paracrystalline structures that are so thin as to comprise very few apatite units.
- With maturation, enamel crystals increase in thickness and width, whereas their length seems to remain constant. This process is particularly conspicuous in enamel, but seems to be a feature of all calcified tissues.
- Changes in crystal size and crystallinity in congenital or acquired diseases and in genetically modified animals may provide clues to the effects of various organic molecules on crystal formation.

References

Abe K, Masatomi Y, Moriwaki Y, Ooshima T (1991) X-ray diffraction analysis and transmission electron microscopic examination of globular dentin. Calcif Tissue Int 48:190–195

Arsenault AL (1988) Crystal-collagen relationships in calcified turkey leg tendons visualized by selected-area dark field electron microscopy. Calcif Tissue Int 43:202–212

Arsenault AL (1989) A comparative electron microscopic study of apatite crystals in collagen fibrils of rat bone, dentin and calcified turkey leg tendons. Bone Miner 6:165–177

Arsenault AL (1990) The ultrastructure of calcified tissues: methods and technical problems. In: Bonucci E, Motta PM (eds) Ultrastructure of skeletal tissues. Kluwer Academic Publishers, Boston, pp 1–18

Arsenault AL, Grynpas MD (1988) Crystals in calcified epiphyseal cartilage and cortical bone of the rat. Calcif Tissue Int 43:219–225

Arsenault AL, Hunziker EB (1988) Electron microscopic analysis of mineral deposits in the calcifying epiphyseal growth plate. Calcif Tissue Int 42:119–126

Ascenzi A, Benedetti EL (1959) An electron microscopic study of the foetal membranous ossification. Acta Anat 37:370–385

Ascenzi A, Bonucci E (1966) The osteon calcification as revealed by the electron microscope. In: Fleisch H, Blackwood HJJ, Owen M (eds) Calcified tissues.Conference proceeding, Springer, Berlin Heidelberg New York, pp 142–146

Ascenzi A, Chiozzotto A (1955) Electron microscopy of the bone ground substance using the pseudo-replica technique. Experientia 11:140

Ascenzi A, François C, Steve Bocciarelli D (1963) On the bone induced by estrogen in birds. J Ultrastruct Res 8:491–505

Ascenzi A, Bonucci E, Ripamonti A, Roveri N (1978) X-ray diffraction and electron microscope study of osteons during calcification. Calcif Tissue Res 25:133–143

Ascenzi A, Bonucci E, Generali P, Ripamonti A, Roveri N (1979) Orientation of apatite in single osteon samples as studied by pole figures. Calcif Tissue Int 29:101–105

Atti E, Gomez S, Wahl SM, Mendelsohn R, Paschalis E, Boskey AL (2002) Effects of transforming growth factor-beta deficiency on bone development: a Fourier tranform-infrared imaging analysis. Bone 31:675–684

Bazylinski DA, Garratt-Reed AJ, Frankel RB (1994) Electron microscopic studies of magnetosomes in magnetotactic bacteria. Microsc Res Tech 27:389–401

Bigi A, Cojazzi G, Panzavolta S, Ripamonti A, Roveri N, Romanello M, Noris Suarez K, Moro L (1997) Chemical and structural characterization of the mineral phase from cortical and trabecular bone. J Inorg Biochem 68:45–51

Bocciarelli DS (1970) Morphology of crystallites in bone. Calcif Tissue Res 5:261–269

Boivin G, Walzer C, Baud CA (1987) Ultrastructural study of the long-term development of two experimental cutaneous calcinoses (topical calciphylaxis and topical calcergy) in the rat. Cell Tissue Res 247:525–532

Bonucci E (1967) Fine structure of early cartilage calcification. J Ultrastruct Res 20:33–50

Bonucci E (1975) The organic-inorganic relationships in calcified organic matrices. Physico-chimie et cristallographie des apatites d'intérêt biologique. Centre National de la Recherche Scientifique, Paris, pp 231–246

Bonucci E, De Matteis A (1968) Aspetti ultrastrutturali dell'ossificazione periostale nell'osteogenesi imperfetta congenita. Ortop Traumatol Appar Motore 36:309–318

Bonucci E, Lozupone E, Silvestrini G, Favia A, Mocetti P (1994) Morphological studies of hypomineralized enamel of rat pups on calcium-deficient diet, and of its changes after return to normal diet. Anat Rec 239:379–395

Boothroyd B (1975) Observations on embryonic chick-bone crystals by high resolution transmission electron microscopy. Clin Orthop Relat Res 106:290–310

Boskey AL, Spevak L, Paschalis E, Doty SB, McKee MD (2002) Osteopontin deficiency increases mineral content and mineral crystallinity in mouse bone. Calcif Tissue Int 71:145–154

Camacho NP, Carroll P, Raggio CL (2003) Fourier transform infrared imaging spectroscopy (FT-IRIS) of mineralization in bisphosphonate-trated *oim/oim* mice. Calcif Tissue Int 72:604–609

Carlström D (1955) X-ray crystallographic studies on apatites and calcified structures. Acta Radiol Suppl. 121:1–59

Carlström D, Finean JB (1954) X-ray diffraction studies on the ultrastructure of bone. Biochim Biophys Acta 13:183–191

Carlström D, Glas J-E (1959) The size and shape of the apatite crystallites in bone as determined from line-broadening measurements on oriented specimens. Biochim Biophys Acta 35:46–53

Clark JH (1931) A study of tendons, bones, and other forms of connective tissue by means of X-ray diffraction patterns. Am J Physiol 98:328–337

Cuisinier F, Bres EF, Hemmerle J, Voegel J-C, Frank RM (1987) Transmission electron microscopy of lattice planes in human alveolar bone apatite crystals. Calcif Tissue Int 40:332–338

Daculsi G, Kerebel B (1978) High-resolution electron microscope study of human enamel crystallites: size, shape, and growth. J Ultrastruct Res 65:163–172

Daculsi G, Kerebel B, Verbaere A (1978) Méthode de mesure des cristaux d'apatite de la dentine humaine en microscopie électronique à transmission de haute résolution. C R Acad Sci Paris 286:1439–1442

Daculsi G, Menanteau J, Kerebel LM, Mitre D (1984) Length and shape of enamel crystals. Calcif Tissue Int 36:550–555

Dauphin Y (2003) Soluble organic matrices of the calcitic prismatic shell layers of two pteriomorphid bivalves. J Biol Chem 278:15168–15177

Dickson GR (1982) Ultrastructure of growth cartilage in the proximal femur of the frog, *Rana temporaria.* J Anat 135:549–564

Eppell SJ, Tong W, Katz JL, Kuhn L, Glimcher MJ (2001) Shape and size of isolated bone mineralites measured using atomic force microscopy. J Orthop Res 19:1027–1034

Fernández-Morán H, Engström A (1957) Electron microscopy and X-ray diffraction of bone. Biochim Biophys Acta 23:260–264

Finean JB, Engström A (1953) The low-angle scatter of X-rays from bone tissue. Biochim Biophys Acta 11:178–189

Fisher LW, Eanes ED, Denholm LJ, Heywood BR, Termine JD (1987) Two bovine models of osteogenesis imperfecta exhibit decreased apatite crystal size. Calcif Tissue Int 40:282–285

Frank RM, Voegel JC (1975) Etude ultrastructurale de la dissolution des cristaux d'apatite au cours de la carie de l'émail dentaire humain. In: Montel G (ed) Physico-chimie et cristallographie des apatites d'intérêt biologique. Centre National de la Recherche Scientiphique, Paris, pp 369–380

Frank RM, Sognnaes RF, Kern R (1967) Calcification of dental tissues with special reference to enamel ultrastructure. In: Sognnaes RF (ed) Calcification in biological systems. American Association Advancement Sciences, Washington, pp 163–202

Fratzl P, Fratzl-Zelman N, Klaushofer K, Vogl G, Koller K (1991) Nucleation and growth of mineral crystals in bone studied by small-angle X-ray scattering. Calcif Tissue Int 48:407–413

Fratzl P, Groschner M, Vogl G, Plenk H Jr, Eschberger J, Fratzl-Zelman N, Koller K, Klaushofer K (1992) Mineral crystals in calcified tissues: a comparative study by SAXS. J Bone Miner Res 7:329–334

Fratzl P, Schreiber S, Klaushofer K (1996a) Bone mineralization as studied by small-angle X-ray scattering. Connect Tissue Res 34:247–254

Fratzl P, Paris O, Klaushofer K, Landis WJ (1996b) Bone mineralization in an osteogenesis imperfecta mouse model studied by small-angle x-ray scattering. J Clin Invest 97:396–402

Frazier PD (1968) Adult human enamel: an electron microscopic study of crystallite size and morphology. J Ultrastruct Res 22:1–11

Garant PR (1970) An electron microscopic study of the crystal-matrix relationship in the teeth of the dogfish *Squalus acanthias* L. J Ultrastruct Res 30:441–449

Gay C, Schraer H (1975) Frozen thin-sections of rapidly forming bone: bone cell ultrastructure. Calcif Tissue Res 19:39–49

Glas J-E, Omnell K-Å (1960) Studies on the ultrastructure of dental enamel 1. Size and shape of the apatite crystallites as deduced from X-ray diffraction data. J Ultrastruct Res 3:334–344

Glick PL (1979) Patterns of enamel maturation. J Dent Res 58(B):883–892

Glick PL, Eisenmann DR (1973) Electron microscopic and microradiographic investigation of a morphologic basis for the mineralization pattern in rat incisor enamel. Anat Rec 176:289–306

Glimcher MJ (1959) Molecular biology of mineralized tissues with particular reference to bone. Rev Modern Phys 31:359–393

Glimcher MJ, Krane SM (1968) The organization and structure of bone, and the mechanism of calcification. In: Gould BS (ed) Biology of collagen. Academic Press, London, pp 67–251

Grove CA, Judd G, Ansell GS (1972) Determination of hydroxyapatite crystallite size in human dental enamel by dark-field electron microscopy. J Dent Res 51:22–29

Gupta HS, Roschger P, Zizak I, Fratzl-Zelman N, Nader A, Klaushofer K, Fratzl P (2003) Mineralized microstructure of calcified avian tendons: a scanning small angle X-ray scattering study. Calcif Tissue Int 72:567–576

Hall DM (1958) Study of the submicroscopic structure of human dental enamel by electron microscopy. J Dent Res 37:243–253

Hayashi Y (1987) Ultrastructure of cementum formation on partially formed teeth in dogs. Acta Anat 129:279–288

Heywood BR, Sparks NH, Shellis RP, Weiner S, Mann S (1990) Ultrastructure, morphology and crystal growth of biogenic and synthetic apatites. Connect Tissue Res 25:103–119

Hoshi K, Ejiri S, Probst W, Seybold V, Kamino T, Yaguchi T, Yamahira N, Ozawa H (2001) Observation of human dentine by focused ion beam and energy-filtering transmission electron microscopy. J Microsc 201:44–49

Hunziker EB, Herrmann W (1990) Ultrastructure of cartilage. In: Bonucci E, Motta PM (eds) Ultrastructure of skeletal tissues. Kluwer Academic Publishers , Boston, pp 79–109

Jackson SA, Cartwright AG, Lewis D (1978) The morphology of bone mineral crystals. Calcif Tissue Res 25:217–222

Johansen E, Parks HF (1960) Electron microscopic observations on the three dimensional morphology of apatite crystallites of human dentine and bone. J Biophys Biochem Cytol 7:743–746

Jongebloed WL, Molenaar I, Arends J (1975) Morphology and size-distribution of sound and acid-treated enamel crystallites. Calcif Tissue Res 19:109–123

Kellenberger E, Rouiller C (1950) Die Knochenstruktur, untersucht mit dem Elektronenmikroskop. Schwiz Z Allgem Pathol Bakteriol 13:783–788

Kelly PG, Oliver PTP, Pautard FGE (1965) The shell of *Lingula unguis*. In: Richelle LJ, Dallemagne MJ (eds) Calcified tissues. Université de Liège, Liège, pp 337–345

Kerebel B, Daculsi G (1977) Ultrastructural study of amelogenesis imperfecta. Calcif Tissue Res 24:191–197

Kerebel B, Daculsi G, Kerebel LM (1979) Ultrastructural studies of enamel crystallites. J Dent Res 58:844–850

Kim H-M, Rey C, Glimcher MJ (1995) Isolation of calcium-phosphate crystals of bone by non-aqueous methods at low temperature. J Bone Miner Res 10:1589–1601

Kinney JH, Pople JA, Marshall GW, Marshall SJ (2001) Collagen orientation and crystallite size in human dentin: a small angle X-ray scattering study. Calcif Tissue Int 69:31–37

Knese K-H, Knoop A-M (1958) Elektronenoptische Untersuchungen über die periostale Osteogenese. Z Zellforsch 48:455–478

Landis WJ, Paine MC, Glimcher MJ (1977a) Electron microscopic observations of bone tissue prepared anhydrously in organic solvents. J Ultrastruct Res 59:1–30

Landis WJ, Hauschka BT, Rogerson CA, Glimcher MJ (1977b) Electron microscopic observations of bone tissue prepared by ultracryomicrotomy. J Ultrastruct Res 59:185–206

Landis WJ, Song MJ, Leith A, McEwen L, McEwen BF (1993) Mineral and organic matrix interaction in normally calcifying tendon visualized in three dimensions by high-voltage electron microscopic tomography and graphic image reconstruction. J Struct Biol 110:39–54

Lees S, Prostak KS, Ingle VK, Kjoller K (1994) The loci of mineral in turkey leg tendon as seen by atomic force microscope and electron microscopy. Calcif Tissue Int 55:180–189

Linde A, Goldberg M (1993) Dentinogenesis. Crit Rev Oral Biol Med 4:679–728

Matsushima N, Akiyama M, Terayama Y, Izumi Y, Miyake Y (1984) The morphology of bone mineral as revealed by small-angle X-ray scattering. Biochim Biophys Acta 801:298–305

McKee MD, Brown JI, Warshawsky H (1988) A simple method for the preparation of isolated enamel crystallites for transmission electron microscopy. J Electron Microsc Techn 8:225–226

Menanteau J, Mitre D, Daculsi G (1984) Aqueous density fractionation of mineralizing tissues: an efficient method applied to the preparation of enamel fractions suitable for crystal and protein studies. Calcif Tissue Int 36:677–681

Molnar Z (1959) Development of the parietal bone of young mice 1. Crystals of bone mineral in frozen-dried preparations. J Ultrastruct Res 3:39–45

Myers HM, Engström A (1965) A note on the organization of hydroxyapatite in calcified tendons. Exp Cell Res 40:182–185

Neuman WF, Neuman MW (1953) The nature of the mineral phase of bone. Chem Rev 53:1–45

Nylen MU, Scott DB, Mosley VM (1960) Mineralization of turkey leg tendon. II. Collagen-mineral relations revealed by electron and X-ray microscopy. In: Sognnaes RF (ed) Calcification in biological systems. American Association for the Advancement of Sciences, Washington, pp 129–142

Nylen MU, Eanes ED, Omnell K-Å (1963) Crystal growth in rat enamel. J Cell Biol 18:109–123

Olson OP, Watabe N (1980) Studies on formation and resorption of fish scales. IV: Ultrastructure of developing scales in newly hatched fry of the sheepshead minnow, *Cyprinodon variegatus* (Atheriniformes: Cyprinodontidae). Cell Tissue Res 211:303–316

Paegle RD (1969) Ultrastructure of calcium deposits in arteriosclerotic human aortas. J Ultrastruct Res 26:412–423

Paine ML, Wang HJ, Luo W, Krebsbach PH, Snead ML (2003) A transgenic animal model resembling amelogenesis imperfecta related to ameloblastin overexpression. J Biol Chem 278:19447–19452

Pautard FGE (1965) Calcification of baleen. In: Richelle LJ, Dallemagne MJ (eds) Calcified tissues. Université de Liège, Liège, pp 347–357

Pautard FGE (1975) The structure and genesis of calcium phosphates in vertebrates and invertebrates. In: Colloque Int.C.N.R.S.N° 230 (ed) Physico-chimie et cristallographie des apatites d'intérêt biologique. Centre National de la Recherche Scientifique, Paris, pp 93–100

Phillips CL, Bradley DA, Schlotzhauer CL, Bergfeld M, Libreros-Minotta C, Gawenis LR, Morris JS, Clarke LL, Hillman LS (2000) Oim mice exhibit altered femur and incisor mineral composition and decreased bone mineral density. Bone 27:219–226

Robinson RA (1952) An electron-microscopic study of the crystalline inorganic component of bone and its relationship to the organic matrix. J Bone Joint Surg 34-A:389–434

Robinson RA, Watson ML (1952) Collagen-crystal relationships in bone as seen in the electron microscope. Anat Rec 114:383–409

Rouiller C, Huber L, Kellenberger E, Rutishauser E (1952a) La structure lamellaire de l'ostéone. Acta Anat 14:9–22

Rouiller C, Huber L, Rutishauser E (1952b) La structure de la dentine. Étude comparée de l'os et de l'ivoire au microscope électronique. Acta Anat 16:16–28

Rönnholm E (1962) The amelogenesis of human teeth as revealed by electron microscopy II. The development of enamel crystallites. J Ultrastruct Res 6:249–303

Rubin MA, Jasiuk I, Taylor J, Rubin J, Ganey T, Apkarian RP (2003) TEM analysis of the nanostructure of normal and osteoporotic human trabecular bone. Bone 33:270–282

Rutishauser E, Huber L, Kellenberger E, Majno G, Rouiller C (1950) Étude de la structure de l'os au microscope électronique. Arch Sci 3:175–180

Saleuddin ASM (1971) Fine structure of normal and regenerated shell of *Helix*. Can J Zool 49:37–41

Schroeder L, Frank RM (1985) High-resolution transmission electron microscopy of adult human peritubular dentine. Cell Tissue Res 242:449–451

Schüler D, Baeuerlein E (1998) Dynamics of iron uptake and Fe_3O_4 biomineralization during aerobic and microaerobic growth of *Magnetospirillum gryphiswaldense*. J Bacteriol 180:159–162

Schwarz W, Pahlke G (1953) Elektronenmikroskopische Untersuchungen an der Interzellularsubstanz des menschlichen Knochengewebes. Z Zellforsch 38:475–487

Scott BL, Pease DC (1956) Electron microscopy of the epiphyseal apparatus. Anat Rec 126:465–495

Selvig KA, Halse A (1972) Mineral content and crystal size in mature rat incisor enamel. A correlated electron microprobe and electron microscope study. J Ultrastruct Res 40:527–531

Sillence DO, Senn AS, Danks DM (1979) Genetic heterogeneity in osteogenesis imperfecta. J Med Genet 16:101–116

Simmelink JW, Nygaard VK, Scott DB (1974) Theory for the sequence of human and rat enamel dissolution by acid and by EDTA: a correlated scanning and transmission electron microscope study. Arch Oral Biol 19:183–197

Siperko LM, Landis WJ (2001) Aspects of mineral structure in normally calcifying avian tendon. J Struct Biol 135:313–320

Sodek KL, Tupy JH, Sodek J, Grynpas MD (2000) Relationships between bone protein and mineral in developing porcine long bone and calvaria. Bone 26:189–198

Stuhler R (1938) Uber den Feinbau des Knochens. Eine Röntgen Feinstruktur Untersuchung. Fortschr Geb Röntgenstr 57:231–234

Su X, Sun K, Cui FZ, Landis WJ (2003) Organization of apatite crystals in human woven bone. Bone 32:150–162

Takuma S (1960) Electron microscopy of the developing cartilaginous epiphysis. Arch Oral Biol 2:111–119

Tong W, Glimcher MJ, Katz JL, Kuhn L, Eppell SJ (2003) Size and shape of mineralites in young bovine bone measured by atomic force microscopy. Calcif Tissue Int 72:592–598

Towe KM, Cifelli R (1967) Wall ultrastructure in the calcareous foraminifera: crystallographic aspects and a model for calcification. J Paleontol 41:742–762

Travis DF (1968a) Comparative ultrastructure and organization of inorganic crystals and organic matrices of mineralized tissues. Biology of the mouth. American Association for the Advancement of Sciences, Washington, pp 237–297

Travis DF (1968b) The structure and organization of, and the relationship between, the inorganic crystals and the organic matrix of the prismatic region of *Mytilus edulis*. J Ultrastruct Res 23:183–215

Travis DF (1970) The comparative ultrastructure and organization of five calcified tissues. In: Schraer H (ed) Biological calcification: cellular and molecular aspects. Appleton-Century-Crofts, New York, pp 203–311

Travis DF, Glimcher MJ (1964) The structure and organization of, and the relationship between the organic matrix and the inorganic crystals of embryonic bovine enamel. J Cell Biol 23:447–497

Travis DF, Gonsalves M (1969) Comparative ultrastructure and organization of the prismatic region of two bivalves and its possible relation to the chemical mechanism of boring. Am Zool 9:635–661

Vetter U, Eanes ED, Kopp JB, Termine JD, Gehron Robey P (1991) Changes in apatite crystal size in bones of patients with osteogenesis imperfecta. Calcif Tissue Int 49:248–250

Warshawsky H (1989) Organization of crystals in enamel. Anat Rec 224:242–262

Warshawsky H, Nanci A (1982) Stereo electron microscopy of enamel crystallites. J Dent Res 61:1504–1514

Watabe N (1963) Decalcification of thin sections for electron microscope studies of crystal-matrix relationships in mollusc shells. J Cell Biol 18:701–703

Weiner S, Price PA (1986) Disaggregation of bone into crystals. Calcif Tissue Int 39:365–375

Weiner S, Traub W (1989) Crystal size and organization in bone. Connect Tissue Res 21:259–265

Weiss MP, Voegel JC, Frank RM (1981) Enamel crystallite growth: width and thickness study related to the possible presence of octocalcium phosphate during amelogenesis. J Ultrastruct Res 76:286–292

Wong V, Saleuddin ASM (1972) Fine structure of normal and regenerated shell of *Helisoma duryi duryi*. Can J Zool 50:1563–1568

Zanini F, Lausi A, Savoia A (1999) The beamlines of ELETTRA and their application to structural biology. Genetica 106:171–180

Ziv V, Weiner S (1994) Bone crystal sizes: a comparison of transmission electron microscopic and X-ray diffraction line width broadening techniques. Connect Tissue Res 30:165–175

Zylberberg L, Nicolas G (1982) Ultrastructure of scales in a teleost (*Carassius auratus* L.) after use of rapid freez-fixation and freeze-substitution. Cell Tissue Res 223:349–367

Zylberberg L, Géraudie J, Meunier F, Sire J-Y (1992) Biomineralization in the integumental skeleton of the living lower vertebrates. In: Hall BK (ed) Bone, volume 4: Bone metabolism and mineralization. CRC Press, Boca Raton, pp 171–224

7 Calcifying Matrices: Bone and Tendons

7.1 Introduction

Studies on the inorganic substance involved with biological calcifications have often treated it as if it were a self-sufficient, independent component of calcified tissues. It is, however, well known by now that these tissues comprise cells and intercellular organic matrix, and that these three components – cells, organic matrix and inorganic substance – show a high degree of interdependence and interrelation, as is confirmed by the frequent finding that abnormal cellular activity may generate an abnormal organic matrix, which, in its turn, may give rise to abnormal calcification. The active role of cells is documented by the knowledge not only that they are responsible for the synthesis of the organic matrix that is destined to calcify, but also that they produce and secrete organic molecules which promote, or, in some cases, interfere with, the calcification process, as is made clear by the observation that alkaline phosphatase – a cellular product – is invariably found wherever calcification occurs (Schajowicz and Cabrini 1954).

These are basic considerations, but this book is mainly dedicated to an analysis of the early stages of the calcification process, so the fine details of cellular behaviour will not be discussed and the origin, structure, function and fate of cells will only be referred to where this will deepen an understanding of the calcification mechanism. As to the morphology, physiology and pathology of calcified tissue cells, readers can consult the many reviews and book chapters available in the literature (Marks and Popoff 1988; Bonucci and Motta 1990; Doty and Schofield 1990; Raisz and Rodan 1990; Poole 1991; Benson and Wilt 1992; Engfeldt and Reinholt 1992; Teti 1993; Deutsch et al. 1995; Marks 1997; Marotti 2000; Nanci 2003).

On this basis, this and the following chapters will mainly be devoted to a detailed consideration of the characteristics and properties of calcifying matrices. If it is true that almost every organic matrix may be, or can be induced to become, a site of calcification, and if studies are extended from vertebrates to invertebrates and unicellular organisms, the numbers of matrices that calcify appear to be extremely high. However, in normal conditions, relatively few matrices can be considered to be representative of the whole range: the matrix of bone, dentin, enamel, cementum, growth cartilage and tendons in vertebrates (Schiffmann et al. 1970); the matrix of shells, scales, tests, spines, spicules and other skeletal or skeletal-like structures in lower organisms and invertebrates (Weiner and Hood 1975; Weiner and Traub 1980; Wheeler et al. 1987; Mann 1988); and endo- and pery-cytoplasmic structures in unicellular organisms (Boyan et al. 1992b). It must be borne in mind, however, that some tissues can calcify as an effect of aging, as may occur in the hyaline cartilage of human ribs (Dearden et al. 1974) and trachea (Bonucci et al. 1974), or in the tendons of turkey legs (Johnson 1960; Likins et al. 1960; Landis 1986; Bigi et al. 1988), and that many other tissues may calcify as a result of diseases (Daculsi et al. 1992). Examples of this possibility are the calcifications brought about by arteriosclerosis in the intima and media of the aorta and other arteries (Bostrom et al. 1995; Campbell and Campbell 2000; Tintut and Demer 2001); those secondary to hypercalcemic/hyperphosphatemic states which may involve various soft tissues, such as the artery wall, myocardium, kidney and lung (Hass et al. 1958; Parfitt 1969; Kuzela et al. 1977); those due to arthritis that occur in the articular cartilage (Ali 1983; Aigner and McKenna 2002); and those that affect several different soft tissues in calciphylaxis and calcergy (Selye 1962; Gabbiani and Tuchweber 1970).

These observations raise a number of questions, most of them still unanswered. The main ones are: what makes such a range of sharply different matrices all susceptible to physiological or pathological calcification?; what is it that triggers off the deposition of inorganic substance in them?; what components, if any, do they have in common?; are there specific structures and/or constituents which induce, regulate and/or inhibit the process?; is there a basic mechanism that is shared by all calcifying matrices? These questions remain unanswered at the time of writing, but a growing array of results converge in pointing to a direct role in calcification for one or more components of the organic matrices that calcify. On this basis, an analysis of their structure and composition becomes mandatory.

The primary focus of attention in this chapter is the organic matrix of bone, which is the main tissue in the vertebrate skeleton and one of the tissues that is most often studied by those whose aim is to explain the mechanism(s) operative in the calcification process. The fact that the organic matrix of tendons is also discussed in this chapter is motivated by the similarities between the two types of tissue and by the fact that tendon calcification has very often been taken as a model for bone calcification.

7.2
The Organic Matrix of Bone: Collagen

The bone matrix consists mainly of a fibrous component – collagen – plus small amounts of non-fibrous, non-collagenous constituents; serum proteins are present in variable amounts. Collagens are structural proteins on which the construction, configuration, integrity and resistance to stress of a large number of living organisms broadly depend. Collagens are quite numerous (van der

Rest 1991; Eyre 2002; Veis 2003); they include fibrillar collagens (types I, II, III, V and XI), reticular collagens (types IV, VIII, X), fibril surface-associated collagens (types IX, XII, XIV), periodic beaded filamentous collagen (type VI), and transmembrane collagenous proteins (types XIII and XVII).

Collagen is the most plentiful protein in the bone matrix. According to Eastoe and Eastoe (1954), the air-dried compact bone of the ox femur diaphysis contains 69.66% by weight of inorganic matter, 18.64% of collagen, 0.24% of mucopolysaccharide-protein complex, 1.2% of resistant protein material, and 8.18% of water. McLean and Urist (1968) estimated that about 90% of the organic matrix of compact bone is collagen, with the remaining 10% made up of non-collagenous proteins. The preponderance of collagen fibrils in bone matrix, as well as their close relationships with the inorganic substance, prompted the conclusion that they play a primary role in bone calcification (reviewed by Glimcher and Krane 1968; Glimcher 1976, 1990, 1992; Veis and Sabsay 1987; Höhling et al. 1990; Bonucci 1992; Veis 2003).

The bone matrix mainly consists of type I collagen (for other types of collagen which are only present in minor amounts in bone, see Sect. 9.2). It is assumed that bone type I collagen does not differ from type I collagen found in non-calcifying mesenchymal tissues, as suggested by the similarity in amino acid composition (Eastoe and Eastoe 1954). Doubts about this equivalence have arisen from to the fact that it is not known which changes, if any, may occur in collagen fibrils before, during or after their calcification. In addition, the close relationship, and the possible biochemical link, between the organic matrix and the inorganic substance is able to affect the former in various ways and may interfere with the biochemical, biophysical and, especially, morphological studies dedicated to it. These, in fact, often involve decalcification to unmask collagen fibrils and other organic matrix components, a procedure that, as already discussed (see Sect. 3.4.4), may alter their structure and organization to varying degrees. For this reason, most of what is known about the structural characteristics of bone collagen is based on studies carried out on the fibrils of uncalcified tendons and other soft tissues (the dermis, for instance) rather than on bone fibrils themselves.

The view that collagen fibrils in bone and those in mesenchymal soft tissues are equivalent receives objective support from the observation that the amino acid composition of the former does not differ greatly from that of the latter (Eastoe and Eastoe 1954; Schiffmann et al. 1970), that mineralized and non-mineralized portions of turkey tendons yield the same amino acids (Likins et al. 1960) and similar X-ray meridional patterns (White et al. 1977), and that the collagen fibrils of still uncalcified osteoid tissue show no specific structural differences with respect to collagen fibrils in skin. This does not, of course, imply that intermolecular aggregation and cross-links in bone collagen are identical with those in type I collagen in other tissues. Although the former is structurally similar to soft tissue collagen, it undergoes different post-translational modifications (Knott and Bailey 1998) and this explains its relative insolubility in organic acid or neutral salt solutions – a feature that differentiates calcified tissue collagen from collagen in uncalcified tissues (Glimcher and Katz 1965). For these reasons, Glimcher and Krane (1968) concluded that "bone collagen, in addition to being impregnated with inorganic crystals in the total biological context, is different from other collagens".

In both bone and non-calcifying mesenchymal tissues, type I collagen consists of fibrils about 78 nm in diameter which are characterized by periodic banding (Fig. 7.1). This is easily recognizable under the transmission electron microscope after either positive or negative staining (Olsen 1963; Bairati et al. 1969; Chapman and Hulmes 1984) and includes two 'bands' – one dense band about 0.4 D in length and another, less dense band about 0.6 D in length. These periodic bands depend on the mutual arrangement of the collagen molecules which make up the fibrils, and on their amino-acid sequences (von der Mark

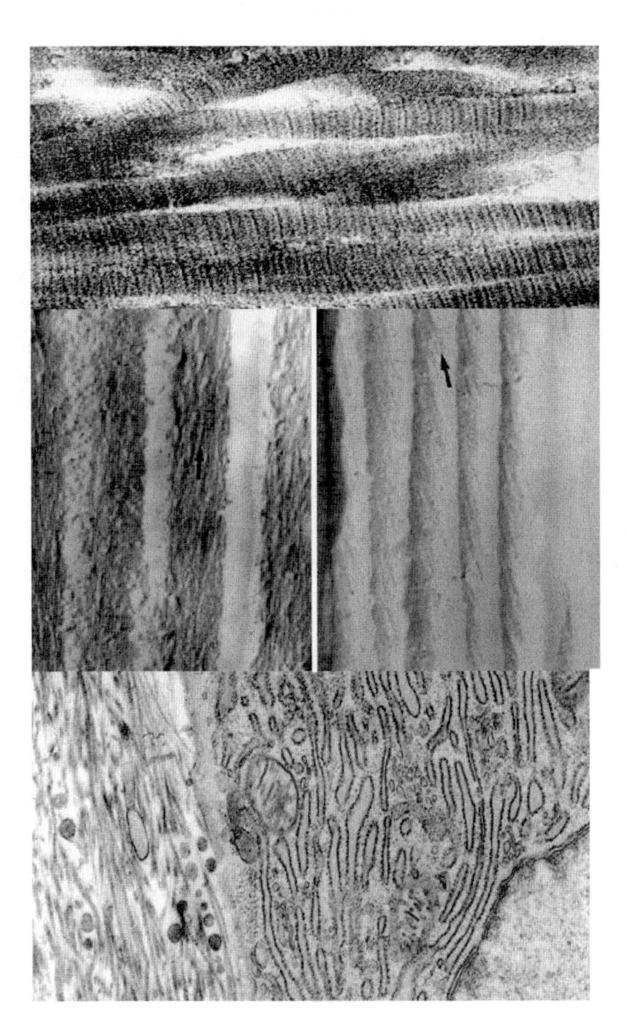

Fig. 7.1. *Above*: type I collagen fibrils; uranyl acetate and lead citrate, ×90,000. *Middle left*: the microfibrils of the tendon collagen fibrils have a straight arrangement, whereas (*right*) those of the derma have a helicoidal arrangement (courtesy of Alessandro Ruggeri, University of Bologna). *Below*: the osteoid tissue of compact bone consists of loosely and irregularly arranged collagen fibrils; matrix vesicles are detectable between them; part of an osteoblast on right, part of its nucleus in the *lower right corner*; uranyl acetate and lead citrate ×24.000

et al. 1970). It should, however, be recalled that the ultrastructural appearance of collagen fibrils partly depends on the treatments to which they have been submitted and on the approach used to allow their visualization (Raspanti et al. 1996).

Collagen molecules are 280–300 nm long, i. e., about 4.48 D, while D ($\sim$ 68–70 nm) is the main low-angle meridional X-ray Bragg spacing in the hydrated state, and 1.4 nm thick. They are formed by three polypeptide chains which aggregate in a left-handed helical configuration (a three-stranded coil coiled structure; Ramachandran and Kartha 1954) and are twisted around a common axis to form a supercoil (for the assembly of procollagen molecules from their three constituent polypeptide chains see Hulmes 2002). About 33% of the amino acid sequence of the polypeptide chains is glycine, which is found at every third position in the triplet gly-x-y, and about 22% consists of prolyne and hydroxyproline (Martin et al. 1963). The amino acid sequence is not the same in all chains, so these can assemble as molecules with different structures. Molecules of type I collagen consist of two equal polypeptide chains, called $\alpha 1(\mathrm{I})$, and of a third, different chain, called $\alpha 2(\mathrm{I})$ (reviewed by van der Rest 1991). Extraction with progressive concentrations of acid yields dimers (β-components) and trimers (γ-components) of α-chains joined by covalent links.

The assemblage of native collagen fibrils is still uncertain (for details see reviews by Cooper and Russell 1969; Steven 1972; Katz and Li 1973a; Piez and Miller 1974; Veis 1975; Miller 1976; Chapman 1984; Chapman and Hulmes 1984; Veis and Sabsay 1987; van der Rest 1991; Wess et al. 1998; Hulmes 2002; Veis 2003). According to the classic model proposed by Hodge and Petruska (1963) and Hodge et al. (1965), the collagen fibril assemblage depends on the spontaneous alignment of collagen molecules according to a quarter-staggered end-to-end overlap array. The molecules are arranged in parallel array but are mutually displaced along their axial plane by integral multiples of the distance D. This arrangement generates regions where the molecules alternately overlap and are separated by gaps which constitute the interval between the 'tail' and 'head' of successive molecules in the same plane. There is a good fit between this model, the ultrastructure of the negatively stained collagen fibrils, and the arrangement of the granular substance in bands (Fig. 7.2).

The dense, or 'overlap', zone is 25–30 nm long, and the low density, or 'hole' zone, is 40–45 nm long; each hole has a diameter equal to that of the collagen molecule (Schiffmann et al. 1970), i. e., about 1.5 nm (Glimcher and Krane 1968). Adjacent molecules are linked by intra- and intermolecular cross-links (Cooper and Russell 1969; Tanzer 1973; Knott and Bailey 1998) which, besides stabilizing the fibril structure, may be responsible for the low degree of solubility of bone collagen (Glimcher and Katz 1965) and may have a role in fibril calcification (Wassen et al. 2000), as further discussed below. The distance between collagen molecules varies to some extent (for a discussion on this topic, see Höhling et al. 1990). According to measurements made by

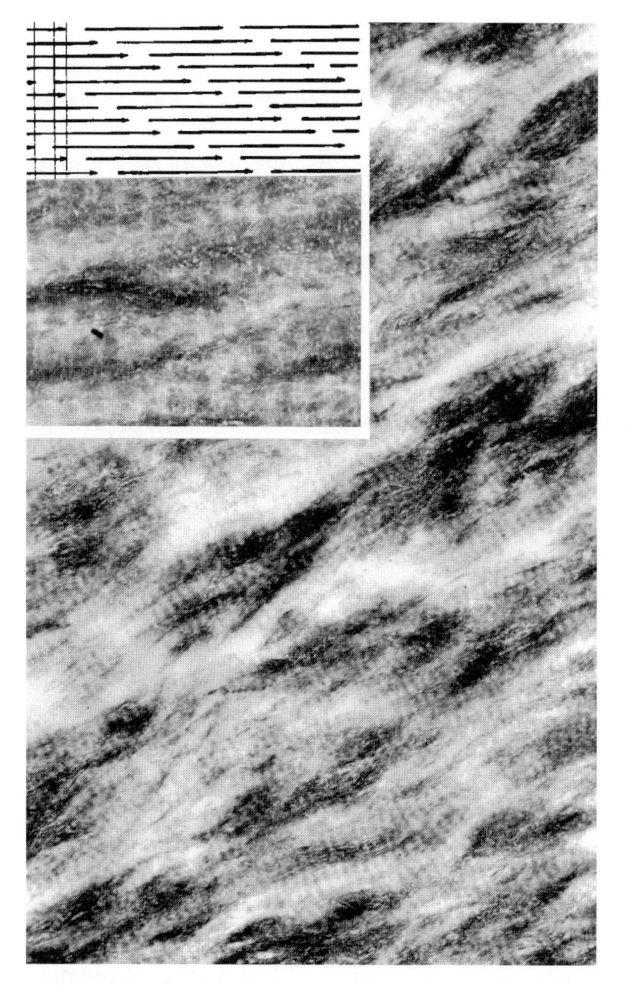

Fig. 7.2. Incompletely calcified osteon: the granular inorganic substance in bands is clearly visible; unstained, ×60,000. *Inset*: according to the model of Hodge-Petruska, the collagen molecules are shifted in such a way as to give rise to holes and overlapping zones; the former contain the granular inorganic substance which appears as periodic bands under the electron microscope. Unstained, ×150,000

Bonar et al. (1985) using X-ray and neutron diffraction analysis of mineralized and demineralized compact bone of the bovine tibia, the collagen equatorial reflections are 1.24 nm and 1.53 nm, respectively, in the wet state, and 1.16 nm and 1.12 nm in the dry state. Using atomic force microscopy and Fourier analysis, the following values have been found: molecular diameter 1.43 nm; intermolecular distance 2.21 nm; periodicity of the threefold screw 1.15 nm; periodicity of a single chain 8.03 nm (Baranauskas et al. 1998).

Type I collagen fibrils are crystal-like structures whose molecules are arranged in a regular three-dimensional lattice (van der Rest 1991). The collagen microfibrils can, however, be organized into a straight or helicoidal texture (Fig. 7.1; Ruggeri et al. 1979; Raspanti et al. 1989), which probably identify different types of collagen fibrils depending on tissue organization and function (Lillie et al. 1977; Raspanti et al. 1989; Ottani et al. 2004). Most of helically arranged fibrils appear to be associated with elastic fibers in highly compliant tissues such as the aortic wall, skin and tendon sheaths, whereas straight fibrils seem to be peculiar to tissues bearing unidirectional forces, such as tendons and some ligaments (Raspanti et al. 1989).

The model suggested by Hodge and Petruska is compatible with a two-dimensional arrangement of molecules, but cannot be adapted easily to a three-dimensional molecular assemblage, because not all lateral contacts between molecules can be quarter-staggered (Smith 1965). As a result, several three-dimensional molecular arrangements have been suggested, such as five-stranded rope, tetrameric or four-stranded rope, hexagonal packing, quasi-hexagonal array, near-hexagonal lattice, two-stranded coiled-coils, five-stranded helical microfibrils, an orthorhombic arrangement of octafibrils, or a multi-helical structure (Miller and Wray 1971; Segrest and Cunningham 1973; Katz and Li 1973a; Hosemann et al. 1974; Piez and Miller 1974; Veis and Yuan 1975; Woodhead-Galloway et al. 1975; Miller 1976; Grynpas 1977; Hulmes and Miller 1979; Trus and Piez 1980; Miller and Tocchetti 1981; Piez 1982; Lees 1981, 1987; Chapman 1984; Raspanti et al. 1989; Katsura et al. 1991; Lee J et al. 1996). Because the equatorial diffraction pattern of type I collagen is diffuse, and for other reasons discussed by Hulmes (2002), a similarity with a two-dimensional fluid, or 'liquid crystal', has been suggested for tendon (Fratzl et al. 1993) and bone (Giraud-Guille 1994; Giraud-Guille et al. 2000) collagen fibrils – a model that, according to Lees (1998), ignores the organization imposed by intermolecular cross-links.

The problem of the assemblage of the fibrils in bone matrix is further complicated by their calcification, which entails that they must contain enough intrafibrillar space to leave room for the inorganic substance. As reported above, the post-translational modifications of bone collagen are different from those of soft tissue collagen (Knott and Bailey 1998), and this is reflected in the average intermolecular gaps which are larger (0.6 nm) in the former than in the latter (0.3 nm; Katz and Li 1972). Weiner and Traub (1986) have suggested that space may also be produced by the lateral, parallel alignment of the gaps which would, on this hypothesis, give rise to grooves and channels within the fibrils.

In reconstituted collagen fibrils, working on the assumption that their molecules are assembled in a quasi-hexagonal configuration (within which collagen molecules are randomly staggered with respect to each other by 0 to 4 D), Katz and Li (1973a, b) have calculated that intrafibrillar spaces have the following dimensions: 0.13 ml/g collagen are attributed to the helical groove of the molecules, 1.01 ml/g is interstitial, 0.66 ml/g is assigned to pores (hexagonally closed packed spaces), and 0.48 ml/g to holes (hexagonal volume defects). These authors conclude that 0.73 ml/g of the intermolecular space is associated with regions where holes are localized, and 0.41 ml/g to regions containing only pores.

If these observations are correct, bone collagen must have more space to accommodate the inorganic substance than has previously been supposed.

Electron microscope and neutral diffraction studies do, however, raise doubts about this assumption. Bonucci (1992) stressed that the study of collagen fibrils in bone implies their decalcification, a process that might profoundly change their molecular array. This has been clearly shown by Bonucci and Reurink (1978), who compared the ultrastructure of the collagen fibrils of bone which had been decalcified before and after embedding. In the first case, the fibrils which had been directly exposed to the decalcifying solution were dissociated, deeply stainable, and showed minute intrafibrillary clefts (Fig. 3.3), probably due to extraction of some organic components together with the inorganic one, and to the breaking of intrafibrillary cross-links; in the second case, where the fibrils were stabilized by the embedding resin during decalcification, they appeared compact, closely associated and lightly stainable, displaying no intrafibrillary clefts (Figs. 3.4, 7.3). These results converge with those of neutron diffraction studies which showed that decalcification had increased the side-to-side spacing of the collagen molecules (from 1.23 to 1.52 nm; Lees and Hukins 1992), and with those of electron microscope studies showing that collagen fibers, which in the osteoid area are loosely and randomly arranged (Fig. 7.1), tend to fuse side by side with increasing calcification, with fusion

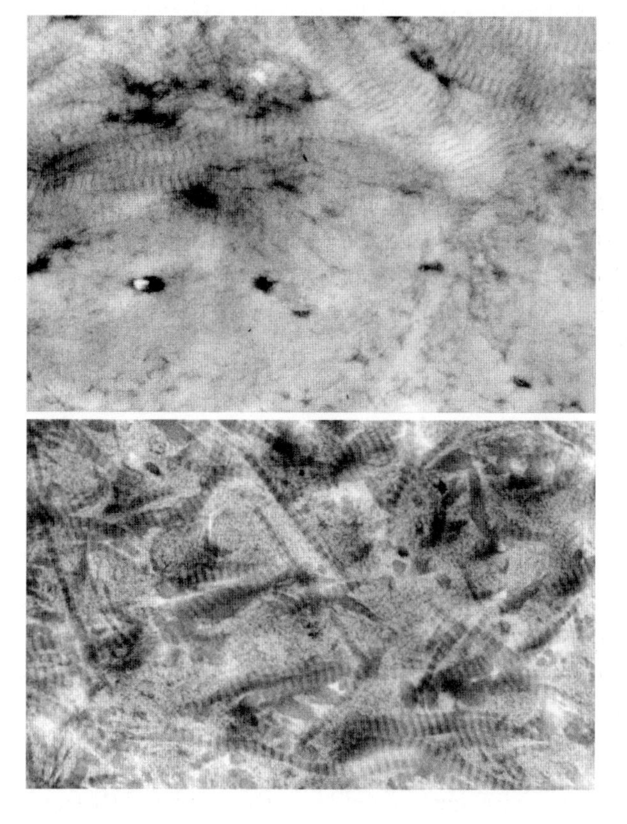

Fig. 7.3. *Above*: organic matrix of a fully calcified osteon decalcified with the PEDS method: note the compactness of the collagen fibrils which are cut longitudinally in the upper part of the figure, and transversally in the lower part. *Below*: organic matrix of the spongy bone of the chick embryo processed as above: plenty of non-collagenous material is located between randomly arranged collagen fibrils. PEDS method, uranyl acetate and lead citrate. ×35,000 and ×30,000

and compactness being at a maximum where calcification is complete (Hoshi et al. 1999). The already reported results of Bonar et al. (1985) showed that collagen in fully calcified bovine bone is very closely packed, with a packing density comparable with that of relatively crystalline collagens such as rat wet tail tendon. These results lead to the conclusion that less space is available between and within collagen fibrils than has previously been assumed.

The density and arrangement of collagen fibrils in bone matrix are not the same in all types of bone, so that amounts of interfibrillary substance differ too (Fig. 7.3). The compact bone of the diaphyses is the tissue where collagen fibril density is at a maximum. Fibrils in the calcified matrix are closely packed in lateral register in this type of bone and their orderly arrangement gives rise to what is called parallel-fibered or lamellar bone. Conversely, fibrils in woven and embryonic bone are randomly oriented, interweave irregularly, and outline rather wide interfibrillary spaces (Bonucci and Silvestrini 1996). This means that woven bone contains greater amounts of non-collagenous material than does lamellar bone. These structural features are particularly conspicuous in the medullary bone of birds, whose matrix shows a unique ultrastructure, with low number of randomly arranged collagen fibers and a high proportion of non-collagenous material lying in wide interfibrillary spaces (Ascenzi et al. 1963; Bonucci and Gherardi 1975). The arrangement of collagen fibrils is important not only because of the different amounts of non-collagenous material that can be fitted into interfibrillary spaces – hence, differences in types and features of inorganic structures – but also because it is largely responsible for the mechanical properties of bone (Ascenzi 1988; Ascenzi et al. 2000).

It is striking that the differences in collagen fibril arrangement that distinguish various types of bone are hard to detect in the osteoid tissue (Fig. 7.1). In the osteoid, collagen fibrils have a low density and are loosely arranged, which makes them resemble the matrix of woven bone, although some degree of packing is detectable in the osteoid of compact bone (Bianco 1992).

7.3
The Organic Matrix of Bone: Non-collagenous Components

Even if they constitute less than 10% of the whole organic matrix, non-collagenous components have an important role within bone (Nusgens et al. 1972; Leaver et al. 1975; Termine et al. 1981a; Butler 1984, 2000; Boskey 1989, 1998; Bianco 1990; Young et al. 1992; Ingram et al. 1993; Robey 1996). Their amounts vary with the type of bone (de Ricqlès et al. 1991; Bianco 1992), generally correlating with the speed of its formation and with the packing density of its collagen fibrils (Nanci 1999). Non-collagenous components are more plentiful in bone tissues with loose collagen fibrils (Fig. 7.3), i. e., with wider interfibrillary spaces, such as the woven bone of the chick embryo (Bonucci and Silvestrini 1996) and the medullary bone of birds (Bonucci and Gherardi

1975). Moreover, their distribution and composition depend on the type of bone, revealing clear differences between woven and lamellar bone (Gorski 1998).

The major non-collagenous components of bone matrix comprise proteoglycans, so-called Gla-proteins, glycoproteins (phosphoproteins) and phospholipids. Moreover, serum-derived proteins, such as albumin and the α_2HS-glycoprotein (Ashton et al. 1976; Triffitt et al. 1978), have been found in the calcified bone matrix, where they may interact with the inorganic substance (Triffitt and Owen 1977) and with other non-collagenous proteins such as osteocalcin and osteopontin (McKee et al. 1993). Alkaline phosphatase (Bonucci et al. 1992) and other enzymes such as matrix metalloproteinases (Breckon et al. 1995) can be found at sites of osteogenesis, where they are often associated with matrix vesicles (Matsuzawa and Anderson 1971; Dean et al. 1992; D'Angelo et al. 2001). Several osteogenic growth factors occur in the calcified matrix (Canalis et al. 1988); their important physiological role in the regulation of bone cell activity falls outside the scope of this book.

7.3.1 Proteoglycans

Proteoglycans have been extracted from various types of bone in vivo and in vitro (Kobayashi 1971; Fisher 1985; Beresford et al. 1987), but they occur in calcified osseous matrices at fairly low concentrations. For this reason, and because, by contrast, they are plentiful in cartilage matrix, their structure and composition are considered in the chapter dedicated to that matrix (see Sect. 9.3.1).

The non-collagenous components of bone include acidic proteoglycans, most of which are entrapped in the calcified matrix, so that their extraction requires decalcification. The extraction of EDTA-decalcified rabbit cortical bone powder yielded two classes of proteoglycans, the first containing chondroitin sulfate, and the second, material resembling keratan sulfate mixed with a smaller amount of chondroitin sulfate (Diamond et al. 1982). The same procedure carried out on the mineralized matrix of ground bovine bone yielded three fractions (Franzén and Heinegård 1984a) of small proteoglycans containing one or probably two chondroitin sulfate chains (Franzén and Heinegård 1984b). At least two classes of chondroitin sulfate-containing proteoglycans of low molecular weight have been extracted from EDTA-decalcified fetal porcine calvaria (Goldberg et al. 1988). Two small proteoglycans have been isolated from the calcified matrix of the developing bone from fetal calves, growing rats and human fetuses (Fisher et al. 1983a). The first of these proteoglycans consisted of one chondroitin sulfate chain attached to a glu/gln-rich core protein; the second, of two chondroitin sulfate chains attached to a leu-rich core protein (Fisher 1985). The first was called decorin (DCN). It is associated with type I collagen in all connective tissues and its core protein can be detected immunohistochemically near the *d* and *e* bands of the collagen fibrils (Pringle and Dodd 1990). It has been found near collagen fibrils in osteoid areas of the embryonic rat calvariae, but its amounts fall in calcifying areas where collagen fibrils fuse after lying side by side (Hoshi et al. 1999).

The role of DCN in regulating collagen fibrillogenesis is shown by the abnormal ultrastructural morphology of collagen fibrils, and by the consequent skin fragility, found in mice that undergo a targeted disruption of the decorin gene (Corsi et al. 2002). The results of in vitro studies on osteoblastic cell clones expressing higher (sense-DCN, S-DCN) and lower (antisense-DCN, As-DCN) levels of decorin testify that this molecule has an inhibitory effect on calcification, which was significantly delayed in S-DCN clones, whereas it was markedly accelerated, with a significant higher number of calcification nodules, in As-DCN clones (Mochida et al. 2003). The removal of DCN, interpreted as an inhibitory substance, and the consequent fusion of collagen fibrils, have been considered a premise to bone matrix calcification (Hoshi et al. 1999). Sugars et al. (2003) have shown that DCN promotes collagen fibrillogenesis and that it can be closely associated with hydroxyapatite, so inhibiting crystal growth. Fischer et al. (2004), on the other hand, have taken the view that DCN actually promotes calcification, because its expression is strongly upregulated in cultures of bovine aortic smooth muscle cells within which calcification had been triggered by the addition of β-glycerophosphate or inorganic phosphate, and because it is associated with calcium deposits in human coronary atherosclerotic lesions.

The second small-sized proteoglycan was called biglycan; it has been found in a range of specialized cell types (Bianco et al. 1990). Biglycan deficiencies lead to structural abnormalities in collagen fibrils in bone, dermis, and tendon, and to a 'subclinical' cutaneous phenotype, with thinning of the dermis but without any overt skin fragility, resembling the rare progeroid variant of the human Ehlers-Danlos syndrome (Corsi et al. 2002).

Not only the localization and features of small proteoglycans, but also those of large hyaluronate-binding proteoglycans have been studied in bone matrix by biochemical and immunohistochemical methods. In the developing mandible of fetal rats, the antibody (Mab)5D5, which specifically recognizes the core proteins of large proteoglycans such as versican and brevican but not aggrecan, has been used with this aim (Lee I et al. 1998); it gave a positive reaction in decalcified calcification nodules, which appear as structures with an electron-dense periphery surrounding fine filamentous and granular material (as this topic is related to those involving crystal ghosts and the calcification mechanism, it will be reconsidered in Sect. 16.2.1). The PCR and immunohistochemical studies by Raouf et al. (2002), carried out on the bone formed in vitro by MC3T3-E1 mouse calvaria osteoprogenitor cells and on mouse embryos, have shown that lumican, a keratan-sulfate-, leucine-rich proteoglycan, is a major bone matrix component secreted by differentiating and mature osteoblasts. Other proteoglycans, such as aggrecan, versican, perlecan and fibromodulin, are poorly represented in bone matrix and will be discussed in the chapter devoted to cartilage proteoglycans (Sect. 9.3.1).

Proteoglycans are not only poorly represented in bone, but their amounts change with the type of bone. They are found at relatively high concentrations in the medullary bone of birds (Stagni et al. 1980), where they are about two to three times more plentiful than in cortical bone (Candlish and Holt 1971); this contains less of them than do embryonic and woven bone. This is substantiated by the observation that the stainability of medullary bone matrix with alcian blue or colloidal iron methods for acid proteoglycans is much higher than that of the surrounding cortical bone (Bonucci and Gherardi 1975). Interestingly, the degree of calcification of medullary bone is, again, higher than that of cortical bone (François 1960; Ascenzi et al. 1963). It must be added that proteoglycans of medullary bone are mainly distinguished by the presence of keratan sulfate, whereas those of cortical bone contain chondroitin sulfate (Candlish and Holt 1971; Fisher and Schraer 1982).

Amounts of proteoglycans not only change with bone type, but may vary in the same kind of bone according to the degree of matrix calcification. Biochemical studies carried out on isolated osteons at different degrees of calcification showed a fall in proteoglycans (evaluated as exosamines) with advancing calcification: the osteoid tissue contained 0.61% of dry weight (8.7 μg/μl) of exosamines vs values of 0.31 (6.2 μg/μl) in osteons at the lowest degree of calcification and 0.28 (6.2 μg/μl) in osteons at the highest degree of calcification (Pugliarello et al. 1970). Electron microprobe analysis also showed that S was high in osteoid, but dropped as calcium concentration increased at the calcification front (Baylink et al. 1972). Biochemical analysis on fractionated bone particles showed that the total amount and molecular size of glycosaminoglycans fell with an increasing degree of mineralization (Engfeldt and Hjerpe 1976), in agreement with ^{35}S studies showing that about 45% of the newly synthesized proteoglycans are removed during calcification (Prince et al. 1983, 1984).

The biochemical findings reported above agree with those obtained by histochemistry (reviewed by Kobayashi 1971). Falling amounts of sulfated glycoconjugates have been found in going from the osteoid border to the transitional zone and the fully calcified matrix of the rat tibia, using the high-iron diamine-thiocarbohydrazide-silver proteinate (HID-TCH-SP) method (Takagi et al. 1983). Osteoid but not mature calcified bone matrix was stained by applying either the alcian blue or the toluidine blue method for acid proteoglycans; the loss of these molecules (mainly chondroitin sulfate, according to the results of the critical electrolyte concentration method) occurred precisely at the calcification front (identified by in vitro lead or procion markers; Baylink et al. 1972).

From personal experience, the degree of proteoglycan staining with cationic dyes is, roughly, inversely correlated with the degree of matrix calcification, so that, in developing bone, the calcified matrix is very lightly stained or not stained at all, the uncalcified osteoid matrix shows a widespread, moderate degree of staining, and the calcification nodules are stained along their periphery but are almost unstained in their central zone. These staining properties are in agreement with the findings reported in the literature.

As regards the fully calcified bone matrix, the immunoreaction for chondroitin sulfate and dermatan sulfate has been found only in the wall of the osteocytic lacunae and canaliculi (Takagi et al. 1991). This has been confirmed by Bonucci and Silvestrini (1992), who also showed chondroitin-4-sulfate to be present at the periphery of calcification nodules (Fig. 7.4). Chondroitin-4-sulfate, dermatan sulfate, and chondroitin 6-sulfate have been reported to be located in the lacuno-canalicular system, and collagen-associated dermatan sulfate proteoglycans have been found to be spread through the calcified matrix (Takagi et al. 1996). Subperiosteal compact bone in rats contained a keratan sulfate-containing glycoconjugate preferentially located in the wall of the lacuno-canalicular system, whereas rabbit bone contained at least two, possibly three types of keratan sulfate-containing glycoconjugates distributed throughout the calcified matrix (Maeno et al. 1992). In the alveolar bone of rats, pigs and men, keratan sulfate could be immunohistochemically and biochemically detected only in rabbit bone, whereas chondroitin sulfate was the predominant glycosaminoglycan in pigs and men (Bartold 1990). Cuprolinic blue-positive proteoglycans, containing chondroitin and/or dermatan sulfate, have been found in the calcified matrix of rat and human lamellar bone (Sauren

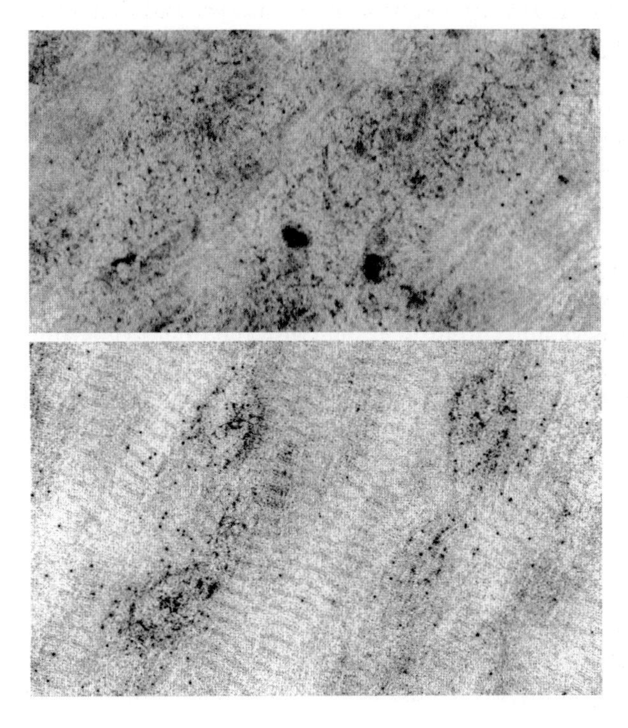

Fig. 7.4. Immunohistochemical detection of chondroitinsulfate in compact bone: at the calcification front (*above*), immunogold particles are located on, and at the periphery of, amorphous, interfibrillary areas corresponding to the matrix of calcification nodules. In fully calcified matrix (*below*), immunogold particles are almost exclusively located in and around osteocyte processes. CS-56 monoclonal antibody, protein A-colloidal gold, ×70,000

et al. 1989, 1992). Dense filamentous reticular patches stained by cuprolinic blue have been reported between collagen fibrils in the calcified matrix of in vivo and in vitro formed bone (Nefussi et al. 1989).

With regard to the osteoid tissue, Baylink et al. (1972) showed that it stained with alcian blue and was metachromatic with toluidine blue and, by means of the critical electrolyte concentration, demonstrated that the major polysaccharide component was chondroitin sulfate. The osteoid tissue of embryonic rat calvaria was shown to be marked by an abundance of cuprolinic blue-stainable proteoglycan granules, whose size fell as calcification proceeded (Hoshi et al. 2001). In uncalcified matrix, cuprolinic blue-positive structures appeared either to be free, with a granular or rod-like appearance, or to be tightly connected to the periphery of the collagen fibers (Nefussi et al. 1989). An inverse correlation has been demonstrated between the histochemical reactions for glycoproteins and those for proteoglycans (Van Den Hooff et al. 1966).

As to calcification nodules, the presence in their matrix of acidic groups, corresponding to glycosaminoglycan sulfate groups, was shown under the electron microscope with colloidal thorium dioxide or ruthenium red (Scherft and Groot 1981; Groot 1982), Thorotrast (Scherft 1968), or colloidal iron (Bonucci 2002). Cuprolinic blue-positive filaments and rods have been found in the calcification nodules of the matrix of the mouse foetus (Sauren et al. 1989). Keratan sulfate proteoglycans have been immuno-localized in the calcification nodules of membranous ossification in the rat calvarium; strikingly, immunoreactivity was strong at the periphery of the nodules but fell in central, fully calcified areas (Nakamura et al. 2001), the same localization for chondroitin sulfate that had previously been found in cartilage calcification nodules (Bonucci and Silvestrini 1992).

An apparently bone-specific proteoglycan which contains keratan sulfate and belongs to the family of leucine-rich repeat proteins of the extracellular matrix has been referred to as osteoadherin because it promotes integrin-mediated cell attachment (Sommarin et al. 1998; Wendel et al. 1998). A recent immunohistochemical study carried out on rat tibial metaphyses and diaphyses and on calvaria has shown that osteoadherin is located in the calcified bone matrix, with the highest label concentration at the border between bone and cartilage remnants – a localization that has been considered similar to that of bone sialoprotein (Ramstad et al. 2003). Studies by Shen et al. (1999) have shown that osteoadherin is primarily expressed by osteoblasts in the rat.

7.3.2 Gla-proteins

Bone matrix includes proteins which have been called Gla-proteins because their molecules are characterized by the presence of γ-carboxyglutamic acid (Gla). They include bone Gla-protein (BGP), otherwise known as osteocalcin (OC), and matrix Gla-protein (MGP).

7.3.2.1
Osteocalcin

Bone osteocalcin (OC, BGP; reviewed by Hauschka 1985; Cole and Hanley 1991; Gundberg 1998) was first described by Hauschka et al. (1975). It is the most plentiful non-collagenous protein restricted to bone matrix (Carlson et al. 1993). Its synthesis is a property of osteoblastic cells (Mark et al. 1987a; Sommer et al. 1996; Bellows et al. 1999) but also of odontoblasts (Bronckers et al. 1989), isolated osteocytes (Aarden et al. 1996), osteoblast-like cells from rat osteosarcoma (Nishimoto and Price 1980) grown in culture and, with some reservations, of hypertrophic chondrocytes (McKee et al. 1992).

Osteocalcin is distinguished by the presence in its small peptide molecule of three residues of vitamin K-dependent Ca^{2+}-binding amino acid γ-carboxyglutamic acid (Gla), which is why it has often been called bone Gla-protein (BGP). Vitamin K is needed for its synthesis, which is inhibited by warfarin. Calcium ions induce changes in osteocalcin molecules, from a random-coil to an α-helical conformation, and the three Gla residues that project from the same face of the helical turns (Dowd et al. 2003) make possible in vitro adsorption to hydroxyapatite (Gundberg 1998). This occurs because the α-helix periodicity is capable of forcing the Gla residues into register, so that they become spaces out at intervals of 0.54 nm, a figure very close to the 0.545 nm Ca-Ca interatomic lattice spacing of the hydroxyapatite (Hauschka 1985). In addition, OC has calcium-binding properties and can interact with hydroxyapatite (Hauschka and Wians 1989), due to the spatial protrusion from its negatively charged surface of five calcium ions which are oriented in a way complementary to that of calcium ions in the crystal lattice of hydroxyapatite (Hoang et al. 2003). Because of this precise interaction with hydroxyapatite, it might inhibit the growth of hydroxyapatite crystals (Menanteau et al. 1982), so controlling their shape and size.

Osteocalcin is bound in vivo to the inorganic phase of bone (Termine 1985; Groot et al. 1986; Hauschka and Wians 1989); this is probably the reason why it persists and can be found in fossil bone (Ulrich et al. 1987). It is localized in bone cells and calcified bone matrix (Bonucci et al. 1986; Boivin et al. 1990; Carlson et al. 1993; Nefussi et al. 1997), through which it is widely distributed (Bianco et al. 1985; Camarda et al. 1987; McKee et al. 1992, 1993; Carlson et al. 1993).Its levels in cortical bone, where it can be detected along the lamellar matrix in fine granular deposits (Vermeulen et al. 1989), appear to be higher than in trabecular bone (Ninomiya et al. 1990; Gorski 1998). An immunohistochemical study on the undecalcified sections of tooth germs and bone in the rat showed osteocalcin immunostaining of young osteoblasts and osteoid (Bronckers et al. 1987); these findings were later confirmed in the bone of developing rat mandible (Ishigaki et al. 2002). Ultrastructural immunohistochemistry has made possible a quite precise localization of osteocalcin during the early stages of calcification: the uncalcified osteoid tissue was not labeled, or was only weakly labeled (Bianco et al. 1985; Camarda et al. 1987); labeling was found in the matrix of the early foci of calcification (Groot et al. 1986), which appeared as 'grey' patches in osteoid tissue (McKee et al. 1990, 1992).

Osteocalcin has long been considered a strong candidate as local regulator of calcification, chiefly because it is synthesized by osteoblasts and odontoblasts during the earliest stage of calcification (but after the synthesis of alkaline phosphatase and other non-collagenous proteins (Bronckers et al. 1987; Mark et al. 1988), and because it is the only non-collagenous protein restricted to bone cells and calcified bone matrix (Carlson et al. 1993). In vivo studies on the effects of osteocalcin have, however, not provided elements capable of clarifying its function. Circulating OC appears to be a marker of osteoblast activity and, in general, its concentration in the serum rises whenever bone turnover rises too, as in primary and secondary hyperparathyroidism (Malluche et al. 1984; Delmas et al. 1986; Giannini et al. 2001), menopause (Garnero and Delmas 1996), and osteoporosis (Delmas et al. 1983), and it falls whenever bone turnover and osteoblast activity fall too, as happens after glucocorticoid treatment (Chavassieux et al. 1993; Delany et al. 1994). Brown et al. (1984), in a study on 35 post-menopausal osteoporotic women, found, however, that the concentration of Gla-protein in serum was normal in 26, rose in 4, and fell in 5. Aerssens et al. (1993) reported that serum OC increased in ovariectomized rats, in which cortical bone density fell significantly, whereas Fiore et al. (1987) found that both OC levels in serum and the bone formation rate had fallen in ovariectomized women.

In vitro studies by Doi et al. (1992) have shown that when OC, as well as ON and DPP, were added to calcium β-glycerophosphate solutions containing catalytic amounts of alkaline phosphatase, the formation of hydroxyapatite needed a higher concentration of inorganic phosphate to induce calcium phosphate precipitation, which was delayed about four times longer by DPP than by OC, which in its turn produced a delay about twice as long as ON; moreover, the proteins tested retarded crystal growth too. The same authors (Doi et al. 1993) showed, however, that the inhibitory effect disappeared when the proteins were immobilized on sepharose beads (for the effects of protein immobilization, see Sect. 8.5.1). Studies by Bronckers et al. (1998) have shown that when OC is added to, or withdrawn from, hamster tooth organ cultures, this does not influence dentinogenesis, and that dentin obtained from two-month-old OC null mutants looks structurally normal. According to Hunter et al. (1996), the effect of OC is essentially that of delaying apatite nucleation.

On the other hand, the inhibition of vitamin K by warfarin treatment induces a fall in bone OC content without causing significant bone changes (Price and Williamson 1981). In vitro studies on the OC effects on lipid-induced hydroxyapatite formation have shown that, by contrast with what happens with the γ carboxyglutamic acid-containing clotting proteins, which display a high affinity for acidic phospholipids, OC does not associate with any of the lipids tested (phosphatidyl serine, phosphatidyl inositol, and the corresponding Ca-acidic phospholipid-phosphate complexes) and has no impact on lipid-induced calcification (Boskey et al. 1985). Osteocalcin-deficient mice develop a phenotype characterized by higher bone mass and bones with improved functional qualities (Ducy et al. 1996), and, in accordance with these results, Fourier transform infrared microspectroscopic analyses of femora in four-week-, six-month-, and nine-month-old OC-knockout mice showed that bone amounts rose with respect to wild-type mice, crystal size and perfection fell, and there was some indication that OC may be needed to stimulate crystal maturation (Boskey et al. 1998). It has, however, been reported that in tooth germ samples of the Hyp mouse – a murine homologue of human X-linked hypophosphatemia, displaying the hypocalcification of bone and dentin – the expression of the OC gene is significantly higher than in wild-type mice, a result confirmed in cultured Hyp mice tooth germ samples (Onishi et al. 2005).

In 1991, Cole and Hanley published an exhaustive review on osteocalcin; after stressing its affinity for hydroxyapatite, and with reference to a paper by P.V. Hauschka et al., they reported that OC may have as many as four functions: inhibition of hydroxyapatite precipitation, mediation of $1{,}25(OH)_2D_3$ action, chemoattraction of monocytes and other cells, and inhibition of leukocyte elastase. In 1998, in a review on the biology, physiology and clinical chemistry of OC, Gundberg reiterated the importance of its chemical structure in the interaction with hydroxyapatite, and stressed its correlation with bone formation, degree of calcification and osteoblast maturation, but concluded that the function of this molecule had never been precisely defined, although it is highly likely it does play a role in bone turnover. The data so far available do not make it possible to determine osteocalcin's role in calcification. One can only conclude with Butler (2000) that OC may intervene as a negative regulator of bone formation, but its exact functional role in this respect has yet to be clarified.

7.3.2.2
Matrix Gla Protein

Matrix Gla protein (MGP; reviewed by Price 1989) is, like osteocalcin, a vitamin K-dependent, γ-carboxyglutamic acid-containing protein purified from urea extracts of decalcified bovine cortical bone (Price et al. 1985). It has enough sequence homology with osteocalcin to make it plausible that both proteins arose by gene duplication, followed by divergent evolution (Price and Williamson 1985); they must, however, be kept distinct, as confirmed by the observation that MGP does not cross-react with BGP (Price et al. 1983), and that they differ in their localizations: like osteocalcin, MGP is found in bone matrix and calcified cartilage matrix, but it also appears in the matrix of other connective tissues (Simes et al. 2003) and has been found in all tested soft tissues (Price 1989). Moreover, MGP shows the same distribution in the bone of newborn and adult rats, whereas BGP proved to have only about 5% of its adult value in newborn rats, rising to 90% of adult levels at 19 days of age (Otawara and Price 1986).

Matrix Gla protein has a negative regulatory role on calcification, as mainly shown by studies on vascular calcifications. Matrix Gla protein-deficient mice do, in fact, develop severe, widespread vascular calcification and die prematurely from aortic rupture (Luo et al. 1997; Boström 2000, 2001; El-Maadawy et al. 2003), a pathological phenotype that can be brought to normality by the restoration of MGP expression in arteries (Murshed et al. 2004). It is striking that the same pathological changes can be noted in osteoprotegerin-deficient (OPG−/−) mice (Bucay et al. 1998); OPG (a cytokin that regulates osteoclast differentiation and activation; Simonet et al. 1997; Lacey et al. 1998; Aubin and Bonnelye 2000; Hofbauer et al. 2000) itself inhibits artery calcification induced by warfarin and by vitamin D (Price et al. 2001). Dhore et al. (2001) have found that MGP occurs in the wall of the normal human aorta, that OPN, BMP-2 and -4, and ON are, conversely, absent, and that all these proteins are upregulated in atherosclerotic, calcified or ossified plaques. They also found that OPG and its ligand RANKL are located in normal aortas and early atherosclerotic lesions, whereas, in advanced calcified lesions, OPG is located in bone structures and RANKL can only be detected around calcium deposits.

The calcification of the vascular wall is mediated by phenotypic changes in smooth muscle cells (discussed in Sect. 14.3.5), but the effect of MGP on these changes is still unknown (discussed by Steitz et al. 2001). According to Boström (2000), its inhibitory activity on calcification might depend on the binding of calcium ions, which prevents calcium phosphate deposition in extracellular fluids near the saturation point, or on the inhibition of a factor, such as the bone morphogenetic protein, that normally helps to regulate cell differentiation in the vascular wall.

7.3.3 Glycoproteins (Phosphoproteins)

Several glycoproteins rich in glutamic, aspartic and sialic acids and, on that basis, known as acidic glycoproteins (Butler et al. 1985), have been extracted from bone matrix using the sequential dissociative extraction procedure. These glycoproteins appear to be buried, or entrapped, in the inorganic component because decalcification is needed for them to become soluble (Termine et al. 1981a; Butler 1984; Fisher and Termine 1985; Termine 1985). For this reason, besides their anionic characteristics and their calcium and hydroxyapatite high affinity binding properties, they are thought to play a role in calcification, by retarding or promoting hydroxyapatite formation (discussed by Fisher and Termine 1985; Veis 1985; Boskey 1989, 1998; Glimcher 1989; Gorski 1992). Alkaline phosphatase, an enzyme long associated with calcification (see Sect. 17.2), is also a glycoprotein occurring not only in the cell membrane and matrix vesicles (Wuthier and Register 1985), but also in calcified matrix (de Bernard et al. 1985, 1986; Bonucci et al. 1992). According to Lowther and Natarajan (1972), glycoproteins, as well as proteoglycans, become firmly bound to collagen fibrils during fibrogenesis in vivo.

Histochemically, it has been shown that the osteoid matrix stains quite intensely when periodic acid-Shiff (PAS) method is used (Caretto 1958; Leblond et al. 1959). This stains the glycoproteins which, because they have vicinal hydroxy groups, are able to form aldehydes when oxidized by periodic acid. Deep PAS staining has also been reported in the calcified matrix of the medullary bone of birds, which is rich in glycoproteins (Bonucci and Gherardi 1975). Lectins have been used to demonstrate the glycoconjugates associated with non-collagenous proteins under the electron microscope (Nanci 1999).

Most bone matrix glycoproteins are highly phosphorylated proteins (phosphoproteins) which contain *o*-phosphoserine and *o*-phosphothreonine in their molecule (Cohen-Solal et al. 1979). They have been isolated from chick bone extracts after the calcified matrix has been decalcified with EDTA or HCl, or has been submitted to periodate solubilization (Spector and Glimcher 1972, 1973; Lee SL and Glimcher 1981; Shuttleworth and Veis 1972). They are not all extracted in the same way: both HCl and formic acid efficiently solubilize osteocalcin and osteonectin, whereas bone sialoprotein is selectively extracted by EDTA (Gerstenfeld et al. 1994).

Phosphoproteins are heterogeneous in weight and composition (Uchiyama et al. 1986); about 20% of them are covalently bound to collagen, mainly to the α2-chains (Cohen-Solal et al. 1979). They are synthesized by osteoblasts (Glimcher et al. 1984) and are exclusively concentrated in the extracellular matrix at the calcification front, essentially without detectable staining in the adjacent unmineralized osteoid matrix (Glimcher 1990; Bruder et al. 1991). Phosphoproteins, however, are not peculiar to bone matrix; apart from those contained in dentin and other calcified tissues (see below), others have been found in non-calcifying tissues.

Bone phosphoproteins include osteonectin, osteopontin, bone sialoprotein, dentin matrix protein 1, matrix extracellular phosphoglycoprotein and acidic glycoprotein-75. Another hydrophobic, glycosilated phosphoprotein, called osteometrin, has been isolated from bovine bone in guanidine HCl after the tissue had been extensively extracted with HCl and EDTA (Zhou et al. 2000). Some of the phosphoproteins are members of the SIBLING (Small Integrin-Binding Ligand, N-linked Glycoprotein) family, which includes genetically related proteins clustered on human chromosome 4 (Fisher et al. 2001; Fisher and Fedarko 2003). These are osteopontin, bone sialoprotein, dentin matrix protein 1, dentin sialophosphoprotein, and matrix extracellular phosphoglycoprotein.The organic bone matrix contains proteins that are synthesized outside bone and accumulate in its matrix, such as α_2HS glycoprotein (or fetuin; see below).

7.3.3.1
Osteonectin

Osteonectin (ON) was first described by Termine et al. (1981a, b). Homologous with BM-40 protein and with SPARC (secreted protein, acid and rich in cysteine) (Dziadek et al. 1986; Mason et al. 1986; Bolander et al. 1988; Domenicucci et al. 1988), it is a calcium-binding glyco-phosphoprotein (reviewed by Tracy and Mann 1991) which binds to both collagen and hydroxyapatite (Termine et al. 1981b; Termine 1985), although purified porcine ON does not seem to share this property (Domenicucci et al. 1988). Osteonectin levels in calcified matrix appear to be heterogeneous: in bovine embryogenesis, its level in primary (woven) bone is about 40% less than that in adult lamellar bone (Termine 1985); in adult human bones, trabecular bone has from 21- to 47-fold higher levels of ON than lamellar bone (Ninomiya et al. 1990). According to Sodek et al. (2000), there is an inverse correlation between the degree of calcification of the bone matrix and its ON content. The close relationship between ON and inorganic substance explains why, as reported above for osteocalcin, it can remain intact and protected for thousands of years in the matrix of fossil bone (Schmidt-Schultz and Schultz 2004).

Osteonectin is synthesized by osteoblasts of several species (Termine 1985). This has been confirmed by immunohistochemistry: immunolabeling for ON has been shown in Golgi vesicles, endoplasmic reticulum cisternae and dense cytoplasmic bodies of the osteoblast, as well as in the deep zones of the osteoid tissue, its intensity increasing at the calcification front (Romanowski et al. 1990). Osteonectin mRNA expression has been found in osteoblasts at the onset of calcification (Sommer et al. 1996). The osteoblast-like cells MC3T3-E1 express ON during the bone matrix formation/maturation period, with maximum expression at day 16 (Choi et al. 1996). Osteonectin has been demonstrated in isolated osteocytes grown in vitro (Aarden et al. 1996), and in osteoblasts and osteoprogenitor cells, as well as in young osteocytes, but not in aged, quiescent osteocytes, which has led to its being considered a marker of osteoblastic functional differentiation (Jundt et al. 1987; Metsäranta et al. 1989). Bianco et al. (1985, 1988) obtained similar results using monoclonal immunohistochemistry and showed strong ON immunoreactivity in the osteoid tissue. In the developing mandible of rat embryos between 15 and 20 days old,ON mRNA expression – as well as that of osteocalcin and bone sialoprotein mRNAs – was first observed in newly differentiated osteoblasts, where ON immunostaining was also positive (Ishigaki et al. 2002). It was again positive in osteoid matrix, but not in calcified matrix, which was, however, immunostained if digested with protease after decalcification with ethanolic trimethylammonium EDTA. This finding was interpreted as indicating that during calcification ON is mainly incorporated into, and becomes a specific component of, the mineralized bone matrix, as also suggested by Western blot analysis of EDTA extracts of fresh rat mandible specimens at embryonic day 15 and 20, which revealed the presence of $\sim$ 50 kDa ON (Ishigaki et al. 2002). Nefussi et al. (1997) reported that ON was distributed uniformly throughout the osteoid tissue and calcified matrix, where it was associated with type I collagen (Nordahl et al. 1995).

In situ hybridization studies of human fetal skeletal tissues have, on the other hand, shown that high ON mRNA levels are expressed not only by osteoblasts, but also by periosteal cells and by hypertrophic chondrocytes, and that weak signals can also be detected in osteocytes, in fibroblasts of tendons and ligaments and in epidermis cells (Metsäranta et al. 1989). Osteonectin has, in fact, been found not only in calcified bone and cartilage, but also in other uncalcified connective tissues (Sage et al. 1989; Chen et al. 1991a). It is expressed by several soft tissues (Carlson et al. 1993), as shown by in situ hybridization studies carried out in the newborn lung, cartilage, thyroid, stomach, skin, heart and peripheral nerve trunk, and in the adult adrenal gland, ovary and testis (Holland et al. 1987), as well as in human decidua and in several carcinomas, where reactivity was confined to basal membranes (Wewer et al. 1988). An osteonectin-like protein has been described in human platelets (Stenner et al. 1986).

In vitro studies have shown that ON is a powerful inhibitor of hydroxyapatite crystal growth (Romberg et al. 1986). This is in line with the fact that comparative studies on the capacity of various non-collagenous proteins to promote or inhibit nucleation have shown that ON – and OPN too – lack nucleation capability even at concentrations as high as 100 μ/ml, whereas BSP and dentin phosphophoryn exhibit nucleation activity at as little as 0.3 and 10 μg/ml, respectively (Hunter et al. 1996). On the other hand, Boskey (1989) has stressed that ON may have a general multifactorial role in regulating calcium-mediated processes. In this connection, Kuboki et al. (1989) have shown that ON undergoes conformational changes once it binds calcium ions, and Maurer et al. (1995) have reported that the C-terminal portion of the molecule is an autonomously folding and crystallizable, calcium-binding domain. In agreement with Bolander et al. (1988), the only conclusion left open appears to be that ON acts as a multifunctional protein with respect to calcium and hydroxyapatite binding sites. Delany et al. (2000) have shown that ON-null mice undergo a curtailment of bone formation, along with a decrease in osteoblast and osteocyte surface areas.

The organic matrix of bone contains fibronectin (Nordahl et al. 1995), a glycoprotein which, because of its specific arginine-glycine-aspartate amino acid sequence, plays a role in cell-cell and cell-matrix attachment.

7.3.3.2
Bone Sialoprotein

Bone sialoprotein (BSP) is a highly glycosylated, sulfated, phosphorylated, sialic acid-rich protein occurring in bone matrix (reviewed by Ganns et al. 1999). Its molecule contains Arg-Gly-Asp motifs, so it can bind to hydroxyapatite and to cell-surface integrins. These properties, and its localization in bone, reveal BSP's analogies with OPN (Chen et al. 1994).

Bone sialoprotein is a prominent component of bone matrix, especially that of woven bone (Gorski 1998), although its content falls with a rising degree of matrix calcification (Sodek et al. 2000). It is expressed by differentiated osteoblastic cells (Chen et al. 1991a, b, 1992a, 1996; Shapiro HS et al. 1993; Zhu et al. 2001); its mRNA has been found in mature bone-forming cells, but not in their immature precursors (Bianco et al. 1991, 1993a; Bellows et al. 1999), and BSP immunolabeling has been reported in osteoblast Golgi and post-Golgi secretory structures (Bianco et al. 1993b). Its transcription by osteoblastic cells is heightened by glucocorticoid and suppressed by vitamin D (Sodek et al. 1995b).

Bone sialoprotein immunostaining of osteoid tissue has been reported to be positive in some studies (Ishigaki et al. 2002) and negative in others (Carlson et al. 1993). In independent studies, discrete, malachite green-stainable sites, which are scattered through the osteoid tissue, have been found to be immunolabeled (Bianco et al. 1993b; Riminucci et al. 1995). These sites resemble those where osteopontin can be demonstrated immunohistochemically (McKee et al. 1990, 1992, 1993; Chen et al. 1994; McKee and Nanci 1996b; Mocetti et al. 2000); like them, they seem to correspond to the decalcified matrix of calcification nodules (Fig. 7.5). They are also similar to the patches of interfibrillary dense material that have been reported several times in bone and other calcifying tissues (Takagi et al. 1983; Ecarot-Charrier et al. 1988; McKee et al. 1990, 1992; McKee and Nanci 1995b; Ayukawa et al. 1998; Bonucci and Silvestrini 1996; Nanci 1999). These findings, which have been confirmed in osteogenic cell cultures derived from fetal rat cranial tissue (Irie et al. 1998), are in line with

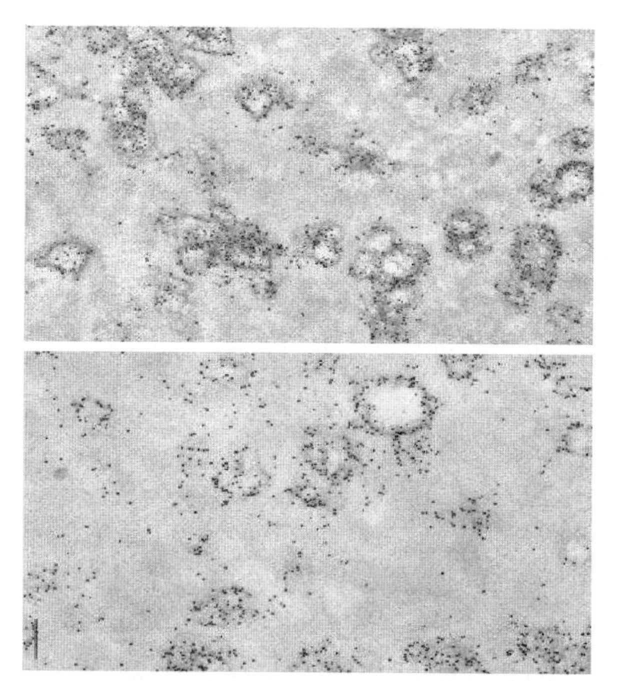

Fig. 7.5. Human bone: immunoreaction for BSP (*above*) and OPN (*below*): note the codistribution of immunogold particles in and around calcification nodules. Not counterstained, ×25,000 (Courtesy of Antonio Nanci, University of Montreal, Canada)

the observation that BSP is associated with clusters of needle-like crystals but not with matrix collagen fibrils (Nefussi et al. 1997), and that the activation of matrix calcification, by adding an organophosphate source to osteoblastic cultures, leads to the localization in specific matrix sites – the later location of crystals – of bone acidic glycoprotein-75 (Midura et al. 2004) and then of BSP (Wang et al. 2000). These sites, which have been called biomineralization foci, appear to correspond to the decalcified matrix of calcification nodules (discussed by Bonucci 2002).

Bone sialoprotein not only builds up in interfibrillary dense patches, but is also found in cement lines and "laminae limitantes", so showing the same distribution as OPN (Nanci 1999). It also codistributes with ON in developing bone trabeculae (Fisher et al. 1983b). According to Hultenby et al. (1994), BSP distribution and synthesis differ from those of OPN, because, in a way divergent to the strong reactivity of OPN at bone surfaces that face cells, no prominent BSP immunoreactivity was observed at these sites. In addition, in the osteogenetic process which follows the bone necrosis induced by colchicine injection, BSP mRNA reaches its maximum expression level at day six, much earlier than OPN, whose maximum level occurs at day ten (Arai et al. 1995). McKee et al. (1993) have shown immunohistochemically that BSP, as well as OPN, interacts with α_2HS-glycoprotein, and Ramstad et al. (2003) have reported that BSP co-distributes with osteoadherin. Bone sialoprotein is synthesized by stimulated osteoclast-like FLG 29.1 cells (Masi et al. 1995), and is expressed in osteoclasts and fetal epiphyseal cartilage cells (Bianco et al. 1991). Unexpectedly, it is also expressed by trophoblastic cells of the placenta (Bianco et al. 1991; Chen et al. 1991a) and by ameloblasts actively synthesizing enamel matrix in rat and hamster incisors and in epithelial cells of the ameloblastoma (Chen et al. 1998). Bone sialoprotein expression has been reported in human breast (Bellahcène et al. 1994; Ibrahim et al. 2000) and prostate (De Pinieux et al. 2001) cancer, in both of which it appears to be correlated with the development of bone metastases (Bellahcène et al. 1996; Waltregny et al. 2000; De Pinieux et al. 2001), and in lung (Bellahcène et al. 1997) and thyroid carcinoma (Veis et al. 1985). BSP, like type I collagen and ON, has been found in in vitro cultures of vascular smooth muscle cells (Severson et al. 1995).

It has been proposed that BSP participates in integrin-mediated cell-matrix interactions (Ross et al. 1993) and plays a role in promoting angiogenesis (Bellahcène et al. 2000) and calcification (Hunter and Goldberg 1993; Ganns et al. 1999; Shibata et al. 2002). This function is indirectly supported by its restriction to cells involved in the formation of calcifying connective matrices (Chen et al. 1991a), and directly supported by its immunohistochemical localization in areas of early calcification, such as calcification nodules and calcification islands (Bianco et al. 1993b; Riminucci et al. 1995; Nanci 1999), as also by the fact that its mRNA expression coincides with the earliest stages of osteogenesis (Chen et al. 1992a; Arai et al. 1995). According to Goldberg et al. (1996), BSP is a powerful nucleator of hydroxyapatite; this would not require intact molecules, but only the presence in them of two glutamic acid-rich sequences in a helical conformation (Tye et al. 2003); their nucleating properties can, in fact, be mimicked by the synthetic homopolymer polyglutamic acid (Hunter and Goldberg 1994). Moreover, hydroxyapatite nucleation appears to be associated with one glutamic acid-rich domain (Harris et al. 2000). These results have been strengthened by in vitro studies, showing that crystal formation can be induced in gel systems containing a metastable solution of calcium and phosphate, as well as BSP, or its synthetic homopolymer analogue: aggregates of rather thick, plate-like crystals were formed, which radiated from a central nidus containing extremely small, needle-shaped crystals and bone sialoprotein (Goldberg et al. 2000). A direct link of BSP with crystals is suggested by the finding that, if extracted from undecalcified samples, protein amounts fall as mineral density rises, whereas they rise if BSP is extracted from decalcified samples (Sodek et al. 2000).

It has been suggested that BSP binds not only to inorganic crystals, but also to collagen fibrils (Nagata et al. 1991). In more precise terms, biotin-labeled BSP preferentially binds to the collagen $\alpha 2$ chain and to peptides derived from that chain and, in reconstituted collagen fibrils, immunogold particles are mainly located in the fibril hole zones, suggesting that the BSP bound to $\alpha 2$ chains could regulate the onset of calcification in the holes (Fujisawa et al. 1995).

7.3.3.3
Osteopontin

Osteopontin (OPN), also called 44-kDa bone phosphoprotein or sialoprotein I, is a phosphorylated, highly glycosilated, sulfated sialoprotein (reviewed by McKee and Nanci 1995a; Gerstenfeld 1999). The OPN molecule contains polyacidic amino acid segments which allow it to bind to hydroxyapatite, and an Arg-Gly-Asp motif which recognizes the vitronectin receptor and can allow cell adhesion through the vitronectin integrin (Chen et al. 1994). Osteopontin is, in fact, found at the bone surfaces that face the cells, especially at the attachment site of the osteoclast (Reinholt et al. 1990; Hultenby et al. 1994). The identification of OPN in isolated rabbit osteoclasts (Tezuka et al. 1992; Dodds et al. 1995) has prompted the suggestion that it may have other functions besides cell attachment (Tezuka et al. 1992).

On the other hand, OPN is one of the predominant non-collagenous proteins in bone matrix (Gotoh et al. 1990; Chen et al. 1994; Boskey 1995), where its concentrations increase with the degree of calcification (Sodek et al. 2000). It is expressed by osteoblasts in developing calvariae, jaw bones and tibial metaphyseal and periosteal bone (Weinreb et al. 1990). It is synthesized by osteoblasts (Mark et al. 1987b; Chen et al. 1991a; Nagata et al. 1991; Chen et al. 1993; Sommer et al. 1996) and is found in their Golgi vesicles (Mark et al. 1987a, b; Hultenby et al. 1991) almost concomitantly with the appearance in them of alkaline phosphatase and before osteoid deposition (Mark et al. 1988). Osteocytes grown in vitro also contain OPN in their Golgi vesicles (Aarden et al. 1996). Sodek et al. (1995a) have reported that osteoblasts express a highly phosphorylated, 44-kDa OPN, whereas preosteoblasts express an only slightly phosphorylated, 55-kDa OPN, and that the former rapidly associates with hydroxyapatite, whereas the synthesis of the latter correlates with the formation of the 'cement' matrix of the lamina limitans. Osteopontin is expressed by periosteal cells and some stromal bone marrow cells (Chen et al. 1991a), by osteoblast and osteoclast progenitors in the bone marrow (Yamate et al. 1997), and by osteoclast precursors grown in vitro from human osteoclastoma (James et al. 1996). It is immunohistochemically demonstrable in the osteoblasts and matrix of woven bone during intramembranous and endochondral calcification, as well as in osteoclasts, so the hypothesis has been put forward that it is osteoclasts, not osteoblasts, that synthesize the cement substance which coats resorption lacunae (Dodds et al. 1995), a hypothesis rejected by McKee and Nancy (1996a). The possibility that the cement line is synthesized by the so-called post-osteoclastic cells during the reversal phase of the basic multicellular unit (BMU) must also be taken into consideration (Bonucci 1990).

Ultrastructural immunohistochemistry clearly shows that OPN is located in preferential bone sites (Fig. 7.5). Immunogold labeling is chiefly found, in fact, in interfibrillary 'dense' or 'grey' patches, which correspond to the matrix of decalcified calcification nodules or calcification islands scattered through a lightly but diffusely labeled osteoid tissue (McKee et al. 1990, 1991, 1992; Chen et al. 1994; McKee and Nanci 1995a, b; Ayukawa et al. 1998; Irie et al. 1998; Nanci 1999), and along the cell- and matrix-matrix interfacial structures (McKee and Nanci 1996b) called 'laminae limitantes' and cementing lines (McKee et al. 1992, 1993; McKee and Nanci 1995a, 1996b, c; Nanci et al. 1996; Ayukawa et al. 1998; Nanci 1999; Mocetti et al. 2000). Interestingly, OPN immunolabeling is concentrated at the periphery of developed calcification nodules (McKee and Nanci 1995b), where crystal ghosts are most prominent (to be discussed in Sect. 16.2.1), a finding confirmed in the bone nodules of primary osteogenic cell cultures (Irie et al. 1998).

Although OPN is a major protein of the bone matrix, it is expressed in other tissues (Carlson et al. 1993). It has, in fact, been found in cells of non-skeletal tissues such as myocardial cells (Graf et al. 1997), mucous and chief cells in the gastric mucosa (Qu-Hong et al. 1997), histiocytes in granulomas of various etiology (McKee and Nanci 1996d; Carlson et al. 1997). Osteopontin mRNA expression has been reported in hypertrophic chondrocytes (Franzén et al. 1989), in the kidney and in the gravid uterus (Chen et al. 1993).

Apart from its cell adhesion, chemoattraction and immunomodulation properties, and its possible role in regulating osteoclast activity and angiogenesis (Reinholt et al. 1990; Chen et al. 1993; McKee and Nanci 1996b, c, d; Shijubo et al. 2000; Asou et al. 2001; Ihara et al. 2001; Razzouk et al. 2002; Shimazu et al. 2002), this highly phosphorylated glycoprotein may have a multifunctional role in the formation and remodeling of bone, and be involved in the early stages of calcification. In this connection, findings by Sodek et al.

(1995a) strengthen the view that the synthesis of a poorly phosphorylated, 55-kDa form of OPN is correlated with the synthesis of a 'cement' matrix, whereas that of a highly phosphorylated, 44-kDa form quickly becomes associated with hydroxyapatite and may well help to regulate crystal growth.

The location of OPN in areas of early bone calcification, such as calcification nodules and calcification islands, that is, specific compartments in forming bone (Nanci et al. 2004), points to its close involvement in the initial phases of the calcification process. These findings should not, however, be taken to imply that OPN directly promotes the deposition of calcium phosphate. Gadeau et al. (2001) have, in fact, shown that in aortic calcification induced by acute vessel wall injury OPN can be detected in eight-day calcified lesions but not in two-day inorganic deposits. Moreover, the expression of its mRNA is low in early bone formation and rises later (Chen et al. 1993), and several studies suggest that OPN quickly becomes bound to already formed hydroxyapatite crystals (Kasugai et al. 1992), besides probably inhibiting calcium phosphate production and crystal formation (Boskey 1995; Goldberg and Hunter 1995; MacNeil et al. 1995; Hunter et al. 1996; Landis et al. 2003). Pampena et al. (2004) synthesized four peptides corresponding to four different phosphoserine-containing sequences in rat OPN; they found that all four raised the nucleation lag time, although they did not significantly impair the subsequent formation of crystals. This is in line with the finding that OPN-deficient mice show no obvious changes in body development and that their bones and teeth appear normal (Rittling et al. 1998). It also seems to be in line with the report by Lomashvili et al. (2004) that the degree of calcification does not increase in cultured normal aortas from OPN-deficient mice. These results do, however, conflict with those of Speer et al. (2002) showing that mice deficient in MGP alone (MGP−/−, OPN+/+) underwent calcification of their arteries as early as two weeks after birth, whereas those deficient in both MGP and OPN had twice as much arterial calcification with respect to MGP(−/−) OPN(+/+) at two weeks, and over three times as much at four weeks; moreover, these mice died significantly earlier (4.4 ± 0.2 weeks) than their MGP(−/−) OPN(+/+) counterparts (6.6 ± 1.0 weeks), as a result of vascular rupture (see review by Giachelli et al. 2005). In vitro studies by the same group of authors showed that OPN inhibits the calcification of vascular smooth muscle cells (Wada et al. 1999), and that smooth muscle cells deficient in OPN display an enhanced susceptibility to calcification (Speer et al. 2005). The view that OPN may have an inhibitory role in crystal formation is further supported by FTIRM analyses of bone in OPN-knockout mice, showing that the relative amounts of inorganic substance in mature areas, as well as crystal size and perfection in all anatomic regions, were significantly higher than in wild-type controls (Boskey et al. 2002). Ohri et al. (2005) detected the significant and rapid (seven days) calcification of glutaraldehyde-fixed bovine pericardium that had been subcutaneously implanted in OPN−/− mice and found that the degree of calcification was reduced by as much as 72% by administering recombinant, histidine-fused OPN, and by up to 91% by its adsorption onto the implant material. It has been suggested by Steitz et al. (2002) that OPN may not only inhibit mineral deposition, but also actively promote its dissolution by physically blocking hydroxyapatite crystal growth and promoting the acidification of the extracellular milieu. On the other hand, Ito et al. (2004) have found that OPN does not induce apatite formation when adsorbed on agarose beads, whereas it promotes calcification when covalently bound to the same substrate.

7.3.3.4 Dentin Matrix Protein 1

Dentin matrix protein 1 (DMP1), first identified from a rat incisor cDNA library (George et al. 1993), was once considered a dentin-specific component. It is now clear that DMP1 mRNA is also expressed by osteoblasts, ameloblasts and cementoblasts (D'Souza et al. 1997; Srinivasan et al. 1999; Thotakura et al. 2000; MacDougall et al. 1998). Butler et al. (2002) reported the presence of 150–200-kDa DMP1 in dentin, and of 57-kDa fragments not only in dentin (predentin, odontoblasts and odontoblast processes), but also in bone (osteocytes and their processes) and in cellular cementum. Qin et al. (2004) have confirmed that 37K and 57K DMP1 fragments are found in bone. The former correspond to the NH2-terminal portion of DMP1 and are less phosphorylated than the latter, which correspond to the COOH-terminal portion. During fracture healing, DMP1 mRNA has been detected in preosteocytes and osteocytes in the osseous callus, but it has not been found in osteoblasts (Toyosawa et al. 2004a). Studies of fetal rat calvaria cells in culture showed that DMP1 first appeared in osteoblasts, but was later strongly expressed in osteocytes (Fen et al. 2002).

These findings suggest that DMP1 is only located in calcified tissues, and the results of a recent study have supported this view (Feng et al. 2003). Both immunohistochemistry and RT-PCR have, however, shown that it is expressed and localized in several soft tissues (Terasawa et al. 2004). Expression of DMP1 mRNA has been found in the jaws and long bone of birds (Toyosawa et al. 2000), which have no teeth; in this case it is expressed in preosteocytes and osteocytes, but not in osteoblasts (Toyosawa et al. 2001). Dentin matrix protein 1 mRNA expression has been detected in phosphaturic mesenchymal tumors that are responsible for oncogenic osteomalacia (Toyosawa et al. 2004b).

One precise function of DMP1 is its activation of MMP-9 (Fedarko et al. 2004). It may also play a role in inducing bone and dentin calcification, as suggested by the finding that its mRNA expression coincides with the onset of crystal formation (Hao et al. 2004), that recombinant DMP1 (rDMP1) has calcium-binding properties (He et al. 2003a), and that, when immobilized on type I collagen fibrils, it facilitates apatite deposition in vitro (He and George 2004). In this connection, the self-assembly of DMP1 into a β-sheet conformation is needed (He et al. 2003b). According to Narayanan et al. (2003), DMP1 has dual functional roles; it is primarily localized in the nucleus of undifferentiated osteoblasts, where it acts as a transcriptional factor for the activation of specific osteoblast genes, and it is then phosphorylated and exported to the matrix, where it may regulate calcification. On the other hand, according to Tartaix et al. (2004), DMP1 in its native form inhibits the calcification process, whereas it promotes it when it is cleaved or dephosphorylated. Hypomineralization and the partial failure of predentin's maturation into dentin are the twin outcomes of the deletion of DMP-1 (Ye et al. 2004). FTIRI studies have shown that the mineral-to-matrix ratio (a parameter of relative mineral content) is significantly lower, whereas the degree of crystallinity (crystal size/perfection) is significantly higher, in DMP-1-knockout mice than in heterozygous and wildtype controls (Ling et al. 2005).

7.3.3.5
Matrix Extracellular Phosphoglycoprotein (MEPE)

Otherwise known as osteoregulin or OF45 (osteoblast/osteocyte Factor 45), matrix extracellular phosphoglycoprotein (MEPE) is an extracellular, bone-tooth matrix glycoprotein that was first detected in tumor-induced osteomalacia and that has been considered a member of a new class of phosphate-regulating factors known as phosphatonins (Schiavi and Kumar 2004). Besides MEPE, these include the fibroblast growth factor 23 (FGF23, expressed by human osteoblast-like cells; Mirams et al. 2004), the stanniocalcin (Blumsohn 2004), and the frizzled-related protein 4 (FRP4) (Fukumoto and Yamashita 2002; Schiavi and Moe 2002; Schiavi and Kumar 2004). Additional studies are, however, required to allow a full understanding of the properties of these recently discovered compounds (Quarles 2003).

The MEPE gene shows major similarities with the group of bone and tooth glycophosphoproteins that comprise OPN, BSP, DSPP and DMP1; all of these contain Arg-Gly-Asp sequence motifs, which are essential for the integrin-receptor interactions (Rowe et al. 2000). They are members of a gene cluster in chromosome 4q21, where genetic linkage studies have identified the critical loci for dentinogenesis imperfecta II and III and for dentin dysplasia type II (MacDougall et al. 2002).

Oncogenic osteomalacia is due to generally small, mesenchymal tumors which have no specific histopathological features but lead to hyperphosphaturia, low serum $1{,}25(OH)_2D_3$ and osteomalacia. MEPE has been found to be associated not only with these phosphaturic tumors, but also with other hypophosphatemic diseases, such as X-linked hypophosphatemic rickets and autosomal dominant hypophosphatemic rickets. In these diseases, it is strongly expressed in osteoblasts, osteocytes and odontoblasts (Rowe et al. 2004) and its murine homologue is expressed by fully differentiated osteoblasts (Argiro et al. 2001). Immunohistochemistry with rabbit polyclonal antibody directed against recombinant human MEPE has, however, revealed that the protein was predominantly expressed by osteocytes rather than osteoblasts: in osteomalacic patients, immunoreaction positivity was found in 87.5% of osteocytes in calcified bone and 7.8% of those in osteoid tissue; in subjects suffering from osteoporosis, 95.3% of osteocytes in calcified bone were MEPE-positive, compared with 4.9% of those in osteoid tissue (Nampei et al. 2004). MEPE expression has been detected by means of RT-PCR not only in cells of tumors that induce osteomalacia, but also in bone marrow and brain, with very low level of expression in cells of the lung, kidney and human placenta (Rowe et al. 2000).

The human-MEPE (Hu-MEPE), besides inducing hypophosphatemia and hyperphosphaturia when administered intraperitoneally in mice, was able to inhibit the BMP-2-mediated calcification of osteoblast-like cells in vitro (Rowe et al. 2004).

According to Hayashibara et al. (2004), a synthetic peptide fragment of MEPE stimulates new bone formation in vitro and in vivo. Studies on mice have shown that it is expressed in bone and teeth in a maturation-dependent way and that its highest levels are reached during calcification; by contrast, in human osteoblastic cells grown in vitro, it attains these high levels in the proliferation and early matrix maturation phases and is strongly suppressed as calcification goes forward (Siggelkow et al. 2004).

7.3.3.6
Acidic Glycoprotein-75

Acidic glycoprotein-75 (BAG-75) is a glycosilated, sialic acid-rich phosphoprotein which is found in bone and dentin calcified matrix in 75 and/or 50 kDa forms (Gorski et al. 1990). It shows limited structural homology with, and is present in primary bone in almost the same amounts as, OPN (Gorski and Shimizu 1988; Gorski et al. 2004). It is enriched in 4 M guanidine HCl-0.5 M EDTA extracts of woven bone from the rat tibial metaphysis, but not in those of lamellar bone from the diaphysis, which contained less than 5% of the amount found in the extracts of calvarial or metaphyseal bone (Gorski 1998). It can self-associate into supramolecular spherical complexes consisting of 10–12 nm thick microfibrils (Gorski et al. 1997).

Acidic glycoprotein-75 complexes can sequester phosphate ions (Gorski et al. 1997) and are influenced by conformational changes induced by calcium (Chen et al. 1992b). Valuable results have been obtained from immunohistochemical studies on bone formation using the rat bone marrow ablation model (Gorski et al. 2004) or the osteogenic UMR 106-01 BSP (Stanford et al. 1995) cell line cultures: BAG-75 is reported not to be directly associated with collagen fibrils, but to be a component of fibrillar scaffolds or vesicle-like structures which are located in the interfibrillary spaces (Midura et al. 2004). As discussed above, both these structures are reminiscent of – and may turn out to be identified with – those that distinguish the decalcified matrix of calcification nodules (reviewed Bonucci 2002) and the interfibrillary matrix of embryonic avian bone (Bonucci and Silvestrini 1996). Double immunohistochemical labeling shows that these scaffolds and vesicles contain BAG-75 and BSP (Midura et al. 2004), an observation in accordance with that of Bianco et al. (1993b) on the location of BSP in calcification nodules.

No precise function has been demonstrated for BAG-75 in calcification, but an important role can be inferred from the following results: BAG-75 is expressed at a very early stage in osteogenic areas; it is localized and concentrated in areas which are about to be calcified; it can self-associate into supramolecular complexes; it delineates vesicle-like, condensed mesenchymal regions where BSP, which is itself thought to be involved in early calcification, will accumulate (Gorski et al. 1997, 2004; Midura et al. 2004).

7.3.3.7 α_2HS Glycoprotein/Fetuin

Otherwise known as α_2-Heremans-Schmid glycoprotein or AHSG, this is a serum protein which is synthesized in the liver and accumulated in bone and dentin (Dickson et al. 1975; Ashton et al. 1976; Triffitt et al. 1978; Pinero et al. 1995). α_2HS glycoprotein is homologous with the 60K glycoprotein of rat bone (Mizuno et al. 1991), and homologous with fetuin, or fetuin-A, the major serum protein in fetal lamb and calf serum (Dziegielewska et al. 1987; Yang et al. 1992), to be distinguished from fetuin-B, a second member of the family in mammals (Olivier et al. 2000). Fetuin-A is a 59-kDa glycoprotein which contains 3.3 mol of protein-bound phosphate when purified from the fetuin-mineral complex (Price et al. 2003); it is a member of the cystatin superfamily of cysteine protease inhibitors.

Fetuin is one of the most plentiful non-collagenous proteins in bone (Ashton et al. 1974, 1976; Quelch et al. 1984), reaching a concentration of about 1 mg/g of tissue (Ashton et al. 1974; Ohnishi et al. 1991); the highest amounts are those found in neonatal bone, with a fall-off in bone from children and adults (Quelch et al. 1984). α_2HS glycoprotein and fetuin have the same immunohistochemical distribution in the brain, bone, kidney, gonads, gastrointestinal tract, and the respiratory and cardiovascular systems, whereas there are notable differences in the liver and thymus (Dziegielewska et al. 1987). They are mainly found in cells or structures undergoing differentiation and transformation (Terkelsen et al. 1998).

Fetuin may regulate osteogenesis by binding to TGF-β and BMP (Demetriou et al. 1996; Binkert et al. 1999; Szweras et al. 2002), and may interfere with the calcification process by inhibiting calcium salt formation (Schinke et al. 1996). This process is accompanied and probably mediated by the formation of a fetuin-mineral complex with a high molecular mass (Price and Lim 2003), consisting of calcium phosphate, fetuin and MGP (Price et al. 2003) in proportions by weight close to 18% mineral, 80% fetuin, and 2% MGP (Price et al. 2002a). The injection of 8 mg etidronate/100 g body weight induces the formation of that complex and raises serum calcium 1.8-fold (to 4.3 mM), phosphate 1.6-fold (to 5.6 mM), and MGP 25-fold (to 12 µg/ml) (Price et al. 2002a). It should be noted that the formation of the fetuin-MGP-calcium phosphate complex in response to etidronate injection is suppressed by treatment with osteoclast activity inhibitors such as calcitonin, osteoprotegerin, or alendronate (Price et al. 2002b).

Several findings show that AHSG/fetuin tend to inhibit calcification. This feature seems to depend on the formation of the AHSG-calcium phosphate complexes discussed above (Price and Lim 2003). These appear to give rise to the transient formation of soluble, colloidal "calciprotein particles", 30–150 nm in diameter, which are initially amorphous and soluble and gradually become more crystalline and insoluble in a time- and temperature-dependent fashion (Heiss et al. 2003). In rats, in which extensive calcification of the artery media had been induced by vitamin D treatment, fetuin-calcium phosphate complexes have been detected in blood; calcification turned out to be correlated with the period during which fetuin-mineral complexes showed their greatest increase in serum, whereas these complexes were not detectable if calcification was inhibited by ibandronate or osteoprotegerin (Price et al. 2004). In vitro studies on vascular smooth muscle cells have shown direct interaction between fetuin-A and matrix vesicle at their moment of release, so inhibiting vascular calcification locally in its early stages (Ketteler 2005). Confocal microscopy and electron microscope immunohistochemistry have shown that fetuin-A is internalized and concentrated in the vesicles of vascular smooth muscle cells, so that it can neutralize their ability to promote calcification when vesicles are released extracellularly from apoptotic or viable cells (Reynolds et al. 2005). In chronic kidney diseases, an inverse correlation has been reported between coronary artery calcification and serum fetuin, with increasing immunostaining for fetuin – and for MGP, too – proportional to the degree of artery wall calcification (Moe et al. 2005). Chronic inflammation and uremia may help to exhaust the release of fetuin-A in the late stages of kidney disease, so favoring vascular calcification and cardiovascular events (Floege and Ketteler 2004; Ketteler 2005; Ketteler et al. 2005). In vitro, the addition of fetuin to bovine vascular smooth muscle cells inhibits calcification (Moe et al. 2005). The inhibitory capacity of AHSG is confirmed by the finding that AHSG-deficient mice, which are phenotypically normal, develop severe calcification of various organs when fed on a mineral and vitamin D rich diet (Schäfer et al. 2003).

7.3.4 Lipids

Lipids are components of the bone matrix, as suggested by the study of Irving (1959, 1960) with pyridine extraction and Sudan black B staining (Fig. 7.6), and as documented by biochemical studies (Shapiro IM 1970a, b; Dirksen and Marinetti 1970). They are intrinsic components of the calcified matrix, as shown by the finding that only a fraction of them – a fraction containing the neutral phospholipids sphingomyelin, lecithin and phosphatidylethanolamine – can be removed from the intact tissue, and that a further quantity, mainly

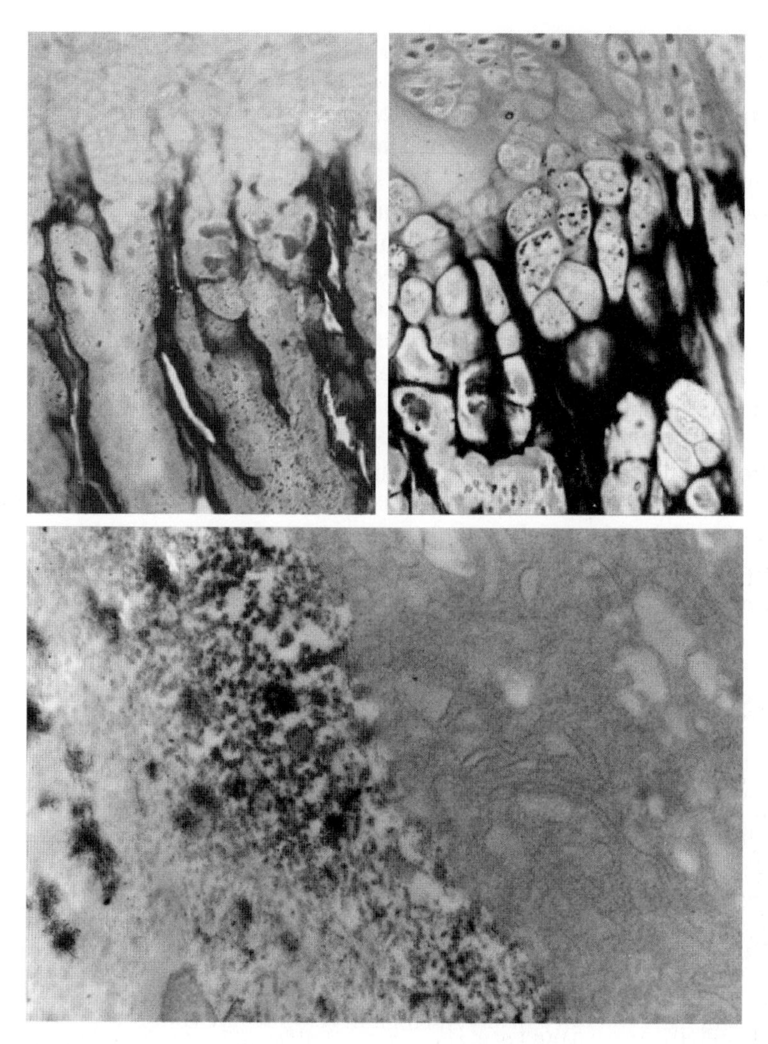

Fig. 7.6. Section of metaphyseal bone extracted with piridine and stained with sudan black B (*upper left*) and PAS (*upper right*): the calcified matrix is stained by both methods; ×50. *Below*: section treated with MC22-33F monoclonal antibody; note the electron density of calcification nodules. Part of an osteoblast on *right*. Immunoperoxidase, ×15,000

comprising acidic phospholipids, can only be extracted after decalcification (Shapiro IM 1970b). Dirksen and Marinetti (1970) reported that total lipids of the bone matrix were 2.005 mg/100 g human bone, and that most of them were triglycerides; Wolinsky and Guggenheim (1970) found that total lipids, triglycerides and phospholipids were 7.50, 1.81 and 1.02 mg/g, respectively.

Histochemical and transmission electron microscope investigations confirmed the presence of lipids in areas of early bone calcification and at the calcification front (Ponlot 1958; Ngoma and Haumont 1974; Bonucci et al.

1978; Nefussi et al. 1992; Bonucci and Silvestrini 1995; Silvestrini et al. 1996; Zini et al. 1996), so supporting the view that acidic lipids may be involved in calcification (discussed by Vogel and Boyan-Salyers 1976; Dziak 1992). Studies carried out with the antibody MC22-33F, which displays a specific reaction with phosphatidylcholine, sphingomyelin and dimethylphosphatidylethanolamine, showed a strong immunoreaction along the osteoblast membrane, in the interfibrillary spaces of the osteoid border, in matrix vesicles and in calcification nodules and their peripheral zone (Fig. 7.6), whereas the fully calcified matrix and the central zone of the nodules were negative (Bonucci et al. 1997). Histochemical studies carried out with Irving's method showed that young osteons are stained by Sudan black at an inverse ratio to their degree of calcification, and that the heavily calcified matrix is left unstained (Haumont and Ngoma 1973; Ngoma and Haumont 1974). In bone matrix a significant proportion of the lipids consists of calcium-phospholipid-phosphate complexes; these constitute a significantly higher proportion of the lipids in young than in mature bone (Boskey and Posner 1976). In chick diaphyseal bone, the calcium-phospholipid-phosphate complexes become more abundant after vitamin D treatment (Boskey and Dickson 1988).

7.3.5
Matrix Vesicles

The term 'matrix vesicles' (MVs) refers to roundish bodies with a diameter of at least 25–250 nm found in the uncalcified matrix of the osteoid tissue and consisting of amorphous, osmiophilic matrix surrounded by a trilaminar membrane. They cannot be considered true constitutive components of the osteoid matrix, but their presence in the interfibrillary spaces of the osteoid tissue implies that they must be discussed in this chapter.

Initially described in epiphyseal cartilage (Bonucci 1967) and in cartilage induced by the injection of amniotic cells into the muscles of cortisone-conditioned mice (Anderson HC 1967), matrix vesicles have aroused great interest because of the finding that they often contain hydroxyapatite crystals and may therefore be the loci of initial calcification (Bonucci 1967). The topic has been reviewed several times (Bonucci 1971, 1975, 1981, 1984; Anderson HC 1976, 1984, 1989, 1995; Boskey 1979; Ali 1992; Boyan et al. 1992a; Sela et al. 1992a). Matrix vesicles were later described in, or isolated from, growth cartilage (Anderson HC 1969; Bonucci 1970), but a number of studies have shown their presence in the long bones of chick embryos and the medullary bone of pigeons (Bonucci 1971), and in intramembranous osteogenesis (Bernard 1969; Bernard and Pease 1969). They were then found in osteogenetic areas of embryonic chick femur (Anderson HC 1973; Anderson HC and Reynolds 1973; Schraer and Gay 1977; Gay et al. 1978), metaphyseal bone (Thyberg et al. 1975), bone from adult rats (Hoshi and Ozawa 2000), bone from thyroparathyroidectomized rats (Weisbrode et al. 1973), medullary bone from pigeons (Bonucci and Gherardi 1975), rat calvaria (Katchburian and Severs 1983), alveolar bone (Bab et al. 1979; Deutsch et al. 1981), tibial bone from normal young rats (Ornoy et al. 1980) and diphosphonate-treated rats (Schenk 1974), bone regenerated after bone marrow removal (Sela et al. 1992b), bone formed in vitro (Marvaso and Bernard 1977; Williams et al. 1980; Ecarot-Charrier et al. 1983; Bhargava et al. 1988) and bone (and cartilage) formed in vivo in diffusion chambers (Bab et al. 1984; Gotoh et al. 1995).

It must be stressed that, although matrix vesicles are present in all types of bone (Sela et al. 1981), compact bone included, their numbers fall with the compactness of the collagen fibrils in the matrix (Bonucci 1981) and few are found in diaphyseal compact bone (Fig. 7.1). This probably explains the view that they do not occur in this type of bone and that they might have a role in the calcification of embryonic bone, but not in that of mature compact bone (Landis et al. 1977).

The features of matrix vesicles do not change greatly in the many, normal and pathological hard tissues where they have been found, and they have mainly been studied in epiphyseal growth cartilage, so their structural and functional properties and their biogenesis will be described and discussed in connection with the components of the organic matrix in cartilage (Sect. 9.4).

7.4 The Organic Matrix of Tendons

Tendons do not normally calcify in humans, whereas calcification almost always occurs with aging in the leg tendons of turkey and other avian species (Abdalla 1979; Vanden Berge and Storer 1995). Calcified turkey leg tendons are therefore used as a model to study the mechanisms of the normal calcification process as it occurs in bone and other tissues (Johnson 1960; Likins et al. 1960; Landis 1986; Bigi et al.1988; Arsenault 1992). As already discussed (Sect. 7.2), no perfect identity exists between the collagen of tendons and bone, but the similarity is so close that the latter can be used as a reliable substitute for the former.

Tendon matrix mainly consists of highly cross-linked (Cannon and Davison 1973; Knott and Bailey 1998) type I collagen fibrils (Landis and Silver 2002) arranged almost exactly in parallel, whose density is almost as high as that of compact bone. Actually, most current knowledge of this type of collagen has come from studies on uncalcified tendons (mainly rat tail tendons) and has then been applied to the collagen of bone matrix. Readers can therefore refer to the discussion on bone collagen (Sect. 7.2) for data on the basic structure and arrangement of type I collagen.

After the pioneering studies of Fitton Jackson (1956) on developing tendons, further ultrastructural studies on various types of uncalcified tendons have shown that their matrix consists of two populations of collagen fibers (Parry 1978) that average 175 and 60 nm in diameter; the larger fibers are connected by interfibrillary filaments (Dyer and Enna 1976). An over tenfold increase in fibril thickness has been reported between fetal to adult life (Scott et al. 1981). These data have been confirmed by applying high voltage electron microscopy to serial thick sections of uncalcified portions of turkey leg tendons; this has shown not only that collagen fibrils may differ in diameter, but also that they may be branched, and be so highly aligned that their hole zones in contiguous arrangement appear as channels and grooves (Landis and Song 1991). This alignment seems to increase the overall volume of available intrafibrillar space, which has been calculated to reach about 60% of total volume (Lees et al. 1994). On the other hand, the freeze-fracture studies of Raspanti et al. (1989) on rat tail tendon collagen fibrils have shown that these were mainly of the larger type and displayed a straight alignment of constitutive helical microfibrils (as opposed to the wavy alignment found in other connective tissues (Fig. 7.1); Ruggeri et al. 1979). The morphological, immunochemical and biochemical studies of Ippolito et al. (1980) have confirmed that collagen fiber diameter rises with age side by side with the fall in proteoglycan and water content.

Proteoglycans are, in fact, constituents of the tendon matrix, which has long been known to contain chondroitin sulfate and dermatansulfate (Meyer et al. 1956; Anderson JC 1975). Ion-exchange chromatography, gel filtration and cesium chloride gradient centrifugation studies of adult bovine deep flexor tendons have demonstrated large proteoglycans in the matrix, whose peptide mass was similar to that of bovine cartilage aggrecan, and shown high levels of aggrecan mRNA in cells (Vogel et al. 1994).

Electron microscope studies of tendon collagen fibers fixed with glutaraldehyde and 1% alcian blue containing 0.3 M $MgCl_2$ (critical electrolyte concentration; Scott and Dorling 1965) have confirmed the previous results of Scott (1980) and Scott et al. (1981), in both cases using cuprolinic blue in a critical electrolyte concentration, that 4–9 nm thick, 180–400 nm long segments adhere to the gap zones of collagen fibrils and that, as they are removed by glycosidase (hyaluronidase, chondroitinase ABC) and stained by cuprolinic blue, they may correspond to proteoglycans (Dell'Orbo et al. 1996). Similar results have been obtained by a tapping-mode atomic force microscope study (Raspanti et al. 1997) which showed transverse ridges decorating the gap zones, and filamentous structures running along the fibril surface, either parallel or perpendicular to the fibril axis; these structures were enhanced by cupromeronic blue and removed by chondroitinase ABC, so the hypothesis has been put forward that the ridges and filaments may well constitute the core protein and the glycosaminoglycan side chains of proteoglycans, respectively. The relationship of proteoglycans to the periodic banding of tendon collagen fibrils has been confirmed by immunohistochemistry: the immunoreaction with monoclonal antibodies for the core protein of decorin has shown that in bovine tendons the immunolabel is associated with the *d* and *e* bands (Pringle and Dodd 1990). Similarly, most fibromodulin immunoreactivity has proved to be preferentially located in the hole region (Hedlund et al. 1994).

The chemical composition of tendons changes with calcification. Normal turkey tendons do not contain osteocalcin, which only appears in them together with, and increases proportionally to, matrix calcification (Glimcher et al. 1979a). Uncalcified samples of turkey tendon contain little or no *o*-phosphoserine, *o*-phosphothreonine and γ-carboxyglutamic acid, which only appear in them at the onset of calcification (Glimcher et al. 1979b).

Matrix vesicles have often been described in calcifying tendons (Yamada 1976; Kubota et al. 1984; Landis 1986; Landis and Arsenault 1989; Arsenault et al. 1991; Landis and Song 1991; Landis et al. 1992; Kohler et al. 1995; Landis and Silver 2002).They do not seem to differ substantially from those in bone.

7.5 Concluding Remarks

The persistence of a number of doubts and inconsistencies related to the structure and composition of bone matrix makes it difficult to draw definitive conclusions from the data reported in the broad literature on these topics. The following points seem worth stressing:

- Collagen is the most abundant matrix protein in bone and tendon, accounting for about 90% of their organic components.
- Bone and tendon contain type I collagen, whose fibrils display a periodic banding of about 68–70 nm.
- The way collagen molecules are assembled into fibrils basically respects the bi-dimensional model suggested by Hodge and Petruska, but it is not exactly known and several three-dimensional models have been proposed.
- The displacement of adjacent molecules along their axial plane generates intrafibrillar regions where molecules alternately overlap, generating 'dense' zones, and are separated by gaps, generating 'hole' zones.
- Collagen molecules are stabilized by intra- and inter-molecular cross-links, which may be responsible for the low degree of solubility of bone collagen.
- The distance between molecules within the fibrils has been calculated as being between 1.12 and 2.21 nm. The arrangement of hole zones in register may give rise to the formation of intrafibrillar grooves and channels.
- The arrangement of the collagen fibrils changes with the type of bone. They are loosely arranged and outline wide interfibrillar spaces in woven bone, above all in embryonic bone and in the medullary bone of birds, whereas they are densely packed with very narrow interfibrillar spaces in compact bone (and in tendons, too).
- Irrespectively of bone type, the collagen fibrils in the osteoid tissue are more loosely and irregularly arranged than those in the calcified bone matrix; they undergo aggregation with calcification.

- Interfibrillar spaces contain non-collagenous material in amounts that are proportional to their width and that therefore vary with bone type; they are greater in woven and medullary bone than in lamellar bone.
- A significant proportion of non-collagenous material is buried, or entrapped, in the calcified matrix and is so closely linked to the inorganic substance that it can only be extracted or stained after decalcification.
- The non-collagenous material consists of proteoglycans, non-collagenous proteins (Gla-proteins and glycoproteins), and lipids, whose molecules show an affinity for calcium ions.
- Proteoglycans are made up of small molecules, such as those of decorin, biglycan, and lumican. Their amounts change with bone type and in the same bone fall as the degree of matrix calcification rises.
- Gla-proteins consist of osteocalcin and matrix Gla-protein. The former is only found in calcified matrix and can be demonstrated immunohistochemically in the matrix of calcification nodules. The latter is located both in bone matrix and in soft-tissue matrix.
- Glycoproteins, besides being located in interfibrillary spaces, may be intrinsic components of collagen fibrils. Most of them are phosphorylated; these include osteonectin, osteopontin, bone sialoprotein, dentin matrix protein 1, matrix extracellular phosphoglycoprotein, acidic glycoprotein-75 and α_2HS-glycoproten/fetuin.
- Alkaline phosphatase is another glycoprotein; it is found in cell membranes, matrix vesicles and bone matrix, both at calcification sites and in fully calcified matrix.
- Bone sialoprotein, osteopontin and acidic glycoprotein-75 are immunohistochemically detected in the matrix of early areas of calcification (calcification nodules and calcification islands, appearing as dense, 'grey' patches) as well as in 'laminae limitantes' and cementing lines.
- Lipids occur in bone matrix as triglycerides and as calcium-phospholipid-phosphate complexes, which constitute a significantly greater proportion in young than in mature bone.
- Matrix vesicles are found in the calcifying matrix of the osteoid tissue and tendons. Their frequency is, however, inversely proportional to the density of collagen fibrils and few are found in compact lamellar bone.
- Non-collagenous proteins may either promote or inhibit the calcification process. Matrix Gla protein, fetuin-A, osteopontin and osteoprotegerin are those with the strongest inhibitory properties.

References

Aarden EM, Wassenaar AM, Alblas MJ, Nijweide PJ (1996) Immunocytochemical demonstration of extracellular matrix proteins in isolated osteocytes. Histochem Cell Biol 106:495–501

Abdalla O (1979) Ossification and mineralization in the tendons of the chicken (*Gallus domesticus*). J Anat 129:351–359

Aerssens J, Van Audekercke R, Geusens P, Schot LPC, Osman AAH, Dequeker J (1993) Mechanical properties, bone mineral content, and bone composition (collagen, osteocalcin, IGF-I) of the rat femur: influence of ovariectomy and nandrolone decanoate (anabolic steroid) treatment. Calcif Tissue Int 53:269–277

Aigner T, McKenna L (2002) Molecular pathology and pathobiology of osteoarthritic cartilage. Cell Mol Life Sci 59:5–18

Ali SY (1983) Calcification of cartilage. In: Hall BK (ed) Cartilage, vol. I. Structure, function and biochemistry. CRC Press, Boca Raton, pp 343–378

Ali SY (1992) Matrix formation and mineralisation in bone. In: Whitehead CC (ed) Bone biology and skeletal disorders. Carfax Pu. Co., Abingdon, pp 19–38

Anderson HC (1967) Electron microscopic studies of induced cartilage development and calcification. J Cell Biol 35:81–101

Anderson HC (1969) Vesicles associated with calcification in the matrix of epiphyseal cartilage. J Cell Biol 41:59–72

Anderson HC (1973) Calcium-accumulating vesicles in the intercellular matrix of bone. In: Ciba Foundation (ed) Hard tissue growth, repair and remineralization. ASP (Elsevier – Excerpta Medica – North-Holland), Amsterdam, pp 213–246

Anderson HC (1976) Matrix vesicles of cartilage and bone. In: Bourne GH (ed) The biochemistry and physiology of bone. Academic Press, New York, pp 135–157

Anderson HC (1984) Mineralization by matrix vesicles. Scanning Electron Microsc 2:953–964

Anderson HC (1989) Mechanism of mineral formation in bone. Lab Invest 60:320–330

Anderson HC (1995) Molecular biology of matrix vesicles. Clin Orthop Relat Res 314:266–280

Anderson HC, Reynolds JJ (1973) Pyrophosphate stimulation of calcium uptake into cultured embryonic bones. Fine structure of matrix vesicles and their role in calcification. Dev Biol 34:211–227

Anderson JC (1975) Isolation of a glycoprotein and proteodermatan sulphate from bovine achilles tendon by affinity cromatography on concanavalin A-Sepharose. Biochim Biophys Acta 379:444–455

Arai N, Ohya K, Kasugai S, Shimokawa H, Ohida S, Ogura H, Amagasa T (1995) Expression of bone sialoprotein mRNA during bone formation and resorption induced by colchicine in rat tibial bone marrow cavity. J Bone Miner Res 10:1209–1217

Argiro L, Desbarats M, Glorieux FH, Ecarot B (2001) Mepe, the gene encoding a tumor-secreted protein in oncogenic hypophosphatemic osteomalacia, is expressed in bone. Genomics 74:342–351

Arsenault AL (1992) Structural and chemical analyses of mineralization using the turkey leg tendon as a model tissue. Bone Miner 17:253–256

Arsenault AL, Frankland BW, Ottensmeyer FP (1991) Vectorial sequence of mineralization in the turkey leg tendon determined by electron microscopic imaging. Calcif Tissue Int 48:46–55

Ascenzi A (1988) The micromechanics versus the macromechanics of cortical bone - A comprehensive presentation. J Biomech Eng 110:357–363

Ascenzi A, François C, Steve Bocciarelli D (1963) On the bone induced by estrogen in birds. J Ultrastruct Res 8:491–505

Ascenzi MG, Benvenuti A, Ascenzi A (2000) Single osteon micromechanical testing. In: An YH, Draughn RA (eds) Mechanical testing of bone and the bone-implant interface. CRC Press, Boca Raton, pp 271–290

Ashton BA, Triffitt JT, Herring GM (1974) Isolation and partial characterization of a glycoprotein from bovine cortical bone. Eur J Biochem 45:525–533

Ashton BA, Höhling H-J, Triffitt JT (1976) Plasma proteins present in human cortical bone: enrichment of the α_2HS-glycoprotein. Calcif Tissue Res 22:27–33

Asou Y, Rittling SR, Yoshitake H, Tsuji K, Shinomiya K, Nifuji A, Denhardt DT, Noda M (2001) Osteopontin facilitates angiogenesis, accumulation of osteoclasts, and resorption in ectopic bone. Endocrinology 142:1325–1332

Aubin JE, Bonnelye E (2000) Osteoprotegerin and its ligand: a new paradigm for regulation of osteoclastogenesis and bone resorption. Osteoporosis Int 11:905–913

Ayukawa Y, Takeshita F, Inoue T, Yoshinari M, Shimono M, Suetsugu T, Tanaka T (1998) An immunoelectron microscopic localization of noncollagenous bone proteins (osteocalcin and osteopontin) at the bone-titanium interface of rat tibiae. J Biomed Mater Res 41:111–119

Bab I, Howlett CR, Ashton BA, Owen ME (1984) Ultrastructure of bone and cartilage formed *in vivo* in diffusion chambers. Clin Orthop Relat Res 187:243–254

Bab IA, Muhlrad A, Sela J (1979) Ultrastructural and biochemical study of extracellular matrix vesicles in normal alveolar bone of rats. Cell Tissue Res 202:1–7

Bairati A, Petruccioli MG, Torri Tarelli L (1969) Studies on the ultrastructure of collagen fibrils 1. Morphological evaluation of the periodic structure. J Submicr Cytol 1:113–141

Baranauskas V, Vidal BC, Parizotto NA (1998) Observation of geometric structure of collagen molecules by atomic force microscopy. Appl Biochem Biotechnol 69:91–97

Bartold PM (1990) A biochemical and immunohistochemical study of the proteoglycans of alveolar bone. J Dent Res 69:7–19

Baylink D, Wergedal J, Thompson E (1972) Loss of proteinpolysaccharides at sites where bone mineralization is initiated. J Histochem Cytochem 20:279–292

Bellahcène A, Merville MP, Castronovo V (1994) Expression of bone sialoprotein, a bone matrix protein, in human breast cancer. Cancer Res 54:2823–2826

Bellahcène A, Kroll M, Liebens F, Castronovo V (1996) Bone sialoprotein expression in primary human breast cancer is associated with bone metastases development. J Bone Miner Res 11:665–670

Bellahcène A, Maloujahmoum N, Fisher LW, Pastorino H, Tagliabue E, Menard S, Castronovo V (1997) Expression of bone sialoprotein in human lung cancer. Calcif Tissue Int 61:183–188

Bellahcène A, Bonjean K, Fohr B, Fedarko NS, Robey FA, Young MF, Fisher LW, Castronovo V (2000) Bone sialoprotein mediates human endothelial cell attachment and migration and promotes angiogenesis. Circ Res 86:885–891

Bellows CG, Reimers S, Heersche JNM (1999) Expression of mRNAs for type-I collagen, bone sialoprotein, osteocalcin, and osteopontin at different stages of osteoblastic differentiation and their regulation by 1,25 dihydroxyvitamin D_3. Cell Tissue Res 297:249–259

Benson SC, Wilt FH (1992) Calcification of spicules in the sea urchin embryo. In: Bonucci E (ed) Calcification in biological systems. CRC Press, Boca Raton, pp 157–178

Beresford JN, Fedarko NS, Fisher LW, Midura RJ, Yanagishita M, Termine JD, Gehron Robey P (1987) Analysis of the proteoglycans synthesized by human bone cells *in vitro*. J Biol Chem 262:17164–17172

Bernard GW (1969) The ultrastructural interface of bone crystals and organic matrix in woven and lamellar endochondral bone. J Dent Res 48:781–788

Bernard GW, Pease DC (1969) An electron microscopic study of initial intramembranous osteogenesis. Am J Anat 125:271–290

Bhargava U, Bar-Lev M, Bellows CG, Aubin JE (1988) Ultrastructural analysis of bone nodules formed in vitro by isolated fetal rat calvaria cells. Bone 9:155–163

Bianco P (1990) Ultrastructural immunohistochemistry of noncollagenous proteins in calcified tissues. In: Bonucci E, Motta PM (eds) Ultrastructure of skeletal tissues. Kluwer Academic Publishers, Boston, pp 63–78

Bianco P (1992) Structure and mineralization of bone. In: Bonucci E (ed) Calcification in biological systems. CRC Press, Boca Raton, pp 243–268

Bianco P, Hayashi Y, Silvestrini G, Termine JD, Bonucci E (1985) Osteonectin and Gla-protein in calf bone: ultrastructural immunohistochemical localization using the protein A-gold method. Calcif Tissue Int 37:684–686

Bianco P, Silvestrini G, Termine JD, Bonucci E (1988) Immunohistochemical localization of osteonectin in developing human and calf bone using monoclonal antibodies. Calcif Tissue Int 43:155–161

Bianco P, Fisher LW, Young MF, Termine JD, Gehron Robey P (1990) Expression and localization of the two small proteoglycans biglycan and decorin in developing human skeletal and non-skeletal tissues. J Histochem Cytochem 38:1549–1563

Bianco P, Fisher LW, Young MF, Termine JD, Gehron Robey P (1991) Expression of bone sialoprotein (BSP) in developing human tissues. Calcif Tissue Int 49:421–426

Bianco P, Riminucci M, Bonucci E, Termine JD, Gehron Robey P (1993a) Bone sialoprotein (BSP) secretion and osteoblast differentiation: relationship to bromodeoxyuridine incorporation, alkaline phosphatase, and matrix deposition. J Histochem Cytochem 41:183–191

Bianco P, Riminucci M, Silvestrini G, Bonucci E, Termine JD, Fisher LW, Gehron-Robey P (1993b) Localization of bone sialoprotein (BSP) to Golgi and post-Golgi secretory structures in osteoblasts and to discrete sites in early bone matrix. J Histochem Cytochem 41:193–203

Bigi A, Ripamonti A, Koch MHJ, Roveri N (1988) Calcified turkey leg tendon as structural model for bone mineralization. Int J Biol Macromol 10:282–286

Binkert C, Demetriou M, Sukhu B, Szweras M, Tenenbaum HC, Dennis JW (1999) Regulation of osteogenesis by fetuin. J Biol Chem 274:28514–28520

Blumsohn A (2004) What have we learnt about the regulation of phosphate metabolism? Curr Opinion Nephrol Hypert 13:397–401

Boivin G, Morel G, Lian JB, Anthoine-Terrier C, Dubois PM, Meunier PJ (1990) Localization of endogenous osteocalcin in neonatal rat bone and its absence in articular cartilage: effect of warfarin treatment. Virchows Arch A Path Anat Histol 417:505–512

Bolander ME, Young MF, Fisher LW, Yamada Y, Termine JD (1988) Osteonectin cDNA sequence reveals potential binding regions for calcium and hydroxyapatite and shows homologies with both a basement membrane protein (SPARC) and a serine proteinase inhibitor (ovomucoid). Proc Natl Acad Sci USA 85:2919–2923

Bonar LC, Lees S, Mook HA (1985) Neutron diffraction studies of collagen in fully mineralized bone. J Mol Biol 181:265–270

Bonucci E (1967) Fine structure of early cartilage calcification. J Ultrastruct Res 20:33–50

Bonucci E (1970) Fine structure and histochemistry of "calcifying globules" in epiphyseal cartilage. Z Zellforsch 103:192–217

Bonucci E (1971) The locus of initial calcification in cartilage and bone. Clin Orthop Relat Res 78:108–139

Bonucci E (1975) The organic-inorganic relationships in calcified organic matrices. Physicochimie et cristallographie des apatites d'intérêt biologique. Centre National de la Recherche Scientifique, Paris, pp 231–246
Bonucci E (1981) The origin of matrix vesicles and their role in the calcification of cartilage and bone. In: Schweiger HG (ed) International Cell Biology 1980–81. Springer, Berlin Heidelberg New York, pp 993–1003
Bonucci E (1984) Matrix vesicles: their role in calcification. In: Linde A (ed) Dentin and dentinogenesis. CRC Press, Boca Raton, pp 135–154
Bonucci E (1990) The basic multicellular unit of bone. It J Miner Electrol Metab 4:115–125
Bonucci E (1992) Role of collagen fibrils in calcification. In: Bonucci E (ed) Calcification in biological systems. CRC Press, Boca Raton, pp 19–39
Bonucci E (2002) Crystal ghosts and biological mineralization: fancy spectres in an old castle, or neglected structures worthy of belief? J Bone Miner Metab 20:249–265
Bonucci E, Gherardi G (1975) Histochemical and electron microscope investigations on medullary bone. Cell Tissue Res 163:81–97
Bonucci E, Motta PM (1990) Ultrastructure of skeletal tissues. Bone and cartilage in health and disease. Kluwer Academic Publishers, Boston
Bonucci E, Reurink J (1978) The fine structure of decalcified cartilage and bone: a comparison between decalcification procedures performed before and after embedding. Calcif Tissue Res 25:179–190
Bonucci E, Silvestrini G (1992) Immunohistochemical investigation on the presence of chondroitin sulfate in calcification nodules of epiphyseal cartilage. Eur J Histochem 36:407–422
Bonucci E, Silvestrini G (1995) Ultrastructure of rat metaphyseal bone after glutaraldehyde-malachite green fixation. It J Miner Electrol Metab 9:15–20
Bonucci E, Silvestrini G (1996) Ultrastructure of the organic matrix of embryonic avian bone after en bloc reaction with various electron-dense 'stains'. Acta Anat 156:22–33
Bonucci E, Cuicchio M, Dearden LC (1974) Investigations of ageing in costal and tracheal cartilage of rats. Z Zellforsch 147:505–527
Bonucci E, Frollà G, Piacentini M, Piantoni L (1978) Presenza di materiale sudanofilo nella matrice ossea e cartilaginea e processo di calcificazione: indagini istochimiche ed ultrastrutturali. Riv Istochim Norm Pathol 22:77–91
Bonucci E, Bianco P, Hayashi Y, Termine JD (1986) Ultrastructural immunohistochemical localization of non-collagenous proteins in bone, cartilage and developing enamel. In: Ali SY (ed) Cell mediated calcification and matrix vesicles. Excerpta Med, Amsterdam, pp 33–38
Bonucci E, Silvestrini G, Bianco P (1992) Extracellular alkaline phosphatase activity in mineralizing matrices of cartilage and bone: ultrastructural localization using a cerium-based method. Histochemistry 97:323–327
Bonucci E, Silvestrini G, Mocetti P (1997) MC22–33F monoclonal antibody shows unmasked polar head groups of choline-containing phospholipids in cartilage and bone. Eur J Histochem 41:177–190
Boskey AL (1979) Models of matrix vesicle calcification. Inorg Perspect Biol Med 2:51–92
Boskey AL (1989) Noncollagenous matrix proteins and their role in mineralization. Bone Miner 6:111–123
Boskey AL (1995) Osteopontin and related phosphorylated sialoproteins: effects on mineralization. Ann N Y Acad Sci 760:249–256
Boskey AL (1998) Biomineralization: conflicts, challenges, and opportunities. J Cell Biochem 30/31:83–91
Boskey AL, Dickson IR (1988) Influence of vitamin D status on the content of complexed acidic phospholipids in chick diaphyseal bone. Bone Miner 4:365–371

Boskey AL, Posner AS (1976) Extraction of calcium-phospholipid-phosphate complex from bone. Calcif Tissue Res 19:273–283

Boskey AL, Wians FH Jr, Hauschka PV (1985) The effect of osteocalcin on in vitro lipid-induced hydroxyapatite formation and seeded hydroxyapatite growth. Calcif Tissue Int 37:57–62

Boskey AL, Gadaleta S, Gundberg C, Doty SB, Ducy P, Karsenty G (1998) Fourier transform infrared microspectroscopic analysis of bones of osteocalcin-deficient mice provides insight into the function of osteocalcin. Bone 23:187–196

Boskey AL, Spevak L, Paschalis E, Doty SB, McKee MD (2002) Osteopontin deficiency increases mineral content and mineral crystallinity in mouse bone. Calcif Tissue Int 71:145–154

Boström KI (2000) Cell differentiation in vascular calcification. Z Kardiol 89:69–74

Boström K (2001) Insights into the mechanism of vascular calcification. Am J Cardiol 88:20E-22E

Boström K, Watson KE, Stanford WP, Demer LL (1995) Atherosclerotic calcification: relation to developmental osteogenesis. Am J Cardiol 75:88B-91B

Boyan BD, Schwartz Z, Swain LD (1992a) Cell maturation-specific autocrine/paracrine regulation of matrix vesicles. Bone Miner 17:263–268

Boyan BD, Swain LD, Everett MM, Schwartz Z (1992b) Mechanisms of microbial mineralization. In: Bonucci E (ed) Calcification in biological systems. CRC Press, Boca Raton, pp 129–156

Breckon JJW, Hembry RM, Reynolds JJ, Meikle MC (1995) Matrix metalloproteinases and TIMP-1 localization at sites of osteogenesis in the craniofacial region of the rabbit embryo. Anat Rec 242:177–187

Bronckers AL, Gay S, Finkelman RD, Butler WT (1987) Developmental appearance of Gla proteins (osteocalcin) and alkaline phosphatase in tooth germs and bones of the rat. Bone Miner 2:361–373

Bronckers ALJJ, Lyaruu DM, Wöltgens JHM (1989) Immunohistochemistry of extracellular matrix proteins during various stages of dentinogenesis. Connect Tissue Res 22:65–70

Bronckers AL, Price PA, Schrijvers A, Bervoets TJ, Karsenty G (1998) Studies of osteocalcin function in dentin formation in rodent teeth. Eur J Oral Sci 106:795–807

Brown JP, Delmas PD, Malaval L, Edouard C, Chapuy MC, Meunier PJ (1984) Serum bone Gla-protein: a specific marker for bone formation in postmenopausal osteoporosis. Lancet 1(8386):1091–1093

Bruder SP, Caplan AI, Gotoh Y, Gerstenfeld LC, Glimcher MJ (1991) Immunohistochemical localization of a ~66 kD glycosylated phosphoprotein during development of the embryonic chick tibia. Calcif Tissue Int 48:429–437

Bucay N, Sarosi I, Dunstan CR, Morony S, Tarpley J, Capparelli C, Scully S, Tan HL, Xu W, Lacey DL, Boyle WJ, Simonet WS (1998) *osteoprotegerin*-deficient mice develop early onset osteoporosis and arterial calcification. Genes Dev 12:1260–1268

Butler W (1984) Matrix macromolecules of bone and dentin. Coll Relat Res 4:297–307

Butler WT (2000) Noncollagenous proteins of bone and dentin: a brief overview. In: Goldberg M, Boskey A, Robinson C (eds) Chemistry and biology of mineralized tissues. American Academy of Orthopaedic Surgeons, Rosemont, IL, pp 137–141

Butler WT, Sato S, Rahemtulla F, Prince CW, Tomana M, Bhown M, Dimuzio MT, Bronckers ALJJ (1985) Glycoproteins of bone and dentin. In: Butler WT (ed) The chemistry and biology of mineralized tissues. EBSCO Media, Birmingham, pp 107–112

Butler WT, Brunn JC, Qin C, McKee MD (2002) Extracellular matrix proteins and the dynamics of dentin formation. Connect Tissue Res 43:301–307

Camarda AJ, Butler WT, Finkelman RD, Nanci A (1987) Immunocytochemical localization of gamma-carboxyglutamic acid-containing proteins (osteocalcin) in rat bone and dentin. Calcif Tissue Int 40:349–355

Campbell GR, Campbell JH (2000) Vascular smooth muscle and arterial calcification. Z Kardiol 89:54–62

Canalis E, McCarthy T, Centrella M (1988) Isolation of growth factors from adult bovine bone. Calcif Tissue Int 43:346–351

Candlish JK, Holt FJ (1971) The proteoglycans of fowl cortical and medullary bone. Comp Biochem Physiol 40B:283–293

Cannon DJ, Davison PF (1973) Cross-linking and aging in rat tendon collagen. Exp Geront 8:51–62

Caretto L (1958) Contributo allo studio della matrice ossea neodeposta. Arch Putti Chir Org Mov 10:211–230

Carlson CS, Tulli HM, Jayo MJ, Loeser RF, Tracy RP, Mann KG, Adams MR (1993) Immunolocalization of noncollagenous bone matrix proteins in lumbar vertebrae from intact and surgically menopausal cynomolgus monkeys. J Bone Miner Res 8:71–81

Carlson I, Tognazzi K, Manseau EJ, Dvorak HF, Brown LF (1997) Osteopontin is strongly expressed by histiocytes in granulomas of diverse etiology. Lab Invest 77:103–108

Chapman JA (1984) Molecular organization in the collagen fibril. In: Hukins DWC (ed) Connective tissue matrix. MacMillan, London, pp 89–132

Chapman JA, Hulmes DJS (1984) Electron microscopy of the collagen fibril. In: Ruggeri A, Motta PM (eds) Ultrastructure of the connective tissue matrix. Martinus Nijhoff Publishers, Boston, pp 1–33

Chavassieux P, Pastoureau P, Chapuy MC, Delmas PD, Meunier PJ (1993) Glucocorticoid-induced inhibition of osteoblastic bone formation in ewes: a biochemical and histomorphometric study. Osteoporosis Int 3:97–102

Chen J, Zhang Q, McCulloch CAG, Sodek J (1991a) Immunohistochemical localization of bone sialoprotein in foetal porcine bone tissues: comparisons with secreted phosphoprotein-1 (SPP-1, osteopontin) and SPARC (osteonectin). Histochem J 23:281–289

Chen J, Shapiro HS, Wrana JL, Reimers S, Heersche JNM, Sodek J (1991b) Localization of bone sialoprotein (BSP) expression to sites of mineralized tissue formation in fetal rat tissues by *in situ* hybridization. Matrix 11:133–143

Chen J, Shapiro HS, Sodek J (1992a) Developmental expression of bone sialoprotein mRNA in rat mineralized connective tissues. J Bone Miner Res 7:987–997

Chen Y, Bal BS, Gorski JP (1992b) Calcium and collagen binding properties of osteopontin, bone sialoprotein, and bone acidic glycoprotein-75 from bone. J Biol Chem 267:24871–24878

Chen J, Singh K, Mukherjee BB, Sodek J (1993) Developmental expression of osteopontin (OPN) mRNA in rat tissues: evidence for a role for OPN in bone formation and resorption. Matrix 13:113–123

Chen J, McKee MD, Nanci A, Sodek J (1994) Bone sialoprotein mRNA expression and ultrastructural localization in fetal porcine calvarial bone: comparisons with osteopontin. Histochem J 26:67–78

Chen J, Thomas HF, Jin H, Jiang H, Sodek J (1996) Expression of rat bone sialoprotein promoter in transgenic mice. J Bone Miner Res 11:654–664

Chen J, Sasaguri K, Sodek J, Aufdemorte TB, Jiang H, Thomas HF (1998) Enamel epithelium expresses bone sialoprotein (BSP). Eur J Oral Sci 106:331–336

Choi JY, Lee BH, Song KB, Park RW, Kim IS, Sohn KY, Jo JS, Ryoo HM (1996) Expression patterns of bone-related proteins during osteoblastic differentiation in MC3T3-E1 cells. J Cell Biochem 15:609–618

Cohen-Solal L, Lian JB, Kossiva D, Glimcher MJ (1979) Identification of organic phosphorus covalently bound to collagen and non-collagenous proteins of chicken-bone matrix. Biochem J 177:81–98

Cole DEC, Hanley DA (1991) Osteocalcin. In: Hall BK (ed) Bone, vol.3: Bone matrix and bone specific products. CRC Press, Boca Raton, pp 239–294

Cooper DR, Russell AE (1969) Intra- and intermolecular crosslinks in collagen in tendon, cartilage and bone. Clin Orthop Relat Res 67:188–209

Corsi A, Xu T, Chen X-D, Boyde A, Liang J, Mankani M, Sommer B, Iozzo RV, Eichstetter I, Robey PG, Bianco P, Young MF (2002) Phenotypic effects of biglycan deficiency are linked to collagen fibril abnormalities, are synergized by decorin deficiency, and mimic Ehlers-Danlos-like changes in bone and other connective tissues. J Bone Miner Res 17:1180–1189

D'Angelo M, Billings PC, Pacifici M, Leboy PS, Kirsch T (2001) Authentic matrix vesicles contain active metalloproteases (mmp). A role for matrix vesicle-associated mmp-13 in activation of transforming growth factor-beta. J Biol Chem 276:11347–11353

D'Souza RN, Cavender A, Sunavala G, Alvarez J, Ohshima T, Kulkarni AB, MacDougall M (1997) Gene expression patterns of murine dentin matrix protein 1 (Dmp1) and dentin sialophosphoprotein (DSPP) suggest distinct developmental functions in vivo. J Bone Miner Res 12:2040–2049

Daculsi G, Pouëzat J, Péru L, Maugars Y, LeGeros RZ (1992) Ectopic calcifications. In: Bonucci E (ed) Calcification in biological systems. CRC Press, Boca Raton, pp 365–397

de Bernard B, Gherardini M, Lunazzi GC, Modricky C, Moro L, Panfili E, Pollesello P, Stagni N, Vittur F (1985) Alkaline phosphatase of matrix vesicles from preosseous cartilage is a Ca^{++} binding glycoprotein. In: Butler WT (ed) The chemistry and biology of mineralized tissues. Ebsco Media, Birmingham, Al, pp 142–145

de Bernard B, Bianco P, Bonucci E, Costantini M, Lunazzi GC, Martinuzzi P, Modricky C, Moro L, Panfili E, Pollesello P, Stagni N, Vittur F (1986) Biochemical and immunohistochemical evidence that in cartilage an alkaline phosphatase is a Ca^{2+}-binding glycoprotein. J Cell Biol 103:1615–1623

De Pinieux G, Flam T, Zerbib M, Taupin P, Bellahcene A, Waltregny D, Vieillefond A, Poupon MF (2001) Bone sialoprotein, bone morphogenetic protein 6 and thymidine phosphorylase expression in localized human prostatic adenocarcinoma as predictors of clinical outcome: a clinicopathological and immunohistochemical study of 43 cases. J Urol 166:1924–1930

de Ricqlès A, Meunier FJ, Castanet J, Francillon-Vieillot H (1991) Comparative microstructure of bone. In: Hall BK (ed) Bone, vol. 3: Bone matrix and bone specific products. CRC Press, Boca Raton, pp 1–78

Dean DD, Schwartz ZVI, Muniz OE, Gomez R, Swain LD, Howell DS, Boyan BD (1992) Matrix vesicles contain metalloproteinases that degrade proteoglycans. Bone and Mineral 17:172–176

Dearden LC, Bonucci E, Cuicchio M (1974) An investigation of ageing in human costal cartilage. Cell Tissue Res 152:305–337

Delany AM, Dong Y, Canalis E (1994) Mechanisms of glucocorticoid action in bone cells. J Cell Biochem 56:295–302

Delany AM, Amling M, Priemel M, Howe C, Baron R, Canalis E (2000) Osteopenia and decreased bone formation in osteonectin-deficient mice. J Clin Invest 105:915–923

Dell'Orbo C, Gioglio L, Quacci D, Soldi C (1996) Evidence of rat tail tendon proteoglycans at emission field scanning electron microscpy. Eur J Histochem 40:125–128

Delmas PD, Wahner HW, Mann KG, Riggs BL (1983) Assessment of bone turnover in postmenopausal osteoporosis by measurement of serum bone Gla-protein. J Lab Clin Med 102:470–476

Delmas PD, Demiaux B, Malaval L, Chapuy MC, Edouard C, Meunier PJ (1986) Serum bone gamma carboxyglutamic acid-containing protein in primary hyperparathyroidism and in malignant hypercalcemia. Comparison with bone histomorphometry. J Clin Invest 77:985–991

Demetriou M, Binkert C, Sukhu B, Tenenbaum HC, Dennis JW (1996) Fetuin/α_2-HS glycoprotein is a transforming growth factor-β type II receptor mimic and cytokine antagonist. J Biol Chem 271:12755–12761

Deutsch D, Bab I, Muhlrad A, Sela J (1981) Purification and further characterization of isolated matrix vesicles from rat alveolar bone. Metab Bone Dis Rel Res 3:209–214

Deutsch D, Catalano-Sherman J, Dafni L, David S, Palmon A (1995) Enamel matrix proteins and ameloblast biology. Connect Tissue Res 32:97–107

Dhore CR, Cleutjens JP, Lutgens E, Cleutjens KB, Geusens PP, Kitslaar PJ, Tordoir JH, Spronk HM, Vermeer C, Daemen MJ (2001) Differential expression of bone matrix regulatory proteins in human atherosclerotic plaques. Arterioscler Thromb Vasc Biol 21:1998–2003

Diamond AG, Triffitt JT, Herring GM (1982) The acid macromolecules in rabbit cortical bone tissue. Arch Oral Biol 27:337–345

Dickson IR, Poole AR, Veis A (1975) Localization of plasma α_2HS glycoprotein in mineralizing human bone. Nature 256:430–432

Dirksen TR, Marinetti GV (1970) Lipids of bovine enamel and dentin and human bone. Calcif Tissue Res 6:1–10

Dodds RA, Connor JR, James IE, Rykaczewski EL, Appelbaum E, Dul E, Gowen M (1995) Human osteoclasts, not osteoblasts, deposit osteopontin onto resorption surfaces: an in vitro and ex vivo study of remodeling bone. J Bone Miner Res 10:1666–1680

Doi Y, Horiguchi T, Kim S-H, Moriwaki Y, Wakamatsu N, Adachi M, Ibaraki K, Moriyama K, Sasaki S, Shimokawa H (1992) Effects of non-collagenous proteins on the formation of apatite in calcium β-glycerophosphate solutions. Arch Oral Biol 37:15–21

Doi Y, Horiguchi T, Kim S-H, Moriwaki Y, Wakamatsu N, Adachi M, Shigeta H, Sasaki S, Shimokawa H (1993) Immobilized DPP and other proteins modify OCP formation. Calcif Tissue Int 52:139–145

Domenicucci C, Goldberg HA, Hofmann T, Isenman D, Wasi S, Sodek J (1988) Characterization of porcine osteonectin extracted from foetal calvariae. Biochem J 253:139–151

Doty SB, Schofield BH (1990) Histochemistry and enzymology of bone-forming cells. In: Hall BK (ed) Bone. Vol. 1: The osteoblast and osteocyte. The Telford Press, Caldwell, pp 71–102

Dowd TL, Rosen JF, Li L, Gundberg CM (2003) The three-dimensional structure of bovine calcium ion-bound osteocalcin using 1H NMR spectroscopy. Biochemistry 42:7769–7779

Ducy P, Desbois C, Boyce B, Pinero G, Story B, Dunstan C, Smith E, Bonadio J, Goldstein S, Gundberg C, Bradley A, Karsenty G (1996) Increased bone formation in osteocalcin-deficient mice. Nature 382:448–452

Dyer RF, Enna CD (1976) Ultrastructural features of adult human tendon. Cell Tissue Res 168:247–259

Dziadek M, Paulsson W, Aumailley M, Timpl R (1986) Purification and tissue distribution of a small protein (BM-40) extracted from a basement membrane tumor. Eur J Biochem 161:455–464

Dziak R (1992) Role of lipids in osteogenesis: cell signaling and matrix calcification. In: Bonucci E (ed) Calcification in biological systems. CRC Press, Boca Raton, pp 59–71

Dziegielewska KM, Mollgard K, Reynolds ML, Saunders NR (1987) A fetuin-related glycoprotein (α_2HS) in human embryonic and fetal development. Cell Tissue Res 248:33–41

Eastoe JE, Eastoe B (1954) The organic constituents of mammalian compact bone. Biochem J 57:453–459

Ecarot-Charrier B, Glorieux FH, van der Rest M, Pireira G (1983) Osteoblasts isolated from mouse calvaria initiate matrix mineralization in culture. J Cell Biol 96:639–643

Ecarot-Charrier B, Shepard N, Charette G, Grynpas M, Glorieux FH (1988) Mineralization in osteoblast cultures: a light and electron microscopic study. Bone 9:147–154

El-Maadawy S, Kaartinen MT, Schinke T, Murshed M, Karsenty G, McKee MD (2003) Cartilage formation and calcification in arteries of mice lacking matrix Gla protein. Connect Tissue Res 44:272–278

Engfeldt B, Hjerpe A (1976) Glycosaminoglycans and proteoglycans of human bone tissue at different stages of mineralization. Acta Pathol Microbiol Scand Sect A 84:95–106

Engfeldt B, Reinholt FP (1992) Structure and calcification of epiphyseal growth cartilage. In: Bonucci E (ed) Calcification in biological systems. CRC Press, Boca Raton, pp 217–241

Eyre D (2002) Collagen of articular cartilage. Arthritis Res 4:30–35

Fedarko NS, Jain A, Karadag A, Fisher LW (2004) Three small integrin binding ligand N-linked glycoproteins (SIBLINGs) bind and activate specific matrix metalloproteinases. FASEB J 18:734–736

Fen JQ, Zhang J, Dallas SL, Lu Y, Chen S, Tan X, Owen M, Harris SE, MacDougall M (2002) Dentin matrix protein 1, a target molecule for Cbfa 1 in bone, is a unique bone marker gene. J Bone Miner Res 17:1822–1831

Feng JQ, Huang H, Lu Y, Ye L, Xie Y, Tsutsui TW, Kunieda T, Castranio T, Scott G, Bonewald LB, Mishina Y (2003) The Dentin matrix protein 1 (Dmp1) is specifically expressed in mineralized, but not soft, tissues during development. J Dent Res 82:776–780

Fiore CE, Falcidia E, Foti R, Caschetto S, Grimaldi DR (1987) Postoophorectomy bone loss is associated with reduced bone Gla-protein serum levels: a possible effect of osteoblastic insufficiency. Calcif Tissue Int 41:303–306

Fischer JW, Steiz S, Johnson P, Burke A, Kolodgie F, Virmani R, Giachelli C, Wight TN (2004) Decorin promotes aortic smooth muscle cell calcification and colocalizes to calcified regions in human atherosclerotic lesions. Arterioscler Thromb Vasc Biol 24:2391–2396

Fisher LW (1985) The nature of the proteoglycans of bone. In: Butler WT (ed) The chemistry and biology of mineralized tissues. Ebsco Media, Inc., Birmingham, pp 188–196

Fisher LW, Fedarko NS (2003) Six genes expressed in bones and teeth encode the current members of the SIBLING family of proteins. Connect Tissue Res 44:33–40

Fisher LW, Schraer H (1982) Keratan sulfate proteoglycan isolated from the estrogen-induced medullary bone in Japanese quail. Comp Biochem Physiol B 72:227–232

Fisher LW, Termine JD (1985) Noncollagenous proteins influencing the local mechanisms of calcification. Clin Orthop Relat Res 200:362–385

Fisher LW, Termine JD, Dejter SW Jr, Whitson SW, Yanagishita M, Kimura JH, Hascall VC, Kleinman HK, Hassell JR, Nilsson B (1983a) Proteoglycans of developing bone. J Biol Chem 258:6588–6594

Fisher LW, Whitson SW, Avioli LV, Termine JD (1983b) Matrix sialoprotein of developing bone. J Biol Chem 258:12723–12727

Fisher LW, Torchia DA, Fohr B, Young MF, Fedarko NS (2001) Flexible Structures of SIBLING Proteins, Bone Sialoprotein, and Osteopontin. Biochem Biophys Res Commun 280:460–465

Fitton Jackson S (1956) The morphogenesis of avian tendon. Proc R Soc London B 144:556–572

Floege J, Ketteler M (2004) Vascular calcification in patients with end-stage renal disease. Nephrol Dial Transplant 19:V59–66

François C (1960) La composition minérale et le degré de minéralisation d'un os jeune: l'os folliculinique du pigeon. Bull Soc Chim Biol 42:259–267

Franzén A, Heinegård D (1984a) Extraction and purification of proteoglycans from mature bovine bone. Biochem J 224:47–58

Franzén A, Heinegård D (1984b) Characterization of proteoglycans from the calcified matrix of bovine bone. Biochem J 224:59–66
Franzén A, Oldberg A, Solursh M (1989) Possible recruitment of osteoblastic precursor cells from hypertrophic chondrocytes during initial osteogenesis in cartilaginous limbs of young rats. Matrix 9:261–265
Fratzl P, Fratzl-Zelman N, Klaushofer K (1993) Collagen packing and mineralization. An X-ray scattering investigation of turkey leg tendon. Biophys J 54:260–266
Fujisawa R, Nodasaka Y, Kuboki Y (1995) Further characterization of interaction between bone sialoprotein (BSP) and collagen. Calcif Tissue Int 56:140–144
Fukumoto S, Yamashita T (2002) Fibroblast growth factor-23 is the phosphaturic factor in tumor-induced osteomalacia and may be phosphatonin. Curr Opin Nephrol Hypertens 11:385–389
Gabbiani G, Tuchweber B (1970) Studies on the mechanism of calcergy. Clin Orthop Relat Res 69:66–74
Gadeau AP, Chaulet H, Daret D, Kockx M, Daniel-Lamaziere JM, Desgranges C (2001) Time course of osteopontin, osteocalcin, and osteonectin accumulation and calcification after acute vessel wall injury. J Exp Med 192
Ganns B, Kim RH, Sodek J (1999) Bone sialoprotein. Crit Rev Oral Biol Med 10:79–98
Garnero P, Delmas PD (1996) New developments in biochemical markers for osteoporosis. Calcif Tissue Int 59: S2-S9
Gay CV, Schraer H, Hargest TE Jr (1978) Ultrastructure of matrix vesicles and mineral in unfixed embryonic bone. Metab Bone Dis Rel Res 1:105–108
George A, Sabsay B, Simonian PA, Veis A (1993) Characterization of a novel dentin matrix acidic phosphoprotein. Implications for induction of biomineralization. J Biol Chem 268:12624–12630
Gerstenfeld LC (1999) Osteopontin in skeletal tissue homeostasis: an emerging picture of the autocrine/paracrine functions of the extracellular matrix. J Bone Miner Res 14:850–855
Gerstenfeld LC, Feng M, Gotoh Y, Glimcher MJ (1994) Selective extractability of noncollagenous proteins from chicken bone. Calcif Tissue Int 55:230–235
Giachelli CM, Speer MY, Li X, Rajachar RM, Yang H (2005) Regulation of vascular calcification. Roles of phosphate and osteopontin. Circ Res 96:717–722
Giannini S, D'Angelo A, Carraro G, Antonello A, Di Landro D, Marchini F, Plebani M, Zaninotto M, Rigotti P, Sartori L, Crepaldi G (2001) Persistently increased bone turnover and low bone density in long-term survivors to kidney transplantation. Clin Nephrol 56:353–363
Giraud-Guille M-M (1994) Liquid crystalline order of biopolymers in cuticles and bones. Microsc Res Tech 27:420–428
Giraud-Guille M-M, Besseau L, Chopin C, Durand P, Herbage D (2000) Structural aspects of fish skin collagen which forms ordered arrays via liquid crystalline states. Biomaterials 21:899–906
Glimcher MJ (1976) Composition, structure, and organization of bone and other mineralized tissues and the mechanism of calcification. In: Greep RO, Astwood EB (eds) Handbook of physiology: endocrinology. American Physiological Society, Washington, pp 25–116
Glimcher MJ (1989) Mechanism of calcification: role of collagen fibrils and collagen-phosphoprotein complexes in vitro and in vivo. Anat Rec 224:139–153
Glimcher MJ (1990) The nature of the mineral component of bone and the mechanism of calcification. In: Avioli LV, Krane SM (eds) Metabolic bone disease and clinically related disorders. W.B. Saunders Company, Philadelphia, pp 42–68
Glimcher MJ (1992) The nature of the mineral component of bone and the mechanism of calcification. In: Coe FL, Favus MJ (eds) Disorders of bone and mineral metabolism. Raven Press, New York, pp 265–286

Glimcher MJ, Katz EP (1965) The organization of collagen in bone: the role of noncovalent bonds in the relative insolubility of bone collagen. J Ultrastruct Res 12:705–729

Glimcher MJ, Krane SM (1968) The organization and structure of bone, and the mechanism of calcification. In: Gould BS (ed) Biology of collagen. Academic Press, London, pp 67–251

Glimcher MJ, Lefteriou B, Kossiva D (1979a) Identification of O-phosphoserine, O-phosphothreonine and γ-carboxyglutamic acid in the non-collagenous proteins of bovine cementum; comparison with dentin, enamel and bone. Calcif Tissue Int 28:83–86

Glimcher MJ, Brickley-Parsons D, Kossiva D (1979b) Phosphopeptides and γ-carboxyglutamic acid-containing peptides in calcified turkey tendon: their absence in uncalcified tendon. Calcif Tissue Int 27:281–284

Glimcher MJ, Kossiva D, Brickley-Parsons D (1984) Phosphoproteins of chicken bone matrix. Proof of synthesis in bone tissue. J Biol Chem 259:290–293

Goldberg HA, Hunter GK (1995) The inhibitory activity of osteopontin on hydroxyapatite formation *in vitro*. Ann N Y Acad Sci 760:305–308

Goldberg HA, Domenicucci C, Pringle GA, Sodek J (1988) Mineral-binding proteoglycans of fetal porcine calvarial bone. J Biol Chem 263:12092–12101

Goldberg HA, Warner KJ, Stillman MJ, Hunter GK (1996) Determination of the hydroxyapatite-nucleating region of bone sialoprotein. Connect Tissue Res 35:385–392

Goldberg HA, Hunter GK, Mundy MA, Warner KJ, McKee MD (2000) Nature of hydroxyapatite crystals formed in the presence of bone sialoprotein, osteopontin and synthetic homopolymer analogues. In: Goldberg M, Boskey A, Robinson C (eds) Chemistry and biology of mineralized tissues. American Academy of Orthopaedic Surgeons, Rosemont, IL, pp 225–228

Gorski JP (1992) Acidic phosphoproteins from bone matrix: a structural rationalization of their role in mineralization. Calcif Tissue Int 50:391–396

Gorski JP (1998) Is all bone the same? Distinctive distributions and properties of non-collagenous matrix proteins in lamellar vs. woven bone imply the existence of different underlying osteogenic mechanisms. Crit Rev Oral Biol Med 9:201–223

Gorski JP, Shimizu K (1988) Isolation of new phosphorylated glycoprotein from mineralized phase of bone that exhibits limited homology to adhesive protein osteopontin. J Biol Chem 263:15938–15945

Gorski JP, Griffin D, Dudley G, Stanford C, Thomas R, Huang C, Lai EL, Karr B, Solursh M (1990) Bone acidic glycoprotein-75 is a major synthetic product of osteoblastic cells and localized as 75- and/or 50-kDa forms in mineralized phases of bone and growth plate and in serum. J Biol Chem 265:14956–14963

Gorski JP, Kremer EA, Chen Y, Ryan S, Fullenkamp C, Delviscio J, Jensen K, McKee MD (1997) Bone acidic glycoprotein-75 self-associates to form macromolecular complexes in vitro and in vivo with the potential to sequester phosphate ions. J Cell Biochem 64:547–564

Gorski JP, Wang A, Lovitch D, Law D, Powell K, Midura RJ (2004) Extracellular bone acidic glycoprotein-75 defines condensed mesenchyme regions to be mineralized and localizes with bone sialoprotein during intramembranous bone formation. J Biol Chem 279:25455–25463

Gotoh Y, Gerstenfeld LC, Glimcher MJ (1990) Identification and characterization of the major bone specific phosphoprotein synthesized by cultured embryonic chick osteoblasts. Eur J Biochem 187:49–58

Gotoh Y, Fujisawa K, Satomura K, Nagayama M (1995) Osteogenesis by human osteoblastic cells in diffusion chamber *in vivo*. Calcif Tissue Int 56:246–251

Graf K, Do YS, Ashizawa N, Meehan WP, Giachelli CM, Marboe CC, Fleck E, Hsueh WA (1997) Myocardial osteopontin expression is associated with left ventricular hypertrophy. Circulation 96:3063–3071

Groot CG (1982) An electron microscopic examination for the presence of acid groups in the organic matrix of mineralization nodules in foetal bone. Metab Bone Dis Rel Res 4:77–84

Groot CG, Danes JK, Blok J, Hoogendijk A, Hauschka PV (1986) Light and electron microscopic demonstration of osteocalcin antigenicity in embryonic and adult rat bone. Bone 7:379–385

Grynpas M (1977) Three-dimensional packing of collagen in bone. Nature 265:381–382

Gundberg CM (1998) Biology, physiology, and clinical chemistry of osteocalcin. J Clin Ligand Assay 21:128–138

Hao J, Zou B, Narayanan K, George A (2004) Differential expression patterns of the dentin matrix proteins during mineralized tissue formation. Bone 34:921–932

Harris NL, Rattray KR, Tye CE, Underhill TM, Somerman MJ, D'Errico JA, Chambers AF, Hunter GK, Goldberg HA (2000) Functional analysis of bone sialoprotein: identification of the hydroxyapatite-nucleating and cell-binding domains by recombinant peptide expression and site-directed mutagenesis. Bone 27:795–802

Hass GM, Trueheart RE, Taylor B, Stumpe M (1958) An experimental histologic study of hypervitaminosis D. Am J Pathol 34:395–431

Haumont S, Ngoma Z (1973) Histochimie des lipides dans l'os compact humain. Acta Orthop Belg 39:460–470

Hauschka PV (1985) Osteocalcin and its functional domains. In: Butler WT (ed) The chemistry and biology of mineralized tissues. Ebsco Media, Birmingham, AL, pp 149–158

Hauschka PV, Wians FH Jr (1989) Osteocalcin-hydroxyapatite interaction in the extracellular organic matrix of bone. Anat Rec 224:180–188

Hauschka PV, Lian JB, Gallop PM (1975) Direct identification of the calcium-binding amino acid, γ-carboxyglutamate, in mineralized tissue. Proc Nat Acad Sci USA 72:3925–3929

Hayashibara T, Hiraga T, Yi B, Nomizu M, Kumagai Y, Nishimura R, Yoneda T (2004) A synthetic peptide fragment of human MEPE stimulates new bone formation *in vitro* and *in vivo*. J Bone Miner Res 19:455–462

He G, George A (2004) Dentin matrix protein 1 immobilized on type I collagen fibrils facilitates apatite deposition *in vitro*. J Biol Chem 279:11649–11656

He G, Dahl T, Veis A, George A (2003a) Dentin matrix protein 1 initiates hydroxyapatite formation *in vitro*. Connect Tissue Res 44:240–245

He G, Dahl T, Veis A, George A (2003b) Nucleation of apatite crystals in vitro by self-assembled dentin matrix protein 1. Nat Mater 2:552–558

Hedlund H, Mengarelli-Widholm S, Heinegård D, Reinholt FP, Svensson O (1994) Fibromodulin distribution and association with collagen. Matrix Biol 14:227–232

Heiss A, DuChesne A, Denecke B, Grötzinger J, Yamamoto K, Renné T, Jahnen-Dechent W (2003) Structural basis of calcification. Inhibition by α_2-HS-glycoprotein/fetuin-A. A formation of colloidal calciprotein particles . J Biol Chem 278:13333–13341

Hoang QQ, Sicheri F, Howard AJ, Yang DS (2003) Bone recognition mechanism of porcine osteocalcin from crystal struture. Nature 425:977–980

Hodge AJ, Petruska JA (1963) Recent studies with the electron microscope on ordered aggregates of the tropocollagen molecules. In: Ramachandran GN (ed) Aspects of protein structure. Academic Press, London, pp 289–300

Hodge AJ, Petruska JA, Bailey AJ (1965) The subunit structure of the tropocollagen macromolecule and its relation to various ordered aggregation states. In: Fitton Jackson S, Harkness RD, Partridge SM, Tristram GR (eds) Structure and function of connective and skeletal tissue. Butterworths, London, pp 31–41

Höhling HJ, Barckhaus RH, Krefting E-R, Althoff J, Quint P (1990) Collagen mineralization: aspects of the structural relationship between collagen and the apatitic crystallites.

In: Bonucci E, Motta PM (eds) Ultrastructure of skeletal tissues. Kluwer Academic Publishers, Boston, pp 41–62

Hofbauer LC, Khosla S, Dunstan CR Lacey DL, Boyle WJ, Riggs BL (2000) The roles of osteoprotegerin and osteoprotegerin ligand in the paracrine regulation of bone resorption. J Bone Miner Res 15:2–12

Holland PWH, Harper SJ, McVaj JH, Hogan BLM (1987) *In vivo* expression of mRNA for the Ca^{++}-binding protein SPARC (osteonectin) revealed by *in situ* hybridization. J Cell Biol 105:473–482

Hosemann R, Dreissig W, Nemetschek T (1974) Schachtelhalm-structure of the octafibrils in collagen. J Mol Biol 83:275–280

Hoshi K, Ozawa H (2000) Matrix vesicle calcification in bones of adult rats. Calcif Tissue Int 66:430–434

Hoshi K, Kemmotsu S, Takeuchi Y, Amizuka N, Ozawa H (1999) The primary calcification in bones follows removal of decorin and fusion of collagen fibrils. J Bone Miner Res 14:273–280

Hoshi K, Ejiri S, Ozawa H (2001) Localizational alterations of calcium, phosphorus, and calcification-related organics such as proteoglycans and alkaline phosphatase during bone calcification. J Bone Miner Res 16:289–298

Hulmes DJS (2002) Building collagen molecules, fibrils, and suprafibrillar structures. J Struct Biol 137:2–10

Hulmes DJS, Miller A (1979) A quasihexagonal model for the molecular packing of collagen. Nature 282:878–880

Hultenby K, Reinholt FP, Oldberg Å, Heinegård D (1991) Ultrastructural immunolocalization of osteopontin in metaphyseal and cortical bone. Matrix 11:206–213

Hultenby K, Reinholt FP, Norgård M, Oldberg Å, Wendel M, Heinegård D (1994) Distribution and synthesis of bone sialoprotein in metaphyseal bone of young rats show a distinctly different pattern from that of osteopontin. Eur J Cell Biol 63:230–239

Hunter GK, Goldberg HA (1993) Nucleation of hydroxyapatite by bone sialoprotein. Proc Natl Acad Sci U S A 90:8562–8565

Hunter GK, Goldberg HA (1994) Modulation of crystal formation by bone phosphoproteins: Role of glutamic acid-rich sequences in the nucleation of hydroxyapatite by bone sialoprotein. Biochem J 302:175–179

Hunter GK, Hauschka PV, Poole AR, Rosenberg LC, Goldberg HA (1996) Nucleation and inhibition of hydroxyapatite formation by mineralized tissue proteins. Biochem J 317:59–64

Ibrahim T, Leong I, Sanchez-Sweatman O, Khokha R, Sodek J, Tenenbaum HC, Ganss B, Cheifetz S (2000) Expression of bone sialoprotein and osteopontin in breast cancer bone metastases. Clin Exp Metastasis 18:253–260

Ihara H, Denhardt DT, Furuya K, Yamashita T, Muguruma Y, Tsuji K, Hruska KA, Higashio K, Enomoto S, Nifuji A, Rittling SR, Noda M (2001) Parathyroid hormone-induced bone resorption does not occur in the absence of osteopontin. J Biol Chem 276:13065–13071

Ingram RT, Clarke BL, Fisher LW, Fitzpatrick LA (1993) Distribution of noncollagenous proteins in the matrix of adult human bone: evidence of anatomic and functional heterogeneity. J Bone Miner Res 8:1019–1029

Ippolito E, Natali PG, Postacchini F, Accinni L, De Martino C (1980) Morphological, immunochemical, and biochemical study of rabbit Achilles tendon at various ages. J Bone Joint Surg 62-A:583–598

Irie K, Zalzal S, Ozawa H, McKee MD, Nanci A (1998) Morphological and immunocytochemical characterization of primary osteogenic cell cultures derived from fetal rat cranial tissue. Anat Rec 252:554–567

Irving JT (1959) A histological staining method for site of calcification in teeth and bone. Arch Oral Biol 1:89–96

Irving JT (1960) Histochemical changes in the early stages of calcification. Clin Orthop Relat Res 17:92–102
Ishigaki R, Takagi M, Igarashi M, Ito K (2002) Gene expression and immunohistochemical localization of osteonectin in association with early bone formation in the developing mandible. Histochem J 34:57–66
Ito S, Saito T, Amano K (2004) In vitro apatite induction by osteopontin: interfacial energy for hydroxyapatite nucleation on osteopontin. J Biomed Mater Res A 69:11–16
James IE, Dodds RA, Lee-Rykaczewski E, Eichman CF, Connor JR, Hart TK, Maleeff BE, Lackman RD, Gowen M (1996) Purification and characterization of fully functional human osteoclast precursors. J Bone Miner Res 11:1608–1618
Johnson LC (1960) Mineralization of turkey leg tendon. I. Histology and histochemistry of mineralization. In: Sognnaes RF (ed) Calcification in biological systems. American Association for the Advancement of Sciences, Washington, pp 117–128
Jundt G, Berghäuser K-H, Termine JD, Schulz A (1987) Osteonectin - a differentiation marker of bone cells. Calcif Tissue Int 248:409–415
Kasugai S, Nagata T, Sodek J (1992) Temporal studies on the tissue compartmentalization of bone sialoprotein (BSP), osteopontin (OPN) and SPARC protein during bone formation in vitro. J Cell Physiol 152:467–477
Katchburian E, Severs NJ (1983) Matrix constituents of early developing bone examined by freeze fracture. Cell Biol Int Rep 7:1063–1070
Katsura N, Tanaka O, Yokoyama M (1991) Three dimensional structure of type I collagen and mineralization. Connect Tissue 22:92–98
Katz EP, Li S-T (1972) The molecular packing of collagen in mineralized and non-mineralized tissues. Biochem Biophys Res Comm 46:1368–1373
Katz EP, Li S-T (1973a) Structure and function of bone collagen fibrils. J Mol Biol 80:1–15
Katz EP, Li ST (1973b) The intermolecular space of reconstituted collagen fibrils. J Mol Biol 73:351–369
Ketteler M (2005) Fetuin-A and extraosseous calcification in uremia. Curr Opinion Nephrol Hypert 14:337–342
Ketteler M, Westenfeld R, Schlieper G, Brandenburg V, Floege J (2005) "Missing" inhibitors of calcification: general aspects and implications in renal failure. Pediatr Nephrol 20:383–388
Knott L, Bailey AJ (1998) Collagen cross-links in mineralizing tissues: a review of their chemistry, function, and clinical relevance. Bone 22:181–187
Kobayashi S (1971) Acid mucopolysaccharides in calcified tissues. Int Rev Cytol 30:257–371
Kohler DM, Crenshaw MA, Arsenault AL (1995) Three-dimensional analysis of mineralizing turkey leg tendon: matrix vesicle-collagen relationships. Matrix Biol 14:543–552
Kuboki Y, Takita H, Komori T, Mizuno M, Furu-uchi E, Taniguchi K (1989) Separation of bone matrix proteins by calcium-induced precipitation. Calcif Tissue Int 44:269–277
Kubota T, Sato K, Kawano H, Yamamoto S, Hirano A, Hashizume Y (1984) Ultrastructure of early calcification in cervical ossification of the posterior longitudinal ligament. J Neurosurg 61:131–135
Kuzela DC, Huffer WE, Conger JD, Winter SD, Hammond WS (1977) Soft tissue calcification in chronic dialysis patients. Am J Pathol 86:403–424
Lacey DL, Tan HL, Lu J, Kaufman S, Van G, Qiu W, Rattan A, Scully S, Fletcher F, Juan T, Kelley M, Burgess TL, Boyle WJ, Polverino AJ (1998) Osteoprotegerin ligand modulates murine osteoclast survival in vitro and in vivo. Am J Pathol 157:435–448
Landis WJ (1986) A study of calcification in the leg tendons from the domestic turkey. J Ultrastruct Mol Struct Res 94:217–238
Landis WJ, Arsenault AL (1989) Vesicle- and collagen-mediated calcification in the turkey leg tendon. Connect Tissue Res 22:35–42

Landis WJ, Silver FH (2002) The structure and function of normally mineralizing avian tendons. Comp Biochem Physiol A Mol Integr Physiol 133:1135–1157

Landis WJ, Song MJ (1991) Early mineral deposition in calcifying tendons characterized by high voltage electron microscopy and three-dimensional graphic imaging. J Struct Biol 107:116–127

Landis WJ, Paine MC, Glimcher MJ (1977) Electron microscopic observations of bone tissue prepared anhydrously in organic solvents. J Ultrastruct Res 59:1–30

Landis WJ, Hodgens KJ, McKee MD, Nanci A, Song MJ, Kiyonaga S, Arena J, McEwen B (1992) Extracellular vesicles of calcifying turkey leg tendon characterized by immunocytochemistry and high voltage electron microscopic tomography and 3-D graphic image reconstruction. Bone Miner 17:237–241

Landis WJ, Jacquet R, Hillyer J, Zhang J (2003) Analysis of osteopontin in mouse growth plate cartilage by application of laser capture microdissection and RT-PCR. Connect Tissue Res 44:28–32

Leaver AG, Triffitt JT, Holbrook IB (1975) Newer knowledge of non-collagenous protein in dentin and cortical bone matrix. Clin Orthop Relat Res 110:269–292

Leblond CP, Lacroix P, Ponlot R, Dhem A (1959) Les stades initiaux de l'ostéogenèse. Nouvelles données histochimique et autoradiographiques. Bull Acad R Med Belgique 24:421–443

Lee I, Ono Y, Lee A, Omiya K, Moriya Y, Takagi M (1998) Immunocytochemical localization and biochemical characterization of large proteoglycans in developing rat bone. J Oral Sci 40:77–87

Lee J, Scheraga HA, Rackovsky S (1996) Computational study of packing a collagen-like molecule: quasi-hexagonal vs "Smith" collagen microfibril model. Biopolymers 40:595–607

Lee SL, Glimcher MJ (1981) Purification, composition, and ^{31}P NMR spectroscopic properties of a noncollagenous phosphoprotein isolated from chicken bone matrix. Calcif Tissue Int 33:385–394

Lees S (1981) A mixed packing model for bone collagen. Calcif Tissue Int 33:591–602

Lees S (1987) Considerations regarding the structure of the mammalian mineralized osteoid from viewpoint of the generalized packing model. Connect Tissue Res 16:281–303

Lees S (1998) Interpreting the equatorial diffraction pattern of collagenous tissues in the light of molecular motion. Biophys J 75:1058–1061

Lees S, Hukins DWL (1992) X-ray diffraction by collagen in the fully mineralized cortical bone of cow tibia. Bone and Mineral 17:59–63

Lees S, Prostak KS, Ingle VK, Kjoller K (1994) The loci of mineral in turkey leg tendon as seen by atomic force microscope and electron microscopy. Calcif Tissue Int 55:180–189

Likins RC, Piez KA, Kunde ML (1960) Mineralization of turkey leg tendon. III. Chemical nature of the protein and mineral phases. In: Sognnaes RF (ed) Calcification in biological systems. American Association for the Advancement of Sciences, Washington, pp 143–149

Lillie JH, MacCallum DK, Scaletta LJ, Occhino JC (1977) Collagen structure: evidence for a helical organization of the collagen fibril. J Ultrastruct Res 58:134–143

Ling Y, Rios HF, Myers ER, Lu Y, Feng JQ, Boskey AL (2005) DMP1 depletion decreases bone mineralization in vivo: an FTIR imaging analysis. J Bone Miner Res 20:2169–2177

Lomashvili KA, Cobbs S, Hennigar RA, Hardcastle KI, O'Neill WC (2004) Phosphate-induced vascular calcification: role of pyrophosphate and osteopontin. J Am Soc Nephrol 15:1392–1401

Lowther DA, Natarajan M (1972) The influence of glycoprotein on collagen fibril formation in the presence of chondroitin sulfate proteoglycan. Biochem J 127:607–608

Luo G, Ducy P, McKee MD, Pinero GJ, Loyer E, Behringer RR, Karsenty G (1997) Spontaneous calcification of arteries and cartilage in mice lacking GLA protein. Nature 386:78–81

MacDougall M, Gu TT, Luan X, Simmons D, Chen J (1998) Identification of a novel isoform of mouse dentin matrix protein 1: spatial expression in mineralized tissues. J Bone Miner Res 13:422–431

MacDougall M, Simmons D, Gu TT, Dong J (2002) MEPE/OF45, a new dentin/bone matrix protein and candidate gene for dentin diseases mapping to chromosome 4q21. Connect Tiss Res 43:320–330

MacNeil RL, Berry J, D'Errico J, Strayhorn C, Piotrowski B, Somerman MJ (1995) Role of two mineral-associated adhesion molecules, osteopontin and bone sialoprotein, during cementogenesis. Connect Tissue Res 33:1–7

Maeno M, Taguchi M, Kosuge K, Otsuka K, Takagi M (1992) Nature and distribution of mineral-binding, keratan sulfate-containing glycoconjugates in rat and rabbit bone. J Histochem Cytochem 40:1779–1788

Malluche HH, Faugere M-C, Fanti P, Price PA (1984) Plasma levels of bone Gla-protein reflect bone formation in patients on chronic maintenance dialysis. Kidney Int 26:869–874

Mann S (1988) Molecular recognition in biomineralization. Nature 332:119–124

Mark MP, Prince CW, Gay S, Austin RL, Bhown M, Finkelman RD, Butler WT (1987a) A comparative immunocytochemical study on the subcellular distributions of 44 kDa bone phosphoprotein and bone γ-carboxyglutamic acid (Gla)-containing protein in osteoblasts. J Bone Miner Res 2:337–346

Mark MP, Prince CW, Oosawa T, Gay S, Bronckers AL, Butler WT (1987b) Immunohistochemical demonstration of a 44-KD phosphoprotein in developing rat bone. J Histochem Cytochem 35:707–715

Mark MP, Butler WT, Prince CW, Finkelman RD, Ruch J-V (1988) Developmental expression of 44-kDa bone phosphoprotein (osteopontin) and bone γ-carboxyglutamic acid (Gla)-containing protein (osteocalcin) in calcifying tissues of rat. Differentiation 37:123–136

Marks SC Jr (1997) The structural basis for bone cell biology. A review. Acta Med Dent Helv 2:141–157

Marks SC Jr, Popoff SN (1988) Bone cell biology: the regulation of development, structure, and function in the skeleton. Am J Anat 183:1–44

Marotti G (2000) The osteocyte as a wiring transmission system. J Muscoloskel Neuron Interact 1:133–136

Martin GR, Piez KA, Lewis MS (1963) The incorporation of [^{14}C] glycine into the subunits of collagens from normal and lathyritic animals. Biochim Biophys Acta 69:472–479

Marvaso V, Bernard GW (1977) Initial intramembraneous osteogenesis in vitro. Am J Anat 149:453–468

Masi L, Brandi ML, Gehron Robey P, Crescioli C, Calvo JC, Bernabei P, Kerr JM, Yanagishita M (1995) Biosynthesis of bone sialoprotein by a human osteoclast-like cell line (FLG 29.1). J Bone Miner Res 10:187–196

Mason IJ, Murphy D, Munke M, Franke U, Elliott RW, Hogan BLM (1986) Developmental and transformation sensitive expression of the SPARC gene on mouse chromosome 11. EMBO J 5:1831–1837

Matsuzawa T, Anderson HC (1971) Phosphatases of epiphyseal cartilage studied by electron microscopic cytochemical methods. J Histochem Cytochem 19:801–809

Maurer P, Hohenadl C, Hohenester E, Göhring W, Timpl R, Engel J (1995) The C-terminal portion of BM-40 (SPARC/osteonectin) is an autonomously folding and crystallisable domain that binds calcium and collagen IV. J Mol Biol 253:347–357

McKee MD, Nanci A (1995a) Osteopontin and the bone remodeling sequence. Colloidal-gold immunocytochemistry of an interfacial extracellular matrix protein. Ann New York Acad Sci 760:177–189

McKee MD, Nanci A (1995b) Postembedding colloidal-gold immunocytochemistry of non-collagenous extracellular matrix proteins in mineralized tissues. Microsc Res Techn 31:44–62

McKee MD, Nanci A (1996a) Osteopontin deposition in remodeling bone: an osteoblast mediated event. J Bone Miner Res 11:873–874

McKee MD, Nanci A (1996b) Osteopontin: an interfacial extracellular matrix protein in mineralized tissues. Connect Tissue Res 35:197–205

McKee MD, Nanci A (1996c) Osteopontin at mineralized tissue interfaces in bone, teeth, and osseointegrated implants: ultrastructural distribution, and implications for mineralized tissue formation, turnover, and repair. Microsc Res Techn 33:141–164

McKee MD, Nanci A (1996d) Secretion of osteopontin by macrophages and its accumulation at tissue surfaces during wound healing in mineralized tissues: a potential requirement for macrophage adhesion and phagocytosis. Anat Rec 245:394–409

McKee MD, Nanci A, Landis WJ, Gotoh Y, Gerstenfeld LC, Glimcher MJ (1990) Developmental appearance and ultrastructural immunolocalization of a major 66kDa phosphoprotein in embryonic and post-natal chicken bone. Anat Rec 228:77–92

McKee MD, Nanci A, Landis WJ, Gotoh Y, Gerstenfeld LC, Glimcher MJ (1991) Effects of fixation and demineralization on the retention of bone phosphoprotein and other matrix components as evaluated by biochemical analyses and quantitative immunocytochemistry. J Bone Miner Res 6:937–945

McKee MD, Glimcher MJ, Nanci A (1992) High-resolution immunolocalization of osteopontin and osteocalcin in bone and cartilage during endochondral ossification in the chicken tibia. Anat Rec 234:479–492

McKee MD, Farach-Carson MC, Butler WT, Hauschka PV, Nanci A (1993) Ultrastructural immunolocalization of noncollagenous (osteopontin and osteocalcin) and plasma (albumin and α_2HS-glycoprotein) proteins in rat bone. J Bone Miner Res 8:485–496

McLean FC, Urist MR (1968) Bone, 3rd edn. The University of Chicago Press, Chicago

Menanteau J, Neuman WF, Neuman MW (1982) A study of bone proteins which can prevent hydroxyapatite formation. Metab Bone Dis Rel Res 4:157–162

Metsäranta M, Young MF, Sandberg M, Termine J, Vuorio E (1989) Localization of osteonectin expression in human fetal skeletal tissues by *in situ* hybridization. Calcif Tissue Int 45:146–152

Meyer K, Davidson E, Linker A, Hoffman P (1956) The acid mucopolysaccharides of connective tissue. Biochim Biophys Acta 21:506–518

Midura RJ, Wang A, Lovitch D, Law D, Powell K, Gorski JP (2004) Bone acidic glycoprotein-75 delineates the extracellular sites of future bone sialoprotein accumulation and apatite nucleation in osteoblastic cultures. J Biol Chem 279:25464–25473

Miller A (1976) Molecular packing in collagen fibrils. In: Ramachandran GN, Reddi AH (eds) Biochemistry of collagen. Plenum Press, New York, pp 85–136

Miller A, Tocchetti D (1981) Calculated X-ray diffraction pattern from a quasi hexagonal model for the molecular arrangement in collagen. Int J Biol Macromol 3:9–18

Miller A, Wray JS (1971) Molecular packing in collagen. Nature 230:437–439

Mirams M, Robinson BG, Mason RS, Nelson AE (2004) Bone as a source of FGF23: regulation by phosphate? Bone 35:1192–1199

Mizuno M, Farach-Carson MC, Pinero GJ, Fujisawa R, Brunn JC, Seyer JM, Bousfield GR, Mark MP, Butler WT (1991) Identification of the rat bone 60K acidic glycoprotein as α_2HS-glycoprotein. Bone Miner 13:1–21

Mocetti P, Ballanti P, Zalzal S, Silvestrini G, Bonucci E, Nanci A (2000) A histomorphometric, structural, and immunocytochemical study of the effects of diet-induced hypocalcemia on bone in growing rats. J Histochem Cytochem 48:1059–1077

Mochida Y, Duarte WR, Tanzawa H, Paschalis EP, Yamauchi M (2003) Decorin modulates matrix mineralization in vitro. Biochem Biophys Res Comm 305:6–9

Moe SM, Reslerova M, Ketteler M, O'Neill K, Duan D, Koczman J, Westenfeld R, Jahnen-Dechent W, Chen NX (2005) Role of calcification inhibitors in the pathogenesis of vascular calcification in chronic kidney disease (CKD). Kidney Int 67:2295–2304

Murshed M, Schinke T, McKee MD, Karsenty G (2004) Extracellular matrix mineralization is regulated locally; different roles of two gla-containing proteins. J Cell Biol 165:625–630

Nagata T, Bellows CG, Kasugai S, Butler WT, Sodek J (1991) Biosynthesis of bone proteins [SPP-1 (secreted phosphoprotein-1, osteopontin), BSP (bone sialoprotein) and SPARC (osteonectin)] in association with mineralized-tissue formation by fetal-rat calvarial cells in culture. Biochem J 274:513–520

Nakamura H, Hirata A, Tsuji T, Yamamoto T (2001) Immunolocalization of keratan sulfate proteoglycan in rat calvaria. Arch Histol Cytol 64:109–118

Nampei A, Hashimoto J, Hayashida K, Tsuboi H, Shi K, Tsuji I, Miyashita H, Yamada T, Matsukawa N, Matsumoto S, Ogihara T, Ochi T, Yoshikawa H (2004) Matrix extracellular phosphoglycoprotein (MEPE) is highly expressed in osteocytes in human bone. J Bone Miner Metab 22:176–184

Nanci A (1999) Content and distribution of noncollagenous matrix proteins in bone and cementum: relationship to speed of formation and collagen packing density. J Struct Biol 126:256–269

Nanci A (2003) Ten Cate's oral histology: Development, structure, and function, 6th edn. Mosby, St. Louis

Nanci A, Zalzal S, Gotoh Y, McKee MD (1996) Ultrastructural characterization and immunolocalization of osteopontin in rat calvarial osteoblast primay culture. Microsc Res Techn 33:214–231

Nanci A, Zalzal S, Gotoh Y, McKee MD (1996) Ultrastructural characterization and immunolocalization of osteopontin in rat calvarial osteoblast primay culture. Microsc Res Techn 33:214–231

Nanci A, Wazen RM, Zalzal SF, Fortin M, Goldberg HA, Hunter GK, Ghitescu D-L (2004) A tracer study with systemically and locally administered dinitrophenylated osteopontin. J Histochem Cytochem 52:1591–1600

Narayanan K, Ramachandran A, Hao J, He G, Park KW, Cho M, George A (2003) Dual functional roles of dentin matrix protein I. Implications in biomineralization and gene transcription by activation of intracellular Ca^{2+} store. J Biol Chem 278:17500–17508

Nefussi J-R, Septier D, Collin P, Goldberg M, Forest N (1989) A comparative ultrahistochemical study of glycosaminoglycans with cuprolinic blue in bone formed *in vivo* and *in vitro*. Calcif Tissue Int 44:11–19

Nefussi J-R, Septier D, Sautier J-M, Forest N, Goldberg M (1992) Localization of malachite green positive lipids in the matrix of bone nodule formed *in vitro*. Calcif Tissue Int 50:273–282

Nefussi JR, Brami G, Modrowski D, Oboeuf M, Forest N (1997) Sequential expression of bone matrix proteins during rat calvaria osteoblast differentiation and bone nodule formation in vitro. J Histochem Cytochem 45:493–503

Ngoma Z, Haumont S (1974) La distribution des lipides dans l'os compact révélée par le noir Soudan B. Acta Histochem 49:220–227

Ninomiya JT, Tracy RP, Calore JD, Gendreau MA, Kelm RJ, Mann KG (1990) Heterogeneity of human bone. J Bone Miner Res 5:933–938

Nishimoto SK, Price PA (1980) Secretion of the vitamin K-dependent protein of bone by rat osteosarcoma cells: evidence for an intracellular precursor. J Biol Chem 255:6579–6583

Nordahl J, Mengarelli-Widholm S, Hultenby K, Reinholt FP (1995) Ultrastructural immunolocalization of fibronectin in epiphyseal and metaphyseal bone of young rats. Calcif Tissue Int 57:442–449

Nusgens B, Chantraine A, Lapiere CM (1972) The protein in the matrix of bone. Clin Orthop Relat Res 88:252–274

Ohnishi T, Arakaki N, Nakamura O, Hirono S, Daikuhara Y (1991) Purification, characterization, and studies on biosynthesis of a 59-kDa bone sialic acid-containing protein (BSP) from rat mandible using a monoclonal antibody. Evidence that 59-kDa BSP may be the rat counterpart of human α_2-HS glycoprotein and is synthesized by both epatocytes and osteoblasts. J Biol Chem 266:14636–14645

Ohri R, Tung E, Rajachar R, Giachelli CM (2005) Mitigation of ectopic calcification in osteopontin-deficient mice by exogenous osteopontin. Calcif Tissue Int 76:307–315

Olivier E, Soury E, Ruminy P, Husson A, Parmentier F, Daveau M, Salier JP (2000) Fetuin-B, a seond member of the fetuin family in mammals. Biochem J 350:589–597

Olsen BR (1963) Electron microscope studies on collagen I.Native collagen fibrils. Z Zellforsch 59:184–198

Onishi T, Ogawa T, Hayashibara T, Hoshino T, Okawa R, Ooshima T (2005) Hyper-expression of osteocalcin mRNA in odontobalsts of Hyp mice. J Dent Res 84:84–88

Ornoy A, Atkin I, Levy J (1980) Ultrastructural studies on the origin and structure of matrix vesicles in bone of young rats. Acta Anat 106:450–461

Otawara Y, Price PA (1986) Developmental appearance of matrix GLA protein during calcification in the rat. J Biol Chem 261:10828–10832

Ottani V, Raspanti M, Ruggeri A (2004) Collagen structure and functional implications. Micron 32:251–260

Pampena DA, Robertson KA, Litvinova O, Lajoie G, Goldberg HA, Hunter GK (2004) Inhibition of hydroxyapatite formation by osteopontin phosphopeptides. Biochem J 378:1083–1087

Parfitt AM (1969) Soft-tissue calcification in uremia. Arch Int Med 124:544–556

Parry DAD (1978) Collagen fibrils and elastic fibers in rat-tail tendon: and electron microscopic study. Biopolymers 17:843–855

Piez KA (1982) Structure and assembly of the native collagen fibril. Connect Tissue Res 10:25–36

Piez KA, Miller A (1974) The structure of collagen fibrils. J Supramol Struct 2:121–137

Pinero GJ, Farach-Carson MC, Devoll RE, Aubin JE, Brunn JC, Butler WT (1995) Bone matrix proteins in osteogenesis and remodelling in the neonatal rat mandible as studied by immunolocalization of osteopontin, bone sialoprotein, α_2HS-glycoprotein and alkaline phosphatase. Arch Oral Biol 40:145–155

Ponlot R (1958) L'intérêt du noir soudan B en histologie des os. Bull Microsc Appl 8:125–127

Poole AR (1991) The growth plate: cellular physiology, cartilage assembly and mineralization. In: Hall B, Newman S (eds) Cartilage: molecular aspects. CRC Press, Boca Raton, pp 179–211

Price PA (1989) Gla-containing proteins of bone. Connect Tissue Res 21:51–60

Price PA, Lim JE (2003) The inhibition of calcium phosphate precipitation by fetuin is accompanied by the formation of a fetuin-mineral complex. J Biol Chem 278:22144–22152

Price PA, Williamson MK (1981) Effects of warfarin on bone. J Biol Chem 256:12754–12759

Price PA, Williamson MK (1985) Primary structure of bovine matrix Gla protein, a new vitamin K-dependent bone protein. J Biol Chem 260:14971–14975

Price PA, Urist MR, Otawara Y (1983) Matrix Gla protein, a new γ-carboxyglutamic acid-containing protein which is associated with the organic matrix of bone. Biochem Biophys Res Comm 117:765–771

Price PA, Williamson MK, Otawara Y (1985) Characterization of matrix Gla protein. A new vitamin K-dependent protein associated with the organic matrix of bone. In: Butler WT (ed) The chemistry and biology of mineralized tissues. Ebsco Media, Birmingham, AL, pp 159–163

Price PA, June HH, Buckley JR, Williamson MK (2001) Osteoprotegerin inhibits artery calcification induced by warfarin and by vitamin D. Arterioscler Thromb Vasc Biol 21:1610–1616
Price PA, Thomas GR, Pardini AW, Figueira WF, Caputo JM, Williamson MK (2002a) Discovery of a high molecular weight complex of calcium, phosphate, fetuin, and matrix γ-carboxyglutamic acid protein in the serum of etidronate-treated rats. J Biol Chem 277:3926–3934
Price PA, Caputo JM, Williamson MK (2002b) Bone origin of the serum complex of calcium, phosphate, fetuin, and matrix Gla protein: biochemical evidence for the cancellous bone-remodeling compartment. J Bone Miner Res 17:1171–1179
Price PA, Nguyen TM, Williamson MK (2003) Biochemical characterization of the serum fetuin-mineral complex. J Biol Chem 278:22153–22160
Price PA, Williamson MK, Nguyen TM, Than TN (2004) Serum levels of the fetuin-mineral complex correlate with artery calcification in the rat. J Biol Chem 279:1594–1600
Prince CW, Rahemtulla F, Butler WT (1983) Metabolism of rat bone proteoglycans *in vivo*. Biochem J 216:589–596
Prince CW, Rahemtulla F, Butler WT (1984) Incorporation of [^{35}S]sulphate into glycosaminoglycans by mineralized tissues in vivo. Biochem J 224:941–945
Pringle GA, Dodd CM (1990) Immunoelectron microscopic localization of the core protein of decorin near the d and e bands of tendon collagen fibrils by use of monoclonal antibodies. J Histochem Cytochem 38:1405–1411
Pugliarello MC, Vittur F, de Bernard B, Bonucci E, Ascenzi A (1970) Chemical modifications in osteones during calcification. Calcif Tissue Res 5:108–114
Qin C, Brunn JC, Cook RG, Orkiszewski RS, Malone JP, Veis A, Butler WT (2004) Evidence for the proteolytic processing of dentin matrix protein 1. Identification and characterization of processed fragments and cleavage sites. J Biol Chem 278:34700–34708
Qu-Hong, Brown LF, Dvorak HF, Dvorak AM (1997) Ultrastructural immunogold localization of osteopontin in human gastric mucosa. J Histochem Cytochem 45:21–33
Quarles LD (2003) FGF23, PHEX, and MEPE regulation of phosphate homeostasis and skeletal mineralization. Am J Physiol Endocrinol Metab 285:E1–9
Quelch KJ, Cole WG, Melick RA (1984) Noncollagenous proteins in normal and pathological human bone. Calcif Tissue Int 36:545–549
Raisz LG, Rodan GA (1990) Cellular basis for bone turnover. In: Avioli LV, Krane SM (eds) Metabolic bone disease and clinically related disorders. W.B. Saunders Company, Philadelphia, pp 1–41
Ramachandran GN, Kartha G (1954) Structure of collagen. Nature 174:269–279
Ramstad VE, Franzen A, Heinegård D, Wendel M, Reinholt FP (2003) Ultrastructural distribution of osteoadherin in rat bone shows a pattern similar to that of bone sialoprotein. Calcif Tissue Int 72:57–64
Raouf A, Ganss B, McMahon C, Vary C, Roughley PJ, Seth A (2002) Lumican is a major proteoglycan component of the bone matrix. Matrix Biol 21:361–367
Raspanti M, Ottani V, Ruggeri A (1989) Different architectures of the collagen fibrils: morphological aspects and functional implications. It J Biol Macromol 11:367–371
Raspanti M, Alessandrini A, Gobbi P, Ruggeri A (1996) Collagen fibril surface: TMAFM, FEG-SEM and freeze-etching observations. Microsc Res Techn 35:87–93
Raspanti M, Alessandrini A, Ottani V, Ruggeri A (1997) Direct visualization of collagen-bound proteoglycans by tapping-mode atomic force microscopy. J Struct Biol 119:118–122
Razzouk S, Brunn JC, Qin C, Tye CE, Goldberg HA, Butler WT (2002) Osteopontin post-translational modifications, possibly phosphorylation, are required for in vitro bone resorption but not osteoclast adhesion. Bone 30:40–47

Reinholt FP, Hultenby K, Oldberg Å, Heinegård D (1990) Osteopontin - a possible anchor of osteoclasts to bone. Proc Natl Acad Sci 87:4473–4475

Reynolds JL, Skepper JN, McNair R, Kasama T, Gupta K, Weissberg PL, Jahnen-Dechent W, Shanahan CM (2005) Multifunctional roles for serum protein fetuin-A in inhibition of human vascular smooth muscle cell calcification. J Am Soc Nephrol 16:2920–2930

Riminucci M, Silvestrini G, Bonucci E, Fisher LW, Gehron Robey P, Bianco P (1995) The anatomy of bone sialoprotein immunoreactive sites in bone as revealed by combined ultrastructural histochemistry and immunohistochemistry. Calcif Tissue Int 57:277–284

Rittling SR, Matsumoto HN, McKee MD, Nanci A, An X-R, Novick KE, Kowalski AJ, Noda M, Denhardt DT (1998) Mice lacking osteopontin show normal development and bone structure but display altered osteoclast formation in vitro. J Bone Miner Res 13:1101–1111

Robey PG (1996) Vertebrate mineralized matrix proteins: structure and function. Connect Tissue Res 35:131–136

Romanowski R, Jundt G, Termine JD, Von der Mark H, Schulz A (1990) Immunoelectron microscopy of osteonectin and type I collagen in osteoblasts and bone matrix. Calcif Tissue Int 46:353–360

Romberg RW, Werness PG, Riggs BL, Mann KG (1986) Inhibition of hydroxyapatite crystal growth by bone-specific and other calcium-binding proteins. Biochemistry 25:1176–1180

Ross FP, Chappel J, Alvarez JI, Sander D, Butler WT, Farach CM, Mintz KA, Robey PG, Teitelbaum SL, Cheresh DA (1993) Interactions between the bone matrix protein osteopontin and bone sialoprotein and the osteoclast integrin alpha beta 3 potentiate bone resorption. J Biol Chem 268:9901–9907

Rowe PS, de Zoysa PA, Dong R, Wang HR, White KE, Econs MJ, Oudet CL (2000) MEPE, a new gene expressed in bone marrow and tumors causing osteomalacia. Genomics 67:54–68

Rowe PS, Kumagai Y, Gutierrez G, Garrett IR, Blacher R, Rosen D, Cundy J, Navvab S, Chen D, Drezner MK, Quarles LD, Mundy GR (2004) MEPE has the properties of an osteoblastic phosphatonin and minhibin. Bone 34:303–319

Ruggeri A, Benazzo F, Reale E (1979) Collagen fibrils with straight and helicoidal microfibrils: a freeze-fracture and thin-sections study. J Ultrastruct Res 68:101–108

Sage H, Vernon RB, Decker J, Funk S, Iruela-Arispe ML (1989) Distribution of the calcium-binding protein SPARC in tissues of embryonic and adult mice. J Histochem Cytochem 37:819–829

Sauren YMHF, Mieremet RHP, Groot CG, Scherft JP (1989) An electron microscopical study on the presence of proteoglycans in the calcified bone matrix by use of cuprolinic blue. Bone 10:287–294

Sauren YMHF, Mieremet RHP, Groot CG, Scherft JP (1992) An electron microscopic study on the presence of proteoglycans in the mineralized matrix of rat and human compact lamellar bone. Anat Rec 232:36–44

Schäfer C, Heiss A, Schwarz A, Westenfeld R, Ketteler M, Floege J, Müller-Esterl W, Schinke T, Jahnen-Dechent W (2003) The serum protein a2-Heremans-Schmid glycoprotein/fetuin-A is a systematically acting inhibitor of ectopic calcification. J Clin Invest 112:357–366

Schajowicz F, Cabrini RL (1954) Histochemical studies of bone in normal and pathological conditions. With special reference to alkaline phosphatase, glycogen and mucopolysaccharides. J Bone Joint Surg 36-B:474–489

Schenk RK (1974) Ultrastruktur des Knochens. Verh Dtsch Ges Path 58:72–83

Scherft JP (1968) The ultrastructure of the organic matrix of calcified cartilage and bone in embryonic mouse radii. J Ultrastruct Res 23:333–343

Scherft JP, Groot CG (1981) The development of matrix vesicles into bone nodules, studied with colloidal thorium dioxide. In: Ascenzi A, Bonucci E, de Bernard B (eds) Matrix vesicles. Wichtig Editore, Milan, pp 173–177
Schiavi SC, Kumar R (2004) The phosphatonin pathway: new insights in phosphate homeostasis. Kidney Int 65:1–14
Schiavi SC, Moe OW (2002) Phosphatonins: a new class of phosphate-regulating proteins. Curr Opin Nephrol Hypertens 11:423–430
Schiffmann E, Martin GR, Miller EJ (1970) Matrices that calcify. In: Schraer H (ed) Biological calcification: cellular and molecular aspects. Appleton-Century-Crofts, New York, pp 27–67
Schinke T, Amendt C, Trindl A, Poschke O, Muller-Esterl W, Jahnen-Dechent W (1996) The serum protein α_2-HS glycoprotein/fetuin inhibits apatite formation *in vitro* and in mineralizing calvaria cells. A possible role in mineralization and calcium homeostasis. J Biol Chem 271:20789–20796
Schmidt-Schultz TH, Schultz M (2004) Bone protects proteins over thousands of years: extraction, analysis, and interpretation of extracellular matrix proteins in archeological skeletal remains. Am J Phys Anthrop 123:30–39
Schraer H, Gay CV (1977) Matrix vesicles in newly synthesizing bone observed after ultracryotomy and ultramicroincineration. Calcif Tissue Res 23:185–188
Scott JE (1980) Collagen-proteoglycan interactions. Localization of proteoglycans in tendon by electron microscopy. Biochem J 187:887–891
Scott JE, Dorling J (1965) Differential staining of acid glycosaminoglycans (mucopolysaccharides) by Alcian blue in salt solutions. Histochemie 5:221–233
Scott JE, Orford CR, Hughes EW (1981) Proteoglycan-collagen arrangements in developing rat tail tendon. An electron-microscopical and biochemical investigation. Biochem J 195:573–581
Segrest JP, Cunningham LW (1973) Unit fibril models derived from the molecular topography of collagen. Biopolymers 12:825–834
Sela J, Bab IA, Muhlrad A (1981) A comparative study on the occurrence and activity of extracellular matrix vesicles in young and adult rat maxillary bone. Calcif Tissue Int 33:129–134
Sela J, Schwartz Z, Swain LD, Boyan BD (1992a) The role of matrix vesicles in calcification. In: Bonucci E (ed) Calcification in biological systems. CRC Press, Boca Raton, pp 73–105
Sela J, Schwartz Z, Amir D, Swain LD, Boyan BD (1992b) The effect of bone injury on ectracellular matrix vesicle proliferation and mineral formation. Bone Miner 17:163–167
Selye H (1962) Calciphylaxis. The University of Chicago Press, Chicago
Severson AR, Ingram RT, Fitzpatrick LA (1995) Matrix proteins associated with bone calcification are present in human vascular smooth muscle cells grown *in vitro*. In vitro Cell Dev Biol - Animal 31:853–857
Shapiro HS, Chen J, Wrana JL, Zhang Q, Blum M, Sodek J (1993) Characterization of porcine bone sialoprotein: primary structure and cellular expression. Matrix 13:431–440
Shapiro IM (1970a) The association of phospholipids with anorganic bone. Calcif Tissue Res 5:13–20
Shapiro IM (1970b) The phospholipids of mineralized tissues I. Mammalian compact bone. Calcif Tissue Res 5:21–29
Shen Z, Gantcheva S, Sommarin Y, Heinegård D (1999) Tissue distribution of a novel cell binding protein, osteoadherin, in the rat. Matrix Biol 18:533–542
Shibata S, Fukada K, Suzuki S, Ogawa T, Yamashita Y (2002) In situ hybridization and immunohistochemistry of bone sialoprotein and secreted phosphoprotein 1 (osteopon-

tin) in the developing mouse mandibular condylar cartilage compared with limb bud cartilage. J Anat 200:309–320

Shijubo N, Uede T, Kon S, Nagata M, Abe S (2000) Vascular endothelial growth factor and osteopontin in tumor biology. Crit Rev Oncog 11:135–146

Shimazu Y, Nanci A, Aoba T (2002) Immunodetection of osteopontin at sites of resorption in the pulp of rat molars. J Histochem Cytochem 50:911–922

Shuttleworth A, Veis A (1972) The isolation of anionic phosphoproteins from bovine cortical bone *via* the periodate solubilization of bone collagen. Biochim Biophys Acta 257:414–420

Siggelkow H, Schmidt E, Hennies B, Hufner M (2004) Evidence of downregulation of matrix extracellular phosphoglycoprotein during terminal differentiation in human osteoblasts. Bone 35:570–576

Silvestrini G, Zini N, Sabatelli P, Mocetti P, Maraldi NM, Bonucci E (1996) Combined use of malachite green fixation and PLA_2-gold complex technique to localize phospholipids in areas of early calcification of rat epiphyseal cartilage and bone. Bone 18:559–565

Simes DC, Williamson MK, Ortiz-Delgado JB, Viegas CS, Price PA, Cancela ML (2003) Purification of matrix Gla protein from a marine teleost fish, *Argyrosomus regius*: calcified cartilage and not bone as the primary site of MGP accumulation in fish. J Bone Miner Res 18:244–259

Simonet WS, Lacey DL, Dunstan CR, Kelley M, Chang M-S, Lüthy R, Nguyen HQ, Wooden S, Bennett L, Boone T, Shimamoto G, DeRose M, Elliott R, Colombero A, Tan H-L, Trail G, Sullivan J, Davy E, Bucay N, Renshaw-Gegg L, Hughes TM, Hill D, Pattison W, Campbell P, Sander S, Van G, Tarpley J, Derby P, Lee R, Amgen Est Program, Boyle WJ (1997) Osteoprotegerin: a novel secreted protein involved in the regulation of bone density. Cell 89:309–319

Smith JW (1965) Packing arrangement of tropocollagen molecules. Nature 205:356–358

Sodek J, Chen J, Nagata T, Kasugai S, Todescan R Jr, Li IW, Kim RH (1995a) Regulation of osteopontin expression in osteoblasts. Ann N Y Acad Sci 760:223–241

Sodek J, Kim RH, Ogata Y, Li J, Yamauchi M, Zhang Q, Freedman LP (1995b) Regulation of bone sialoprotein gene transcription by steroid hormones. Connect Tissue Res 32:209–217

Sodek KL, Tupy JH, Sodek J, Grynpas MD (2000) Relationships between bone protein and mineral in developing porcine long bone and calvaria. Bone 26:189–198

Sommarin Y, Wendel M, Shen Z, Hellman U, Heinegård D (1998) Osteoadherin, a cell-binding keratan sulfate proteoglycan in bone, belongs to the family of leucine-rich repeat proteins of the extracellular matrix. J Biol Chem 273:16723–16729

Sommer B, Bickel M, Hofstetter W, Wetterwald A (1996) Expression of matrix proteins during the development of mineralized tissues. Bone 19:371–380

Spector AR, Glimcher MJ (1972) The extraction and characterization of soluble anionic phosphoproteins from bone. Biochim Biophys Acta 263:593–603

Spector AR, Glimcher MJ (1973) The identification of *o*-phosphoserine in the soluble anionic phosphoproteins of bone. Biochim Biophys Acta 303:360–362

Speer MY, McKee MD, Guldberg RE, Liaw L, Yang HY, Tung E, Karsenty G, Giachelli CM (2002) Inactivation of the osteopontin gene enhances vascular calcification of matrix Gla protein-deficient mice: evidence for osteopontin as an inducible inhibitor of vascular calcification in vivo. J Exp Med 196:1047–1055

Speer MY, Chien YC, Quan M, Yang HY, Vali H, McKee MD, Giachelli CM (2005) Smooth muscle cells deficient in osteopontin have enhanced susceptibility to calcification *in vitro*. Cardiovasc Res 66:324–333

Srinivasan R, Chen B, Gorski JP, George A (1999) Recombinant expression and characterization of dentin matrix protein 1. Connect Tissue Res 40:251–258

Stagni N, de Bernard B, Liut GF, Vittur F, Zanetti M (1980) Ca^{2+}-binding glycoprotein in avian bone induced by estrogen. Connect Tissue Res 7:121–125

Stanford CM, Jacobson PA, Eanes ED, Lembke LA, Midura RJ (1995) Rapidly forming apatitic mineral in an osteoblastic cell line (UMR 106–01 BSP). J Biol Chem 270:9420–9428

Steitz SA, Speer MY, Curinga G, Yang HY, Haynes P, Aebersold R, Schinke T, Karsenty G, Giachelli CM (2001) Smooth muscle cell phenotypic transition associated with calcification: upregulation of Cbfa1 and downregulation of smooth muscle lineage markers. Circ Res 89:1147–1154

Steitz SA, Speer MY, McKee MD, Liaw L, Almeida M, Yang H, Giachelli CM (2002) Osteopontin inhibits mineral deposition and promotes regression of ectopic calcification. Am J Pathol 161:2035–2046

Stenner DD, Tracy RP, Riggs BL, Mann KG (1986) Human platelets contain and secrete osteonectin, a major protein of mineralized bone. Proc Natl Acad Sci 83:6892–6896

Steven FS (1972) Current concepts of collagen structure. Clin Orthop Relat Res 85:257–274

Sugars RV, Milan AM, Brown JO, Waddington RJ, Hall RC, Embery G (2003) Molecular interaction of recombinant decorin and biglycan with type I collagen influences crystal growth. Connect Tissue Res 44:189–195

Szweras M, Liu D, Partridge EA, Pawling J, Sukhu B, Clokie C, Jahnen-Dechent W, Tenenbaum HC, Swallow CJ, Grynpas MD, Dennis JW (2002) α_2-HS glycoprotein/fetuin, a transforming growth factor-β/bone morphogenetic protein antagonist, regulates postnatal bone growth and remodeling. J Biol Chem 277:19991–19997

Takagi M, Parmley RT, Toda Y, Denys FR (1983) Ultrastructural cytochemistry of complex carbohydrates in osteoblasts, osteoid, and bone matrix. Calcif Tissue Int 35:309–319

Takagi M, Maeno M, Kagami A, Takahashi Y, Otsuka K (1991) Biochemical and immunocytochemical characterization of mineral binding proteoglycans in rat bone. J Histochem Cytochem 39:41–50

Takagi M, Maeno M, Yamada T, Miyashita K, Otsuka K (1996) Nature and distribution of chondroitin sulphate and dermatan sulphate proteoglycans in rabbit alveolar bone. Histochem J 28:341–351

Tanzer ML (1973) Cross-linking of collagen. Endogenous aldehydes in collagen react in several ways to form a variety of unique covalent cross-links. Science 180:561–566

Tartaix PH, Doulaverakis M, George A, Fisher LW, Butler WT, Qin C, Salih E, Tan M, Fujimoto Y, Spevak L, Boskey AL (2004) *In vitro* effects of dentin matrix protein-1 on hydroxyapatite formation provide insights into *in vivo* functions. J Biol Chem 279:18115–18120

Terasawa M, Shimokawa R, Terashima T, Ohya K, Takagi Y (2004) Expression of dentin matrix protein 1 (DMP1) in nonmineralized tissues. J Bone Miner Metab 22:430–438

Terkelsen OB, Jahnen-Dechent W, Nielsen H, Moos T, Fink E, Nawratil P, Muller-Esterl W, Mollgard K (1998) Rat fetuin: distribution of protein and mRNA in embryonic and neonatal rat tissues. Anat Embryol (Berl) 197:125–133

Termine JD (1985) The tissue specific proteins of the bone matrix. In: Butler WT (ed) The chemistry and biology of mineralized tissues. EBSCO Media, Birmingham,AL, pp 94–97

Termine JD, Belcourt AB, Conn KM, Kleinman HK (1981a) Mineral and collagen-binding proteins of fetal calf bone. J Biol Chem 256:10403–10408

Termine JD, Kleinman HK, Whitson SW, Conn KM, McGarvey ML, Martin GR (1981b) Osteonectin, a bone-specific protein linking mineral to collagen. Cell 26:99–105

Teti A (1993) Biology of osteoclast and molecular mechanisms of bone resorption. It J Miner Electrol Metab 7:123–133

Tezuka K, Sato T, Kamioka H, Nijweide PJ, Tanaka K, Matsuo T, Ohta M, Kurihara N, Hakeda Y, Kumegawa M (1992) Identification of osteopontin in isolated rabbit osteoclasts. Biochem Biophys Res Comm 186:911–917

Thotakura SR, Karthikeyan N, Smith T, Liu K, George A (2000) Cloning and characterization of rat dentin matrix protein 1 (*DMP1*) gene and its 5′-upstream region. J Biol Chem 275:10272–10277

Thyberg J, Nilsson S, Friberg U (1975) Electron microscopic and enzyme cytochemical studies on the guinea pig metaphysis with special reference to the lysosomal system of different cell types. Cell Tissue Res 156:273–299

Tintut Y, Demer LL (2001) Recent advances in multifactorial regulation of vascular calcification. Curr Opin Lipidol 12:555–560

Toyosawa S, Sato A, O'hUigin C, Tichy H, Klein J (2000) Expression of the dentin matrix protein 1 gene in birds. J Mol Evol 50:31–38

Toyosawa S, Shintani S, Fujiwara T, Ooshima T, Sato A, Ijuhin N, Komori T (2001) Dentin matrix protein 1 is predominantly expressed in chicken and rat osteocytes but not in osteoblasts. J Bone Miner Res 16:2017–2026

Toyosawa S, Kanatani N, Shintani S, Kobata M, Yuki M, Kishino M, Ijuhin N, Komori T (2004a) Expression of dentin matrix protein 1 (DMP1) during fracture healing. Bone 35:553–561

Toyosawa S, Tomita Y, Kishino M, Hashimoto J, Ueda T, Tsujimura T, Aozasa K, Ijuhin N, Komori T (2004b) Expression of dentin matrix protein 1 in tumors causing oncogenic osteomalacia. Mod Pathol 17:573–578

Tracy RP, Mann KG (1991) Osteonectin. In: Hall BK (ed) Bone, vol. 3: Bone matrix and bone specific products. CRC Press, Boca Raton, pp 295–319

Triffitt JT, Owen M (1977) Preliminary studies on the binding of plasma albumin to bone tissue. Calcif Tissue Res 23:303–305

Triffitt JT, Owen ME, Ashton BA, Wilson JM (1978) Plasma disappearance of rabbit α_2HS-glycoprotein and its uptake by bone tissue. Calcif Tissue Res 26:155–161

Trus BL, Piez KA (1980) Compressed microfibril models of the native collagen fibril. Nature 286:300–301

Tye CE, Rattray KR, Warner KJ, Gordon JAR, Sodek J, Hunter GK, Goldberg HA (2003) Delineation of the hydroxyapatite-nucleating domains of bone sialoprotein. J Biol Chem 278:7949–7955

Uchiyama A, Suzuki M, Lefteriou B, Glimcher MJ (1986) Isolation and chemical characterization of the phosphoproteins of chicken bone matrix: heterogeneity in molecular weight and composition. Biochemistry 25:7572–7583

Ulrich MMW, Perizonius WRK, Spoor CF, Sanberg P, Vermeer C (1987) Extraction of osteocalcin from fossil bones and teeth. Biochem Biophys Res Comm 149:712–719

Van Den Hooff A, Van Nie CJ, Buitenweg DW (1966) Histology and polysaccharide histochemistry of heterotopic bone formation. Pathol Microbiol 29:17–28

van der Rest M (1991) The collagens of bone. In: Hall BK (ed) Bone. Vol. 3: Bone matrix and bone specific products. CRC Press, Boca Raton, pp 187–237

Vanden Berge JC, Storer RW (1995) Intratendinous ossification in birds: a review. J Morphol 226:47–77

Veis A (1975) The biochemistry of collagen. Ann Clin Lab Sci 5:123–131

Veis A (1985) Phosphoproteins of dentin and bone. Do they have a role in matrix mineralization. In: Butler WT (ed) The chemistry and biology of mineralized tissues. EBSCO Media, Birmingham,AL, pp 170–176

Veis A (2003) Mineralization in organic matrix frameworks. Rev Miner Geochem 54:249–289

Veis A, Sabsay B (1987) The collagen of mineralized matrices. In: Peck WA (ed) Bone and mineral research/5. Elsevier Science Publishers, Amsterdam, pp 1–63

Veis A, Yuan L (1975) Structure of the collagen microfibril. A four-strand overlap model. Biopolymers 14:895–900

Veis A, Tsay T-G, Kanwar Y (1985) An immunological study of the localization of dentin phosphophoryns in the tooth. In: Belcourt AB, Ruch J-V (eds) Morphogenèse et différenciation dentaires, Coll. INSERM 125, 1984. INSERM, Paris, pp 223–232
Vermeulen AHM, Vermeer C, Bosman FT (1989) Histochemical detection of osteocalcin in normal and pathological human bone. J Histochem Cytochem 37:1503–1508
Vogel JJ, Boyan-Salyers BD (1976) Acidic lipids associated with the local mechanism of calcification. A review. Clin Orthop Relat Res 118:230–241
Vogel KG, Sandy JD, Pogany G, Robbins JR (1994) Aggrecan in bovine tendon. Matrix Biol 14:171–179
von der Mark K, Wendt P, Rexrodt F, Kühn K (1970) Direct evidence for a correlation between amino acid sequence and cross striation pattern of collagen. FEBS Lett 11:105–108
Wada T, McKee MD, Steitz S, Giachelli CM (1999) Calcification of vascular smooth muscle cell cultures: inhibition by osteopontin. Circ Res 84:166–178
Waltregny D, Bellahcène A, De Leval X, Florkin B, Weidle U, Castronovo V (2000) Increased expression of bone sialoprotein in bone metastases compared with visceral metastases in human breast and prostate cancer. J Bone Miner Res 15:834–843
Wang A, Martin JA, Lembke LA, Midura RJ (2000) Reversible suppression of in vitro biomineralization by activation of protein kinase A. J Biol Chem 275:11082–11091
Wassen MHM, Lammens J, Tekoppele JM, Sakkers RJB, Liu Z, Verbout AJ, Bank RA (2000) Collagen structure regulates fibril mineralization in osteogenesis as revealed by cross-link patterns in calcifying callus. J Bone Miner Res 15:1776–1785
Weiner S, Hood L (1975) Soluble protein of the organic matrix of mollusk shells: a potential template for shell formation. Science 190:987–989
Weiner S, Traub W (1980) X-ray diffraction study of the insoluble organic matrix of mollusk shells. FEBS Lett 111:311–316
Weiner S, Traub W (1986) Organization of hydroxyapatite crystals within collagen fibrils. Feder Eur Bioch Soc 206:262–266
Weinreb M, Shinar D, Rodan GA (1990) Different pattern of alkaline phosphatase, osteopontin, and osteocalcin expression in developing rat bone visualized by in situ hybridization. J Bone Miner Res 5:831–842
Weisbrode SE, Capen CC, Nagode LA (1973) Fine structural and enzymatic evaluation of bone in thyroparathyroidectomized rats receiving various levels of vitamin D. Lab Invest 28:29–37
Wendel M, Sommarin Y, Heinegård D (1998) Bone matrix protein: isolation and characterization of a novel cell-binding keratan sulfate proteoglycan (osteoadherin) from bovine bone. J Cell Biol 141:839–847
Wess TJ, Hammersley AP, Wess L, Miller A (1998) A consensus model for molecular packing of type I collagen. J Struct Biol 122:92–100
Wewer UM, Albrechtsen R, Fischer LW, Young MF, Termine JD (1988) Osteonectin/SPARC/BM-40 in human decidua and carcinoma tissues characterized by *de novo* formation of basement membrane. Am J Pathol 132:345–355
Wheeler AP, Rusenko KW, George JW, Sikes CS (1987) Evaluation of calcium binding by molluscan shell organic matrix and its relevance to biomineralization. Comp Biochem Physiol 87B:953–960
White SW, Hulmes DJS, Miller A, Timmins PA (1977) Collagen-mineral axial relationship in calcified turkey leg tendon by X-ray and neutron diffraction. Nature 266:421–425
Williams DC, Boder GB, Toomey RE, Paul DC, Hillman CC Jr, King KL, Van Frank RM, Johnston CC Jr (1980) Mineralization and metabolic response in serially passaged adult rat bone cells. Calcif Tissue Int 30:233–246
Wolinsky I, Guggenheim K (1970) Lipid metabolism of chick epiphyseal bone and cartilage. Calcif Tissue Res 6:113–119

Woodhead-Galloway J, Hukins DWL, Wray JS (1975) Closest packing of two-stranded coiled-coils as a model for the collagen fibril. Biochem Biophys Res Comm 64:1237–1244

Wuthier RE, Register TC (1985) Role of alkaline phosphatase, a polyfunctional enzyme, in mineralizing tissues. In: Butler WT (ed) The chemistry and biology of mineralized tissues. Ebsco Media, Inc., Birmingham, pp 113–124

Yamada M (1976) Ultrastructural and cytochemical studies on the calcification of the tendon-bone joint. Arch Histol Jpn 39:347–378

Yamate T, Mocharla H, Taguchi Y, Igietseme JU, Manolagas SC, Abe E (1997) Osteopontin expression by osteoclast and osteoblast progenitors in the murine bone marrow: demonstration of its requirement for osteoclastogenesis and its increase after ovariectomy. Endocrinology 138:3047–3055

Yang F, Chen Z-I, Bergeron JM, Cupples RL, Friedrichs WE (1992) Human α_2-HS-glycoprotein/bovine fetuin homologue in mice: identification and developmental regulation of the gene. Biochim Biophys Acta 1130:149–156

Ye L, MacDougall M, Zhang S, Xie Y, Zhang J, Li Z, Lu Y, Mishina Y, Feng JQ (2004) Deletion of dentin matrix protein-1 leads to a partial failure of maturation of predentin into dentin, hypomineralization, and expanded cavities of pulp and root canal during postnatal tooth development. J Biol Chem 279:19141–19148

Young MF, Kerr JM, Ibaraki K, Heegaard A-M, Robey PG (1992) Structure, expression, and regulation of the major noncollagenous matrix proteins of bone. Clin Orthop Relat Res 281:275–294

Zhou H-Y, Glimcher MJ, Wang J, Salih E (2000) A novel glycosylated phosphoprotein: osteometrin. In: Goldberg M, Boskey A, Robinson C (eds) Chemistry and biology of mineralized tissues. American Academy of Orthopaedic Surgeons, Rosemont, IL, pp 185–188

Zhu XL, Ganss B, Goldberg HA, Sodek J (2001) Synthesis and processing of bone sialoproteins during de novo bone formation in vitro. Biochem Cell Biol 79:737–746

Zini N, Sabatelli P, Silvestrini G, Bonucci E, Maraldi NM (1996) Influence of specimen preparation on the identification of phospholipids by the phospholipase A_2-gold method in mineralizing cartilage and bone. Histochem Cell Biol 105:283–296

8 Calcifying Matrices: Dentin and Cementum

8.1 Introduction

Much of the skeleton of vertebrates consists of bone. Other calcified tissues are, however, present both in the skeleton itself (epiphyseal growth cartilage) and in extraskeletal tissues (dentin, enamel, cementum). Some tissues which do not normally calcify may become calcified as a result of aging (tendons, hyaline cartilage) or pathological conditions. While the bone matrix consists mainly of collagen, calcifying non-osseous matrices have varying percentages of this component or may contain no collagen at all. Evaluation of the structure and composition of these matrices may offer important clues to an understanding of the early stages of the calcification process. This chapter deals with dentin, a tissues which shows a large area of overlap with bone.

The organic matrices of dentin and bone share the presence of collagen fibrils (Fig. 8.1), non-collagenous proteins, and lipids (reviewed by Jones and Boyde 1984; Linde 1992; Linde and Goldberg 1993; Butler 1995; Goldberg et al. 1995a). Moreover, just as the osteoid tissue appears as the thin border of uncalcified organic matrix separating osteoblasts from the calcified matrix in bone, so predentin forms a thin uncalcified layer separating odontoblasts and pulp soft tissue from the already calcified dentin matrix (Fig. 8.1). The organic matrix of dentin and predentin are fundamentally similar, but do show some important differences (Linde 1985; described below). A third, 0.5–2.5 μm thick, transition zone, known as metadentin, has been described between dentin and predentin (Goldberg and Septier 1996). Metadentin accumulates lanthanum both in vivo and in vitro, a finding considered to be due either to the high porosity of the zone, or to La^{3+} binding to Ca^{2+}-binding sites (Goldberg et al. 2000).

8.2 Collagen of Dentin

The fibrous components of the organic matrix of predentin and dentin resemble those of bone: they consist, in fact, almost exclusively of type I collagen (Fig. 8.1; reviewed by Butler 1984a; for structure and composition of type I collagen fibers see Sect. 7.2). Type III, IV, V and VI (pro)collagens have been detected

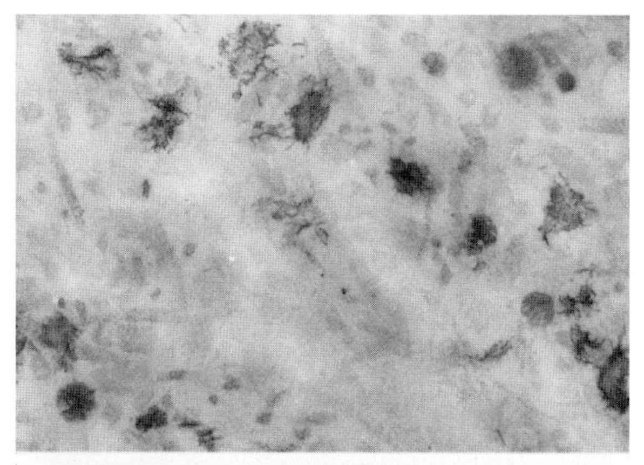

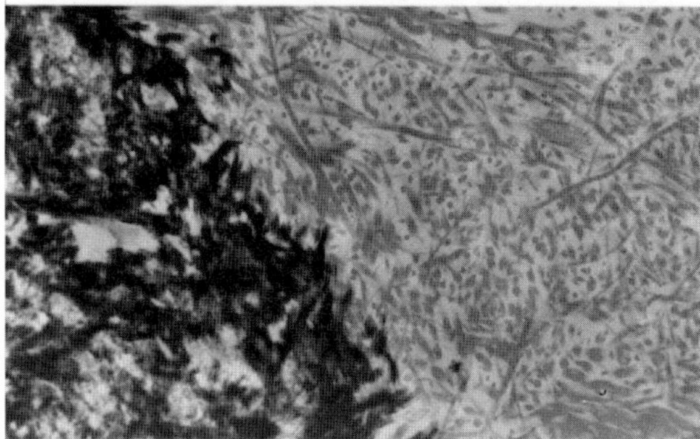

Fig. 8.1. *Above*: rat mantle predentin; besides loosely arranged collagen fibrils, matrix vesicles are present which are still uncalcified (*upper right corner*) or show an initial degree of calcification (*lower left corner*); calcification nodules at various degrees of calcification can be detected, too. Uranyl acetate and lead citrate, ×70,000. *Below*: circumpulpal predentin: collagen fibrils are loosely and irregularly arranged; no matrix vesicles are present. Uranyl acetate and lead citrate, ×24,000

immunohistochemically in odontoblasts (Bronckers et al. 1989), but they are virtually absent in dentin matrix, even if type III collagen has been shown in the dentin of mice (Nagata et al. 1992).

Collagen fibers, which are synthesized and then released by odontoblasts (Weinstock M and Leblond 1974), form a loose network located close to the cell surface; their thickness and density increase in going from this to the calcification front (Beniash et al. 2000). They are more closely interwoven than in compact bone (Kramer 1951), and their thickness and arrangement vary with the type of dentin (discussed by Jones and Boyde 1984). In mantle dentin – the earliest dentin to be formed peripherally – they are rather coarse, are grouped in bundles, and run almost parallel to odontoblast processes, whereas in circumpulpal dentin – the dentin located between the mantle dentin and the dental pulp – they are thinner, and lie within the plane of the developing surface. A study by atomic force microscopy (tapping mode) of human molar dentin partly decalcified with citric acid and treated with aqueous solutions of NaOCl, which probably removes non-collagenous proteins, showed fibrils distinguished by three specific values for their diameters (83, 91 and 100 nm) and an axial repeat distance of 67 nm in the hydrated state; after dehydration, the values of the diameter ranged between 75 and 105 nm, and that of the axial repeat was divided into three groups at 57, 62 and 67 nm (Habelitz et al. 2002).

The presence and distribution of collagen fibrils in peritubular dentin deserve to be taken into separate consideration, because of their structural peculiarities (reviewed by Jones and Boyde 1984), which may have an impact on their calcification. Peritubular dentin is the dentin which makes up the wall of the dentinal tubules and forms a collar around the odontoblast processes. It is reported to be highly calcified, as shown by the backscattered electron image of the human dentin surface (Jones and Boyde 1984); its mineral content, although variable, may exceed that of intertubular dentin by 9% (Miller et al. 1971) or more. Its collagen content is disputed, in spite of the various studies dedicated to this topic (Johansen and Parks 1962; Takuma and Eda 1966; Sundström et al. 1970; Kodaka et al. 1992). Doubts persist on this point, but it seems probable that, unlike intertubular dentin, peritubular dentin contains few collagen fibrils (Takuma and Eda 1966; Weiner et al. 1999) and that fine filamentous structures are found in their place (Takuma 1960; Sundström et al. 1970; Goldberg et al. 1980). Moreover, peritubular dentin contains proteins which are soluble after EDTA decalcification (Weiner et al. 1999). This has led to its being called a 'hypercalcified ground substance' by Rouiller et al. (1952); for the same reason, according to Jones and Boyde (1984), the fracture surface of 'afibrillar' peritubular dentin differs greatly from that of 'fibrillar' intertubular dentin and is easily recognizable (see also Weiner et al. 1999). Vose and Baylink (1970) did, however, stress the point that the increased X-ray absorption of peritubular matrix may depend on the orientation of their calcified collagen fibrils, and Magne et al. (2002), on the basis of FTIR investigations, reported that peritubular and intertubular horse dentins have a very similar inorganic and organic composition, and that, in both, collagen is the main protein component. By contrast, morphological and biochemical investigations have shown that the organic components of sedimented fragments of peritubular dentin – obtained by density fractionation of human dentin powder and identified on the basis of the presence of fractured cylindrical tubes with empty cores – consist of quite strongly acidic proteins which are made soluble by decalcification, stain with Stains All, and have a higher molecular weight than, and a different amino acid composition from, intertubular dentin phosphoproteins (Weiner et al. 1999).

8.3 Dentin Non-collagenous Proteins: Proteoglycans

As in bone (to which readers can refer for further information), non-collagenous proteins of dentin comprise proteoglycans, glycoproteins (Gla-proteins and phosphoproteins), lipids, and serum proteins such as albumin and α_2HS glycoprotein (reviewed by Linde 1984, 1992; Fisher and Termine 1985; Butler and Ritchie 1995; Goldberg et al. 1995a; Ritchie et al. 1998); as in bone, most of these components are entrapped in the calcified matrix and can only be extracted after decalcification (Butler 1984b); as in bone, they are located in interfibrillary spaces and are associated with collagen fibrils (Fujisawa and Kuboki 1992).

It has been shown several times that predentin and dentin contain proteoglycans. Engfeldt and Hjerpe (1972) reported the presence of glycosaminoglycans in predentin and dentin, the former containing relatively more chondroitin-6-sulfate than the latter. This finding was confirmed by Linde (1973) and Jontell and Linde (1983), who detected glycosaminoglycans in predentin that had been isolated – together with an odontoblast fraction – from porcine teeth using microdissection. Jones and Leaver (1974) reported that chondrotin-4-sulfate is the major glycosaminoglycan of the dentin matrix and that chondroitin-6-sulfate, hyaluronic acid, dermatan sulfate and a non-sulfated galactosaminoglycan are present in smaller quantities. Studies by Hjerpe and Engfeldt (1976) have shown that the glycosaminoglycans extracted from predentin, calcification front and dentin, which had been separated by the density gradient fractionation procedure, migrated with proteins, giving rise to polydispersed proteoglycans; the amino acid composition of the three fractions showed that amounts (expressed in numbers of amino acids per 1000 residues) of Asp and Ser rose in moving from predentin to the calcification front and from that to dentin (167–247–350 and 170–277–342, respectively), whereas amounts of Glu and Gly fell (134–117–62 and 179–122–82, respectively). Linde (1992) reported that a class of proteoglycans in predentin has a molecular size greater than that of the proteoglycans in dentin.

The extraction of calcified dentin with 4 M guanidine hydrochloride releases a proteoglycan fraction which is not representative of the whole proteoglycan content of the matrix, because a second fraction can be extracted, together with other non-collagenous components, if the extraction procedure is carried out after EDTA decalcification. This important observation (Linde et al. 1980) showed that the proteoglycans of dentin, like those of bone (see Sect. 7.3.3), are entrapped by the inorganic substance. Rahemtulla et al. (1984) extracted proteoglycans from ^{35}S-labeled predentin and EDTA decalcified dentin: the first predentin extraction yielded two proteoglycan populations – a minor population, of larger molecular size, with larger glycosaminoglycan chains consisting of chondroitin 4- and 6-sulfate isomers, and a major population, of smaller hydrodynamic size and with smaller glycosaminoglycan chains, consisting entirely of chondroitin 4-sulfate; the second dentin extraction yielded only one fraction resembling the second one extracted from predentin. Lormée et al. (1996) found two distinct groups of ^{35}S-labeled proteoglycans, one exclusively related to predentin and disappearing over time, the other stably located in dentin; this did not derive from proteoglycans in predentin and first appeared at the calcification front. Steinfort et al. (1994), who also labeled dentin proteoglycans with ^{35}S, described five fractions whose size varied from 100 to 400 kDa; in all of them, the glycosaminoglycans mainly consisted of chondroitin-4-sulfate.

Actually, the predominant proteoglycans in predentin and dentin are their chondroitin-sulfate-rich decorin and biglycan and their keratan-sulfate-, leucine-rich fibromodulin and lumican (Embery et al. 2001), the latter occurring throughout the predentin (Hall et al. 1997; Goldberg et al. 2003) and, in lower concentrations, in the dentinal tubules (Yamauchi et al. 2000). The glycosaminoglycans are 68% chondroitin sulfate and 32% dermatan sulfate in predentin, 92% chondroitin sulfate and 8% dermatan sulfate in the predentin-dentin interface, and 100% chondroitin sulfate in dentin (Milan et al. 2005). A falling gradient of chondroitin sulfate from the pulpal aspect toward the calcification front, and an opposite, rising gradient of keratan sulfate and decorin, have been documented by immunohistochemistry (Embery et al. 2001). Similarly, immunoreaction with the 2B6 antibody showed a developmentally regulated chondroitin sulfate/dermatan sulfate gradient (Septier et al. 1998) which fell from the proximal to the distal predentin, in contrast with a constant biglycan concentration and a sharp increase in decorin found in the distal predentin, near the transition of predentin into dentin (Goldberg et al. 2003).

Ultrastructural HID-TCH-SP histochemical studies showed that predentin contains three times the amount of iron precipitates, i. e., sulfated glycosaminoglycans, compared with the area of the calcification front (Takagi et al. 1981). Light microscope immunohistochemistry also showed not only that there was a definite difference in proteoglycans between predentin and dentin (more plentiful in the former), but also that a selective decrease in their concentrations took place during dentinogenesis and calcification: proteoglycans were distributed homogeneously in predentin and their glycosaminoglycans were mainly chondroitin-4-sulfate, chondroitin-6-sulfate, keratan sulfate and dermatan sulfate, whereas, in dentin, chondroitin-4-sulfate and dermatan sulfate were the prevailing glycosaminoglycans and staining was mostly limited to dentinal tubules surrounding odontoblastic processes with only weak staining in the rest of the matrix (Takagi et al. 1990a). Using colloidal gold coated with poly-L-lysine, Goldberg and Lécolle (1995) confirmed that the amount of glycosaminoglycans in dentin is about a half that in predentin. Intense decorin immunoreaction extended along the calcification front and dentinal tubules, whereas none was detected in predentine matrix (Yoshiba et al. 1996). Using antibodies against the proforms and fully processed forms of decorin and biglycan, Septier et al. (2001) found that the former were mostly found in odontoblasts and, to a lower extent, in their processes, as well as in extracellular matrix, where they were uniformly distributed throughout predentin, with higher labeling for pro-biglycan than for pro-decorin; the processed, secreted forms of biglycan were found at constant levels in predentin and dentin, whereas decorin was chiefly located at distal predentin.

Using electron microscopy and ruthenium hexammine trichloride as a stain for acidic groups, proteoglycans were detected in the predentin as granules 10–15 nm in diameter and as filaments, and non-aggregating proteoglycans were observed in the spaces between collagen fibers as an amorphous substance

(Goldberg and Septier 1986). Using the same method, two different staining patterns emerged: the interfibrillary spaces in the circumpulpal dentin contained a homogeneous, electron-dense material, whereas the predentin matrix was dotted by dense granules, some of them related to collagen periodic banding (Kogaya et al. 1987). Using analogous methods, Goldberg and Takagi (1993) showed the presence of amorphous ground substance, homogeneously stained with alcian Blue, located in the predentin between the collagen fibers. After fixation with cetylpyridinium chloride in glutaraldehyde, Chardin et al. (1990) found that the proteoglycans of predentin appear as interfibrillar, filamentous, crystal-like structures 300–500 nm long and 7–14 nm thick, and that in decalcified dentin they appeared as needle-like structures at the periphery of unstained globular structures. The same authors observed that glycosaminoglycans were probably bound to lipids, because they were removed by treatment with a chloroform/methanol mixture. After fixation with glutaraldehyde-cuprolinic blue, Tenorio et al. (1990) showed that the proteoglycans of early developing rat dentin appeared as ribbon-like precipitates associated with collagen fibrils. Using the same method, Everts et al. (1995) found that proteoglycans in the enamel-related predentin of mice appeared as long, slender, often star-shaped precipitates, whereas in the cementum-related dentin they were thick and short, suggesting the presence of a variety of proteoglycan types. In their review on dentinogenesis, Linde and Goldberg (1993) reported that mantle dentin is easy to recognize because of its high stainability with cationic dyes, such as alcian blue, a result of its richness in proteoglycans.

Osteoadherin, a proteoglycan that contains keratan sulfate, previously described in bone (see Sect. 7.3.1), has been detected in developing teeth (Buchaille et al. 2000), where it is located in predentin, dentin, and enamel matrix close to the apical pole of ameloblasts (Couble et al. 2004). It has also been found in extracts of bovine and rat calcified dentin matrix (Petersson et al. 2003).

8.4 Dentin Non-collagenous Proteins: Gla-proteins

Gla-proteins have been detected in dentin matrix, from which they can be extracted and separated by anion-exchange chromatography in four different, closely related fractions (Linde et al. 1982).

8.4.1 Osteocalcin

Osteocalcin has been found in much higher concentrations in bone than in dentin matrix (Finkelman et al. 1990), where it has been reported to occur in moderate amounts (Camarda et al. 1987; Bronckers et al. 1989), to a very low degree (Bronckers et al. 1989), or to be virtually absent (Gorter de Vries et al. 1988a). Discrepancies have, in fact, been found between reports on the distribution of OC in dentin, at least partly due to differences in composition in different species (Linde 1988), or even to technical reasons (Bronckers et al. 1985, 1987a; Camarda et al. 1987).

Immunohistochemical investigations on the growth regions of rat tooth germs have shown that OC is present in young odontoblasts and in early predentin, right in the neighbourhood of positive odontoblast processes (Bronckers et al. 1987b; Gorter de Vries et al. 1988b), but not throughout the predentin (Bronckers et al. 1985, 1987a; Camarda et al. 1987), as if it was transported through the odontoblast processes directly into calcified dentin (Bronckers et al. 1985). This has been reported to be immunolabeled in the rat by OC antibodies (Gorter de Vries et al. 1988b; Bronckers et al. 1994; McKee and Nanci 1995), whose amounts are higher in mantle than in circumpulpal dentin (Gorter de Vries et al. 1987; McKee et al. 1996). Immunolabeling, on the other hand, has not been detected in the dentin of human and bovine teeth (Gorter de Vries et al. 1988a; Papagerakis et al. 2002), where a weak positive reaction was only found at the dentin-enamel junction (Gorter de Vries et al. 1988a). In human teeth, Papagerakis et al. (2002) found that OC is expressed at late stages of odontoblast differentiation, but is not localized extracellularly in predentin or intertubular dentin. An in vitro culture of rat molars showed that odontoblasts synthesize OC concurrently with the elaboration of predentin matrix, but independently of inorganic substance deposition (Finkelman and Butler 1985).

8.4.2 Dentin Matrix Gla-protein

Dentin matrix Gla-protein has been reported in dentin, where it was demonstrated using specific antibodies and radioimmunoassay (Price et al. 1985).

8.5 Dentin Non-collagenous Proteins: Phosphoproteins

On the basis of a histochemical study of dentin extracellular proteins, the following non-collagenous proteins were reported in 1989 by Bronckers et al. (1989): dentinophosphophoryn, osteocalcin, osteopontin and 95-kDa glycoprotein. In the book *Ten Cate's Oral Histology* published 14 years later, the list of dentin non-collagenous proteins is longer and comprises proteoglycans, dentin phosphoproteins/phosphophoryn, dentin sialophosphoprotein, dentin matrix protein, dentin matrix protein I, osteonectin, osteocalcin, bone sialoprotein, osteopontin and matrix extracellular phosphoglycoprotein (Nanci 2003). From this list – which will very probably become longer in the future – it can already be inferred that dentin shares some non-collagenous proteins with bone.

The presence of glycosilated non-collagenous proteins in dentin has long been known, and has been demonstrated using both biochemical (Holbrook and Leaver 1976) and histochemical techniques (Fullmer and Alpher 1958). ^{3}H-fucose is incorporated into glycoproteins by odontoblasts and is deposited at the calcification front in dentin (Weinstock A et al. 1972). Vicinal glycol-containing glycoconjugates have been shown with Thiéry's electron microscope method to be located on and between collagen fibrils at the calcification front (Takagi et al. 1981), where they are quickly transported after their synthesis by odontoblasts (Weinstock A et al. 1972).

Most dentin glycoproteins are phosphorylated. Phosphoproteins (DPPs) are found in dentin in two major molecular classes, but other minor forms are possible (Linde et al. 1980). Depending on the species and the degree of phosphorylation, Goldberg et al. (1995a) have listed three phosphoprotein groups: highly phosphorylated (containing 46% phosphoserine and 45% aspartic acid in bovine teeth and over 40% phosphoserine in rat incisors), moderately phosphorylated (25% phosphoserine), and weakly phosphorylated (5–7% phosphoserine) proteins. Actually, the phosphoprotein content does not appear to be homogeneous throughout the dentin: the portion of dentin facing the enamel contains about twice the organic phosphate, about 4 times the highly phosphorylated phosphoproteins, and about 1.4 times the amount of proteins found in the portion of dentin facing the cementum (Steinfort et al. 2004). Besides this, the phosphoproteins in peritubular dentin differ in their amino acid composition from that of the phosphophoryn in intertubular dentin (Weiner et al. 1999).

Phosphoproteins bind large numbers of calcium ions (Lee et al. 1977; Zanetti et al. 1981); a high proportion of them bind to collagen (Carmichael et al. 1971; Cohen-Solal et al. 1979; Veis 1993), and most are closely linked to the inorganic substance, so they can only be extracted after decalcification (Linde et al. 1980). Predentin does not contain DPPs, despite containing non-collagenous glycoproteins, as shown by microdissection and chemical analysis (Carmichael et al. 1975) and by autoradiography (Weinstock M and Leblond 1973). Radioautographic studies using ^{3}H-serine show that rat incisor phosphoproteins are related to secretory granules and are secreted by odontoblasts at the mineralization front, but are later incorporated in the calcified dentin (Inage and Toda 1988). The highly phosphorylated proteins have been named phosphophoryns (Dimuzio and Veis 1978a), a name that has been considered synonymous with all other phosphoproteins.

8.5.1 Phosphophoryn

Phosphophoryn is the most representative phosphoprotein in dentin (Dimuzio and Veis 1978a). It is a tooth-specific, calcium-binding member of a complex family of proteins which are highly phosphorylated – although to varying degrees – and are therefore highly acidic, their anionic character mostly depending on the phosphate groups in phosphoserine, and carboxyl groups in aspartic acid (about 50 and 40% of the amino acid residues, respectively; DiMuzio and Veis 1978a, b; Takagi and Veis 1984; Linde and Goldberg 1993; Chang et al. 1996; Gu et al. 2000). Even though phosphophoryn does not occur in predentin (Dimuzio and Veis 1978b; Jontell and Linde 1983; Nakamura et al. 1985; Veis et al. 1985; Gorter de Vries et al. 1986; Rahima et al. 1988), it is exclusively synthesized by odontoblasts (Bronckers et al. 1989) and can be detected in their cytoplasm and cytoplasmic processes, as well as in dentin (Takagi et al. 1986; Veis et al. 1985; Nakamura et al. 1985; Rahima et al. 1988). This suggests that phosphophoryn is directly secreted in, or quickly transported to, the calcification front, where collagen-phosphophoryn conjugates are assembled from precursors which follow different secretory pathways (Maier et al. 1983). Phosphophoryn immunoreaction is not detected in mantle dentin, whereas it is in circumpulpal dentin, where the degree of staining gradually falls from the predentin-dentin junction toward the enamel (Nakamura et al. 1985), and it is directly related to the distribution of the inorganic substance (Rahima et al. 1988). In vitro studies by Satoyoshi et al. (1995) have shown that after secretion by odontoblasts, phosphophoryn is accumulated in calcification nodules.

A number of observations have shown that phosphophoryn binds to collagen fibrils (Butler 1995). Beniash et al. (2000) were able to localize phosphophoryn at the boundary between the gap and overlap zones of the collagen fibrils nearest the calcification front. Turkey tendon collagen fibrils treated with rat incisor phosphophoryn in the presence of calcium ions were characterized by the formation of positively or negatively stained globular particles – considered to be phosphophoryn aggregates – predominantly at the *e* band of the fibril gap regions (Traub et al. 1992). By mixing purified rat incisor phosphophoryn and native, monomeric lathyritic rat skin collagen, a specific protein–protein interaction was shown under the electron microscope by the appearance of 15 nm thick phosphophoryn globules centered about 210 nm from the N-terminus of the 270 nm long collagen filaments (Dahl et al. 1998).

Because of its structural and constitutive properties, phosphophoryn can exert a profound influence on the calcification process, as instanced by its property of binding Ca^{2+} ions with a strong affinity (Zanetti et al. 1981; Stetler-Stevenson and Veis 1987). He et al. (2005) have reported that phosphorylation is essential to phosphophoryn's characteristic functions – dentin matrix protein 2, a non-phosphorylated, reconstituted portion of phosphophoryn, possesses, for example, a much lower calcium-binding capability than the native protein. Marsh (1989) has shown that rat phosphophoryn almost exclusively binds PO_4^{3-} ions and that for each bound phosphate ion, the phosphoprotein binds about 1.5 Ca^{2+} ions.

In solution, phosphophoryn forms calcium complexes that reduce the net ion activity coefficient, so inhibiting calcification; by contrast, when immobilized on collagen fibrils or other substrates, it promotes hydroxyapatite formation even when present in very small amounts (Lussi et al. 1988; Linde et al. 1989; Lussi and Linde 1993), a property it shares with other proteins (Addadi and Weiner 1985; Doi et al. 1993; Lussi and Linde 1993). Proteins immobilized on crystal surfaces, through their adhesion to stereospecific crystal parts, can either inhibit further ion apposition and crystal growth, or can act as templates able to induce further inorganic ion apposition. In these cases, the organic-inorganic molecular complementarity might be regulated by cation-induced conformational folding-unfolding processes, i. e., transitions from a 'random coil' state to a 'predominantly' β-sheet structure, within the proteins themselves (Evans and Chan 1994).

In vitro studies relying on a gelatin gel diffusion system have shown that the effect of this protein on calcium phosphate deposition is concentration-dependent: at low concentrations (0.01–1.0 μg/ml), phosphophoryn promotes hydroxyapatite formation, whereas at higher ones (100 μg/ml) it inhibits hydroxyapatite growth and induces a fall in numbers of mineral clusters pointing to an inhibition of secondary nucleation (Boskey et al. 1990). According to George et al. (1996), the effects of phosphophoryn on calcification depend on the presence at its carboxy-terminal domain of triplet amino acid repeat sequences which form ordered carboxyl-phosphate interaction ridges; these may be involved either in the binding of phosphophoryn to specific collagen fibril sites, so favoring the onset of crystal nucleation, or in regulating crystal growth.

8.5.2
Osteonectin

Osteonectin (ON) is a component of the organic matrix in fetal calf dentin (Termine et al. 1981). It appears to be barely detectable (Salonen et al. 1990) or absent in rat dentin, but present in porcine dentin (Goldberg et al. 1995a) and in the predentin of developing human teeth (Papagerakis et al. 2002). It has been found in odontoblasts – as SPARC – using in situ hybridization (Holland et al. 1987). It has been observed immunohistochemically in the dentin of unerupted porcine teeth and in the associated alveolar bone, whereas in erupted teeth ON immunostaining was concentrated around dentinal tubules; both odontoblasts and ameloblasts in fetal tissues stained for ON and type III collagen (Tung et al. 1985). In human teeth, odontoblasts, their processes and predentin all show an intense immunological reaction (Reichert et al. 1992).

8.5.3
Osteopontin

Osteopontin (OPN) has been shown immunohistochemically to be located in small calcification nodules in mantle predentin (McKee et al. 1996; Bosshardt and Nanci 2000) and in interfibrillar reticular patches in decalcified mantle dentin (Bosshardt and Nanci 2000), whereas its presence in circumpulpar dentin has not been completely clarified (Linde 1992). Weak OPN antigenicity was found in predentin by Helder et al. (1993), but no OPN transcripts could be detected in odontoblasts, either by in situ hybridization or Northern blotting, whereas strong positive signals were detected in the osteoblasts of alveolar bone. Bronckers et al. (1989, 1994) did, however, report that, in advanced stages of dentinogenesis, OPN could be detected immunologically in odontoblasts and in dentin matrix. In a comparative SDS-PAGE and Western immunoblot study of sialic acid-rich proteins in rat dentin and bone, it was found that the OPN level in the former is less than one-seventieth of that in the latter (Qin et al. 2001; Butler et al. 2003).

8.5.4 Dentin Sialoprotein

Dentin sialoprotein (DSP), also known as 95k glycoprotein (Butler et al. 1985; Dimuzio et al. 1985) or 53-kDa DSP (Butler et al. 1992), is a dentin-specific glycoprotein which includes 30% carbohydrates, about 9% of which are sialic acid (Butler et al. 1992). It is a member of the SIBLING family (Fisher et al. 2001), which includes, besides bone osteopontin and bone sialoprotein, dentin sialoprotein, dentin matrix protein 1 and matrix extracellular phosphoglycoprotein (Qin et al. 2004); all of these are distinguished by their relatively large amounts of sialic acid and phosphate. According to Yamakoshi et al. (2005), porcine DSP is a proteoglycan with glycosaminoglycan chains containing chondroitin 6-sulfate. Its amino acids comprise aspartic acid, serine, glutamic acid and glycine (Butler 1998). DSP, however, does not contain the Arg-Gly-Asp sequence which is found in BSP and OPN and which promotes cell attachment activity (Butler et al. 1992; Ritchie et al. 1994). The total phosphorylation level of DSP is not known exactly; it has been considered a non-phosphorylated sialoprotein (Goldberg et al. 1995a), but 6.2 phosphates per molecule have been reported as a feature of DSP in the rat (Qin et al. 2003a). Very acidic, high molecular weight isoforms of DSP (designated HMW-DSP) have been isolated by antibody affinity chromatography (Qin et al. 2003a).

Dentin sialoprotein is expressed by odontoblasts and, transitorily, by pre-secretory ameloblasts (Papagerakis et al. 2002); it occurs in early predentin (D'Souza et al. 1992). Dentin sialoprotein mRNA has been reported to be expressed by young and mature odontoblasts and, transiently, by pre-ameloblasts, whereas no DSP mRNA expression has been observed in a number of soft tissues (Ritchie et al. 1997). Interestingly, DSP and DPP are expressed as a single RNA transcript, which codes for the large precursor protein dentin sialophosphoprotein (DSPP), whose proteolytic cleavage liberates DSP and DPP (MacDougall et al. 1997; Butler et al. 2002, 2003). The dssp gene is not tooth-specific and is expressed in calvaria (Qin et al. 2002), even if at a much lower level than in odontoblasts; the level of DSP in bone is, in fact, about 1/400 of its level in dentin (Qin et al. 2003b).

In developing rat molars and incisors, early immunostaining for DSP has been found in newly differentiated odontoblasts; the staining level later gradually rose in predentin, odontoblasts and dentin (Bronckers et al. 1993). A strong DSP immunoreaction has been detected in predentin and in the dentin tubules of developing rat molars (Baba et al. 2004). In advanced stages of rat dentinogenesis, both crown and root odontoblasts and dentin were immunolabeled for DSP (Bronckers et al. 1994).

Little is known about the effects of DSP on calcification (Ritchie et al. 1998). Using an in vitro gelatin gel diffusion system, Boskey et al. (2000) found that at low concentrations (< 25 μg/ml) DSP slightly raised the yield of hydroxyapatite formed at 3.5 and 5 days, whereas at higher concentrations (50–100 μg/ml) it slightly inhibited accumulation; their overall conclusion was that the role of DSP in dentin is not primarily that of a calcification regulator.

Bone sialoprotein is detected in small amounts in dentin: Chen et al. (1993) reported BSP immunoreactivity in odontoblasts and their processes and in the peritubular dentin of developing porcine teeth. It has been identified in small calcification nodules in the mantle dentin (McKee et al. 1996).

8.5.5 Acidic Glycoprotein-75

Acidic glycoprotein-75 is similar to the BAG-75 protein already mentioned in bone. It was detected in dentin (as well as in bone) as an 83-kDa band, distinct from that of DMP1 (Qin et al. 2001).

8.5.6 Dentin Matrix Protein-1

Dentin matric protein-1 (DMP1) is a very acidic, phosphorylated matrix glycoprotein (George et al. 1993, 1995) which was considered to be specific for dentin and other calcified tissues (Feng et al. 2003), but is now recognized as being expressed not only by odontoblasts, but also by osteoblasts, ameloblasts and cementoblasts (D'Souza et al. 1997; Srinivasan et al. 1999; Thotakura et al. 2000; MacDougall et al. 1998; Toyosawa et al. 2004), as well as non-calcified tissues and organs such as the liver, brain, muscle, pancreas and kidney (Terasawa et al. 2004).

In situ hybridization studies on rat molars showed that DMP1 mRNA was expressed in root odontoblasts in parallel with dentin calcification, but the level of expression fell near the coronal part and disappeared in coronal odontoblasts (Toyosawa et al. 2004). Immunohistochemical studies carried out in parallel with those just mentioned showed that DMP1 protein was localized in the tubules of calcified root dentin, but shifted towards their base as dentin formation progressed. Immunoreaction for DMP1 was again detected in dentinal tubules by Baba et al. (2004), who found a positive DMP1 immunoreaction in the predentin and odontoblasts of molars in rats older than two weeks, and a strong and widely distributed immunoreaction in the odontoblasts and predentin of five- and eight-week-old rats, the same localization as DSP. The presence in dentin of 150–200 kDa DMP1 was reported by Butler et al. (2002). According to He and George (2004), DMP1 is localized in the gap region of the collagen fibrils, to which it binds specifically through two acidic clusters. Immunocytochemistry studies by Massa et al. (2005) detected DMP1 around calcification nodules when calcification spreads from matrix vesicles into the surrounding matrix.

8.5.7 Matrix Extracellular Phosphoglycoprotein

Matrix extracellular phosphoglycoprotein (MEPE) is expressed not only in bone (see Sect. 7.3.3), but also in dental tissues, in particular odontoblasts (MacDougall et al. 2002). The MEPE gene has been reported to be located in a segment of the human chromosome 4q21, which also includes osteopontin, bone sialoprotein, dentin sialophosphoprotein and dentin matrix protein 1, where genetic linkage studies have identified the critical loci for dentinogenesis imperfecta II and III and for dentin dysplasia type II (MacDougall 2003; MacDougall et al. 2002).

8.5.8 α_2HS Glycoprotein

α_2HS glycoprotein is found in the dentin matrix at concentrations higher than those detected in the bone matrix. Immunohistochemical studies have, however, failed to show the epitope in dentin, whereas a positive reaction was recorded in both bovine and human peritubular dentin – but not in intertubular dentin – when the immunoreaction was carried out after etching dentin surface with diluted HCl (Takagi et al. 1990b). Positive results have also been obtained in dentin by post-embedding immunohistochemistry (McKee and Nanci 1995).

8.6 Dentin Lipids

The possibility that lipids are components of the dentin matrix was initially suggested by Irving (1959), who, after pyridine extraction, showed the presence of sudanophilic material at the calcification front between predentin and dentin. Lipids were then extracted from the decalcified dentin, of which they are minor components (20.28 mg/100 g dentin compared with 2005 mg/100 g human bone; Dirksen and Marinetti 1970). They include cholesterol, cholesterol esters, triacylglycerols and phospholipids, which consist of phosphatidylcholine and small amounts of phosphatidylethanolamine, sphyngomyelin and diphosphatidylglycerol (Wuthier 1984). Phospholipids mainly associated with cell membranes can be extracted from undecalcified dentin by lipid solvents; another phospholipid fraction, probably linked to the mineral phase and mainly consisting of acidic phospholipids (above all, phosphatidylserine), can be extracted after decalcification (Dirksen and Marinetti 1970; Goldberg and Septier 2002). Ellingson et al. (1977) showed that bovine predentin contained a significantly lower percentage of phosphatidylserine and sphingomyelin than dental pulp, and that the major phospholipid of undecalcified dentin was phosphatidylcholine (52–56% of total phospholipids). In murine molar dentin, Dunglas et al. (1999) found a gradual accumulation of phospholipids in the extracellular matrix at an early stage of tooth germ and, later, as predentin/dentin and enamel components that participate in the calcification process; phospholipids containing phosphatidylethanolamine, phosphatidylserine and sphingomyelin were found in the extracellular matrix, where they constituted 68% of the decalcified lipid extract.

Using electron microscopy and malachite green-glutaraldehyde fixation (Goldberg and Septier 1985), rapid-freezing and malachite green-acrolein-osmium tetroxide freeze-substitution fixation (Goldberg and Escaig 1987), osmium fixation-*p*-phenylenediamine staining (Goldberg et al. 1985), iodoplatinate (Vermelin et al. 1994), or imidazole-osmium tetroxide fixation (Goldberg and Septier 2002), phospholipids have been demonstrated in interfibrillar spaces in predentin and as needle-like structures along the collagen fibrils and at the surface of calcified areas in dentin (Goldberg et al. 1995b). They appear to be combined with filamentous proteoglycan structures (Chardin et al. 1990).

8.7 Dentin Matrix Vesicles

Matrix vesicles are components of the mantle predentin organic matrix (reviewed by Bonucci 1984), where vesicles similar to those found in bone have been described several time (Bernard 1972; Eisenmann and Glick 1972; Sisca and Provenza 1972; Katchburian 1973a, b; Katchburian and Severs 1982; Hayashi 1983; Kitamura et al. 1991; Diekwisch et al. 1995; Sasaki and Garant 1996; Stratmann et al. 1996). They have been found in mantle dentin (Fig. 8.1) but not in circumpulpal dentin (Katchburian 1973a; Slavkin et al. 1976; Larsson 1973, 1974; Larsson and Bloom 1973; Almuddaris and Dougherty 1979; Takano et al. 1986, 2000; Stratmann et al. 1997). They are also present in the reparative dentin of healing pulp in rats and dogs (Sela et al. 1981; Hayashi 1982; Yamada et al. 1987) and were detected in the mineralized tissues developed within cell multilayers on day 28 of a culture of dental pulp cells (clone RCP-K) isolated from rat incisors (Hayashi et al. 1993). Their origin seems to be no different from that of matrix vesicles in bone and cartilage, i. e., they arise by budding off from the odontoblast processes and membrane (Almuddaris and Dougherty

1979; Katchburian and Severs 1982); their presence around degenerate and necrotic cells in culture was reported by Hayashi et al. (1993).

Matrix vesicles in predentin are closely associated with proteoglycans (Tenorio et al. 1990). They are stained by ZIO (zink-iodide-osmium), a method which stains lipidic substances (Uchida and Ozawa 1984), and react with the MC22-33F antibody for coline phospholipids (Tsuji et al. 1994). According to Dunglas et al. (1999), matrix vesicles in dentin are the source of some of the lipids detected in the matrix.

8.8 Cementum

Cementum is the least known calcified tissue in vertebrates (Bosshardt and Schroeder 1996) and several questions about cementogenesis have not yet been answered. This partly depends on the thinness of the cementum layer, which makes its separation from the other components of the dental root and, therefore, study of its biochemistry, quite difficult. As a result, some particular cementum-related structures – such as the dentino-cemental junction and the relationship between cellular and acellular cementum – are best studied by morphological methods.

Cementum comprises acellular, or primary, cementum, which covers the upper, or cervical, portion of the root and contains the fibers of the periodontal ligament which connect bone to cementum (Sharpey's fibers), and cellular, or secondary, cementum, which, besides the fibrils of the periodontal ligament, contains cementocytes and covers the lower, apical portion of the root. One can also distinguish between acellular extrinsic fiber cementum, which consists of a thin layer of calcified matrix and a fringe of collagen fibrils perpendicular to the root surface; acellular afibrillar cementum, which is located over enamel and dentin near the cemento-enamel junction; cellular intrinsic fiber cementum, whose fiber bundles run parallel to the root surface; and mixed (intrinsic and extrinsic) fiber cementum, which constitutes the bulk of secondary cementum (for details see Nanci 2003). It is striking that the degree of calcification of afibrillar cementum exceeds that of fibrillar cementum (Kodaka and Debari 2002). A thin layer of uncalcified cementum (cementoid) separates cementoblasts from the calcified matrix during cementogenesis.

The nature of the cementum-forming cells (cementoblasts) is one of the still unraveled questions; these are considered phenotypically different from bone-forming cells by some investigators (e.g., Grzesik et al. 2000) and phenotypically similar to osteoblasts by others (discussed by Bosshardt, Nanci and Somerman in Nanci 2003, pp 250–251). In any case, the organic components of the cementum matrix, especially those of cellular intrinsic fiber cementum, resemble those of bone matrix (Bosshardt and Schroeder 1992, 1993; D'Errico et al. 1995). Type I collagen, the major constituent of the organic matrix of bone, is also the major constituent of the cementum matrix (Birkedal-Hansen et al. 1977; Christner et al. 1977); as in bone, the cementum interfibrillary spaces contain glycosaminoglycans, such as protein-bound hyaluronic acid, chondroitin sulfate and dermatan sulfate (Bartold et al. 1988, 1990). Two families of chondroitin sulfate proteoglycans have been detected, mainly in the uncalcified fraction of cementum (Cheng et al. 1999): the first contained only chondroitin-4-sulfate and corresponded to decorin and biglycan, the second contained chondroitin-4-sulfate and chondroitin-6-sulfate and corresponded to versican. Immunostaining for decorin was mainly associated with collagen fibers in the periodontal ligament and, to a slight degree, in cementum matrix, that for biglycan with cementoblasts and precementum, and that for versican with the lacunae of cementocytes. Keratan sulfate proteoglycans such as lumican and fibromodulin have been found almost exclusively in non-calcified precementum and pericementocyte areas (Cheng et al. 1996). Osteoadherin was found in matrix of cementum by Petersson et al. (2003) using immunohistochemistry.

As in bone, the interfibrillary spaces contain non-collagenous proteins, mostly phosphoproteins (Glimcher and Lefteriou 1989), such as osteocalcin, osteonectin, osteopontin and bone sialoprotein (see Table I in Nanci 2000). In bovine cementum, the non-collagenous fraction of the organic matrix contains *o*-phosphoserine, *o*-phosphothreonine and γ-carboxyglutamic acid; the content of this last is less than that in bone (Glimcher et al. 1979). Collagen fibrils thinner than those of type I collagen have been described by Jande and Bélanger (1970).

Immunohistochemical studies (reviewed by Goldberg et al. 1995a) confirm that the composition of the cementum matrix is similar to that of bone, with a few minor differences. In this connection, acellular extrinsic fiber cementum has wider interfibrillary spaces, and contains more non-collagenous proteins, than cellular intrinsic fiber cementum (Bosshardt et al. 1998), so recalling the relationship between collagen fibrils and non-collagenous proteins that occurs in woven as compared with compact bone (see Sect. 7.2). As in bone, the interfibrillary spaces of the cementum matrix contain glycosilated phosphoproteins (Glimcher and Lefteriou 1989) which, with minor variations, resemble those of the bone matrix. So too, during early tooth development, an OPN-containing cement line – like those found in bone – may occasionally appear at the interface between dentin and cementum (McKee et al. 1996), and BSP may be immunologically localized in the cemental root surface, coinciding with the initiation of root formation and cementogenesis (MacNeil et al. 1995a, b). Acellular cementum – which is synthesized by cementoblasts expressing osteocalcin (Kagayama et al. 1997) – immunostains for OPN (Bronckers et al. 1994), which could, however, at least partly derive from circulating blood (Vandenbos et al. 1999), whereas it does not stain for osteocalcin or DSP (Bronckers et al. 1994). Cellular cementum, on the other hand, stains for osteocalcin and OPN and not for DSP (Bronckers et al. 1994) and only lightly stains for ON (Tung et al. 1985). Acellular afibrillar cementum contains no DSP, but is immunoreactive for BSP, OPN, osteocalcin and α_2HS-glycoprotein (Bosshardt and Nanci 1997), which, according to McKee et al. (1996), are prominent organic constituents of both cellular and acellular cementum. OPN and BSP essentially show the same distribution and predominate in interfibrillary regions (Bosshardt and Nanci 2000), where the protein A-gold complexes indicative of positive immunoreaction are found over an extensive reticular network of fine filaments (McKee and Nanci 1995). As in bone and dentin, so in the cementum decalcified matrix, OPN and BSP labeling associates with isolated or interconnected, interfibrillar patches of reticulated organic matrix (McKee et al. 1996; Nanci 1999).

Light microscopy immunohistochemistry and in situ hybridization studies have shown that BSP is localized in the cementum root surface, coinciding with the initiation of root formation and cementogenesis, and that OPN is distributed in a non-specific fashion throughout the periodontal ligament and the eruption pathway of the forming tooth (MacNeil et al. 1995a).

Two polypeptides have been isolated from cementum – the cementum-derived growth factor, or CGF, and the cementum attachment protein, or CAP, which promote the growth and attachment of periodontal cells, respectively (Narayanan et al. 1995). Proteinaceous chemotactic factors, which have been detected in the cementum, root dentin and alveolar bone, appear to regulate the migration and orientation of the cells of the periodontium (Ogata et al. 1997).

Matrix vesicles have been described by Hayashi (1985, 1987) in the apical cementum of cats, and by Takano et al. (2000) in that of rats.

8.9 Concluding Remarks

Dentin and cementum are fibrous tissues dedicated to supporting strong mechanical functions. They share a number of structural and constitutive properties with bone.

- The organic matrix of dentin and of cementum – as well as that of bone – mainly consists of type I collagen fibrils. These are irregularly oriented in dentin, so resembling woven bone.
- Like the osteoid border that separates osteoblasts from the calcified bone matrix during osteogenesis, predentin separates odontoblasts from calcified dentin during dentinogenesis and the cementoid separates cementoblasts from calcified cementum during cementogenesis. In all cases, a more or less developed calcification front is interposed between the calcified and uncalcified matrix.
- Very few collagen fibrils are found in peritubular dentin and none occur in acellular afibrillar cementum. In both cases, the matrix consists of a framework of thin filaments.

- The degree of calcification in peritubular dentin is higher than that of intertubular dentin, while that of acellular afibrillar cementum is higher than that of fibrillar cementum.
- The interfibrillary spaces of both tissues contain non-collagenous material, which comprises proteoglycans, Gla-proteins, glycoproteins and lipids. Some of these are buried in the inorganic substance and can only be completely extracted after decalcification.
- Proteoglycans in predentin and dentin are rich in chondroitin sulfate. The predominant ones are the chondroitin-sulphate-rich decorin and biglycan, and the keratan-sulphate-, leucine-rich lumican.
- Amounts of proteoglycans fall with increasing calcification.
- Dentin matrix contains fewer Gla-proteins than bone matrix; its osteocalcin content is variable, depending on the species.
- As in bone, most glycoproteins are acidic and phosphorylated.
- Dentin contains a unique phosphoprotein, called phosphophoryn, which does not occur in predentin. It is bound to collagen fibrils.
- Phosphophoryn is a phosphorylated, highly acidic, calcium-binding protein.
- Besides phosphophoryn, dentin contains phosphoproteins common to bone and cementum; these comprise osteonectin, osteopontin, bone sialoprotein and varying amounts of dentin sialoprotein, dentin matrix protein-1, acidic glycoprotein-75, matrix extracellular phosphoglycoprotein and α_2HS glycoprotein.
- Osteopontin and bone sialoprotein generally codistribute in dentin. They have been demonstrated by immunohistochemistry in cement lines, laminae limitantes and reticular, fine filamentous, interfibrillar material. At the calcification front, this material forms "grey" patches which correspond to the decalcified matrix of the calcification nodules.
- Lipids are present in dentin in smaller amounts than in bone; some of them can only be extracted after decalcification.
- Matrix vesicles are found in both dentin and cementum, but in the former they have only been detected in mantle dentin, not in circumpulpal dentin.

References

Addadi L, Weiner S (1985) Interactions between acidic proteins and crystals: stereochemical requirements in biomineralization. Proc Natl Acad Sci USA 82:4110–4114

Almuddaris MF, Dougherty WJ (1979) The association of amorphous mineral deposits with the plasma membrane of pre- and young odontoblasts and their relationship to the origin of dentinal matrix vesicles in rat incisor teeth. Am J Anat 155:223–244

Baba O, Qin C, Brunn JC, Wygant JN, McIntyre BW, Butler WT (2004) Colocalization of dentin matrix protein 1 and dentin sialoprotein at late stages of rat molar development. Matrix Biol 23:371–379

Bartold PM, Miki Y, McAllister B, Narayanan AS, Page RC (1988) Glycosaminoglycans of human cementum. J Periodont Res 23:13–17

Bartold PM, Reinboth B, Nakae H, Narayanan AS, Page RC (1990) Proteoglycans of bovine cementum: isolation and characterization. Matrix 10:10–19

Beniash E, Traub W, Veis A, Weiner S (2000) A transmission electron microscope study using vitrified ice sections of predentin: structural changes in the dentin collagenous matrix prior to mineralization. J Struct Biol 132:212–225

Bernard GW (1972) Ultrastructural observations of initial calcification in dentine and enamel. J Ultrastruct Res 41:1–17

Birkedal-Hansen H, Butler WT, Taylor RC (1977) Ultrastructural observations of initial calcification in dentine and enamel. Calcif Tissue Res 23:39–44

Bonucci E (1984) Matrix vesicles: their role in calcification. In: Linde A (ed) Dentin and dentinogenesis, 1st vol. CRC Press, Boca Raton, pp 135–154

Boskey AL, Maresca M, Doty S, Sabsay B, Veis A (1990) Concentration-dependent effects of dentin phosphophoryn in the regulation of in vitro hydroxyapatite formation and growth. Bone Miner 11:55–65

Boskey A, Spevak L, Tan M, Doty SB, Butler WT (2000) Dentin sialoprotein (DSP) has limited effects on in vitro apatite formation and growth. Calcif Tissue Int 67:472–478

Bosshardt DD, Nanci A (1997) Immunodetection of enamel- and cementum-related (bone) proteins at the enamel-free area and cervical portion of the tooth in rat molars. J Bone Miner Res 12:367–379

Bosshardt DD, Nanci A (2000) The pattern of expression of collagen determines the concentration and distribution of non-collagenous proteins along the forming root. In: Goldberg M, Boskey A, Robinson C (eds) Chemistry and biology of mineralized tissues. American Academy of Orthopaedic Surgeons, Rosemont, IL, pp 129–136

Bosshardt DD, Schroeder HE (1992) Initial formation of cellular intrinsic fiber cementum in developing human teeth. A light- and electron-microscopic study. Cell Tissue Res 267:321–335

Bosshardt DD, Schroeder HE (1993) Attempts to label matrix synthesis of human root cementum in vitro. Cell Tissue Res 274:343–352

Bosshardt DD, Schroeder HE (1996) Cementogenesis reviewed: a comparison between human premolars and rodent molars. Anat Rec 245:267–292

Bosshardt DD, Zalzal S, McKee MD, Nanci A (1998) Developmental appearance and distribution of bone sialoprotein and osteopontin in human and rat cementum. Anat Rec 250:1–21

Bronckers ALJJ, Gay S, Dimuzio MT, Butler WT (1985) Immunolocalization of γ-carboxyglutamic acid-containing proteins in developing molar tooth germs of the rat. Collagen Rel Res 5:17–22

Bronckers ALJJ, Gay S, Finkelman RD, Butler WT (1987a) Immunolocalization of Gla proteins (osteocalcin) in rat tooth germs: comparison between indirect immunofluorescence, peroxidase-antiperoxidase, avidin-biotin-peroxidase complex, and avidin-biotin-gold complex with silver enhancement. J Histochem Cytochem 35:825–830

Bronckers AL, Gay S, Finkelman RD, Butler WT (1987b) Developmental appearance of Gla proteins (osteocalcin) and alkaline phosphatase in tooth germs and bones of the rat. Bone Miner 2:361–373

Bronckers ALJJ, Lyaruu DM, Wöltgens JHM (1989) Immunohistochemistry of extracellular matrix proteins during various stages of dentinogenesis. Connect Tissue Res 22:65–70

Bronckers AL, D'Souza RN, Butler WT, Lyaruu DM, Van Dijk S, Gay S, Woltgens JH (1993) Dentin sialoprotein: biosynthesis and developmental appearance in rat tooth germs in comparison with amelogenins, osteocalcin and collagen type-I. Cell Tissue Res 272:237–247

Bronckers AL, Farach-Carson MC, Van Waveren E, Butler WT (1994) Immunolocalization of osteopontin, osteocalcin, and dentin sialoprotein during dental root formation and early cementogenesis in the rat. J Bone Miner Res 9:833–841

Buchaille R, Couble M-L, Magloire H, Bleicher F (2000) Expression of the small leucine-rich proteoglycan osteoadherin/osteomodulin in human dental pulp and developing rat teeth. Bone 27:265–270

Butler WT (1984a) Dentin collagen: chemical structure and role in mineralization. In: Linde A (ed) Dentin and dentinogenesis, 2nd vol. CRC Press, Boca Raton, pp 37–54

Butler WT (1984b) Matrix macromolecules of bone and dentin. Collagen Rel Res 4:297–307

Butler WT (1995) Dentin matrix proteins and dentinogenesis. Connect Tissue Int 33:59–65

Butler WT (1998) Dentin matrix protein. Eur J Oral Sci 106:204–210

Butler WT, Ritchie H (1995) The nature and functional significance of dentin extracellular matrix proteins. Int J Dev Biol 39:169–179

Butler WT, Sato S, Rahemtulla F, Prince CW, Tomana M, Bhown M, Dimuzio MT, Bronckers ALJJ (1985) Glycoproteins of bone and dentin. In: Butler WT (ed) The chemistry and biology of mineralized tissues. EBSCO Media, Birmingham, pp 107–112

Butler WT, Bhown M, Brunn JC, D'Souza RN, Farach-Carson MC, Happonen R-P, Schrohenloher RE, Seyer JM, Somerman MJ, Foster RA, Tomana M, Van Dijk S (1992) Isolation, characterization and immunolocalization of a 53-kDal dentin sialoprotein (DSP). Matrix 12:343–351

Butler WT, Brunn JC, Qin C, McKee MD (2002) Extracellular matrix proteins and the dynamics of dentin formation. Connect Tissue Res 43:301–307

Butler WT, Brunn JC, Qin C (2003) Dentin extracellular matrix (ECM) proteins: comparison to bone ECM and contribution to dynamics of dentinogenesis. Connect Tissue Res 44:171–178

Camarda AJ, Butler WT, Finkelman RD, Nanci A (1987) Immunocytochemical localization of gamma-carboxyglutamic acid-containing proteins (osteocalcin) in rat bone and dentin. Calcif Tissue Int 40:349–355

Carmichael DJ, Veis A, Wang ET (1971) Dentin matrix collagen: evidence for a covalently linked phosphoprotein attachment. Calcif Tissue Res 7:331–344

Carmichael DJ, Chovelon A, Pearson CH (1975) The composition of the insoluble collagenous matrix of bovine predentine. Calcif Tissue Res 17:263–271

Chang SR, Chiego D Jr, Clarkson BH (1996) Characterization and identification of a human dentin phosphophoryn. Calcif Tissue Int 59:149–153

Chardin H, Septier D, Goldberg M (1990) Visualization of glycosaminoglycans in rat incisor predentin and dentin with cetylpyridinium chloride-glutaraldehyde as fixative. J Histochem Cytochem 38:885–894

Chen J, McCulloch CAG, Sodek J (1993) Bone sialoprotein in developing porcine dental tissues: Cellular expression and comparison of tissue localization with osteopontin and osteonectin. Arch Oral Biol 38:241–249

Cheng H, Caterson B, Neame PJ, Lester GE, Yamauchi M (1996) Differential distribution of lumican and fibromodulin in tooth cementum. Connect Tissue Res 34:87–96

Cheng H, Caterson B, Yamauchi M (1999) Identification and immunolocalization of chondroitin sulfate proteoglycans in tooth cementum. Connect Tissue Res 40:37–47

Christner P, Robinson P, Clark CC (1977) A preliminary characterization of human cementum collagen. Calcif Tissue Res 23:147–150

Cohen-Solal L, Lian JB, Kossiva D, Glimcher MJ (1979) Identification of organic phosphorus covalently bound to collagen and non-collagenous proteins of chicken-bone matrix. Biochem J 177:81–98

Couble M-L, Bleicher F, Farges JC, Peyrol S, Lucchini M, Magloire H, Staquet MJ (2004) Immunodetection of osteoadherin in murine tooth extracellular matrix. Histochem Cell Biol 121:47–53

D'Errico JA, MacNeil RL, Strayhorn CL, Piotrowski BT, Somerman MJ (1995) Models for the study of cementogenesis. Connect Tissue Res 33:9–17

D'Souza RN, Bronckers AL, Happonen R-P, Doga DA, Farach-Carson MC, Butler WT (1992) Developmental expression of a 53 KD dentin sialoprotein in rat tooth organs. J Histochem Cytochem 40:359–366

D'Souza RN, Cavender A, Sunavala G, Alvarez J, Ohshima T, Kulkarni AB, MacDougall M (1997) Gene expression patterns of murine dentin matrix protein 1 (Dmp1) and dentin sialophosphoprotein (DSPP) suggest distinct developmental functions in vivo. J Bone Miner Res 12:2040–2049

Dahl T, Sabsay B, Veis A (1998) Type I collagen-phosphophoryn interactions: specificity of the monomer-monomer binding. J Struct Biol 123:162–168

Diekwisch TGH, Berman BJ, Gentner S, Slavkin HC (1995) Initial enamel crystals are not spatially associated with mineralized dentine. Cell Tissue Res 279:149–167

Dimuzio MT, Veis A (1978a) Phosphophoryns - Major noncollagenous proteins of rat incisor dentin. Calcif Tissue Res 25:169–178

Dimuzio MT, Veis A (1978b) The biosynthesis of phosphophoryns and dentin collagen in the continuously erupting rat incisor. J Biol Chem 253:6845–6852

Dimuzio MT, Bhown M, Walton RK, Butler WT (1985) Bone and dentin organ cultures. Non collagenous protein biosynthesis. In: Butler WT (ed) The chemistry and biology of mineralized tissues. EBSCO Media, Birmingham, pp 296–302

Dirksen TR, Marinetti GV (1970) Lipids of bovine enamel and dentin and human bone. Calcif Tissue Res 6:1–10

Doi Y, Horiguchi T, Kim S-H, Moriwaki Y, Wakamatsu N, Adachi M, Shigeta H, Sasaki S, Shimokawa H (1993) Immobilized DPP and other proteins modify OCP formation. Calcif Tissue Int 52:139–145

Dunglas C, Septier D, Carreau JP, Goldberg M (1999) Developmentally regulated changes in phospholipid composition in murine molar tooth. Histochem J 31:535–540

Eisenmann DR, Glick PL (1972) Ultrastructure of initial crystal formation in dentin. J Ultrastruct Res 41:18–28

Ellingson JS, Smith M, Larson LR (1977) Phospholipid composition and fatty acid profiles of the phospholipids in bovine predentin. Calcif Tissue Res 24:127–133

Embery G, Hall R, Waddington R, Septier D, Goldberg M (2001) Proteoglycans in dentinogenesis. Crit Rev Oral Biol Med 12:331–349

Engfeldt B, Hjerpe A (1972) Glycosaminoglycans of dentine and predentine. Calcif Tissue Res 10:152–159

Evans JS, Chan SI (1994) Phosphophoryn, an "acidic" biomineralization regulatory protein: conformational folding in the presence of Cd(II). Biopolymers 34:1359–1375

Everts V, Schutter M, Niehof A (1995) Proteoglycans in cementum- and enamel-related predentin of young mouse incisors as visualized by cuprolinic blue. Tissue Cell 27:55–60

Feng JQ, Huang H, Lu Y, Ye L, Xie Y, Tsutsui TW, Kunieda T, Castranio T, Scott G, Bonewald LB, Mishina Y (2003) The Dentin matrix protein 1 (Dmp1) is specifically expressed in mineralized, but not soft, tissues during development. J Dent Res 82:776–780

Finkelman RD, Butler WT (1985) Appearance of dentin γ-carboxyglutamic acid-containing proteins in developing rat molars *in vitro*. J Dent Res 64:1008–1015

Finkelman RD, Mohan S, Jennings JC, Taylor AK, Jepsen S, Baylink DJ (1990) Quantitation of growth factors IGF-I,SGF/IGF-II, and TGF-β in human dentin. J Bone Miner Res 5:717–723

Fisher LW, Termine JD (1985) Noncollagenous proteins influencing the local mechanisms of calcification. Clin Orthop Relat Res 200:362–385

Fisher LW, Torchia DA, Fohr B, Young MF, Fedarko NS (2001) Flexible Structures of SIBLING Proteins, Bone Sialoprotein, and Osteopontin. Biochem Biophys Res Commun 280:460–465

Fujisawa R, Kuboki Y (1992) Affinity of bone sialoprotein and several other bone and dentin acidic proteins to collagen fibrils. Calcif Tissue Int 51:438–442

Fullmer HM, Alpher N (1958) Histochemical polysaccharide reactions in human developing teeth. Lab Invest 7:163–170

George A, Sabsay B, Simonian PA, Veis A (1993) Characterization of a novel dentin matrix acidic phosphoprotein. Implications for induction of biomineralization. J Biol Chem 268:12624–12630

George A, Silberstein R, Veis A (1995) *In situ* hibridization shows Dmp1 (AG1) to be a developmentally regulated dentin-specific protein produced by mature odontoblasts. Connect Tissue Res 33:67–72

George A, Bannon L, Sabsay B, Dillon JW, Malone J, Veis A, Jenkins NA, Gilbert DJ, Copeland NG (1996) The carboxyl-terminal domain of phosphophoryn contains unique extended triplet amino acid repeat sequences forming ordered carboxyl-phosphate interaction ridges that may be essential in the biomineralization process. J Biol Chem 271:32869–32873

Glimcher MJ, Lefteriou B (1989) Soluble glycosylated phosphoproteins of cementum. Calcif Tissue Int 45:165–172

Glimcher MJ, Lefteriou B, Kossiva D (1979) Identification of O-phosphoserine, O-phosphothreonine and γ-carboxyglutamic acid in the non-collagenous proteins of bovine cementum; comparison with dentin, enamel and bone. Calcif Tissue Int 28:83–86

Goldberg M, Escaig F (1987) Rapid-freezing and malachite green-acrolein-osmium tetroxide freeze-substitution fixation improve visualization of extracellular lipids in rat incisor pre-dentin and dentin. J Histochem Cytochem 35:427–433

Goldberg M, Lécolle S (1995) Poly-L-lysine-gold complexes used at different pH are probes for differential detection of glycosaminoglycans and phosphoproteins in the predentine and dentine of rat incisor. Histochem J 27:401–410

Goldberg M, Septier D (1985) Improved lipid preservation by malachite green-glutaraldehyde fixation in rat incisor predentine and dentine. Archs Oral Biol 30:717–726

Goldberg M, Septier D (1986) Visualization of proteoglycans and membrane-associated components in rat incisor predentine and dentine using ruthenium hexammine trichloride. Arch Oral Biol 31:205–212

Goldberg M, Septier D (1996) A comparative study of the transition between predentin and dentin, using various preparative procedures in the rat. Eur J Oral Sci 104:269–277

Goldberg M, Septier D (2002) Phospholipids in amelogenesis and dentinogenesis. Crit Rev Oral Biol Med 13:276–290

Goldberg M, Takagi M (1993) Dentine proteoglycans: composition, ultrastructure and functions. Histochem J 25:781–806

Goldberg M, Noblot MM, Septier D (1980) Effets de deux méthodes de démineralisation sur la préservation des glycoprotéines et des protéoglycanes dans les dentines intercanaliculaires et péricanaliculaires chez le cheval. J Biol Buccale 8:315–330

Goldberg M, Escaig F, Septier D (1985) Distribution of lipids in odontoblasts and at the predentine-dentine junction of rat incisor. In: Belcourt AB, Ruch J-V (eds) Morphogenèse et différenciation dentaire, Coll. INSERM 125, 1984. INSERM, Paris, pp 199–207

Goldberg M, Septier D, Lécolle S, Chardin H, Quintana MA, Acevedo AC, Gafni G, Dillouya D, Vermelin L, Thonemann B, Schmalz G, Bissila-Mapahou P, Carreau JP (1995a) Dental mineralization. Int J Dev Biol 39:93–110

Goldberg M, Septier D, Lécolle S, Vermelin L, Bissila-Mapahou P, Carreau JP, Gritli A, Bloch-Zupan A (1995b) Lipids in predentine and dentine. Connect Tissue Res 33:105–114

Goldberg M, Septier D, Torres-Quintana M-A, Lécolle S, Hall R, Gafni G, Menachi S, Embery G (2000) New insights on the dynamics of dentin formation. In: Goldberg M, Boskey A, Robinson C (eds) Chemistry and biology of mineralized tissues. American Academy of Ortopaedic Surgeons, Rosemont, pp 297–303

Goldberg M, Rapoport O, Septier D, Palmier K, Hall R, Embery G, Young M, Ameye L (2003) Proteoglycans in predentin: the last 15 micrometers before mineralization. Connect Tissue Res 44:184–188

Gorter de Vries I, Quartier E, Van Steirteghem A., Boute P, Coomans D, Wisse E (1986) Characterization and immunocytochemical localization of dentine phosphoprotein in rat and bovine teeth. Arch Oral Biol 31:57–66

Gorter de Vries I, Quartier E, Boute P, Wisse E, Coomans D (1987) Immunocytochemical localization of osteocalcin in developing rat teeth. J Dent Res 66:784–790

Gorter de Vries I, Coomans D, Wisse E (1988a) Immunocytochemical localization of osteocalcin in human and bovine teeth. Calcif Tissue Int 43:128–130

Gorter de Vries I, Coomans D, Wisse E (1988b) Ultrastructural localization of osteocalcin in rat tooth germs by immunogold staining. Histochemistry 89:509–514

Grzesik WJ, Cheng H, Oh JS, Kuznetsov SA, Mankani MH, Uzawa K, Robey PG, Yamauchi M (2000) Cementum-forming cells are phenotypically distinct from bone-forming cells. J Bone Miner Res 15:52–59

Gu K, Chang S, Ritchie HH, Clarkson BH, Rutherford RB (2000) Molecular cloning of a human dentin sialophosphoprotein gene. Eur J Oral Sci 108:35–42

Habelitz S, Balooch M, Marshall SJ, Balooch G, Marshall GW Jr (2002) In situ atomic force microscopy of partially demineralized human dentin collagen fibrils. J Struct Biol 138:227–236

Hall RC, Embery G, Lloyd D (1997) Immunochemical localization of the small leucine-rich proteoglycan lumican in human predentine and dentine. Arch Oral Biol 42:783–786

Hayashi Y (1982) Ultrastructure of initial calcification in wound healing following pulpotomy. J Oral Pathol 11:174–180

Hayashi Y (1983) Crystal growth in matrix vesicles of permanent tooth germs in kittens. Acta Anat 116:62–68

Hayashi Y (1985) Ultrastructural characterization of extracellular matrix vesicles in the mineralizing fronts of apical cementum in cats. Arch Oral Biol 30:445–449

Hayashi Y (1987) Ultrastructure of cementum formation on partially formed teeth in dogs. Acta Anat 129:279–288

Hayashi Y, Imai M, Goto Y, Murakami N (1993) Pathological mineralization in a serially passaged cell line from rat pulp. J Oral Pathol Med 22:175–179

He G, George A (2004) Dentin matrix protein 1 immobilized on type I collagen fibrils facilitates apatite deposition *in vitro*. J Biol Chem 279:11649–11656

He G, Ramachandran A, Dahl T, George S, Schultz D, Cookson D, Veis A, George A (2005) Phosphorylation of phosphophoryn is crucial for its function as a mediator of biomineralization. J Biol Chem 280:33109–33114

Helder MN, Bronckers AL, Woltgens JH (1993) Dissimilar expression patterns for the extracellular matrix proteins osteopontin (OPN) and collagen type I in dental tissues and alveolar bone of the neonatal rat. Matrix 13:415–425

Hjerpe A, Engfeldt B (1976) Proteoglycans of dentine and predentine. Calcif Tissue Res 22:173–182

Holbrook IB, Leaver AG (1976) Glycoproteins in the dentine of human deciduous teeth. Archs Oral Biol 21:509–512

Holland PWH, Harper SJ, McVaj JH, Hogan BLM (1987) *In vivo* expression of mRNA for the Ca^{++}-binding protein SPARC (osteonectin) revealed by *in situ* hybridization. J Cell Biol 105:473–482

Inage T, Toda Y (1988) Phosphoprotein synthesis and secretion by odontoblasts in rat incisors as revealed by electron microscopic radioautography. Am J Anat 182:369–380

Irving JT (1959) A histological staining method for site of calcification in teeth and bone. Arch Oral Biol 1:89–96

Jande SS, Bélanger LF (1970) Fine structural study of rat molar cementum. Anat Record 167:439–464

Johansen E, Parks HF (1962) Electron-microscopic observations on sound human dentine. Arch Oral Biol 7:185–193

Jones IL, Leaver AG (1974) Glycosaminoglycans of human dentine. Calcif Tissue Res 16:37–44

Jones SJ, Boyde A (1984) Ultrastructure of dentin and dentinogenesis. In: Linde A (ed) Dentin and dentinogenesis, 1st vol. CRC Press, Boca Raton, pp 81–134

Jontell M, Linde A (1983) Non-collagenous proteins of predentine from dentinogenically active bovine teeth. Biochem J 214:769–776

Kagayama M, Li HC, Zhu J, Sasano Y, Hatakeyama Y, Mizoguchi I (1997) Expression of osteocalcin in cementoblasts forming acellular cementum. J Periodont Res 32:273–278

Katchburian E (1973a) Membrane-bound bodies as initiators of mineralization of dentine. J Anat 116:285–302

Katchburian E (1973b) Role of extracellular bodies in calcification of dentine. J Anat 115:139–158

Katchburian E, Severs NJ (1982) Membranes of matrix-vesicles in early developing dentine. A freeze fracture study. Cell Biol Int Rep 6:941–950

Kitamura C, Terashita M, Noguchi T (1991) Presence of vesicles containing lactate dehydrogenase in the dentin of bovine tooth germs. J Dent Res 70:1444–1446

Kodaka T, Debari K (2002) Scanning electron microscopy and energy-dispersive X-ray microanalysis studies of afibrillar cementum and cementicle-like structures in human teeth. J Electron Microsc 51:327–335

Kodaka T, Hirayama A, Abe M, Miake K (1992) Organic structures of the hypercalcified peritubular matrix in horse dentine. Acta Anat 145:181–188

Kogaya Y, Kato T, Furuhashi K (1987) Ultrastructural visualization of proteoglycans in dentine forming sites of developing rat molar tooth germs by use of ruthenium hexammine trichloride (RHT). J Electron Microsc 36:40–44

Kramer IRH (1951) The distribution of collagen fibrils in the dentin matrix. Br Dent J 91:1–7

Larsson Å (1973) Studies on dentinogenesis in the rat. Ultrastructural observations on early dentin formation with special reference to "dentinal globules" and alkaline phosphatase activity. Z Anat Entwickl-Gesch 142:103–115

Larsson Å (1974) Studies on dentinogenesis in the rat. The interaction between lead-pyrophosphate solutions and dentinal globules. Calcif Tissue Res 16:93–107

Larsson Å, Bloom GD (1973) Studies on dentinogenesis in the rat. Fine structure of developing odontoblasts and predentin in relation to the mineralization process. Z Anat Entwickl-Gesch 139:227–246

Lee SL, Veis A, Glonek T (1977) Dentin phosphoprotein: an extracellular calcium-binding protein. Biochemistry 16:2971–2979

Linde A (1973) Glycosaminoglycans of the odontoblast-predentine layer in dentinogenically active porcine teeth. Calcif Tissue Res 12:281–294

Linde A (1984) Non-collagenous proteins and proteoglycans in dentinogenesis. In: Linde A (ed) Dentin and dentinogenesis, 2nd vol. CRC Press, Boca Raton, pp 55–92

Linde A (1985) Predentin; biochemical and functional aspects. In: Belcourt AB, Ruch J-V (eds) Morphogenèse et différènciation dentaires. INSERM, Paris, pp 175–189
Linde A (1988) Differences between non-collagenous protein content of rat incisor and permanent bovine dentin. Scand J Dent Res 96:188–198
Linde A (1992) Structure and calcification of dentin. In: Bonucci E (ed) Calcification in biological systems. CRC Press, Boca Raton, pp 269–311
Linde A, Goldberg M (1993) Dentinogenesis. Crit Rev Oral Biol Med 4:679–728
Linde A, Bhown M, Butler WT (1980) Noncollagenous proteins of dentin. A re-examination of proteins from rat incisor dentin utilizing techniques to avoid artifacts. J Biol Chem 255:5931–5942
Linde A, Bhown M, Cothran WC, Höglund A, Butler WT (1982) Evidence for several γ-carboxyglutamic acid-containing proteins in dentin. Biochim Biophys Acta 704:235–239
Linde A, Lussi A, Crenshaw MA (1989) Mineral induction by immobilized polyanionic proteins. Calcif Tissue Int 44:286–295
Lormée P, Septier D, Lécolle S, Baudoin C, Goldberg M (1996) Dual incorporation of (^{35}S)sulfate into dentin proteoglycans acting as mineralization promotors in rat molars and predentin proteoglycans. Calcif Tissue Int 58:368–375
Lussi A, Linde A (1993) Mineral induction in vivo by dentine proteins. Caries Res 27:241–248
Lussi A, Crenshaw MA, Linde A (1988) Induction and inhibition of hydroxyapatite formation by rat dentine phosphoprotein *in vitro*. Archs Oral Biol 33:685–691
MacDougall M (2003) Dental structural diseases mapping to human chomosome 4q21. Connect Tissue Res 44:285–291
MacDougall M, Simmons D, Luan X, Nydegger J, Feng J, Gu TT (1997) Dentin phosphoprotein and dentin sialoprotein are cleavage products expressed from a single transcript coded by a gene on human chromosome 4. Dentin phosphoprotein DNA sequence determination. J Biol Chem 272:835–842
MacDougall M, Gu TT, Luan X, Simmons D, Chen J (1998) Identification of a novel isoform of mouse dentin matrix protein 1: spatial expression in mineralized tissues. J Bone Miner Res 13:422–431
MacDougall M, Simmons D, Gu TT, Dong J (2002) MEPE/OF45, a new dentin/bone matrix protein and candidate gene for dentin diseases mapping to chromosome 4q21. Connect Tissue Res 43:320–330
MacNeil RL, Berry J, D'Errico J, Strayhorn C, Piotrowski B, Somerman MJ (1995a) Role of two mineral-associated adhesion molecules, osteopontin and bone sialoprotein, during cementogenesis. Connect Tissue Res 33:1–7
MacNeil RL, Berry J, D'Errico J, Strayhorn C, Somerman MJ (1995b) Localization and expression of osteopontin in mineralized and nonmineralized tissues of the periodontium. Ann N Y Acad Sci 760:166–176
Magne D, Guicheux J, Weiss P, Pilet P, Daculsi G (2002) Fourier transform infrared microspectroscopic investigation of the organic and mineral constituents of peritubular dentin: a horse study. Calcif Tissue Int 71:179–185
Maier GD, Lechner JH, Veis A (1983) The dynamics of formation of a collagen-phosphophoryn conjugate in relation to the passage of the mineralization front in rat incisor dentin. J Biol Chem 258:1450–1455
Marsh ME (1989) Binding of calcium and phosphate ions to dentin phosphophoryn. Biochemistry 28:346–352
Massa LF, Ramachandran A, George A, Arana-Chavez VE (2005) Developmental appearance of dentin matrix protein 1 during the early dentinogenesis in rat molars as identified by high-resolution immunocytochemistry. Histochem Cell Biol 124: 197–205
McKee MD, Nanci A (1995) Postembedding colloidal-gold immunocytochemistry of noncollagenous extracellular matrix proteins in mineralized tissues. Microsc Res Techn 31:44–62

McKee MD, Zalzal S, Nanci A (1996) Extracellular matrix in tooth cementum and mantle dentin: localization of osteopontin and other noncollagenous proteins, plasma proteins, and glycoconjugates by electron microscopy. Anat Rec 245:293–312

Milan AM, Sugars RV, Embery G, Waddington RJ (2005) Modulation of collagen fibrillogenesis by dentinal proteoglycans. Calcif Tiss Int 76:127–135

Miller WA, Eick JD, Neiders ME (1971) Inorganic components of the peritubular dentin in young human permanent teeth. Caries Res 5:264–269

Nagata K, Huang YH, Ohsaki Y, Kukita T, Nakata M, Kurisu K (1992) Demonstration of type III collagen in the dentin of mice. Matrix 12:448–455

Nakamura O, Gohda E, Ozawa M, Senba I, Miyazaki H, Murakami T, Daikuhara Y (1985) Immunohistochemical studies with a monoclonal antibody on the distribution of phosphophoryn in predentin and dentin. Calcif Tissue Int 37:491–500

Nanci A (1999) Content and distribution of noncollagenous matrix proteins in bone and cementum: relationship to speed of formation and collagen packing density. J Struct Biol 126:256–269

Nanci A (2000) Matrix mediated mineralization in enamel and the collagen-based hard tissues. In: Goldberg M, Boskey A, Robinson C (eds) Chemistry and biology of mineralized tissues. American Academy of Orthopaedic Surgeons, Rosemont,IL, pp. 217–224

Nanci A (2003) Ten Cate's oral histology: Development, structure, and function, 6th edn. Mosby, St. Louis

Narayanan AS, Ikezawa K, Wu D, Pitaru S (1995) Cementum specific components which influence periodontal connective tissue cells. Connect Tissue Res 33:19–21

Ogata Y, Niisato N, Moriwaki K, Yokota Y, Furuyama S, Sugiya H (1997) Cementum, root dentin and bone extracts stimulate chemotactic behaviour in cells from periodontal tissue. Comp Biochem Physiol 116B:359–365

Papagerakis P, Berdal A, Mesbah M, Peuchmaur M, Malaval L, Nydegger J, Simmer J, MacDougall M (2002) Investigation of osteocalcin, osteonectin, and dentin sialophosphoprotein in developing human teeth. Bone 30:377–385

Petersson U, Hultenby K, Wendel M (2003) Identification, distribution and expression of osteoadherin during tooth formation. Eur J Oral Sci 111:128–136

Price PA, Williamson MK, Otawara Y (1985) Characterization of matrix Gla protein. A new vitamin K-dependent protein associated with the organic matrix of bone. In: Butler WT (ed) The chemistry and biology of mineralized tissues. EBSCO Media, Birmingham, AL, pp 159–163

Qin C, Brunn JC, Jones J, George A, Ramachandran A, Gorski JP, Butler WT (2001) A comparative study of sialic acid-rich proteins in rat bone and dentin. Eur J Oral Sci 109:133–141

Qin C, Brunn JC, Cadena E, Ridall A, Tsujigiwa H, Nagatsuka H, Nagai N, Butler WT (2002) The expression of dentin sialophosphoprotein gene in bone. J Dent Res 81:392–394

Qin C, Brunn JC, Baba O, Wygant JN, McIntyre BW, Butler WT (2003a) Dentin sialoprotein isoforms: detection and characterization of a high molecular weight dentin sialoprotein. Eur J Oral Sci 111:235–242

Qin C, Brunn JC, Cadena E, Ridall A, Butler WT (2003b) Dentin sialoprotein in bone and dentin sialophosphoprotein gene expressed by osteoblasts. Connect Tissue Res 44:179–183

Qin C, Brunn JC, Cook RG, Orkiszewski RS, Malone JP, Veis A, Butler WT (2004) Evidence for the proteolytic processing of dentin matrix protein 1. Identification and characterization of processed fragments and cleavage sites. J Biol Chem 278:34700–34708

Rahemtulla F, Prince CW, Butler WT (1984) Isolation and partial characterization of proteoglycans from rat incisors. Biochem J 218:877–885

Rahima M, Tsay T-G, Andujar M, Veis A (1988) Localization of phosphophoryn in rat incisor dentin using immunocytochemical techniques. J Histochem Cytochem 36:153–157

Reichert T, Storkel S, Becker K, Fisher LW (1992) The role of osteonectin in human tooth development: an immunological study. Calcif Tissue Int 50:468–472

Ritchie HH, Hou H, Veis A, Butler WT (1994) Cloning and sequence determination of rat dentin sialoprotein, a novel dentin protein. J Biol Chem 269:3698–3702

Ritchie HH, Berry JE, Somerman MJ, Hanks CT, Bronckers ALJJ, Hotton D, Papagerakis P, Berdal A, Butler WT (1997) Dentin sialoprotein (DSP) transcripts: developmentally-sustained expression in odontoblasts and transient expression in pre-ameloblasts. Eur J Oral Sci 105:405–413

Ritchie HH, Ritchie DG, Wang L-H (1998) Six decades of dentinogenesis research. Historical and prospective views on phosphophoryn and dentin sialoprotein. Eur J Oral Sci 106:211–220

Rouiller C, Huber L, Rutishauser E (1952) La structure de la dentine. Étude comparée de l'os et de l'ivoire au microscope électronique. Acta Anat 16:16–28

Salonen J, Domenicucci C, Goldberg HA, Sodek J (1990) Immunohistochemical localization of SPARC (osteonectin) and denatured collagen and their relationship to remodelling in rat dental tissues. Arch Oral Biol 35:337–346

Sasaki T, Garant PR (1996) Structure and organization of odontoblasts. Anat Rec 245:235–249

Satoyoshi M, Koizumi T, Teranaka T, Iwamoto T, Takita H, Kuboki Y, Saito S, Mikuni-Takagaki Y (1995) Extracellular processing of dentin matrix protein in the mineralizing odontoblast culture. Calcif Tissue Int 57:237–241

Sela J, Tamari I, Hirschfeld Z, Bab I (1981) Transmision electron microscopy of reparative dentin in rat molar pulps. Primary mineralization via extracellular matrix vesicles. Acta Anat 109:247–251

Septier D, Hall RC, Lloyd D, Embery G, Goldberg M (1998) Quantitative immunohistochemical evidence of a functional gradient of chondroitin 4-sulphate/dermatan sulphate, developmentally regulated in the predentine of rat incisor. Histochem J 30:275–284

Septier D, Hall RC, Embery G, Goldberg M (2001) Immunoelectron microscopic visualization of pro- and secreted forms of decorin and biglycan in the predentin and during dentin formation in the rat incisor. Calcif Tissue Int 69:38–45

Sisca RF, Provenza DV (1972) Initial dentin formation in human deciduous teeth. An electron microscope study. Calcif Tissue Res 9:1–16

Slavkin HC, Croissant RD, Bringas P, Matosian P, Wilson P, Mino W, Guenther H (1976) Matrix vesicles heterogeneity: possible morphogenetic functions for matrix vesicles. Fed Proc 35:127–134

Srinivasan R, Chen B, Gorski JP, George A (1999) Recombinant expression and characterization of dentin matrix protein 1. Connect Tissue Res 40:251–258

Steinfort J, van de Stadt R, Beertsen W (1994) Identification of new rat dentin proteoglycans utilizing C18 chromatography. J Biol Chem 269:22397–22404

Steinfort J, van den Bos T, Beertsen W (2004) Differences between enamel-related and cementum-related dentin in the rat incisor with special emphasis on the phosphoproteins. J Biol Chem 264:2840–2845

Stetler-Stevenson WG, Veis A (1987) Bovine dentin phosphophoryn: calcium ion binding properties of a high molecular weight preparation. Calcif Tissue Int 40:97–102

Stratmann U, Schaarschmidt K, Wiesmann HP, Plate U, Höhling HJ (1996) Mineralization during matrix-vesicle-mediated mantle dentine formation in molars of albino rats: a microanalytical and ultrastructural study. Cell Tissue Res 284:223–230

Stratmann U, Schaarschmidt K, Wiesmann HP, Plate U, Höhling HJ, Szuwart T (1997) The mineralization of mantle dentine and of circumpulpal dentine in the rat: an ultrastructural and element-amalytical study. Anat Embryol 195:289–297

Sundström B, Takuma S, Nagai N (1970) Ultrastructural aspects of human dentine decalcified with chromium sulphate. Calcif Tissue Res 4:305–313

Takagi Y, Veis A (1984) Isolation of phosphophoryn from human dentin organic matrix. Calcif Tissue Int 36:259–265

Takagi M, Parmley RT, Denys FR (1981) Ultrastructural localization of complex carbohydrates in odontoblasts, predentin, and dentin. J Histochem Cytochem 29:747–758

Takagi Y, Fujisawa R, Sasaki S (1986) Identification of dentin phosphophoryn localization by histochemical stainings. Connect Tissue Res 14:279–292

Takagi M, Hishikawa H, Hosokawa Y, Kagami A, Rahemtulla F (1990a) Immunohistochemical localization of glycosaminoglycans and proteoglycans in predentin and dentin of rat incisors. J Histochem Cytochem 38:319–324

Takagi Y, Shimokawa H, Suzuki M, Nagai H, Sasaki S (1990b) Immunohistochemical localization of α_2HS glycoprotein in dentin. Calcif Tissue Int 47:40–45

Takano Y, Ozawa H, Crenshaw MA (1986) Ca-ATPase and ALPase activities at the initial calcification sites of dentin and enamel in the rat incisor. Cell Tissue Res 243:91–99

Takano Y, Sakai H, Baba O, Terashima T (2000) Differential involvement of matrix vesicles during the initial and appositonal mineralization processes in bone, dentin, and cementum. Bone 26:333–339

Takuma S (1960) Electron microscopy of the structure around the dentinal tubules. J Dent Res 39:973–981

Takuma S, Eda S (1966) Structure and development of the peritubular matrix in dentine. J Dent Res 45:683–692

Tenorio D, Reid AR, Katchburian E (1990) Ultrastructural visualisation of proteoglycans in early unmineralised dentin of rat tooth germs stained with cuprolinic blue. J Anat 169:257–264

Terasawa M, Shimokawa R, Terashima T, Ohya K, Takagi Y (2004) Expression of dentin matrix protein 1 (DMP1) in nonmineralized tissues. J Bone Miner Metab 22:430–438

Termine JD, Kleinman HK, Whitson SW, Conn KM, McGarvey ML, Martin GR (1981) Osteonectin, a bone-specific protein linking mineral to collagen. Cell 26:99–105

Thotakura SR, Karthikeyan N, Smith T, Liu K, George A (2000) Cloning and characterization of rat dentin matrix protein 1 (*DMP1*) gene and its 5′-upstream region. J Biol Chem 275:10272–10277

Toyosawa S, Okabayashi K, Komori T, Ijuhin N (2004) mRNA expression and protein localization of dentin matrix protein 1 during dental root formation. Bone 34:124–133

Traub W, Jodaikin A, Arad T, Veis A, Sabsay B (1992) Dentin phosphophoryn binding to collagen fibrils. Matrix 12:197–201

Tsuji T, Mark MP, Ruch J-V (1994) Immunocytochemical localization of choline-phospholipids in postnatal mouse molars. Archs Oral Biol 39:81–86

Tung PS, Domenicucci C, Wasi S, Sodek J (1985) Specific immunohistochemical localization of osteonectin and collagen types I and III in fetal and adult porcine dental tissues. J Histochem Cytochem 33:531–540

Uchida T, Ozawa H (1984) ZIO staining of matrix vesicles in the epiphyseal cartilage and mineralizing dentin. In: Cohn DV, Potts JT, Fujita T (eds) Endocrine control of bone and calcium metabolism. Elsevier, Amsterdam, pp 441–443

Vandenbos T, Bronckers AL, Goldberg HA, Beertsen W (1999) Blood circulation as source for osteopontin in acellular extrinsic fiber cementum and other mineralizing tissues. J Dent Res 78:1688–1695

Veis A (1993) Mineral-matrix interactions in bone and dentin. J Bone Miner Res 8:S493-S497

Veis A, Tsay T-G, Kanwar Y (1985) An immunological study of the localization of dentin phosphophoryns in the tooth. In: Belcourt AB, Ruch J-V (eds) Morphogenèse et différenciation dentaires, Coll. INSERM 125, 1984. INSERM, Paris, pp 223–232

Vermelin L, Septier D, Goldberg M (1994) Iodoplatinate visualization of phospholipids in rat incisor predentine and dentine, compared with malachite green aldehyde. Histochemistry 101:63–72

Vose GP, Baylink DJ (1970) Effect of fibrillar structure of pericanalicular and intercanalicular bone on X-ray absorption. Anat Rec 166:239–246

Weiner S, Veis A, Beniash E, Arad T, Dillon JW, Sabsay B, Siddiqui F (1999) Peritubular dentin formation: crystal organization and the macromolecular constituents in human teeth. J Struct Biol 126:27–41

Weinstock A, Weinstock M, Leblond CP (1972) Autoradiographic detection of ^{3}H-fucose incorporation into glycoprotein by odontoblasts and its deposition at the site of the calcification front in dentin. Calcif Tissue Res 8:181–189

Weinstock M, Leblond CP (1973) Radioautographic visualization of the deposition of a phosphoprotein at the mineralization front in the dentin of the rat incisor. J Cell Biol 56:838–845

Weinstock M, Leblond CP (1974) Synthesis, migration, and release of precursor collagen by odontobalsts as visualized by radioautography after [^{3}H]proline administration. J Cell Biol 60:92–127

Wuthier RE (1984) Lipids in dentinogenesis. In: Linde A (ed) Dentin and dentinogenesis, 2nd vol. CRC Press, Boca Raton, pp 93–106

Yamada M, Hirayama A, Miake K, Kasugai S, Ogura H (1987) An ultrastructural study of reparative dentinogenesis in the rat incisor after colchicine administration. J Electron Microsc 36:398–407

Yamakoshi Y, Hu JC-C, Fukae M, Iwata T, Kim J-W, Zhang H, Simmer JP (2005) Porcine dentin sialoprotein is a proteoglycan with glycosaminoglycan chains containing chondroitin 6-sulfate. J Cell Biol 280:1552–1560

Yamauchi M, Uzawa K, Katz EP, Lopes MM, Verdelis K, Cheng H (2000) Distribution of small keratan sulfate proteoglycans in predentin and dentin. In: Goldberg M, Boskey A, Robinson C (eds) Chemistry and biology of mineralized tissues. American Academy of Orthopaedic Surgeons, Rosemont, pp 305–310

Yoshiba N, Yoshiba K, Iwaku M, Ozawa H (1996) Immunolocalization of the small proteoglycan decorin in human teeth. Archs Oral Biol 41:351–357

Zanetti M, de Bernard B, Jontell M, Linde A (1981) Ca^{2+}-binding studies of the phosphoprotein from rat-incisor dentine. Eur J Biochem 113:541–545

9 Calcifying Matrices: Cartilage

9.1 Introduction

There are basic similarities between the organic matrix of vertebrate cartilage and that of bone; both consist of a collagen fibril network whose meshes are rich in non-collagenous material. However, not only is the collagen of the cartilage matrix mainly of type II, instead of the type I collagen that distinguishes the bone matrix, but its content of non-collagenous material far exceeds that of bone. This chapter deals with the structure and composition of the vertebrate cartilage matrix in areas of endochondral calcification, especially the areas where the process has been studied most, such as the growth plates of the epiphyses and the calcification centers of the cartilaginous skeletal anlages. Calcification takes place in these areas after distinctive, easily recognizable 'differentiation' and 'maturation' changes have occurred in cells and matrix, so allowing investigation of the role that the various components and structures have in this process. That is why the growth cartilage of the epiphyses will be taken as paradigmatic of all types of calcifying cartilage (reviewed by Heinegård et al. 1988; Poole et al. 1989; Hunziker and Herrmann 1990; Poole 1991; Engfeldt and Reinholt 1992). Articular cartilage, other types of hyaline cartilage that normally do not calcify, as well as elastic and fibrous cartilage, will not be considered in any detail.

The growth cartilage is easy to distinguish from the abutting, non-calcifying cartilage known as resting or reserve cartilage, as this is marked out by a few roundish chondrocytes scattered through an abundant intercellular matrix. By contrast, chondrocytes in the growth cartilage overlap, forming almost regular cellular columns which are oriented longitudinally with respect to the metaphyseal plate and are separated by the interposition of longitudinal septa. The single cells in a column are separated from each other by thin transverse septa. This arrangement permits subdivision of the growth plate into compartments both vertically, i. e., parallel to the axis of the longitudinal septa, in which case it is based on cell morphology along the columns, and in a transverse direction, i. e., across the matrix of the longitudinal intercellular septa, in which case it is based on the distance from chondrocytes.

According to the first form of subdivision, the growth cartilage can be divided into separate zones according to the types of chondrocytes that appear along characteristic cell columns (Fig. 9.1, upper left). The top of each column

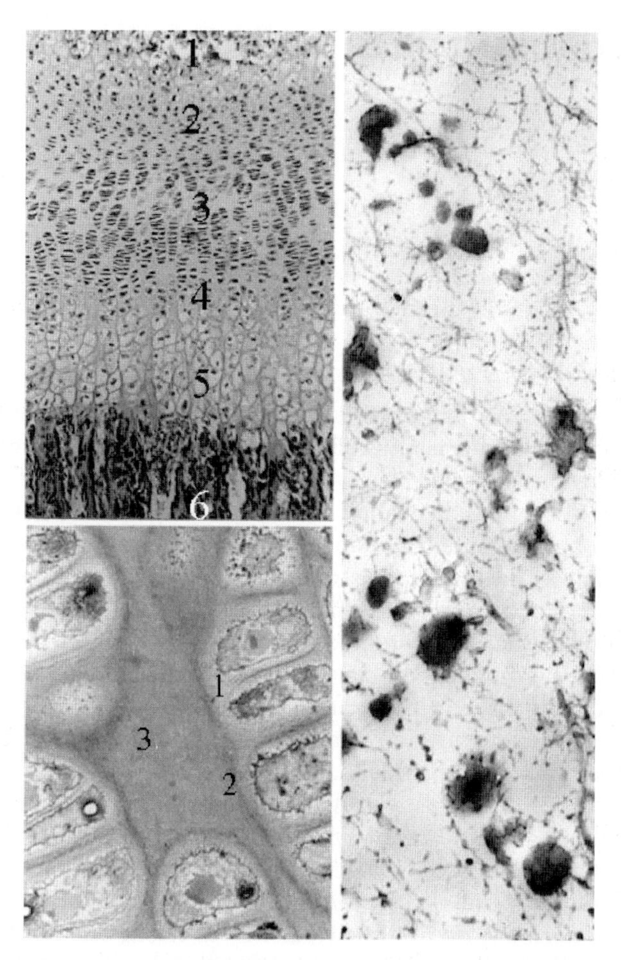

Fig. 9.1. *Upper left*: growth cartilage; from above to below: 1) epiphyseal nucleation center, 2) resting cartilage, 3) proliferating cartilage, 4) maturing cartilage, 5) hypertrophic-degenerating cartilage and, 6) metaphyseal trabeculae. Azure II-Methylene blue, ×50. *Lower left*: the intercolumnar matrix usually comprises 1) the pericellular matrix, which appears as a thin, clear halo around chondrocytes, 2) the territorial matrix, which is often demarcated by a thin, dense border, and 3) the interterritorial matrix. Azure II-Methylene blue, ×220. *Right*: three matrix components characterize the growth cartilage matrix under the electron microscope: thin, loosely and irregularly arranged type II collagen fibrils, proteoglycan granules that are often attached to them, and matrix vesicles. Uranyl acetate and lead citrate, ×36,000

is in contact with the resting cartilage, which is actually a type of cartilage that may be more active than is usually supposed (Abad et al. 2002). The first zone corresponds to the upper end of the columns and is called the proliferation or proliferative zone, because its chondrocytes undergo mitosis and proliferate, so keeping the length of the columns stable by compensating for the cell degeneration and death that occur at the other end. The proliferation zone is marked by the presence of a few flat chondrocytes which are superimposed on each other and separated only by thin transverse septa. These chondrocytes become larger, take on a round shape and give rise to the maturation zone. Within that zone, chondrocytes further increase in volume (from five- to tenfold; Buckwalter et al. 1986) and become hypertrophic chondrocytes, which end up located in the hypertrophic zone, which is subdivided into an upper and a lower hypertrophic zone. Hypertrophic chondrocytes are the terminal differentiation stage of the cell line that originates in the proliferation zone (Pacifici et al. 1990a) and are often thought of as degenerating cells which are about to die (Brighton 1978; Poole 1991), a fate they share with the many (up to 44%) post-mitotic chondrocytes of the growth plate which undergo apoptosis (Ohyama et al. 1997) and are mainly found in the hypertrophic zone (Aizawa et al. 1997). A range of data validate the alternative hypothesis that they are viable, metabolically active cells (discussed by Hunziker 1988) which could differentiate into osteoblast- like cells (Descalzi Cancedda et al. 1992; Roach 1997; Bianco et al. 1998). Leaving these questions open, there is no doubt that most hypertrophic cells ultimately disappear, leaving room for penetrating capillary vessels and associated mesenchymal cells. On one hand these developments make up for the proliferation activity occurring at the other end of the columns, on the other they permit bone matrix deposition on the surface of the calcified longitudinal septa. Concomitantly with the maturation and hypertrophy of chondrocytes, in fact, the thickness of the longitudinal septa decreases, matrix vesicles appear in them, and their matrix calcifies. Taken together, the proliferation of chondrocytes at one end of the columns, and their death at the other, constitute a dynamic process which leads to elongation of the skeletal segment (Frost and Schönau 2001).

The second form of subdivision allows a territorial and an interterritorial matrix to be distinguished, the former corresponding to the wall of the chondrocyte lacunae and a thin halo of the matrix around them, the latter to the remaining portion of the longitudinal intercellular septa (Fig. 9.1, lower left). Initially this distinction was based on differences in the histochemical properties of the matrix, which, however, vary with the experimental conditions (Szirmai 1963) and do not constitute a valid or reproducible method of differentiation (Hunziker and Herrmann 1990). As a result, ultrastructural parameters have come to be preferred, although the division of cartilage matrix into sharply demarcated compartments can hardly be achieved on that basis alone.

When cartilage is examined under the electron microscope, a pericellular empty space is often detected; this points to the presence of a lacuna. The pericellular space is due to water extraction, and to the consequent retraction and shrinkage of the chondrocyte within a relatively rigid matrix, which inevitably occur during the usual preparation of specimens for electron microscopy. This type of artifact can be precluded if the cartilage is preserved by fixation with cationic dyes or by freezing methods such as high-pressure freezing, freeze substitution and low-temperature embedding (Hunziker and Schenk 1984; Maitland and Arsenault 1989).

In any case, three concentric compartments – the pericellular matrix, the territorial matrix, and the interterritorial matrix, each gradually merging into the next – can be made out, extending outward in a radial pattern from a chondrocyte into the surrounding matrix. The pericellular matrix can be clearly recognized after using fixation methods based on cationic dyes, above all ruthenium red, which react with the anionic molecules of proteoglycans. A pericellular, 1 2μm thick matrix layer becomes evident; it is rich in acidic proteoglycans and devoid of fibrillar collagen (Hunziker et al. 1983). Proceeding from the pericellular matrix outwards, a territorial matrix is encountered next. It consists of acid proteoglycans and collagen fibrils (Fig. 9.1, right), which are initially rather haphazardly arranged, but acquire a more parallel alignment as the interterritorial matrix is approached (Eggli et al. 1985). Proceeding further outward, the territorial matrix gradually merges into the interterritorial matrix. This also consists of acid proteoglycans and collagen fibrils, mostly lying roughly parallel to the longitudinal axis of the longitudinal intercellular septa; most matrix vesicles are reported to lie in this interterritorial compartment (Eggli et al. 1985) and, to some extent, in personal experience, in the transitional zone between territorial and interterritorial matrix (Fig. 9.1, right).

9.2 Collagen

As in bone, dentin and cementum, collagen is the major fibrillar component of the extracellular cartilage matrix. Structural differences are, however, evident between the collagen found in the matrix of those tissues and that found in the hyaline cartilage.

The most abundant collagen type in the growth cartilage matrix is type II (see reviews by Miller 1973; Mayne and Von der Mark 1983), but its concentrations are not homogeneous through the growth plate. They rise during the evolution from proliferating to hypertrophic chondrocytes, but then, due to resorption, fall as chondrocytes become increasingly hypertrophic (Horton and Machado 1988; Mwale et al. 2000, 2002), probably in association with the unwinding of the triple helix (Alini et al. 1992). Fibrils of type II collagen are much thinner than those of the type I collagen found in bone, dentin and cementum; they are also more loosely arranged, and they outline wider interfibrillar spaces containing plenty of non-collagenous material. This is probably the reason why early ultrastructural studies failed to show a periodic binding in these fibrils (Bonucci 1971b; Stark et al. 1972), or only occasionally showed a 22.0 nm periodic pattern (Takuma 1960). In reality, type II collagen molecules are homotrimers made of $\alpha 1$(II) chains (Mayne and Von der Mark 1983; van der Rest 1991) that are arranged in fibrils according to a periodic D-band pattern similar to, but not identical with, that of type I collagen fibrils (Ortolani and Marchini 1995); in particular, a more prominent b_1 band is apparent in native type II than in type I collagen fibrils (Ortolani et al. 2000).

A protein called chondrocalcin, isolated by Choi et al. (1983) from fetal epiphyseal cartilage, where it is associated with calcification (Poole et al. 1984), was later recognized as the C-propeptide of type II collagen (van der Rest et al. 1986). Its characteristics and properties have been reviewed by Poole et al. (1989). It is a calcium-binding protein which is concentrated in the matrix of calcification nodules (Poole et al. 1984; Lee et al. 1996).

Nascent type II collagen fibrils are heteropolymers because they are covalently linked with type IX and type XI collagens. Besides type II collagen, other types of fibrillar and non-fibrillar collagens are, in fact, expressed in vertebrate tissues, including the hyaline cartilage of the growth plate (Eyre 2004). Some of these collagens belong to the so-called FACIT subfamily.

The term FACIT, the acronym for fibril-associated collagen with interrupted triple helices, refers to collagens whose molecules consist of two or three triple-helical domains (COL 1, 2, 3) interrupted by non-triple-helical domains (NC1, 2, 3, 4) which permit the molecules to stay flexible (Gordon and Olsen 1990; Shaw and Olsen 1991), to associate with one another, and to be stabilized by covalent cross-links, so forming molecular bridges which may stabilize the intercellular matrix. Different tissues express different FACIT molecules (Olsen 1989); type IX, XI, XII and XIV collagens are cross-linked with collagen type II in the hyaline cartilage.

Collagen type IX has been reported as located at the surface of type II collagen fibrils (discussed by Mwale et al. 2002) and at their intersections (Müller-Glauser et al. 1986), and to be covalently linked with them (Eyre 2002, 2004; Fernandes et al. 2003; Eyre et al. 2004). It may be distributed periodically along the fibrils, or may copolymerize with them (Vaughan et al. 1988). It is synthesized as a heterotrimer comprising $\alpha1$(IX)-, $\alpha2$(IX)- and $\alpha3$(IX)-chains (van der Rest et al. 1985); the $\alpha2$(IX) chain may include a glycosaminoglycan chain (Roughley and Lee 1994), so that type IX collagen can be structured either as a proteoglycan or as a non-proteoglycan (Ayad et al. 1991). According to Mwale et al. (2002), as chondrocytes hypertrophy and the Ca^{2+}/Pi molar ratio increases, a matrix remodeling occurs which involves a progressive, selective removal of type II and IX collagens, with the retention of the PG aggrecan.

Type XI collagen is a heterotrimer consisting of $\alpha1$(XI), $\alpha2$(XI) and $\alpha3$(XI) chains, the latter having the same primary sequence as that of $\alpha1$(II)B, the splicing variant of the type II collagen gene (Eyre 2002). It copolymerizes with type II collagen and its molecules are cross-linked to each other (Wu and Eyre 1995).

Type XII and XIV collagens have been reported as present around cartilage canals, the articular surface and the perichondrium, as occurring in small amounts in cartilage matrix, and as absent from the growth plate (Watt et al. 1992). Type XII collagen was initially thought to be located only in bone; it was later detected by RT-PCR and Northern blot analysis in human fetal epiphyseal cartilage, as well as in the heart, placenta, lung, skeletal muscle and pancreas (Dharmavaram et al. 1998). It has also been shown in articular cartilage and in the longitudinal and transverse septa of the growth cartilage in young rats (Gregory et al. 2001), and has been immunodetected in fetal dense connective tissues such as tendons, ligaments, perichondrium and periostium (Sugrue et al. 1989). Its molecule is partially homologous with type IX collagen, and has a unique structure which, under the electron microscope (rotary shadowing image), shows a cross-like shape with a 75-nm tail, a central globule, and three 60-nm arms each ending in a small globule (Gordon et al. 1989). It is possible that it interacts with type I collagen fibrils through lateral association with the tail COL1 domain, and that the arms project into the perifibrillar space, where they may interact with other matrix fibrils or components (Gordon and Olsen 1990).

Another collagen of the FACIT family is type XX collagen, found in embryonic sternal cartilage, skin, cornea and tendons (Koch et al. 2001). Many other types of collagen have been described (at least 27, so far), some of them located in epiphyseal cartilage. Type III collagen has been consistently found in normal and osteoarthritic human articular cartilage (Wotton and Duance 1994; Young et al. 1995), where it colocalizes in the same fibrils with type II collagen (Young et al. 2000). In the growth cartilage, type III collagen has been found in the lacunae of hypertrophic chondrocytes in embryonic chick tibia which had been invaded by bone marrow-derived cells, just before the synthesis of type I collagen and the formation of an osteoid border (Von der Mark and Von der Mark 1977).

Type V collagen has been detected in the growth cartilage of brachymorphic mice (Wikström et al. 1984), while type VI collagen, unlike V, has been reported in normal cartilage (Ayad et al. 1984) and, focally, around individual cells in a series (53 cases) of chondroblastomas, in whose chondroid matrix, as also in fetal cartilage, a strong to moderate type VI collagen immunoreaction has been detected (Edel et al. 1992). Similar results were reported in a series of benign and malignant cartilaginous tumors, which, however, showed the presence of type V collagen in the matrix of the most atypical, grade III, chondrosarcoma (Ueda et al. 1990). Minor amounts of types IX, XI and VI are found in the matrix of the proliferative zone in normal growth cartilage (Engfeldt and Reinholt 1992).

Type X collagen appears to be of interest because of its typical localization in the growth cartilage and its pattern of expression during chondrocyte differentiation: it is, in fact, specifically expressed by hypertrophic chondrocytes (reviewed by Linsenmayer et al. 1988). None is found in the proliferative zone; it first appears in the hypertrophic zone of the growth cartilage (Schmid and Linsenmayer 1985; Poole and Pidoux 1989), and its synthesis precedes calcification both in vivo and in vitro (Gibson and Flint 1985; Silbermann and Von der Mark 1990; Iyama et al. 1991; Schmid et al. 1991; Kirsch and Von der Mark

1992; Claassen and Kirsch 1994; Ohashi et al. 1997; Kergosien et al. 1998). It has, however, been reported that, at least in culture, there is no correlation linking chondrocyte size with the expression of type X collagen (Ekanayake and Hall 1994). This is associated with type II collagen fibrils (Linsenmayer et al. 1988; Poole and Pidoux 1989; Poole et al. 1989; Chen et al. 1992) and, in avian embryonic cartilage, it occurs in two forms, one of which is pericellular (Lu Valle et al. 1992), consisting of thin (about 5 nm thick) filaments, while the other is associated with type II collagen fibrils (Schmid and Linsenmayer 1990). It binds calcium in a specific, dose-dependent manner, and its synthesis precedes calcium deposition in the nodules of fetal human chondrocytes that develop in cell culture (Kirsch and Von der Mark 1991). For these reasons, and because of its localization in the hypertrophic cartilage, type X collagen has been thought to play a role in calcification.

9.3 Non-collagenous Components

Besides collagen, the cartilage matrix comprises a conspicuous fraction of non-collagenous components (Fig. 9.2). These are mostly proteoglycan molecules, a name derived from the fact that they consist of a core protein which links glycosaminoglycan chains. Both the core protein and the chains linked to it can show a variety of different compositions and arrangements, that give rise to a number of different proteoglycans (Hardingham and Fosang 1992; Iozzo 1998). These implement very different functions, including regulation of the calcification process. According to Iozzo (1998), they can act as tissue organizers and can control cell growth, influence tissue maturation, behave as biological filters, regulate collagen fibrillogenesis, and implement mechanical functions. Some are components of cellular structures, whereas others are secreted into the pericellular matrix; only the latter will be discussed below, on account of their possibly direct role in cartilage calcification.

9.3.1 Proteoglycans

It has long been known that the matrix of epiphyseal cartilage contains plenty of proteoglycans (for reviews see Kobayashi 1971; Buckwalter 1983; Lash and Vasan 1983; de Bernard et al. 1977; Hunziker and Schenk 1987; Takagi 1990; Yamada et al. 1991; Engfeldt and Reinholt 1992; Roughley and Lee 1994). They can be divided in two fractions, one of which is extractable with 4 M guanidinium chloride, while the other is resistant to extraction (Campo 1974, 1981). The distribution of these fractions changes with age and species (Campo 1976). The extractable pool, which is more abundant in the lower hypertrophic zone, appears to be localized between collagen fibrils, whereas the non-extractable

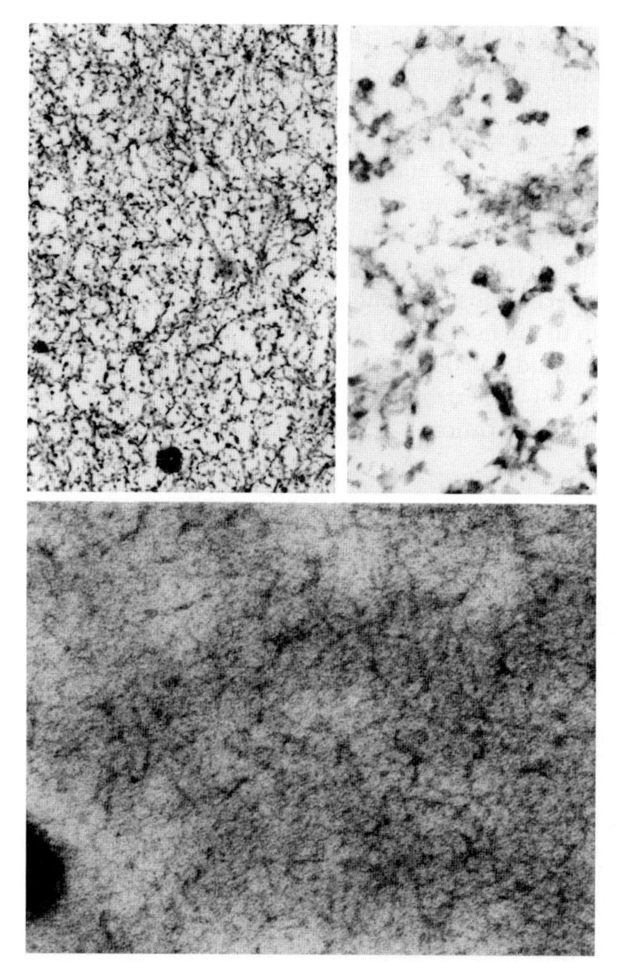

Fig. 9.2. *Above*: cartilage matrix processed by glutaraldehyde-osmium fixation and Araldite embedding; the acid proteoglycans have collapsed, due to the effects of fixation and dehydration, and appear as small granules; note the loose network of thin collagen fibrils. ×15,000 and ×90,000. *Below*: cartilage matrix processed by cryofixation and Lowicryl embedding: the acid proteoglycans maintain their extended structure; ×90,000. All sections stained with uranyl acetate and lead citrate

pool appears to be linked to the fibrils (Eisenstein et al. 1973). The proteoglycans of the calcified matrix can only be totally extracted after decalcification (Diamond et al. 1982; Franzén and Heinegård 1984).

Examination of tissue sections under the electron microscope shows that the shape of the acid proteoglycans is radically modified by preparatory procedures (Fig. 9.2): while their native molecules are elongated and bottlebrush-like, after fixation and dehydration appear as small granules (Campo and Phillips 1973; Thyberg et al. 1973a; Newbrey and Banks 1975; Thyberg 1977; Hascall 1980). Both histochemical and immunohistochemical studies have shown that the granules are proteoglycan monomers (Takagi et al. 1982; Thyberg et al. 1973b); in particular, they are digested by enzymes like chondroitinases and proteases. A direct relationship has been reported between size and numbers of granules and content in sulfated proteoglycans (Matukas et al. 1967), the largest granules being those in the territorial matrix and the smallest those in the interterritorial matrix (Campo and Phillips 1973). In agreement with biochemical (Lohmander and Hjerpe 1975; Vittur et al. 1979) and immunochemical (Hirschman and Dziewiatkowski 1966) data showing loss of protein-polysaccharide during calcification, the size and numbers of granules fall during this process (Matukas and Krikos 1968; Takagi et al. 1983; Campo and Romano 1986; Buckwalter et al. 1987a), in parallel with a fall in their sulfur content (Althoff et al. 1982) and a rise in unsulfated residues (Byers et al. 1997). There are, however, conflicting findings on this topic: Poole et al. (1982) reported that there is no net loss of proteoglycans during calcification and that, more precisely, they become entombed in the calcified cartilage.

The granular appearance shown by proteoglycans under the electron microscope is due to the collapse and shrinkage their molecules undergo when the tissue is chemically processed prior to its embedding (Smith 1970; Serafini-Fracassini and Smith 1974; Hascall 1980).The ultrastructural morphology of proteoglycans is, in fact, largely dependent on the tissue preparation methods used (Arsenault 1990): with cryogenic methods of tissue preparation, followed by freeze substitution, freeze drying and ultracryomicrotomy, or with anhydrous technique, proteoglycan molecules retain their extended state (Fig. 9.2) and appear as thin filaments that are interconnected in a dense network (Hunziker and Schenk 1984; Arsenault et al. 1988); when the usual electron microscopy procedure for fixation and dehydration is applied, the proteoglycans not only collapse into granules (Matukas et al. 1967), but are also partly lost (Engfeldt and Hjertquist 1968). The leakage and shrinkage of proteoglycan molecules can be almost completely avoided by adding cationic dyes (alcian blue, colloidal thorium, cupromeronic blue, ruthenium hexammine trichloride, ruthenium red; for a complete list see Takagi 1990) to the fixatives (usually aldehyde and/or osmium solutions). By reacting with the proteoglycan anionic groups, these cationic dyes stabilize proteoglycan molecules and prevent their solubilization and collapse. This occurs with the cationic molecules just listed as well as with the low-molecular-weight safranin O or acridine orange (Sheppard and Mitchell 1981; Shepard 1992; Engfeldt et al. 1994), with substances like bismuth nitrate (Serafini-Fracassini and Smith 1966; Smith 1970), or with treatment with *N*-*N*-dimethylformamide after aldehyde fixation and Lowicryl or Spurr embedding (Kagami et al. 1990); all these reactants prevent the shrinkage of proteoglycans and preserve their extended state almost to the same degree as that seen after cryogenic methods.

Some of the cationic molecules mentioned above can be used as 'stains' to reveal the presence of acid proteoglycans under the electron microscope (Spicer and Schulte 1982). To achieve this, one of the substances most often used is colloidal iron (Fig. 9.3), which yields a good concordance with biochemical findings (Joel et al. 1956; Curran et al. 1965; Matukas et al. 1967). If this procedure is carried out at low pH (about 2.8), a satisfactory specificity is obtained, as also shown by the disappearance of iron-positive granules after protease or hyaluronidase digestion (Matukas et al. 1967; Bonucci 1971a).

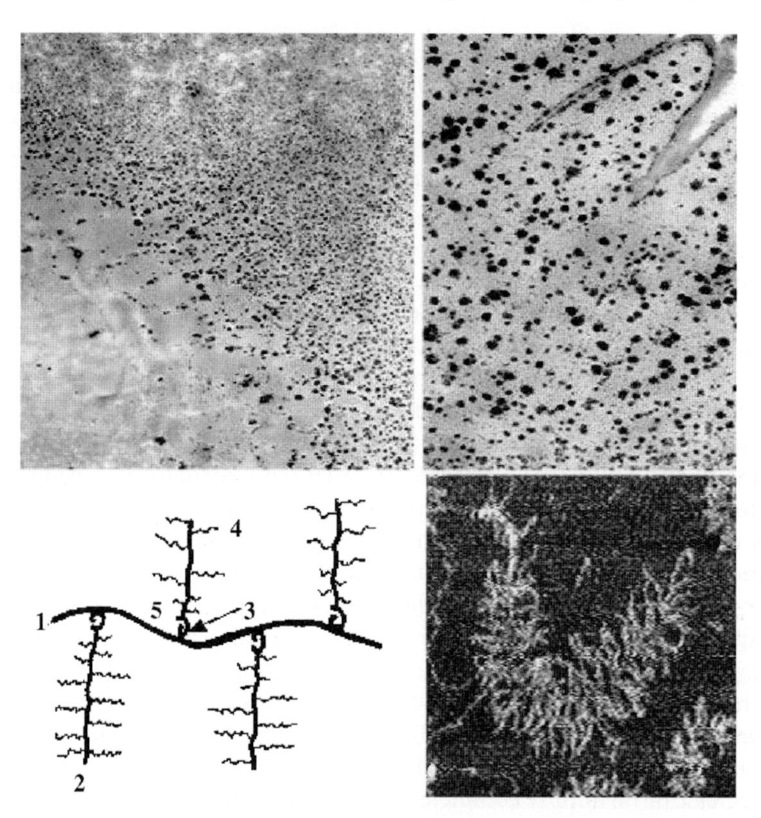

Fig. 9.3. *Above*: epiphyseal cartilage processed by glutaraldehyde-osmium fixation and Araldite embedding; sections stained with colloidal iron, pH 2.8. Plenty of granules are stained, showing that they correspond to collapsed acid proteoglycans. Part of a chondrocyte lacuna on *lower left*; part of chondrocyte processes on *upper right*. ×12,000 and ×45,000. *Lower left*: diagram of proteoglycan aggregate: 1) hyaluronic acid; 2) core protein; 3) link protein; 4) chondroitin sulfate chains; 5) keratan sulfate chains. *Lower right*: image of aggrecan visualized via atomic force microscopy; note the glycosaminoglycan chains and (*upper left corner*) the core protein (courtesy of Laurel Ng, Alan Grodzinsky, and Christine Ortiz, Massachusetts Institute of Technology, Cambridge, MA)

Under the electron microscope, the extended molecules of native proteoglycans appear as bottlebrush-like structures (i. e., they appear to consist of an axial filament with attached lateral chains) when aggregates are isolated and spread (Hascall 1980; Morgelin et al. 1994; Ng et al. 2003; Yeh and Luo 2004). Aggrecan is the most representative proteoglycan in the cartilage matrix (Roughley and Lee 1994; Hardingham and Fosang 1995; Knudson and Knudson 2001). It consists of many negatively charged glycosaminoglycan chains which are covalently linked to a core protein (Fig. 9.3, diagram), so that it shows a typical bottlebrush-like shape (Fig. 9.3, lower right). Once exported into the extracellular matrix (Luo et al. 2000), aggrecan molecules aggregate by binding to hyaluronan (hyaluronic acid), which in its turn is bound to a globular link protein (Hardingham et al. 1994), so becoming stable and forming wide macromolecular aggregates. The glycosaminoglycan side-chains mainly comprise chondroitin sulfate (each aggrecan contains approximately 100 chondroitin sulfate chains, about 20 kDa each) and smaller amounts of keratan sulfate (up to 60 chains, 5–15 kDa), but their pattern of sulfation, as well as their number and length, can vary, so resulting in different sulfate chain sequences that have specific functions (Hardingham and Fosang 1992). In spite of their variability, all these chains have a net negative charge, mainly due to the presence of sulfate groups in them; this explains why they can be stained with cationic dyes such as alcian blue or colloidal iron even at a pH below 2.8. The charged anionic groups are also responsible for the load-bearing properties of the cartilage (discussed in Kiani et al. 2002). They do, in fact, induce an imbalance in ion concentrations between the immobile proteoglycans and the surrounding structures, which is compensated by mobile Na^+ counterions. This induces an osmotic imbalance, which is compensated by water permeation of the matrix. This, in its turn, causes swelling and expansion of the proteoglycan-rich matrix and the formation of hydrated gel-like, visco-elastic structures which are responsible for the load-bearing properties of cartilage.

Aggrecan shows variations in the different zones of the growth cartilage. The rate both of its synthesis and its turnover is higher in the resting and proliferative zone than in the upper or lower hypertrophic zones (Shapses et al. 1994). In bovine tibial and costo-chondral growth plates an increase in the hydrodynamic size of chondroitin sulfate chains occurs between the resting and hypertrophic zones, probably because a proportion of the aggrecan in the resting zone is removed and replaced by molecules rich in chondroitin sulfate (Deutsch AJ et al. 1995). In the ovine growth plate, the proliferative zone contains proteoglycan monomers larger than those of the resting cartilage; these monomers increase further in size through an increase in chondroitin sulfate chains in the maturing zone and, above all, in the hypertrophic zone, where the aggrecan molecules contain a higher proportion of unsulfated residues (Byers et al. 1997). Morphometric analysis revealed that in bovine fetal cartilage the content of large aggregating proteoglycans reaches a maximum in the extracellular matrix of the hypertrophic zone (Matsui et al. 1991). Using antibodies to chondroitin 4- and 6-sulfate and keratan sulfate along with alcian blue staining of sulfated proteoglycans, Farquharson et al. (1994) showed a similar glycosaminoglycan content in the proliferative and transitional zones of the chick growth plate; in the hypertrophic zone, chondroitin 4- and 6-sulfate were slightly lower (13 and 18%, respectively), keratan sulfate was markedly lower (58%) and, compared with the proliferative zone, the alcian blue staining of sulfated glycosaminoglycans was markedly lower in both the transitional (46%) and hypertrophic (22%) zones.

Closely related core proteins give rise to other members of the aggrecan family, such as versican, neurocan and brevican (Schwartz NB et al. 1999). Versican has been detected in loose connective tissues, in fibrous, articular and elastic cartilages, in the central and peripheral nervous system, in the epidermis, and in the wall of veins and arteries (Lebaron 1996). Its expression and immunostaining rapidly disappear from tibial cartilage during fetal development, and are replaced by the expression and formation of aggrecan (Shibata et al. 2001, 2003). Neurocan and brevican are largely restricted to nervous tissue.

Besides chondroitin and keratan sulfate, the cartilage of the growth plate contains heparan sulfate, which, to a slight degree (0.1%), enters into the structure of aggrecan but makes up as much as 25% of that of perlecan (Govindraj et al. 2002; Hassell et al. 2003). A proteoglycan that is found in all basement membranes, perlecan is a prominent pericellular component of mature hypertrophic chondrocytes in two- to five-day-old sheep (Melrose et al. 2002), and accumulates impressively during cartilage development; it continues to be the major heparan sulfate proteoglycan in adult cartilage (Gomes et al. 2004). Syndecans are a four-member family of heparan sulfate proteoglycans that provide matrix binding sites and cell-surface receptors for growth factors (Hardingham and Fosang 1992). Syndecan-3 has been immunodetected in the proliferative zone of the growth cartilage (Shimazu et al. 1996).

Decorin and biglycan are small proteoglycans that have been reported in resting cartilage, but not in the proliferative, maturing, or hypertrophic zones of the growth cartilage (Poole et al. 1986). In human fetal epiphyses, biglycan core protein was almost exclusively localized in the articular cartilage, whereas decorin was found in the resting cartilage, where biglycan staining was weak; moreover, both biglycan and decorin were detected in the upper proliferative zone, where the former was restricted to the territorial matrix and the latter to the interterritorial matrix (Bianco et al. 1990). According to Takagi et al. (2000), the expression of mRNA and the core protein of biglycan is prominent in hypertrophic and degenerative chondrocytes, but is almost absent from the rest of the epiphyseal cartilage; moreover, the chondroitinase ABC-digested, EDTA extract contains biglycan core proteins, showing that it is present in the calcified matrix (Takagi et al. 2000).

9.3.2 Glycoproteins

Non-collagenous, non-proteoglycan macromolecules are components of the cartilage matrix (reviewed by Neame et al. 1999). Some of them are glycoproteins containing vicinal glycol groups, as can be shown by the PAS reaction. The cartilage matrix is, in fact, deeply stained by this method (de Bernard et al. 1977).

The two major phosphorylated glycoproteins in bone, osteopontin and bone sialoprotein, are also expressed by hypertrophic chondrocytes, as shown by the detection in them of the mRNA for both molecules (Shibata et al. 2002). The immunohistochemical studies of Bianco et al. (1991) showed BSP mRNA expression in human fetal cartilage cells, especially in hypertrophic chondrocytes of the growth plate. Chen et al. (1991) showed that OPN, as well as ON, occur in cartilage, including the hypertrophic chondrocytes, and that BSP also occurs in the calcified cartilage and associated cells of the epiphysis, but not in the hypertrophic zone. Copray et al. (1989) reported the presence of ON in the cartilage of rat mandible condyle, and Pacifici et al. (1990b) detected very low levels of ON in the chondrocytes of the resting, proliferating, and early hypertrophic zones, and large amounts in the matrix of the calcifying zone. Analyses carried out by Landis et al. (2003) on chondrocytes, isolated from mouse growth plate cartilage by laser capture microdissection and examined by RT-PCR, showed that OPN expression in the youngest cartilage was steady throughout the microdissected zones, whereas within the old cartilage it was qualitatively greatest in resting, and lowest in hypertrophic, regions. In the chicken tibia growth plate studied with high resolution immunocytochemistry, McKee et al. (1992) found that OPN was mostly associated with *laminae limitantes* containing condensed, filamentous structures at the periphery of small nodules and large masses of calcified cartilage, that moderate labeling could be detected throughout the interior of the calcified cartilage, and that OC labeling was associated with filamentous structures throughout the calcified cartilage.

9.3.3 Lipids

Lipids occur in the calcified cartilage, from which they can only be totally extracted once the tissue has been decalcified (reviewed by Boyan et al. 1989). They are particularly plentiful in areas of early calcification, as shown by ^{32}P-orthophosphate incorporation into phospholipids: these accumulate at the calcification front and show a turnover pattern different from that of the phospholipids which can be extracted before decalcification (Eisenberg et al. 1970). Lipid accumulation in calcified cartilage areas is also shown by staining with sudan black after pyridine extraction (Bonucci et al. 1978), and using the MC22-33F antibody, which displays a specific reaction with phosphatidylcholine, sphingomyelin and dimethylphosphatidylethanolamine: the immunoreaction was found in the cytoplasm of maturing, hypertrophic and, to a lesser extent, proliferating and degenerating chondrocytes, at the periphery of calcification nodules, and at the periphery of the calcified matrix (Bonucci et al. 1997). The presence of phospholipids in epiphyseal cartilage was confirmed by combining malachite green fixation with phospholipase A_2-gold complex, a method which localizes phospholipids: the labeling intensity was higher at the periphery of the calcification nodules than in the central, fully calcified matrix (Silvestrini et al. 1996; Zini et al. 1996). The proteolipid/dry weight and proteolipid/total lipid ratios were greater in the lower than in the upper (resting) part of the epiphyseal cartilage (Boyan and Ritter 1984).

The amounts of total lipids, triglycerides and phospholipids in cartilage were 5.20, 1.46 and 0.53 mg/g, respectively (Wolinsky and Guggenheim 1970). The phospholipids in the cartilage mostly belong to matrix vesicles, as discussed below.

9.4 Matrix Vesicles

Matrix vesicles are roundish bodies with a diameter between 25 and 250 nm, or even more, and consist of amorphous, osmiophilic matrix surrounded by a trilaminar membrane (Figs. 9.1 and 9.4). They are found in several calcifying matrices, but are specially abundant in the intercellular matrix of the growth cartilage, which probably accounts for the fact that they were first discovered in this tissue (Bonucci 1967). The recurrent observation that the earliest aggregates of crystals recognizable in the growth cartilage are found in matrix vesicles (Fig. 9.4) prompted the concept that these structures are the earliest areas of calcification in that tissue and that, therefore, they possess all that is needed to induce the formation of inorganic substance in cartilage.

This hypothesis attracted a great deal of interest, giving rise to a number of studies on these structures, which were called by a variety of names: 'dense bodies' (Bonucci 1967; Sisca and Provenza 1972), 'osteoblast extrusions' (Bernard and Pease 1969), 'calcifying globules' (Bonucci 1970a; Silbermann and Frommer 1974), 'osmiophilic bodies' (Hall et al. 1971), 'osmiophilic spherules' (Kim and Huang 1971), 'membrane-bound bodies' (Katchburian 1973a), 'extracellular bodies' (Katchburian 1973b), and 'dentinal globules' (Larsson 1973; Larsson and Bloom 1973). The term 'matrix vesicle', introduced by Anderson (1969), is, by now, universally used, even if it has been judged "pathetically inaccurate" (Ghadially 2001). The many studies carried out on matrix vesicles have been reviewed several times (Bonucci 1971b, 1984; Alcock 1972; Anderson 1973, 1976, 1984, 1995, 2003; Ali 1976, 1983; Wuthier 1982; Sela et al. 1992; Hoshi et al. 2000).

Recently, the topic has been subjected to severe, but unjustified, criticism by Ghadially (2001), who claimed that "there is nothing special or wonderful about matrix vesicles". In his view, they are no more than the "matrical lipidic debris" he had described in articular cartilage, two years before the discovery of matrix vesicle (Ghadially et al. 1965), as structures deriving from the in situ necrosis of chondrocytes or from extruded chondrocyte processes, i. e., from a physiological, continuous process of cell remodeling. In his criticisms, Ghadially stresses that calcium precipitation is common to degenerate or necrotic cells and tissues and that, therefore, it is hardly surprising that the same occurs in cell-derived debris.

It is true that cellular debris was already known from studies by several investigators, Ghadially among them; as early as 1928 Wolf and Cerný (translated and summarized in Wolf and Cerný 1928) had reported globular structures

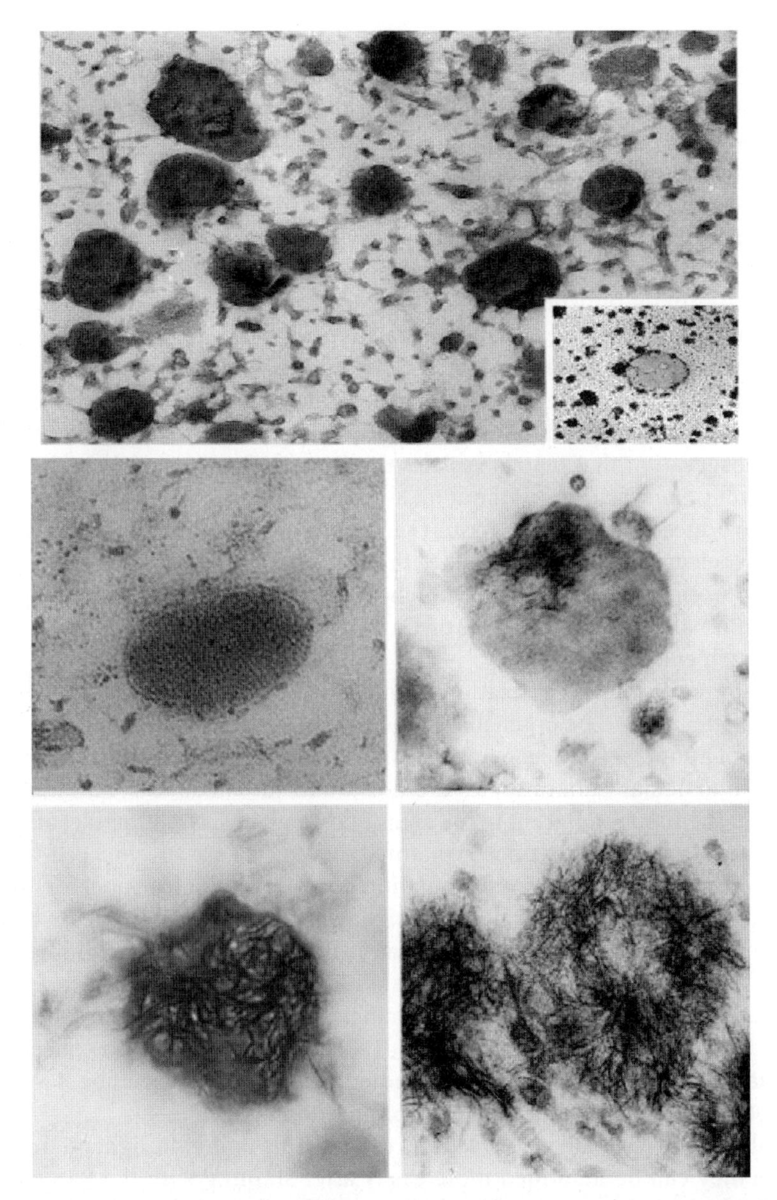

Fig. 9.4. Maturing zone of growth cartilage: matrix vesicles correspond to roundish bodies with homogeneous matrix surrounded by a membrane and connected to collagen fibrils. Uranyl acetate and lead citrate. ×70,000. *Inset*: in sections treated with colloidal iron at pH 2.8, acid proteoglycans appear as electron-dense, iron-stained granules; note that they coat the membrane of a matrix vesicle. ×47,000. *Middle and below*: series of matrix vesicles at different degrees of calcification. *Middle left*: uncalcified (note peripheral membrane; treated with uranium and lead); *middle right*: partly calcified (untreated); *lower left*: almost fully calcified, with a few crystal-like structures protruding into the perivesicular matrix (untreated); *lower right*: calcification nodules; vesicles no longer recognizable (uranium and lead). ×90,000

surrounding chondrocytes and involved in calcification, while in 1930 (Schaffer 1930) had described a process of disappearance of chondrocytes with residual matrical dense bodies. The nature and biogenesis of matrix vesicles are discussed below. It must be pointed out here that, even if the cartilage matrix vesicles resemble the lipidic debris described by Ghadially in articular cartilage, and may show a degree of overlap with that type of structure, their morphological similarities do not appear sufficient to justify the claim that matrix vesicles are simply lipidic cellular debris, although it is well known that they contain phospholipids. The demonstration that matrix vesicles have reproducible, distinctive morphological, biochemical and histochemical properties (discussed below) makes them structures with specific characteristics and help to distance them from the generic, largely misleading, concept that they are secondary products of cell degeneration. This is supported by the in vitro observation that matrix vesicles produced by non-calcifying tissue have a different composition (without alkaline phosphatase and annexin V) from those produced by calcifying tissue (Kirsch et al. 1997a) and that, unlike matrix vesicles released from chondrocytes that have been treated with retinoic acid, the apoptotic bodies released from chondrocytes treated with staurosporine contain no alkaline phosphatase or Ca^{2+} channel-forming annexins II, V and VI (Kirsch et al. 2003).

In any case, the identification of matrix vesicles with cell degeneration products appears to be untenable in the case of matrix vesicles in bone, where figures of apoptosis or degeneration comparable with those described by Ghadially in articular chondrocytes are not found in osteoblasts, although they occur in the tissues adjacent to the areas of osteogenesis (Palumbo et al. 2003). The introduction of the concept that matrix vesicles are specific, cell-derived structures was not just the reformulation of a morphological label for already known structures. The crucial new insight was that matrix particles, which up till then had been known generically as "cell debris" or "spherical microparticles", have a precise identity, above all that they can be identified as the structures responsible for the early calcification process in cartilage, bone and other biological tissues (Bonucci 1967, 1971b; Anderson 1969), a concept that at that time clashed with the deeply rooted belief that collagen fibers were the only structures able to nucleate hydroxyapatite in bone.

Only the biggest matrix vesicles are recognizable under the light microscope, where they can be identified more easily after staining with the PAS method (Bonucci 1967) or by periodic acid-silver methenamine (Bonucci 1970a; Dearden and Espinosa 1974). Their staining with these methods shows, incidentally, that they contain glycoproteic material. Moreover, the outer surface of vesicle membranes is coated with acid proteoglycans, as shown by staining with colloidal iron (Fig. 9.4, inset) or colloidal thorium (Bonucci 1970a; Dearden and Espinosa 1974; Scherft and Groot 1981). Spikes of proteoglycan material extend into the surrounding matrix and connect matrix vesicles with collagen fibrils (Hunziker et al. 1981; Scherft and Groot 1981; Kogaya and Furuhashi

1985). Matrix vesicles do, in fact, interact strongly with matrix collagen fibrils and with proteoglycans (Wu et al. 1989, 1991), the association being mediated by several matrix vesicle proteins such as alkaline phosphatase, annexins V and VI, proteoglycan link protein and the hyaluronate binding region (Wu et al. 1992). Keratansulfate has been immunolocalized within and around matrix vesicles in rat calvaria (Nakamura et al. 2001).

According to Buckwalter et al. (1987b), cartilage matrix vesicles are particularly plentiful in the lower proliferative and upper hypertrophic zone, but become less frequent as mineralization proceeds in the lower hypertrophic zone. According to Reinholt et al. (1983), their concentrations are highest in the resting zone, are low in the proliferative zone, are high again in the hypertrophic zone, and rapidly falls with mineralization in the lower hypertrophic zone. In the growth cartilage, they are mainly found in longitudinal septa (Anderson 1969; Reinholt and Wernerson 1988), where they are often located in a matrix zone corresponding to the territorial matrix (Eggli et al. 1985) or, more precisely, to a band at the boundary between the territorial and interterritorial matrix (unpublished personal observation); very few can be found in the transverse septa (Reinholt and Wernerson 1988). Most are scattered irregularly through these zones, but some, mainly in the proliferative and maturing zones, may be clustered in groups whose arrangement suggests that they are confined to the area of degenerate chondrocytes (Bonucci 1970a, b). This raises the problem of the origin of matrix vesicles.

The presence of a peripheral membrane around matrix vesicles unequivocally shows that they are cellular structures, so much so that, soon after their first description, they were taken to be cross-sections of chondrocyte processes. Serial sections have shown that they are true spherical bodies – although they are sometimes irregular in shape – which have no direct connection with chondrocytes (Bonucci 1970a, b, 1978); in this regard, serial sections also show that early crystals are not only produced on the membrane of matrix vesicles (Hunziker et al. 1981), but inside them, too, deep in their matrix. Similar results have been found in dentin, where matrix vesicles show no direct contact with odontoblasts or their processes (Katchburian 1973a). The fact that matrix vesicles can be isolated by matrix collagenase digestion, cell pellet removal, and differential centrifugation, or else by matrix homogenization and density gradient fractionation, both from cartilage (Ali 1976; Ali and Evans 1973; Ali et al. 1970, 1971; Wuthier et al. 1978, 1985; Watkins et al. 1980; Warner et al. 1983; Kakuta et al. 1985) and from bone (Deutsch D et al. 1981; Bab et al. 1983) and dentin (Slavkin et al. 1972), further confirms that they are true matrix bodies, without any direct link with cells.

By now it is generally accepted that most cartilage matrix vesicles are formed by budding from the cell membrane or the tip of cell processes, as shown by transmission (Bonucci 1970a, b, 1978; Newbrey and Banks 1975; Cecil and Anderson 1978; Dickson 1981a, b) and scanning (Ornoy and Langer 1978; Ornoy et al. 1979) electron microscopy and by freeze-fracture studies (Cecil and Anderson 1978; Akisaka and Gay 1985a). The same formation mechanism has been shown in bone (Ornoy et al. 1980) and dentin (Almuddaris and Dougherty 1979; Katchburian and Severs 1982). Budding is not yet fully understood: it might be due to a membrane degradation process triggered by Ca-activated neutral proteases (Pollesello et al. 1990), a perturbation of the filament network that supports membrane microvilli (Hale and Wuthier 1987), a rise in intracellular calcium concentrations alone (Iannotti et al. 1994), or a rise in concentrations of membrane-bound calcium and of chondroitin sulfate (Takagi et al. 1989). Histochemical studies with K-pyroantimonate-osmium fixation have, however, shown that matrix vesicle biogenesis is not associated with an increase in intracellular calcium complexes, and that matrix vesicles may, rather, function as carriers of calcium from the plasma membrane to the surrounding matrix (Lewinson and Silbermann 1981). Using the same histochemical method to demonstrate calcium, Morris and Appleton (1980) described the presence of pyroantimonate precipitates in vesicular structures closely adjacent to the chondrocyte membrane and apparently extruded from the cells into the pericellular matrix, confirming their calcium content by energy-dispersive X-ray analysis. The same authors used lanthanum to study the cartilage calcium-binding sites, on the assumption that it links to the same sites as calcium ions; they described coarse deposits of lanthanum at and along the hypertrophic chondrocyte membrane and in the closely adjacent matrix, and showed that these became fewer and smaller in vitamin D-deprived animals (Morris and Appleton 1984). Gomez et al. (1996) obtained the same results using the same method and interpreted the deposits as corresponding to a specific type of matrix vesicles which have the intrinsic capacity to precipitate lanthanum as soon as they originate from the hypertrophic chondrocytes. These results confirm that calcium is transported from the cells into the matrix, but doubts persist about whether the structures found close to the outer surface of the chondrocyte membrane are truly vesicular, and about the significance of the precipitates connected with them.

According to Wuthier et al. (1977), matrix vesicles are formed by a rapid, metabolically active process, not by degenerative changes in the cell membrane. On the other hand, the frequent report that matrix vesicles are located around degenerating chondrocytes (Hayashi et al. 1993) or can be found collected in groups reminiscent of chondrocyte lacunae suggests that they may also derive from the degeneration and fragmentation of chondrocytes (Bonucci 1970a, 1978; Dearden 1974; Bonucci and Dearden 1976; Takagi et al. 1979; Akisaka and Shigenaga 1983; Akisaka and Gay 1985a; Reinholt et al. 1982, 1984). This process has long been known as "Verdämmern der Zellen" ('vanishing of cells'; Schaffer 1930) and is now thought to be a possible expression of cellular apoptosis (Kim 1976; Kardos and Hubbard 1982; Anderson 1995; Hashimoto et al. 1998; Lotz et al. 1999), a process that has been induced experimentally in articular cartilage (Hashimoto et al. 1998) and that seems to characterize the terminal phase of hypertrophic chondrocytes (Rauterberg and Becker 1970;

Kim 1995; Zenmyo et al. 1996; Gibson et al. 1997; Szuwart et al. 1998; Magne et al. 2003). However, as reported above, the composition of apoptotic bodies differs to some extent from that of matrix vesicles (Kirsch et al. 2003).

While budding from cells or cell processes and cell fragmentation are now generally recognized to be the main processes of matrix vesicle generation, other mechanisms that had been suggested at the outset (Rabinovitch and Anderson 1976), such as the assembly of secreted subunits or the extrusion of preformed structures (if any, only in bone), have not been well documented. It is worth noting that matrix vesicle formation is enhanced by the administration of glucocorticoids (Silbermann et al. 1980; Dearden et al. 1981).

One of the most distinctive properties of matrix vesicles is that they show alkaline phosphatase activity (Ali et al. 1970, 1971; Matsuzawa and Anderson 1971; Larsson 1973; Salomon 1974; Majeska and Wuthier 1975; Bernard 1978; Fortuna et al. 1978, 1980; Kahn et al. 1978; Takagi and Toda 1979; Väänänen and Korhonen 1980; Watkins et al. 1980; Lewinson et al. 1982; Morris et al. 1986, 1990, 1992a; Bonucci et al. 1986; de Bernard et al. 1986; Miller and DeMarzo 1988; Henson et al. 1995; Herbert et al. 1997; Kirsch and Claassen 2000; Hoshi et al. 2001). Light microscope studies on epiphyseal cartilage, as well as on bone and other calcifying tissues (Fig. 9.5), have shown that enzyme activity first appears on the membrane of osteoblasts and chondrocytes and, as globule-like precipitates, around and outside the cell membrane of lower maturing chondrocytes and hypertrophic chondrocytes (Morris et al. 1986; Akisaka and Gay 1985b; Bianco 1990). Under the electron microscope, dense precipitates indicative of an enzymatic reaction were found at the chondrocyte plasma membrane and in association with spherical or ovoid structures which were 80–120 nm in diameter and were mainly located in the territorial matrix (Matsuzawa and Anderson 1971; Göthlin and Ericsson 1973; Salomon 1974; Thyberg et al. 1975; Meikle 1976; Bernard 1978; Takagi and Toda 1979; Bonucci et al. 1986; de Bernard et al. 1986; Morris et al. 1986, 1990, 1992a; Ecarot-Charrier et al. 1988). These results were confirmed by alkaline phosphatase immunohistochemistry (Fig. 9.5, inset) (de Bernard et al. 1986; Bonucci et al. 1986; Morris et al. 1986, 1992a; Bianco 1990; Masuhara et al. 1992) and by the presence of alkaline phosphatase activity in isolated matrix vesicles (Ali et al. 1970, 1971; Majeska and Wuthier 1975; Fortuna et al. 1978; Kahn et al. 1978; Väänänen and Korhonen 1980; Watkins et al. 1980).

Alkaline phosphatase-negative vesicles have been reported in epiglottis elastic cartilage (Nielsen 1978) and in the resting zone of epiphyseal cartilage (Ohashi-Takeuchi et al. 1990). According to Genge et al. (1988), 65–70% of alkaline phosphatase activity is lost as calcium is accumulated in matrix vesicles, probably because of denaturation of the enzyme due to the loss of Zn and Mg from its active sites or to a reaction with the inorganic substance (de Bernard et al. 1986). According to Watkins et al. (1980), the alkaline phosphatase-rich, low-density fractions obtained from chick cartilage by a non-enzymatic method (Wuthier et al. 1978) were not identical with matrix vesicles isolated

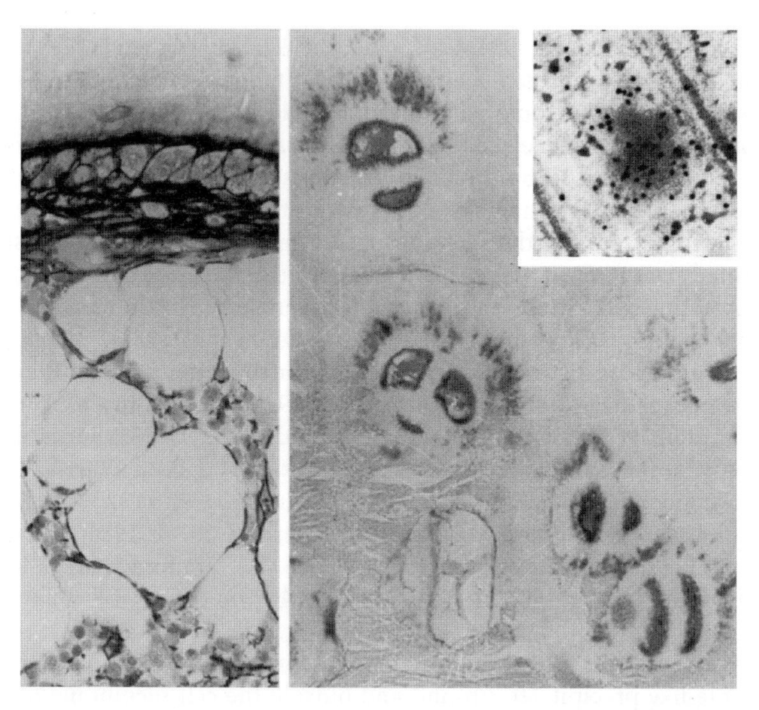

Fig. 9.5. Histochemical demonstration of alkaline phosphatase activity. *Left*, osteogenic area in bone: the products of the enzymatic reaction (*black*) can be seen outlining the membrane of the osteoblasts; it also marks the osteoid border which separates the osteoblasts from the calcified matrix (*above*), a few elongated cells located below the osteoblasts, and thin processes of reticular cells in the bone marrow; ×220. *Right*, epiphyseal cartilage: the enzymatic reaction has occurred on the membrane of the chondrocytes and on small granular structures forming a halo around them and corresponding to matrix vesicles; calcified matrix (unstained) on *lower left*; ×500. *Inset*: alkaline phosphatase immunolocalized on a cartilage matrix vesicle; ×48,000

by enzymatic digestion but consisted primarily of plasma membranes; two of the phospholipids characteristic of matrix vesicles were enriched in these fractions but, unlike matrix vesicles, there was no loss of phosphatidylcholine or increase in lysophospholipids.

Besides alkaline phosphatase activity, pyrophosphatase, ATPase, 5′AMPase, lactate dehydrogenase and carbonic anhydrase activities have also been demonstrated in matrix vesicles (Ali et al. 1970; Larsson 1974; Majeska and Wuthier 1975; Thyberg et al. 1975; Felix and Fleisch 1976; Bab et al. 1979; Väänänen and Korhonen 1979, 1980; Hsu 1983; Takano et al. 1986; Hosokawa et al. 1988; Kitamura et al. 1991; Stechschulte et al. 1992; Hsu and Anderson 1996; Maki et al. 2000). According to Hsu and Anderson (1984), alkaline phosphatase is not the major factor in CaP deposition in vitro by matrix vesicles; instead, it appears that the nucleoside triphosphate pyrophosphohydrolase is directly involved in the early stages of deposition. Matrix vesicles contain metalloproteinases such MMP-2, -9 and -13 (Katsura and Yamada 1986; D'Angelo et al. 2001), stromelysin-1 and 72-kDa gelatinase (Schmitz et al. 1996). According to Dean et al. (1992, 1994), matrix vesicles are selectively enriched by the presence of enzymes which degrade proteoglycans, and the highest concentrations of these enzymes are found in matrix vesicles produced by growth zone chondrocytes.

Besides enzymatic activities, matrix vesicles contain substantial amounts of lipids. According to Peress et al. (1974), the sphingomyelin and phosphatidylserine content, and the ratio between moles of cholesterol and moles of total phospholipids, are higher in the matrix vesicle fraction than in the cellular fraction, a finding consistent with the origin of matrix vesicles in the chondrocyte plasma membrane. Majeska et al. (1979) localized phosphatidylserine in isolated chick epiphyseal cartilage matrix vesicles. According to Wuthier (1975, 1976), the lipids in matrix vesicles show a composition similar to that of the cellular plasma membrane: they are rich in cholesterol, free fatty acids, sphingomyelin, glycolipids, lysophospholipids, and phosphatidylserine, and poor in phosphatidylcholine and phosphatidylethanolamine. The immunoreaction with CM22-33F, an antibody specific for phosphatidylcholine, sphingomyelin and dimethylphosphatidylethanolamine, already demonstrated in matrix vesicles of dentin (Tsuji et al. 1994), has been confirmed in those of cartilage (Bonucci et al. 1997).

Among the major constituents of matrix vesicles there are acidic phospholipid-dependent proteins of 33 and 36 kDa, corresponding to annexin V and annexin II, respectively (Genge et al. 1991, 1992a, b), and of 67 kDa, corresponding to annexin VI (Cao et al. 1993). When inserted in membranes, annexins display voltage-dependent Ca^{2+}-ion channel activity and a calcium-binding property; they could therefore be involved in intravesicular calcium uptake (Rojas et al. 1992; Arispe et al. 1996; Kirsch et al. 1997b).

Other substances which have been proposed as constituents of matrix vesicles are actin (Muhlrad et al. 1982; Morris et al. 1992b) and calbindin-D9k. The latter is a vitamin-D-dependent, calcium-binding protein localized in mature chondrocytes, osteoblasts and osteocytes; it is co-localized with calbindin-D28k in ameloblasts (Balmain 1991; Berdal et al. 1991a, b), but is also found in matrix vesicles (Balmain et al. 1989, 1991; Balmain 1991, 1992). Personal immunohistochemical studies on cartilage calcification (unpublished) showed that matrix vesicles do not contain BSP, which can only be detected in them with the earliest intravesicular crystals (Fig. 9.6).

Matrix vesicles, mainly those found in growth cartilage, may be mixed with, or may show the same shape as, the cell debris described in articular cartilage by Ghadially et al. (1965), and may therefore be confused with them. This could explain the often reported heterogeneity of matrix vesicles. Two fractions of matrix vesicles have been isolated by means of the equilibrium density fractionation method: a heavy fraction, rich in enzymatic activities, and a light fraction (Bab et al. 1983). Two types of matrix vesicles have been

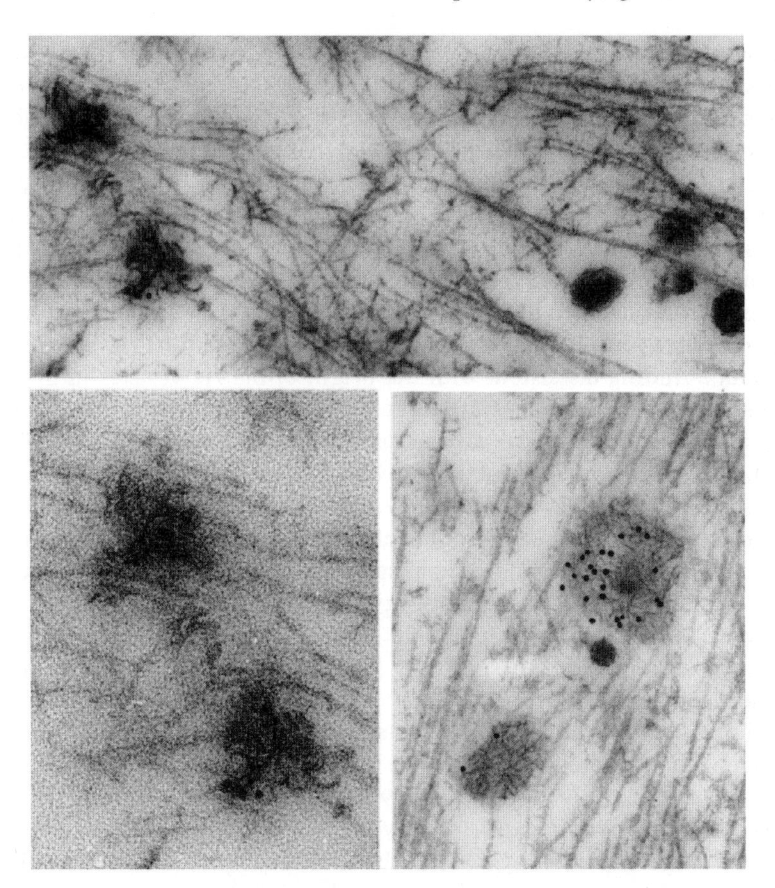

Fig. 9.6. Specimen of growth cartilage embedded in LR-White and decalcified and stained according to the PEDS method; BSP immunolabeling. *Above*: immunogold particles are not recognizable in not yet calcified matrix vesicles (*right*), whereas they can be detected in those (*left*) which contain the earliest crystals (and appear in this figure as earliest aggregates of crystal ghosts); ×45,000. *Lower left*: enlargement of the crystal ghost aggregates shown above; ×70,000. *Lower right*: BSP immunogold particles in two early areas of calcification; note negativity of the interposed matrix vesicle. ×30,000

described too, on the basis of their ultrastructure and enzymatic properties: a rare type I, comprising vesicles with lysosome-like properties, and a frequent type II, showing alkaline phosphatase activity (Thyberg 1972; Thyberg et al. 1975). Meikle (1976) drew a distinction between lysosome-like bodies and matrix vesicles. On the basis of their reactivity with lanthanum chloride, two types of vesicles – dense and light – have been described in the epiphyseal cartilage of the young rat (Gomez et al. 1996); the dense vesicles have an intrinsic capacity to precipitate LaP intravesicularly, whereas the light vesicles seem to lack this feature. According to Kirsch et al. (1997a), chondrocytes could release two types of matrix vesicles, which are morphologically similar but differ in their capacity to induce calcification. Another type of differentiation between two types of vesicles focuses on density-gradient fractionation in different regions of the growth cartilage: a P-I, low-density, non-calcifiable fraction, and a P-II, high-density, readily calcifiable fraction (Warner et al. 1983). On the basis of the density and distribution of particles observed in freeze-fracture images of the P-face of the vesicular membrane, three types of matrix vesicles have been distinguished in the hypertrophic zone of epiphyseal cartilage, with those in calcifying areas (type I) showing numbers of particles higher than those (type III) in the resting cartilage (Takagi et al. 1979). Using electron microscopy and quantitative computerized morphometry, four types of matrix vesicles have been described in healing bone, which seem to be different morphological expressions of the same underlying type of vesicle: 'empty' (23.2%), which have an electron-lucent content; 'amorphous' (74%), which have an amorphous, electron-dense content; 'crystal' (2.5%), which contain crystalline structures; and 'rupture' (0.3%), which release crystals into the extracellular matrix (Schwartz Z et al. 1987; Sela et al. 1987a; Amir et al. 1988a, b). According to Schwartz Z et al. (1987), the relative frequencies of these vesicle types are 10, 31, 51 and 8%, respectively, whereas they are 23.2, 74.0, 2.5 and 0.3% according to Amir et al. (1988a). A time-dependent decrease in numbers of empty and amorphous types which correlated with an increase in crystal and rupture types, and an increase in the rupture type close to the calcification front, were noted (Schwartz Z et al. 1989). In this connection, the hypothesis has been put forward that cells produce empty vesicles, that these become amorphous vesicles as they accumulate Ca and P_i, that these amorphous vesicles turn into crystal vesicles as crystals are formed inside them, and that crystal vesicles finally become rupture vesicles, rupturing and releasing crystals into the matrix (Sela et al. 1987b, 1992).

9.5 Concluding Remarks

The cartilage in the growth plate displays characteristic zones, allowing different phases of cell differentiation to be studied with reference to matrix calcification. The findings reported above justify the following concluding remarks.

- Cartilage matrix consists of collagen and interfibrillary substance; type II collagen is the major fibrillar component. Its distribution is not homogeneous – its concentrations rise as proliferating chondrocytes evolve into hypertrophic ones and then fall as chondrocytes continue to hypertrophy.
- Other types of collagen occur in the cartilage matrix and can form heteropolymers with type II collagen. Type X collagen is only produced by hypertrophic chondrocytes and precedes matrix calcification.

- The C-propeptide of type II collagen (chondrocalcin) is a calcium-binding protein which is concentrated in the matrix of calcification nodules.
- Aggrecan is the most representative proteoglycan in cartilage matrix. Chondroitin and keratan sulfate chains are the glycosaminoglycans linked to its core protein. Aggrecan molecules form wide aggregates by binding to hyaluronan, which in its turn binds to a globular link protein, so forming wide macromolecular, highly negatively charged hydrophilic aggregates.
- Besides aggrecan, other proteoglycans occur in the cartilage matrix, including decorin, biglycan, and other members of the aggrecan family.
- The ultrastructure of cartilage proteoglycans changes with the preparation procedure: native aggrecan molecules have a bottlebrush-like conformation. They have an extended, filament-like shape after cryofixation and cryosectioning or after adding of cationic dyes to the fixative solutions used; they collapse and appear as small granules if the usual methods for preparing tissues for electron microscopy are used.
- Concentrations of acid proteoglycans fall, and their structure and composition may well change, before or during matrix calcification.
- Acid proteoglycans are embedded in the calcified matrix and can only be totally extracted after decalcification.
- Some of the phosphoglycoproteins of the bone matrix have been found in cartilage matrix (specifically, osteopontin and bone sialoprotein); BSP has not been found in uncalcified matrix vesicles.
- Phospholipids are present in the calcified cartilage, and can only be totally extracted after tissue decalcification.
- Plenty of matrix vesicles can be detected in the intercellular matrix of the growth cartilage. They function as areas of initial calcification.
- Matrix vesicles originate by budding from the chondrocyte membrane or the tip of cellular processes, or by fragmentation of whole chondrocytes, a process that has been compared to apoptosis.
- Alkaline phosphatase activity is the most typical histochemical marker of matrix vesicles. Pyrophosphatase, ATPase, 5′ AMPase, lactate dehydrogenase and carbonic anhydrase activities have also been demonstrated in them.
- Matrix vesicles contain substantial amounts of phospholipids and acidic phospholipid-dependent proteins such as annexins II, V and VI, which, once inserted in the chondrocyte membrane, display voltage-dependent Ca^{2+}-ion channel activity and a calcium-binding property.

References

Abad V, Meyers JL, Weise M, Gafni RI, Barnes KM, Nilsson O, Bacher JD, Baron J (2002) The role of the resting zone in growth plate chondrogenesis. Endocrinology 143:1851–1857

Aizawa T, Kokubun S, Tanaka Y (1997) Apoptosis and proliferation of growth plate chondrocytes in rabbits. J Bone Joint Surg 79-B:483–486

Akisaka T, Gay CV (1985a) The plasma membrane and matrix vesicles of mouse growth plate chondrocytes during differentiation as revealed in freeze-fracture replicas. Am J Anat 173:269–286

Akisaka T, Gay CV (1985b) Ultrastructural localization of calcium-activated adenosine triphosphatase (Ca2+-ATPase) in growth-plate cartilage. J Histochem Cytochem 33:925–932

Akisaka T, Shigenaga Y (1983) Ultrastructure of growing epiphyseal cartilage processed by rapid freezing and freeze-substitution. J Electron Microsc 32:305–320

Alcock NW (1972) Calcification of cartilage. Clin Orthop Relat Res 86:287–311

Ali SY (1976) Analysis of matrix vesicles and their role in the calcification of epiphyseal cartilage. Fed Proc 35:135–143

Ali SY (1983) Calcification of cartilage. In: Hall BK (ed) Cartilage, vol. I. Structure, function and biochemistry. CRC Press, Boca Raton, pp 343–378

Ali SY, Evans L (1973) The uptake of [^{45}Ca]calcium ions by matrix vesicles isolated from calcifying cartilage. Biochem J 134:647–650

Ali SY, Sajdera SW, Anderson HC (1970) Isolation and characterization of calcifying matrix vesicles from epiphyseal cartilage. Proc Nat Acad Sci 67:1513–1520

Ali SY, Anderson HC, Sajdera SW (1971) Enzymic and electron-microscopic analysis of extracellular matrix vesicles associated with calcification in cartilage. Biochem J 122:56

Alini M, Matsui Y, Dodge GR, Poole AR (1992) The extracellular matrix of cartilage in the growth plate before and during calcification: changes in composition and degradation of type II collagen. Calcif Tissue Int 50:327–335

Almuddaris MF, Dougherty WJ (1979) The association of amorphous mineral deposits with the plasma membrane of pre- and young odontoblasts and their relationship to the origin of dentinal matrix vesicles in rat incisor teeth. Am J Anat 155:223–244

Althoff J, Quint P, Krefting E-R, Höhling HJ (1982) Morphological studies on the epiphyseal growth plate combined with biochemical and X-ray microprobe analyses. Histochemistry 74:541–552

Amir D, Schwartz Z, Weinberg H, Sela J (1988a) The distribution of extracellular matrix vesicles in healing of rat tibial bone three days after intramedullary injury. Arch Orthop Traumat Surg 107:1–6

Amir D, Schwartz Z, Sela J, Weinberg H (1988b) The relationship between extracellular matrix vesicles and the calcifying front on the 21st day after injury to rat tibial bone. Clin Orthop Relat Res 230:289–295

Anderson HC (1969) Vesicles associated with calcification in the matrix of epiphyseal cartilage. J Cell Biol 41:59–72

Anderson HC (1973) Calcium-accumulating vesicles in the intercellular matrix of bone. In: Ciba Foundation (ed) Hard tissue growth, repair and remineralization. ASP (Elsevier-Excerpta Medica-North-Holland), Amsterdam, pp 213–246

Anderson HC (1976) Matrix vesicles of cartilage and bone. In: Bourne GH (ed) The biochemistry and physiology of bone. Academic Press, New York, pp 135–157

Anderson HC (1984) Mineralization by matrix vesicles. Scanning Electron Microsc 2: 953–964

Anderson HC (1995) Molecular biology of matrix vesicles. Clin Orthop Relat Res 314:266–280

Anderson HC (2003) Matrix vesicles and calcification. Curr Rheumatol Rep 5:222–226
Arispe N, Rojas E, Genge BR, Wu LNY, Wuthier RE (1996) Similarity in calcium channel activity of annexin V and matrix vesicles in planar lipid bilayers. Biophys J 71:1764–1775
Arsenault AL (1990) The ultrastructure of calcified tissues: methods and technical problems. In: Bonucci E, Motta PM (eds) Ultrastructure of skeletal tissues. Kluwer Academic Publishers, Boston, pp 1–18
Arsenault AL, Ottensmeyer FP, Heath IB (1988) An electron microscopic and spectroscopic study of murine epiphyseal cartilage: analysis of fine structure and matrix vesicles preserved by slam freezing and freeze substitution. J Ultrastruct Mol Struct Res 98:32–47
Ayad S, Evans H, Weiss JB, Holt L (1984) Type VI collagen but not type V collagen is present in cartilage. Collagen Rel Res 4:165–168
Ayad S, Marriott A, Brierley VH, Grant ME (1991) Mammalian cartilage synthesizes both proteoglycan and non-proteoglycan forms of type IX collagen. Biochem J 278:441–445
Bab IA, Muhlrad A, Sela J (1979) Ultrastructural and biochemical study of extracellular matrix vesicles in normal alveolar bone of rats. Cell Tissue Res 202:1–7
Bab I, Deutsch D, Schwartz Z, Muhlrad A, Sela J (1983) Correlative morphometric and biochemical analysis of purified extracellular matrix vesicles from rat alveolar bone. Calcif Tissue Int 35:320–326
Balmain N (1991) Calbindin-D_{9k}. A vitamin-D-dependent, calcium-binding protein in mineralized tissues. Clin Orthop Relat Res 265:265–276
Balmain N (1992) Identification of calbindin-D9k in matrix vesicles. Bone Miner 17:197–201
Balmain N, Hotton D, Cuisinier-Gleizes P, Mathieu H (1989) Immunoreactive calbindin-D_{9k} localization in matrix vesicle-initiated calcification in rat epiphyseal cartilage: an immunoelectron microscope study. J Bone Miner Res 4:565–575
Balmain N, Hotton D, Cuisinier-Gleizes P, Mathieu H (1991) Immunoreactive calbindin-D_{9k} in bone matrix vesicles. Histochemistry 95:459–469
Berdal A, Hotton D, Kamyab S, Cuisinier-Gleizes P, Mathieu H (1991a) Subcellular colocalization and co-variations of two vitamin D-dependent proteins in rat ameloblasts. Arch Oral Biol 36:715–725
Berdal A, Nanci A, Smith CE, Ahluwalia JP, Thomasset M, Cuisinier-Gleizes P, Mathieu H (1991b) Differential expression of calbindin-D 28 kDa in rat incisor ameloblasts throughout enamel development. Anat Rec 230:149–163
Bernard GW (1978) Ultrastructural localization of alkaline phosphatase in initial intramembranous osteogenesis. Clin Orthop Relat Res 135:218–225
Bernard GW, Pease DC (1969) An electron microscopic study of initial intramembranous osteogenesis. Am J Anat 125:271–290
Bianco P (1990) Ultrastructural immunohistochemistry of noncollagenous proteins in calcified tissues. In: Bonucci E, Motta PM (eds) Ultrastructure of skeletal tissues. Kluwer Academic Publishers, Boston, pp 63–78
Bianco P, Fisher LW, Young MF, Termine JD, Gehron Robey P (1990) Expression and localization of the two small proteoglycans biglycan and decorin in developing human skeletal and non-skeletal tissues. J Histochem Cytochem 38:1549–1563
Bianco P, Fisher LW, Young MF, Termine JD, Gehron Robey P (1991) Expression of bone sialoprotein (BSP) in developing human tissues. Calcif Tissue Int 49:421–426
Bianco P, Descalzi Cancedda F, Riminucci M, Cancedda R (1998) Bone formation via cartilage models: the "borderline" chondrocyte. Matrix Biol 17:185–192
Bonucci E (1967) Fine structure of early cartilage calcification. J Ultrastruct Res 20:33–50
Bonucci E (1970a) Fine structure and histochemistry of "calcifying globules" in epiphyseal cartilage. Z Zellforsch 103:192–217

Bonucci E (1970b) Fine structure of epiphyseal cartilage in experimental scurvy. J Pathol 102:219–227
Bonucci E (1971a) Problemi attuali attinenti all'istochimica di talune matrici calcificanti normali e patologiche. Riv Istochim Norm Patol 17:153–234
Bonucci E (1971b) The locus of initial calcification in cartilage and bone. Clin Orthop Relat Res 78:108–139
Bonucci E (1978) Matrix vesicle formation in cartilage of scorbutic guinea pigs: electron microscope study of serial sections. Metab Bone Dis Rel Res 1:205–212
Bonucci E (1984) Matrix vesicles: their role in calcification. In: Linde A (ed) Dentin and dentinogenesis. CRC Press, Boca Raton, pp 135–154
Bonucci E, Dearden LC (1976) Matrix vesicles in aging cartilage. Fed Proc 35:163–168
Bonucci E, Frollà G, Piacentini M, Piantoni L (1978) Presenza di materiale sudanofilo nella matrice ossea e cartilaginea e processo di calcificazione: indagini istochimiche ed ultrastrutturali. Riv Istochim Norm Pathol 22:77–91
Bonucci E, Bianco P, Hayashi Y, Termine JD (1986) Ultrastructural immunohistochemical localization of non-collagenous proteins in bone, cartilage and developing enamel. In: Ali SY (ed) Cell mediated calcification and matrix vesicles. Excerpta Medica, Amsterdam, pp 33–38
Bonucci E, Silvestrini G, Mocetti P (1997) MC22–33F monoclonal antibody shows unmasked polar head groups of choline-containing phospholipids in cartilage and bone. Eur J Histochem 41:177–190
Boyan BD, Ritter NM (1984) Proteolipid-lipid relationships in normal and vitamin D-deficient chick cartilage. Calcif Tiss Int 36:332–337
Boyan BD, Schwartz Z, Swain LD, Khare A (1989) Role of lipids in calcification of cartilage. Anat Rec 224:211–219
Brighton CT (1978) Structure and function of the growth plate. Clin Orthop Relat Res 136:22–32
Buckwalter JA (1983) Proteoglycan structure in calcifying cartilage. Clin Orthop Relat Res 172:207–232
Buckwalter JA, Mower D, Ungar R, Schaeffer J, Ginsberg B (1986) Morphometric analysis of chondrocyte hypertrophy. J Bone Joint Surg 68A:243–255
Buckwalter JA, Rosenberg LC, Ungar R (1987a) Changes in proteoglycan aggregates during cartilage mineralization. Calcif Tissue Int 41:228–236
Buckwalter JA, Mower D, Schaeffer J (1987b) Differences in matrix vesicle concentration among growth plate zones. J Orthop Res 5:157–163
Byers S, van Rooden JC, Foster BK (1997) Structural changes in the large proteoglycan, aggrecan, in different zones of the ovine growth plate. Calcif Tissue Int 60:71–78
Campo RD (1974) Soluble and resistant proteoglycans in epiphyseal plate cartilage. Calcif Tissue Res 14:105–119
Campo RD (1976) Resistant proteoglycans in epiphyseal plate cartilage. Variations in their distribution in relationship to age and species. Calcif Tissue Res 22:99–115
Campo RD (1981) Studies on extractable and resistant proteoglycans from metaphyseal and cortical bone and cartilage. Calcif Tissue Int 33:89–99
Campo RD, Phillips J (1973) Electron microscopic visualization of proteoglycans and collagen in bovine costal cartilage. Calcif Tissue Res 13:83–92
Campo RD, Romano JE (1986) Changes in cartilage proteoglycans associated with calcification. Calcif Tissue Int 39:175–184
Cao X, Genge BR, Wu LNY, Buzzi WR, Showman RM, Wuthier RE (1993) Characterization, cloning and expression of the 67-kDa annexin from chick growth plate cartilage matrix vesicles. Biochem Biophys Res Comm 197:556–561

Cecil RNA, Anderson HC (1978) Freeze-fracture studies of matrix vesicle calcification in epiphyseal growth plate. Metab Bone Dis Rel Res 1:89–95
Cerný J, Wolf J (1928) Contribution a l'étude de la calcification du cartilage. C R Assoc Anat (April 2–4, 1928)
Chen J, Zhang Q, McCulloch CAG, Sodek J (1991) Immunohistochemical localization of bone sialoprotein in foetal porcine bone tissues: comparisons with secreted phosphoprotein-1 (SPP-1, osteopontin) and SPARC (osteonectin). Histochem J 23:281–289
Chen Q, Fitch JM, Linsenmayer C, Linsenmayer TF (1992) Type X collagen: covalent crosslinking to hypertrophic cartilage-collagen fibrils. Bone Miner 17:223–227
Choi HU, Tang L-H, Johnson TL, Pal S, Rosenberg C, Reiner A, Poole AR (1983) Isolation and characterization of a 35,000 molecular weight subunit fetal cartilage matrix protein. J Biol Chem 258:655–661
Claassen H, Kirsch T (1994) Immunolocalization of type X collagen before and after mineralization of human thyroid cartilage. Histochemistry 101:27–32
Copray JCVM, Johnson PM, Decker JD, Hall SH (1989) Presence of osteonectin/SPARC in mandibular condylar cartilage of the rat. J Anat 162:43–51
Curran RC, Clark AE, Lovell D (1965) Acid mucopolysaccharides in electron microscopy. The use of the colloidal iron method. J Anat 99:427–434
D'Angelo M, Billings PC, Pacifici M, Leboy PS, Kirsch T (2001) Authentic matrix vesicles contain active metalloproteases (mmp). A role for matrix vesicle-associated mmp-13 in activation of transforming growth factor-beta. J Biol Chem 276:11347–11353
de Bernard B, Stagni N, Colautti I, Vittur F, Bonucci E (1977) Glycosaminoglycans and endochondral calcification. Clin Orthop Relat Res 126:285–291
de Bernard B, Bianco P, Bonucci E, Costantini M, Lunazzi GC, Martinuzzi P, Modricky C, Moro L, Panfili E, Pollesello P, Stagni N, Vittur F (1986) Biochemical and immunohistochemical evidence that in cartilage an alkaline phosphatase is a Ca^{2+}-binding glycoprotein. J Cell Biol 103:1615–1623
Dean DD, Schwartz Z, Muniz OE, Gomez R, Swain LD, Howell DS, Boyan BD (1992) Matrix vesicles are enriched in metalloproteinases that degrade proteoglycans. Calcif Tissue Int 50:342–349
Dean DD, Schwartz Z, Bonewald L, Muniz OE, Morales S, Gomez R, Brooks BP, Qiao M, Howell DS, Boyan BD (1994) Matrix vesicles produced by osteoblast-like cells in culture become significantly enriched in proteoglycan-degrading metalloproteinases after addition of β-glycerophosphate and ascorbic acid. Calcif Tissue Int 54:399–408
Dearden LC (1974) Enhanced mineralization of the tibial epiphyseal plate in the rat following propylthiouracil treatment: a histochemical, light, and electron microscopic study. Anat Rec 178:671–690
Dearden LC, Espinosa T (1974) Comparison of mineralization of the tibial epiphyseal plate in immature rats following treatment with cortisone, propylthiouracyl or after fasting. Calcif Tissue Res 15:93–110
Dearden LC, Mosier HD Jr, Jaffe NR (1981) The effect of glucocorticoids on calcification of the fetal rat skeleton. In: Ascenzi A, Bonucci E, de Bernard B (eds) Matrix vesicles. Wichtig Editore, Milan, pp 135–140
Descalzi Cancedda F, Gentili C, Manduca P, Cancedda R (1992) Hypertrophic chondrocytes undergo further differentiation in culture. J Cell Biol 117:427–435
Deutsch AJ, Midura RJ, Plaas AHK (1995) Structure of chondroitin sulfate on aggrecan isolated from bovine tibial and costochondral growth plates. J Orthop Res 13:230–239
Deutsch D, Bab I, Muhlrad A, Sela J (1981) Purification and further characterization of isolated matrix vesicles from rat alveolar bone. Metab Bone Dis Rel Res 3:209–214
Dharmavaram RM, Huynh AI, Jimenez SA (1998) Characterization of human chondrocyte and fibroblast type XII collagen cDNAs. Matrix Biol 16:343–348

Diamond AG, Triffitt JT, Herring GM (1982) The acid macromolecules in rabbit cortical bone tissue. Arch Oral Biol 27:337–345

Dickson GR (1981a) The origin of matrix vesicle membranes in cartilage. IRCS Med Sci 9:1104

Dickson GR (1981b) The role of matrix vesicles in the calcification of amphibian growth cartilage. In: Ascenzi A, Bonucci E, de Bernard B (eds) Matrix vesicles. Wichtig Editore, Milan, pp 191–196

Ecarot-Charrier B, Shepard N, Charette G, Grynpas M, Glorieux FH (1988) Mineralization in osteoblast cultures: a light and electron microscopic study. Bone 9:147–154

Edel G, Ueda Y, Nakanishi J, Brinker KH, Roessner A, Blasius S, Vestring T, Müller-Miny H, Erlemann R, Wuisman P (1992) Chondroblastoma of bone. A clinical, radiological, light and immunohistochemical study. Virchows Arch A Pathol Anat 421:355–366

Eggli PS, Herrmann W, Hunziker EB, Schenk RK (1985) Matrix compartments in the growth plate of the proximal tibia of rats. Anat Rec 211:246–257

Eisenberg E, Wuthier RE, Frank RB, Irving JT (1970) Time study of *in vivo* incorporationof ^{32}P orthophosphate into phospholipids of chicken epiphyseal tissues. Calcif Tissue Res 6:32–48

Eisenstein R, Larsson S-E, Sorgente N, Kuettner KE (1973) Collagen-proteoglycan relationships in epiphyseal cartilage. Am J Pathol 73:443–456

Ekanayake S, Hall BK (1994) Hypertrophy is not a prerequisite for type X collagen expression or mineralization of chondrocytes derived from cultured chick mandibular ectomesenchyme. Int J Dev Biol 38:683–694

Engfeldt B, Hjertquist S-O (1968) Studies on the epiphysial growth zone. I. The preservation of acid glycosaminoglycans in tissues in some histochemical procedures for electron microscopy. Virchows Arch B Cell Path 1:222–229

Engfeldt B, Reinholt FP (1992) Structure and calcification of epiphyseal growth cartilage. In: Bonucci E (ed) Calcification in biological systems. CRC Press, Boca Raton, pp 217–241

Engfeldt B, Reinholt FP, Hultenby K, Widholm SM, Müller M (1994) Ultrastructure of hypertrophic cartilage: histochemical procedures compared with high pressure freezing and freeze substitution. Calcif Tissue Int 55:274–280

Eyre D (2002) Collagen of articular cartilage. Arthritis Res 4:30–35

Eyre DR (2004) Collagens and cartilage matrix homeostasis. Clin Orthop Relat Res 427:S118–122

Eyre DR, Pietka T, Weis MA, Wu J-J (2004) Covalent cross-linking of the NC1 domain of collagen type IX to collagen type II in cartilage. J Biol Chem 279:2568–2574

Farquharson C, Whitehead CC, Loveridge N (1994) Alterations in glycosaminoglycan concentration and sulfation during chondrocyte maturation. Calcif Tissue Int 54:296–303

Felix R, Fleisch H (1976) Pyrophosphatase and ATPase of isolated cartilage matrix vesicles. Calcif Tissue Res 22:1–7

Fernandes RJ, Schmid TM, Eyre DR (2003) Assembly of collagen types II, IX and XI into nascent hetero-fibrils by a rat chondrocyte cell line. Eur J Biochem 270:3243–3250

Fortuna R, Anderson HC, Carty RP, Sajdera SW (1978) The purification and molecular characterization of alkaline phosphatases from chondrocytes and matrix vesicles of bovine fetal epiphyseal cartilage. Metab Bone Dis Rel Res 1:161–168

Fortuna R, Anderson HC, Carty RP, Sajdera SW (1980) Enzymatic characterization of the matrix vesicle alkaline phosphatase isolated from bovine fetal epiphyseal cartilage. Calcif Tissue Int 30:217–225

Franzén A, Heinegård D (1984) Extraction and purification of proteoglycans from mature bovine bone. Biochem J 224:47–58

Frost HM, Schönau E (2001) On longitudinal bone growth, short stature, and related matters: insights about cartilage physiology from the Utah paradigm. J Ped Endocrinol Metab 14:481–496

Genge BR, Sauer GR, Wu LNY, McLean FM, Wuthier RE (1988) Correlation betwen loss of alkaline phosphatase activity and accumulation of calcium during matrix vesicle-mediated mineralization. J Biol Chem 263:18513–18519

Genge BR, Wu LNY, Adkisson HDI, Wuthier RE (1991) Matrix vesicle annexins exhibit proteolipid-like properties. Selective partitioning into lipophilic solvents under acidic conditions. J Biol Chem 266:10678–10685

Genge B, Cao X, Wu LNY, Wuthier RE (1992a) Establishment of the primary structure of the two major matrix vesicle annexins by peptide and DNA sequencing. Bone Miner 17:202–208

Genge BR, Cao X, Wu LN, Buzzi WR, Showman RW, Arsenault AL, Ishikawa Y, Wuthier RE (1992b) Establishment of the primary structure of the major lipid-dependent Ca2+ binding proteins of chicken growth plate cartilage matrix vesicles: identity with anchorin CII (annexin V) and annexin II. J Bone Miner Res 7:807–819

Ghadially FN (2001) As you like it, part 3: a critique and historical review of calcification as seen with the electron microscope. Ultrastruct Pathol 25:243–267

Ghadially FN, Meachim G, Collins DH (1965) Extracellular lipid in the matrix of human articular cartilage. Ann Rheum Dis 24:136–145

Gibson GJ, Flint MH (1985) Type X collagen synthesis by chick sternal cartilage and its relationship to endochondral development. J Cell Biol 101:277–284

Gibson G, Lin D-L, Roque M (1997) Apoptosis of terminally differentiated chondrocytes in culture. Exp Cell Res 233:372–382

Göthlin G, Ericsson JLE (1973) Fine structural localization of alkaline phosphatase in the fracture callus of the rat. Histochemie 36:225–236

Gomes RR Jr, Farach-Carson MC, Carson DD (2004) Perlecan functions in chondrogenesis: insights from *in vitro* and *in vivo* models. Cells Tissues Organs 176:79–86

Gomez S, Lopez-Cepero JM, Silvestrini G, Mocetti P, Bonucci E (1996) Matrix vesicles and focal proteoglycan aggregates are the nucleation sites revealed by the lanthanum incubation method: a correlated study on the hypertrophic zone of the rat epiphyseal cartilage. Calcif Tissue Int 58:273–282

Gordon MK, Olsen BR (1990) The contribution of collagenous proteins to tissue-specific matrix assemblies. Curr Opin Cell Biol 2:833–838

Gordon MK, Gerecke DR, Dublet B, van der Rest M, Olsen BR (1989) Type XII collagen. A large multidomain molecule with partial homology to type IX collagen. J Biol Chem 264:19772–19778

Govindraj P, West L, Koob TJ, Neame P, Doege K, Hassell JR (2002) Isolation and identification of the major heparan sulfate proteoglycans in the developing bovine rib growth plate. J Biol Chem 277:19461–19469

Gregory KE, Keene DR, Tufa SF, Lunstrum GP, Morris NP (2001) Developmental distribution of collagen type XII in cartilage: association with articular cartilage and the growth plate. J Bone Miner Res 16:2005–2016

Hale JE, Wuthier RE (1987) The mechanism of matrix vesicle formation. Studies on the composition of chondrocyte microvilli and on the effects of microfilament-perturbing agents on cellular vesiculation. J Biol Chem 262:1916–1925

Hall TA, Höhling HJ, Bonucci E (1971) Electron probe X-ray analysis of osmiophilic globules as possible sites of early mineralization in cartilage. Nature 231:535–536

Hardingham TE, Fosang AJ (1992) Proteoglycans: many forms and many functions. FASEB J 6:861–870

Hardingham TE, Fosang AJ (1995) The structure of aggrecan and its turnover in cartilage. J Rheumatol Suppl 43:86–90
Hardingham TE, Fosang AJ, Dudhia J (1994) The structure, function and turnover of aggrecan, the large aggregating proteoglycan from cartilage. Eur J Clin Chem Clin Biochem 32:249–257
Hascall GK (1980) Cartilage proteoglycans: comparison of sectioned and spread whole molecoles. J Ultrastruct Res 70:369–375
Hashimoto S, Ochs RL, Rosen F, Quach J, McCabe G, Solan J, Seegmiller JE, Terkeltaub R, Lotz M (1998) Condrocyte-derived apoptotic bodies and calcification of articular cartilage. Proc Natl Acad Sci USA 95:3094–3099
Hassell J, Yamada Y, Arikawa-Hirasawa E (2003) Role of perlecan in skeletal development and diseases. Glycoconj J 19:263–267
Hayashi Y, Imai M, Goto Y, Murakami N (1993) Pathological mineralization in a serially passaged cell line from rat pulp. J Oral Pathol Med 22:175–179
Heinegård D, Antonsson P, Hedbom E, Larsson T, Oldberg Å, Sommarin Y, Wndel M (1988) Noncollagenous matrix constituents of cartilage. Path Immunopathol Res 7:27–31
Henson FMD, Davies ME, Skepper JN, Jeffcott LB (1995) Localisation of alkaline phosphatase in equine growth cartilage. J Anat 187:151–159
Herbert B, Lecouturier A, Masquelier D, Hauser N, Remacle C (1997) Ultrastructure and cytochemical detection of alkaline phosphatase in long-term cultures of osteoblast-like cells from rat calvaria. Calcif Tissue Int 60:216–223
Hirschman A, Dziewiatkowski DD (1966) Protein-polysaccharide loss during endochondral ossification: immunochemical evidence. Science 154:393–395
Horton WA, Machado MM (1988) Extracellular matrix alterations during endochondral ossification in humans. J Orthop Res 6:793–803
Hoshi K, Ejiri S, Ozawa H (2000) Ultrastructural, cytochemical, and biophysical aspects of mechanisms of bone matrix calcification. Acta Anat Nippon 75:457–465
Hoshi K, Ejiri S, Ozawa H (2001) Localizational alterations of calcium, phosphorus, and calcification-related organics such as proteoglycans and alkaline phosphatase during bone calcification. J Bone Miner Res 16:289–298
Hosokawa R, Uchida Y, Fujiwara S, Noguchi T (1988) Lactate dehydrogenase isoenzymes are present in matrix vesicles. J Biol Chem 263:10045–10047
Hsu HHT (1983) Purification and partial characterization of ATP pyrophosphohydrolase from fetal bovine epiphyseal cartilage. J Biol Chem 258:3463–3468
Hsu HHT, Anderson HC (1984) The deposition of calcium pyrophosphate and phosphate by matrix vesicles isolated from fetal bovine epiphyseal cartilage. Calcif Tissue Int 36:615–621
Hsu HHT, Anderson HC (1996) Evidence of the presence of a specific ATPase responsible for ATP-initiated calcification by matrix vesicles isolated from cartilage and bone. J Biol Chem 271:26383–26388
Hunziker EB (1988) Growth plate structure and function. Pathol Immunopathol Res 7:9–13
Hunziker EB, Herrmann W (1990) Ultrastructure of cartilage. In: Bonucci E, Motta PM (eds) Ultrastructure of skeletal tissues. Kluwer Academic Publishers, Boston, pp. 79–109
Hunziker EB, Schenk RK (1984) Cartilage ultrastructure after high pressure freezing, freeze substitution, and low temperature embedding. II. Intercellular matrix ultrastructure – preservation of proteoglycans in their native state. J Cell Biol 98:277–282
Hunziker EB, Schenk RK (1987) Structural organization of proteoglycans in cartilage. In: Wight TN, Mecham P (eds) Biology of proteoglycans. Academic Press, Orlando (FL), pp 155–185
Hunziker EB, Herrmann W, Schenk RK, Marti T, Müller M, Moor H (1981) Structural integration of matrix vesicles in calcifying cartilage after cryofixation and freeze sub-

stitution. In: Ascenzi A, Bonucci E, de Bernard B (eds) Matrix vesicles. Wichtig Editore, Milan, pp 25–31

Hunziker EB, Herrmann W, Schenk RK (1983) Ruthenium hexammine trichloride (RHT)-mediated interaction between plasmalemmal components and pericellular matrix proteoglycans is responsible for the preservation of chondrocytic plasma membranes in situ during cartilage fixation. J Histochem Cytochem 31:717–727

Iannotti JP, Naidu S, Noguchi Y, Hunt RM, Brighton CT (1994) Growth plate matrix vesicle biogenesis. The role of intracellular calcium. Clin Orthop Relat Res 306:222–229

Iozzo RV (1998) Matrix proteoglycans: from molecular design to cellular function. Ann Rev Biochem 67:609–652

Iyama K, Ninomiya Y, Olsen BR, Linsenmayer TF, Trelstad RL, Hayashi M (1991) Spatiotemporal pattern of type X collagen gene expression and collagen deposition in embryonic chick vertebrae undergoing endochondral ossification. Anat Rec 229:462–472

Joel W, Masters YF, Shetlar MR (1956) Comparison of histochemical and biochemical methods for the polysaccharides of cartilage. J Histochem Cytochem 4:476–478

Kagami A, Takagi M, Hirama M, Sagami Y, Shimada T (1990) Enhanced ultrastructural preservation of cartilage proteoglycans in the extended state. J Histochem Cytochem 38:901–906

Kahn SE, Jafri AM, Lewis NJ, Arsenis C (1978) Purification of alkaline phosphatase from extracellular vesicles of fracture callus cartilage. Calcif Tissue Res 25:85–92

Kakuta S, Malamud D, Golub EE, Shapiro IM (1985) Isolation of matrix vesicles by isoelectric focusing in Pevikon-Sephadex. Bone 6:187–191

Kardos TB, Hubbard MJ (1982) Are matrix vesicles apoptotic bodies? Prog Clin Biol Res 101:45–60

Katchburian E (1973a) Membrane-bound bodies as initiators of mineralization of dentine. J Anat 116:285–302

Katchburian E (1973b) Role of extracellular bodies in calcification of dentine. J Anat 115:139–158

Katchburian E, Severs NJ (1982) Membranes of matrix-vesicles in early developing dentine. A freeze fracture study. Cell Biol Int Rep 6:941–950

Katsura N, Yamada K (1986) Isolation and characterization of a metalloprotease associated with chicken epiphyseal cartilage matrix vesicles. Bone 7:137–143

Kergosien N, Sautier J-M, Forest N (1998) Gene and protein expression during differentiation and matrix mineralization in a chondrocyte cell culture system. Calcif Tissue Int 62:114–121

Kiani C, Chen L, Wu YJ, Yee AJ, Yang BB (2002) Structure and function of aggrecan. Cell Res 12:19–32

Kim KM (1976) Calcification of matrix vesicles in human aortic valve and aortic media. Fed Proc 35:156–162

Kim KM (1995) Apoptosis and calcification. Scanning Microsc 9:1137–1178

Kim KM, Huang S (1971) Ultrastructural study of calcification of human aortic valve. Lab Invest 25:357–366

Kirsch T, Claassen H (2000) Matrix vesicles mediate mineralization of human thyroid cartilage. Calcif Tissue Int 66:292–297

Kirsch T, Von der Mark H (1991) Ca^{2+} binding properties of type X collagen. Feder Eur Bioch Soc 294:149–152

Kirsch T, Von der Mark H (1992) Remodelling of collagen types I, II and X and calcification of human fetal cartilage. Bone Miner 18:107–117

Kirsch T, Nah H-D, Shapiro IM, Pacifici M (1997a) Regulated production of mineralization-competent matrix vesicles in hypertrophic chondrocytes. J Cell Biol 137:1149–1160

Kirsch T, Nah H-D, Demuth DR, Harrison G, Golub EE, Adams SL, Pacifici M (1997b) Annexin V-mediated calcium flux across membranes is dependent on the lipid composition: implications for cartilage mineralization. Biochemistry 36:3359–3367

Kirsch T, Wang W, Pfander D (2003) Functional differences between growth plate apoptotic bodies and matrix vesicles. J Bone Miner Res 18:1872–1881

Kitamura C, Terashita M, Noguchi T (1991) Presence of vesicles containing lactate dehydrogenase in the dentin of bovine tooth germs. J Dent Res 70:1444–1446

Knudson CB, Knudson W (2001) Cartilage proteoglycans. Semin Cell Dev Biol 12:69–78

Kobayashi S (1971) Acid mucopolysaccharides in calcified tissues. Int Rev Cytol 30:257–371

Koch M, Foley JE, Hahn R, Zhou P, Burgeson RE, Gerecke DR, Gordon MK (2001) $\alpha 1$(XX) collagen, a new member of the collagen subfamily, fibril-associated collagens with interrupted triple helices. J Biol Chem 276:23120–23126

Kogaya Y, Furuhashi K (1985) Ultrastructural distribution of acidic glycosaminoglycans associated with matrix vesicle-mediated calcification in mouse progenitor predentine. Calcif Tissue Int 37:36–41

Landis WJ, Jacquet R, Hillyer J, Zhang J (2003) Analysis of osteopontin in mouse growth plate cartilage by application of laser capture microdissection and RT-PCR. Connect Tissue Res 44:28–32

Larsson Å (1973) Studies on dentinogenesis in the rat. Ultrastructural observations on early dentin formation with special reference to "dentinal globules" and alkaline phosphatase activity. Z Anat Entwickl -Gesch 142:103–115

Larsson Å (1974) Studies on dentinogenesis in the rat. The interaction between lead-pyrophosphate solutions and dentinal globules. Calcif Tiss Res 16:93–107

Larsson Å, Bloom GD (1973) Studies on dentinogenesis in the rat. Fine structure of developing odontoblasts and predentin in relation to the mineralization process. Z Anat Entwickl -Gesch 139:227–246

Lash JW, Vasan NS (1983) Glycosaminoglycans of cartilage. In: Hall BK (ed) Cartilage. Structure, function and biochemistry. Academic Press, New York, pp 215–251

Lebaron RG (1996) Versican. Perspect Dev Neurobiol 3:261–271

Lee ER, Smith CE, Poole AR (1996) Ultrastructural localization of the C-propeptide released from type II procollagen in fetal bovine growth plate cartilage. J Histochem Cytochem 44:433–443

Lewinson D, Silbermann M (1981) Second thoughts on the correlative relationship between: chondrocyte metabolic state, formation of matrix vesicles and cartilage mineralization, using calcium complexes as ultrastructural markers. In: Ascenzi A, Bonucci E, de Bernard B (eds) Matrix vesicles. Wichtig Editore, Milan, pp 85–90

Lewinson D, Toister Z, Silbermann M (1982) Quantitative and distributional changes of alkaline phosphatase during the maturation of cartilage. J Histochem Cytochem 30:261–269

Linsenmayer TF, Eavey RD, Schmid TM (1988) Type X collagen: a hypertrophic cartilage-specific molecule. Pathol Immunopathol Res 7:14–19

Lohmander S, Hjerpe A (1975) Proteoglycans of mineralizing rib and epiphyseal cartilage. Biochim Biophys Acta 404:93–109

Lotz M, Hashimoto S, Kuhn K (1999) Mechanisms of chondrocyte apoptosis. Osteoarthritis Cartilage 7:389–391

Lu Valle P, Daniels K, Hay ED, Olsen BR (1992) Type X collagen is transcriptionally activated and specifically localized during sternal cartilage maturation. Matrix 12:404–413

Luo W, Guo C, Zheng J, Chen T-L, Wang PY, Vertel BM, Tanzer ML (2000) Aggrecan from start to finish. J Bone Miner Metab 18:51–56

Magne D, Bluteau G, Faucheux C, Palmer G, Vignes-Colombeix C, Pilet P, Rouillon T, Caverzasio J, Weiss P, Daculsi G, Guicheux J (2003) Phosphate is a specific signal for

ATDC5 chondrocyte maturation and apoptosis-associated mineralization: possible implication of apoptosis in the regulation of endochondral ossification. J Bone Miner Res 18:1430–1442

Maitland ME, Arsenault AL (1989) Freeze-substitution staining of rat growth plate cartilage with Alcian blue for electron microscopic study of proteoglycans. J Histochem Cytochem 37:383–387

Majeska RJ, Wuthier RE (1975) Studies on matrix vesicles isolated from chick epiphyseal cartilage. Association of pyrophosphatase and ATPase activities with alkaline phosphatase. Biochim Biophys Acta 391:51–60

Majeska RJ, Holwerda DL, Wuthier RE (1979) Localization of phosphatidylserine in isolated chick epiphyseal cartilage matrix vesicles with trinitrobenzenesulfonate. Calcif Tissue Int 27:41–46

Maki K, Hayashi S, Nishioka T, Kimura M, Noguch T (2000) A new type of matrix vesicles is found in fetal bovine tracheal cartilage. Connect Tissue Res 41:109–115

Masuhara K, Suzuki S, Yoshikawa H, Tsuda T, Takaoka K, Ono K, Morris DC, Hsu HH, Anderson HC (1992) Development of a monoclonal antibody specific for human bone alkaline phosphatase. Bone Miner 17:182–186

Matsui Y, Alini M, Webber C, Poole AR (1991) Characterization of aggregating proteoglycans from the proliferative, maturing, hypertrophic, and calcifying zones of the cartilaginous physis. J Bone Joint Surg 73-A:1064–1074

Matsuzawa T, Anderson HC (1971) Phosphatases of epiphyseal cartilage studied by electron microscopic cytochemical methods. J Histochem Cytochem 19:801–809

Matukas VJ, Krikos GA (1968) Evidence for changes in protein polysaccharide associated with the onset of calcification in cartilage. J Cell Biol 39:43–48

Matukas VJ, Panner BJ, Orbison JL (1967) Studies on ultrastructural identification and distribution of protein-polysaccharide in cartilage matrix. J Cell Biol 32:365–377

Mayne R, Von der Mark K (1983) Collagens of cartilage. In: Hall BK (ed) Cartilage. Structure, function and biochemistry. Academic Press, New York, pp 181–214

McKee MD, Glimcher MJ, Nanci A (1992) High-resolution immunolocalization of osteopontin and osteocalcin in bone and cartilage during endochondral ossification in the chicken tibia. Anat Rec 234:479–492

Meikle MC (1976) The mineralization of condylar cartilage in the rat mandible: an electron microscopic enzyme histochemical study. Arch Oral Biol 21:33–43

Melrose J, Smith S, Knox S, Whitelock J (2002) Perlecan, the multidomain HS-proteoglycan of basement membranes, is a prominent pericellular component of ovine hypertrophic vertebral growth plate and cartilaginous endplate chondrocytes. Histochem Cell Biol 118:269–280

Miller EJ (1973) A review of biochemical studies on the genetically distinct collagens of the skeletal system. Clin Orthop Relat Res 92:260–280

Miller GJ, DeMarzo AM (1988) Ultrastructural localization of matrix vesicles and alkaline phosphatase in the Swarm rat chondrosarcoma: their role in cartilage calcification. Bone 9:235–241

Morgelin M, Heinegård D, Engel J, Paulsson M (1994) The cartilage proteoglycan aggregate: assembly through combined protein-carbohydrate and protein-protein interactions. Biophys Chem 50:113–128

Morris DC, Appleton J (1980) Ultrastructural localization of calcium in the mandibular condylar growth cartilage of the rat. Calcif Tissue Int 30:27–34

Morris DC, Appleton J (1984) The effects of lanthanum on the ultrastructure of hypertrophic chondrocytes and the localization of lanthanum precipitates in condylar cartilages of rats fed on normal and rachitogenic diets. J Histochem Cytochem 32:239–247

Morris DC, Väänänen HK, Munoz P, Anderson HC (1986) Light and electron microscopic immunolocalization of alkaline phosphatase in bovine growth plate cartilage. In: Ali SY (ed) Cell mediated calcification and matrix vesicles. Excerpta Medica, Amsterdam, pp 21–26

Morris DC, Randall JC, Stechschulte DJ Jr, Zeiger S, Mansur DB, Anderson HC (1990) Enzyme cytochemical localization of alkaline phosphatase in cultures of chondrocytes derived from normal and rachitic rats. Bone 11:345–352

Morris DC, Masuhara K, Takaoka K, Ono K, Anderson HC (1992a) Immunolocalization of alkaline phosphatase in osteoblasts and matrix vesicles of human fetal bone. Bone Miner 19:287–298

Morris DC, Moylan PE, Anderson HC (1992b) Immunochemical and immunocytochemical identification of matrix vesicle proteins. Bone Miner 17:209–213

Müller-Glauser B, Humbel B, Glatt M, Strauli P, Winterhalter KH, Bruckner P (1986) On the role of type IX collagen in the extracellular matrix of cartilage: type IX collagen is localized to intersections of collagen fibrils. J Cell Biol 102:1931–1939

Muhlrad A, Bab IA, Deutsch D, Sela J (1982) Occurrence of actin-like protein in extracellular matrix vesicles. Calcif Tissue Int 34:376–381

Mwale F, Billinghurst C, Wu W, Alini M, Webber C, Reiner A, Ionescu M, Poole J, Poole AR (2000) Selective assembly and remodelling of collagens II and IX associated with expression of the chondrocyte hypertrophic phenotype. Dev Dyn 218:648–662

Mwale F, Tchetina E, Wu CW, Poole AR (2002) The assembly and remodeling of the extracellular matrix in the growth plate in relationship to mineral deposition and cellular hypertrophy: an in situ study of collagens II and IX and proteoglycan. J Bone Miner Res 17:275–283

Nakamura H, Hirata A, Tsuji T, Yamamoto T (2001) Immunolocalization of keratan sulfate proteoglycan in rat calvaria. Arch Histol Cytol 64:109–118

Neame PJ, Tapp H, Azizan A (1999) Noncollagenous, nonproteoglycan macromolecules of cartilage. Cell Mol Life Sci 55:1327–1340

Newbrey JW, Banks WJ (1975) Characterization of developing antler cartilage matrix II. An ultrastructural study. Calcif Tissue Res 17:289–302

Ng L, Grodzinsky AJ, Patwari P, Sandy J, Plaas A, Ortiz C (2003) Individual cartilage aggrecan macromolecules and their constituent glycosaminoglycans visualized via atomic force microscopy. J Struct Biol 143:242–257

Nielsen EH (1978) Ultrahistochemistry of matrix vesicles in elastic cartilage. Acta Anat 100:268–272

Ohashi N, Ejiri S, Hanada K, Ozawa H (1997) Changes in type I, II, and X collagen immunoreactivity of the mandibular condylar cartilage in a naturally aging rat model. J Bone Miner Metab 15:77–83

Ohashi-Takeuchi H, Yamada N, Hosokawa R, Noguchi T (1990) Vesicles with lactate dehydrogenase and without alkaline phosphatase present in the resting zone of epiphyseal cartilage. Biochem J 266:309–312

Ohyama K, Farquharson C, Whitehead CC, Shapiro IM (1997) Further observations on programmed cell death in the epiphyseal growth plate: comparison of normal and dyschondroplastic epiphyses. J Bone Miner Res 12:1647–1656

Olsen BR (1989) The next frontier: molecular biology of extracellular matrix. Connect Tissue Res 23:115–121

Ornoy A, Langer Y (1978) Scanning electron microscopy studies on the origin and structure of matrix vesicles in epiphyseal cartilage from young rats. Israel J Med Sci 14:745–752

Ornoy A, Levy J, Atkin I, Salamon J (1979) Scanning and trasmission electron microscopic observations on the origin and structure of matrix vesicles in normal and papain-digested epiphyseal cartilage of young rats. Israel J Med Sci 15:928–936

Ornoy A, Atkin I, Levy J (1980) Ultrastructural studies on the origin and structure of matrix vesicles in bone of young rats. Acta Anat 106:450–461

Ortolani F, Marchini M (1995) Cartilage type II collagen fibrils show distinctive negative-staining band patterns differences between type II and type I unfixed or glutaraldehyde-fixed collagen fibrils. J Electron Microsc 44:365–375

Ortolani F, Giordano M, Marchini M (2000) A model for type II collagen fibrils: distinctive D-band patterns in native and reconstituted fibrils compared with sequence data for helix and telopeptide domains. Biopolymers 54:448–463

Pacifici M, Golden EB, Oshima O, Shapiro IM, Leboy PS, Adams SL (1990a) Hypertrophic chondrocytes. The terminal stage of differentiation in the chondrogenic cell lineage? Ann N Y Acad Sci 599:45–57

Pacifici M, Oshima O, Fisher LW, Young MF, Shapiro IM, Leboy PS (1990b) Changes in osteonectin distribution and levels are associated with mineralization of the chicken tibial growth cartilage. Calcif Tissue Int 47:51–61

Palumbo C, Ferretti M, De Pol A (2003) Apoptosis during intramembranous ossification. J Anat 203:589–598

Peress NS, Anderson HC, Sajdera SW (1974) The lipids of matrix vesicles from bovine fetal epiphyseal cartilage. Calcif Tissue Res 14:275–281

Pollesello P, D'Andrea P, Martina M, de Bernard B, Vittur F (1990) Modification of plasma membrane of differentiating preosseous chondrocytes: evidence for a degradative process in the mechanism of matrix vesicle formation. Exp Cell Res 188:214–218

Poole AR (1991) The growth plate: cellular physiology, cartilage assembly and mineralization. In: Hall B, Newman S (eds) Cartilage: molecular aspects. CRC Press, Boca Raton, pp 179–211

Poole AR, Pidoux I (1989) Immunoelectron microscopic studies of type X collagen in endochondral ossification. J Cell Biol 109:2547–2554

Poole AR, Pidoux I, Rosenberg L (1982) Role of proteoglycans in endochondral ossification: immunofluorescent localization of link protein and proteoglycan monomer in bovine fetal epiphyseal growth plate. J Cell Biol 92:249–260

Poole AR, Pidoux I, Reiner A, Choi H, Rosenberg LC (1984) Association of an extracellular protein (chondrocalcin) with the calcification of cartilage in endochondral bone formation. J Cell Biol 98:54–65

Poole AR, Webber C, Pidoux I, Choi H, Rosenberg LC (1986) Localization of a dermatan sulfate proteoglycan (DS-PGII) in cartilage and the presence of an immunologically related species in other tissues. J Histochem Cytochem 34:619–625

Poole AR, Matsui Y, Hinek A, Lee ER (1989) Cartilage macromolecules and the calcification of cartilage matrix. Anat Rec 224:167–179

Rabinovitch AL, Anderson HC (1976) Biogenesis of matrix vesicles in cartilage growth plates. Fed Proc 35:112–116

Rauterberg K, Becker W (1970) Das Problem der Knorpelmineralisation am Beispiel des verkalkenden freien Gelenkkörpers. Arch Orthop Unfall-Chir 69:12–34

Reinholt FP, Wernerson A (1988) Septal distribution and the relationship of matrix vesicle size to cartilage mineralization. Bone Miner 4:63–71

Reinholt FP, Engfeldt B, Hjerpe A, Jansson K (1982) Stereological studies on the epiphyseal growth plate with special reference to the distribution of matrix vesicles. J Ultrastruct Res 80:270–279

Reinholt FP, Hjerpe A, Jansson K, Engfeldt B (1983) Stereological studies on matrix vesicle distribution in the epiphyseal growth plate during healing of low phosphate, vitamin D deficiency rickets. Virchows Arch Cell Pathol 44:257–266

Reinholt FP, Hjerpe A, Jansson K, Engfeldt B (1984) Stereological studies on the epiphyseal growth plate in low phosphate, vitamin D-deficiency rickets with special reference to the distribution of matrix vesicles. Calcif Tissue Int 36:95–101

Roach HI (1997) New aspects of endochondral ossification in the chick: chondrocyte apoptosis, bone formation by former chondrocytes, and acid phosphatase activity in the endochondral bone matrix. J Bone Miner Res 12:795–805

Rojas E, Arispe N, Haigler HT, Burns AL, Pollard HB (1992) Identification of annexins as calcium channels in biological membranes. Bone Miner 17:214–218

Roughley PJ, Lee ER (1994) Cartilage proteoglycans: structure and potential functions. Microsc Res Tech 28:385–397

Salomon CD (1974) A fine structural study on the extracellular activity of alkaline phosphatase and its role in calcification. Calcif Tissue Res 15:201–212

Schaffer J (1930) Die Stützgewebe. In: von Möllendorff W (ed) Handbuch der mikroskopischen Anatomie des Manschen. Springer, Berlin,

Scherft JP, Groot CG (1981) The development of matrix vesicles into bone nodules, studied with colloidal thorium dioxide. In: Ascenzi A, Bonucci E, de Bernard B (eds) Matrix vesicles. Wichtig Editore, Milan, pp 173–177

Schmid TM, Linsenmayer TF (1985) Immunohistochemical localization of short chain cartilage collagen (type X) in avian tissue. J Cell Biol 100:598–605

Schmid TM, Linsenmayer TF (1990) Immunoelectron microscopy of type X collagen: supramolecular forms within embryonic chick cartilage. Dev Biol 138:53–62

Schmid TM, Bonen DK, Luchene L, Linsenmayer TF (1991) Late events in chondroyte differentiation: hypertrophy, type X collagen synthesis and matrix calcification. In Vivo 5:533–540

Schmitz JP, Dean DD, Schwartz Z, Cochran DL, Grant GM, Klebe RJ, Nakaya H, Boyan BD (1996) Chondrocyte cultures express matrix metalloproteinase mRNA and immunoreactive protein; stromelysin-1 and 72 kDa gelatinase are localized in extracellular matrix vesicles. J Cell Biochem 61:375–391

Schwartz NB, Pirok EWI, Mensch JR Jr, Domowicz MS (1999) Domain organization, genomic structure, evolution, and regulation of expression of the aggrecan gene family. Progr Nucleic Acid Res Mol Biol 62:177–225

Schwartz Z, Amir D, Weinberg H, Sela J (1987) Extracellular matrix vesicle distribution in primary mineralization two weeks after injury to rat tibial bone (ultrastructural tissue morphometry). Eur J Cell Biol 45:97–101

Schwartz Z, Sela J, Ramirez V, Amir D, Boyan BD (1989) Changes in extracellular matrix vesicles during healing of rat tibial bone: a morphometric and biochemical study. Bone 10:53–60

Sela J, Amir D, Schwartz Z, Weinberg H (1987a) Ultrastructural tissue morphometry of the distribution of extracellular matrix vesicles in remodeling rat tibial bone six days after injury. Acta Anat 128:295–300

Sela J, Amir D, Schwartz Z, Weinberg H (1987b) Changes in the distribution of extracellular matrix vesicles during healing of rat tibial bone (computerized morphometry and electron microscopy). Bone 8:245–250

Sela J, Schwartz Z, Swain LD, Boyan BD (1992) The role of matrix vesicles in calcification. In: Bonucci E (ed) Calcification in biological systems. CRC Press, Boca Raton, pp 73–105

Serafini-Fracassini A, Smith JW (1966) Observations on the morphology of the protein-polysaccharide complex of bovine nasal cartilage and its relationship to collagen. Proc R Soc London B 165:440–449

Serafini-Fracassini A, Smith JW (1974) The structure and biochemistry of cartilage. Churchill Livingstone, Edinburgh

Shapses SA, Sandell LJ, Ratcliffe A (1994) Differential rates of aggrecan synthesis and breackdown in different zones of the bovine growth plate. Matrix Biol 14:77–86
Shaw LM, Olsen BR (1991) FACIT collagens: diverse molecular bridges in extracellular matrices. Trends Biochem Sci 16:191–194
Shepard N (1992) Role of proteoglycans in calcification. In: Bonucci E (ed) Calcification in biological systems. CRC Press, Boca Raton, pp 41–58
Shepard N, Mitchell N (1981) Acridine orange stabilization of glycosaminoglycans in beginning endochondral ossification. A comparative light and electron microscopic study. Histochemistry 70:107–114
Shibata S, Fukada K, Suzuki S, Ogawa T, Yamashita Y (2001) Histochemical localisation of versican, aggrecan and hyaluronan in the developing condylar cartilage of the fetal rat mandible. J Anat 198:129–135
Shibata S, Fukada K, Suzuki S, Ogawa T, Yamashita Y (2002) In situ hybridization and immunohistochemistry of bone sialoprotein and secreted phosphoprotein 1 (osteopontin) in the developing mouse mandibular condylar cartilage compared with limb bud cartilage. J Anat 200:309–320
Shibata S, Fukada K, Imai H, Abe T, Yamashita Y (2003) In situ hybridization and immunohistochemistry of versican, aggrecan and link protein, and histochemistry of hyaluronan in the developing mouse limb bud cartilage. J Anat 203:425–432
Shimazu A, Nah H-D, Kirsch T, Koyama E, Leatherman JL, Golden EB, Kosher RA, Pacifici M (1996) Syndecan-3 and the control of chondrocyte proliferation during endochondral ossification. Exp Cell Res 229:126–136
Silbermann M, Frommer J (1974) Initial locus of calcification in chondrocytes. Clin Orthop Relat Res 98:288–293
Silbermann M, Von der Mark H (1990) An immunohistochemical study of the distribution of matrical proteins in the mandibular condyle of neonatal mice. I. Collagens. J Anat 170:11–22
Silbermann M, Lewinson D, Toister Z (1980) Early cartilage response to systemic glucocorticoid administration: an ultrastructural study. Metab Bone Dis Rel Res 2:267–279
Silvestrini G, Zini N, Sabatelli P, Mocetti P, Maraldi NM, Bonucci E (1996) Combined use of malachite green fixation and PLA_2-gold complex technique to localize phospholipids in areas of early calcification of rat epiphyseal cartilage and bone. Bone 18:559–565
Sisca RF, Provenza DV (1972) Initial dentin formation in human deciduous teeth. An electron microscope study. Calcif Tissue Res 9:1–16
Slavkin HC, Croissant R, Bringas P Jr (1972) Epithelial-mesenchymal interactions during odontogenesis. III. A simple method for the isolation of matrix vesicles. J Cell Biol 53:841–849
Smith JW (1970) The disposition of protein-polysaccharide in the epiphyseal plate cartilage of the young rabbit. J Cell Sci 6:843–864
Spicer SS, Schulte BA (1982) Ultrastructural methods for localizing complex carbohydrates. Human Pathol 13:343–354
Stark M, Miller EJ, Kühn K (1972) Comparative electron-microscope studies on the collagens extracted from cartilage, bone, and skin. Eur J Biochem 27:192–196
Stechschulte DJ Jr, Morris DC, Silverton SF, Anderson HC, Väänänen HK (1992) Presence and specific concentration of carbonic anhydrase II in matrix vesicles. Bone Miner 17:187–191
Sugrue SP, Gordon MK, Seyer J, Dublet B, van der Rest M, Olsen BR (1989) Immunoidentification of type XII collagen in embryonic tissues. J Cell Biol 109:309–345
Szirmai JA (1963) Quantitative approaches in the histochemistry of mucopolysaccharides. J Histochem Cytochem 11:24–34

Szuwart T, Kierdorf H, Kierdorf U, Clemen G (1998) Ultrastructural aspects of cartilage formation, mineralization, and degeneration during primary antler growth in fallow deer (Dama dama). Anat Anz 180:501–510

Takagi M (1990) Ultrastructural cytochemistry of cartilage proteoglycans and their relation to the calcification process. In: Bonucci E, Motta PM (eds) Ultrastructure of skeletal tissues. Kluwer Academic Publishers, Boston, pp 111–127

Takagi M, Toda Y (1979) Electron microscopic study of the intercellular activity of alkaline phosphatase in rat epiphyseal cartilage. J Electron Microsc 28:117–127

Takagi M, Kasahara Y, Takagi H, Toda Y (1979) Freeze-fracture images of matrix vesicles of epiphyseal cartilage and non-calcified tracheal cartilage. J Electron Microsc 28:165–175

Takagi M, Parmley RT, Toda Y, Austin RL (1982) Ultrastructural cytochemistry and immunocytochemistry of sulfated glycosamiglycans in epiphyseal cartilage. J Histochem Cytochem 30:1179–1185

Takagi M, Parmley RT, Denys FR (1983) Ultrastructural cytochemistry and immunocytochemistry of proteoglycans associated with epiphyseal cartilage calcification. J Histochem Cytochem 31:1089–1100

Takagi M, Sasaki T, Kagami A, Komiyama K (1989) Ultrastructural demonstration of increased sulfated proteoglycans and calcium associated with chondrocyte cytoplasmic processes and matrix vesicles in rat growth plate cartilage. J Histochem Cytochem 37:1025–1033

Takagi M, Kamiya N, Urushizaki T, Tada Y, Tanaka H (2000) Gene expression and immunohistochemical localization of biglycan in association with mineralization in the matrix of epiphyseal cartilage. Histochemistry J 32:175–186

Takano Y, Ozawa H, Crenshaw MA (1986) Ca-ATPase and ALPase activities at the initial calcification sites of dentin and enamel in the rat incisor. Cell Tissue Res 243:91–99

Takuma S (1960) Electron microscopy of the developing cartilaginous epiphysis. Arch Oral Biol 2:111–119

Thyberg J (1972) Ultrastructural localization of aryl sulfatase activity in the epiphyseal plate. J Ultrastruct Res 38:332–342

Thyberg J (1977) Electron microscopy of cartilage proteoglycans. Histochem J 9:259–266

Thyberg J, Lohmander S, Friberg U (1973a) Electron microscopic demonstration of proteoglycans in guinea pig epiphyseal cartilage. J Ultrastruct Res 45:407–427

Thyberg J, Nilsson S, Friberg U (1973b) Electron microscopic studies on guinea pig rib cartilage. Structural heterogeneity and effects of extraction with guanidine-HCl. Z Zellforsch 146:83–102

Thyberg J, Nilsson S, Friberg U (1975) Electron microscopic and enzyme cytochemical studies on the guinea pig metaphysis with special reference to the lysosomal system of different cell types. Cell Tissue Res 156:273–299

Tsuji T, Mark MP, Ruch J-V (1994) Immunocytochemical localization of choline-phospholipids in postnatal mouse molars. Archs Oral Biol 39:81–86

Ueda Y, Oda Y, Tsuchiya H, Tomita K, Nakanishi I (1990) Immunohistological study on collagenous proteins of benign and malignant human cartilaginous tumours of bone. Virchows Arch A Path Anat Histopathol 417:291–297

van der Rest M (1991) The collagens of bone. In: Hall BK (ed) Bone. Vol. 3: Bone matrix and bone specific products. CRC Press, Boca Raton, pp 187–237

van der Rest M, Mayne R, Ninomiya Y, Seidah NG, Chretien M, Olsen BR (1985) The structure of type IX collagen. J Biol Chem 260:220–225

van der Rest M, Rosenberg LC, Olsen BR, Poole AR (1986) Chondrocalcin is identical with the C-propeptide of type II procollagen. Biochem J 237:923–925

Vaughan L, Mendler M, Huter S, Bruckner P, Winterhalter K, Irwin M, Mayne R (1988) D-periodic distribution of collagen type IX along cartilage fibrils. J Cell Biol 106:991–997

Väänänen HK, Korhonen LK (1979) Matrix vesicles in chicken epiphyseal cartilage. Separation from lysosomes and the distribution of inorganic pyrophosphatase activity. Calcif Tissue Int 28:65–72

Väänänen HK, Korhonen LK (1980) Purification of matrix-vesicle alkaline phosphatase from chicken epiphyseal cartilage and experiments on its ATP-hydrolyzing properties. Clin Orthop Relat Res 148:291–296

Vittur F, Zanetti M, Stagni N, de Bernard B (1979) Further evidence for the participation of glycoproteins to the process of calcification. Perspect Inherit Metab Dis 2:13–30

Von der Mark K, Von der Mark H (1977) The role of three genetically distinct collagen types in endochondral ossification and calcification of cartilage. J Bone Joint Surg 59-B:458–464

Warner GP, Hubbard HL, Lloyd GC, Wuthier RE (1983) ^{32}Pi- and ^{45}Ca-metabolism by matrix vesicle-enriched microsomes prepared from chicken epiphyseal cartilage by isosmotic percoll density-gradient fractionation. Calcif Tissue Int 35:327–338

Watkins EL, Stillo JV, Wuthier RE (1980) Subcellular fractionation of epiphyseal cartilage. Isolation of matrix vesicles and profiles of enzymes, phospholipids, calcium and phosphate. Biochim Biophys Acta 631:289–304

Watt SL, Lunstrum GP, McDonough AM, Keene DR, Burgeson RE, Morris NP (1992) Characterization of collagen types XII and XIV from fetal bovine cartilage. J Biol Chem 267:20093–20099

Wikström B, Gay R, Gay S, Hjerpe A, Mengarelli S, Reinholt FP, Engfeldt B (1984) Morphological studies of the epiphyseal growth zone in the brachymorphic (bm/bm) mouse. Virchows Arch (Cell Pathol) 47:167–176

Wolf J, Cerný J (1928) Príspêvek k otázce zvápenatêníchrupavky (Contribution a l'étude de la calcification du cartilage). Rozpravy II Tridy Ceske Akademie 37:1–13

Wolinsky I, Guggenheim K (1970) Lipid metabolism of chick epiphyseal bone and cartilage. Calcif Tissue Res 6:113–119

Wotton SF, Duance VC (1994) Type III collagen in normal human articular cartilage. Histochem J 26:412–416

Wu J-J, Eyre DR (1995) Structural analysis of cross-linking domains in cartilage type XI collagen. Insights on polymeric assembly. J Biol Chem 270:18865–18870

Wu LN, Sauer GR, Genge BR, Wuthier RE (1989) Induction of mineral deposition by primary cultures of chicken growth plate chondrocytes in ascorbate-containing media. Evidence of an association between matrix vesicles and collagen. J Biol Chem 264:21346–21355

Wu LNY, Genge BR, Wuthier RE (1991) Association between proteoglycans and matrix vesicles in the extracellular matrix of growth plate cartilage. J Biol Chem 266:1187–1194

Wu LNY, Genge BR, Wuthier RE (1992) Evidence for specific interactions between matrix vesicle proteins and the connective tissue matrix. Bone Miner 17:247–252

Wuthier RE (1975) Lipid composition of isolated epiphyseal cartilage cells, membranes and matrix vesicles. Biochim Biophys Acta 409:128–143

Wuthier RE (1976) Lipids of matrix vesicles. Fed Proc 35:117–121

Wuthier RE (1982) A review of the primary mechanism of endochondral calcification with special emphasis on the role of cells, mitochondria and matrix vesicles. Clin Orthop Relat Res 169:219–242

Wuthier RE, Majeska RJ, Collins GM (1977) Biosynthesis of matrix vesicles in epiphyseal cartilage I. In vivo incorporation of ^{32}P orthophosphate into phospholipids of chondrocyte, membrane, and matrix vesicle fraction. Calcif Tiss Res 23:135–139

Wuthier RE, Linder RE, Warner GP, Gore ST, Borg TK (1978) Non-enzymatic isolation of matrix vesicles: characterization and initial studies on ^{45}Ca and ^{32}P-orthophosphate metabolism. Metab Bone Dis Rel Res 1:125–136

Wuthier RE, Chin JE, Hale JE, Register TC, Hale LV, Ishikawa Y (1985) Isolation and characterization of calcium-accumulating matrix vesicles from chondrocytes of chicken epiphyseal growth plate cartilage in primary culture. J Biol Chem 260:15972–15979

Yamada Y, Horton W, Miyashita T, Savagner P, Hassell J, Doege K (1991) Expression and structure of cartilage proteins. J Craniofac Genet Dev Biol 11:350–356

Yeh ML, Luo ZP (2004) The structure of proteoglycan aggregate determined by atomic force microscopy. Scanning 26:273–276

Young RD, Lawrence PA, Duance VC, Aigner T, Monaghan P (1995) Immunolocalization of type III collagen in human articular cartilage prepared by high-pressure cryofixation, freeze-substitution, and low-temperature embedding. J Histochem Cytochem 43:421–427

Young RD, Lawrence PA, Duance VC, Aigner T, Monaghan P (2000) Immunolocalization of collagen type II and III in single fibrils of human articular cartilage. J Histochem Cytochem 48:424–432

Zenmyo M, Komiya S, Kawabata R, Sasaguri Y, Inoue A, Morimatsu M (1996) Morphological and biochemical evidence for apoptosis in the terminal hypertrophic chondrocytes of the growth plate. J Pathol 180:430–433

Zini N, Sabatelli P, Silvestrini G, Bonucci E, Maraldi NM (1996) Influence of specimen preparation on the identification of phospholipids by the phospholipase A_2-gold method in mineralizing cartilage and bone. Histochem Cell Biol 105:283–296

10 Calcifying Matrices: Enamel

10.1 Introduction

Among vertebrate calcified tissues, enamel has a unique profile. It is of epithelial origin; it is acellular; it is exposed to the oral environment; once formed, it is never renewed or remodeled; it contains no collagen; its organic matrix is almost completely reabsorbed during calcification; its degree of calcification is at the top of the range. Moreover, enamel builds up through a complex process that reveals morphological and functional changes both in cells (ameloblasts) and matrix. Because of its complexity, the amelogenetic process is briefly summarized below (for fuller accounts, see Nanci and Smith 1992; Nanci 2003).

Amelogenesis is usually divided into three consecutive periods: a presecretory stage, during which ameloblasts differentiate and acquire their definitive phenotype, a secretory stage, marked out by the secretion of the entire thickness of the enamel, and a maturation stage during which the crystal structure reaches maturity. Enamel deposition begins soon after the earliest formation of dentin: the differentiating ameloblasts push cytoplasmic processes through the basal membrane which separates them from the dentin; by so doing they break up the membrane. Early enamel secretion then follows; this marks the beginning of the secretory stage. The differentiated, secretory ameloblasts located at the dentin-enamel junction appear as a single layer of elongated cells with a polarized nucleus, a developed Golgi complex, a wide endoplasmic reticulum, and two series of junctional complexes, one at their proximal (that is, far from enamel), another at their distal extremity. This junction marks the limit between the body of the cytoplasm and the so-called Tomes' processes, whose distal part penetrates into the secreted, partially calcified enamel layer. A flux of secretory granules progresses from the Golgi complex through the Tomes' processes and is then released into the extracellular space. Matrix secretion occurs both around the distal portion of the Tomes' processes and at their tip. In the first case, a thin wall is formed around the processes, which therefore come to be contained in elongated pits; in the second, enamel is secreted within the tips, which gradually fill up. As a result, the Tomes' processes are compressed and thinned, and eventually withdraw, so leaving an apparently split space filled with organic material. In this way, two different arrangements emerge: the first involves the presence of interrod enamel, which forms around Tomes' processes (i. e., the wall of the cylindrical pits), and the second the presence of rod enamel, which forms within the pits. These two forms of enamel, rod and interrod, have the same composition but differ in crystal orientation. The organic material which comes to fills each cavity after the Tomes' processes have withdrawn, forms the rod sheath. The ameloblasts that synthesize the initial and final enamel have no Tomes' processes, so no rod or interrod structures are formed.

The secretory stage is followed by a short transitory stage, during which the volume of the ameloblasts and their organelle content decrease, and then by a maturation stage: the growth of its crystals makes the enamel harder and, simultaneously, the proteins that constitute the matrix are almost completely removed. The morphology of ameloblasts evolves: their height is reduced, the Golgi complex shrinks, and the endoplasmic reticulum loses some of its cisternae; significant numbers of ameloblasts undergo apoptosis. At this stage, the most distinctive changes in ameloblasts are those that feature in 'modulation', a term used to label cyclical changes in their membrane, which shows an alternation between smoothness and multiple folds, so giving rise to a ruffled border. Modulation is probably related to the transport of calcium and/or organic material.

The many special features which make the enamel organic matrix unique among calcified tissues have been reviewed several times (Eastoe 1968; Nanci and Smith 1992; Brookes et al. 1995; Deutsch et al. 1995a; Goldberg et al. 1995; Robinson et al. 1995, 1998; Simmer and Fincham 1995; Fincham et al. 1999). Because the enamel matrix contains no fibrous structures comparable with those of bone, cartilage, dentin or other hard tissues, it has been considered to be a thixotropic gel (a gel whose viscosity changes with mechanical stress) or a relatively "structureless" sol (Eastoe 1968). As reported above, the characteristics of the enamel matrix change with the stage of amelogenesis and the degree of calcification. As a result, studies on the enamel organic matrix must be carried out bearing in mind that the composition, structure and amount of organic components change continuously from the beginning of enamel formation to the end of crystal maturation. The difficulties are greater in the case of morphological studies, because these imply decalcification, which can cause extraction or other changes in the organic components. In addition, the calcification of enamel matrix occurs as soon as it has been secreted by ameloblasts, so no uncalcified pre-enamel border comparable with predentin or the bone osteoid border is available. In this connection, a brief comment is necessary about the so-called 'stippled material'.

The term 'stippled material' (Watson 1960) refers to a granular (Fearnhead 1961) or coarse-textured (Reith 1967; Kallenbach 1971) material found closely adjacent to the ameloblast membrane and in ameloblast secretory vesicles. The picture suggests that this stippled material is a still uncalcified product of ameloblast secretion and constitutes a microenvironment for enamel crystal nucleation (Diekwisch et al. 1995). It has been proposed that particles of the stippled material are monomers of the most abundant enamel protein, i.e., amelogenin, or intermediate structures in its assembly (Fincham et al. 1998), or even amelogenin nanospheres (Moradian-Oldak et al. 1995; Dunglas et al. 2002), and it has been supposed that it is produced before calcification begins (Yamada et al. 1980).

It has, however, been reported that no stippled material is present in well-fixed developing teeth, that its ultrastructural morphology is due to poor tissue fixation, that, rather than being a precursor to mineralized enamel, it is an artifactual breakdown product (Nanci and Warshawsky 1984), and that the organic phase of well-fixed developing enamel appears homogeneous under the electron microscope (Nanci et al. 1998). Lyaruu et al. (1984) have, in fact, shown that stippled material occurs in specimens that have been fixed at 0–4 °C, and that, when fixation is carried out at 37 °C, it is substituted by a dense material which has the same electron density as ameloblast secretory granules and forms a reticular network containing enamel crystals. These authors suppose that the stippled material consists mainly of amelogenin and that its variable morphology is due to the temperature-dependent aggregation-disaggregation properties of this molecule. By contrast, Chen and Eisenmann (1984) failed to detect any relationship between the formation of stippled material and the ultrastructural preservation of cells; they found that fluoride administration induced an accumulation of stippled material between the calcification front and secretory ameloblasts. Stippled material has been found between the proximal ends of ameloblasts in vitro, where it has been considered a possible result of local disturbances in matrix secretion (Gorter de Vries et al. 1986). Dense patches of granular or amorphous material that are immunoreactive to enamel proteins (Nanci et al. 1984; Inage et al. 1989), amelogenin (Herold et al. 1987; Nanci et al. 1989a; Uchida et al. 1989; Nanci and Smith 1992), enamelins (Herold and Rosenbloom 1990), and ameloblastin (Nanci et al. 1998) have been detected in the apices of presecretory stage ameloblasts and laterally between secretory stage ameloblasts. They do not seem to be related to stippled material.

Whether the morphology of the stippled material is as described by Watson (1960) or is that of the amorphous or finely granular patches described for well-fixed teeth, is a largely formal question, because, in either case, it appears to correspond to secreted enamel organic matrix, as shown by the just discussed positive immunoreactions for amelogenin and other enamel proteins. This is supported by results obtained by Dunglas et al. (2002) in mice bearing a transgene that disrupts the amelogenin self-assembly domains: two different transgenic mouse lines were bred, following deletions either of the amino terminal (A-domain deletions) or the carboxyl region (B-domain deletions). In the A-domain-deleted newborn mice the formation of the initial layer of aprismatic enamel was delayed and this led to accumulation of stippled material through the entire thickness of the forming enamel. Similar results were obtained by Sakakura (1986), who showed that in agar chamber cultures of mouse molar teeth there is an accumulation of stippled material whose fibrils are continuous with developing enamel crystals. Accumulation of glycoproteic organic material, immunoreactive for amelogenin and ameloblastin, or its breakdown products, has been reported at the interface between ameloblasts and enamel in rats fed with a calcium-free diet (Nanci et al. 2000).

It must be added that Kallenbach (1971) observed that patches of coarse-textured material similar to stippled material are found in dentin when the ameloblast basal membrane is still present, so that they seem to have an odontoblast origin, and that not all the earliest enamel crystals are found over the fine-textured stippled material located at the tip of the ameloblast processes. This finding was developed by Nanci et al. (1989a), who reported that amelogenin is present over the developing dentin matrix before the disappearance of the basement membrane which separates ameloblasts from odontoblasts, and that patches of electron-dense granular material can be detected near the apex of the ameloblasts. A detailed review of the enamel protein secretion by ameloblasts has been reported by Nanci and Smith (1992).

The role, and the existence itself, of this stippled material remain an unresolved issue, but it must be taken as proved that the enamel matrix normally calcifies as soon as it is secreted, so that enamel crystals are in contact with the ameloblast membrane. A pre-calcification material, comparable to the osteoid border in bone or to predentin, does not exist in enamel and the stippled material, if any, seems to be the outcome of inadequate preparative procedures or of pathological conditions.

10.2 Morphology of Decalcified Matrix

Light microscopy has produced plenty of information on the microscopic architecture of enamel, whereas what is known about the morphological appearance of the enamel matrix is mainly due to electron microscope investigations. This means that results are conditioned by the technical procedures and, above all, by the need to unmask the calcified organic components. This can only be achieved by decalcification, which, besides removing the inorganic material, causes dissolution of organic components. This artifactual removal, and the fact that early studies made no distinction between immature and mature enamel, which contains very few proteins, explain the longstanding belief that enamel has no organic matrix. Using perfusion fixation with 2.55% glutaraldehyde, decalcification in isotonic, neutral disodium EDTA, and post-osmication, Warshawsky (1971) was able to describe the organic matrix of nearly mature enamel in rat incisors as consisting of rod profiles and inter-rod material comprising variously oriented, tubular and filamentous, subunits 25 nm in diameter.

More satisfactory results were obtained when the unmasking of the organic components was achieved by decalcifying already embedded tissues using a PEDS-like method (see Sect. 3.4), i.e., by treating ultrathin sections with phosphotungstic acid (PTA), which at one and the same time decalcified and 'stained', or by using EDTA or formic acid and an electron-dense 'stain'. Nylen and Omnell (1962) and Rönnholm (1962) were probably the first to apply these techniques to the study of enamel; they described organic structures showing the same shape and arrangement as dissolved crystals, or outlining spaces the same width as crystals, so that the tissue appeared to be undecalcified. Structures with a fibrillar, filamentous, lamellar or helical, crystal-like morphology, or outlining compartments or tubules with the same width as crystals, have then been reported as the organic constituents of the organic matrix of immature enamel (Travis and Glimcher 1964; Travis 1968; Garant 1970; Decker 1973; Kemp and Park 1974; Smales 1975; Frank 1979; Nylen 1979; Bishop and Warshawsky 1982; Nanci et al. 1983; Bai and Warshawsky 1985; Hayashi et al. 1986; Hayashi 1989; Bonucci et al. 1994; Bonucci 1995; Nanci

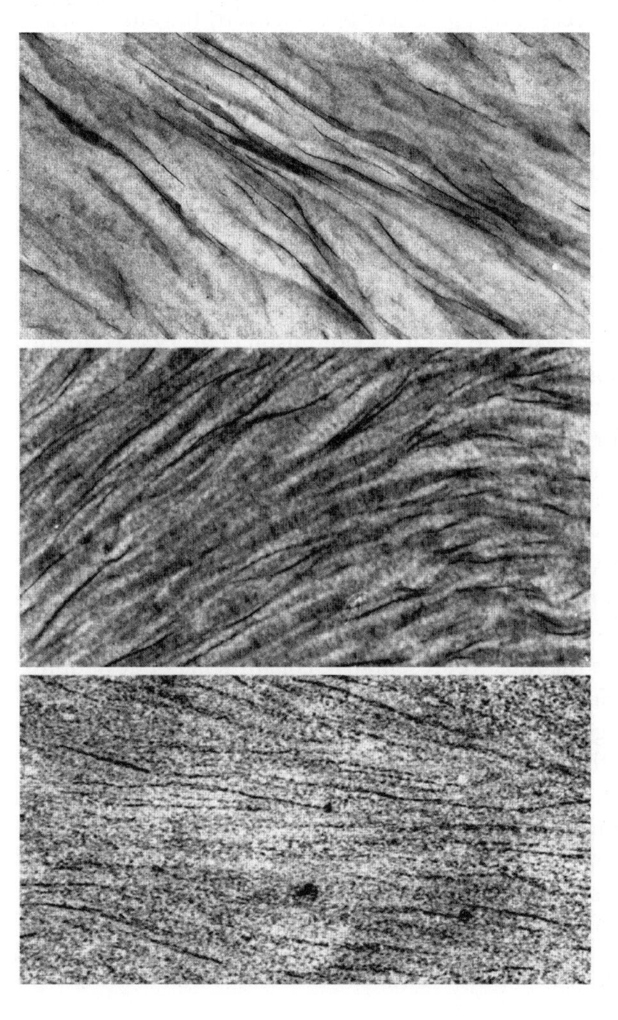

Fig. 10.1. Crystals of immature enamel. *Above*: in untreated sections, they appear as long, filament- or ribbon-like structures with interposed, apparently amorphous material. *Middle*: in sections stained with uranyl acetate and lead citrate, their shape is not significantly changed; ribbon-like structures are easily recognizable. *Below*: after the PEDS method (decalcification after embedding, staining with uranyl acetate and lead citrate), they are replaced by crystal ghosts. ×70,000

et al. 2000). Decalcification by PEDS method and uranyl acetate and lead citrate staining unequivocally show filament- and ribbon-like organic structures with the same shape as that of untreated crystals (crystal ghosts; Fig. 10.1). In spite of the many data provided by these investigations, the question of the structure of enamel organic matrix has not yet been definitively settled. Because of its pertinence to the calcification process, the topic will be further discussed below (see Sect. 16.4).

10.3 Matrix Components

Studies on the components of the enamel matrix have encountered a series of difficulties chiefly arising from the heterogeneous composition of the matrix, which is partly due to the secretion of a variety of proteins by the ameloblasts, and partly to the post-secretory sequential degradation of these proteins (Robinson et al. 1998; Smith et al. 1992). Probably because it is secreted by ameloblasts, that is, by ectodermal cells, and also because of its high sulfur content, it has been suggested that the organic matrix of immature enamel could consist of keratin-related proteins (Eastoe 1963). Actually, an extracellular organic component sharing antigenic determinants with keratins has been found in enamel, as well as in dentin and bone (Lesot et al. 1988). In this connection, tuft protein, an extremely insoluble enamel protein lying near the dentine in mature teeth (to be discussed), contains ameloblastic secretory products which seem to be related to keratins (Robinson et al. 1989a). In any case, the high amount of proline – higher than the combined content of proline and hydroxyproline in collagen – and the virtual absence of cystine sharply differentiate the enamel matrix from keratins, in the same way as the poor glycine content and the absence of hydroxyproline and hydroxylysine distinguish it from collagen (Eastoe 1968).

The organic matrix in enamel is heterogeneous. Glycoconjugates and lipids make up only a minor proportion of its components. Although early mention to them can be found in the 1950s (Sobel and Burger 1954) and early 1960s (Irving 1963), relatively few studies have been dedicated to these substances (Smith 1985). Proteins are the most representative constituents of developing enamel and much attention has been devoted to them. They fall into in two major classes: the amelogenins, which represent about 90% of the matrix in developing enamel, and the non-amelogenins, also called enamelins (Robinson and Kirkham 1985; Zeichner-David et al. 1997). The latter include other ameloblastic proteins (ameloblastin, tuftelin) and non-ameloblastic proteins (albumin, α_2-HS glycoprotein, salivary proteins). Enzymes with proteinase activity are also components of the matrix. No matrix vesicles or collagen fibrils have ever been found; recently, low amounts of proline and hydroxyproline have been detected in enamel matrix (Açil et al. 2005).

10.3.1 Proteoglycans

Most of the enamel carbohydrates associated with proteins take the form of glycoproteins; some may appear in developing enamel as glycosaminoglycans or proteoglycans (Goldberg et al. 1995). Ultrastructural studies on carbohydrates in developing rat enamel using soybean agglutinin-gold complexes have shown the presence of *N*-acetyl-D-galactosamine in the crystal-associated organic material (Hayashi 1989). Immunohistochemical and lectin-binding studies have shown that most of the lectins had not become bound to porcine amelogenins, while a large number of non-amelogenins were strongly stained by conjugated WGA, Con A, HPL and MPA lectins (Nanci et al. 1989b; Akita et al. 1992). On this basis, porcine non-amelogenins were divided into two groups: the 60–90-kDa glycoproteins, which reacted with WGA and Con A, and the 13–17-kDa glycoproteins, which reacted with MPA. The former were concentrated in the outer region next to ameloblasts, and disappeared (through degradation) in the underlying inner secretory enamel; the latter were found in all zones of the secretory enamel and their quantity remained relatively constant (Nanci et al. 1989b; Akita et al. 1992).

The presence of glycosaminoglycans in enamel was probably first mentioned by Sobel and Burger (1954), who found chondroitin sulfate to be a component of the organic matrix. This is in agreement with studies showing that the organic matrix of almost mature enamel is stained by alcian blue, pH 2.6, and bismuth nitrate, pH 1.2 (Goldberg et al. 1978; Goldberg and Septier 1986). Similar results have been reported in rat molar tooth germs: the organic matrix is stained by alcian blue at lower concentrations of $MgCl_2$, probably due to a reaction of the cationic dye with carboxyl and phosphate groups of proteins bearing traces of sulfated glycosaminoglycans (Smith 1985); in the same study, *N*-acetylgalactosamine and glucose were found as matrix macromolecule terminal sugars. Histochemical studies carried out using the high iron diamine, thiocarbohydrazide silver proteinate method showed the presence of sulfated glyconjugates, which resisted digestion with heparitinase, testicular hyaluronidase and chondroitinase ABC, in the interdigitating cell membrane of Tomes' processes, inside some secretory granules in ameloblasts, on the lateral cell membrane of the stratum intermedium, in the basal membrane associated with the outer enamel epithelium and endothelial cells of capillary vessels, as well as in enamel matrix near the future enamel-cement junction (Kogaya and Furuhashi 1988). Chardin et al. (1990), after using the hyaluronidase-gold method, reported finding hyaluronic acid and/or chondroitin sulfate during the earliest stage of organic matrix secretion, and their disappearance as enamel becomes thicker. The rapid incorporation of $^{35}SO_4$ in a $\sim$ 65-kDa, short-lived protein, which was secreted into the enamel within 6–7.5 min and rapidly destroyed, has been reported by Smith et al. (1995). On the other hand, intense ^{35}S labeling but very slight alcian blue staining were detected in developing enamel, whereas moderate ^{35}S labeling, and high alcian blue positivity of labeled sites, were found in mature enamel (Suga et al. 1970). According to Reith and Cotty (1967), sulfur, or sulfate compound, were eliminated by ameloblasts.

10.3.2 Lipids

The presence of lipids in the enamel organic matrix was first suggested on the basis of the observation that the developing enamel is intensely sudanophilic after pyridine extraction (Irving 1963). Further investigations showed, however, that sudan-stainable material is not found in immature enamel matrix undergoing calcification but does occur in mature enamel, and that immature enamel is stained after lipid extraction, whereas mature enamel is not, suggesting that the former contains bound lipids that are not available for staining (Beynon 1976). More recently, it has been argued that, on the basis of Irving's method, no definitive conclusions can be reached about whether lipids are present (Frederiks 1977), and that sudanophilia is more probably due to the stain's reaction with hydrophobic matrix proteins (Beynon 1976), but the early results led to other histochemical and biochemical investigations.

In an attempt to demonstrate enamel lipids under the electron microscope, Goldberg et al. (1983) showed an osmiophilic material whose nature could not be determined but which could be extracted in chloroform-methanol solution and lay within tubule-like structures in the extracellular matrix. In a review on the structure and composition of mature enamel, Frank (1979) reported that the organic components, estimated to be less than 0.3 wt %, consist primarily of proteins (58%) and secondarily of lipids (42%), which comprise triglycerides (30%), cholesterol (20%), cholesterol esters (20%), lecithins (10%), and neutral and complex lipids (20%). He stressed that it is hard to extract lipids completely, and that some polar lipids can only be extracted after decalcification. Fincham et al. (1972) confirmed that, prior to demineralization of bovine dental enamel, only 44% of the total neutral lipid complement could be removed; within this, cholesterol esters, triglycerides and free fatty acids were predominant. After enamel decalcification, residual lipids were freed, showing a profile like that of the first fraction. Shapiro et al. (1966) reported a phospholipid content of 0.06% by weight of tissue, only about 40% of which could be extracted without prior decalcification. Structures identifiable with phospholipids have been reported by Goldberg and Septier (2002) in electron microscope studies of enamel treated with various fixation methods. The results of Goldberg et al. (1998) seem to show that fragments of distal Tomes' processes may be released during enamel formation, and that their membranes might be degraded enzymatically, so producing phospholipids, which go on to form structures resembling crystal ghosts (see Nefussi et al. 1992). Using confocal microscopy and Nile red fluorescence, lipid material has been detected in enamel cross-striations, the lines of Retzius, the Hunter-Screger bands, and interprismatic spaces (Girija and Stephen 2003).

10.3.3 Amelogenins

Amelogenins are a group of glycosylated and phosphorylated, highly conserved proteins characterized by high concentrations of proline, glutamic acid, leucine and histidine (for an exhaustive description, see Fincham et al. 1999). Amelogenin molecules are hydrophobic, except for a 13–15 amino acid hydrophilic C-terminal domain (Beniash et al. 2005). They are relatively small; their molecular weight (SDS-PAGE) ranges between about 5 and 30 kDa (Seyer and Glimcher 1977; Nanci et al. 1989b; Aoba and Moreno 1991; Fincham et al. 1994a; Brookes et al. 1995; Fincham and Moradian-Oldak 1995; Goldberg et al. 1995; Smith and Nanci 1996). This wide range is, to a small extent, due to genetic factors such as sexual dimorphism (Gibson et al. 1995; Fincham et al. 1991a) or alternative splicing (Lau et al. 1992; Yamakoshi et al. 1994; Simmer et al. 1994; Simmer 1995; Veis 2003), and mainly due to the enzymatic degradation processing which takes place after amelogenins are secreted extracellularly (Seyer and Glimcher 1977; Fincham et al. 1991b; Aoba et al. 1992; Bronckers et al. 1994; Moradian-Oldak et al. 1994a; Robinson et al. 1998).

Amelogenins represent the primary (about 90%), distinctive molecular species of developing enamel matrix, from which they can be extracted by 4 M guanidinium HCl, pH 7.4, without dissolution of the crystals (Termine et al. 1980). Amelogenins are expressed by cells of the ameloblast lineage (Inage et al. 1989; Nanci et al. 1989a, 1992, 1994; Uchida et al. 1989; Inai et al. 1991; Wurtz et al. 1996; Karg et al. 1997; Hu et al. 2001). Their earliest secretion occurs in enamel's presecretory stage, when ameloblasts and odontoblasts are not fully differentiated, mantle predentin is absent, and enamel crystals are not detected (Nanci et al. 1989a); this has suggested that amelogenins may initially participate in epithelial-mesenchymal interactions (Nanci and Smith 1992; Sawada and Nanci 1995). Amelogenin expression continues through the secretory stage, and ends in the early maturation stage (Wakida et al. 1999). It cannot be excluded that odontoblasts may express various alternatively spliced amelogenin transcripts during mantle dentin deposition (Papagerakis et al. 2003).

Extracellular amelogenins are localized immunohistochemically throughout the enamel (Figs. 10.2 and 10.3): the immunoreaction is weak near the surface of the secretory stage ameloblasts, where it is found over enamel and over patches of material at the baso-lateral surfaces of these cells (Nanci et al. 1992); it then increases over a distance of about 1.25 μm (Nanci et al. 1998) and gradually declines in the direction of the dentin-enamel junction (Uchida et al. 1989), the lowest degree of immunoreactivity being found over the midportion of the enamel layer (Orsini et al. 2001). The immunoreaction occurs

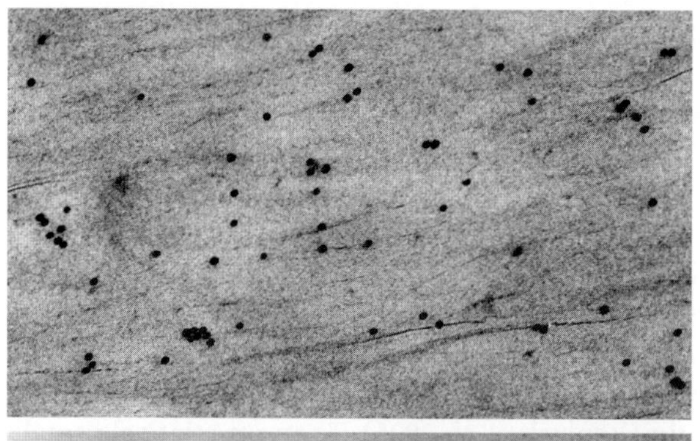

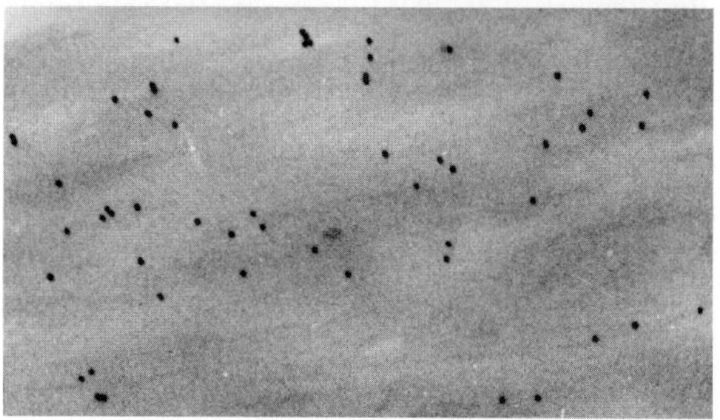

Fig. 10.2. *Above*: antiamelogenin immunoreaction in immature enamel; undecalcified, ×40,000. *Below*: the same immunoreaction carried out on a section decalcified with EDTA for 10 min; not counterstained, ×40,000

in undecalcified sections (Fig. 10.2), showing that the epitope is not masked by the mineral substance (Hayashi et al. 1986), a finding that converges with results of immunohistochemical reactions for amelogenin indicating that protein A-gold particles were located between enamel crystals (Herold et al. 1987; Inage et al. 1989).

Amelogenin molecules undergo self-assembly and form supramolecular aggregates (Limeback and Simic 1990; Moradian-Oldak et al. 1995, 1998, 2000; Brookes et al. 2000; Moradian-Oldak 2001; Wen et al. 2001), initially consisting of intermediate structures with a 4–5-nm hydrodynamic radius (Fincham et al. 1998), and then of 20-nm diameter spherical structures which have been called "nanospheres" on account of their shape (Fincham et al. 1994b, 1995, 1998). These are aligned with and between the developing enamel crystals (Fincham et al. 1991b, 1995; Moradian-Oldak et al. 1994b) and can further assemble to give rise to a variety of higher order assemblies (Wen et al. 1999, 2001), depending on temperature and pH (Moradian-Oldak et al. 1998). The supramolecular assembling properties of amelogenin have been confirmed by in vitro researches showing that amelogenin nanospheres gave rise to linear aggregates which, in their turn, developed into birefringent microribbons 200–1,200 μm long, 29 μm wide and 2.7 μm thick (Du et al. 2005). The immersion of these microribbons into a metastable calcium phosphate solution resulted in the formation of ordered, oriented crystals which ran parallel to the ribbons.

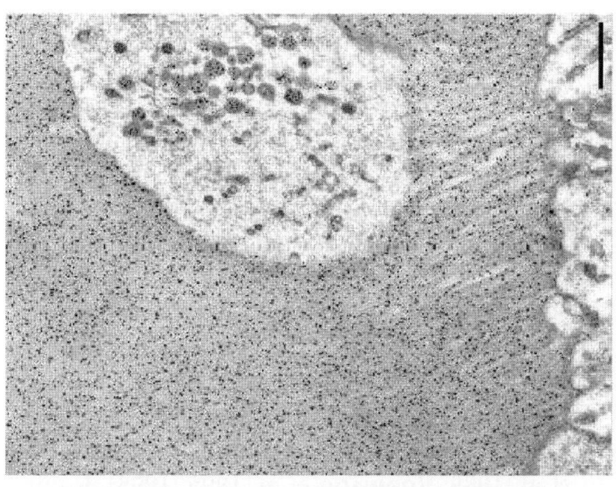

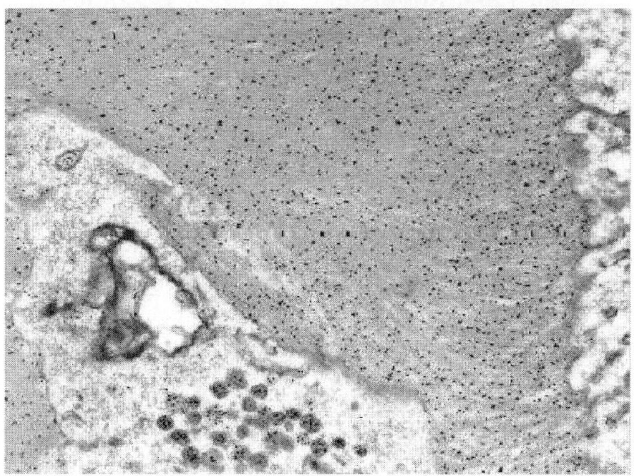

Fig. 10.3. Immature enamel: immunoreaction for amelogenin (*above*) and for ameloblastin (*below*; antibody recognizing both entire ameloblastin molecule and its degradation products); ×24,000 (courtesy of Antonio Nanci, University of Montreal, Montreal, Canada)

The mechanism of amelogenin self-assembly is still uncertain and is probably dependent on the amino-terminal and carboxyl-terminal residues (reviewed by Paine ML and Snead 2005). The leucine-rich amelogenin peptide (LRAP), which results from alternate splicing of the amelogenin mRNA, and the amino-terminus tyrosine-rich amelogenin peptide (TRAP), may both have a role in regulating the formation of amelogenin nanosphere.

The relationship between amelogenin nanospheres and enamel crystals could be the key to a better understanding of how these can develop into such long structures. Amelogenins could be selectively adsorbed on to crystal surfaces, so curtailing their growth in thickness (Aoba et al. 1987; Aoba and Moreno 1991); the interposition of amelogenins would, in fact, regulate crystal growth and thickness and preclude crystal fusion, so allowing crystals to grow preferentially in length (reviewed by Robinson et al. 1998; Fincham et al.1999; Moradian-Oldak 2001). As noted above, this process is thought to depend on the self-assembly of amelogenin monomers into 15–19 nm thick 'nanospheres'

(Moradian-Oldak et al. 1994b), which form 'beaded rows' aligned with, and located between and around, immature crystals at the initial stage of enamel formation, so determining their spacing, inhibiting their fusion, controlling their shape and creating anionic channels for ion transport within the calcified matrix (Fincham et al. 1994b, 1995). According to Beniash et al. (2005), amelogenin's limitation of crystal growth is due to its hydrophobic portion, whose C-terminal domain is essential for the alignment of crystals into parallel arrays.

As enamel matures, amelogenins are degraded (reviewed by Brookes et al. 1995; Aoba 1996). They are cleft by proteinases soon after their secretion, so that whole molecules can only be found at the superficial enamel layer, while their cleavage products are detected in the inner, more mature enamel (Uchida et al. 1991a; Tanabe et al. 1992; Fukae et al. 1993). Eventually, amelogenins almost completely disappear (Smith et al. 1989; Robinson et al. 1995, 1998), leaving two main residual polypeptides, one rich in tyrosine, the other in leucine. This proteolytic process allows crystals to grow in thickness and reach maturation. Several proteinases may occur in enamel (reviewed by Bartlett and Simmer 1999); among them, enamelysin, otherwise known as MMP-20 (Bartlett et al. 1996), and enamel matrix serine proteinase-1, often called EMSP1 (Simmer et al. 1998) (see Sect. 10.4).

10.3.4 Non-amelogenins

Non-amelogenin proteins were first described by Termine et al. (1980), who used the already mentioned sequential dissociative extraction method to isolate them: they found that the treatment of fetal teeth with buffered guanidine hydrochloride released most of the enamel matrix proteins (amelogenin), and that further treatment with the same solution with EDTA added to it released a fraction of acidic, phosphorylated, glycoproteic proteins rich in aspartic, serine and glycine amino acid residues. They called these proteins 'enamelins'. Lyaruu et al. (1982) preferred the name "enamel crystal proteins", because they could only be extracted after crystal dissolution by decalcification. A consensus has now emerged in favor of 'non-amelogenin proteins' (Robinson and Kirkham 1985), to allow a clear distinction from the enamelins, which include enamelin proper, ameloblastin (amelin, sheathlin) and tuftelin.

The non-amelogenin proteins (Bronckers et al. 1988) are highly acidic proteins, rich in aspartic, glutamic, and sialic acids. They are highly heterogeneous molecules whose most consistently occurring molecular weights are 55–65, 60–65 and 70–72 kDa in various species (see Table I in Robinson et al. 1989b), and between 25 and 155 kDa in porcine immature enamel (Fukae and Tanabe 1987; Hu et al. 1997a); they reach a maximum value of 130 kDa in bovine immature enamel (Fukae et al. 1993). As already reported for amelogenins, this heterogeneity may be due to mRNA alternative splicing, or to different genes leading to coding for different proteins (Ogata et al. 1988), or else to a rapid, almost complete protein degradation during enamel maturation (Belcourt et al. 1982; Tanabe et al. 1994).

It must be stressed that non-amelogenin proteins partly consist of albumin and other serum proteins which have been detected in developing porcine (Limeback et al. 1989), rat (Chen et al. 1995) and bovine (Strawich and Glimcher 1989, 1990; Strawich et al. 1993) enamel. In vitro studies of Robinson et al. (1996) have shown albumin degradation from the transition to the maturation stage, where it is ultimately removed.

The non-amelogenin protein fraction has been detected in the extracellular matrix since the earliest stages of enamel formation (Slavkin et al. 1988). Enamelin antibodies, having no cross-reactivity with amelogenin, showed that immunolabeling was located in almost exactly the same cell organelles as amelogenin (Inage et al. 1989) and that, in developing enamel, it was mainly attached to the enamel crystals. Concentrations of enamelins are relatively high in developing enamel and decrease with enamel maturation, but they, or their fragments, can be detected in enamel throughout the secretory and maturation stages (Belcourt et al. 1982; Dohi et al. 1998).

X-ray and electron diffraction studies on fixed and decalcified enamel proteins from rat and hippopotamus teeth have shown a strong 0.47-nm reflection, indicative of a β pleated-sheet conformation, and a crystalline pattern matching the dimensions of the *hhl* plane of hydroxyapatite; both have been ascribed to the enamelin fraction (Traub et al. 1985; Jodaikin et al. 1986).

Specific non-amelogenin proteins have now been cloned and characterized. The best known are enamelin, ameloblastin (also called sheathlin or amelin), and tuftelin (Deutsch et al. 1995b; Fukae et al. 1996; Krebsbach et al. 1996; Hu et al. 1997b). A glycosilated and sulfated non-amelogenin with a short life and a molecular weight of about 65 kDa has been described in rat enamel by Smith et al. (1995) and Smith and Nanci (1996).

10.3.4.1 Enamelin

Enamelin is the name given to an acidic, glycosilated and phosphorylated protein (Uchida et al. 1991a), whose molecular weight changes with its processing, ranging from 140–150 kDa in its native state through 89 kDa and 65 kDa to 32 kDa in the deepest regions of enamel (Ogata et al. 1988; Fukae and Tanabe 1987; Uchida et al. 1991b; Fukae et al. 1993, 1996; Hu et al. 1997a, 1998). Although its amount is probably less than initially thought, and its very presence in enamel has been called into question, several studies have provided evidence that enamelin can be found both in enamel and ameloblasts (Rosenbloom et al. 1986; Inage et al. 1989; Fukae et al. 1993; Hu et al. 1998) and that it, and its cleavage products, are the main constituents of non-amelogenin proteins (Fukae and Tanabe 1987; Uchida et al. 1991a; Fukae et al. 1993, 1996; Hu et al. 1997a, 1998). The intact molecule has been detected only at the calcification front, while its cleavage products accumulate in the rod and interrod enamel and have a high affinity for the mineral (Simmer and Hu 2002; Hu et al. 1997a, 1998; Brookes et al. 2002). The onset of its expression almost coincides with that of amelogenin and is synchronous with the initial predentin formation; both proteins are later expressed by ameloblasts throughout the secretory, transitory, and early maturation stages (Hu et al. 2001).

Immunohistochemical and immunochemical enamelin studies carried out in porcine tooth germs by Dohi et al. (1998) using antibodies against 89-kDa enamelin and its N-terminal synthetic peptide, respectively, showed that at the presecretory stage the 89-kDa fraction occurred in enamel islands in dentin, while the N-terminus occurred both in enamel islands and in stippled material; at the matrix secretion stage, both antibodies stained enamel prisms located in the outer layer, but the immunoreaction disappeared at the end of the transition stage and in the early maturation stage; in the inner layer, the 89-kDa antibody stained enamel matrix moderately and homogeneously, whereas the N antibody stained the prism sheath. These results have been assumed to demonstrate that enamelin expression continues from the late presecretory to the transitory stage and that the cleavage of the enamelin N-terminal region occurs soon after secretion.

The finding that, unlike amelogenin, enamelin can only be extracted from enamel after decalcification (Termine et al. 1980; Belcourt et al. 1982; Lyaruu et al. 1982; Bronckers et al. 1988), shows that an intimate relationships exists between this protein and the inorganic substance. This has led to the suggestion that enamelin acts as crystal template for the initiation and growth of crystals (Bronckers et al. 1988). This perspective appears to be in agreement with its acidic nature and time of secretion (Deutsch et al. 1995a), and is supported by ultrastructural immunohistochemical findings showing that enamelin immunoreactivity is mainly linked to the crystals (Inage et al. 1989; Herold and Rosenbloom 1990), and that while amelogenin, which is located between crystals, is detectable in undecalcified sections, decalcification is needed to display enamelin immunoreactivity (Fig. 10.4) (Hayashi et al. 1986). Similar results have been obtained by Bai and Warshawsky (1985): intercrystalline particulate materials and crystal ghosts could be detected in decalcified and stained sections, whereas only crystal ghosts could be found in 4 M guanidine extracted enamel. The authors' conclusion was that the particulate material corresponded to amelogenins and the crystal ghosts to enamelins, seen as the integral template proteins which initially permit the elongation of enamel crystallites. Bouropoulos and Moradian-Oldak (2004), using a 10% gelatin gel diffusion device, found that the induction time for apatite precipitation was delayed when the gel was loaded with 0.75 wt % and 1.5 wt % native porcine amelogenins, whereas it was accelerated in a dose-dependent manner when 18 and 80 μg/mL of 32-kDa enamelin were added to gel containing 1.5% amelogenin. The authors concluded that amelogenins and 32-kDa enamelin cooperate to promote crystal formation.

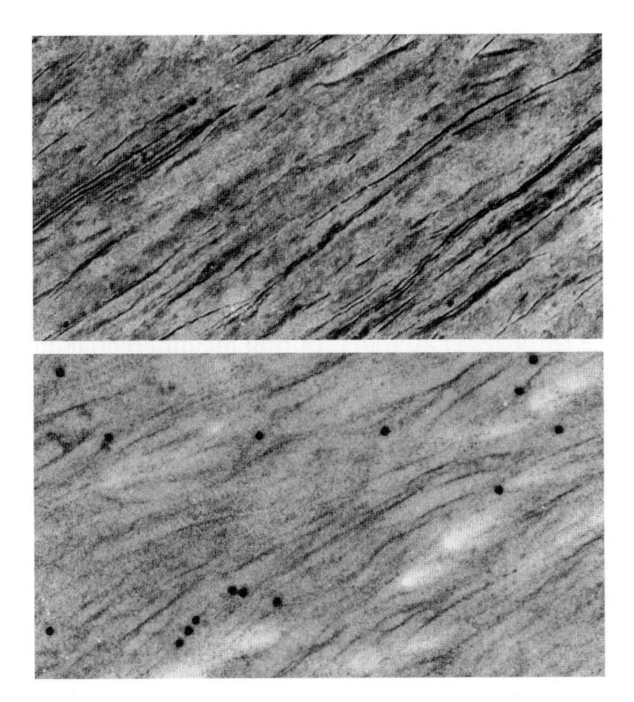

Fig. 10.4. *Above*: antienamelin immunoreaction in undecalcified immature enamel; only a few, occasional immunogold particles can be detected. *Below*: the same immunoreaction carried out on a section decalcified according to the PEDS method: several immunogold particles can be detected. PEDS method; uranyl acetate and lead citrate counterstaining; ×48,000

10.3.4.2 Ameloblastin (Amelin, Sheathlin)

Ameloblastin (Krebsbach et al. 1996), otherwise called amelin (Fong et al. 1996a) or sheathlin (Hu et al. 1997b; Uchida et al. 1998), is a glycosylated protein which is known to have two alternatively spliced variants (1 and 2) (Hu et al. 1997b; Lee et al. 2003). The nascent molecule is distributed in the superficial, secretory- and transition-stage enamel (Fig. 10.3) and appears to migrate at 62–68 kDa on SDS gels (Uchida et al. 1995; Hu et al. 1997b; Brookes et al. 2001). Other bands, that are immunoreactive at 52, 40, 37, 19, 17, 16, 15, 14 and 13 kDa, have been reported; these can be assumed to be ameloblastin processing products (Brookes et al. 2001). Of these, the 13–17 kDa have been immunolocalized throughout the layer of developing, immature enamel, where it takes on a honeycomb appearance because of the localization of the ameloblastin molecule at the prism periphery, i. e., in the interprismatic space, or "sheath", which incompletely separates rod and interrod enamel (Uchida et al. 1991b, 1995; Hu et al. 1997b). A fraction of the 17- and 16-kDa ameloblastins could not be extracted with simulated enamel fluid, but these two ameloblastins were desorbed from the enamel crystal surface with phosphate buffer, showing that they were mineral-bound; the remainder of these fractions, as well as the 15- and 14-kDa amelins, could only be extracted with SDS, suggesting that they were present in vivo as insoluble aggregates (Brookes et al. 2001). According to Uchida et al. (1997), ameloblastin is synthesized as a 55-kD core protein and is post-translationally changed into the 65-kD secretory form, whose cleavage generates N- and C-terminal polypeptides; some of the former concentrate in the prism sheath, whereas the latter are degraded and lost. Unlike amelogenin, ameloblastin does not self-aggregate (Paine ML et al. 1998).

Ameloblastin has been reported to be sequentially expressed by ameloblasts from the enamel presecretory to the late maturation stage (Fong et al. 1996b; Lee et al. 1996). Proteins immunoreactive to the anti-ameloblastin antibodies have, however, been detected during the secretory and transitory stages, but not during the maturation stage (Brookes et al. 2001). During the differentiation of inner epithelium, ameloblastin was found to be expressed in the Tomes' processes, while only modest amounts could be detected in enamel matrix (Krebsbach et al. 1996).

Ameloblastin mRNA expression, like that of amelogenin mRNA, is intense in secretory ameloblasts and remains so in post-secretory ameloblasts (Fong et al. 1996b). Ameloblastin-1 mRNA expression has been reported to occur in pulp mesenchymal cells, preodontoblasts and young odontoblasts, whereas ameloblastin-1 has been localized immunohistochemically in the enamel matrix adjacent to secretory ameloblasts and, in the maturation stage, at the dentino-enamel junction (Fong et al. 1998). Immunohistochemistry, in combination with immunoblotting and in situ hybridization, have shown that, unlike amelogenin, there is no correlation between mRNA signals and the sites of ameloblastin secretion during the early presecretory and the mid- to late maturation stages, and that, while amelogenin immunoreactivity is weak near secretory surfaces, that of ameloblastin shows an inverse pattern, reaching high levels near secretory surfaces and low ones deep in the enamel layer (Nanci et al. 1998). The same study showed that ameloblastin immunoreactivity, which is high at the rod growth site, where enamel crystals elongate, becomes practically negative after the inhibition of protein secretion by brefeldin A or cycloheximide, which suggests that ameloblastin is very rapidly broken down into fragments with lower molecular weights. A study of the ameloblastin temporal expression pattern in mouse mandibular molars has shown that it is initially expressed at E-14, one day after the initial expression of tuftelin and one day before that of amelogenin (Simmons et al. 1998). A specific interaction of ameloblastin with amelogenin has been reported (Ravindranath et al. 2004).

10.3.4.3 Tuftelin

Tuftelin, which was first characterized by Deutsch et al. (1991), is an acidic enamel protein which has been considered to play a major role in enamel calcification (see review by Deutsch et al. 1995b, 1997, 1998, 2002). Its identity has been confirmed using several methods (Bashir et al. 1997, 1998; MacDougall et al. 1998; Mao et al. 2001). Tuftelin should be clearly distinguished from "tuft" proteins, a term used to indicate extremely insoluble proteins or fragments of proteins secreted during early amelogenesis, left in the calcified matrix during enamel maturation, and identified with the 'tufts', the sinuous bundles of organic material which can be recognized histologically at the dentin-enamel junction (Robinson et al. 1989c, 2000; Farge et al. 1991; Amizuka et al. 1992). Tuftelin and tuft proteins do, however, have similar amino acid compositions. The composition of the latter may represent a residual, perhaps degraded, form of the former (Deutsch et al. 1995b).

Tuftelin is not enamel-specific (see review by Deutsch et al. 2002): alternatively spliced tuftelin mRNAs have been found in several non-calcifying tissues of the mouse (MacDougall et al. 1998; Mao et al. 2001). Tuftelin has been demonstrated immunohistochemically within the odontoblast processes during early stages of tooth organ development, and in the portion of enamel adjacent to the dentin-enamel junction during later stages (Robinson et al. 1989a; Deutsch et al. 1991; Simmer and Fincham 1995). In developing tooth organs from the mouse mandible, tuftelin was detectable in odontoblast processes at day E19 and the first postnatal day, whereas it was later (postnatal days 3–11) recognized in an enamel layer adjacent to the dentino-enamel junction (Diekwisch et al. 1997).

Tuftelin, like amelogenin, has been considered a self-assembling protein by Paine ML et al. (1998); these authors went on to identify tuftelin interacting proteins (TIPs), one of which, called TIP39 because of its molecular mass of approximately 39 kDa, has been found in the apical portion of the secretory ameloblasts and has been considered a membrane-bound protein which might participate in the secretion of cell-specific proteins (Paine CT et al. 2000).

10.4 Proteinases

It has been reported above (Sect. 10.3.3) that the organic components of enamel matrix are processed extracellularly during enamel formation and that their heterogeneity and variable molecular weights are at least partly due to this process. In their reviews of amelogenin, Brookes et al. (1995) and Robinson et al. (1998) report, for instance, the extracellular processing pathway of amelogenin in porcine enamel: the nascent, 25-kDa, amelogenin is rapidly processed to a 20-kDa amelogenin-like protein and then to an insoluble, tyrosine-rich amelogenin peptide (TRAP) and to highly soluble 11–13-kDa fragments which are, presumably, further processed and lost from the tissue (see also Tanabe et al. 1992; Moradian-Oldak 2001). Molecular processing may, in fact, lead either to the assembly of proteins into supramolecular aggregates or to their degradation to progressively smaller fragments which are removed from the tissue (Smith et al. 1989). This is why a conspicuous fall in protein concentration occurs during amelogenesis (Blumen and Merzel 1972, 1982; Blumen 1979; Robinson et al. 1990; Fukae and Tanabe 1998; Moradian-Oldak et al. 2002); the protein content falls from approximate values of 15–20 wt % in immature enamel to values of 0.1% or less in the most mature enamel (Glimcher et al. 1977). This process, which is essential to enamel maturation, relies on the extracellular activity of proteolytic enzymes (for reviews, see Bartlett and Simmer 1999 and Simmer and Hu 2002).

Proteolytic enzymes have often been reported as components of the enamel matrix (Shimizu et al. 1979; Moe and Birkedal-Hansen 1979; Overall and Limeback 1988; Carter et al. 1989; DenBensten and Heffeman 1989). They belong to two main classes, the metalloproteinase and the serine proteinase class (Overall and Limeback 1988). These are active in a biphasic process: the former processes amelogenin soon after enamel formation – but its activity extends to enamelin, too (Tanabe et al. 1994) – and it continues to be active through the secretory and maturation stages; the latter almost completely demolishes the residual organic components (Brookes et al. 1998). Other proteinases – as well as their inhibitors – may be present. This is true of gelatinase A (MMP-2), which has been found in developing tooth tissues and has been reported to cleave recombinant amelogenin (Caron et al. 2001), of MMP-3, which is similar to the metalloproteinase stromelysin (DenBesten et al. 1989), of ameloprotease-1, a serine proteinase capable of amelogenin degradation in maturing enamel (Moradian-Oldak et al. 1996a), and of non-specific esterases, which may be involved in the degradation of matrix proteins (Moe and Kirkeby 1990). In vitro experiments have shown that serine proteinases, as well as alkaline phosphatase – whose activity has been reported in enamel with a maximum at the maturation stage – are able to facilitate crystal growth (Robinson et al. 1990). Alkaline phosphatase has been found in the stratum intermedium, in the proximal borders of ameloblasts and within Tomes' processes in ameloblasts, but not in enamel matrix (Orams 1981). According to Takano et al. (1986) it is not present in ameloblasts.

Enamelysin (Bartlett et al. 1996; Llano et al. 1997; Fukae et al. 1998; Caterina et al. 2000; Ryu et al. 2000), or matrix metalloproteinase-20 (MMP-20), is the most representative processing enzyme in the first class mentioned above; serine proteinase 1, otherwise known as kallikrein 4 (KLK 4), or EMSP1, is the principal degradation enzyme in the second (Robinson et al. 1998; Simmer and Hu 2002). Enamelysin is a proteolytic enzyme which degrades amelogenin by cleaving its C-terminal region (Moradian-Oldak et al. 2001, 2002), so triggering amelogenin to amelogenin interactions and self-assembly (Moradian-Oldak et al. 2000). It is expressed by ameloblasts before dentin mineralization and is active throughout the secretory amelogenesis stage (Ryu et al. 1999), when amelogenin and other enamel proteins are processed. Kallikrein 4 is active during enamel maturation (Fukae et al. 2002), when it clears the degraded enamel proteins from enamel (Simmer and Hu 2002). Using recombinant mouse amelogenin M179 as a substrate, two classes of enamel proteinases were identified in vitro, the first corresponding to a high molecular weight, Ca-dependent metalloproteinase which cleaves the M179 C-terminal segment, the second to a low molecular weight serine proteinase which causes further degradation of amelogenin by removing the TRAP sequence (Moradian-Oldak et al. 1996b). All these results go to strengthen the concept that enamel proteolysis is a highly controlled process (Moradian-Oldak et al. 1996b) and that the temporal distribution and physical state of enamel proteinases change with the stage of amelogenesis. In particular, enamelysin activity is prevalent during the secretory stage, and KLK-4 activity during the maturation stage (Robinson et al. 1998). According to Bartlett and Simmer (1999), amelogenin, enamelin and ameloblastin are enzymatically processed soon after they are secreted, so that the intact proteins are only found at the enamel surface, while their cleavage products may accumulate in the mature enamel. In agreement with these findings, it has been reported that a progressive disappearance of 25-kDa amelogenin occurs within 40 μm from the enamel surface (Tanabe et al. 1992).

10.5 Concluding Remarks

The structure and composition of enamel organic matrix make it a unique tissue. Its most distinctive features can be summarized as follows:

- The enamel organic matrix calcifies as soon as it is secreted by ameloblasts.
- The stippled material, if any, has been considered a just secreted, not yet calcified organic material which can accumulate in some circumstances and whose ultrastructural morphology depends on technical procedures.
- The organic matrix lacks a true fibrous supporting framework comparable to collagen fibrils in bone; enamel contains no collagen fibrils at all.
- The morphology of the organic matrix can only be studied after decalcification. Organic structures with the same shape and/or width as inorganic crystals (crystal ghosts) can be recognized in immature enamel using the PEDS method.
- The organic matrix consists of glycoproteins, proteoglycans, lipids, and two groups of specific proteins, amelogenins and non-amelogenins.
- The structure and composition of the organic matrix change with the degree of enamel maturation; amounts of proteins dwindle from the early secretory to the final maturative stages, reaching values of 0.1 wt % or less in mature enamel.
- Most of the organic matrix proteins in developing enamel are amelogenins.
- Amelogenins are glycosilated, phosphorylated, hydrophobic proteins which undergo self-assembly and, in developing enamel, form supramolecular aggregates which have been called 'nanospheres' on account of their shape.

- Changes in the molecular weight of amelogenins mainly derive from the enzymatic degradation processes that break proteins down in the secretory and maturative stages.
- Amelogenin can be extracted, or immunodetected, without prior decalcification, suggesting that it is located between crystals and is not masked by the inorganic substance.
- Non-amelogenin proteins, known generically as enamelins, comprise enamelin, ameloblastin (amelin, sheathlin), tuftelin and glycosilated and sulfated, 65-kDa protein.
- Enamelin is an acidic, glycosilated and phosphorylated protein whose molecular weight changes with its processing; the intact molecule is found only at the calcification front, while its cleavage products accumulate in the rod and interrod enamel and have a high affinity for the mineral.
- Enamelin can only be extracted, and is preferentially immunodetected, after decalcification, which suggests it has a close relationship with the inorganic crystals.
- Ameloblastin (also known as amelin or sheathlin) is a glycosylated protein whose molecular weight changes with molecule processing; a small cleavage product has been found localized in the interprismatic space, or ‘sheath’.
- One fraction of ameloblastin is mineral-bound, and another can only be extracted with sodium dodecyl sulfate, suggesting that it is present in vivo as insoluble aggregates.
- Tuftelin is an acidic protein which has been found in several non-calcifying tissues and is not specific to enamel.
- Proteolytic enzymes are components of the enamel matrix; they belong to two main classes, the metalloproteinases and the serine proteinases. The former interact with amelogenin from the stage of enamel formation to that of maturation, the latter remove the residual organic components. Other proteinases – as well as their inhibitors – may be present, too.
- Enamelysin, or matrix metalloproteinase-20 (MMP-20), is the most representative processing enzyme in the first class; serine proteinase 1, or kallikrein 4 (KLK4, or EMSP1), is the main enzyme in the second.
- Enamelysin is a proteolytic enzyme which degrades amelogenin by cleaving its C-terminal region, so triggering amelogenin interactions and self-assembly. It is highly active throughout the secretory amelogenesis stage.
- Kallikrein 4 is active during enamel maturation, when it clears the degraded proteins from enamel.

References

Açil Y, Mobasseri AE, Warnke PH, Terheyden H, Wiltfang J, Springer I (2005) Detection of mature collagen in human dental enamel. Calcif Tissue Int 76:121–126

Akita H, Fukae M, Shimoda S, Aoba T (1992) Localization of glycosylated matrix proteins in secretory porcine enamel and their possible functional roles in enamel mineralization. Archs Oral Biol 37:953–962

Amizuka N, Uchida T, Fukae M, Yamada M, Ozawa H (1992) Ultrastructural and immunocytochemical studies of enamel tufts in human permanent teeth. Arch Histol Cytol 55:179–190

Aoba T (1996) Recent observations on enamel crystal formation during mammalian amelogenesis. Anat Rec 245:208–218

Aoba T, Moreno EC (1991) Structural relationship of amelogenin proteins to their regulatory function of enamel mineralization. In: Sikes CS, Wheeler AP (eds) Surface reactive peptides and polymers. American Chemical Society, pp 85–106

Aoba T, Fukae M, Tanabe T, Shimizu M, Moreno EC (1987) Selective adsorption of porcine-amelogenins onto hydroxyapatite and their inhibitory activity on hydroxyapatite growth in supersaturated solutions. Calcif Tissue Int 41:281–289

Aoba T, Shimoda S, Shimokawa H, Inage T (1992) Common epitopes of mammalian amelogenins at the C-terminus and possible functional roles of the corresponding domain in enamel mineralization. Calcif Tissue Int 51:85–91

Bai P, Warshawsky H (1985) Morphological studies on the distribution of enamel matrix proteins using routine electron microscopy and freeze-fracture replicas in the rat incisor. Anat Rec 212:1–16

Bartlett JD, Simmer JP (1999) Proteinases in developing dental enamel. Crit Rev Oral Biol Med 10:425–441

Bartlett JD, Simmer JP, Xue J, Margolis HC, Moreno EC (1996) Molecular cloning and mRNA tissue distribution of a novel matrix metalloproteinase isolated from porcine enamel organ. Gene 183:123–128

Bashir MM, Abrams WR, Rosenbloom J (1997) Molecular cloning and characterization of the bovine tuftelin gene. Arch Oral Biol 42:489–496

Bashir MM, Abrams WR, Tucker T, Sellinger B, Budarf M, Emanuel B, Rosenbloom J (1998) Molecular cloning and characterization of the bovine and human tuftelin genes. Connect Tissue Res 39:13–24

Belcourt AB, Fincham AG, Termine JD (1982) Acid-soluble bovine fetal enamelins. J Dent Res 61:1031–1032

Beniash E, Simmer JP, Margolis HC (2005) The effect of recombinant mouse amelogenins on the formation and organization of hydroxyapatite crystals *in vitro*. J Struct Biol 149:182–190

Beynon AD (1976) A reappraisal of Sudan black staining of enamel matrix in mouse molar. Arch Oral Biol 21:83–90

Bishop MA, Warshawsky H (1982) Electron microscopic studies on the potential loss of crystallites from routinely processed sections of young enamel in the rat incisor. Anat Rec 202:177–186

Blumen G (1979) Preliminary autohistoradiographic studies about the possible participation of the secretory ameloblasts during the enamel maturation. Cell Mol Biol 24:143–150

Blumen G, Merzel J (1972) The decrease in the concentration of organic material in the course of formation of the enamel matrix. Experientia 28:545–548

Blumen G, Merzel J (1982) New evidences for the role of secretory ameloblasts in the removal of proline labelled proteins from young enamel as visualized by autoradiography. J Biol Buccale 10:73–83

Bonucci E (1995) Ultrastructural organic-inorganic relationships in calcified tissues: cartilage and bone vs. enamel. Connect Tissue Res 33:157–162

Bonucci E, Lozupone E, Silvestrini G, Favia A, Mocetti P (1994) Morphological studies of hypomineralized enamel of rat pups on calcium-deficient diet, and of its changes after return to normal diet. Anat Rec 239:379–395

Bouropoulos N, Moradian-Oldak J (2004) Induction of apatite by the cooperative effect of amelogenin and the 32-kDa enamelin. J Dent Res 83:278–282

Bronckers ALJJ, Lyaruu DM, Bervoets TJM, Wöltgens JHM (1988) Autoradiographic, ultrastructural and biosynthetic study of the effect of colchicine on enamel matrix secretion and enamel mineralization in hamster tooth germs *in vitro*. Arch Oral Biol 33:7–16

Bronckers ALJJ, Bervoets TJM, Lyaruu DM, Wöltgens JHM (1994) Degradation of hamster amelogenins during secretory stage enamel formation in organ culture. Matrix Biol 14:553–541

Brookes SJ, Robinson C, Kirkham J, Bonass WA (1995) Biochemistry and molecular biology of amelogenin proteins of developing dental enamel. Arch Oral Biol 40:1–14

Brookes SJ, Kirkham J, Shore RC, Bonass WA, Robinson C (1998) Enzyme compartmentalization during bophasic enamel matrix processing. Connect Tissue Res 39:89–99

Brookes SJ, Kirkham J, Lyngstadaas SP, Shore RC, Wood SR, Robinson C (2000) Spatially related amelogenin interactions in developing rat enamel as revealed by molecular cross-linking studies. Arch Oral Biol 45:937–943

Brookes SJ, Kirkham J, Shore RC, Wood SR, Slaby I, Robinson C (2001) Amelin extracellular processing and aggregation during rat incisor amelogenesis. Arch Oral Biol 46:201–208

Brookes SJ, Lyngstadaas SP, Robinson C, Shore RC, Kirkham J (2002) Enamelin compartmentalization in developing porcine enamel. Connect Tissue Res 43:477–481

Caron C, Xue J, Sun X, Simmer JP, Bartlett JD (2001) Gelatinase A (MMP-2) in developing tooth tissues and amelogenin hydrolysis. J Dent Res 80:1660–1664

Carter J, Smillie AA, Shepherd MJ (1989) Purification and properties of a protease from developing porcine dental enamel. Arch Oral Biol 34:399–404

Caterina J, Shi J, Sun X, Qian Q, Yamada S, Liu Y, Krakora S, Bartlett JD, Yamada Y, Engler JA, Birkedal-Hansen H, Simmer JP (2000) Cloning, characterization, and expression analysis of mouse enamelysin. J Dent Res 79:1697–1703

Chardin H, Londono I, Goldberg M (1990) Visualization of glycosaminoglycans in rat incisor extracellular matrix using a hyaluronidase-gold complex. Histochem J 22:588–594

Chen S, Eisenmann DR (1984) Ultrastructural study of the effects of fixation and fluoride injection on stippled material during amelogenesis in the rat. Arch Oral Biol 29:681–686

Chen W-Y, Nanci A, Smith CE (1995) Immunoblotting studies of artifactual contamination of enamel homogenates by albumin and other proteins. Calcif Tissue Int 57:145–151

Decker JD (1973) Fixation effects on the fine structure of enamel crystal-matrix relationships. J Ultrastruct Res 44:58–74

DenBensten PK, Heffeman LM (1989) Separation by polyacrylamide gel electrophoresis of multiple proteases in rat and bovine enamel. Arch Oral Biol 34:399–404

DenBesten PK, Heffernan LM, Treadwell BV, Awbrey BJ (1989) The presence and possible functions of the matrix metalloproteinase collagenase activator protein in developing enamel matrix. Biochem J 264:917–920

Deutsch D, Palmon A, Fisher LW, Kolodny N, Termine JD, Young MF (1991) Sequencing of bovine enamelin ("tuftelin") a novel acidic enamel protein. J Biol Chem 266:16021–16028

Deutsch D, Catalano-Sherman J, Dafni L, David S, Palmon A (1995a) Enamel matrix proteins and ameloblast biology. Connect Tissue Res 32:97–107

Deutsch D, Palmon A, Dafni L, Catalano-Sherman J, Young MF, Fisher LW (1995b) The enamelin (tuftelin) gene. Int J Dev Biol 39:135–143

Deutsch D, Dafni L, Palmon A, Hekmati M, Young MF, Fischer LW (1997) Tuftelin: enamel mineralization and amelogenesis imperfecta. Ciba Found Symp 205:135–147
Deutsch D, Palmon A, Dafni L, Mao Z, Leytin V, Young M, Fisher LW (1998) Tuftelin - aspects of protein and gene structure. Eur J Oral Sci 106:315–323
Deutsch D, Leiser Y, Shay B, Fermon E, Taylor A, Rosenfeld E, Dafni L, Charuvi K, Cohen Y, Haze A, Fuks A, Mao Z (2002) The human tuftelin gene and the expression of tuftelin in mineralizing and nonmineralizing tissues. Connect Tissue Res 43:425–434
Diekwisch TGH, Berman BJ, Gentner S, Slavkin HC (1995) Initial enamel crystals are not spatially associated with mineralized dentine. Cell Tissue Res 279:149–167
Diekwisch TGH, Ware J, Fincham AG, Zeichner-David M (1997) Immunohistochemical similarities and differences between amelogenin and tuftelin gene products during tooth development. J Histochem Cytochem 45:859–866
Dohi N, Murakami C, Tanabe T, Yamakoshi Y, Fukae M, Yamamoto Y, Wakida K, Shimizu M, Simmer JP, Kurihara H, Uchida T (1998) Immunocytochemical and immunochemical study of enamelins, using antibodies against porcine 89-kDa enamelin and its N-terminal synthetic peptide, in porcine tooth germs. Cell Tissue Res 293:313–325
Du C, Falini G, Fermani S, Abbott C, Moradian-Oldak J (2005) Supramolecular assembly of amelogenin nanospheres into birefringent microribbons. Science 307:1450–1454
Dunglas C, Septier D, Paine ML, Zhu DH, Snead ML, Goldberg M (2002) Ultrastructure of forming enamel in mouse bearing a transgene that disrupts the amelogenin self-assembly domains. Calcif Tissue Int 71:155–166
Eastoe JE (1963) The amino-acid composition of proteins from the oral tissues – II. The matrix proteins in dentine and enamel from developing human decisuous teeth. Arch Oral Biol 8:633–652
Eastoe JE (1968) Chemical aspects of the matrix concept in calcified tissue organisation. Calcif Tissue Res 2:1–19
Farge P, Ricard-Blum S, Joffre A, Ville G, Magloire H (1991) Immunoblotting and cytochemical characterization of human enamel proteins. Arch Oral Biol 36:89–94
Fearnhead RW (1961) Electron microscopy of forming enamel. Arch Oral Biol a:24–28
Fincham AG, Moradian-Oldak J (1995) Recent advances in amelogenin biochemistry. Connect Tissue Res 32:119–124
Fincham AG, Burkland GA, Shapiro IM (1972) Lipophilia of enamel matrix. A chemical investigation of the neutral lipids and lipophilic proteins of enamel. Calcif Tissue Res 9:247–259
Fincham AG, Bessem CC, Lau EC, Pavlova Z, Shuler C (1991a) Human developing enamel proteins exhibit a sex-linked dimorphism. Calcif Tissue Int 48:288–290
Fincham AG, Hu CC, Lau EC, Slavkin HC, Snead ML (1991b) Amelogenin post-secretory processing during biocalcification in the postnatal mouse molar tooth. Arch Oral Biol 36:305–317
Fincham AG, Moradian-Oldak J, Sarte PE (1994a) Mass-spectrographic analysis of a porcine amelogenin identifies a single phosphorylated locus. Calcif Tissue Int 55:398–400
Fincham AG, Moradian-Oldak J, Simmer JP, Sarte P, Lau P, Diekwisch T, Slavkin HC (1994b) Self-assembly of a recombinant amelogenin protein generates supramolecular structures. J Struct Biol 112:103–109
Fincham AG, Moradian-Oldak J, Diekwisch TGH, Lyaruu DM, Wright JT, Bringas P Jr, Slavkin HC (1995) Evidence for amelogenin "nanospheres" as functional components of secretory-stage enamel. J Struct Biol 115:50–59
Fincham AG, Leung W, Tan J, Moradian-Oldak J (1998) Does amelogenin nanosphere assembly proceed through intermediary-sized structures? Connect Tissue Res 38:237–240

Fincham AG, Moradian-Oldak J, Simmer JP (1999) The structural biology of the developing dental enamel matrix. J Struct Biol 126:270–299
Fong CD, Slaby I, Hammarström L (1996a) Amelin: an enamel-related protein, transcribed in the cells of epithelial root sheath. J Bone Miner Res 11:892–898
Fong CD, Hammarström L, Lundmark C, Wurtz T, Slaby I (1996b) Expression patterns of RNAs for amelin and amelogenin in developing rat molars and incisors. Adv Dent Res 10:195–200
Fong CD, Cerny R, Hammarström L, Slaby I (1998) Sequential expression of an amelin gene in mesenchymal and epithelial cells during odontogenesis in rats. Eur J Oral Sci 106:324–330
Frank RM (1979) Tooth enamel: current state of the art. J Dent Res 58:684–693
Frederiks WM (1977) Some aspects of the value of Sudan black B in lipid histochemistry. Histochemistry 54:27–37
Fukae M, Tanabe T (1987) Nonamelogenin components of porcine enamel in the protein fraction free from the enamel crystals. Calcif Tissue Int 40:286–293
Fukae M, Tanabe T (1998) Degradation of enamel matrix proteins in porcine secretory enamel. Connect Tissue Res 39:123–129
Fukae M, Tanabe T, Uchida T, Yamakoshi Y, Shimizu M (1993) Enamelins in the newly formed bovine enamel. Calcif Tissue Int 53:257–261
Fukae M, Tanabe T, Marukami C, Dohi N, Uchida T, Shimizu M (1996) Primary structure of the porcine 89-kDa enamelin. Adv Dent Res 10:111–118
Fukae M, Tanabe T, Uchida T, Lee SK, Ryu OH, Murakami C, Wakida K, Simmer JP, Yamada Y, Bartlett JD (1998) Enamelysin (matrix metalloproteinase-20): localization in the developing tooth and effects of pH and calcium on amelogenin hydrolysis. J Dent Res 77:1580–1588
Fukae M, Tanabe T, Nagano T, Ando H, Yamakoshi Y, Yamada M, Simmer JP, Oida S (2002) Odontoblasts enhance the maturation of enamel crystals by secreting EMSP1 at the enamel-dentin junction. J Dent Res 81:668–672
Garant PR (1970) An electron microscopic study of the crystal-matrix relationship in the teeth of the dogfish *Squalus acanthias* L. J Ultrastruct Res 30:441–449
Gibson CW, Kucick U, Collier P, Shen G, Decker S, Bashir M, Rosenbloom J (1995) Analysis of amelogenin proteins using monospecific antibodies to define sequences. Connect Tissue Res 32:109–114
Girija V, Stephen HC (2003) Characterization of lipid in mature enamel using confocal laser scanning microscopy. J Dent 31:303–311
Glimcher MJ, Brickley-Parsons D, Levine PT (1977) Studies of enamel proteins during maturation. Calcif Tissue Res 24:259–270
Goldberg M, Septier D (1986) Ultrastructural location of complex carbohydrates in developing rat incisor enamel. Anat Rec 216:181–190
Goldberg M, Septier D (2002) Phospholipids in amelogenesis and dentinogenesis. Crit Rev Oral Biol Med 13:276–290
Goldberg M, Genotelle-Septier D, Molon-Noblot M, Weill R (1978) Ultrastructural study of the proteoglycans in enamel from rat incisors during late enamel maturation. Arch Oral Biol 23:1007–1011
Goldberg M, Lelous M, Escaig F, Boudin M (1983) Lipids in the developing enamel of the rat incisor. Parallel histochemical and biochemical investigations. Histochemistry 78:145–156
Goldberg M, Septier D, Lécolle S, Chardin H, Quintana MA, Acevedo AC, Gafni G, Dillouya D, Vermelin L, Thonemann B, Schmalz G, Bissila-Mapahou P, Carreau JP (1995) Dental mineralization. Int J Dev Biol 39:93–110

Goldberg M, Vermelin L, Mostermans P, Lécolle S, Septier D, Godeau G, LeGeros RZ (1998) Fragmentation of the distal portion of Tomes' processes of secretory ameloblasts in the forming enamel of rat incisors. Connect Tissue Res 38:159–169

Gorter de Vries I, Ameloot PC, Coomans D, Wisse E (1986) An ultrastructural study of dentinogenesis and amelogenesis in rat molar tooth germs cultured in vitro. Cell Tissue Res 246:623–634

Hayashi Y (1989) Ultrastructural demonstration of the carbohydrate in developing rat enamel using soybean agglutinin-gold complexes. Arch Oral Biol 34:517–522

Hayashi Y, Bianco P, Shimokawa H, Termine JD, Bonucci E (1986) Organic-inorganic relationships, and immunohistochemical localization of amelogenins and enamelins in developing enamel. Basic Appl Histochem 30:291–299

Herold RC, Rosenbloom J (1990) Immunocytochemical localization of enamelin proteins in developing bovine teeth. Arch Oral Biol 35:373–379

Herold RC, Boyde A, Rosenbloom J, Lally ET (1987) Monoclonal antibody and immunogold cytochemical localization of amelogenins in bovine secretory amelogenesis. Arch Oral Biol 32:439–444

Hu C-C, Fukae M, Uchida T, Qian Q, Zhang CH, Ryu OH, Tanabe T, Yamakoshi Y, Murakami C, Dohi N, Shimizu M, Simmer JP (1997a) Cloning and characterization of porcine enamelin mRNAs. J Dent Res 76:1720–1729

Hu C-C, Fukae M, Uchida T, Qian Q, Zhang CH, Ryu OH, Tanabe T, Yamakoshi Y, Murakami C, Dohi N, Shimizu M, Simmer JP (1997b) Sheathlin: cloning, cDNA/polypeptide sequences, and immunolocalization of porcine enamel sheath proteins. J Dent Res 76:648–657

Hu C-C, Simmer JP, Bartlett JD, Qian Q, Zhang C, Ryu OH, Xue J, Fukae M, Uchida T, MacDougall M (1998) Murine enamelin: cDNA and derived protein sequences. Connect Tissue Res 39:47–61

Hu JC, Sun X, Zhang C, Simmer JP (2001) A comparison of enamelin and amelogenin expression in developing mouse molars. Eur J Oral Sci 109:125–132

Inage T, Shimokawa H, Teranishi Y, Iwase T, Toda Y, Moro I (1989) Immunocytochemical demonstration of amelogenins and enamelins secreted by ameloblasts during the secretory and maturation stages. Arch Histol Cytol 52:213–229

Inai T, Kukita T, Ohsaki Y, Nagata K, Kukita A, Kurisu K (1991) Immunohistochemical demonstration of amelogenin penetration toward the dental pulp in the early stages of ameloblast development in rat molar tooth germs. Anat Rec 229:259–270

Irving JT (1963) The sudanophil material at sites of calcification. Arch Oral Biol 8:735–745

Jodaikin A, Traub W, Weiner S (1986) Protein conformation in rat tooth enamel. Arch Oral Biol 31:685–689

Kallenbach E (1971) Electron microscopy of the differentiating rat incisor ameloblast. J Ultrastruct Res 35:508–531

Karg HA, Burger EH, Lyaruu DM, Wöltgens JHM, Bronckers ALJJ (1997) Gene expression and immunolocalisation of amelogenins in developing embryonic and neonatal hamster teeth. Cell Tissue Res 288:545–555

Kemp NE, Park JH (1974) Ultrastructure of the enamel layer in developing teeth of the shark *Carcharhinus menisorrah*. Arch Oral Biol 19:633–644

Kogaya Y, Furuhashi K (1988) Sulfated glycoconjugates in rat incisor secretory ameloblasts and developing enamel matrix. Calcif Tissue Int 43:307–318

Krebsbach PH, Lee SK, Matsuki Y, Kozak CA, Yamada KM, Yamada Y (1996) Full-length sequence, localization, and chromosomal mapping of ameloblastin. A novel tooth-specific gene. J Biol Chem 271:4431–4435

Lau EC, Simmer JP, Bringas P, Hu CC, Zeichner-David M, Thiemann F (1992) Alternative splicing of the mouse amelogenin primary mRNA transcript contributes to amelogenin heterogeneity. Biochem Biophys Res Comm 188:1253–1260

Lee SK, Krebsbach PH, Matsuki Y, Nanci A, Yamada KM, Yamada Y (1996) Ameloblastin expression in rat incisors and human tooth germs. Int J Dev Biol 40:1141–1150

Lee SK, Kim SM, Lee YJ, Yamada KM, Chi JG (2003) The structure of the rat ameloblastin gene and its expression in amelogenesis. Mol Cells 15:216–225

Lesot H, Smith AJ, Matthews JB, Ruch J-V (1988) An extracellular matrix protein of dentine, enamel, and bone shares common antigenic determinants with keratins. Calcif Tissue Int 42:53–57

Limeback H, Simic A (1990) Biochemical characterization of stable high molecular-weight aggregates of amelogenins formed during porcine enamel development. Arch Oral Biol 35:459–468

Limeback H, Sakarya H, Chu W, Mackinnon M (1989) Serum albumin and its acid hydrolysis peptides dominate preparations of mineral-bound enamel proteins. J Bone Miner Res 4:235–241

Llano E, Pendas AM, Knauper V, Sorsa T, Salo T, Salido E, Murphy G, Simmer JP, Bartlett JD, Lopez-Otin C (1997) Identification and structural and functional characterization of human enamelysin (MMP-20). Biochemistry 36:15101–15108

Lyaruu DM, Belcourt A, Fincham AG, Termine JD (1982) Neonatal hamster molar tooth development: extraction and characterization of amelogenins, enamelins, and soluble dental proteins. Calcif Tissue Int 34:86–96

Lyaruu DM, Bronckers ALJJ, Wöltgens JHM (1984) The Tomes' process: is stippled material a reality? The effect of fixative temperature on the structure of stippled material in enamel. In: Belcourt AB, Ruch J-V (eds) Morphogenèse et différenciation dentaires, Coll. INSERM 125, 1984. INSERM, Paris, pp 257–272

MacDougall M, Simmons D, Dodds A, Knight C, Luan X, Zeichner-David M, Zhang C, Ryu OH, Qian Q, Simmer JP, Hu CC (1998) Cloning, characterization, and tissue expression pattern of mouse tuftelin cDNA. J Dent Res 77:1970–1978

Mao Z, Shay B, Hekmati M, Fermon E, Taylor A, Dafni L, Heikinheimo K, Lustmann J, Fisher LW, Young MF, Deutsch D (2001) The human tuftelin gene: cloning and characterization. Gene 279:181–196

Moe D, Birkedal-Hansen H (1979) Proteolytic acctivity in developing bovine enamel. J Dent Res 58 B:1012–1013

Moe D, Kirkeby S (1990) Non-specific esterases in partly mineralized bovine enamel. Acta Odontol Scand 48:327–332

Moradian-Oldak J (2001) Amelogenins: assembly, processing and control of crystal morphology. Matrix Biol 20:293–305

Moradian-Oldak J, Simmer JP, Sarte PE, Zeichner-David M, Fincham AG (1994a) Specific cleavage of a recombinant murine amelogenin at the carboxy-terminal region by a proteinase fraction isolated from developing bovine tooth enamel. Arch Oral Biol 39:647–656

Moradian-Oldak J, Simmer JP, Lau EC, Sarte PE, Slavkin HC, Fincham AG (1994b) Detection of monodisperse aggregates of a recombinant amelogenin by dynamic light scattering. Biopolymers 34:1339–1347

Moradian-Oldak J, Simmer JP, Lau EC, Diekwisch T, Slavkin HC, Fincham AG (1995) A review of the aggregation properties of a recombinant amelogenin. Connect Tissue Res 32:125–130

Moradian-Oldak J, Leung W, Simmer JP, Zeichner-David M, Fincham AG (1996a) Identification of a novel proteinase (ameloprotease-I) responsible for the complete degradation of amelogenin during enamel maturation. Biochem J 318:1015–1021

Moradian-Oldak J, Sarte PE, Fincham AG (1996b) Description of two classes of proteinases from enamel extracellular matrix cleaving a recombinant amelogenin. Connect Tissue Res 35:231–238
Moradian-Oldak J, Leung W, Fincham AG (1998) Temperature and pH-dependent supramolecular self-assembly of amelogenin molecules: a dynamic light-scattering analysis. J Struct Biol 122:320–327
Moradian-Oldak J, Paine ML, Lei YP, Fincham AG, Snead ML (2000) Self-assembly properties of recombinant engineered amelogenin proteins analyzed by dynamic light scattering and atomic force microscopy. J Struct Biol 131:27–37
Moradian-Oldak J, Jimenez J, Maltby D, Fincham AG (2001) Controlled proteolysis of amelogenins reveals exposure of both carboxy- and amino-terminal regions. Biopolymers 58:606–616
Moradian-Oldak J, Gharakhanian N, Jimenez J (2002) Limited proteolysis of amelogenin: toward understanding the proteolytic processes in enamel extracellular matrix. Connect Tissue Res 43:450–455
Nanci A (2003) Ten Cate's oral histology: development, structure, and function, 6th edn. Mosby, St. Louis
Nanci A, Smith CE (1992) Development and calcification of enamel. In: Bonucci E (ed) Calcification in biological systems. CRC Press, Boca Raton, pp 313–343
Nanci A, Warshawsky H (1984) Relationship between the quality of fixation and the presence of stippled material in newly formed enamel of the rat incisor. Anat Rec 208:15–31
Nanci A, Bai P, Warshawsky H (1983) The effect of osmium postfixation and uranyl and lead staining on the ultrastructure of young enamel in the rat incisor. Anat Rec 207:1–16
Nanci A, Bendayan M, Bringas P, Slavkin HC (1984) High resolution immunocytochemical localization of enamel proteins in mouse ameloblasts *in situ* and in culture. In: Belcourt AB, Ruch J-V (eds) Morphogenèse et différenciation dentaires. INSERM, vol 125, Paris, pp 333–339
Nanci A, Ahluwalia JP, Pompura JR, Smith CE (1989a) Biosynthesis and secretion of enamel proteins in the rat incisor. Anat Rec 224:277–291
Nanci A, Ahluwalia JP, Zalzal S, Smith CE (1989b) Cytochemical and biochemical characterization of glycoproteins in forming and maturing enamel of the rat incisor. J Histochem Cytochem 37:1619–1633
Nanci A, McKee MD, Smith CE (1992) Immunolocalization of enamel proteins during amelogenesis in the cat. Anat Rec 233:335–349
Nanci A, Kawaguchi H, Kogaya Y (1994) Ultrastructural studies and immunolocalization of enamel proteins in rodent secretory stage ameloblasts processed by various cryofixation methods. Anat Rec 238:425–436
Nanci A, Zalzal S, Lavoie P, Kunikata M, Chen W-Y, Krebsbach PH, Yamada Y, Hammarström L, Simmer JP, Fincham AG, Snead ML, Smith CE (1998) Comparative immunochemical analyses of the developmental expression and distribution of ameloblastin and amelogenin in rat incisors. J Histochem Cytochem 46:911–934
Nanci A, Mocetti P, Sakamoto Y, Kunikata M, Lozupone E, Bonucci E (2000) Morphological and immunocytochemical analyses on the effects of diet-induced hypocalcemia on enamel maturation in the rat incisor. J Histochem Cytochem 48:1043–1057
Nefussi J-R, Septier D, Sautier J-M, Forest N, Goldberg M (1992) Localization of malachite green positive lipids in the matrix of bone nodule formed *in vitro*. Calcif Tissue Int 50:273–282
Nylen MU (1979) Matrix-mineral relationships - a morphologist's viewpoint. J Dent Res 58:922–926

Nylen MU, Omnell K-Å (1962) The relationship between the apatite crystals and the organic matrix of rat enamel. Fifth International Congress for Electron Microscopy, Academic Press, New York, p QQ-4

Ogata Y, Shimokawa H, Sasaki S (1988) Purification, characterization, and biosynthesis of bovine enamelins. Calcif Tissue Int 43:389–399

Orams H (1981) Ultrastructural sites of alkaline phosphatase activity during amelogenesis in forming rodent incisors. In: Veis A (ed) The chemistry and biology of mineralized connective tissues. Elsevier North Holland, Amsterdam, pp 571–575

Orsini G, Lavoie P, Smith CE, Nanci A (2001) Immunochemical characterization of a chicken egg yolk antibody to secretory forms of rat incisor amelogenin. J Histochem Cytochem 49:285–292

Overall CM, Limeback H (1988) Identification and characterization of enamel proteinases isolated from developing enamel. Amelogeninolytic serine proteinases are associated with enamel maturation in pig. Biochem J 256:965–972

Paine CT, Paine ML, Luo W, Okamoto T, Lyngstadaas SP, Snead ML (2000) A tuftelin-interacting protein (TIP39) localizes to the apical secretory pole of mouse ameloblasts. J Biol Chem 275:22284–22292

Paine ML, Snead ML (2005) Ultrastructure of forming enamel in mouse bearing a transgene that disrupts the amelogenin self-assembly domains. Orthodont Craniofac Res 8:239–251

Paine ML, Krebsbach PH, Chen LS, Paine CT, Yamada Y, Deutsch D, Snead ML (1998) Protein-to-protein interactions: criteria defining the assembly of the enamel organic matrix. J Dent Res 77:496–502

Papagerakis P, MacDougall M, Hotton D, Bailleul-Forestier I, Oboeuf M, Berdal A (2003) Expression of amelogenin in odontoblasts. Bone 32:228–240

Ravindranath HH, Chen L-S, Zeichner-David M, Ishima R, Ravindranath RM (2004) Interaction between the enamel matrix proteins amelogenin and ameloblastin. Biochem Biophys Res Comm 323:1075–1083

Reith EJ (1967) The early stage of amelogenesis as observed in molar teeth of young rats. J Ultrastruct Res 17:503–526

Reith EJ, Cotty VF (1967) The absorptive activity of ameloblasts during the maturation of enamel. Anat Rec 157:577–588

Robinson C, Kirkham J (1985) Dynamics of amelogenesis as revealed by protein compositional studies. In: Butler WT (ed) The chemistry and biology of mineralized tissues. Ebsco Media, Inc., Birmingham, AL, pp 248–263

Robinson C, Shore RC, Kirkham J (1989a) Tuft protein: its relationship with the keratins and the developing enamel matrix. Calcif Tissue Int 44:393–398

Robinson C, Kirkham J, Fincham A (1989b) The enamelin/non-amelogenin problem. A brief review. Connect Tissue Res 22:93–100

Robinson C, Kirkham J, Storehouse NJ, Shore RC (1989c) Extracellular processing of enamel matrix and origin and function of Tuft protein. In: Fearnhead RW (ed) Tooth enamel V. Florence Publishers, Yokohama, pp 59–63

Robinson C, Shore RC, Kirkham J, Stonehouse NJ (1990) Extracellular processing of enamel matrix proteins and the control of crystal growth. J Biol Buccale 18:355–361

Robinson C, Kirkham J, Brookes SJ, Bonass WA, Shore RC (1995) The chemistry of enamel development. Int J Dev Biol 39:145–152

Robinson C, Brookes SJ, Kirkham J, Bonass WA, Shore RC (1996) Crystal growth in dental enamel: the role of amelogenins and albumin. Adv Dent Res 10:173–179

Robinson C, Brookes SJ, Shore RC, Kirkham J (1998) The developing enamel matrix: nature and function. Eur J Oral Sci 106:282–291

Robinson C, Kirkham J, Shore RC, Brookes SJ, Wood SR (2000) Enamel matrix function and the tuft enigma: a role in directing enamel tissue-architecture: a partial sequence of human ameloblastin. In: Goldberg M, Boskey A, Robinson C (eds) Chemistry and biology of mineralized tissues. American Academy of Orthopaedic Surgeons, Rosemont, IL, pp 209–213
Rönnholm E (1962) The structure of the organic stroma of human enamel during amelogenesis. J Ultrastruct Res 3:368–389
Rosenbloom J, Lally E, Dixon M, Spencer A, Herold R (1986) Production of monoclonal antibody to enamelins which does not cross-react with amelogenins. Calcif Tissue Int 39:412–415
Ryu OH, Fincham AG, Hu C-C, Zhang C, Qian Q, Bartlett JD, Simmer JP (1999) Characterization of recombinant pig enamelysin activity and cleavage of recombinant pig and mouse amelogenins. J Dent Res 78:743–750
Ryu OH, Hsiung DT, Hu C-C, Sun X, Cao X, Bartlett JD, Simmer JP (2000) The structure and function of enamelysin (MMP-20). In: Goldberg M, Boskey A, Robinson C (eds) Chemistry and biology of mineralized tissues. American Academy of Orthopaedic Surgeons, Rosemont, IL, pp 363–367
Sakakura Y (1986) A new culture method assuring the three-dimensional development of the mouse embryonic molar tooth *in vitro*. Calcif Tissue Int 39:271–278
Sawada T, Nanci A (1995) Spatial distribution of enamel proteins and fibronectin at early stages of rat incisor tooth formation. Arch Oral Biol 40:1029–1038
Seyer JM, Glimcher MJ (1977) Evidence for the presence of numerous protein components in immature bovine dental enamel. Calcif Tissue Res 24:253–257
Shapiro IM, Wuthier RE, Irving JT (1966) A study of the phospholipids of bovine dental tissues. I. Enamel matrix and dentine. Arch Oral Biol 11:501–512
Shimizu M, Tanabe T, Fukae M (1979) Proteolytic enzyme in porcine immature enamel. J Dent Res 58B:782–789
Simmer JP (1995) Alternative splicing of amelogenins. Connect Tissue Res 32:131–136
Simmer JP, Fincham AG (1995) Molecular mechanisms of dental enamel formation. Crit Rev Oral Biol Med 6:84–108
Simmer JP, Hu JC (2002) Expression, structure, and function of enamel proteinases. Connect Tissue Res 43:441–449
Simmer JP, Hu CC, Sarte P, Slavkin HC, Fincham AG (1994) Alternative splicing of the mouse amelogenin primary RNA transcript. Calcif Tissue Int 55:302–310
Simmer JP, Fukae M, Tanabe T, Yamakoshi Y, Uchida T, Xue J, Margolis HC, Shimizu M, DeHart BC, Hu CC, Bartlett JD (1998) Purification, characterization, and cloning of enamel matrix serine proteinase 1. J Dent Res 77:377–386
Simmons D, Gu TT, Krebsbach PH, Yamada Y, MacDougall M (1998) Identification and characterization of a cDNA for mouse ameloblastin. Connect Tissue Res 39:3–12
Slavkin HC, Bessem C, Bringas P, Zeichner-David M, Nanci A, Snead ML (1988) Sequential expression and differential function of multiple enamel proteins during fetal neonatal and early postnatal stages of mouse molar organogenesis. Differentiation 37:26–39
Smales FC (1975) Structural subunit in prisms of immature rat enamel. Nature 258:772–774
Smith AJ (1985) Histochemistry of enamel. In: Belcourt AB, Ruch J-R (eds) Morphogenèse et différenciation dentaires, Coll. INSERM 125, 1984. INSERM, Paris, pp 307–314
Smith CE, Nanci A (1996) Protein dynamics of amelogenesis. Anat Rec 245:186–207
Smith CE, Pompura JR, Borenstein S, Fazel A, Nanci A (1989) Degradation and loss of matrix proteins from developing enamel. Anat Rec 224:292–316
Smith CE, Dahan S, Fazel A, Lai W, Nanci A (1992) Correlated biochemical and radioautographic studies of protein turnover in developin rat incisor enamel following pulse-chase labeling with L-[35S]- and L-[methyl-3H] methionine. Anat Rec 232:1–14

Smith CE, Chen W-Y, Issid M, Fazel A (1995) Enamel matrix protein turnover during amelogenesis: basic biochemical properties of short-lived sulfated enamel proteins. Calcif Tissue Int 57:133–144

Sobel AE, Burger M (1954) Calcification XIV. Investigation of the role of chondroitin sulfate in the calcifying mechanism. Proc Soc Exper Biol Med 87:7–13

Strawich E, Glimcher MJ (1989) Major "enamelin" protein in enamel of developing bovine teeth is albumin. Connect Tissue Res 22:111–121

Strawich E, Glimcher MJ (1990) Tooth 'enamelins' identified mainly as serum proteins. Major 'enamelin' is albumin. Eur J Biochem 191:47–56

Strawich E, Seyer J, Glimcher MJ (1993) Immuno-identification of two non-amelogenin proteins of developing bovine enamel isolated by affinity chromatography. Further proof that tooth "enamelins" are mainly serum proteins. Connect Tissue Res 29:163–169

Suga S, Murayama Y, Musashi T (1970) A study of the mineralization process in the developing enamel of guinea pig. Arch Oral Biol 15:597–612

Takano Y, Ozawa H, Crenshaw MA (1986) Ca-ATPase and ALPase activities at the initial calcification sites of dentin and enamel in the rat incisor. Cell Tissue Res 243:91–99

Tanabe T, Fukae M, Uchida T, Shimizu M (1992) The localization and characterization of proteinases for the initial cleavage of porcine amelogenin. Calcif Tissue Int 51:213–217

Tanabe T, Fukae M, Shimizu M (1994) Degradation of enamelins by proteinases found in porcine secretory enamel *in vitro*. Archs Oral Biol 39:277–281

Termine JD, Belcourt AB, Christner PJ, Conn KM, Nylen MU (1980) Properties of dissociatively extracted fetal tooth matrix proteins I. Principal molecular species in developing bovine enamel. J Biol Chem 255:9760–9768

Traub W, Jodaikin A, Weiner S (1985) Diffraction studies of enamel protein-mineral structural relations. In: Butler WT (ed) The chemistry and biology of mineralized tissues. Ebsco Media, Inc., Birmingham, pp 221–225

Travis DF (1968) Comparative ultrastructure and organization of inorganic crystals and organic matrices of mineralized tissues. Biology of the mouth. American Association for the Advancement of Sciences, Washington, pp 237–297

Travis DF, Glimcher MJ (1964) The structure and organization of, and the relationship between the organic matrix and the inorganic crystals of embryonic bovine enamel. J Cell Biol 23:447–497

Uchida T, Tanabe T, Fukae M (1989) Immunocytochemical localization of amelogenins in the deciduous tooth germs of the human fetus. Arch Histol Cytol 52:543–552

Uchida T, Tanabe T, Fukae M, Shimizu M, Yamada M, Miake K, Kobayashi S (1991a) Immunochemical and immunohistochemical studies, using antisera against porcine 25 kDa amelogenins, 89 kDa enamelin and the 13–17 kDa nonamelogenins, on immature enamel of the pig and rat. Histochemistry 96:129–138

Uchida T, Tanabe F, Fukae M, Shimizu M (1991b) Immunocytochemical and immunochemical detection of a 32 kDa nonamelogenin and related proteins in porcine tooth germs. Arch Histol Cytol 54:527–538

Uchida T, Fukae M, Tanabe T, Yamakoshi Y, Satoda T, Murakami C, Takahashi O, Shimizu M (1995) Immunochemical and immunocytochemical study of the 15kDA non-amelogenin and related proteins in the porcine immature enamel: proposal for a new group of enamel-proteins "sheath proteins". Biomed Res 16:131–140

UchidaT, Muratami C, Dohi N, Wakida K, Satoda T, Takahashi O (1997) Synthesis, secretion, degradation, and fate of ameloblastin during the matrix formation stage of the rat incisor as shown by immunocytochemistry and immunochemistry using region-specific antibodies. J Histochem Cytochem 45:1329–1340

Uchida T, Murakami C, Wakida K, Dohi N, Iwai Y, Simmer JP, Fukae M, Satoda T, Takahashi O (1998) Sheath proteins: synthesis, secretion, degradation and fate in forming enamel. Eur J Oral Sci 106:308–314

Veis A (2003) Amelogenin gene splice products: potential signaling molecules. Cell Mol Life Sci 60:38–55

Wakida K, Amizuka N, Murakami C, Satoda T, Fukae M, Simmer JP, Ozawa H, Uchida T (1999) Maturation ameloblasts of the porcine tooth germ do not express amelogenin. J Histochem Cytochem 111:297–303

Warshawsky H (1971) A light and electron microscopic study of the nearly mature enamel of rat incisors. Anat Rec 169:559–584

Watson ML (1960) The extracellular nature of enamel in the rat. J Biophys Biochem Cytol 7:489–492

Wen HB, Moradian-Oldak J, Leung W, Bringas P Jr, Fincham AG (1999) Microstructures of an amelogenin gel matrix. J Struct Biol 126:42–51

Wen HB, Fincham AG, Moradian-Oldak J (2001) Progressive accretion of amelogenin molecules during nanospheres assembly revealed by atomic force microscopy. Matrix Biol 20:387–395

Wurtz T, Lundmark C, Christersson C, Bawden JW, Slaby I, Hammarström L (1996) Expression of amelogenin mRNA sequences during development of rat molars. J Bone Miner Res 11:125–131

Yamada M, Bringas P Jr, Grodin M, MacDougall M, Slavkin HC (1980) Developmental comparisons of murine secretory amelogenesis in vivo, as xenografts on the chick chorio-allantoic membrane, and in vitro. Calcif Tissue Int 31:161–171

Yamakoshi Y, Tanabe T, Fukae M, Shimizu M (1994) Porcine amelogenesis. Calcif Tissue Int 54:69–75

Zeichner-David M, Vo H, Tan H, Diekwisch T, Berman B, Thiemann F, Alcocer D, Hsu P, Wang T, Eyna J, Caton J, Slavkin HC, MacDougall M (1997) Timing of the expression of enamel gene products during mouse tooth development. Int J Dev Biol 41:27–38

11 Calcifying Matrices: Lower Vertebrates

11.1 Introduction

Lower vertebrates have a double type of calcifying matrix, that of mesodermal origin pertaining to the endoskeleton, and that of ecto-mesodermal origin pertaining to the integumental or dermal skeleton. There are a few points of contact between the matrix of the integumental skeleton and that of mammalian teeth, and these may offer a basis for the study of the evolution of mineralized tissues in vertebrates (Kawasaki et al. 2004).

11.2 Endoskeleton

The endoskeleton of lower vertebrates is highly distinctive with regard to its first (anatomical level) and second (histological level) order structures, whereas its third (cytological level) and fourth (submicroscopic level) order structures appear similar to those of the highest vertebrates (for a review on the comparative microstructure of bone, and for a discussion on the four levels of integration, see de Ricqlès et al. 1991). For the purposes of this book, the structures of major interest belong to the fourth order and attention will be mainly directed to them.

The organic matrix of the endoskeleton of lower vertebrates mainly consists of collagen fibrils (reviewed by Kimura 1985), which are fundamentally homologous with the two types of collagen – types I and II – most often detectable in the skeletons of higher vertebrates (Bigi et al. 2000): type II collagen is the main constituent of cartilaginous skeletal segments and type I collagen that of the osseous structures. Type I collagen is also one of the main components of lower vertebrate skin (Kimura et al. 1981).

As in higher vertebrates, collagen fibrils may be arranged in a disorderly way (woven bone) or in almost parallel bundles (parallel-fibered bone, lamellar bone). Periosteal ossification and endochondral ossification can, clearly, give rise to distinct calcified tissues. Vascular canals may be absent, acellular bone (or aspidin) can be found in most advanced teleosts, chondroid bone, i. e., tissue whose properties are to some extent intermediate between bone and cartilage, is no rarity, and during periosteal ossification non-ossified spaces containing loose connective tissue may remain entrapped in cancellous bone. As a result, different skeletal segments from the same individual may display quite distinct morphological arrangements. They do, however, show an extensive homoplasy (parallelism or convergence) at the bone tissue level of organization, close similarity in their fine structure, and their third and fourth order structures appear to be very stable right through their evolution (de Ricqlès et al. 1991).

The calcified matrix of the endoskeleton of lower vertebrates can be considered similar to that of the highest vertebrates, at least as far as the fourth order structures are concerned: it mainly consists of collagen fibrils of different densities and of non-collagenous material, and there are good reasons to believe that its calcification occurs in the same way as in the skeletons of the highest vertebrates. The same seems to be true of dental structures, as suggested by the fact that reptilian and amphibian amelogenins have been cloned and sequenced (Wang et al. 2005), and that amelogenin has been immunodetected in the secretory enamel of developing tooth germs in the snake, *Elaphe quadrivirgata* (Ishiyama et al. 1998).

11.3 Dermal (or Integumental) Skeleton

The dermal skeleton of lower vertebrates shows a high species variability and is marked by a number of polymorphic structures which are the outcome of a long evolutionary process, comprising scales, scutes, fin rays, spines and osteoderms, to which teeth and odontodes – superficial tooth-like structures – can be added (for a review see Zylberberg et al. 1992). The high degree of polymorphism of these structures impedes detailed analysis of the organic matrix organization in each of them; on the other hand, the relative simplicity of their constituents – essentially, collagenous and non-collagenous material – allows a simplification to be made by limiting description to generalized models. Fish scales can be taken as paradigmatic; a few calcified tissues typical of the dermal skeleton and unknown in the endoskeleton will be discussed in some detail below.

The main calcifiable components of fish scales (elasmoid scales) are a basal fibrillary layer with a highly variable degree of calcification, which is able to reveal the histological and histochemical characteristics of lamellar bone (known as 'isopedine'), and a superficial layer consisting of woven bone. This contains acid proteoglycans and can be variably stained with PAS, alcian blue at low pH, and ruthenium red; it is slightly metachromatic with toluidine blue (Zylberberg and Nicolas 1982). In teleosts, this layer can be divided into two parts, the "external layer", consisting of a loose network of thin collagen fibrils and interfibrillar proteoglycans (Zylberberg 1985), and the "outer limiting layer", which corresponds to the outer surface of the scale, is devoid of collagen fibrils and is rich in proteoglycans (Schönbörner et al. 1979; Zylberberg and Nicolas 1982). Bundles of thin (under 30 nm in diameter) collagen fibrils –

so-called 'TC fibres' (Onozato and Watabe 1979) – rise from this layer, run perpendicularly through the fibrillary plate and connect the scale to the dermis. These fibrils calcify early than those that form the fibrillary plate (Onozato and Watabe 1979; Schönbörner et al. 1979).

Matrix vesicles have been reported in ossification areas in the rainbow trout and in salmon scales (Kobayashi et al. 1972), in the external layer (Kobayashi et al. 1972; Schönbörner et al. 1979; Zylberberg and Nicolas 1982; Zylberberg et al. 1992), and in regenerating scales (Sire and Géraudie 1984), whereas they have not been found in the scales of *Fundulus heteroclitus* (Yamada and Watabe 1979) or in the developing scales of the newly hatched fry of the sheepshead minnow, *Cyprinodon variegatus* (Olson and Watabe 1980). Many matrix vesicles have been reported in the dentin of tooth germs in the killifish *Oryzias latipes* (Yamada and Ozawa 1978).

The basal fibrillary plate varies in its degree of calcification. In some cases it is structurally similar to lamellar bone (isopedine), but in others is completely uncalcified (Meunier and François 1980), with a range of intermediate degrees of calcification (Onozato and Watabe 1979; Schönbörner et al. 1979). The lamellae in isopedine consist of collagen fibrils arranged in parallel, whose thickness changes with species (see Table 2 in Zylberberg et al. 1992) and whose direction too changes from one lamella to the next, so that they show a plywood-like arrangement (Brown and Wellings 1969; Meunier and Géraudie 1980; Meunier 1984; Bigi et al. 2001) like that found in osteons (those termed alternate, type 2, osteons; Ascenzi and Bonucci 1968). Varying amounts of non-collagenous material are interspersed with collagen bundles.

Collagen fibrils and non-collagenous organic material, in various arrangements, form the bulk of the calcifiable matrices in other structures (scutes, spines, fin rays, and so on) in the dermal skeleton. Their proportions vary quite widely, and collagen fibrils are sometimes completely absent. In some lizards, for instance, the superficial calcified layer of osteoderms – the calcified dermal plates of the integument in Anura and Reptilia – contains no collagen fibrils, only a proteoglycan microfibrillar meshwork (Levrat-Calviac and Zylberberg 1986).

In the most primitive Actinopterygii, the superficial layer of scales may be substituted by dental structures, that is, either a central dentine-like layer topped by a superficial enamel-like layer, or only a non-collagenous, enamel-like layer (ganoine). Ganoine is devoid of collagen fibrils (Sire et al. 1987) but contains microfibrils, in this respect resembling the organic matrix of enamel (Sire et al. 1987; Sire 1994, 1995). The homology between ganoine and enamel is strengthened by the immunodetection in its matrix of molecules which cross-react with mammalian amelogenin proteins (Zylberberg et al. 1997).

Dental structures can also be found in the so-called odontodes, which in primitive lower vertebrates may appear as ornaments of scales or dermal plaques. They stick out of the skin and have the same organization as teeth, comprising a basal osseous plate, a middle layer of dentin, and an outer hypermineralized layer of enamel or enamel-like material. This is known as 'enameloid'.

11.3.1 Enameloid

As already reported (see Sect. 4.3.1), enameloid is an enamel-like, highly calcified tissue which, in the form of a thin cap, covers the outer surface of teeth and odontodes in lower vertebrates (Fincham et al. 1999). In some cases, it has been considered to have the same ectodermal origin and the same structure and composition, including immuno-histochemical reactivity (Krejsa et al. 1985), as the enamel of high vertebrates – a feature that is certainly true as far as its high degree of calcification is concerned; in other cases, enameloid has been considered to be a mesodermal, or ecto-mesodermal, tissue containing collagen fibrils and matrix vesicles and, therefore, similar to dentin in several respects (Kemp and Park 1974; Sasagawa 1998; Sasagawa and Ishiyama 1988). Doubts are probably heightened by the fact that the enameloid tissue is not homogeneous, that its structure and composition may change according to species, and that the poorly defined boundary between enameloid and dentin may lead to an attribution to the former of structures belonging to the latter.

In their review of the evolution of tooth developmental biology, Slavkin and Diekwisch (1996) reported that the enameloid organic matrix comprises collagen fibrils and epithelial proteins, including amelogenin-like proteins, and that it is formed prior to dentin mineralization by molecules deriving from the inner enamel epithelium, and from adjacent odontoblasts. Prostak et al. (1993) found that secretory stage, inner dental epithelial cells secreted procollagen granules into the enameloid collagen matrix.

The histochemical investigations of Kogaya (1989) on the developing enameloid of the fish *Polypterus senegalus* showed collagen fibrils and a closely associated HID-TCH-SP stained material, consisting of chondroitin 4- and 6-sulfate, which tended to disappear as calcification went forward; the same author (Kogaya 1999) reported that, in larval urodele teeth, chondroitin sulfates and type I collagen fibrils are elaborated as enameloid inside the dental basement membrane, that they are then deposited below the membrane as initial dentin matrix, and that enamel-like matrix, without collagen fibrils or sulfated glycoconjugates but with amelogenins, is deposited on the dentin. The author concluded that enameloid begins its evolution as a dentin-like tissue. Inage (1975) reported that the formation of fish enameloid matrix begins with the appearance of non-striated fibers 14–18 nm thick, followed by fibers with 64-nm periodicity, and that the entire enameloid matrix consists of these fibers. Collagen fibrils have been reported in the enameloid matrix of teleost fish during the secretory stage, but not in mature enameloid (Prostak and Skobe 1986). Sasagawa (1995) described relatively dispersed, thin collagen fibrils at the early and middle stages of enameloid matrix formation in a teleost fish, with thicker (about 30 nm in diameter) fibrils appearing at a late stage, when the interwoven bundles that are characteristic of teleost cap enameloid are formed. The same author reported the presence of an interfibrillar material, which could consist of glycosaminoglycans or proteoglycans, and proposed that a considerable proportion of enameloid, including collagen fibrils, is produced by odontoblasts.

A transmission and scanning electron microscope study on the teeth of teleost fish showed on one hand that the enameloid cap (about 50 μm thick) was much more highly calcified than collar enameloid (about 1 μm thick), and on the other that cap enameloid differed with respect to dentin in several respects, which could be summed up as follows: greater electron density in TEM; structural arrangement perpendicular to dentin; higher F content; lower Mg content (Kerebel and Le Cabellec 1980). Another electron microscope study showed that the organic matrix of the outer tooth layer of *Squalus acanthias* has a membranous organization in the form of saccular compartments and that it is devoid of collagen fibrils, which, by contrast, are present in the inner layer, which resembles dentin (Garant 1970). Similar results were reported by Kemp and Park (1974) in developing teeth of the shark *Carcharhinus menisorrah*; tubular vesicles limited by a unit membrane and forming the bulk of the elasmobranch enameloid matrix during the stage of enameloid matrix formation were reported in elasmobranch fish by Sasagawa (2002) and in teleost fish by Sasagawa and Ferguson (1990).

Immunohistochemical studies on the presence of enamel proteins in the enameloid matrix have not solved the problem of its nature. Ishiyama et al. (2001) studied the teeth of the actinopterygian fish *Lepisosteus oculatus*, which feature the presence of both enamel and enameloid, the former covering the tooth shaft, and the latter the tooth cap. They found that amelogenin can be demonstrated immunohistochemically in enamel, but not in enameloid, and that, during early enameloid formation, this molecule was expressed by the inner enamel epithelial cells but was not secreted extracellularly. Similarly, biochemical studies of the proteins extracted from placoid scales and dermal denticles of the blue shark *Prionace glauca* L. failed to show any amelogenin-like material (Kawasaki et al. 1980). On the other hand, amelogenin was immunodetected in the enameloid of teeth and dermal denticles of Chondrichthyes, in the enameloid of Teleostei and Amphibia – as well as in the enamel of Reptilia – using antisera obtained from preparations of bovine enamel matrix (Herold et al. 1980), and in Selachian teeth using indirect immunofluorescence with rabbit polyclonal antibodies acting against purified murine amelogenins (Slavkin et al. 1983). Amelogenin genes have been identified and defined in some monotremes (platypus and echidna), reptiles (caiman), and amphibians (African clawed toad) (Toyosawa et al. 1998). Immunohistochemical studies with specific monoclonal antibodies acting against bovine amelogenin and bovine enamelins showed that only the latter could be detected in teeth and dermal denticles of the shark, bony fish and larval amphibian, whereas both could be found in higher vertebrates, suggesting an evolutionary appearance of enamelins at an earlier stage than amelogenins (Herold et al. 1989). Enamelin-like proteins have been reported in the enameloid of *Squalus acanthias* (Graham 1984) and the blue shark *Prionace glauca* (Graham 1985).

Many of the findings reported above contain contradictory elements, but the concept that fish enameloid is distinct from mammalian enamel seems to prevail (Kawasaki et al. 2004). Another finding, however, fixes a few points of contact between the two tissues: in the elasmobranch *Heterodontus japonicus* most of the organic matrix disappears at the maturation stage of enameloid, when this is occupied by large numbers of crystals (Sasagawa 1999), a process similar to that found in mammalian enamel.

11.4 Concluding Remarks

- The organic matrix of the calcified structures of lower vertebrates does not appear to differ substantially from those of higher vertebrates, especially as far as the endoskeleton is concerned. However, some distinctive features should be mentioned.
- The structures that make up the integumental skeletons (scales, scutes, fin rays, spines, osteoderms and odontodes) basically consist of varying amounts of collagen fibrils and interfibrillary glycoproteic and proteoglycanic material.
- Taking elasmoid scales as model, two layers can be described, a superficial layer consisting of woven bone, and a deep layer consisting of collagen fibrils displaying a plywood-like arrangement.
- The superficial, osseous layer comprises an outer, limiting stratum, devoid of collagen fibrils, and an underlying stratum, consisting of randomly oriented collagen fibrils, some of which (the TC fibrils) connect the scales to the dermis.
- An enamel-like tissue (ganoine) can be found in place of this superficial matrix in structures like odontodes. Ganoine has no collagen fibrils and its matrix contains amelogenin.
- Enameloid, which is another enamel-like tissue – at least as far as its high degree of calcification is concerned –, covers, in the form of a thin cap, the outer tooth surface in several lower vertebrates.
- The nature of enameloid has still not been exactly defined; its organic matrix has been reported to comprise collagen fibrils and matrix vesicles, so displaying a dentin-like organization, or to be afibrillar, in this case consisting of organic structures outlining vesicular compartments.

- Immunohistochemistry has provided mixed results: some studies have shown that the enameloid organic matrix comprises amelogenin, or enamelin, or both, whereas other studies have failed to show these enamel-specific proteins.

References

Ascenzi A, Bonucci E (1968) The compressive properties of single osteons. Anat Rec 161:377–392

Bigi A, Koch MH, Panzavolta S, Roveri N, Rubini K (2000) Structural aspects of the calcification process of lower vertebrate collagen. Connect Tissue Res 41:37–43

Bigi A, Burghammer M, Falconi R, Koch MH, Panzavolta S, Riekel C (2001) Twisted plywood pattern of collagen fibrils in teleost scales: an X-ray diffraction investigation. J Struct Biol 136:137–143

Brown GA, Wellings SR (1969) Collagen formation and calcification in teleost scales. Z Zellforsch 93:571–582

de Ricqlès A, Meunier FJ, Castanet J, Francillon-Vieillot H (1991) Comparative microstructure of bone. In: Hall BK (ed) Bone, vol 3. Bone matrix and bone specific products. CRC Press, Boca Raton, pp 1–78

Fincham AG, Moradian-Oldak J, Simmer JP (1999) The structural biology of the developing dental enamel matrix. J Struct Biol 126:270–299

Garant PR (1970) An electron microscopic study of the crystal-matrix relationship in the teeth of the dogfish *Squalus acanthias* L. J Ultrastruct Res 30:441–449

Graham EE (1984) Protein biosynthesis during spiny dogfish (*Squalus acanthias*) enameloid formation. Arch Oral Biol 29:821–825

Graham EE (1985) Isolation of enamelinlike proteins from blue shark (*Prionace glauca*) enameloid. J Exp Zool 234:185–191

Herold RC, Graver HT, Christner P (1980) Immunohistochemical localization of amelogenins in enameloid of lower vertebrate teeth. Science 207:1357–1358

Herold R, Rosenbloom J, Granovsky M (1989) Phylogenetic distribution of enamel proteins: immunohistochemical localization with monoclonal antibodies indicates the evolutionary appearance of enamelins prior to amelogenins. Calcif Tissue Int 45:88–94

Inage T (1975) Electron microscopic study of early formation of the tooth enameloid of a fish (*Hoplognathus fasciatus*). I. Odontoblasts and matrix fibers. Arch Histol Jpn 38:209–227

Ishiyama M, Mikami M, Shimokawa H, Oida S (1998) Amelogenin protein in tooth germs of the snake *Elaphe quadrivirgata*, immunohistochemistry, cloning and cDNA sequence. Arch Histol Cytol 61:467–474

Ishiyama M, Inage T, Shimokawa H (2001) Abortive secretion of an enamel matrix in the inner enamel epithelial cells during an enameloid formation in the gar-pike, Lepisosteus oculatus (Holostei, Actinopterygii). Arch Histol Cytol 64:99–107

Kawasaki H, Kawaguchi T, Yano T, Fujimura S, Yago M (1980) Chemical nature of proteins in the placoid scale of the blue shark, *Prionace glauca* L. Arch Oral Biol 25:313–320

Kawasaki K, Suzuki T, Weiss KM (2004) Genetic basis for the evolution of vertebrate mineralized tissue. Proc Natl Acad Sci U S A 101:11356–11361

Kemp NE, Park JH (1974) Ultrastructure of the enamel layer in developing teeth of the shark *Carcharhinus menisorrah*. Arch Oral Biol 19:633–644

Kerebel L-M, Le Cabellec MT (1980) Enameloid in the teleost fish Lophius. An ultrastructural study. Cell Tissue Res 206:211–223

Kimura S (1985) The interstitial collagens of fish. In: Bairati A, Garrone R (eds) Biology of invertebrate and lower vertebrate collagens. Plenum Press, New York, pp 397–408

Kimura S, Kamimura T, Takema Y, Kubota M (1981) Lower vertebrate collagen. Evidence for type I-like collagen in the skin of lamprey and shark. Biochim Biophys Acta 669:251–257
Kobayashi S, Yamada J, Maekawa K, Ouchi K (1972) Calcification and nucleation in fish-scales. Biomineralization 6:84–90
Kogaya Y (1989) Histochemical properties of sulfated glycoconjugates in developing enameloid matrix of the fish *Polypterus senegalus*. Histochemistry 91:185–190
Kogaya Y (1999) Immunohistochemical localisation of amelogenin-like proteins and type I collagen and histochemical demonstration of sulphated glycoconjugates in developing enameloid and enamel matrices of the larval urodele (*Triturus pyrrhogaster*) teeth. J Anat 195:455–464
Krejsa RJ, Samuel N, Bessem C, Slavkin HC (1985) Immunogenetic and phylogenetic comparisons between teleost scale and dental enameloid with mammalian enamel antigens. In: Belcourt AB, Ruch J-V (eds) Morphogenèse et différenciation dentaires. Coll. INSERM 125. INSERM, Paris, pp 369–376
Levrat-Calviac V, Zylberberg L (1986) The structure of the osteoderms in the gekko: *Tarentola mauritanica*. Am J Anat 176:437–446
Meunier FJ (1984) Spatial organization and mineralization of the basal plate of elasmoid scales in Osteichthyans. Am Zoolog 24:953–964
Meunier FJ, François Y (1980) L'organisation spatiale des fibres collagènes et la minéralisation des écailles des Dipneustes actuels. Bull Soc Zool Fr 105:215–226
Meunier FJ, Géraudie J (1980) Les structures en contre-plaqué du derme et des écailles des vertébrés inférieurs. Ann Biol 19:1–18
Olson OP, Watabe N (1980) Studies on formation and resorption of fish scales. IV: Ultrastructure of developing scales in newly hatched fry of the sheepshead minnow, *Cyprinodon variegatus* (Atheriniformes: Cyprinodontidae). Cell Tissue Res 211:303–316
Onozato H, Watabe N (1979) Studies on fish scale formation and resorption III. Fine structure and calcification of the fibrillary plates of the scales in *Carassius auratus* (Cypriniformes: cyprinidae). Cell Tissue Res 201:409–422
Prostak K, Skobe Z (1986) Ultrastructure of the dental epithelium during enameloid mineralization in a teleost fish, *Cichlasoma cyanoguttatum*. Arch Oral Biol 31:73–85
Prostak KS, Seifert P, Skobe Z (1993) Enameloid formation in two tetraodontiform fish species with high and low fluoride contents in enameloid. Arch Oral Biol 38:1031–1044
Sasagawa I (1995) Fine structure of tooth germs during the formation of enameloid matrix in *Tilapa nilotica*, a teleost fish. Arch Oral Biol 40:801–814
Sasagawa I (1998) Mechanisms of mineralization in the enameloid of elasmobranchs and teleosts. Connect Tissue Res 39:207–214
Sasagawa I (1999) Fine structure of dental epithelial cells and the enameloid during the enameloid formation stages in an elasmobranch, *Heterodontus japonicus*. Anat Embryol (Berl) 200:477–486
Sasagawa I (2002) Mineralization patterns in elasmobranch fish. Microsc Res Tech 59:396–407
Sasagawa I, Ferguson MW (1990) Fine structure of the organic matrix remaining in the mature cap enameloid in *Halichoeres poecilopterus*, teleost. Arch Oral Biol 35:765–770
Sasagawa I, Ishiyama M (1988) The structure and development of the collar enameloid in two teleost fishes, *Halichoeres poecilopterus* and *Pagrus major*. Anat Embryol (Berl) 178:499–511
Schönbörner AA, Boivin G, Baud CA (1979) The mineralization processes in teleost fish scales. Cell Tissue Res 202:203–212
Sire J-Y (1994) Light and TEM study of nonregenerated and experimentally regenerated scales of *Lepisosteus oculatus* (Holostei) with particular attention to ganoine formation. Anat Rec 240:189–207

Sire J-Y (1995) Ganoine formation in the scales of primitive actinopterygian fishes, lepisosteids and polypterides. Connect Tissue Res 33:213–222
Sire J-Y, Géraudie J (1984) Fine structure of regenerating scales and their associated cells in the cichlid Hemichromis bimaculatus (Gill). Cell Tissue Res 237:537–547
Sire J-Y, Géraudie J, Meunier FJ, Zylberberg L (1987) On the origin of ganoine: histological and ultrastructural data on the experimental regeneration of the scales of *Calamoichthys calabaricus* (Osteichthyes, Brachyopterygii, Polypteridae). Am J Anat 180:391–402
Slavkin HC, Diekwisch T (1996) Evolution in tooth developmental biology: of morphology and molecules. Anat Rec 245:131–150
Slavkin HC, Samuel N, Bringas P Jr, Nanci A, Santos V (1983) Selachian tooth development: II. Immunolocalization of amelogenin polypeptides in epithelium during secretory amelogenesis in *Squalus acanthias*. J Craniofac Genet Dev Biol 3:43–52
Toyosawa S, O'hUigin C, Figueroa F, Tichy H, Klein J (1998) Identification and characterization of amelogenin genes in monotremes, reptiles, and amphibians. Proc Natl Acad Sci USA 95:13056–13061
Wang X, Ito Y, Luan X, Yamane A, Diekwisch TG (2005) Amelogenin sequence and enamel biomineralization in *Rana pipiens*. J Exp Zool B Mol Dev Evol 304:177–186
Yamada M, Ozawa H (1978) Ultrastructural and cytochemical studies on the matrix vesicle calcification in the teeth of the killifish, *Oryzias latipes*. Arch Histol Jpn 41:309–323
Yamada J, Watabe N (1979) Studies on fish scale formation and resorption I. Fine structure and calcification of the scales in *Fundulus heteroclitus* (Atheriniformes: cyprinodontidae). J Morphol 159:49–66
Zylberberg L (1985) Collagen and mineralization in the elasmoid scales. In: Bairati A, Garrone R (eds) Biology of invertebrate and lower vertebrate collagens. Plenum Press, New York, pp 457–463
Zylberberg L, Nicolas G (1982) Ultrastructure of scales in a teleost (*Carassius auratus* L.) after use of rapid freez-fixation and freeze-substitution. Cell Tissue Res 223:349–367
Zylberberg L, Géraudie J, Meunier F, Sire J-Y (1992) Biomineralization in the integumental skeleton of the living lower vertebrates. In: Hall BK (ed) Bone, vol 4: Bone metabolism and mineralization. CRC Press, Boca Raton, pp 171–224
Zylberberg L, Sire J-Y, Nanci A (1997) Immunodetection of amelogenin-like proteins in the ganoine of experimentally regenerating scales of *Calamoichthys calabaricus*, a primitive actinopterygian. Anat Rec 249:86–95

12 Calcifying Matrices: Invertebrates

12.1 Introduction

The calcified structures of invertebrates are ranged along a spectrum of sharply differing morphological expressions. These comprise intracellular inorganic granules, the outer coverings of certain protozoa, single polymorphous crystals in sponges, calcified tubes formed by worms, skeletal spines and other calcified structures in echinoderms, calcified exoskeletons in arthropods, shells of mollusks and skeletal deposits of corals (see Wilbur 1976). This array of different forms implies that the organic matrix in each should be considered in detail, a demanding task in many respects. In practice, studies have mainly been focused on the organic matrix of the invertebrate calcified structures which, from time to time, seemed to be of the greatest biological interest, those most accessible to investigation, or those easiest to locate. As a result, research has been preferentially concentrated on some categories of calcified structures, such as the shells of mollusks (reviewed by Wheeler 1992; Wilt et al. 2003), the echinoderm endoskeleton – more specifically, the spicules of the sea urchin embryo (reviewed by Decker and Lennarz 1988; Benson and Wilt 1992; Wilt 1999, 2002; Wilt et al. 2003), the carapace of Crustacea (reviewed by Wilt et al. 2003), and the exoskeleton of corals and other cnidarians (reviewed by Marshall 2002). Amounts of organic material in all these structures are very low, but the possibility that organic molecules may be enzymatically degraded and lost during calcification – a process parallel to that found in enamel – seems to have been overlooked.

12.2 Mollusk Shells

The organic matrix of mollusk shells has been defined as the organic material which is an integral constituent of the calcified structures, pervading their main parts; it is located both around and within crystals (Wilbur 1976). It is less than 5 wt % and consists of proteins and glycoproteins which can be collected in two fractions by utilizing their solubility properties (Crenshaw 1972; Wilbur 1976): shell proteins treated with an almost neutral solution of EDTA yield two fractions, one of soluble, highly anionic proteins or glycoproteins, another of insoluble, hydrophobic proteins. There is, however, no clear-cut separation between these fractions, especially as regards their amino acid composition and the charge of their constituents (Wheeler 1992). On that account, and because organic components displaying many of the features of the insoluble fraction can be extracted in ultra-pure water without previous decalcification (Pereira-Mouriès et al. 2002), the idea of a sharp division into soluble and insoluble fractions may call for reconsideration.

In broad terms, the soluble fraction, which is similar in different species, consists of proteins rich in aspartic acid molecules (Weiner 1979), which form sequences separated either by glycine or by serine (Weiner and Hood 1975). This fraction also contains sulfate proteoglycans (for a review of papers published before 1970 on this topic, see Kobayashi 1971). Marxen et al. (1998) have reported acid proteoglycans, glycosaminoglycans and glycoproteins in the shell matrix of the snail *Biomphalatia glabrata*, and Marxen and Becker (1997) have found that the soluble protein fraction from the nacre layer of the same snail contains neutral sulfate sugar, amino sugar, small amounts of uronic acid and phosphate, besides glycosaminoglycans and proteoglycans of undetermined size, which may be calcium-binding constituents of the organic matrix (Marxen and Becker 2000). Dauphin (2003) and Dauphin et al. (2003) quoted and confirmed the results of Wada, showing that the acid proteoglycan fraction corresponds to chondroitin sulfate. A strong staining reaction with alcian blue, considered to point to the presence of sulfated proteoglycans, has been reported by Marin et al. (1994). Scanning X-ray microscopy of the shell prismatic layer of *Pinna nobilis* and *Pinctada margaritifera* has shown that the spectra, both of the interprismatic wall and of the intraprismatic area, are characterized by a 2.482-keV peak, indicative of sulfated sugars – either chondroitin or keratan sulfate (Dauphin et al. 2003).

The insoluble fraction, whose composition varies in different species, is rich in glycine, alanine, phenylalanine and thyrosine (Meenakshi et al. 1971). The ordered protein components of this fraction take up an antiparallel β-sheet conformation (Weiner and Traub 1980; Choi and Kim 2000) and the polysaccharide phase, when present, has been named as chitin (Peters 1972; Weiner and Traub 1980). Characteristic chitin diagrams have been obtained from the decalcified shells of *Lingula unguis* (Kelly et al. 1965), where both the organic and the decalcified layer show the X-ray diffraction pattern of β-chitin (Iijima and Moriwaki 1990). In other calcified tissues, such as the cuticle of crustaceans, chitin is in the α-form (see Falini et al. 2003). Chitin is closely bound to less ordered protein (Zentz et al. 2001), whose diffraction pattern and amino acid composition resemble those of silk fibroin (Weiner et al. 1983). In vitro studies on chitin/silk fibroin interactions have shown that silk fibroin is inserted between the molecular planes of chitin by means of the acetyl groups in the latter (Falini et al. 2003).

Ca-binding glycoproteins have been found to be components of these fractions (Crenshaw 1972; Krampitz et al. 1976). Interestingly, Ca-binding proteins have the tendency to polymerize and form needle-like crystals which lie parallel when polymerization is allowed on a plastic surface; once polymerized, they cannot be dissolved (Krampitz et al. 1976).

The organic components reported above interact to generate biopolymeric complexes which differ structurally in the prismatic and nacreous layers, even if they favor or facilitate calcification in both (Ravindranath and Rajeswari Ravindranath 1974). The morphology of these biopolymers has been studied using several techniques, including X-ray diffraction and electron microscopy. The first step in crystal formation in the prismatic and nacre layers of mollusk shells has been reported to be the "polymerization" of part of the pallial fluid, leading to the formation of very thin organic lamellae on, and parallel to, the surface of the mantle epithelium (Bevelander and Nakahara 1969; Nakahara and Bevelander 1971). These lamellae form compartments that enclose a modified pallial fluid and allow crystal formation. Organic material (conchiolin) acting as a "mortar" between and around the tabular aragonite crystals in the mother-of-pearl of several species was observed by Grégoire (1957). According to Towe and Hamilton (1968), who used the replica method to study developing and mature nacre in bivalve mollusks, the uniform aragonite tablets are separated by very thin (about 20 nm thick) boundaries of organic matrix. The same findings were reported by Watabe (1963, 1965) and Saleuddin (1971).

It is now accepted that aragonite crystals are surrounded by an organic sheet consisting of several laminae which include a chitin core, itself sandwiched between two layers of water-insoluble proteins on which water-soluble proteins are adsorbed (Weiner 1979; Weiner and Traub 1980; Weiss et al. 2000). With reference to the structure of nacre, which is the most widely studied of the shell layers, Nakahara et al. (1982) and Weiner and Traub (1984) proposed a five-layered model (discussed by Mann S 1988) of the sheets of organic matrix which surround the aragonite crystals on all sides: a hydrophobic core of β-chitin (Nakahara et al. 1982) sandwiched between two layers of silk-fibroin-like proteins (Weiner et al. 1983), both sides of which are in contact with layers of EDTA-soluble, hydrophilic, acidic components aligned with the underlying anti-parallel β-plated sheet polypeptide chains. To some extent, this proposed arrangement has been modified (for diagram see Levi-Kalisman et al. 2001) on the basis of the cryo-transmission electron microscope (cryo-TEM) observation that in the bivalve *Atrina* the sheets that lie between aragonite lamellae mainly consist of highly ordered, aligned β-chitin layers between which a randomly oriented organic material, probably corresponding to a silk-fibroin gel and Asp-rich glycoproteins, is located (Levi-Kalisman et al. 2001; Pereira-Mouriès et al. 2002). The artificial assembly of β-chitin from fins of the squid *Loligo*, silk fibroin from cocoons of the silkworm *Bombyx mori*, and acidic, soluble macromolecules from aragonite or calcite shell layers, which had been shown to promote aragonite or calcite formation, respectively (Falini et al. 1996), fails to change the ultrastructure of β-chitin, so suggesting that this protein alone is responsible for the lamellar ultrastructural appearance of decalcified aragonite layers (Levi-Kalisman et al. 2001). Chitin is found in a variety of different calcified biological systems and plays crucial role in controlling the calcification process (Falini and Fermani 2004).

The cryo-TEM imaging of vitrified, decalcified *Atrina* nacreous organic matrix reported by Levi-Kalisman et al. (2001) appears to be the most direct illustration made available so far of the ordered arrangement of β-chitin lamellae in nacre. An apparently disordered architecture of interconnecting bundles, outlining pores and channels, has been reported in a shell nacreous layer decalcified with 5% hydrochloric acid solution by Tong et al. (2002). The same authors observed that the organic matrix of the decalcified prismatic layer consisted of ordered prismatic cavities whose shape and size were nearly the same as those of the columnar crystals. They also showed that partly decalcified crystals containing intracrystalline organic material could be detected in the transition zone between fully decalcified and not decalcified matrix. Blank et al. (2003) carried out an atomic force microscope study of *Haliota laevigata* nacre after this had been cleft parallel to the aragonite tablet (i. e., between the tablet layer and the protein layer) and had been decalcified: organic filamentous structures – interpreted as collagen – were detected; they outlined honeycomb-like spaces whose diameter was similar to that of aragonite tablets and appeared to be their imprints. High resolution imaging of the area containing these spaces revealed tightly packed, globular protein molecules only a few nanometers thick. Wierzbicki et al. (1994), also using the atomic force microscope, observed calcite bound oyster shell proteins on the surface of calcite crystals; they had a globular appearance.

In spite of the investigations reported above, no direct morphological evidence has been obtained on the overall arrangement of proteins in the aragonite or calcite layers, probably because of the difficulties inherent in the microscopic study of heavily calcified, hard structures such as mollusk shells. In addition, even if these structures reveal a gradient of increasing calcification (Kelly et al. 1965), no border or sheath of uncalcified organic matrix comparable with the bone osteoid border can be detected, while the zone of incipient calcification shows only a few lamellae (Towe and Hamilton 1968).

Identification of the components of the organic material (Tables 12.1 and 12.2) may be fundamental to an understanding of the molecular mechanism of shell calcification, as suggested by the findings that aragonite crystallization requires the presence of a protein complex (Matsushiro et al. 2003), that macromolecules from aragonitic shell layers induce aragonite formation in vitro when preadsorbed on a substrate of β-chitin and silk fibroin (Falini et al. 1996), and that the formation of calcite and aragonite crystals depends on the synthesis of polyanionic molecules whose interactions with calcium ions control deposition of the inorganic phase (Belcher et al. 1996). Acidic proteins are, in fact, thought to be the matrix components that trigger and regulate shell calcification (Weiner and Addadi 1991).

Table 12.1. Proteins isolated from the invertebrate prismatic layer and extrapallial fluid (EP)

Name	Source	Invertebrate	pI	Reference
MSI31	Prism	Oyster, *Pinctada fucata*	a	Sudo et al. (1997)
MSP-1	Prism	Scallop, *Patinopecten yessoensis*	3.20	Sarashina and Endo (1998)
EP-A	EP	Mussel, *Mytilus edulis*	4.43	Hattan et al. (2001)
Aspein	Prism	Oyster, *Pinctada fucata*	1.45	Tsukamoto et al. (2004)
Prismalin-14	Prism	Oyster, *Pinctada fucata*	–	Suzuki et al. (2004)
Asprich	Prism	Bivalve, *Atrina rigida*	b	Gotliv et al. (2005)

pI: Isoelectric point
a A large acidic region
b Very acidic protein

Table 12.2. Proteins isolated from the mollusk nacreous layer and pearl organic matrix

Name	Source	Invertebrate	Reference
Nacrein	Pearl (nacre)	Oyster, *Pinctada fucata*	Miyamoto et al. (1996)
N66	Nacre	Oyster, *Pictada maxima*	Kono et al. (2000)
MSI60	Nacre	Oyster, *Pinctada fucata*	Sudo et al. (1997)
Lustrin A	Nacre	Abalone, *Haliotis rufescens*	Shen et al. (1997)
N16	Nacre	Oyster, *Pinctada fucata*	Samata et al. (1999)
N14	Nacre	Oyster, *Pinctada fucata*	Kono et al. (2000)
Pearlin	Pearl (nacre)	Oyster, *Pinctada fucata*	Miyashita et al. (2000)
Pearlin-keratin	Pearl & nacre	Oyster, *Pinctada fucata*	Matsushiro et al. (2003)
Mucoperlin	Nacre	Mussel, *Pinna nobilis*	Marin et al. (2000)
Perlucin	Shell	Abalone, *Halotis laevigata*	Weiss et al. (2000)
Perlustrin	Nacre	Abalone, *Halotis laevigata*	Chen and Boskey (1986)
AP7–AP24	Nacre	Abalone, *Haliotis rufescens*	Michenfelder et al. (2003)

A significant proportion of the soluble proteins in the organic matrix of the mollusk shell consists of a repeating sequence of aspartic acid separated by either glycine or serine (Weiner and Hood 1975), and several polyanionic proteins have been isolated from mollusk shells (Weiner and Hood 1975; Nakahara et al. 1982; Cariolou and Morse 1988). Highly acidic, soluble organic components have been extracted from the shells of the Antarctic scallop, *Adamussium colbecki*; they consisted of 1.5 wt % carbohydrate and 12.8 wt % phosphate, contained approximately 31% Asx (Asp+Asn), 29% Ser and 18% Gly, and could be separated by high-performance liquid chromatography into six fractions (RP-1 through RP-6), the first of which was found to inhibit calcium carbonate crystal nucleation in vitro (Halloran and Donachy 1995). The extrapallial fluid, which fills the cavity between the mantle and the external shell (on the structure and properties of mantle and extrapallial cavity, see Wheeler 1992), is considered to play a fundamental role in shell calcification and to contain the heterogeneous macromolecules that compose the mature shell. Hattan et al. (2001) have shown that 56% of the total protein in the extrapallial fluid of the mollusk *Mytilus edulis* consists of an acidic (isoelectric point, 4.43), calcium-binding glycoprotein, here indicated as EP(A), which is a dimer of 28,340 Da monomers rich in histidine (11.14%), as well as in Asx (Asp+Asn) and Glx (Glu+Gln) residues. According to Yin et al. (2005), it might have multiple functions, contributing to calcium transport, heavy metal detoxification, and shell construction.

A major water-soluble protein, known as Mollusk Shell Protein 1 (MSP-1), has been extracted from the calcite layer of the scallop *Patinopecten yessoensis* (Sarashina and Endo 1998). MSP-1 is an acidic glycoprotein (isoelectric point 3.2) with a calculated molecular mass of 74.5 kDa and a deduced amino acid sequence which includes a high proportion of Ser (32%), Gly (25%), and Asp (20%). A sequential unit between 158 and 177 amino acids long recurs four times in the mid-section of the protein. It comprises three domains: the SG domain, consisting almost exclusively of Ser and Gly residues; the D domain, which is rich in Asp residues and exhibits some repeated motifs, such as Asp-Gly-Ser-Asp and Asp-Ser-Asp; and the K domain, which is rich in Gly residues (Sarashina and Kazuyoshi 2001).

Another matrix protein has been purified from the acid-insoluble fraction of the shell prismatic layer of the oyster *Pinctada fucata* (Suzuki et al. 2004). Named Prismalin-14, it consists of 105 amino acid residues, including a Gly/Tyr-rich region and N- and C-terminal Asp-rich regions.

Unusually acidic (predicted isoelectric point 1.45) matrix protein, called Aspein, has been identified and characterized in the organic matrix of the pearl oyster *Pinctada fucata* (Tsukamoto et al. 2004). The amino acid sequence of Aspein includes 60.4% Asp, 16.0% Gly, and 13.2% Ser. It is expressed at the outer edge of the mantle, which corresponds to the prismatic layer, but not in its inner portion, which corresponds to the nacreous layer. Ten different proteins have been isolated by Gotliv et al. (2005) by screening a cDNA library made up of the mRNA of the shell-forming cells of the bivalve *Atrina rigida*, using probes for short Asp-containing repeat sequences. Seven proteins were then identified using more specific probes. This subfamily of unusually acidic, aspartic acid-rich proteins has been named Asprich. Polyclonal antibodies raised against a synthetic peptide of the conserved acidic domain of these proteins reacted with components of the calcite prismatic layer but not with the nacre layer, showing that Asprich proteins are constituents of the prismatic shell layer matrix. A conserved domain rich in aspartic acid, comprising a sequence very similar to the calcium-binding domain of Calsequestrin, has been identified, along with another domain rich in aspartic acid that shows variation between the seven sequences.

The study of acidic macromolecules like Asprich is fundamental, because they appear to be the agents that most actively control calcification (Weiner and Addadi 1991). It is hard to characterize them, precisely because many of them are polyanions (Weiner and Hood 1975) with a high negative charge. In spite of this problem, several shell proteins from the nacreous layer have now been identified.

Taking pearls from the oyster *Pinctada fucata* as a material equivalent to shell nacre, Miyamoto et al. (1996) extracted a soluble, 60-kDa protein, that they called Nacrein, which was specifically expressed in mantle pallial. Analysis of the amino acid sequence of Nacrein showed two functional domains: the first, similar to carbonic anhydrase, inhibited calcium carbonate precipitation from a saturated solution (Miyamoto et al. 2005), and the second was a Gly-Xaa-Asn repeat domain, where Xaa is either Asp, Asn, or Glu. The former domain had been split into two subdomains by the insertion of the latter between them. Kono et al. (2000) have isolated a protein named N66 from the nacreous layer of *Pinctada maxima*; it consisted of carbonic anhydrase-like and repeat domains, as described for Nacrein.

Two insoluble proteins from the decalcified nacreous layer of the *Pinctada fucata* shell have been isolated; they were called MSI60 and MSI31 because of their molecular weight, of 60,000 and 31,000, respectively (Sudo et al. 1997). Their amino acid composition corresponds with those previously found for insoluble proteins from the nacreous and prismatic layers, respectively, by Samata (quoted by Sudo et al. 1997).

Another multidomain protein has been extracted from the nacreous layer of the shell and pearl produced by the marine gastropod abalone, *Haliotis rufescens* (Shen et al. 1997). Called Lustrin A, it is a member of the lustrin superfamily, a unique group of proteins localized between, and probably connecting, as an elastomer, the layers of aragonite plates (Zhang et al. 2002). The deduced amino acid sequence of Lustrin A reveals a highly modular structure with 16% Ser, 14% Pro, 13% Gly, and 9% Cys (Shen et al. 1997).

Another class of EDTA-insoluble protein, called Pearlin, has been extracted with 0.3 M EDTA/8 M urea from powdered pearl decalcified with 0.3 M EDTA (Miyashita et al. 2000). Pearlin is expressed specifically in mantle epithelium, has a molecular weight of about 15 kDa, contains a sulphated proteoglycan, and comprises 129 amino acids, with a high proportion of Gly (10.8%), Tyr (10.0%), Cys (8.5%), Asn (7.7%), Asp (7.7%) and Arg (7.7%). A protein complex, considered to be a prerequisite for aragonite crystallization, has been isolated from the nacreous layer of pearl beads and oyster shell by Matsushiro et al. (2003). It consists of a 15-kDa protein (Pearlin) and a 20-kDa protein showing homologies with keratin (pearl keratin). A matrix protein family specific to the nacreous layer of the pearl oyster, *Pinctada fucata*, and homologous with Pearlin, has been described and named N16 by Samata et al. (1999) and N14 by Kono et al. (2000).

Two proteins, called Perlucin and Perlustrin (molecular weights 717,000 and 13,000, respectively) have been isolated from the shell of the abalone *Halotis laevigata* after decalcification of the shell in 10% acetic acid (Weiss et al. 2000). The sequence of the first 130 amino acids of Perlucin showed similarities with the C-type carbohydrate-recognition domains of the group of C-type lectins (Mann K et al. 2000), the same as those found in other proteins isolated from $CaCO_3$ biominerals such as litostathine, the pancreatic stone protein, and ovocleidin 17, the eggshell protein. Perlustrin was the minor component of the protein mixture and showed some similarities with a portion of the protein Lustrin A. Its 84-amino-acid-long sequence was homologous with the N-terminal domain of mammalian insulin-like growth factor binding proteins (IGFBPs), with a 40% sequential identity (Weiss et al. 2001). Experiments by Blank et al. (2003) have not only shown that purified nacre Perlustrin is a $CaCO_3$-nucleating protein (see also Weiss et al. 2001), but also that it becomes incorporated in $CaCO_3$ crystals, so forming a protein-mineral composite (see Sect. 16.6).

Another protein, called Mucoperlin, has been identified by screening a cDNA library made from the mantle tissue mRNA of the mussel *Pinna nobilis* using antibodies raised against the acetic acid-soluble shell matrix of the same species (Marin et al. 2000). Mucoperlin has a molecular mass of 66.7 kDa and an isoelectric point of 4.8, is rich in serine and proline residues, and exhibits some mucin-like characteristics. Immunohistochemistry has shown that it is specifically localized in the nacre, where it surrounds the aragonite polygonal elements.

Two proteins, known as AP7 and AP24 (aragonite proteins of molecular weight 7 kDa and 24 kDa, respectively) have been isolated from the nacre layer of the abalone, *Haliotis rufescens*, and peptides have been generated that correspond to the first 30 residues of the N-terminal sequences of both proteins, and include $CaCO_3$-interacting domains (Michenfelder et al. 2003). The peptide from AP7, called AP7-1, exists as an unfolded sequence and is able to inhibit the in vitro growth of $CaCO_3$ crystals. Solid-state NMR magic angle spinning studies have shown that AP7-1 is bound to calcite fragments from which it cannot be completely displaced using competitive Ca-washing (Kim et al. 2004). AP24-1 has a random, coil-like structure and there is evidence that it has a turn-like, bend, or loop conformation resembling the structures of Lustrin A (Wustman et al. 2004).

A matrix protein of 19.6 kDa has been isolated by preparative electrophoresis from the shell of the freshwater snail *Biomphalaria glabrata* (Marxen et al. 2003). Its sequence of 148 amino acids showed a 32% and 34–37% sequence identity with mammalian and invertebrate dermatopontin sequences, respectively.

12.3 Echinoderm Skeleton

Studies on the endoskeleton of echinoderms are usually carried out on the calcified structures of the sea urchin embryo. Rather than dealing with the skeleton of adult echinoids, whose ossicles consist of a rather complex tridimensional network of calcified trabeculae (known as 'a stereom'), research has mainly been directed to the spicules of the larval skeleton. These consist of single or slightly ramified trabeculae whose structure appears to be simpler –

though this concept could prove to be deceptive (Wilt 1999) – and, therefore, much easier to study, than that of stereoms, or many other calcified tissues. Spicules offer two further advantages: they occur in a developmental context, and can be grown in vitro. This can be achieved quite easily by isolating the mesenchymal micromeres that are formed at the embryonic vegetal pole of the fourth cell division and growing them in culture (Okazaki 1975; Urry et al. 2000): they give rise to primary mesenchyme cells that then differentiate to form needle-shaped spicules (for reviews on the embryology of spicule formation see Decker et al. 1987; Decker and Lennarz 1988; Benson and Wilt 1992; Wilt 1999). Probably because they are harder to study than spicules, sea urchin teeth have only rarely been examined (Bonfield and Scandola 1979; Weiner 1985; Veis DJ et al. 1986; Wang et al. 1997; Ameye et al. 1999, 2000; Veis A et al. 2002).

Several proteins have been isolated from the sea urchin embryo spicules (Table 12.3). Most of them (about 90%) are soluble in dilute aqueous buffers and consist of acidic proteins (Benson and Wilt 1992). Radio-iodination of the decalcified matrix of spicules from *Strongylocentrotus purpuratus* embryos showed six bands on SDS protein gels, some of them corresponding to N-linked glycoproteins (Venkatesan and Simpson 1986). Using two-dimensional gel electrophoresis of [^{35}S]methionine-labeled proteins of spicules from the same type

Table 12.3. Sea urchin proteins involved in skeletogenesis and calcification

Name	Source	pI	Reference
SM50	Sp,Stp	Alkaline	Benson et al. (1987)
LSM34	Sp,Lyp	Alkaline	Livingston et al. (1991)
PM27	Sp, WE,Syp	Alkaline	Harkey et al. (1995)
SM37	Sp,Stp	Alkaline	Lee et al. (1999)
SM30	Sp,Stp	Acidic	George et al. (1991)
SM30-A	WE,Stp	Acidic	Killian and Wilt (1996)
SM30-B	WE, AS,Stp	Acidic	Killian and Wilt (1996)
SM30-C	AS,Stp	Acidic	Killian and Wilt (1996)
MSP130	CS,Stp	Acidic	Leaf et al. (1987)
MSP130-related-1	Sp,Stp	–	Illies et al. (2002)
MSP130-related-2	Sp,Stp	–	Illies et al. (2002)
SM50-related	Sp,Stp	Alkaline	Illies et al. (2002)
SpC-lectin	Sp,Stp	Alkaline	Illies et al. (2002)
SpSM29	Sp,Stp	Alkaline	Illies et al. (2002)
SpSM32	Sp,Stp	Alkaline	Illies et al. (2002)
SpP16	CS,Stp	Acidic	Illies et al. (2002)
SpP19	CS,Stp	–	Illies et al. (2002)

Sp: embryo spicule; WE: whole embryo; AS: adult spine; CS: cell surface;
Stp: *Strongylocentrotus purpurats*; Lyp: *Lytechynus pictus*

of sea urchin, 12 strongly labeled proteins and approximately three dozen less strongly labeled proteins were identified (their molecular mass ranged between 20 and over 100 kDa); most have acidic isoelectric points (Killian and Wilt 1996). Four major components (with molecular weights of 64, 57, 50 and 47 kDa) have been detected by SDS-PAGE electrophoresis and immunoblotting among spicule proteins (Benson et al. 1986). The collagen fibrils and hydroxyproline reported by Pucci-Minafra et al. (1980) in *Paracentrotus lividus* have not been confirmed and were probably due to contamination (Benson et al. 1986).

Among those just mentioned, a few individual integral matrix proteins of the spicule matrix have been isolated and characterized. Antisera produced against the bulk of proteins isolated from spicules of the sea urchin *Strongylocentrotus purpuratus* have been used to screen sea urchin cDNA libraries. The first cDNA isolated encoded a 50-kDa (more precisely, a 48-kDa; Killian and Wilt 1996) spicule protein that has been called SM50 (Benson et al. 1987; Sucov et al. 1987; Killian and Wilt 1989; Katoh-Fukui et al. 1992). It is a non-glycosylated protein with the most alkaline isoelectric point of all (Killian and Wilt 1996). The genomic DNA region that includes the SM50 gene comprises a second gene, which encodes a skeletal matrix protein called SM37 on the basis of its molecular weight (Lee et al. 1999). SM37 has an overall structure very similar to SM50, although the two proteins are only about 30% identical in their amino acid sequences. A protein orthologous with SM50, known as LSM34, has been reported to be a component of the organic matrix of the spicules of the sea urchin *Lytechinus pictus* (Livingston et al. 1991). LSM34 has been found to be essential to calcification of sea urchin spicules (Peled-Kamar et al. 2002). Another protein, showing some sequence similarity with SM50, has been named PM27; it mostly builds up at the advancing mineralizing surface of the spicule tips (Harkey et al. 1995). A "glycine-loop"-like coiled structure has been identified in its 34 AA (AA = amino acid) Pro-Gly-Met repeat domain, which may constitute a putative elastic or mobile domain (Wustman et al. 2002). All these proteins are predicted to be basic proteins comprising a signal sequence, a C-type lectin domain, and a region with a variable number of proline-rich repeats (discussed by Illies et al. 2002).

Another integral matrix protein of sea urchin spicules has been named SM30 (George et al. 1991): characterized by an acidic isoelectric point, it is probably the main occluded protein (Wilt 1999). Like SM50, SM30 contains a signal sequence and a C-type lectin domain but has no proline-rich repeats. Moreover, while SM50 expression precedes the appearance of spicules, SM30 expression can first be detected around the onset of spicule formation, and then increases rapidly as spicules grow (Kitajima et al. 1996). Using western blotting analysis and the anti-SM30 antiserum, three distinct forms of SM30 have been detected, two in embryonic spicules (called SM30-A and SM30-B; their apparent molecular weights are 43 and 46 kDa) and a third (called SM30-C; its apparent molecular weight is 49 kDa) in the adult spines of *Strongylocentrotus purpuratus* (Killian and Wilt 1996). Akasaka et al. (1994) have shown that SM30 protein is encoded by a multigene family of two and four members.

Other proteins possibly involved in sea urchin skeletogenesis have been identified. One such protein, known as MSP130, is a cell surface, acidic protein expressed by primary mesenchyme cells in both embryonic and adult echinoids (Anstrom et al. 1987; Leaf et al. 1987; Drager et al. 1989; Harkey et al. 1992). It contains an N-terminal signal peptide, two glycine-rich domains, and a C-terminal glycophospholipid anchor (Parr et al. 1990). A full-length analysis of mRNA expression by these mesenchymal cells led to the identification of partial sequences of two putative new spicule matrix proteins (SM50-related and C-lectin) and of an MSP130-related protein (MSP130-related-1) (Zhu et al. 2001). The complete sequences and patterns of expression of these products have been identified in an investigation in which two other spicule matrix proteins have been discovered: SpSM29 and SpSM32, which are characterized by highly alkaline isoelectric points and characteristic proline-rich repeat regions, so that they appear to be members of the same family as SM50, SM37 and PM27. Another MSP130-related protein (MSP130-related-2), and two abundant gene products (SpP16, SpP19) specific to the primary mesenchyme cells, have been identified (Illies et al. 2002). A calcium-binding protein has been detected in primary mesenchyme cells and the spicular syncytium and has been named CBP180 (Iwata and Nakano 1986).

12.4 Crustacean Cuticle

Most arthropods are characterized by their molt process, during which the external cuticle is lost and then renewed to allow the growing organism greater scope for expansion. Crustaceans have a calcified cuticle (the carapace), which is also cyclically lost and then reconstructed. This implies the cyclical storage of calcium and its deposition in the cuticle, a process which offers an excellent way of studying calcification. The molting cycle varies to some extent with the crustacean species; in basic terms, it can be subdivided into a premolt stage, which ends with the ecdysis, during which the two outer layers of the exoskeleton (the epicuticle and the exocuticle) are synthesized and calcium is stored in specialized organelles, a postmolt stage, during which calcium carbonate is deposited in the cuticle, and an intermediate stage which ends with the beginning of a new molting cycle. The calcium required partly comes from the old carapace, partly from food, and partly from specialized storage organelles (gastroliths in decapods, ceca of the midgut in terrestrial crustaceans, sternal deposits in terrestrial isopods, etc.) which have often been the object of investigation, arguably because they can be easily dissected from the surrounding tissues.

The matrix of gastroliths is organized into concentric fibrous or filamentous lamellar sheets which consist of chitin in an α-configuration and a high proportion of soluble proteins in a fully extended β-configuration, whose amino acid constituents are mostly Ala, Gly, Asp, Ser and Pro (Travis et al. 1967). A cDNA has been cloned that encodes an insoluble matrix protein from crayfish gastroliths (Tsutsui et al. 1999). Called GAMP (Gastrolith Associated Matrix Protein), it is rich in glutamine, aspartic acid and alanine. It has been shown by immunohistochemistry that it is localized in gastroliths and the calcified exocuticle of crayfish (Takagi et al. 2000).

Proteins involved in skeletogenesis and calcification have been isolated from crustacea (Table 12.4). A cDNA that encodes an acidic protein (pI 4.4) has been characterized from the storage organelles (ceca of the midgut) of the terrestrial crustacea *Orchestia cavimana*, which explains its name orchestin (Testenière et al. 2002). The calcium storage structures are concretions appearing as very small spherules that tend to associate with one another; they consist of amorphous calcium carbonate (Raz et al. 2002). The protein has a calculated molecular weight of 12.4 kDa, whereas the native protein extracted from the $CaCO_3$ concretions migrates as a 23-kDa band on SDS-PAGE, the difference probably being due to the high amounts of acidic amino acids (approx. 30%) in the latter. Orchestin is a phosphorylated calcium-binding protein (Luquet et al. 2000; Hecker et al. 2004); phosphorylation occurs on serine and tyrosine residues, whereas calcium binding only occurs via the phosphoserine residues (Hecker et al. 2003). Orchestin takes part in two successive calcification phases; in the first it is expressed during the premolt stage by becoming a component of the storage spherules, in the second it is resynthesized during the postmolt stage, becoming a component of the resorption spherulites (Hecker et al. 2004).

In terrestrial isopods, calcium storage occurs, as it does in terrestrial crustacea, through the formation of calcareous spherules which accumulate in various storage sites, the most common being a sternal site. The isopods are characterized by a biphasic molt, the first phase affecting the cuticle of the posterior part of the body, the second that of the anterior part (Hawkes and Schraer 1973). During premolt, $CaCO_3$ is resorbed from the posterior cuticle and stored as spherulites between the cuticle and epithelium of the first

Table 12.4. Proteins of crustacea involved in skeletogenesis and calcification

Name	Source	Crustacean	Reference
GAMP	Gastrolith, exocuticle	*Procambarus clarkii*	Takagi et al. (2000)
CAP-1[b,c]	Exoskeleton	*Procambarus clarkii*	Inoue et al. (2001)
Orchestin[a]	Ceca, midgut	*Orchestia cavimana*	Testenière et al. (2002)
Crustocalcin[b]	Epidermid	*Penaeus japonicus*	Endo et al. (2004)

[a] Acidic
[b] Acidic, phosphorylated
[c] Ca-binding

four anterior sternites. The spherulites are then degraded and calcium used for calcification of the posterior part (for details, see Fabritius and Ziegler 2003). In spherulites, $CaCO_3$ is associated with an organic matrix consisting of radial and concentric elements (Ziegler 1994). The formation of spherules appears to be initiated by the formation of nanoparticles comprising organic and inorganic components (Fabritius and Ziegler 2003).

The various types of storage have the main aim of ensuring a quick supply of calcium and, probably, carbonate to the newly formed, postmolt cuticle, which calcifies after ecdysis. This process most likely involves one or more organic components of the cuticle, but information about proteins associated with the inorganic substance is hard to come by. Glycosaminoglycans have been described in initial calcification sites (the interprismatic septa) of the crab exocuticle, which also proved to contain carbonic anhydrase (Giraud-Guille 1984). Tweedie et al. (2004) have reported I. Yano's finding that acid mucopolysaccharide-protein complexes are detectable in the exocuticle, especially in its less calcified portion. Postecdysial cuticle alteration in the blue crab, *Callinectes sapidus*, involves the disappearance of 66 and 79 kDa glycoproteins at the onset of calcification (Shafer et al. 1995). Further investigations have shown that a heavily glycosylated soluble protein can be isolated from a soluble extract of the uncalcified dorsal cuticle of the same type of crab, that it appears to be the same protein species previously described as disappearing with calcification, and that, while it can be demonstrated immunohistochemically through the exocuticle at ecdysis, the immunoreaction decreases at the interprismatic septa as little as 2 h post-molt (Tweedie et al. 2004). Proteins have been detected by polyacrylamide gel electrophoresis of water soluble extracts from the uncalcified premolt cuticle of the blue crab *Collinectes sapidus*, but not from the calcified postmolt cuticle, so that the role of being calcification inhibitors has been attributed to them (Burgess et al. 1992; Burgess and Oxendine 1995).

In an attempt to identify proteins capable of regulating calcification in the crustacean cuticle, research on genes expressed during calcification in the postmolt stage has led to the identification of cDNA, named DD4, in the epidermal cells underlying the exoskeleton of the prawn *Penaeus japonicus* (Endo et al. 2000). The deduced protein was acidic and proline rich, and possessed a Ca^{2+}-binding site. Further investigations have led to the identification of another gene, called DD9A, encoding a putative protein precursor which showed similarities with a group of crustacean and insect cuticular proteins and might have a potential role in calcifying the exoskeleton (Watanabe et al. 2000). A DD5 gene encoding a putative exoskeletal protein was then identified (Ikeya et al. 2001). An additional open reading frame of 289 amino acids was later identified in addition to that of DD4; it was found that the newly identified part of the encoded protein included a region that was rich in glutamate residues and contained Ca^{2+}-binding sites (Endo et al. 2004). This protein has been named Crustocalcin (CCN). It has been detected in the calcified regions of the endocuticular layer of the exoskeleton and its expression does not go beyond the early postmolt period.

A phosphorylated peptide, named CAP-1 (calcification associated peptide-1), has been extracted and purified from the exoskeleton of the crayfish *Procambarus clarkii* (Inoue et al. 2001). CAP-1 has been found to be rich in acidic amino acids residues, to inhibit calcium carbonate precipitation in vitro, and to have a chitin-binding ability. The mRNA which encodes CAP-1 was expressed in the epidermal tissue during the postmolt stage in the same places and at the same time as calcification (Inoue et al. 2003). Another peptide, known as CAP-2, showing the same properties as CAP-1, has subsequently been described in the exoskeleton of the same type of crayfish.

12.5 Corals

In his review of calcified tissue proteoglycans, Kobayashi (1971) reported the longstanding findings of Goreau showing that coral calcification occurs outside the calicoblastic epidermis, which secretes an organic matrix rich in acid mucopolysaccharides, and the results of Wainwright that indicate chitin as the major organic component of the skeleton. By now it is certain that calcified areas in coral contain organic material at the nanometre scale (Cuif and Dauphin, 2005), but its composition and its effect on calcification are still uncertain. An important study by Allemand et al. (1998) has, in any case, clearly shown that the synthesis of organic material and the occurrence of calcification are directly linked.

The amounts of organic material present in calcified segments of corals are very low. In one study, the organic components extracted from the exoskeleton of the coral *Galaxea fascicularis* were examined by SDS-PAGE, which showed one major glycosylated protein with an apparent molecular weight of 53 kDa and a few other minor bands; a cDNA encoding the major protein (named galaxin) could be cloned (Fukuda et al. 2003; Watanabe et al. 2003). In the same type of coral, 26 nm thick organic fibrils were found located between calicoblastic ectodermal cells and the underlying $CaCO_3$ skeleton; small nodular structures (37 nm in diameter) were observed on this fibrillar, S-rich organic material, probably corresponding to localized Ca-rich regions and nascent crystals of $CaCO_3$ (Clode and Marshall 2003). Analysis of the organic matrix in the endoskeletal spicules from the alcyonaria coral *Synularia polydactyla* showed the presence of proteins that could be separated by SDS-PAGE into seven bands, whose apparent molecular mass was 109, 83, 70, 63, 41, 30 and 22 kDa; of these, the 109 and 63 kDa were found to bind ^{45}Ca and the 83 and 63 kDa were glycosylated (Rahman et al. 2005). Puverel et al. (2005) extracted three proteins from *Stylophora pistillata* and at least five from *Pavona cactus*, some of them possessing calcium binding properties; one sequence of the former was unusual because it contained a long polyaspartate domain, like that found in proteins belonging to the calsequestrin family and in some mollusk proteins, indicative of a possible role in regulating the calcification process.

Phospholipids that were capable of binding calcium ions were found in the coral skeleton by Isa and Okazaki (1987). Histochemical and ultrastructural studies by Goldberg (2001) showed that acid proteoglycans were contained in the skeletal matrix of the coral *Mycetophyllia reesi*, and suggested the presence in it of a hyluronan-like substance.

12.6 Concluding Remarks

Although the calcified structures of invertebrates are highly polymorphic, both in terms of molecular organization and types of mineral, some general conclusions can be formulated:

- All calcified structures of invertebrates contain very low amounts of organic material; it is not known whether this material is quantitatively stable during the calcification process.
- The organic material found in calcified areas is located both between the crystals and within them: in the second case, it is occluded by inorganic material.
- Most of the organic molecules contain high proportions of aspartic acid and are strongly acidic. They comprise glycoproteins and proteoglycans.
- Chitin is present in mollusc shells. It is closely associated with less ordered, silk fibroin-like proteins, which lie between the chitin molecular planes.
- Several specific proteins have been isolated from the organic material that can be extracted from the mollusk calcified shells. Some of them (MSP-1, aspein, asprich) have an unusually acidic character, others (Nacrein or N66, MS160, Lustrin A, N16 and N14, Pearlin, Pearlin-keratin, Mucoperlin, Perlucin, Perlustrin, AP7-AP24) have so far only been extracted from the nacre aragonitic layer.
- Individual proteins have been isolated and characterized from the organic matrix occluded in the calcified spines of the sea urchin embryo. Some of these proteins (SM50, SM50-related, LSM34, PM27, SM37, SpC-lectin, SpSM29, SpSM32) share a basic character and show molecular homologies, others (SM30, SM30A, B, C, MSP130) are acidic.
- In crustacea, the calcified matrix of storage organelles and carapaces contains proteins, some of which (GAMP, CAP-1, Orchestin, Crustocalcin) have been isolated and characterized.
- Coral exoskeletons contain occluded organic material whose major protein is galaxin; seven other proteins have been separated by SDS-PAGE.

References

Akasaka K, Frudakis TN, Killian CE, George NC, Yamasu K, Khaner O, Wilt FH (1994) Genomic organization of a gene encoding the spicule matrix protein SM30 in the sea urchin *Strongylocentrotus purpuratus*. J Biol Chem 269:20592–20598

Allemand D, Tambutté È, Girard JP, Jaubert J (1998) Organic matrix synthesis in the scleractinian coral *Stylophora pistillata*: role in biomineralization and potential target of the organotin tributyltin. J Exp Biol 201:2001–2009

Ameye L, Hermann R, Killian C, Wilt F, Dubois P (1999) Ultrastructural localization of proteins involved in sea urchin biomineralization. J Histochem Cytochem 47:1189–1200

Ameye L, Hermann R, Dubois P (2000) Ultrastructure of sea urchin calcified tissues after high-pressure freezing and freeze substitution. J Struct Biol 131:116–125

Anstrom JA, Chin JE, Leaf DS, Parks AL, Raff RA (1987) Localization and expression of smp130, a primary mesenchyme lineage-specific cell surface protein in the sea urchin embryo. Development 101:255–265

Belcher AM, Wu XH, Christensen RJ, Hansma PK, Stucky GD, Morse DE (1996) Control of crystal phase switching and orientation by soluble mollusc-shell proteins. Nature 381:56–58

Benson SC, Wilt FH (1992) Calcification of spicules in the sea urchin embryo. In: Bonucci E (ed) Calcification in biological systems. CRC Press, Boca Raton, pp 157–178

Benson SC, Benson NC, Wilt F (1986) The organic matrix of the skeletal spicule of sea urchin embryos. J Cell Biol 102:1878–1886

Benson S, Sucov H, Stephens L, Davidson E, Wilt F (1987) A lineage-specific gene encoding a major matrix protein of the sea urchin embryo spicule. I. Authentication of the cloned gene and its developmental expression. Dev Biol 120:499–506

Bevelander G, Nakahara H (1969) An electron microscope study of the formation of the nacreous layer in the shell of certain bivalve molluscs. Calcif Tissue Res 3:84–92

Blank S, Arnoldi M, Khoshnavaz S, Treccani L, Kuntz M, Mann K, Grathwohl G, Fritz M (2003) The nacre protein perlucin nucleates growth of calcium carbonate crystals. J Microsc 212:280–291

Bonfield W, Scandola M (1979) Natural carbonate-reinforced composite materials Part 1 Morphology of sea urchin teeth. J Mater Sci 14:2865–2871

Burgess SK, Oxendine SL (1995) A comparison of calcium binding in *Callinectes sapidus* premolt and postmolt cuticle homogenates: implications for regulation of biomineralization. J Protein Chem 14:655–664

Burgess SK, Carey DM, Oxendine SL (1992) Novel protein inhibits in vitro precipitation of calcium carbonate. Arch Biochem Biophys 297:383–387

Cariolou MA, Morse DE (1988) Purification and characterization of calcium-binding conchiolin shell peptides from the mollusc, *Haliotis rufescens*, as a function of development. J Comp Physiol 157B:717–729

Chen CC, Boskey AL (1986) The effects of proteoglycans from different cartilage types on *in vitro* hydroxyapatite proliferation. Calcif Tissue Int 39:324–327

Choi CS, Kim YW (2000) A study of the correlation between organic matrices and nanocomposite materials in oyster shell formation. Biomaterials 21:213–222

Clode PL, Marshall AT (2003) Calcium associated with a fibrillar organic matrix in the scleractinian coral *Galaxea fascicularis*. Protoplasma 220:153–161

Crenshaw MA (1972) The soluble matrix from *Mercenaria mercenaria* shell. Biomineralization 6:6–11

Cuif J-P, Dauphin Y (2005) The two-step mode of growth in the scleractinian coral skeleton from the micrometre to the overall scale. J Struct Biol 150:319–331

Dauphin Y (2003) Soluble organic matrices of the calcitic prismatic shell layers of two Pteriomorphid bivalves. *Pinna nobilis* and *Pinctada margaritifera.* J Biol Chem 278:15168–15177

Dauphin Y, Cuif J-P, Doucet J, Salomé M, Susini J, Willams CT (2003) In situ chemical speciation of sulfur in calcitic biominerals and the simple prism concept. J Struct Biol 142:272–280

Decker G, Lennarz WJ (1988) Skeletogenesis in the sea urchin embryo. Development 103:231–247

Decker GL, Morrill JB, Lennarz WJ (1987) Characterization of sea urchin primary mesenchyme cells and spicules during biomineralization in vitro. Development 101:297–312

Drager BJ, Harkey MA, Iwata M, Whiteley AH (1989) The expression of embryonic primary mesenchyme genes of the sea urchin, *Strongylocentrotus purpuratus*, in the adult skeletogenic tissues of this and other species of echinoderms. Dev Biol 133:14–23

Endo H, Persson P, Watanabe T (2000) Molecular cloning of the crustacean DD4 cDNA encoding a Ca(2+)-binding protein. Biochem Biophys Res Comm 276:286–291

Endo H, Takagi Y, Ozaki N, Kogure T, Watanabe T (2004) A crustacean Ca^{2+}-binding protein with a glutamate-rich sequence promotes CaCO3 crystallization. Biochem J 384:159–167

Fabritius H, Ziegler A (2003) Analysis of $CaCO_3$deposit formation and degradation during the molt cycle of the terrestrial isopod *Porcellio scaber* (Crustacea, Isopoda). J Struct Biol 142:281–291

Falini G, Fermani S (2004) Chitin mineralization. Tissue Eng 10:1–6

Falini G, Albeck S, Weiner S, Addadi L (1996) Control of aragonite or calcite polymorphism by mollusk shell macromolecules. Science 271:67–69

Falini G, Weiner S, Addadi L (2003) Chitin-silk fibroin interactions: relevance to calcium carbonate formation in invertebrates. Calcif Tissue Int 72:548–554

Fukuda I, Ooki S, Fujita T, Murayama E, Nagasawa H, Isa Y, Watanabe T (2003) Molecular cloning of a cDNA encoding a soluble protein in the coral exoskeleton. Biochem Biophys Res Comm 304:11–17

George NC, Killian CE, Wilt FH (1991) Characterization and expression of a gene encoding a 30.6-kDa *Strongylocentrotus purpuratus* spicule matrix protein. Dev Biol 147:334–342

Giraud-Guille M-M (1984) Calcification initiation sites in the crab cuticle: the interprismatic septa. An ultrastructural cytochemical study. Cell Tissue Res 236:413–420

Goldberg WM (2001) Acid polysaccharides in the skeletal matrix and calicoblastic epithelium of the stony coral *Mycetophyllia reesi.* Tissue Cell 33:376–387

Gotliv BA, Kessler N, Sumerel JL, Morse DE, Tuross N, Addadi L, Weiner S (2005) Asprich: a novel aspartic acid-rich protein family from the prismatic shell matrix of the bivalve *Atrina rigida.* Chembiochem 6:304–314

Grégoire C (1957) Topography of the organic components in mother-of-pearl. J Biophys Biochem Cytol 3:797–806

Halloran BA, Donachy JE (1995) Characterization of organic matrix macromolecules from the shells of the Antarctic scallop, *Adamussium colbecki.* Comp Biochem Physiol B Biochem Mol Biol 111:221–231

Harkey MA, Whiteley HR, Whiteley AH (1992) Differential expression of the msp30 gene among skeletal lineage cells in the sea urchin embryo: a three dimensional *in situ* hybridization analysis. Mech Dev 37:173–184

Harkey MA, Klueg K, Sheppard P, Raff RA (1995) Structure, expression, and extracellular targeting of PM27, a skeletal protein associated specifically with growth of the sea urchin larval spicule. Dev Biol 168:549–566

Hattan SJ, Laue TM, Chasteen ND (2001) Purification and characterization of a novel calcium-binding protein from the extrapallial fluid of the mollusc, *Mytilus edulis*. J Biol Chem 276:4461–4468

Hawkes JW, Schraer H (1973) Mineralization during the molt cycle in *Lirceus brachyurus* (Isopoda: Crustacea). I. Chemistry and light microscopy. Calcif Tissue Res 12:125–136

Hecker A, Testenière O, Marin F, Luquet G (2003) Phosphorylation of serine residues is fundamental for the calcium-binding ability of Orchestin, a soluble matrix protein from crustacean calcium storage structures. FEBS Lett 535:49–54

Hecker A, Quennedey B, Testenière O, Quennedey A, Graf F, Luquet G (2004) Orchestin, a calcium-binding phosphoprotein, is a matrix component of two successive transitory calcified biomineralizations cyclically elaborated by a terrestrial crustacean. J Struct Biol 146:310–324

Iijima M, Moriwaki Y (1990) Orientation of apatite and organic matrix in *Lingula unguis* shell. Calcif Tissue Int 47:237–242

Ikeya T, Persson P, Kono M, Watanabe T (2001) The *DD5* gene of the decapod crustacean *Penaeus japonicus* encodes a putative exoskeletal protein with a novel tandem repeat structure. Comp Biochem Physiol B Biochem Mol Biol 128:379–388

Illies MR, Peeler MT, Dechtiaruk AM, Ettensohn CA (2002) Identification and developmental expression of new biomineralization proteins in the sea urchin *Strongylocentrotus purpuratus*. Dev Genes Evol 212:419–431

Inoue H, Ozaki N, Nagasawa H (2001) Purification and structural determination of a phosphorylated peptide with anti-calcification and chitin-binding activities in the exoskeleton of the crayfish, *Procambarus clarkii*. Biosci Biotech Biochem 65:1840–1848

Inoue H, Ohira T, Ozaki N, Nagasawa H (2003) Cloning and expression of a cDNA encoding a matrix peptide associated with calcification in the exoskeleton of the crayfish. Comp Biochem Physiol B Biochem Mol Biol 136:755–765

Isa Y, Okazaki M (1987) Some observations on the Ca^{2+}-binding phospholipids from scleractinian coral skeletons. Comp Biochem Physiol 87B:507–512

Iwata M, Nakano E (1986) A large calcium-binding protein associated with the larval spicules of the sea urchin embryo. Cell Differ 19:229–236

Katoh-Fukui Y, Noce T, Ueda T, Fujiwara Y, Hashimoto N, Tanaka S, Higashinakagawa T (1992) Isolation and characterization of cDNA encoding a spicule matrix protein in *Hemicentrotus pulcherrimus* micromeres. Int J Dev Biol 36:353–361

Kelly PG, Oliver PTP, Pautard FGE (1965) The shell of *Lingula unguis*. In: Richelle LJ, Dallemagne MJ (eds) Calcified tissues. Université de Liège, Liège, pp 337–345

Killian CE, Wilt FH (1989) The accumulation and translation of a spicule matrix protein mRNA during sea urchin embryo development. Dev Biol 133:148–156

Killian CE, Wilt FH (1996) Characterization of the proteins comprising the integral matrix of *Strongylocentrotus purpuratus* embryonic spicules. J Biol Chem 271:9150–9159

Kim IW, Morse DE, Evans JS (2004) Molecular characterization of the 30-AA N-terminal mineral interaction domain of the biomineralization protein AP7. Langmuir 20:11664–11673

Kitajima T, Tomita M, Killian CE, Akasaka K, Wilt FH (1996) Expression of spicule matrix protein gene SM30 in embryonic and adult mineralized tissues of sea urchin *Hemicentrotus pulcherrimus*. Develop Growth Differ 38:687–695

Kobayashi S (1971) Acid mucopolysaccharides in calcified tissues. Int Rev Cytol 30:257–371

Kono M, Hayashi N, Samata T (2000) Molecular mechanism of the nacreous layer formation in Pinctada maxima. Biochem Biophys Res Commun 269:213–218

Krampitz G, Engels J, Cazaux C (1976) Biochemical studies of water-soluble proteins and related components of gastropod shells. In: Watabe N, Wilbur KM (eds) The mechanisms of mineralization in the invertebrates and plants. The University of South Carolina Press, Columbia, SC, pp 155–173

Leaf DS, Anstrom JA, Chin JE, Harkey MA, Showman RM, Raff RA (1987) Antibodies to a fusion protein identify a cDNA clone encoding msp130, a primary mesenchyme-specific cell surface protein of the sea urchin embryo. Dev Biol 121:29–40

Lee Y-O, Britten RJ, Davidson EH (1999) *SM37*, a skeletogenic gene of the sea urchin embryo linked to the *SM50* gene. Develop Growth Differ 41:303–312

Levi-Kalisman Y, Falini G, Addadi L, Weiner S (2001) Structure of the nacreous organic matrix of a bivalve mollusk shell examined in the hydrated state using cryo-TEM. J Struct Biol 135:8–17

Livingston BT, Shaw R, Bailey A, Wilt F (1991) Characterization of a cDNA encoding a protein involved in the formation of the skeleton during development of the sea urchin *Lytechinus pictus*. Dev Biol 148:473–480

Luquet G, Testenière O, Graf F (2000) *Orchestia cavimana* as a model to study the hormonal regulation of a calcium storage process. In: Goldberg M, Boskey A, Robinson C (eds) Chemistry and biology of mineralized tissues. American Academy of Orthopaedic Surgeons, Rosemont, IL, pp 7–12

Mann K, Weiss IM, Andre S, Gabius HJ, Fritz M (2000) The amino-acid sequence of the abalone (*Haliotis laevigata*) nacre protein perlucin. Detection of a functional C-type lectin domain with galactose/mannose specificity. Eur J Biochem 267:5257–5264

Mann S (1988) Molecular recognition in biomineralization. Nature 332:119–124

Marin F, Muyzer G, Dauphin Y (1994) Caractérisations électrophorétique et immunologique des matrices organiques solubles des tests de deux bivalves ptériomorphes actuels, *Pinna nobilis* L. et *Pinctada margaritifera* (L.). C R Acad Sci Paris 318:1653–1659

Marin F, Corstjens P, de Gaulejac B, de Vrind-De Jong E, Westbroek P (2000) Mucins and molluscan calcification. Molecular characterization of mucoperlin, a novel mucin-like protein from the nacreous shell layer of the fan mussel *Pinna nobilis (Bivalvia, Pteriomorphia)*. J Biol Chem 275:20667–20675

Marshall AT (2002) Occurrence, distribution, and localisation of metals in cnidarians. Microsc Res Tech 56:341–357

Marxen JC, Becker W (1997) The organic shell matrix of the freshwater snail *Biomphalaria glabrata*. Comp Biochem Physiol 118B:23–33

Marxen JC, Becker W (2000) Calcium binding constituents of the organic shell matrix from the freshwater snail *Biomphalaria glabrata*. Comp Biochem Physiol B 127:235–242

Marxen JC, Hammer M, Gehrke T, Becker W (1998) Carbohydrates of the organic shell matrix and the shell-forming tissue of the snail *Biomphalaria glabrata* (say). Biol Bull 194:231–240

Marxen JC, Nimtz M, Becker W, Mann K (2003) The major soluble 19.6 kDa protein of the organic shell matrix of the freshwater snail *Biomphalaria glabrata* is an N-glycosylated dermatopontin. Biochim Biophys Acta 1650:92–98

Matsushiro A, Miyashita T, Miyamoto H, Morimoto K, Tonomura B, Tanaka A, Sato K (2003) Presence of protein complex is prerequisite for aragonite crystallization in the nacreous layer. Mar Biotechnol 5:37–44

Meenakshi VR, Hare PE, Wilbur KM (1971) Amino acids of the organic matrix of neogastropod shells. Comp Biochem Physiol 40B:1037–1043

Michenfelder M, Fu G, Lawrence C, Weaver JC, Wustman BA, Taranto L, Evans JS, Morse DE (2003) Characterization of two molluscan crystal-modulating biomineralization proteins and identification of putative mineral binding domains. Biopolymers 70:522–533

Miyamoto H, Miyashita T, Okushima M, Nakano S, Morita T, Matsushiro A (1996) A carbonic anhydrase from the nacreous layer in oyster pearls. Proc Natl Acad Sci U S A 93:9657–9660

Miyamoto H, Miyoshi F, Kohno J (2005) The carbonic anhydrase domain protein nacrein is expressed in the epithelial cells of the mantle and acts as a negative regulator in calcification in the mollusc *Pinctada fucata*. Zool Sci 22:311–315

Miyashita T, Takagi R, Okushima M, Nakano S, Miyamoto H, Nishikawa E, Matsushiro A (2000) Complementary DNA cloning and characterization of Pearlin, a new class of matrix protein in the nacreous layer of oyster pearls. Mar Biotechnol 2:409–418

Nakahara H, Bevelander G (1971) The formation and growth of the prismatic layer of *Pinctada radiata*. Calcif Tissue Res 7:31–45

Nakahara H, Bevelander G, Kakei M (1982) Electron microscopic and amino acid studies on the outer and inner shell layer of *Haliotis rufescens*. Venus, Jpn J Malacol 41:33–46

Okazaki K (1975) Spicule formation by isolated micromeres of the sea urchin embryo. Am Zool 15:567–582

Parr BA, Parks AL, Raff RA (1990) Promoter structure and protein sequence of smp130, a lipid-anchored sea urchin glycoprotein. J Biol Chem 265:1408–1413

Peled-Kamar M, Hamilton P, Wilt FH (2002) Spicule matrix protein LSM34 is essential for biomineralization of the sea urchin spicule. Exp Cell Res 272:56–61

Pereira-Mouriès L, Almeida MJ, Ribeiro C, Peduzzi J, Barthélemy M, Milet C, Lopez E (2002) Soluble silk-like organic matrix in the nacreous layer of the bivalve *Pinctada maxima*. Eur J Biochem 269:4994–5003

Peters W (1972) Occurrence of chitin in mollusca. Comp Biochem Physiol 41B:541–550

Pucci-Minafra I, Fanara M, Minafra S (1980) Chemical and physical changes in the organic matrix of mineralized tissues from embryo to adult of *Paracentrotus lividus*. J Submicr Cytol 12:267–273

Puverel S, Tambutté É, Pereira-Mouries L, Zoccola D, Allemand D, Tambutté S (2005) Soluble organic matrix of two Scleractinian corals: partial and comparative analysis. Comp Biochem Physiol B Biochem Mol Biol 141:480–487

Rahman MA, Isa Y, Uehara T (2005) Proteins of calcified endoskeleton: II Partial amino acid sequences of endoskeletal proteins and the characterization of proteinaceous organic matrix of spicules from the alcyonarian, *Synularia polydactyla*. Proteomics 5:885–893

Ravindranath MH, Rajeswari Ravindranath MH (1974) The chemical nature of the shell of molluscs: I. Prismatic and nacreous layers of a bivalve *Lamellidans marginalis* (Unionidae). Acta Histochem 48:26–41

Raz S, Testenière O, Hecker A, Weiner S, Luquet G (2002) Stable amorphous calcium carbonate is the main component of the calcium storage structures of the crustacean *Orchestia cavimana*. Biol Bull 203:269–274

Saleuddin ASM (1971) Fine structure of normal and regenerated shell of *Helix*. Can J Zool 49:37–41

Samata T, Hayashi N, Kono M, Hasegawa K, Horita C, Akera S (1999) A new matrix protein family related to the nacreous layer formation of Pinctada fucata. FEBS Lett 462:225–229

Sarashina I, Endo K (1998) Primary structure of a soluble matrix protein of scallop shell: implications for calcium carbonate biomineralization. Am Mineral 83:1510–1515

Sarashina I, Kazuyoshi E (2001) The complete primary structure of molluscan shell protein 1 (MSP-1), an acidic glycoprotein in the shell matrix of the scallop *Patinopecten yessoensis*. Mar Biotechnol 3:362–369

Shafer TH, Roer RD, Midgette-Luther C, Brookins TA (1995) Postecdysial cuticle alteration in the blue crab, *Callinectes sapidus*: synchronous changes in glycoproteins and mineral nucleation. J Exp Zool 271:171–182

Shen X, Belcher AM, Hansma PK, Stucky GD, Morse DE (1997) Molecular cloning and characterization of Lustrin A, a matrix protein from shell and pearl nacre of *Haliotis rufescens*. J Biol Chem 272:32472–32481

Sucov HM, Benson S, Robinson JJ, Britten RJ, Wilt F, Davidson EH (1987) A lineage-specific gene encoding a major matrix protein of the sea urchin embryo spicule. II. Structure of the gene and derived sequence of the protein. Dev Biol 120:507–519

Sudo S, Fujikawa T, Nakagura T, Ohkubo T, Sakaguchi K, Tanaka M, Nakashima K, Takahashi T (1997) Structures of mollusc shell framework proteins. Nature 387:563–564

Suzuki M, Murayama E, Inoue H, Ozaki N, Tohse H, Kogure T, Nagasawa H (2004) Characterization of Prismalin-14, a novel matrix protein from the prismatic layer of the Japanese pearl oyster (*Pinctada fucata*). Biochem J 382(1):205–213

Takagi Y, Ishii K, Ozaki N, Nagasawa N (2000) Immunolocalization of gastrolith matrix protein (GAMP) in the gastroliths and exoskeleton of cryfish, *Procambarus clarkii.* Zool Sci 17:179–184

Testenière O, Hecker A, Le Gurun S, Quennedey B, Graf F, Luquet G (2002) Characterization and spatiotemporal expression of orchestin, a gene encoding an ecdysone-inducible protein from a crustacean organic matrix. Biochem J 361(2):327–335

Tong H, Hu J, Ma W, Zhong G, Yao S, Cao N (2002) In situ analysis of the organic framework in the prismatic layer of mollusc shell. Biomaterials 23:2593–2598

Towe KM, Hamilton GH (1968) Ultrastructure and inferred calcification of the mature and developing nacre in bivalve mollusks. Calcif Tissue Res 1:306–318

Travis DF, François CJ, Bonar LC, Glimcher MJ (1967) Comparative studies of the organic matrices of invertebrate mineralized tissues. J Ultrastruct Res 18:519–550

Tsukamoto D, Sarashina I, Endo K (2004) Structure and expression of an unusually acidic matrix protein of pearl oyster shells. Biochem Biophys Res Commun 320:1175–1180

Tsutsui N, Ishii K, Takagi Y, Watanabe T, Nagasawa N (1999) Cloning and expression of a cDNA encoding an insoluble matris protein in the gastroliths of a crayfish, *Procambarus clarkii.* Zool Sci 16:619–628

Tweedie EP, Coblentz FE, Shafer TH (2004) Purification of a soluble glycoprotein from the undecalcified ecdysial cuticle of the blue crab *Callinectes sapidus* and its possible role in initial mineralization. J Exp Biol 207:2589–2598

Urry LA, Hamilton PC, Killian CE, Wilf FH (2000) Expression of spicule matrix proteins in the sea urchin embryo during normal and experimentally altered spiculogenesis. Dev Biol 225:201–213

Veis A, Barss J, Dahl T, Rahima M, Stock S (2002) Mineral-related proteins of sea urchin teeth: *Lytechinus variegatus.* Microsc Res Tech 59:342–351

Veis DJ, Albinger TM, Clohisy J, Rahima M, Sabsay B, Veis A (1986) Matrix proteins of the teeth of the sea urchin *Lytechinus variegatus.* J Exp Zool 240:35–46

Venkatesan M, Simpson RT (1986) Isolation and characterization of spicule proteins from *Strongylocentrotus purpuratus.* Exp Cell Res 166:259–264

Wang RZ, Addadi L, Weiner S (1997) Designed strategies of sea urchin teeth: structure, composition and micromechanical relations to function. Philos Trans R Soc London B Biol Sci 352:469–480

Watabe N (1963) Decalcification of thin sections for electron microscope studies of crystal-matrix relationships in mollusc shells. J Cell Biol 18:701–703

Watabe N (1965) Studies on shell formation. XI. Crystal-matrix relationships in the inner layer of mollusk shells. J Ultrastruct Res 12:351–370

Watanabe T, Persson P, Endo H, Kono M (2000) Molecular analysis of two genes, *DD9A* and *B*, which are expressed during the postmolt stage in the decapod crustacean *Penaeus japonicus.* Comp Biochem Physiol B Biochem Mol Biol 125:127–136

Watanabe T, Fukuda I, China K, Isa Y (2003) Molecular analyses of protein components of the organic matrix in the exoskeleton of two scleractinian coral species. Comp Biochem Physiol B Biochem Mol Biol 136:767–774

Weiner S (1979) Aspartic acid-rich proteins: major components of the soluble organic matrix of mollusk shells. Calcif Tissue Int 29:163–167

Weiner S (1985) Organic matrixlike macromolecules associated with the mineral phase of sea urchin skeletal plates and teeth. J Exp Zool 234:7–15
Weiner S, Addadi L (1991) Acidic macromolecules of mineralized tissues: the controllers of crystal formation. Trends Biochem Sci 16:252–256
Weiner S, Hood L (1975) Soluble protein of the organic matrix of mollusk shells: a potential template for shell formation. Science 190:987–989
Weiner S, Traub W (1980) X-ray diffraction study of the insoluble organic matrix of mollusk shells. FEBS Lett 111:311–316
Weiner S, Traub W (1984) Macromolecules in mollusc shells and their function in biomineralization. Phil Trans R Soc London 304 B:425–434
Weiner S, Talmon Y, Traub W (1983) Electron diffraction of mollusk shell organic matrices and their relationship to the mineral phase. Int J Biol Macromol 5:325–328
Weiss IM, Kaufman S, Mann K, Fritz M (2000) Purification and characterization of perlucin and perlustrin, two new proteins from the shell of the mollusc *Haliotis Laevigata.* Biochem Biophys Res Comm 267:17–21
Weiss IM, Gohring W, Fritz M, Mann K (2001) Perlustrin, a *Haliotis laevigata* (abalone) nacre protein, is homologous to the insulin-like growth factor binding protein N-terminal module of vertebrates. Biochem Biophys Res Comm 285:244–249
Wheeler AP (1992) Mechanisms of molluscan shell formation. In: Bonucci E (ed) Calcification in biological systems. CRC Press, Boca Raton, pp 179–216
Wierzbicki A, Sikes CS, Madura JD, Drake B (1994) Atomic force microscopy and molecular modeling of protein and peptide binding to calcite. Calcif Tissue Int 54:133–141
Wilbur KM (1976) Recent studies of invertebrate mineralization. In: Watabe N, Wilbur KM (eds) The mechanisms of mineralization in invertebrates and plants. University of South Carolina Press, Columbia, S.C., pp 79–108
Wilt FH (1999) Matrix and mineral in the sea urchin larval skeleton. J Struct Biol 126:216–226
Wilt FH (2002) Biomineralization of the spicules of sea urchin embryos. Zool Sci 19:253–261
Wilt FH, Killian CE, Livingston BT (2003) Development of calcareous skeletal elements in invertebrates. Differentiation 71:237–250
Wustman BA, Santos R, Zhang B, Evans JS (2002) Identification of a "glycine-loop"-like coiled structure in the 34 AA Pro,Gly,Met repeat domain of the biomineral-associated protein, PM27. Biopolymers 65:362–372
Wustman BA, Morse DE, Evans JS (2004) Structural characterization of the N-terminal mineral modification domains from the molluscan crystal-modulating biomineralization proteins, AP7 and AP24. Biopolymers 74:363–376
Yin Y, Huang J, Paine ML, Reinhold VN, Chasteen ND (2005) Structural characterization of the major extrapallial fluid protein of the mollusc *Mytilus edulis*: implications for function. Biochemistry 44:10720–10731
Zentz F, Bédouet L, Almeida MJ, Milet C, Lopez E, Giraud M (2001) Characterization and quantification of chitosan extracted from nacre of the abalone *Haliotis tuberculata* and the oyster *Pinctada maxima.* Mar Biotechnol 3:36–44
Zhang B, Wustman BA, Morse D, Evans JS (2002) Model peptide studies of sequence regions in the elastomeric biomineralization protein, Lustrin A. I. The C-domain consensus-PG-, -NVNCT-motif. Biopolymers 63:358–369
Zhu X, Mahairas G, Illies M, Cameron RA, Davidson EH, Ettensohn CA (2001) A large-scale analysis of mRNAs expressed by primary mesenchyme cells of the sea urchin embryo. Development 128:2615–2627
Ziegler A (1994) Ultrastructure and electron spectroscopic diffraction analysis of the sternal calcium deposits of *Porcellio scaber* Latr. (Crustacea) during the moult cycle. J Struct Biol 112:110–116

13 Calcifying Matrices: Non-skeletal Structures

13.1 Introduction

A number of non-skeletal tissues normally undergo calcification, which may follow the same general rules that hold in skeletal tissues. Calcification can also occur in unicellular organisms, in some of them constituting structures with skeletal functions. Theoretically, they should be excluded from this chapter, which deals with non-skeletal calcified tissues, but they have been included here to allow continuity in the discussion of all the topics related to calcification in unicellular organisms.

13.2 Otoliths and Otoconia

Otoliths, otherwise called statoliths, are large calcified structures of the fish inner ear which are located in the endolymph of the sacs and canals of the membranous labyrinth, each of them lying over a macula. They show analogies with the otoconia, or statoconia, of vertebrates, even if these consist of hundreds of very small organic-inorganic structures. The fact that otoliths and otoconia calcify increases the stimulus exerted by the force of gravity on sensory cells, so facilitating its transduction into neuronal signals. They consist of calcium carbonate (either aragonite, calcite or, in rare cases, vaterite, mainly depending on the species) and of organic material which, in any case, makes up less than 10% of the total. Kobayashi (1971) has cited several histochemical studies showing that the otolith organic matrix contains glycoproteins and acid proteoglycans, and a paper by Degens et al., who had named the matrix otolin, arguing that it is probably a keratin-like substance. Borelli et al. (2003) have reported a proteoglycan concentration of 0.7–1.4 μg/mg $CaCO_3$ in turbot (*Psetta maxima*) otoliths.

Several investigations have shown that otolith calcification occurs after appropriate organic matrix is secreted (for a brief description of the way otoliths are formed, see Murayama et al. 2004), and that this matrix actually comprises glycoproteins and acid proteoglycans. A set of proteins (Table 13.1), collectively known as otoconins, which characterize each of the three calcium carbonate polymorphs, has been found in otoconia (Pote and Ross 1991). Among them,

Table 13.1. Proteins of otoconia and otoliths

Name	Properties	Source	Reference
		Otoconia	
Otoconin-22	Homol. phospholipase A2	*Xenopus laevis*	Pote et al. (1993)
Otoconin-90	Homol. phospholipase A2	Mouse, guinea pig	Wang Y et al. (1998)
Otoconin-95	Glycoprotein	Mouse	Verpy et al. (1999)
		Otolith	
Starmaker	Phosphorylated, acidic	Zebra fish	Söllner et al. (2003)
OMP-1	Glycoprotein	Rainbow trout	Murayama et al. (2000)
Otolin-1	Collagenous (type VIII,X)	Chum salmon	Murayama et al. (2002)

OMP-1: Otolith matrix protein-1

the name otoconin-22 refers to a protein that has been extracted from the aragonitic otoconia of *Xenopus laevis*; it has been found to be homologous with phospholipase A_2 (Pote and Ross 1991; Pote et al. 1993). Two other proteins homologous with phospholipase A_2 have been described and named asotoconin-90 (OC90), the main aorganic component of otoconia (Wang Y et al. 1998), and otoconin-95, a 95-kDa glycoprotein, which is the major protein of the utricular and saccular otoconia and may be present in multiple isoforms as a result of differential splicing (Verpy et al. 1999).

The gene *starmaker* has recently been identified in teleost fish otoliths; it encodes a protein, itself called starmaker, which is hydrophilic, contains a number of phosphorylation sites, and comprises a high proportion (35% in all) of acidic residues (Söllner et al. 2003). Two additional proteins have been isolated from the otoliths of the rainbow trout, *Oncorhynchus mykiss*, and the chum salmon, *Onchorhynchus keta*: otolith matrix protein-1 (OMP-1) and otolin-1, major components of the EDTA-soluble and EDTA-insoluble otolith organic fraction, respectively (Murayama et al. 2000, 2002). Otolin-1 has been shown to be a collagenous protein belonging to types VIII and X of the collagen family. Histochemically, both OMP-1 and otolin-1 were found to be co-localized along the organic and inorganic otolith rings which form daily during trout embryonic and larval stages (Murayama et al. 2004). Otolith formation and inner-ear development need two other proteins: otopetrin 1 (Hughes et al. 2004) and pendrin, which has been named after the observation that targeted disruption of mouse *Pds* provides insight about the inner-ear defects encountered in Pendred syndrome (Everett et al. 2001).

13.3 Pineal Gland

The pineal gland contains a number of calcified concretions, called 'brain sand' or 'acervuli' (reviewed by Vigh et al. 1998), which consist of calcium carbonate (calcite), in this respect sharing some similarities with otoconia (Baconnier et al. 2002). Their numbers increase with age; they may be one outcome of the process of senescence, although they are already found in small numbers in children, and do not seem to be caused by specific pathological processes. They are formed within intracellular vesicles, vacuoles, lipid droplets and mitochondria, and may appear as roundish, amorphous concretions with concentric layers of different density (calcospherites), or as aggregates of thin, needle-shaped crystals. Their enlargement and coalescence may induce the breakdown of pinealocytes, which allows these concretions to migrate to the extracellular space, as shown by their association with cellular debris (Humbert and Pevet 1995). It has, however, been reported that in the pineal gland of the pig the calcified concretions are formed by the leptomeningeal tissue, not by pinealocytes (Lewczuk et al. 1994). The demonstration that the calcified structures contain sulfur (Kodaka et al. 1994) suggests the presence of an organic matrix, whose composition is still unknown.

13.4 Avian Eggshell

Avian eggshell is a biocomposite material consisting of organic matrix and inorganic material (reviewed by Arias et al. 1993), which are arranged in four layers. Proceeding from the egg white outwards, the following layers are met:

1. An inner layer, the shell membrane, which consists of two (inner and outer) sublayers; these are strongly adherent, but become separate at the obtuse pole of the egg, where they form the wall of the air chamber.
2. The mammillary layer, whose name is due to the presence of mammillae, or calcium reserve bodies, which are knob-like, organic-inorganic structures; their base is in close contact with the fibers of the outer shell membranes and their upper part corresponds to the base of the columns of the next palisade layer. These mammillary knobs consist of the calcium reserve assembly (CRA), and the crown region. The former is made up of a dense, flocculent material, and contains easy available calcium. It is capped by the calcium reserve body sac, which contains many electron-dense spherical vesicles.
3. The palisade, or calcified layer, the thickest layer of the eggshell, which is characterized by the presence of plate-like crystals of calcium carbonate (calcite in avian eggs) and of gas vesicles between them. These crystals are arranged in columns perpendicular to the egg surface; single, unbranched

channels, the pores, penetrate the palisade from the outside and end in clefts between mammillary knobs.

4. The cuticle, the outermost eggshell layer, which comprises glycoproteins and pigments. A vertical matrix layer, which contains hydroxyapatite crystals, has been described between the palisade and the cuticle (Dennis et al. 1996).

This complex architecture is built as the egg moves through the oviduct: roughly speaking, the inner and outer shell membranes are completed while in transit through the isthmus, the mammillae are initiated in the distal part of this site (red isthmus or tubular shell gland) and develop in the uterus (or shell gland), and the bulk of calcite is deposited in the uterus. The whole process is completed in about 22 h (Fernandez et al. 1997), calcium ions being provided by intestinal absorption and, to a large extent (25–40% according to Comar and Driggers 1949), by resorption of medullary bone (Bloom et al. 1942).

The organization of avian eggshell has been studied under transmission and scanning electron microscopes (Quintana and Sandoz 1978; Dennis et al. 1996). Its subdivision into four layers has been confirmed, and sublayers have been described; a detailed analysis of the findings of these and other studies does not appear to be needed at this point, and only the main morphological data will be reported, together with biochemical results where this is appropriate.

The inner and outer shell membranes show a fibrillary structure whose fibrils consist of a central, dense collagenous core surrounded by a less dense, amorphous material (the mantle), which is rich in polyanions. The fibrils of the inner layer are thinner than those of the outer layer, some of which penetrate the mammillary layer (Quintana and Sandoz 1978). The early mammillary knobs consist of organic, flocculent material, which can also be demonstrated in the palisade layer after decalcification with 5% acetic acid. It forms a continuous layer around the egg (Arias et al. 1993) and can be stained by methylene blue (Terepka 1963). The organic material in the palisade zone displays electron-transparent microvesicles outlined by a border showing rising densities.

Eggshell is a polymer-ceramic composite whose organic component consists of proteins and polysaccharides. Kobayashi (1971) has reported the results of Baker and Balk showing that, in fact, polysaccharides and glycosaminoglycans account for about 11%, and proteins for about 70% of the organic material of the shell. Arias et al. (1993) have reviewed the findings, showing that the shell membrane does not, as hypothesized, consist of keratin-like material; it is collagenous in nature, while acid proteoglycans are contained in the mammillary layer. Type I and type X collagens, keratan sulfate (mammillan), and dermatan sulfate (ovoglycan), have been described in the egg membranes (Carrino et al. 1996; Fernandez et al. 1997, 2001); the expression of type X collagen has been reported in the tubular gland cells of the isthmus (Wang X et al. 2002). Further studies have shown that both the membranes and the eggshell matrix contain type X collagen (Carrino et al. 1996; Arias et al. 1997), glycosaminoglycans and sulfate proteoglycans (Picard et al. 1973; Nakano et al. 2001). According to

Arias et al. (1997), boundaries are established by type X collagen non-helical domain which protect from calcification and must be removed to facilitate the calcification of the egg shell.

Uronic acid has been found both in the inner and outer membranes and, at a concentration five times greater, in the shell organic matrix, where sialic acid has also been detected (Nakano et al. 2003). Moreover, it has been shown that the water-soluble material extracted from EDTA-decalcified eggshell has a high hexosamine and uronic acid content (Abatangelo et al. 1978), that the organic matrix of eggshell contains chondroitinase-sensitive proteoglycans comprising only a small proportion of dermatan sulfate and proteins which cannot be separated from proteoglycans by second anion exchange chromatography (Carrino et al. 1997), and that a purified preparation of the extracted glycosaminoglycan corresponds to hyaluronic acid (Heaney and Robinson 1976). Dermatan sulfate proteoglycans, which are able to affect the growth of calcite crystals, have been extracted from the eggshell (Wu et al. 1995). Galactosaminoglycan uronic acid, mostly linked to chondroitin sulfate-dermatan sulfate copolymers, has also been detected (Nakano et al. 2001, 2002). A number of different proteoglycans have been found in discrete regions within the eggshell (Arias et al. 1992): dermatan sulfate was observed in the matrix of the eggshell in all species except quail, which contained chondroitin-6-sulfate, and keratan sulfate was found in the mammillary bodies of hen, quail, pheasant and turkey, while chondroitin sulfate was also detected in guinea-fowl and duck (Panhéleux et al. 1999). Western blotting and immunogold labeling with the antibody Epi2 successfully detected molecules with the apparent molecular weight of the eggshell dermatan sulfate proteoglycan, which were localized in the secretory vesicles of non-ciliated epithelial cells and in the eggshell matrix (Dennis et al. 2000).

Studies have been carried out to identify specific proteins in the eggshell but, as this is built through the organization of calcium carbonate and organic components from the uterine fluid, research has been done not only on the matrix of the shell, but also on the native proteins that may be present in the fluid (Table 13.2). This has led to the chromatographic identification of calcium-binding proteins whose properties change according to the phase of eggshell formation (Gautron et al. 1997). Further analysis of the fluid has permitted additional soluble protein microsequencing into 15, 32, 36 and 80-kDa fractions, and the identification of the 15 and 80-kDa fractions as the egg-white proteins lysozyme and ovotransferrin, respectively.

The complex array of proteins present in the uterine fluid and eggshell matrix has been divided into three groups (see Reyes-Grajeda et al. 2004): lysozyme (14.4-kDa), ovotransferrin (78-kDa), and ovalbumin (43-kDa), which correspond to proteins of egg-white; osteopontin, which is similar to the homonymous bone protein; ovocleidins and ovocalyxins, which are specifically located in the uterus and in eggshell.

Lysozyme and ovotransferrin are present in the uterine fluid, in eggshell, and in cells of the isthmus and uterus, which also express their mRNA (Gautron

Table 13.2. Proteins of the oviduct and eggshell

Name	Properties	Location	Reference
Lysozyme	Egg-white protein	SMa,SMe	Hincke et al. (2000)
Ovotransferrin	Egg-white protein	CI,CU,Mam,SMe	Gautron et al. (2001a)
Ovalbumin	Egg-white protein	SMa	Hincke (1995)
Osteopontin	See bone matrix	CI,CU,SMe,Mam,SM	Pines et al. (1995)
Ovocleidin-17[a]	C-type lectin domain	SMa	Hincke et al. (1995)
Ovocleidin-23[b]	C-type lectin domain	SMa	Mann (1999)
Ovocleidin-116	Core protein DerSulf	Mam,SMa	Hincke et al. (1999)
Ovocalyxin-32	80-kDa protein	UF,outer SMa, Cut	Gautron et al. (2000)
Ansocalcin	C-type lectin domain	SMa	Lakshminarayanan et al. (2003)
Struthiocalcin-1	C-type lectin domain	SMa	Mann and Siedler (2004)
Struthiocalcin-2	C-type lectin domain	SMa	Mann and Siedler (2004)
Clusterin	Glycoprotein	UF,Sme,Mam,SMa	Mann et al. (2003)

CI: cells of the isthmus; CU: cells of the uterus; Cut: cuticle; DerSulf: dermatan sulfate; Mam: mammillae; SMa: shell matrix; SMe: shell membrane; UF: uterine fluid

[a] Two phosphorylated serines

[b] One phosphorylated serine

et al. 2000). More specifically, lysozyme protein is abundant in the innermost layer of the shell membranes and in the matrix of the calcified shell (Hincke et al. 2000), and ovotransferrin protein has been localized in the eggshell membranes and mammillae, and has been found, by northern blotting and RT-PCR, to be expressed in the proximal oviduct (magnum and white isthmus) and, to a lower degree, in the distal oviduct (red isthmus and uterus; Gautron et al. 2001a). Ovalbumin, or at least a protein whose molecular weight and chromatographic properties resemble those of purified egg ovalbumin, has been detected in eggshell (Hincke 1995).

The non-collagenous bone protein osteopontin is synthesized by the epithelial cells of the eggshell gland in the laying chicken; once secreted out of cells, it accumulates in eggshell and becomes part of its organic matrix (Pines et al. 1995). In a comparative study on several birds, OPN has been observed in laying and breeder hens and quail (Panhéleux et al. 1999). Immunohistochemical studies have shown that it is localized, concomitantly with the presence of the egg, in the ciliated epithelial cells of the isthmus and non-ciliated cells of the uterus; it has, moreover, been found in the core of non-calcified shell membrane fibers, at the base of the mammillae, and in the outermost part of the palisade (Fernandez et al. 2003).

The third group of eggshell proteins comprises ovocleidin-17 (OC-17), which is secreted during shell formation and is a major component of the shell matrix (Hincke et al. 1995), showing uniform distribution throughout; in the shell matrix, it is mainly associated with the sheets of flocculent material that permeate the palisade layer (Hincke et al. 1999). It belongs to a heterogeneous group of soluble proteins consisting of a single C-type lectin domain. Its amino-acid sequence contains 142 amino acids comprising two phosphorylated serines (Mann and Siedler 1999); a minor glycosylated form with an identical sequence, but only one phosphorylated serine (OC-23), has been described, too (Mann 1999). The crystal structure of monomeric OC-17 has been determined by Reyes-Grajeda et al. (2004).

Another eggshell protein, ovocleidin-116, has been detected by electron microscopy immunohistochemistry in non-ciliated (granular) cells of the shell gland epithelium, in the organic matrix of the eggshell, in small vesicles found throughout the calcified palisade layer and in the CRA of the mammillary layer (Hincke et al. 1999). OC-116 contains two N-glycosylation sites (Nimtz et al. 2004); it is considered to be a major component of the chicken eggshell matrix and to correspond to the core protein of an eggshell dermatan sulfate proteoglycan, which has still only been partly characterized (Hincke et al. 1999; Mann et al. 2002).

Two additional proteins have been identified in the eggshell matrix. One is a 32-kDa protein named ovocalyxin-32 (Gautron et al. 2000); it is predominantly localized in the outer eggshell and is found in the uterine fluid, mainly in the terminal phase of shell calcification (Hincke et al. 2003). Immunohistochemical studies have identified ovocalyxin-32 in the outer palisade layer, the vertical crystal layer and the cuticle of eggshell (Gautron et al. 2001b). Another, 15-kDa protein, known as ansocalcin, is a major constituent of goose eggshell extract; its complete amino acid sequence shows homology with ovocleidin-17 and C-type lectins (Lakshminarayanan et al. 2002, 2003); its central region is distinguished by multiplets of charged amino acids (Ajikumar et al. 2003). Two different C-type lektin-like proteins, named struthiocalcin-1 (SCA-1) and struthiocalcin-2 (SCA-2), have been reported to be major components of the ostrich eggshell matrix: SCA-1 is the ortholog of goose eggshell ansocalcin; SCA-2 shows features both of ansocalcin and chicken OC-17 (Mann and Siedler 2004).

The chicken eggshell matrix contains clusterin, a glycoprotein found in the uterine fluid as a disulfide-bonded heterodimer; it has been detected in the shell membranes and in mammillary and palisade layers using immunofluorescence and colloidal gold immunocytochemistry (Mann et al. 2003).

13.5 Unicellular Organisms

The immense population of unicellular organisms offers plenty of examples of intra-, peri- and extra-cellular calcification with the formation of a number of different inorganic salts (Pautard 1970). It appears to be reasonable to hypothesize that the inorganic substance, whatever its composition, must be closely related to some organic components, as it is in higher organisms.

The main difference is that in these organisms the process mainly takes place in the extracellular matrix, whereas in unicellular organisms it occurs within cells or close to their outer membrane. So, calcification is not controlled by a cohort of cells – osteoblasts, odontoblasts, ameloblasts, and so on – but the micro-organism itself supervises its own calcification (it is worth mentioning that pathological intracellular calcifications are not considered here; they are discussed in the next chapter). The organic components related to the calcification process must, therefore, correspond to substances or structures that are proper to the cytoplasm. This should theoretically simplify the problem: even so, though plenty of studies have been carried out to detect and characterize those components, gaps in knowledge are still exceedingly numerous.

Calcification in unicellular organisms has chiefly been studied in oral and marine bacteria, Foraminifera and Coccolithophorids. Diatomaceae, which have a silica shell, have not been considered in this volume.

13.5.1 Bacteria

Bacteria characterized by the deposition of inorganic substance (calcium phosphate or carbonate) in their cytoplasm are good examples of biological calcification and can provide useful cues to clarification of the roles of the organic structures that take part in the process, mainly because they can easily be grown in vitro under widely differing experimental conditions. They cannot be reviewed in any detail, both because they are exceedingly numerous and because differences between and within species are frequent but finely drawn, so much so that closely related bacterial strains may show different patterns of calcification (Lie and Selvig 1974a). For this reason, and because of their easy availability and their role in inducing the formation of dental calculus and secondary oral diseases, studies have mainly been directed to oral bacteria (Fig. 4.5). In this connection, the microorganism *Corynebacterium matruchotii*, also known as *Bacterionema matruchotii*, which is found in tooth calcified deposits, has received a great deal of attention. An excellent review on microbial calcification is available (Boyan et al. 1992).

The calcification of *C. matruchotii*, like that of many other oral bacteria (Ennever 1960), occurs intracellularly, but electron microscopy of dental plaque during its formation has clearly shown the presence of needle-shaped crystals both within and between bacteria, as well as on their peripheral membrane (Gonzales and Sognnaes 1960; Takazoe et al. 1963; Höhling et al. 1969; Vogel and Ennever 1971; Theilade et al. 1976; Lo Storto et al. 1990), the predominance of one site or another probably depending on type of bacteria and nutritional factors (Lie and Selvig 1974b).

It has been reported that calcification of the bacterial plaque begins in the intermicrobial spaces (Zander et al. 1960). Vesicles associated with calcification, to some extent resembling cartilage matrix vesicles, have been described as being released by *C. matruchotii* (Lie and Selvig 1974a) and as being found in the extracellular space in supragingival calculus (Lo Storto et al. 1990), where they could be produced by the degeneration and death of bacteria. Since these vesicles display alkaline phosphatase activity (Lo Storto et al. 1992), they might be responsible for early calcium phosphate deposition. Alkaline phosphatase activity can, however, been found in association with the bacterial membrane (Fig. 4.5), independently of extracellular vesicles. It must be recalled that a dental calculus can occur in germ-free rats (Fitzgerald and McDaniel 1960). On the other hand, it has been shown that there is a perfect coherence of lattice fringes between dental calculus crystals and those of enamel (Hayashi 1993), as if the former were nucleated by the latter.

Several reports have linked bacterial calcification with the nucleating effect of lipids (Boyan et al. 1984), a possibility supported by the often reported close relationship between bacterial membranes and inorganic substance (Vogel and Ennever 1971) and by the extraction of calcium-phospholipid-phosphate complexes and proteoilipids from calcifying *Bacterionema matruchotii* (Boyan and Boskey 1984).

13.5.1.1 Magnetosomal Bacteria

Besides oral bacteria, other microorganisms capable of accumulating inorganic material in their cytoplasm have been studied. Of these, magnetosomal bacteria (Fig. 4.6) have attracted attention because of the unique nature of their iron-containing magnetosomes, the mechanism of their formation, and the important implications that can be drawn from the organic-inorganic relationships that develop in them. Organic material is, in fact, closely related to iron nanoparticles (reviewed by Matsunaga and Okamura 2003; Schüler 2004). It surrounds the magnetosome (Balkwill et al. 1980; Gorby et al. 1988; Spring et al. 1998) and, in its turn, is enclosed by a trilaminar membrane probably formed by invagination of the cytoplasmic membrane (Matsunaga et al. 2000). It consists of phospholipids and proteins, some of which appear to be specific constituents (Balkwill et al. 1980; Gorby et al. 1988). Phosphatidylethanolamine and phosphatidylglycerol are the most abundant polar lipids; some proteins are tightly bound to the iron crystals, others are loosely bound and easily solubilized. Five proteins that are specifically expressed on the magnetosome membrane of *Magnetospirillum magneticum* AMB-1 have been reported by Okamura et al. (2000); their molecular weight (SDS-PAGE) was 12.0, 16.0, 24.8, 35.6, and 66.2 kDa, the 16 kDa being the most plentiful in the membrane. The 12.0 and 66.2-kDa proteins have not been isolated; the 16.0-kDa, also called Mms16, is a GTPase, the 24.8-kDa, also called Mam22, is a tetratricopeptide, and the 35.6-kDa (also called MpsA) is an acyl-CoA transferase (Okamura et al. 2001).

Several other proteins have been detected in the magnetosome organic material. Their current names are rather confusing; in some cases, different names may indicate the same protein. Thus, Matsunaga and Okamura (2003) have reported that a 24-kDa protein, isolated from *Magnetospirillum magneticum* and called Mms24, corresponds to Mam22 from strain MS-1 and to MamA from strain MSR-1. A list of 18 proteins in the magnetosome organic material, identified in *Magnetospirillium gryphiswaldense* (Grünberg et al. 2004), has been reported by Schüler (2004), who has summarized the characteristics and properties of some of them: MamA is a receptor that interacts with cytoplasmic proteins or is involved in the assembly of multiprotein complexes; MamB and MamM are cation transporters; MamE and MamO display sequence similarity to HtrA-serine proteases; MamC and MamF are unique components of magnetosome; MamD, MamG and Mms6 show the presence in their sequence of repetitive motifs showing similarities with repetitive sequences found in fibroin, mollusc shell proteins, elastin and some cartilage proteins; MamJ is rich in repeats of the acidic amino acid residues glutamate and aspartate. Several low molecular mass proteins tightly bound to magnetite have been obtained from magnetosomes of *Magnetospirillum magneticum*; four of these (Mms5, Mms6, Mms7, and Mms13) have carboxyl and hydroxyl groups that bind iron ions (Arakaki et al. 2002). Mms7 is homologous with the C-terminal region of MamD, and Mms13 to Man C (Matsunaga and Okamura 2003). Komeili et al. (2004) have reported that MamA is required for the formation of functional magnetosome vesicles, which may exist in the absence of magnetite, as shown by their presence in *Magnetospirillum* sp. AMB-1 grown in an iron-deprived medium.

13.5.2 Foraminifera

Although it has long been known that the calcified testes of Foraminifera comprise both organic and inorganic material, knowledge of the former is rather limited. Towe and Cifelli (1967) have reported Moss' view that the organic matrix is secreted before calcification and that it consists of two different materials, which are defined by them as being 'active' and 'passive' with respect to calcification. The former (Moss's organized or oriented phase) would then be a protein, the latter (Moss's unorganized ground substance) a chitin-like polysaccharide. The nature of the calcifiable organic matrix in Foraminifera is largely unknown.

13.5.3 Coccolithophorids

As in other unicellular organisms, the calcified structures, or coccoliths, which form the skeleton of the planctonic algae called Coccolithophorids, consist of a mixture of organic and inorganic material. The early phase of their development occurs in vesicles of the Golgi apparatus (for clear illustrations of coccoliths, and their development and calcification in *Pleurochrysis carterae*, one of the most actively studied species, see van der Wal et al. 1983, Marsh 1994, Young et al. 1999 and Marsh et al. 2002). Coccolith development is initiated by the formation of oval organic base-plate scales which are synthesized in the medial- and trans-Golgi vesicles. They can remain uncalcified until they are exocytosed and incorporated in the inner layer of the coccosphere (see below). Alternatively, they fuse with Golgi vesicles containing 25 nm thick, calcium-rich particles or coccolithosomes (Outka and Williams 1971), so forming coccolith saccules, otherwise known as mineralizing vesicles. In these vesicles, calcium-rich particles and acid polysaccharides mediate the deposition of a rim of calcite crystals around the base-plates. This is followed by crystal growth; at the same time, the vesicles dilate and an organic material can be detected which coats and connects crystals. Once coccoliths have completed their development, they are exocytosed: calcified scales and coccoliths assemble into an extracellular layer which, together with underlying layers of uncalcified scales, form the wall of the alga, or coccosphere.

Possibly because of its high acid polysaccharide content, the organic material which envelops and connects crystals is particularly evident when cells are fixed in the presence of cetylpyridinium chloride (van der Wal et al. 1983). This is in agreement with the finding that the coccolithosome-containing Golgi vesicles are preferentially stained by Thiery's periodic acid-thiocarbohydrazide-silver proteinate method for glycoproteins. The presence of polysaccharides in coccoliths has been confirmed by the isolation of a polysaccharide from the coccoliths of *Emiliania huxleyi* (Van Emburg et al. 1986) and by the detection of three polysaccharides from coccoliths of *Pleurochrysis cartera* (Marsh et al. 1992) and from the surface of *Pleurochrysis haptonemofera* (Hirokawa et al. 2005). These are acidic polysaccharides that have been called PS1, PS2 and PS3. PS1 makes up about 22% of the polyanion fraction and is predominantly a polymer of galacturonic and glucuronic acids. PS2 is the most abundant polyanion (accounting for about 77% of all polyanions); it displays an unusual repetitive sequence of d-glucuronic, meso-tartaric, and glyoxylic acid residues, which provide a high density of negatively charged groups. PS3, which makes up less than 2% of the coccolith polyanions, is predominantly a sulphated polymer of galacturonic acid (Marsh et al. 1992). PS1 and PS2 show some analogies with dentin phosphoprotein (Marsh et al. 1992). On the basis of immunohistochemical ultrastructural studies, it has been shown that these polysaccharides are synthesized in medial Golgi cisternae and then follow the developmental pathway that leads to coccolith formation; it has been reported that they become part of the amorphous organic coat that surrounds and links mature crystals, disappearing after the maturation of coccoliths which, however, contain and are coated by the PS1 and PS2 polysaccharides (Marsh 1994). PS3 is a sulphated galacturonomannan which is considered to be necessary for the formation and growth of the definitive coccolith (Marsh et al. 2002; Marsh 2003). A polysaccharide with anti-calcification activity, known as CNAP, has been isolated from

EDTA-soluble organic materials extracted from the coccoliths of *Pleurochrysis cartera* (Ozaki et al. 2001). Young et al. (1999) have reported the results of Corstjens et al. showing that a calcium-binding protein can be found in the crystal-coating material in coccoliths of *Emiliania huxleyi*.

13.6 Concluding Remarks

Calcified non-skeletal matrices and unicellular organisms provide important, reliable models of biological calcification. The following features deserve special consideration:

- The calcified tissues, cells and structures considered in this chapter are analogous with skeletal tissues as far as the presence of organic material in calcified areas is concerned.
- Polysaccharides and proteins are the components most often described in this material; proteolipids and calcium-acidic phospholipid-phosphate complexes do, however, characterize the calcification of oral bacteria.
- Many of these substances are acidic or strongly acidic, and may be phosphorylated.
- Several specific proteins have been isolated and characterized; their descriptive, at least partly subjective names point to the uncertainty of their nature and function.
- Intracellular or extracellular vesicles or compartments have been described as sites of initial calcification.

References

Abatangelo G, Daga-Gordini D, Castellani I, Cortivo R (1978) Some observations on the calcium ion binding of the eggshell matrix. Calcif Tissue Int 26:247–252

Ajikumar PK, Lakshminarayanan R, Ong BT, Valiyaveettil S, Kini RM (2003) Eggshell matrix protein mimics: designer peptides to induce the nucleation of calcite crystal aggregates in solution. Biomacromolecules 4:1321–1326

Arakaki A, Webb J, Matsunaga T (2002) A novel protein tightly bound to bacterial magnetic particles in *Magnetospirillum magneticum* strain AMB-1. J Biol Chem 278:8745–8750

Arias JL, Carrino DA, Fernandez MS, Rodriguez JP, Dennis JE, Caplan AI (1992) Partial biochemical and immunochemical characterization of avian eggshell extracellular matrices. Arch Biochem Biophys 298:293–302

Arias JL, Fink DJ, Xiao S-Q, Heuer AH, Caplan AI (1993) Biomineralization and eggshells: cell-mediated acellular compartments of mineralized extracellular matrix. Int Rev Cytol 145:217–250

Arias JL, Nakamura O, Fernández MS, Wu J-J, Knigge P, Eyre DR, Caplan AI (1997) Role of type X collagen on experimental mineralization of eggshell membranes. Connect Tiss Res 36:21–33

Baconnier S, Lang SB, Polomska M, Hilczer B, Berkovic G, Meshulam G (2002) Calcite microcrystals in the pineal gland of the human brain: first physical and chemical studies. Bioelectromagnetics 23:488–495

Balkwill DL, Maratea D, Blakemore RP (1980) Ultrastructure of a magnetotactic spirillum. J Bacteriol 141:1399–1408

Bloom W, Bloom MA, McLean FC (1942) Calcification and ossification. Medullary bone changes in the reproductive cycle of female pigeons. Anat Rec 83:443–466

Borelli G, Mayer-Gostan N, Merle PL, De Pontual H, Boeuf G, Allemand D, Payan P (2003) Composition of biomineral organic matrices with special emphasis on turbot (*Psetta maxima*) otolith and endolimph. Calcif Tissue Int 72:717–725

Boyan BD, Boskey AL (1984) Co-isolation of proteolipids and calcium-phospholipid-phosphate complexes. Calcif Tissue Int 36:214–218

Boyan BD, Landis WJ, Knight J, Dereszewski G, Zeagler J (1984) Microbial hydroxyapatite formation as a model of proteolipid-dependent membrane-mediated calcification. Scann Electr Microsc 4:1793–1800

Boyan BD, Swain LD, Everett MM, Schwartz Z (1992) Mechanisms of microbial mineralization. In: Bonucci E (ed) Calcification in biological systems. CRC Press, Boca Raton, pp 129–156

Carrino DA, Dennis JE, Wu T-M, Arias JL, Fernandez MS, Rodriguez JP, Fink DJ, Heuer AH, Caplan AI (1996) The avian eggshell extracellular matrix as a model for biomineralization. Connect Tissue Res 35:325–329

Carrino DA, Rodriguez JP, Caplan AI (1997) Dermatan sulfate proteoglycans from the mineralized matrix of the avian eggshell. Connect Tissue Res 36:175–193

Comar CL, Driggers JC (1949) Secretion of radioactive calcium in the hen's egg. Science 109:282

Dennis JE, Xiao S-Q, Agarwal M, Fink DJ, Heuer AH, Caplan AI (1996) Microstructure of matrix and mineral components of eggshells from white leghorn chickens (*Gallus gallus*). J Morphol 228:287–306

Dennis JE, Carrino DA, Yamashita K, Caplan AI (2000) Monoclonal antibodies to mineralized matrix molecules of the avian eggshell. Matrix Biol 19:683–692

Ennever J (1960) Intracellular calcification by oral filamentous microrganisms. J Periodontol 31:304–307

Everett LA, Belyantseva IA, Noben-Trauth K, Cantos R, Chen A, Thakkar SI, Hoogstraten-Miller SL, Kachar B, Wu DK, Green ED (2001) Targeted disruption of mouse *Pds* provides insight about the inner-ear defects encountered in Pendred syndrome. Hum Mol Genet 10:153–161

Fernandez MS, Araya M, Arias JL (1997) Eggshells are shaped by a precise spatio-temporal arrangement of sequentially deposited macromolecules. Matrix Biol 16:13–20

Fernandez MS, Moya A, Lopez L, Arias JL (2001) Secretion pattern, ultrastructural localization and function of extracellular matrix molecules involved in eggshell formation. Matrix Biol 19:793–803

Fernandez MS, Escobar C, Lavelin I, Pines M, Arias JL (2003) Localization of osteopontin in oviduct tissue and eggshell during different stages of the avian egg laying cycle. J Struct Biol 143:171–180

Fitzgerald RJ, McDaniel EG (1960) Dental calculus in the germ-free rat. Arch Oral Biol 2:239–240

Gautron J, Hincke MT, Nys Y (1997) Precursor matrix proteins in the uterine fluid change with stage of eggshell formation in hens. Connect Tissue Res 36:195–210

Gautron J, Hincke MT, Panhéleux M, Garcia-Ruiz J, Nys Y (2000) Identification and characterization of matrix proteins from hen's eggshell. In: Goldberg M, Boskey A, Robinson C (eds) Chemistry and biology of mineralized tissues. American Academy of Orthopaedic Surgeons, Rosemont, pp 19–23

Gautron J, Hincke MT, Panheleux M, Garcia-Ruiz JM, Boldicke T, Nys Y (2001a) Ovotransferrin is a matrix protein of the hen eggshell membranes and basal calcified layer. Connect Tissue Res 42:255–267

Gautron J, Hincke MT, Mann K, Panhéleux M, Bain M, McKee MD, Solomon SE, Nys Y (2001b) Ovocalyxin-32, a novel chicken eggshell matrix protein. Isolation, amino acid sequencing, cloning, and immunocytochemical localization. J Biol Chem 276:39243–39252

Gonzales HA, Sognnaes RF (1960) Electron microscopy of dental calculus. Science 131:156–158

Gorby YA, Beveridge TJ, Blakemore RP (1988) Characterization of the bacterial magnetosome membrane. J Bacteriol 170:834–841

Grünberg K, Müller EC, Otto A, Reszka R, Linder D, Kube M, Reinhardt R, Schüler D (2004) Biochemical and proteomic analysis of the magnetosome membrane in *Magnetospirillum gryphiswaldense*. Appl Environ Microbiol 70:1040–1050

Hayashi Y (1993) High resolution electron microscopy of the junction between enamel and dental calculus. Scanning Microsc 7:973–978

Heaney RK, Robinson DS (1976) The isolation and characterization of hyaluronic acid in egg shell. Biochim Biophys Acta 451:133–142

Hincke MT (1995) Ovalbumin is a component of the chicken eggshell matrix. Connect Tissue Res 31:227–233

Hincke MT, Tsang CPW, Courtney M, Hill V, Narbaitz R (1995) Purification and immunohistochemistry of a soluble matrix protein of the chicken eggshell (Ovocleidin 17). Calcif Tissue Int 56:578–583

Hincke MT, Gautron J, Tsang CPW, McKee MD, Nys Y (1999) Molecular cloning and ultrastructural localization of the core protein of an eggshell matrix proteoglycan, ovocleidin-116. J Biol Chem 274:32915–32923

Hincke MT, Gautron J, Panhéleux M, Garcia-Ruiz J, McKee MD, Nys Y (2000) Identification and localization of lysozyme as a component of eggshell membranes and eggshell matrix. Matrix Biol 19:443–453

Hincke MT, Gautron J, Mann K, Panheleux M, McKee MD, Bain M, Solomon SE, Nys Y (2003) Purification of ovocalyxin-32, a novel chicken eggshell matrix protein. Connect Tissue Res 44:16–19

Hirokawa Y, Fujiwara S, Tsuzuki M (2005) Three types of acidic polysaccharides associated with coccolith of *Pleurochrysis haptonemofera*: comparison with *Pleurochrysis carterae* and analysis using fluorescein-isothiocyanate-labeled lectins. Mar Biotechnol (NY) 7:634–644

Höhling HJ, Pfefferkorn G, Radicke J, Vahl J (1969) Elektronenmikroskopische Untersuchungen zur organischen Matrix und Kristalbildung in meschlichen Speichelsteinen. Deut Zahnärztl Z 24:663–670

Hughes I, Blasiole B, Huss D, Warchol ME, Rath NP, Hurle B, Ignatova E, Dickman JD, Thalmann R, Levenson R, Ornitz DM (2004) Otopetrin 1 is required for otolith formation in the zebrafish *Danio rerio*. Dev Biol 276:391–402

Humbert P, Pevet P (1995) Calcium concretions in the pineal gland of aged rats: an ultrastructural and microanalytical study of their biogenesis. Cell Tissue Res 279:565–573

Kobayashi S (1971) Acid mucopolysaccharides in calcified tissues. Int Rev Cytol 30:257–371

Kodaka T, Mori R, Debari K, Yamada M (1994) Scanning electron microscopy and electron probe microanalysis studies of human pineal concretions. J Electron Microsc (Tokyo) 43:307–317

Komeili A, Vali H, Beveridge TJ, Newman DK (2004) Magnetosome vesicles are present before magnetite formation, and MamA is required for their activation. Proc Natl Acad Sci U S A 101:3839–3844

Lakshminarayanan R, Kini RM, Valiyaveettil S (2002) Investigation of the role of ansocalcin in the biomineralization in goose eggshell matrix. Proc Natl Acad Sci U S A 99:5155–5159

Lakshminarayanan R, Valiyaveettil S, Rao VS, Kini RM (2003) Purification, characterization, and *in vitro* mineralization studies of a novel goose eggshell matrix protein, ansocalcin. J Biol Chem 278:2928–2936

Lewczuk B, Przybylska B, Wyrzykowski Z (1994) Distribution of calcified concretions and calcium ions in the pig pineal gland. Folia Histochem Cytobiol 32:243–249

Lie T, Selvig KA (1974a) Calcification of oral bacteria: an ultrastructural study of two strains of Bacterionema matruchotii. Scand J Dent Res 82:8–18

Lie T, Selvig KA (1974b) Effect of salivary proteins on calcification of oral bacteria. Scand J Dent Res 82:135–143

Lo Storto S, Di Grezia R, Silvestrini G, Cattabriga M, Bonucci E (1990) Studio morfologico ultrastrutturale di tartaro sopragengivale. Min Stomat 39:83–89

Lo Storto S, Silvestrini G, Bonucci E (1992) Ultrastructural localization of alkaline and acid phosphatase activities in dental plaque. J Periodont Res 27:161–166

Mann K (1999) Isolation of a glycosylated form of the chicken eggshell protein ovocleidin and determination of the glycosylation site. Alternative glycosylation/phosphorylation at an N-glycosylation sequon. FEBS Lett 463:12–14

Mann K, Siedler F (1999) The amino acid sequence of ovocleidin 17, a major protein of the avian eggshell calcified layer. Biochem Mol Biol Int 47:997–1007

Mann K, Siedler F (2004) Ostrich (*Struthio camelus*) eggshell matrix contains two different C-type lectin-like proteins. Isolation, amino acid sequence, and posttranslational modifications. Biochim Biophys Acta 1696:41–50

Mann K, Hincke MT, Nys Y (2002) Isolation of ovocleidin-116 from chicken eggshells, correction of its amino acid sequence and identification of disulfide bonds and glycosylated Asn. Matrix Biol 21:383–387

Mann K, Gautron J, Nys Y, McKee MD, Bajari T, Schneider WJ, Hincke MT (2003) Disulfide-linked heterodimeric clusterin is a component of the chicken eggshell matrix and egg white. Matrix Biol 22:397–407

Marsh ME (1994) Polyanion-mediated mineralization - assembly and reorganization of acidic polysaccharides in the Golgi system of a coccolithophorid alga during mineral deposition. Protoplasma 177:108–122

Marsh ME (2003) Regulation of $CaCO_3$ formation in coccolithophores. Comp Biochem Physiol B Biochem Mol Biol 136:743–754

Marsh ME, Chang D-K, King GC (1992) Isolation and characterization of a novel acidic polysaccharide containing tartrate and glyoxylate residues from the mineralized scales of a unicellular coccolithophorid alga *Pleurochrysis carterae*. J Biol Chem 267:20507–20512

Marsh ME, Ridall AL, Azadi P, Duke PJ (2002) Galacturonomannan and Golgi-derived membrane linked to growth and shaping of biogenic calcite. J Struct Biol 139:39–45

Matsunaga T, Okamura Y (2003) Genes and proteins involved in bacterial magnetic particle formation. Trends Microbiol 11:536–541

Matsunaga T, Tsujimura N, Okamura Y, Takeyama H (2000) Cloning and characterization of a gene, *mpsA*, encoding a protein associated with intracellular magnetic particles from *Magnetospirillum* sp. strain AMB-1. Biochem Biophys Res Comm 268:932–937

Murayama E, Okuno A, Ohira T, Takagi Y, Nagasawa H (2000) Molecular cloning and expression of an otolith matrix protein cDNA from the rainbow trout, *Oncorhynchus mykiss*. Comp Biochem Physiol 126B:511–520

Murayama E, Takagi Y, Ohira T, Davis JG, Greene MI, Nagasawa H (2002) Fish otolith contains a unique structural protein, otolin-1. Eur J Biochem 269:688–696
Murayama E, Takagi Y, Nagasawa H (2004) Immunohistochemical localization of two otolith matrix proteins in the otolith and inner ear of the rainbow trout, *Oncorhynchus mykiss*: comparative aspects between the adult inner ear and embryonic otocysts. Histochem Cell Biol 121:155–166
Nakano T, Ikawa N, Ozimek L (2001) Extraction of glycosaminoglycans from chicken eggshell. Poult Sci 80:681–684
Nakano T, Ikawa N, Ozimek L (2002) Galactosaminoglycan composition in chicken eggshell. Poult Sci 81:709–714
Nakano T, Ikawa NI, Ozimek L (2003) Chemical composition of chicken eggshell and shell membranes. Poult Sci 82:510–514
Nimtz M, Conradt HS, Mann K (2004) LacdiNAc (GalNAcβ1–4GlcNAc) is a mjor motif in N-glycan structures of the chicken eggshell protein ovocleidin-116. Biochim Biophys Acta 1675:71–80
Okamura Y, Takeyama H, Matsunaga T (2000) Two-dimensional analysis of proteins specific to the bacterial magnetic particle membrane from *Magnetospirillum* sp. AMB-1. Appl Biochem Biotechnol 84–86:441–446
Okamura Y, Takeyama H, Matsunaga T (2001) A magnetosome-specific GTPase from the magnetic bacterium *Magnetospirillum magneticum* AMB-1. J Biol Chem 276:48183–48188
Outka DE, Williams DC (1971) Sequential coccolith morphoegensis in *Hymenomonas carterae*. J Protozool 18:285–297
Ozaki N, Sakuda S, Nagasawa H (2001) Isolation and some characterization of an acidic polysaccharide with anti-calcification activity from coccoliths of a marine alga, *Pleurochrysis carterae*. Biosci Biotech Biochem 65:2330–2333
Panhéleux M, Bain M, Fernandez MS, Morales I, Gautron J, Arias JL, Solomon SE, Hincke M, Nys Y (1999) Organic matrix composition and ultrastructure of eggshell: a comparative study. Br Poult Sci 40:240–252
Pautard FGE (1970) Calcification in unicellular organisms. In: Schraer H (ed) Biological calcification: cellular and molecular aspects. Appleton-Century-Crofts, New York, pp 105–201
Picard J, Paul-Gardais A, Vedel M (1973) Glycoprotéines sulfates des membranes de l'oeuf de poule et de l'oviducte. Isolement et caractérization de glycopeptides sulfates. Biochim Biophys Acta 320:427–441
Pines M, Knopov V, Bar A (1995) Involvement of osteopontin in egg shell formation in the laying chicken. Matrix Biol 14:765–771
Pote KG, Ross MD (1991) Each otoconia polymorph has a protein unique to that polymorph. Comp Biochem Physiol B 98:287–295
Pote KG, Hauer CR III, Michel H, Shabanowitz J, Hunt DF, Kretsinger RH (1993) Otoconin-22, the major protein of aragonitic frog otoconia, is a homolog of phospholipase A_2. Biochemistry 32:5017–5024
Quintana C, Sandoz D (1978) Coquille de l'oeuf de caille: étude ultrastructurale et cristallographique. Calcif Tissue Res 25:145–159
Reyes-Grajeda JP, Moreno A, Romero A (2004) Crystal structure of Ovocleidin-17, a major protein of the calcified *Gallus gallus* eggshell. J Biol Chem 279:40876–40881
Schüler D (2004) Molecular analysis of a subcellular compartment: the magnetosome membrane in *Magnetospirillum gryphiswaldense*. Arch Microbiol 181:1–7
Söllner C, Burghammer M, Busch-Nentwich E, Berger J, Schwarz H, Riekel C, Nicolson T (2003) Control of crystal size and lattice formation by starmaker in otolith biomineralization. Science 302:282–286

Spring S, Lins U, Amann R, Schleifer K-H, Ferreira LCS, Esquivel DMS, Farina M (1998) Phylogenetic affiliation and ultrastructure of uncultured magnetotactic bacteria with unusually large magnetosome. Arch Microbiol 169:136–147

Takazoe I, Kurahashi Y, Takuma S (1963) Electron microscopy of intracellular mineralization of oral filamentous microrganisms *in vitro*. J Dent Res 42:681–685

Terepka AR (1963) Organic-inorganic interrelationships in avian egg shell. Exper Cell Res 30:183–192

Theilade J, Fejerskov O, Horsted M (1976) A transmission electron microscopic study of 7-day old bacterial plaque in human tooth fissures. Arch Oral Biol 21:587–598

Towe KM, Cifelli R (1967) Wall ultrastructure in the calcareous foraminifera: crystallographic aspects and a model for calcification. J Paleontol 41:742–762

van der Wal P, de Jong EW, Westbroek P, de Bruijn WC, Mulder-Stapel AA (1983) Polysaccharide localization, coccolith formation, and Golgi dynamics in the coccolithophorid *Hymenomonas carterae*. J Ultrastruct Res 85:139–158

Van Emburg PR, de Jong EW, Daems WT (1986) Immunochemical localization of a polysaccharide from biomineral structures (coccoliths) of *Emiliania huxleyi*. J Ultrastruct Molec Struct Res 94:246–259

Verpy E, Leibovici M, Petit C (1999) Characterization of otoconin-95, the major protein of murine otoconia, provides insights into the formation of these inner ear biominerals. Proc Natl Acad Sci U S A 96:529–534

Vigh B, Szel A, Debreceni K, Fejer Z, Manzano e Silva MJ, Vigh-Teichmann I (1998) Comparative histology of pineal calcification. Histol Histopathol 13:851–870

Vogel JJ, Ennever J (1971) The role of a lipoprotein in the intracellular hydroxyapatite formation in *Bacterionema matruchotii*. Clin Orthop Relat Res 78:218–222

Wang X, Ford BC, Praul CA, Leach RM Jr (2002) Collagen X expression in oviduct tissue during the different stages of the egg laying cycle. Poult Sci 81:805–808

Wang Y, Kowalski PE, Thalmann I, Ornitz DM, Mager DL, Thalmann R (1998) Otoconin-90, the mammalian otoconial matrix protein, contains two domains of homology to secretory phospholipase A_2. Proc Natl Acad Sci USA 95:15345–15350

Wu T-M, Rodriguez JP, Fink DJ, Carrino DA, Blackwell J, Caplan AI, Heuer AH (1995) Crystallization studies on avian eggshell membranes: implications for the molecular factors controlling eggshell formation. Matrix Biol 14:507–513

Young JR, Davis SA, Bown PR, Mann S (1999) Coccolith ultrastructure and biomineralisation. J Struct Biol 126:195–215

Zander HA, Hazen SP, Scott DB (1960) Mineralization of dental calculus. Proc Soc Exp Biol (NY) 103:257–260

14 Calcifying Matrices: Pathological Calcifications

14.1 Introduction

There are so many examples of pathological calcification that a detailed review of the organic matrix of each of them is not feasible. In this chapter the only pathological calcifications that will be discussed are those that have most attracted the attention of scientists or that, because of their specific biological topic, are the most pertinent to the aims of this book.

14.2 Urinary Tract Stones

The organic matrix associated with calcific urinary tract stones constitutes about 2–3% of their weight (Sugimoto et al. 1985; Khan 1992). Histologically, it appears more prominent than is suggested by this low percentage (Khan and Hackett 1993): it may have a fibrillar and/or amorphous structure, or appear stratified with concentric laminations, or be mixed with membranous cellular degradation products (Khan and Hackett 1987). On the whole, the organic components of the stone matrix appear to be closely associated with the crystalline inorganic fraction (Khan and Hackett 1984, 1987, 1993; Iwata et al. 1986; Boevé et al. 1993; Khan et al. 1996; Ryall et al. 2001; Fleming et al. 2003; Walton et al. 2003). It has been supposed that the closeness of this association may arise from the fact that the organic matrix is sandwiched between the growing crystals (Ogbuji and Finlayson 1981), or from the adsorption of urinary proteins on the surface of oxalate crystals (Leal and Finlayson 1977; Khan et al. 1983; Fleming et al. 2001). Electron microscope studies on the crystals of calcium phosphate stones, carried out using a PEDS-like method, consisting of the simultaneous fixation and decalcification of stone specimens after embedding them in 1% agar, have shown that the spherulitic aggregates of needle- or plate-like crystals that can be found in untreated sections appear after fixation-decalcification as spherulitic aggregates of organic structures that keep the shape of the original crystals (Khan et al. 1996). These structures have been called crystal ghosts (see Sect. 16.2.1) to stress their very close relationship with these crystals; they have been thought to be linked to the crystal surface (Khan et al. 1996), but the same types of crystal-like structure have been detected in stone crystals after treatment with NaOH, which completely removes the superficial organic material (Ryall et al. 2001). The possibility that organic material is occluded within, and is an intrinsic component of, the crystals of urinary stones is now supported by several findings (Iwata et al. 1986; Doyle et al. 1991; Boevé et al. 1993; Khan and Hackett 1993; Ryall et al. 2000, 2001; Fleming et al. 2003). This topic will be discussed further in Sect. 16.6.

Histochemically, the organic matrix of kidney stones is positively stained by the PAS method for glycoproteins and by alcian blue or colloidal iron for acid proteoglycans (Khan and Hackett 1986; Khan et al. 1996). Glycosaminoglycans have been demonstrated histochemically in calcium-containing urinary calculi; those characterized by calcium phosphate crystals contain greater amounts of glycoproteic material than those containing calcium oxalate (Szabó and Módis 1980). The major components of the organic matrix found in 22 stones were identified as glycoproteins and/or proteoglycans (Sugimoto et al. 1985). Heparan sulfate has been isolated from the organic matrix of calcium oxalate crystals precipitated from the whole urine of healthy subjects (Suzuki et al. 1994). Different types of glycosaminoglycans have been isolated from stones showing different forms of inorganic composition: the main matrix glycosaminoglycans turned out to be hyaluronic acid in apatite and struvite stones, heparan sulfate in calcium oxalate monohydrate and uric acid stones, and hyaluronic acid and heparan sulfate in calcium oxalate dihydrate stones (Nishio et al. 1985). According to Khan (1992), all stone matrices contain uronic and sialic acids.

Kidney stones with different types of inorganic composition (calcium oxalate, struvite, or uric acid) all contain lipid material (Khan et al. 1988), whose amounts change with the stone type. Lipids have, in fact, been reported to make up 80% of the organic matrix in calcium oxalate stones, 67% in calcium phosphate stones, 26% in struvite stones and 25% in uric acid stones (Khan et al. 2002). The fact that the organic crystal-like structures and cellular membranes located in decalcified stone matrix react with malachite green-osmium shows that they contain lipid material (Khan et al. 1996).

Proteins have been extracted from the organic matrix of kidney stones, their relative concentrations ranging from 20 to 75%. Their amino acids change with stone composition, but aspartic and glutamic acids are those best represented (Malagodi and Moye 1981). The matrices of all stones contain a mucoprotein, known as matrix substance A, which is rich in aspartic and glutamic acid and contains galactose, mannose and methylpentose, with glucosamine as the predominant hexosamine (Khan 1992). Considering five different types of stone, the following proteins have been found: human serum albumin, which is the main protein component in all stones; α1-acid glycoprotein; α1-microglobulin; immunoglobulins; apolipoprotein A1; transferrin; α1-antitrypsin; retinol-binding protein; renal lithostatine; the Tamm-Horsfall protein (a glycoprotein normally secreted in the ascending limb of Henle's loop and in distal tubules) was not detected (Dussol et al. 1995). Proteins have been extracted from the calcium oxalate and calcium phosphate crystals generated in urine from normal subjects or stone-forming patients by adding an oxalate or phosphate load: when formed in control urine, oxalate crystals were mainly associated with protrombin-related proteins (osteocalcin) and with minor amounts of albumin and osteopontin, while phosphate crystals were mainly associated with the Tamm-Horsfall protein and minor amounts of albumin, OC and OPN; when formed in the urine of stone-forming patients, crystals were mainly associated with albumin in both types of crystal (Atmani et al. 1998; Atmani and Khan 2002). Similar results had been obtained in calcium oxalate crystals (a mixture of calcium oxalate mono- and dihydrate) induced in vitro in normal human urine, and in calcium oxalate dihydrated crystals produced in rat urine: human crystals were associated with albumin, OC and OPN; only albumin and OPN, with trace amounts of OC, were found in rat crystals (Atmani et al. 1996). Osteopontin has been reported in calcium oxalate-containing stones (Kleinman et al. 2004), and both OPN and OC have been documented immunohistochemically in the stone matrix (McKee et al. 1995). Protein-bound γ-carboxyglutamic acid has been detected in calcium-containing renal calculi (Lian et al. 1977).

14.3 Ectopic Calcifications

Ectopic calcifications are those that occur in extraskeletal soft tissues which do not normally calcify. A variety of matrices may be involved in the process, ranging from intracellular organelles to extracellular matrix components. Moreover, all these matrices may have become altered for pathological reasons. Depending on the site where ectopic calcifications are initiated, a rough subdivision can be made between intra- and extracellular calcifications; both can be found in the same tissue, and one may gradually merge with the other, so that, when advanced, calcification may be both intra- and extracellular (e.g., Boivin 1975). Moreover, ectopic calcifications can arise spontaneously, often as an effect of hypercalcemia and/or hyperphosphatemia (Kuzela et al. 1977; Ahmed et al. 2001; Giachelli et al. 2001, 2005) and/or of a deficiency of inhibitors, such as matrix Gla protein (Shanahan et al. 1998) and fetuin-A (Schäfer et al. 2003), or can be induced experimentally, by administration of vitamin D (Hass et al. 1958; Kent et al. 1958), dihydrotachysterol (Selye et al. 1960; Altenähr et al. 1970; Bonucci and Sadun 1972, 1973, 1975; Brandt and Bässler 1972), parathyroid hormone (Berry 1970), calcium gluconate (Duffy et al. 1971), or by calciphylaxis and calcergy (Selye 1962). This chapter deals with the ectopic calcifications which appear most pertinent to the aims of this book, independently of their etiology.

14.3.1 Calcification of the Kidney

Besides urinary stones, calcifications may occur in the renal parenchyma, where their formation is probably made easier by the function of the kidney in regulating calcium and phosphate serum concentration. Most of them are intracellular and first involve mitochondria, in some cases spreading to other cell components and extracellular structures.

Calcifications of mitochondria in the kidneys seem to develop through the same process as in other pathologically calcified soft tissues and to involve the same type of matrix (reviewed by Bonucci et al. 1973; Ghadially 2001). They largely consist of the inorganic granules described above (see Sect. 4.6.1) or of needle-shaped crystals: the former can be released from calcium loaded-mitochondria by the disruption of the mitochondrial membranes in alkaline detergent solutions; they have been found to contain not only inorganic material but proteins, too (Weinbach and von Brand 1965). For this reason, sections treated with the PEDS method and examined with the transmission electron microscope show organic structures (Fig. 14.1), which have the same morphology as the untreated inorganic granules (Bonucci et al. 1973). The

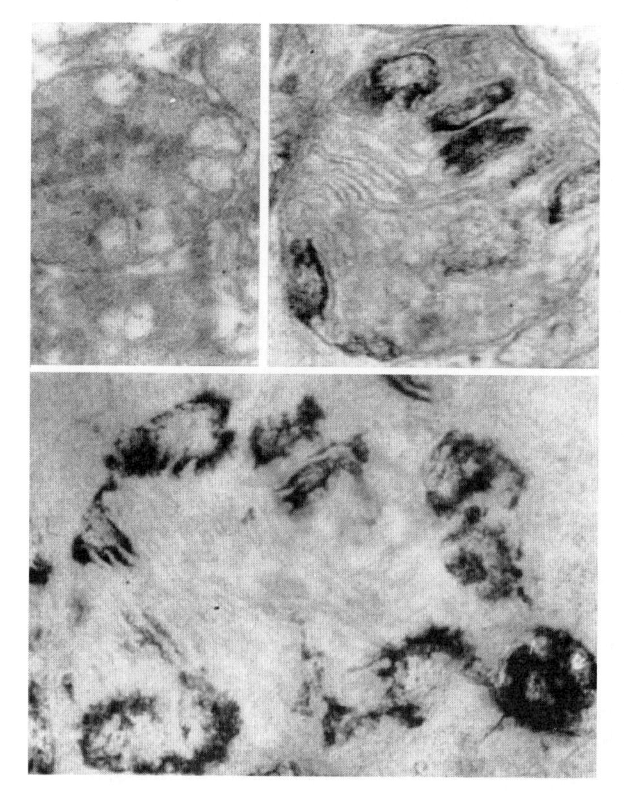

Fig. 14.1. Calcified mitochondria of carbon tetrachloride-intoxicated rats. *Upper left*: decalcified after embedding and unstained (first phase of the PEDS method); the calcified areas now appear as empty spaces; ×90,000. *Upper right*: decalcified after embedding and stained with uranium and lead (PEDS method): the previously empty spaces now appear to contain an amorphous, electron-dense substance; ×90,000. *Below*: decalcified after embedding and stained with acidic PTA (PEDS method): the calcified areas are deeply stained, showing that they contain glycoproteins; ×110,000

needle-shaped crystals, on the other hand, have been reported to be closely related to the membranes of the mitochondrial cristae (Bonucci et al. 1973). Studies by Thomas and Greenawalt (1968) have also shown that organic and inorganic material are intimately associated constituents of the granules that develop in isolated mitochondria of the rat liver after their incubation in calcium phosphate.

The properties and characteristics of the inorganic granules in renal mitochondria do not seem to differ from those in the mitochondria of other soft tissues. It has long been debated whether the matrix of these inorganic granules is the same as that of the 'dense' or 'native' granules, a hypothesis supported by the finding that the latter often disappear just before, or during, the formation of the former (Reynolds 1965). The native granules have been reported to contain phospholipids, glycoproteins, calcium precipitable lipoproteins, and cytochrome C oxidase (Jacob et al. 1994). In addition, they are stained by methods that reveal glycoproteins (see Bonucci et al. 1973), such as PAS (light microscopy), acidic phosphotungstic acid, or periodic acid-silver nitrate-methenamine (electron microscopy), and by methods for the detection of acid proteoglycans, such as ruthenium red (Allenspach and Babiarz 1975). The matrix of inorganic granules in renal mitochondria, as in other soft tissues, also seems to contain lipid molecules, as suggested by its osmiophilia.

Mitochondria can be more or less completely occupied by needle-shaped crystals. These also appear to be related to lipid material, because they are closely connected to the membranes of the mitochondrial cristae.

Renal epithelial cells with mitochondrial calcification may undergo degenerative changes and necrosis, and, when they are shed in the glomeruli or tubules, they may give rise to the formation of casts. Moreover, calcification may occur on the basal membranes (Berry 1970), where the electron microscope reveals small electron-dense particles (Duffy et al. 1971; Scarpelli 1965) and/or lamellar bodies (Caulfield and Schrag 1964). Under the light microscope, the von Kossa method for calcium phosphate shows positive linear profiles outlining basal membranes (Duffy et al. 1971). In this case, the inorganic substance might be related to a glycoproteic material, since the basal membranes are known to be strongly PAS-positive. Especially when advanced, renal calcifications may be found located in the interstice (Altenähr et al. 1970).

14.3.2
Calcification of the Myocardium and Skeletal Muscles

The calcification of skeletal muscles is not a frequent event in human pathology, but it is certainly not exceptional (Walton 1969): cases of myositis ossificans progressiva (Maxwell et al. 1977; Povysil and Matejovsky 1979; Ogilvie-Harris and Fornasier 1980) or of myositis ossificans traumatica (Pintér et al. 1980) are well known, and a list of degenerative muscle calcifications that may occur in several pathological conditions can be added. So too, the calcification of myocardium is a common outcome of cell necrosis or other pathological lesions (Gore and Arons 1949). Besides the calcifications that may occur spontaneously in the course of specific pathologies, others can be induced experimentally in voluntary muscles or in the myocardium, for instance, by intoxicating rat with dihydrotachysterol (Bonucci and Sadun 1972, 1973). In both these tissues, calcium phosphate deposits are found in mitochondria, either as roundish granular aggregates or filament-like crystals, the former located close to, and between, the cristae, the latter closely adhering to them and sometimes apparently lying on them. In voluntary muscles early calcification occurs in the vesicles of the sarcoplasmic reticulum and only comes to involve mitochondria and myofibrils later; in the myocardium, by contrast, mitochondria appear to be the early sites of calcification, which may then spread to myofibrils as cells degenerate, while the structures of the sarcoplasmic reticulum seem to be unaffected by this process. This is in line with the observation of Patriarca and Carafoli (1968) that Ca^{2+} movements in heart mitochondria are more active than those in the sarcoplasmic reticulum. Similar results were found in the muscle fibres of patients with tetanus (Agostini 1967).

In the early stages of calcification the morphology of mitochondria does not appear to change greatly, and their component structures are easily recognizable. There may, however, be so many granular aggregates and/or filament-like crystals that they completely fill the mitochondria, a finding also reported in the myocardium of rats with experimental magnesium deficiency (Heggtveit et al. 1964). Similar intramitochondrial calcifications have been reported in myocardial cells after the administration of 9α-fluorocortisol and sodium phosphate (D'Agostino 1964), injection of isoproterenol (Bloom and Davis 1972; Kutsuna 1972), perfusion with various ionic solutions (Legato et al. 1968), prolonged ventricular fibrillation (Ghidoni et al. 1969) and cardiac surgery (D'Agostino and Chiga 1970).

One possibility, when the calcium-binding capacity of the vesicles of the sarcoplasmic reticulum in voluntary muscles, or that of mitochondria in the myocardium, is exhausted, is that calcification may spread to other cellular components. In this case, filament-like crystals may be found outside the mitochondria, in a close relationship with myofilaments (Bonucci and Sadun 1972, 1973). The accumulation of calcium in the mitochondria of voluntary muscles and the myocardium seems to occur in a similar way, and appears to involve the same type of lipid and glycoproteic matrix components as those discussed above in other tissues (see Sects. 4.6.1 and 5.6.1). The filament-like crystals are related to other types of structure (mitochondrial cristae and cellular myofibrils), but in this case too phospholipids and glycoproteins seem to be involved, as shown by osmiophilia, and by the PAS-, alcian blue- and colloidal iron-positivity of the calcified areas (Bonucci and Sadun 1973).

14.3.3 Calcification of the Skin

Calcification of the skin may occur in spontaneous dermatologic diseases or in experimentally induced calcinoses. Among the former, 'familial tumoral calcinosis' – non-familial cases are known, too (McGregor et al. 1995) – is a disorderly reparative process which develops in the periarticular derma and is characterized by the formation of tumor-like calcified masses sometimes containing a milky, gritty material (for clinical symptoms, presentation and course, see Slavin et al. 1993; Pakasa and Kalengayi 1997). Hydroxypatite crystals related to membranous and cellular debris develop both in the cytoplasmic vacuoles of osteoclast-like giant cells and in interstitial cystic cavities (Slavin et al. 1993; McGregor et al. 1995). The chemical analysis of calcified deposits in three cases of tumoral calcinosis showed that the total lipid content was 1–10% of the decalcified dry weight; the content of the calcium-acidic phospholipid phosphate was 1–14 μg/mg dry weight, that of the uronic acid was 1–6.4 μg/mg dry weight, and that of osteocalcin was very low (Boskey et al. 1983).

Another cutaneous disease characterized by calcification is pseudoxanthoma elasticum, a systemic, autosomal, recessive, inheritable disorder characterized by abnormalities and, possibly, calcification of the elastic components of the skin, sometimes involving eyes and the cardiovascular system (reviewed by Ringpfeil et al. 2001). Hydroxyapatite crystals develop in a close relationship with elastic fibrils (Otkjaer-Nielsen et al. 1977), which appear to be structurally abnormal and to consist of granulo-filamentous material, possibly corresponding to abnormal elastic precursors of degenerated and disorganized elastic units (.Huang et al. 1967). Some doubts about the direct role of elastin in calcification have, however, been raised by the observation that non-crystalline inorganic substance shares the localization sites within elastic fibers with hyaluronidase-resistant polyanions (Martinez-Hernandez and Huffer 1974), and that matrix proteins with a high calcium affinity (osteopontin, bone sialoprotein, vitronectin, alkaline phosphatase) may be associated with calcification within elastic fibers (Baccarani Contri et al. 1996).

The presence of γ-carboxyglutamic acid in proteins associated with ectopic calcifications has been reported in scleroderma and dermatomyositis (Lian et al. 1976).

Experimental calcification of the skin can be induced by various methods (McClure and Gardner 1976), the most effective of which appear to be those related to topical calciphylaxis (sensitization of the animal with dihydrotachysterol and induction of calcification by subcutaneous injection of $CaFe_2$) or to calcergy (injection of $KMnO_4$ without previous sensitization), as suggested by Selye (1962). An extensive study of calcification induced in the rat skin by these methods has been carried out by Boivin (1975) and Boivin et al. (1987). In topical calciphylaxis, calcification was again first detected in mitochondria, followed soon after by extracellular calcification occurring in matrix vesicles, then around and between collagen fibrils, and finally within collagen fibrils. In topical calcergy, mitochondrial granules and matrix vesicles were the early sites of calcification (Walzer et al. 1980). Probably as a result of the denaturing effect of permanganate, other initial dermal changes in calcergy have, however, been described, including loosening of fibrous tissue, swelling of fibroblasts, partial lysis of fibrils, proliferation of fibroblasts, increase in acid proteoglycans and alkaline phosphatase activity (Seifert and v.Hentig 1967; Seifert 1970); these are followed by the formation of crystals. Crystals are reported to develop in connection with collagen fibrils and with structures derived from the cells (Placková 1975), and to be located around and between collagen fibrils (Boivin 1975; Boivin et al. 1987; Walzer et al. 1980; Vandeputte et al. 1990), in association with the interfibrillary ground substance (Seifert and v.Hentig 1967). An increase in osteocalcin content proportional to the degree of calcification has been shown in skin calcification due to calcergy and calciphylaxis (Lian et al. 1983).

14.3.4 Calcification of Other Soft Tissues

Spontaneous calcifications may develop in a number of tissues and organs, and experimental calcifications can be induced in all of them. In his review on experimental calcification, Seifert (1970) mentions calciphylaxis-induced calcification of the skin, the lung, the pancreas and islets of Langerhans, the myocardium, the thymus, arteries, salivary glands, and parathyroid glands. Several other tissues could be added to this list, e. g., voluntary muscles (Bonucci and Sadun 1972), the vagus nerve (Robinson et al. 1968), the cornea and conjunctiva (Obenberger et al. 1969; Parfitt 1969), the gallbladder (Arsenault et al. 1991), the brain (Bignami and Appicciutoli 1964; Fujita et al. 2003), the mammary gland (Brandt and Bässler 1972), the placenta (Brandt 1972; Varma and Kim 1985). Moreover, calcifications are often found in neoplastic tissues (Anderson 1983); this occurs mainly in the case of breast cancer, where calcifications are so common that they have acquired a diagnostic value: they are detected by preoperative mammograms in 48%, and by histological examinations in 63%, of patients with breast carcinoma, whereas calcium concretions are only revealed in 20% of mammograms of patients with benign breast diseases (Millis et al. 1976). An association of needle-like crystals with what is defined simply as amorphous material has been found in human breast carcinoma (Ahmed 1975). Mitochondrial granules have been reported as early sites of calcification in the mammary gland of rats treated with dihydrotachysterol (Brandt and Bässler 1972). Mitochondrial inclusions, partly corresponding to calcium-containing granules linked to a glycoproteic material, and partly to uncalcified phospholipids, have been described in human cancer cells of the gastrointestinal tract (Marinozzi et al. 1977) and of the lung (Wang and Steele 1979).

Lung calcification may occur as a complication of various systemic diseases and especially of renal failure (Firooznia et al. 1977; Justrabo et al. 1979; Bestetti-Bosisio et al. 1984) or can be induced by the administration of dihydrotachysterol (Bonucci et al. 1994); calcium deposition mainly occurs in the alveolar wall and, in this context, in the basal membrane (Eggermann and Kapanci 1971). The earliest deposits of calcium are, however, found in mitochondria of type II pneumocytes and other stromal cells, and in collagen fibrils of the interstitium, whereas calcification of the basal membrane and elastic fibrils occurs at a later stage (Bonucci et al. 1994). The calcified portions of the alveolar walls are stained by PAS for glycoproteins and acidic colloidal iron for acid proteoglycans (Bonucci et al. 1994). Matrix vesicles have been described as early calcification sites in alveolar walls in cases of alveolar microlithiasis (Bab et al. 1981).

Calcification is common in human placentas and is a recurrent process during their maturation and aging. The earliest crystalline deposits have been reported along the trophoblastic basal membrane of the chorionic microvilli undergoing fibrinoid degeneration, but early deposits have also been found within small, membrane-bound vesicles derived from degenerated cells and resembling matrix vesicles (Varma and Kim 1985).

The calcification of cornea and conjunctiva has been reported as a specific aspect of the wide spectrum of ectopic calcifications which may occur in the course of chronic renal failure (Parfitt 1969; Porter and Crombie 1973; Michaud 1978; De Graaf et al. 1980) or as a result of dihydrotachysterol administration (Obenberger et al. 1969). Roundish structures resembling matrix vesicles and containing needle-shaped crystals have been described in the cornea of dihydrotachysterol-sensitized rats after perfusion of the anterior eye chamber with dilute permanganate (Valouch et al. 1974). In a similar experiment in rabbits, a conspicuous edema developed in the cornea, which was stained by histochemical methods for acid proteoglycans (alcian blue at pH 2.5, acidic colloidal iron, toluidine blue at pH 4.5), while biochemical analyses showed that amounts of hexosamines and glucuronic acid were decreased (Obenberger et al. 1970).

The calcification of articular cartilage may occur as a consequence of osteoarthritis (Ali 1985; Ali et al. 1992). It has been shown by Ali and Griffiths (1981) that in this pathological condition there is an increase in the number of matrix vesicles and in their alkaline phosphatase activity. According to Hashimoto et al. (1998), this increase might be due to chondrocyte apoptosis, with the accumulation of apoptotic bodies around chondrocytes.

Calcium pyrophosphate dihydrate crystal deposition disease (tophaceous pseudogout) is a rare joint disease characterized by the deposition of calcium pyrophosphate crystals in the synovial tissue and sinovial liquid, and in periarticular soft tissues (reviewed by Dijkgraaf et al. 1995). Matrix vesicles from calcified areas produce hydroxyapatite crystals in vitro (Derfus et al. 1998), but they also seem to play a key role in calcium pyrophosphate dihydrate crystal formation (Derfus et al. 1995).

14.3.5 Calcification of Arteries

Although calcification of the artery wall is a common pathological change in humans, especially as a result of arteriosclerosis and/or aging, and is certainly one of the most life-threatening (Chen and Moe 2003, 2004; Moe and Chen 2004), the organic components that are associated with the inorganic substance are still poorly known. The arterial elastic fibers have long been considered the earliest structures where deposition of inorganic substance occurs (Seifert and Dreesbach 1966), a hypothesis supported by the results of histological, histochemical and ultrastructural studies on experimentally induced calcification of the rat aorta (Fig. 14.2; Bonucci and Sadun 1975). The same studies also showed a close relationship between crystalline inorganic structures and thin filaments probably belonging to elastin. Elastic fibers certainly play a direct role in artery calcification in some cases (Meyer and Stelzig 1967; Paule et al. 1992), but all the structures in the media may eventually do so (Eisenstein and Zeruolis 1964). According to Tanimura et al. (1986a), calcification in atherosclerotic intima is not closely associated with elastic fibers, whereas in the media this association may be very close. According to Proudfoot and Shanahan (2001), elastic fibers are involved in the calcification of the media, whereas the calcification of intima, such as that which occurs in atherosclerosis, is independent of them. In several cases in which the elastic fibers appear to be calcified, the inorganic substance is actually related to hyaluronidase-resistant polyanions (Martinez-Hernandez and Huffer 1974), a finding supported by the extraction from calcified human aorta of an acidic glycoprotein, which is associated with an organic phosphorus moiety, cholesterol and cholesterol esters (Keeley 1977), and by the observation that, in the experimentally calcified aorta of the rat, the calcified areas are stained by the PAS method for glycoproteins and by alcian blue for acid proteoglycans (Bonucci and Sadun 1975). Calcification has been reported to be closely associated with elastic coating material rather than directly with mature elastic lamellae in aortic medial calcification (Gardner and Blankenhorn 1968), and a connection between inorganic substance and lipids, presumably phospholipids, has been shown by oil red and Nile blue histochemistry in human calcified aortic valves (Kim and Huang 1971) and by silver precipitation and malachite green uptake in calcified porcine aortic valve leaflets implanted sub-cutis (Ortolani et al. 2003a, b). Calcium-acidic phospholipid-phosphate complexes have been detected in atherosclerotic plaques and adjacent zones in human aortas (Dmitrovsky and Boskey 1985). Calcium deposits mixed with cellular debris and contained in intra- and extracellular vesicles have been reported by Kim and Huang (1971).

These and other findings have raised a number of doubts about the old concept that vascular calcification is due to a passive deposition of inorganic substance on elastic fibers which, for some reason, had acquired an excessive affinity for calcium, and on the theory that the calcification of atherosclerotic

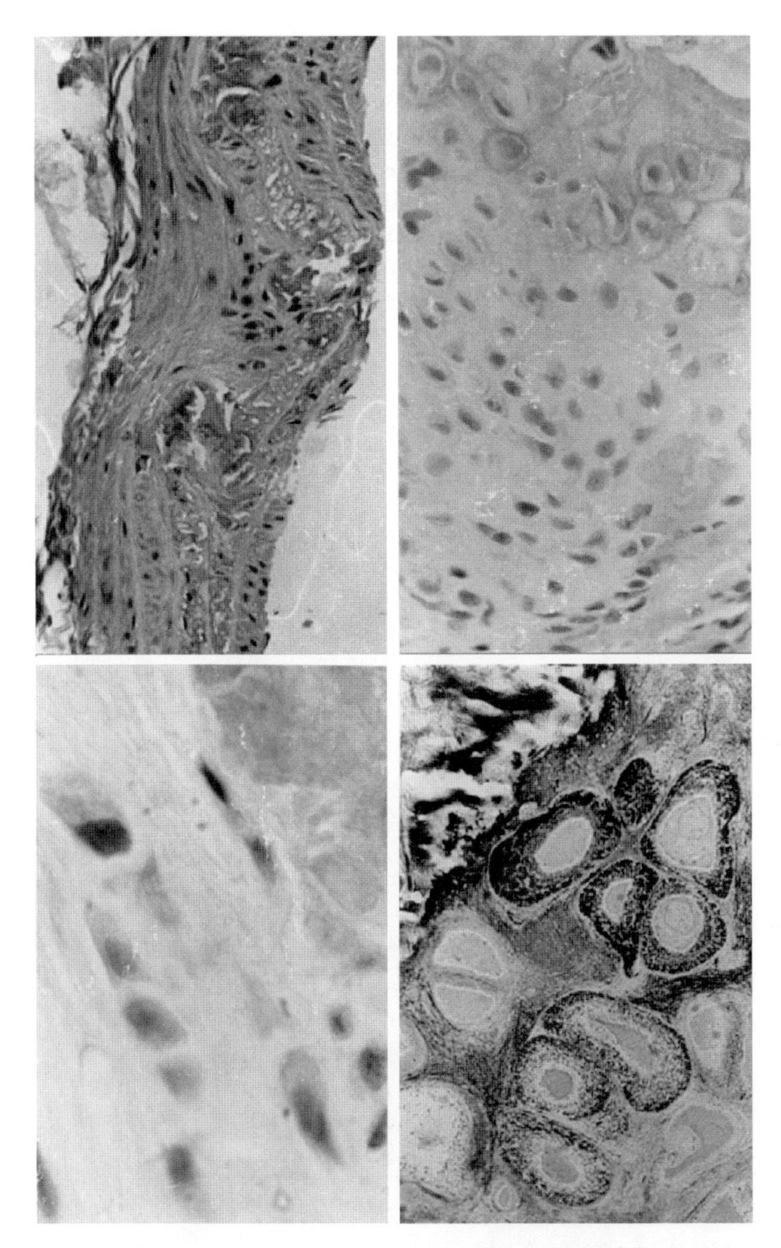

Fig. 14.2. Calcified aorta of dihydrotachysterol-treated rat. *Upper left*: distortion of the microarchitecture and replacement of normal cells by basophilic cells (center of figure); haematoxylin-eosin, ×100. *Lower left*: detail of the basophilic, osteoblast-like cells which replace the normal cells of the aorta; haematoxylin-eosin, ×500. *Upper right*: pseudocartilaginous area; haematoxylin-eosin, ×250. *Lower right*: pseudocartilaginous area stained with PASNM: glycoproteins (*black*) form a halo around chondrocyte-like cells. PASNM, ×500

plaques was a dystrophic process occurring in the context of a necrotic tissue. Although pathological calcification may occur by passive, cell-autonomous mechanisms (reviewed by Doherty et al. 2003), there is now a consensus (depicted in Figs. 14.2 and 14.3) that cells belonging to the artery wall – either a stem cell population or a smooth muscle cell subpopulation (Campbell and Campbell 2000), or even pericytes (Collett and Canfield 2005) – may differentiate into osteoblast-like cells (Rajamannan et al. 2003), that these cells, called calcifying vascular cells, or CVCs, produce a calcifying matrix, and that vascular calcification is therefore an active, biologically regulated process which shares some features with the process of osteogenesis (reviewed by Doherty

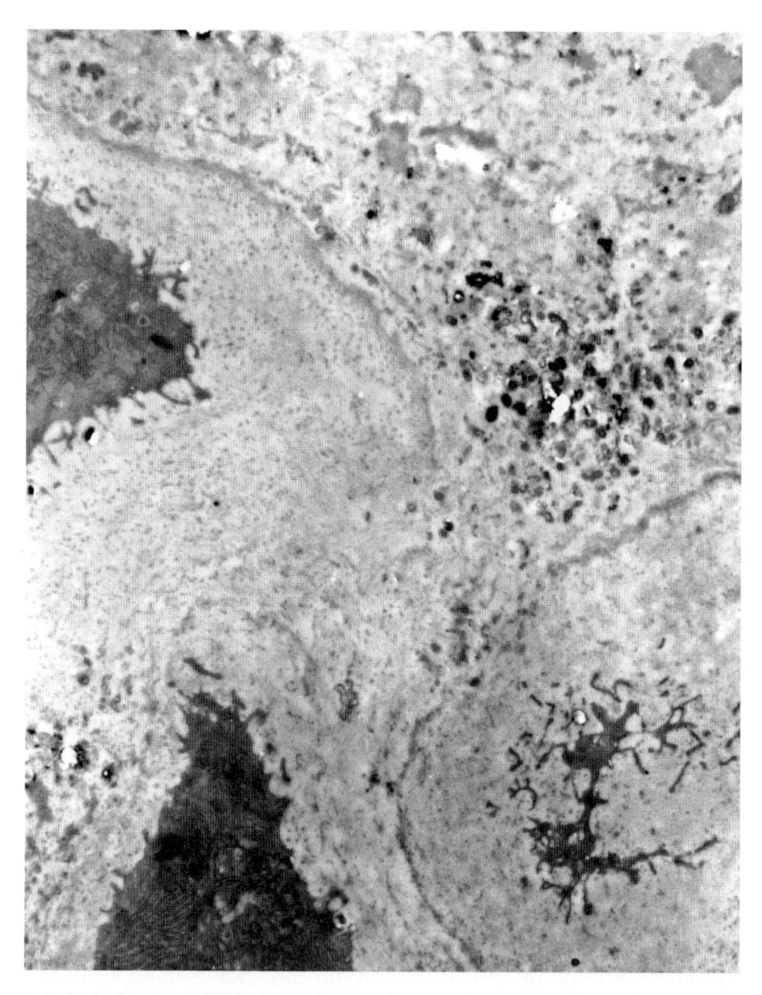

Fig. 14.3. Calcified aorta of dihydrotachysterol-treated rat; pseudocartilaginous area. The cells show a chondrocyte phenotype; plenty of matrix vesicles, some of them calcified, are visible in the intercellular matrix. Uranyl acetate and lead citrate, ×9,700

and Detrano 1994; Boström and Demer 2000; Jakoby and Semenkovich 2000; Boström 2001; Proudfoot and Shanahan 2001; Demer et al. 2002; Chen and Moe 2003, 2004; Doherty et al. 2003; Reslerova and Moe 2003; Giachelli 2004; Vattikuti and Towler 2004; Giachelli et al. 2005).

More than one stimulus may be responsible for the development of CVCs (reviewed by Boström 2000). In discussing the mechanisms of vascular calcification, Abedin et al. (2004) have stressed the fact that artery cells grown in culture become osteogenic under the influence of inflammatory and atherogenic stimuli, and that oxidized lipids may represent the inflammatory stimulus in atherosclerosis. Demer et al. (2002) accepted the concept that oxidized lipids and inflammatory mediators can promote the osteoblastic differentiation of vascular cells, but emphasized that this process, and vascular calcification as a whole, are regulated by a wide variety of mechanisms. Proudfoot et al. (2002) have observed that modified lipoproteins enhance the osteogenic differentiation of vascular smooth muscle cells. In a review of arterial calcification in diabetes and in end-stage kidney disease, Chen and Moe (2003, 2004) and Moe et al. (2003), although concluding that the pathogenesis of vascular calcification is not completely understood, have suggested that an important role in CVC differentiation, besides those of high blood glucose, hypoxia and hyperphosphatemia, may be played by the osteoblast differentiation factor (Cbfa 1; transcription core binding factor $\alpha 1$, also called Runx 2, which is essential to the development of osteoblasts), whose expression in cultured vascular smooth cells is up-regulated by uremic serum, which also accelerates the degree of calcification in cultured vascular smooth muscle cells (Moe and Chen 2004).

The concept that vascular calcification may share features with osteogenesis is confirmed by several findings (Figs. 14.2 and 14.3): the appearance of cartilaginous areas in the aorta of dihydrotachysterol-treated rats (Bonucci and Sadun 1975); the presence of chondrocyte-like cells in sclerotic aortic valves (Seemayer et al. 1973) and in areas surrounding human atherosclerotic necrotic plaques (Qiao et al. 2003; Bobryshev 2005); the development of cartilage in the calcified vessels of matrix Gla protein-deficient mice (MGP–/– mice) (Luo et al. 1997; Boström 2000, 2001; El-Maadawy et al. 2003), and of osteoblast- and osteoclast-like cells and osteoid tissue in the arteries of DHT-treated rats (Fig. 14.2) and of patients with chronic renal failure (Boström et al. 1995; Demer et al. 2002; Doherty et al. 2002; Chen and Moe 2004); the marked co-localization of calcium deposits and decorin (Fischer et al. 2004); and the presence near sites of carotid calcification of lipid-filled vascular smooth muscle cells expressing alkaline phosphatase, OC and BSP (Proudfoot et al. 2002). The similarity between the calcification of blood vessels and that of cartilage and bone is accentuated by the observations (Fig. 14.3) that calcifying matrix vesicles deriving from chondrocytes, and membrane-bound vesicles deriving from the destruction of chondrocytes, or from the apoptosis or degeneration of smooth muscle cells, have been detected in the atherosclerotic plaques of humans (Kim 1976; Ejerblad et al. 1979; Tanimura et al. 1983, 1986a; Hsu and Camacho 1999; Bobryshev 2005) and of atherogenic diet-administered rabbits (Tanimura et al. 1983, 1986b; Hsu et al. 2000, 2002, 2004) and chickens (Tanimura et al. 1986b), as well as in the experimental aneurysms of sheep (Martin et al. 1992) and rabbits (Rogers and Stehbens 1986), in the organ cultures of fresh and freeze thawed rat aorta (Kim 1984), in those of calcifying bovine aortic smooth muscle cells (Wada et al. 1999), and in cultures of human vascular smooth muscle cells which were able to form calcified nodules spontaneously (Proudfoot et al. 2001; Reynolds et al. 2004).

As summarized in Table 14.1, several bone-associated proteins and enzymes (OPN, BSP, OC, alkaline phosphatase), which may help to promote calcification or regulate it, as well as proteins with an inhibitory activity (MGP, fetuin-A, OPG), can be detected in areas of vascular calcification (Shanahan et al. 2000; Canfield et al. 2002; Chen and Moe 2004; Moe and Chen 2004; Price et al. 2004) and may be down-regulated in pathological conditions (Floege and Ketteler 2004; Ketteler et al. 2005).

One example of this is the extraction of OPN from calcified human aortic valves by Mohler et al. (1997). Nitta et al. (2001), using immunohistochemistry, demonstrated OPN in the calcified areas and in the surrounding macrophages of the aorta of hemodialysis patients. Chen et al. (2002) observed that phosphorus and uremic serum up-regulated OPN expression in cultures of vascular smooth muscle cells. Gadeau et al. (2001) found OPN and OC in established (from 8 to 14 day-old) but not in early (2 day) aortic calcifications induced in rabbits by acute vessel wall injury, and Levy et al. (1980) detected OC in calcified aortic stenosis. Fitzpatrick et al. (1994) demonstrated the presence of OPN immunohistochemically in the outer margin of the calcified plaques in human coronary arteries, Ahmed et al. (2001) found OPN in calcified small arteries of patients suffering from calciphylaxis, and Canfield et al. (2002), using the same method, detected OPN and MGP at sites of calcification within atherosclerotic lesions, as well as in microvessels in renal failure calciphylaxis, a disease characterized by the extensive calcification of cutaneous and subcutaneous arterioles. Bini et al. (1999) found that ON, OPN, and OC co-localize with calcium deposits in arteriosclerotic carotid arteries, Rajamannan et al.

Table 14.1. Bone matrix proteins in vascular wall

Protein	Material	Pathology	Method	Reference
OPN	Hum., coronary	Atheromasia	Ih	Fitzpatrick et al.(1994)
OPN	Hum., aortic valv.	Stenosis	PAGE,Wb	Mohler et al. (1997)
OPN	Hum., aorta, VSMC	In vitro	Nb	Proudfoot et al. (1998)
OPN	Hum., carotid	Atheromasia	Ih	Bini et al. (1999)
OPN	Hum., small arter.	Calciphylax.	Ih	Ahmed et al. (2001)
OPN	Hum., aorta	Atheromasia	Wb,Ich,Ish	Dhore et al. (2001)

Table 14.1. (continued)

Protein	Material	Pathology	Method	Reference
OPN	Rabbit, aorta	Traumatic	Ih	Gadeau et al. (2001)
OPN	Hum., aorta	Haemodial..	Ih	Nitta et al. (2001)
OPN	Hum., aorta	Atheromasia	Ih	Canfield et al. (2002)
OPN	Hum., small arter.	Calciphylax.	Ih	Canfield et al. (2002)
OPN	Bovine, VSMC	In vitro	Wb	Chen et al. (2002)
OPN	Hum., epigastr.art.	Dialysis	Ih	Moe et al. (2002)
OPN	Hum., epigastr.art.	Dialysis	Ih	Moe et al. (2003)
OPN	Hum., aortic valv.	Stenosis	RP	Rajamannan et al. (2003)
ON	Hum., carotid	Atheromasia	Ih	Bini et al. (1999)
ON	Hum., aorta	Atheromasia	Wb,Ich,Ish	Dhore et al. (2001)
ON	Hum., aorta, carotid	Atheromasia	Ih,RP,Ish	Tyson et al. (2003)
OC	Hum., aortic valv.	Stenosis	Ca	Levy et al. (1980)
OC	Hum., carotid	Atheromasia	Ih	Bini et al. (1999)
OC	Rabbit, aorta	Traumatic	Ih	Gadeau et al. (2001)
OC	Hum., aorta	Atheromasia	Wb,Ih,Ish	Dhore et al. (2001)
OC	Hum., aortic valv.	Stenosis	RP	Rajamannan et al. (2003)
OC	Hum., aorta, carotid	Atheromasia	Ih,RP,Ish	Tyson et al. (2003)
BSP	Hum., aortic valv.	Stenosis	Ih	Kaden et al. (2004)
BSP	Hum., aorta	Atheromasia	Wb,Ih,Ish	Dhore et al. (2001)
BSP	Hum., epigastr.art.	Dialysis	Ih.	Moe et al. (2002)
BSP	Hum., aortic valv.	Stenosis	RP	Rajamannan et al. (2003)
BSP	Hum., aorta, carotid	Atheromasia	Ih,RP,Ish	Tyson et al. (2003)
MGP	Hum., aorta, VSMC	In vitro	Nb	Proudfoot et al. (1998)
MGP	Hum., aorta	Atheromasia	Wb,Ih,Ish	Dhore et al. (2001)
MGP	Hum., aorta	Atheromasia	Ih	Canfield et al. (2002)
MGP	Hum., artery, VSMC	Iv, atherom.	Ih,Nb,Ish	Engelse et al. (2001)
MGP	Hum., rat	Aorta	Ih	Spronk et al. (2001)
MGP	Rats, aorta	Aging	Ih	Sweatt et al. (2003)
BMP-2	Hum., carotid	Atheromasia	Ih	Boström et al. (1993)
BMP-2	Hum., aorta	Atheromasia	Wb,Ih,Ish	Dhore et al. (2001)
BMP-2	Hum., aortic valv.	Stenosis	Ih	Kaden et al. (2004)
BMP-4	Hum., aorta	Atheromasia	Wb,Ih,Ish	Dhore et al. (2001)
BMP-4	Hum., small arter.	Calciphylax.	Ih	Griethe et al. (2003)
BMP-6	Hum., carotid	Atheromasia	Ih	Jeziorska et al. (1998)
OPG	Hum., aorta	Atheromasia	Wb,Ih,Ish	Dhore et al. (2001)
OPG	Hum., aorta, carotid	Atheromasia	Ih,RP,Ish	Tyson et al. (2003)
Cbfa 1	Hum., VSMC, artery	Iv, atherom.	Ih,Nb,Ish	Engelse et al. (2001)
Cbfa 1	Bovine, VSMC	In vitro	RP,Nb,Ih	Steitz et al. (2001)
Cbfa 1	Hum.,epigastr.art.	Dialysis	Ih	Moe et al. (2002)
Cbfa 1	Hum.,epigastr.art.	Dialysis	Ish,RP,Ih	Moe et al. (2003)
Cbfa 1	Hum., aorta, carotid	Atheromasia	Ih,RP,Ish	Tyson et al. (2003)
Cbfa 1	Hum., aortic valv.	Stenosis	RP	Rajamannan et al. (2003)

Iv: in vitro; Ih: immunohistochemistry; PAGE: polyacrylamide gel electrophoresis; Wb: Western blotting; Nb: Northern blotting; RP: RT-PCR; Ca: chemical analysis; Ish: In situ hybridization

(2003) detected increased mRNA levels of OPN, OC, BSP and Cbfa 1 in calcified aortic valves, and Moe et al. (2002) found that the degree of calcification in inferior epigastric arteries of dialysis patients at the time of kidney transplants was proportional to the expression of OPN, BSP, alkaline phosphatase and Cbfa 1. Spronk et al. (2001) reported that high amounts of MGP were demonstrable in the extracellular matrix at borders of media and intimal calcification in both human and rat arteries. Proudfoot et al. (1998) found that the calcification of human vascular smooth muscle cells grown in vitro expressed low levels of OPN and high levels of MGP. Sweatt et al. (2003) reported the important finding that, consistently with a role for MGP as inhibitor of BMP-2 (Boström 2000; Boström and Demer 2000; Spronk et al. 2001; El-Maadawy et al. 2003), BMP-2/MGP complexes existed in vivo – e.g., in the glandular structures of the nasal mucosa of the rat – and that the calcified aortic wall of the aging rat contained high concentrations of MGP which was, however, poorly γ-carboxylated and did not bind to BMP-2. Engelse et al. (2001) found that cultured endothelial cells and smooth muscle cells expressed MGP, that this was high on the luminal side of the normal vascular wall but fell towards the center of the media, and that it was low or absent in calcified areas; they also detected that these areas, but not the normal wall, contained Cbfa 1. Steitz et al. (2001) showed a gain in Cbfa 1 and OPN as cells grown in vitro lost their smooth muscle markers (SM22α and SM α-actin) and underwent a phenotypic transition into osteoblast-like cells. Hao et al. (2004) found that MGP and ON were expressed in the normal aorta of newborn and young subjects (up to ten years) and that their expression decreased as atherosclerosis developed in adults. A protein (named atherocalcin) containing γ-carboxyglutamic acid has been extracted from atherosclerotic plaque (Levy et al. 1979). Kaden et al. (2004) observed significant increases in BSP and BMP-2 immunostaining in calcified human aortic valves. Bonström et al. (1993) described fully formed bone tissue, including marrow, in human carotid endarterectomy specimens and found that the calcified plaques contained BMP-2, which was also expressed by cells cultured from the aortic wall. Jeziorska et al. (1998), who also described areas of ossification, or 'osteosomes', within atherosclerotic, calcified plaques of human carotids, reported the presence of BMP-6 around the ossified structures. Griethe et al. (2003) found BMP-4 in intradermal cells at the border of arterioles of patients with uremic calciphylaxis. Studies by Dhore et al. (2001) showed the constitutive immunoreactivity of MGP, OC and BSP, and the absence of OPN, BMP-2, BMP-4 and ON, in normal human aortas; the calcification or ossification of atherosclerotic plaques were associated with up-regulation of the latter (OPN, BMP-2, BMP-4 and ON,) and sustained immunoreactivity of the former (MGP, OC and BSP). Moreover, the same authors found that osteoprotegerin (OPG), an osteoclastogenesis inhibitory factor, and its ligand (RANKL), were detectable in normal aortas and early atherosclerotic lesions, and that, in advanced calcified plaques, OPG was present in bone structures and RANKL was only detectable around calcified deposits. In this connection, it has been reported by Bucay et al. (1998) that OPG−/− mice exhibited osteoporosis, as expected, but unexpectedly developed widespread calcification of the aortas and renal arteries, suggesting that OPG inhibits artery calcification, as demonstrated later (Price et al. 2001). These findings, which suggest a link between the local factors that regulate osteoclastic bone resorption, the immune system and calcification of the arterial wall (Schoppet et al. 2002; Collin-Osdoby 2004), have found support in the observation that the inhibition of osteoclast activity by SB 242784 – a selective inhibitor of the osteoclast V-H^+-ATPase – prevents arterial calcification in vitamin D-intoxicated rats (Price et al. 2002). Jian et al. (2003) have reported the presence of TGF-β, and Jian et al. (2001) that of matrix metalloproteinase-2 and tenascin-C, in calcific, stenotic aortic cusps.

In examining a range of calcified and non-calcified arteries, Tyson et al. (2003) have shown that MGP, OPG, ON and aggrecan are basically expressed by normal vascular smooth muscle cells and are down-regulated in calcified arteries, whereas BSP, OC, alkaline phosphatase and type II collagen are predominantly expressed in calcified arteries, together with transcription factors (Cbfa 1, Msx2, Sox9) that regulate the expression of osteoblast and chondrocyte markers.

The suggested presence of nanobacteria in calcified human arteries (Miller et al. 2004), along with their role in calcification (Kajander and Çiftçioglu 1998), require further supporting data before being accepted (Cisar et al. 2000).

14.3.6
Calcification of Implanted Cardiac Bioprosthetic Valves

Thrombo-embolism and occlusion are mostly responsible for an adverse evolution of mechano-prosthetic valves, whereas calcification is the most frequent cause of late failure of implanted bioprosthetic valves – i.e., valves derived from porcine aortic valves or bovine pericardium (Schoen and Hobson 1985; Schoen and Levy 1994; Levy et al. 1991). Calcification of degenerative type does, in fact, affect over 50% of glutaraldehyde-treated porcine aortic valves implanted in young patients, and from 5–10% of those implanted in adults, 5–10 years after implantation (Levy et al. 1985). Calcification initiates in association with fragments and/or membranes of residual, devitalized cells (Schoen et al. 1985, 1986; Kim 2002) and is largely conditioned by the young age of the recipient, mechanical stress, and, above all, valve treatment before implantation (reviewed by Schoen and Levy 2005). Obviously, this potentially highly dangerous process has repeatedly been studied to determine the role of the host metabolism, the effect of the mechanical forces involved and – the question most pertinent to the aims of this book – which of the valve components is locally responsible for the calcification process. In addition, because bioprosthetic valves are fixed with glutaraldehyde to improve their durability and make them less biodegradable, a number of studies have been carried out to detect the role this type of fixation may have in promoting valve calcification

(Grabenwoger et al. 1996; Nimni et al. 1997), its effects on intracellular calcium and phosphate concentration (Kim et al. 1999; Kim 2001), and the changes it may induce in the molecular tissue configuration (Girardot et al. 1995).

Attention was initially drawn to the components of the implanted valves. Paule et al. (1992) subdermally implanted small pieces of rabbit aorta which had been cross-linked with glutaraldehyde. After an observation period of one month or longer, they found many areas of calcification mainly associated with elastic fibers. Bailey et al. (2003) made a similar study by implanting heart valves subdermally after these had been extracted using trypsin/DNAase/RNAase, cyanogen bromide, and sodium hydroxide; they showed that the removal of cells and collagen did not significantly reduce aortic valve calcification, whereas partial degradation of elastic fibers significantly increased elastin-oriented calcification. Pretreatment with formaldehyde or glutaraldehyde promotes the calcification of collagen sponge implanted subcutaneously in weanling rats, although not proportionally to the degree of cross-linking (Levy et al. 1986). Subcutaneous implantation in the rat of collagen-elastin matrices, chosen in such a way that amounts of elastin increased from 10 to 50 and then 80%, showed that the total concentration of accumulated calcium increased with the increase in elastin (Singla and Lee 2002).

Electron microscope investigations of explanted porcine valves have shown three main features: intracytoplasmic and interstitial calcospherulae; calcified collagen fibrils; platelet-like calcium deposits upon amorphous material (Levy et al. 1985; Valente et al. 1985; Schoen et al. 1994). The first phase of valve calcification therefore appears to be a true dystrophic process and it is possible that phospholipids of degenerated cell membranes play a primary role in allowing linkage with the inorganic substance. This is supported by the observation that the extraction of phospholipids by valve pre-treatment with dodecyl sulfate (Hirsch et al. 1993), ethanol (Lee et al. 1998; Vyavahare et al. 1998, 2000), or ethanol-aluminum chloride (Levy et al. 2003), induces significant, but only partly effective, inhibition of aortic valve calcification. This suggests that cell activity or cell products are not necessary for bioprosthetic valve calcification, which is basically of dystrophic type. Several findings do, however, point to a cellular function. Schoen and Levy (1992) have observed that concomitantly with the onset of bioprosthetic tissue calcification, the content of alkaline phosphatase increases sharply, and Srivatsa et al. (1997) have reported that the matrix proteins that regulate skeletal calcification (OPN, OC, ON) may also regulate calcification in normal and bioprosthetic valves. Levy et al. (1980) have detected OC in three calcified porcine xenograft valves and in a calcified homograft valve. Grabenwoger et al. (1998) have evaluated cells, collagen fibrils and matrix proteins in bovine pericardium implanted subcutaneously into rats after various different treatments; they found that calcification was at a maximum in twice glutaraldehyde-fixed specimens, in which calcium deposition occurred together with the accumulation of glycoproteins, identified as sialoglycans. Gura et al. (1997) found non-collagenous proteins tightly bound to the mineral phase in subdermal implants of glutaraldehyde-preserved bovine pericardium with marked accumulation of highly acidic, phosphorylated proteins; they reported that ON increased within three days after implantation but its duration decreased by two weeks, and that BAG-75 and OPN uptake could be detected in initial inorganic deposits and increased with calcification.

14.4 Concluding Remarks

Although the findings reported above have been obtained from very different tissues in different spontaneous or experimental pathological conditions, and are therefore heterogeneous, it is possible to draw some concluding remarks which are valid for all of them:

- Ectopic calcifications can occur spontaneously or because of calciphylaxis or calcergy in various organic matrices in a number of very different tissues.
- Intracellular calcification begins in mitochondria, which may contain roundish aggregates of granular inorganic material or needle-shaped crystals, both of which are related to a lipidic, glycoproteic material.
- Matrix vesicles, possibly produced by cell apoptosis, are found in the intercellular space, where they may be mixed with cellular membranes and debris. They are sites of calcification, which may spread from them into the surrounding tissue.
- In all extracellular calcified areas, the inorganic substance is closely related to an organic matrix which may be the pre-existing, altered matrix peculiar to the tissue, or a newly synthesized matrix.
- Among the organic structures specific to the tissue, the first to be calcified may be altered elastic fibrils. Collagen fibrils are involved at a later stage, with the deposition of crystals on and between them.
- As shown both by histochemical methods and biochemical analyses, the calcified organic matrix contains glycoproteins, acid proteoglycans and lipids, some of which correspond to those already present in the tissue.
- Non collagenous proteins are also found; these include osteopontin, osteocalcin and bone sialoprotein. Alkaline phosphatase has been detected in many cases.
- Some proteins may be occluded within the crystals.
- Arterial calcification is not a dystrophic or metastatic process; it shares many features with the process of osteogenesis.
- Smooth muscle cells, or stem cells, in the artery wall can differentiate under appropriate stimuli and acquire an osteoblastic phenotype, so synthesizing

an organic matrix whose structure and composition permit, or promote, calcification.

- Vascular smooth muscle cells express bone morphogenetic proteins (BMP-2, -4 and -6); they also express Matrix Gla protein, fetuin-A, and osteoprotegerin. MGP surrounds calcified atherosclerotic plaques.

References

Abedin M, Tintut Y, Demer LL (2004) Vascular calcification: mechanisms and clinical ramifications. Arterioscler Thromb Vasc Biol 24:1161–1170

Agostini B (1967) Licht- und elektronenmikroskopische Untersuchungen an den Muskelfasern und an den Endplatten bei Tetanus. Beitr Path Anat 135:250–275

Ahmed A (1975) Calcification in human breast carcinomas: ultrastructural observations. J Path 117:247–251

Ahmed S, O'Neill KD, Hood AF, Evan AP, Moe SM (2001) Calciphylaxis is associated with hyperphosphatemia and increased osteopontin expression by vascular smooth muscle cells. Am J Kidney Dis 37:1267–1276

Ali SY (1985) Apatite-type crystal deposition in arthritic cartilage. Scanning Electron Microsc 4:1555–1566

Ali SY, Griffiths S (1981) Matrix vesicles and apatite deposition in osteoarthritis. In: Ascenzi A, Bonucci E, de Bernard B (eds) Matrix vesicles. Wichtig Editore, Milan, pp 241–247

Ali SY, Rees JA, Scotchford CA (1992) Microcrystal deposition in cartilage and in osteoarthritis. Bone Miner 17:115–118

Allenspach AL, Babiarz BS (1975) Intramitochondrial binding of ruthenium red in degenerating chondroblasts. J Ultrastruct Res 51:348–353

Altenähr E, Schäfer H-J, Wulff CA (1970) Zur Wirkung von Thyreocalcitonin auf die experimentelle Weichteilverkalkung. Virchows Arch Abt A Path Anat 349:124–137

Anderson HC (1983) Calcific diseases. A concept. Arch Path Lab Med 107:341–348

Arsenault AL, Jevon GP, Riddell RH, Simon GT (1991) Apatite determination and distribution in an early stage of gallbladder calcification. Ultrastruct Pathol 15:175–183

Atmani F, Khan SR (2002) Quantification of proteins extracted from calcium oxalate and calcium phosphate crystals induced *in vitro* in the urine of healthy controls and stone-forming patients. Urol Int 68:54–59

Atmani F, Opalko FJ, Khan SR (1996) Association of urinary macromolecules with calcium oxalate crystals induced in vitro in normal human and rat urine. Urol Res 24:45–50

Atmani F, Glenton PA, Khan SR (1998) Identification of proteins extracted from calcium oxalate and calcium phosphate crystals induced in the urine of healthy and stone forming subjects. Urol Res 26:201–207

Bab I, Rosenmann E, Neleman Z, Sela J (1981) The occurrence of extracellular matrix vesicles in pulmonary alveolar microlithiasis. Virchows Arch [Pathol Anat] 391:357–361

Baccarani Contri M, Boraldi F, Taparelli F, De Paepe A, Pasquali Ronchetti I (1996) Matrix proteins with high affinity for calcium ions are associated with mineralization within the elastic fibers of pseudoxanthoma elasticum dermis. Am J Pathol 148:569–577

Bailey MT, Pillarisetti S, Xiao H, Vyavahare NR (2003) Role of elastin in pathologic calcification of xenograft heart valves. J Biomed Mater Res 66A:93–102

Berry JP (1970) Néphrocalcinose éxperimentale par injection de parathormone. Etude au microanalyseur à sonde électronique. Nephron 7:97–116

Bestetti-Bosisio M, Cotelli F, Schiaffino E, Sorgato G, Schmid C (1984) Lung calcification in long-term dialysed patients: a light and electronmicroscopic study. Histopathology 8:69–79

Bignami A, Appicciutoli L (1964) Micropolygyria and cerebral calcification in cytomegalic inclusion disease. Acta Neuropathol 4:127–137

Bini A, Mann KG, Kudryk BJ, Schoen FJ (1999) Noncollagenous bone matrix proteins, calcification, and thrombosis in carotid artery atherosclerosis. Arterioscler Thromb Vasc Biol 19:1852–1861

Bloom S, Davis DL (1972) Calcium as mediator of isoproterenol-induced myocardial necrosis. Am J Pathol 69:459–470

Bobryshev YV (2005) Transdifferentiation of smooth muscle cells into chondrocytes in atherosclerotic arteries *in situ*: implications for diffuse intimal calcification. J Path 205:641–650

Boevé ER, Ketelaars GA, Vermeij M, Cao LC, Schröder FH, de Bruijn WC (1993) An ultrastructural study of experimentally induced microliths in rat proximal and distal tubules. J Urol 149:893–899

Boivin G (1975) Étude chez le rat d'une calcinose cutanée induite par calciphylaxie locale I. – Aspects ultrastructuraux. Arch Anat Microsc Morphol Expér 64:183–205

Boivin G, Walzer C, Baud CA (1987) Ultrastructural study of the long-term development of two experimental cutaneous calcinoses (topical calciphylaxis and topical calcergy) in the rat. Cell Tissue Res 247:525–532

Bonucci E, Sadun R (1972) An electron microscope study on experimental calcification of skeletal muscle. Clin Orthop Relat Res 88:197–217

Bonucci E, Sadun R (1973) Experimental calcification of the myocardium. Ultrastructural and histochemical investigations. Am J Pathol 71:167–192

Bonucci E, Sadun R (1975) Dihydrotachysterol-induced aortic calcification. A histochemical and ultrastructural investigation. Clin Orthop Relat Res 107:283–294

Bonucci E, Derenzini M, Marinozzi V (1973) The organic-inorganic relationship in calcified mitochondria. J Cell Biol 59:185–211

Bonucci E, Silvestrini G, Ballanti P, Della Rocca C, Mocetti P (1994) Dihydrotachysterol-induced lung calcification in the rat. It J Miner Electrol Metab 8:12–22

Boskey AL, Vigorita VJ, Sencer O, Stuchin SA, Lane JM (1983) Chemical, microscopic, and ultrastructural characterization of the mineral deposits in tumoral calcinosis. Clin Orthop Relat Res 178:258–269

Boström KI (2000) Cell differentiation in vascular calcification. Z Kardiol 89:69–74

Boström K (2001) Insights into the mechanism of vascular calcification. Am J Cardiol 88:20E–22E

Boström K, Demer LL (2000) Regulatory mechanisms in vascular calcification. Crit Rev Eukaryot Gene Expr 10:151–158

Boström K, Watson KE, Horn S, Wortham C, Herman IM, Demer LL (1993) Bone morphogenetic protein expression in human atherosclerotic lesions. J Clin Invest 91:1800–1809

Boström K, Watson KE, Stanford WP, Demer LL (1995) Atherosclerotic calcification: relation to developmental osteogenesis. Am J Cardiol 75:88B–91B

Brandt G (1972) Experimentelle Pathomorphogenese placentarer Verkalkungen. Morphologische und analytische Untersuchungen mit dem Modell der Calciphylaxie. Virchows Arch Abt A Path Anat 356:235–248

Brandt G, Bässler R (1972) Die Wirkung der experimentellen Hypercalcämie durch Dihydrotachysterin auf Drüsenfunktion und Verkalkungsmuster der Mamma. Licht-, elektronenmikroskopische und chemisch-analytische Untersuchungen. Virchows Arch Abt A Path Anat 356:155–172

Bucay N, Sarosi I, Dunstan CR, Morony S, Tarpley J, Capparelli C, Scully S, Tan HL, Xu W, Lacey DL, Boyle WJ, Simonet WS (1998) *osteoprotegerin*-deficient mice develop early onset osteoporosis and arterial calcification. Genes Dev 12:1260–1268
Campbell GR, Campbell JH (2000) Vascular smooth muscle and arterial calcification. Z Kardiol 89:54–62
Canfield AE, Farrington C, Dziobon MD, Boot-Handford RP, Heagerty AM, Kumar SN, Roberts IS (2002) The involvement of matrix glycoproteins in vascular calcification and fibrosis: an immunohistochemical study. J Pathol 196:228–234
Caulfield JB, Schrag PE (1964) Electron microscopic study of renal calcification. Am J Pathol 44:365–381
Chen NX, Moe SM (2003) Arterial calcification in diabetes. Curr Diab Rep 3:28–32
Chen NX, Moe SM (2004) Vascular calcification in chronic kidney disease. Semin Nephrol 24:61–68
Chen NX, O'Neill KD, Duan D, Moe SM (2002) Phosphorus and uremic serum up-regulate osteopontin expression in vascular smooth muscle cells. Kidney Int 62:1724–1731
Cisar JO, Xu D-Q, Thompson J, Swaim W, Hu L, Kopecko DJ (2000) An alternative interpretation of nanobacteria-induced biomineralization. Proc Natl Acad Sci USA 97:11511–11515
Collett GDM, Canfield AE (2005) Angiogenesis and pericytes in the initiation of ectopic calcification. Circ Res 96:930–938
Collin-Osdoby P (2004) Regulation of vascular calcification by osteoclast regulatory factors RANKL and osteoprotegerin. Circ Res 95:1046–1057
D'Agostino AN (1964) An electron microscopic study of cardiac necrosis produced by 9α-fluorocortisol and sodium phosphate. Am J Pathol 45:633–644
D'Agostino AN, Chiga M (1970) Mitochondrial mineralization in human myocardium. Am J Clin Path 53:820–824
De Graaf P, Polak BCP, De Wolff D, Schicht IM, Oosterhuis JA, De Graef J (1980) Ocular calcification in dialysis patients. Metab Pediat Ophtalmol 4:73–77
Demer LL, Tintut Y, Parhami F (2002) Novel mechanisms in accelerated vascular calcification in renal disease patients. Curr Opin Nephrol Hypertens 11:437–443
Derfus B, Steinberg M, Mandel N, Buday M, Daft L, Ryan L (1995) Characterization of an additional articular cartilage vesicle fraction that generates calcium pyrophosphate dihydrate crystals *in vitro*. J Rheumat 22:1514–1519
Derfus B, Kranendonk S, Camacho N, Mandel N, Kushnaryov V, Lynch K, Ryan L (1998) Human osteoarthritic cartilage matrix vesicles generate both calcium pyrophosphate dihydrate and apatite *in vitro*. Calcif Tissue Int 63:258–262
Dhore CR, Cleutjens JP, Lutgens E, Cleutjens KB, Geusens PP, Kitslaar PJ, Tordoir JH, Spronk HM, Vermeer C, Daemen MJ (2001) Differential expression of bone matrix regulatory proteins in human atherosclerotic plaques. Arterioscler Thromb Vasc Biol 21:1998–2003
Dijkgraaf LC, Liem RS, de Bont LG, Boering G (1995) Calcium pyrophosphate dihydrate crystal deposition disease: a review of the literature and a light and electron microscopic study of a case of the temporomandibular joint with numerous intracellular crystals in the chondrocytes. Osteoarthritis Cartilage 3:35–45
Dmitrovsky E, Boskey AL (1985) Calcium-acidic phospholipid-phosphate complexes in human atherosclerotic aortas. Calcif Tissue Int 37:121–125
Doherty TM, Detrano RC (1994) Coronary arterial calcification as an active process: a new perspective on an old problem. Calcif Tissue Int 54:224–230
Doherty TM, Uzui H, Fitzpatrick LA, Tripathi PV, Dunstan CR, Asotra K, Rajavashisth TB (2002) Rationale for the role of osteoclast-like cells in arterial calcification. FASEB J 16:577–582

Doherty TM, Asotra K, Fitzpatrick LA, Qiao J-H, Wilkin DJ, Detrano RC, Dunstan CR, Shah PK, Rajavashisth TB (2003) Calcification in atherosclerosis: bone biology and chronic inflammation at the arterial crossroads. Proc Natl Acad Sci U S A 100:11201–11206

Doyle IR, Ryall RL, Marshall VR (1991) Inclusion of proteins into calcium oxalate crystals precipitated from human urine: a highly selective phenomenon. Clin Chem 37:1589–1594

Duffy JL, Meadow E, Suzuki Y, Churg J (1971) Acute calcium nephropathy. Early proximal tubular changes in the rat kidney. Arch Pathol 91:340–350

Dussol B, Geider S, Lilova A, Leonetti F, Dupuy P, Daudon M, Berland Y, Dagorn JC, Verdier J-M (1995) Analysis of the soluble organic matrix of five morphologically different kidney stones. Evidence for a specific role of albumin in the constitution of the stone protein matrix. Urol Res 23:45–51

Eggermann J, Kapanci Y (1971) Experimental pulmonary calcinosis in the rat. Ultrastructural and morphometric study. Lab Invest 24:469–482

Eisenstein R, Zeruolis L (1964) Vitamin-D induced aortic calcification. Arch Path 77:27–35

Ejerblad S, Ericsson JL, Eriksson I (1979) Arterial lesions of the radial artery in uraemic patients. Acta Chir Scand 145:415–428

El-Maadawy S, Kaartinen MT, Schinke T, Murshed M, Karsenty G, McKee MD (2003) Cartilage formation and calcification in arteries of mice lacking matrix Gla protein. Connect Tissue Res 44:272–278

Engelse MA, Neele JM, Bronckers AL, Pannekoek H, de Vries CJ (2001) Vascular calcification: expression patterns of the osteoblast-specific gene core binding factor $\alpha 1$ and the protective factor matrix gla protein in human atherogenesis. Cardiovasc Res 52:281–289

Firooznia H, Pudlowski R, Golimbu C, Rafii M, McCauley D (1977) Diffuse interstitial calcification of the lungs in chronic renal failure mimicking pulmonary edema. Am J Roentgenol 129:1103–1105

Fischer JW, Steiz S, Johnson P, Burke A, Kolodgie F, Virmani R, Giachelli C, Wight TN (2004) Decorin promotes aortic smooth muscle cell calcification and colocalizes to calcified regions in human atherosclerotic lesions. Arterioscler Thromb Vasc Biol 24:2391–2396

Fitzpatrick LA, Severson A, Edwards WD, Ingram RT (1994) Diffuse calcification in human coronary arteries. Association of osteopontin with atherosclerosis. J Clin Invest 94:1597–1604

Fleming DE, van Bronswijk W, Ryall RL (2001) A comparative study of the adsorption of amino acids on to calcium minerals found in renal calculi. Clin Sci 101:159–168

Fleming DE, Van Riessen A, Chauvet MC, Grover PK, Hunter B, van Bronswijk W, Ryall RL (2003) Intracrystalline proteins and urolithiasis: a synchrotron X-ray diffraction study of calcium oxalate monohydrate. J Bone Miner Res 18:1282–1291

Floege J, Ketteler M (2004) Vascular calcification in patients with end-stage renal disease. Nephrol Dial Transplant 19:V59–66

Fujita D, Terada S, Ishizu H, Yokota O, Nakashima H, Ishihara T, Kuroda S (2003) Immunohistochemical examination on intracranial calcification in neurodegenerative diseases. Acta Neuropathol (Berl) 105:259–264

Gadeau AP, Chaulet H, Daret D, Kockx M, Daniel-Lamaziere JM, Desgranges C (2001) Time course of osteopontin, osteocalcin, and osteonectin accumulation and calcification after acute vessel wall injury. J Exp Med 192:

Gardner MB, Blankenhorn DH (1968) Aortic medial calcification. An ultrastructural study. Arch Pathol 85:397–403

Ghadially FN (2001) As you like it, part 3: a critique and historical review of calcification as seen with the electron microscope. Ultrastruct Pathol 25:243–267

Ghidoni JJ, Liotta D, Thomas H (1969) Massive subendocardial damage accompanying prolonged ventricular fibrillation. Am J Pathol 56:15–29

Giachelli CM (2004) Vascular calcification mechanisms. J Am Soc Nephrol 15:2959–2964

Giachelli CM, Jono S, Shioi A, Nishizawa Y, Mori K, Morii H (2001) Vascular calcification and inorganic phosphate. Am J Kidney Dis 38:S34–37

Giachelli CM, Speer MY, Li X, Rajachar RM, Yang H (2005) Regulation of vascular calcification. Roles of phosphate and osteopontin. Circ Res 96:717–722

Girardot M-N, Torrianni M, Dillehay D, Girardot J-M (1995) Role of glutaraldehyde in calcification of porcine heart valves: comparing cusp and wall. J Biomed Mater Res 29:793–801

Gore I, Arons W (1949) Calcification of the myocardium. A pathologic study of thirteen cases. Arch Path 48:1–12

Grabenwoger M, Sider J, Fitzal F, Zelenka C, Windberger U, Grimm M, Moritz A, Bock P, Wolner E (1996) Impact of glutaraldehyde on calcification of pericardial bioprosthetic heart valve material. Ann Thorac Surg 62:772–777

Grabenwoger M, Bock P, Fitzal F, Schmidberger A, Grimm M, Laufer G, Bergmeister H, Wolner E (1998) Acidic glycoproteins accumulate in calcified areas of bioprosthetic tissue. J Heart Valve Dis 7:229–234

Griethe W, Schmitt R, Jurgensen JS, Bachmann S, Eckardt KU, Schindler R (2003) Bone morphogenetic protein-4 expression in vascular lesions of calciphylaxis. J Nephrol 16:728–732

Gura TA, Wright KL, Veis A, Webb CL (1997) Identification of specific calcium-binding noncollagenous proteins associated with glutaraldehyde-preserved bovine pericardium in the rat subdermal model. J Biomed Mater Res 35:483–495

Hao H, Hirota S, Ishibashi-Ueda H, Kushiro T, Kanmatsuse K, Yutani C (2004) Expression of matrix Gla protein and osteonectin mRNA by human aortic smooth muscle cells. Cardiovasc Pathol 13:195–202

Hashimoto S, Ochs RL, Rosen F, Quach J, McCabe G, Solan J, Seegmiller JE, Terkeltaub R, Lotz M (1998) Condrocyte-derived apoptotic bodies and calcification of articular cartilage. Proc Natl Acad Sci USA 95:3094–3099

Hass GM, Trueheart RE, Taylor B, Stumpe M (1958) An experimental histologic study of hypervitaminosis D. Am J Pathol 34:395–431

Heggtveit HA, Herman L, Mishra RK (1964) Cardiac necrosis and calcification in experimental magnesium deficiency. A light and electron microscopic study. Am J Pathol 45:757–782

Hirsch D, Drader J, Thomas TJ, Schoen FJ, Levy JT, Levy RJ (1993) Inhibition of calcification of glutaraldehyde pretreated porcine aortic valve cusps with sodium dodecyl sulfate: preincubation and controlled release studies. J Biomed Mater Res 27:1477–1484

Hsu HH, Camacho NP (1999) Isolation of calcifiable vesicles from human atherosclerotic aortas. Atherosclerosis 143:353–362

Hsu HH, Camacho NP, Sun F, Tawfik O, Aono H (2000) Isolation of calcifiable vesicles from aortas of rabbits fed with high cholesterol diets. Atherosclerosis 153:337–348

Hsu HH, Camacho NC, Tawfik O, Sun F (2002) Induction of calcification in rabbit aortas by high cholesterol diets: roles of calcifiable vesicles in dystrophic calcification. Atherosclerosis 161:85–94

Hsu HH, Tawfik O, Sun F (2004) Mechanism of dystrophic calcification in rabbit aortas: temporal and spatial distributions of calcifying vesicles and calcification-related structural proteins. Cardiovasc Pathol 13:3–10

Huang S-N, Steele HD, Kumar G, Parker JO (1967) Ultrastructural changes of elastic fibers in pseudoxanthoma elasticum. Arch Path 83:108–113

Iwata H, Abe Y, Nishio S, Wakatsuki A, Ochi K, Takeuchi M (1986) Crystal-matrix interrelations in brushite and uric acid calculi. J Urol 135:397–401
Jacob WA, Bakker A, Hertsens RC, Biermans W (1994) Mitochondrial matrix granules: their behavior during changing metabolic situations and their relationship to contact sites between inner and outer mitochondrial membranes. Microsc Res Tech 27:307–318
Jakoby MG4, Semenkovich CF (2000) The role of osteoprogenitors in vascular calcification. Curr Opin Nephrol Hypertens 9:11–15
Jeziorska M, McCollum C, Woolley DE (1998) Observations on bone formation and remodelling in advanced atherosclerotic lesions of human carotid arteries. Virchows Arch 433:559–565
Jian B, Jones PL, Li Q, Mohler ER3, Schoen FJ, Levy RJ (2001) Matrix metalloproteinase-2 is associated with tenascin-C in calcific aortic stenosis. Am J Pathol 159:321–327
Jian B, Narula N, Li QY, Mohler ER3, Levy RJ (2003) Progression of aortic stenosis: TGF-β1 is present in calcific aortic valve cusps and promotes aortic valve interstitial cell calcification via apoptosis. Ann Thorac Surg 75:457–465
Justrabo E, Genin R, Rifle G (1979) Pulmonary metastatic calcification with respiratory insufficiency in patients on maintenance haemodialysis. Thorax 34:384–388
Kaden JJ, Bickelhaupt S, Grobholz R, Vahl CF, Hagl S, Brueckmann M, Haase KK, Dempfle CE, Borggrefe M (2004) Expression of bone sialoprotein and bone morphogenetic protein-2 in calcific aortic stenosis. J Heart Valve Dis 13:560–566
Kajander EO, Çiftçioglu N (1998) Nanobacteria: an alternative mechanism for pathogenic intra- and extracellular calcification and stone formation. Proc Natl Acad Sci U S A 95:8274–8279
Keeley FW (1977) The extraction and partial characterization of proteins released by decalcification from calcified human aortic plaques. Biochim Biophys Acta 494:384–394
Kent SP, Vawter GF, Dowben RM, Benson RE (1958) Hypervitaminosis D in monkeys; a clinical and pathologic study. Am J Pathol 34:37–59
Ketteler M, Westenfeld R, Schlieper G, Brandenburg V, Floege J (2005) "Missing" inhibitors of calcification: general aspects and implications in renal failure. Pediatr Nephrol 20:383–388
Khan SR (1992) Structure and development of calcific urinary stones. In: Bonucci E (ed) Calcification in biological systems. CRC Press, Boca Raton, pp 345–363
Khan SR, Hackett RL (1984) Microstructure of decalcified human calcium oxalate urinary stones. Scanning Electron Microsc Pt. 2:935–941
Khan SR, Hackett RL (1986) Histochemistry of colloidal iron stained crystal associated material in urinary stones and experimentally induced intrarenal deposits in rats. Scanning Electron Microsc Pt. 2:761–765
Khan SR, Hackett RL (1987) Crystal-matrix relationships in experimentally induced urinary calcium oxalate monohydrate crystals, an ultrastructural study. Calcif Tissue Int 41:157–163
Khan SR, Hackett RL (1993) Role of organic matrix in urinary stone formation: an ultrastructural study of crystal matrix interface of calcium oxalate monohydrate stones. J Urol 150:239–245
Khan SR, Finlayson B, Hackett RL (1983) Stone matrix as proteins adsorbed on crystal surfaces: a microscopic study. Scanning Electron Microsc 1:379–385
Khan SR, Shevock PN, Hackett RL (1988) Presence of lipids in urinary stones: results of preliminary studies. Calcif Tissue Int 42:91–96
Khan SR, Atmani F, Glenton P, Hou Z-C, Talham DR, Khurshid M (1996) Lipids and membranes in the organic matrix of urinary calcific crystals and stones. Calcif Tissue Int 59:357–365

Khan SR, Glenton PA, Backov R, Talham DR (2002) Presence of lipids in urine, crystals and stones: implications for the formation of kidney stones. Kidney Int 62:2062–2072

Kim KM (1976) Calcification of matrix vesicles in human aortic valve and aortic media. Feder Proc 35:156–162

Kim KM (1984) Cell injury and calcification of rat aorta in vitro. Scanning Electron Microsc Pt 4:1809–1818

Kim KM (2001) Cellular mechanism of calcification and its prevention in glutaraldehyde treated vascular tissue. Z Kardiol 90:99–105

Kim KM (2002) Cells, rather than extracellular matrix, nucleate apatite in glutaraldehyde-treated vascular tissue. J Biomed Mater Res 59:639–645

Kim KM, Huang S (1971) Ultrastructural study of calcification of human aortic valve. Lab Invest 25:357–366

Kim KM, Herrera GA, Battarbee HD (1999) Role of glutaraldehyde in calcification of porcine aortic valve fibroblasts. Am J Pathol 154:843–852

Kleinman JG, Wesson JA, Hughes J (2004) Osteopontin and calcium stone formation. Nephron Physiol 98:43–47

Kutsuna F (1972) Electron microscopic studies on isoproterenol-induced myocardial lesion in rats. Jpn Heart J 13:168–175

Kuzela DC, Huffer WE, Conger JD, Winter SD, Hammond WS (1977) Soft tissue calcification in chronic dialysis patients. Am J Pathol 86:403–424

Leal JJ, Finlayson B (1977) Adsorption of naturally occurring polymers onto calcium oxalate crystal surface. Invest Urol 14:278–283

Lee CH, Vyavahare N, Zand R, Kruth H, Schoen FJ, Bianco R, Levy RJ (1998) Inhibition of aortic wall calcification in bioprosthetic heart valves by ethanol pretreatment: biochemical and biophysical mechanisms. J Biomed Mater Res 42:30–37

Legato MJ, Spiro D, Langer GA (1968) Ultrastructural alterations produced in mammalian myocardium by variation in perfusate ionic composition. J Cell Biol 37:1–12

Levy RJ, Lian JB, Gallop P (1979) Atherocalcin, a γ-carboxyglutamic acid containing protein from atherosclerotic plaque. Biochem Biophys Res Comm 14:41–49

Levy RJ, Zenker JA, Lian JB (1980) Vitamin K-dependent calcium binding proteins in aortic valve calcification. J Clin Invest 65:563–566

Levy RJ, Schoen FJ, Langer R, Levy JT, Wolfrum J, Hawley MA, Lund SA, Carnes DL (1985) Bioprosthetic heart valve calcification. In: Butler WT (ed) The chemistry and biology of mineralized tissues. Ebsco Media, Inc., Birmingham, AL, pp 375–380

Levy RJ, Schoen FJ, Sherman FS, Nichols J, Hawley MA, Lund SA (1986) Calcification of subcutaneously implanted type I collagen sponges. Effects of formaldehyde and glutaraldehyde pretreatments. Am J Pathol 122:71–82

Levy RJ, Schoen FJ, Anderson HC, Harasaki H, Koch TH, Brown W, Lian JB, Cumming R, Gavin JB (1991) Cardiovascular implant calcification: a survey and update. Biomaterials 12:707–714

Levy RJ, Vyavahare N, Ogle M, Ashworth P, Bianco R, Schoen FJ (2003) Inhibition of cusp and aortic wall calcification in ethanol- and aluminum-treated bioprosthetic heart valves in sheep: background, mechanisms, and synergism. J Heart Valve Dis 12:209–216

Lian JB, Skinner M, Glimcher MJ, Gallop P (1976) The presence of γ-carboxyglutamic acid in the proteins associated with ectopic calcification. Biochem Biophys Res Comm 73:349–355

Lian JB, Prien EL Jr, Glimcher MJ, Gallop PM (1977) The presence of protein-bound γ-carboxyglutamic acid in calcium-containing renal calculi. J Clin Invest 59:1151–1157

Lian JB, Boivin G, Patterson-Allen P, Grynpas M, Walzer C (1983) Calcergy and calciphylaxis: timed appearance of γ-carboxyglutamic acid and osteocalcin in mineral deposits. Calcif Tissue Int 35:555–561

Luo G, Ducy P, McKee MD, Pinero GJ, Loyer E, Behringer RR, Karsenty G (1997) Spontaneous calcification of arteries and cartilage in mice lacking GLA protein. Nature 386:78–81
Malagodi MH, Moye HA (1981) Physical and chemical characteristics of renal stone matrix. Urol Surv 31:87–91
Marinozzi V, Derenzini M, Nardi F, Gallo P (1977) Mitochondrial inclusions in human cancer of the gastrointestinal tract. Cancer Res 37:1556–1563
Martin BJ, Thomas S, Greenhill NS, Ryan PA, Davis PF, Stehbens WE (1992) Isolation and purification of extracellular matrix vesicles from blood vessels. Prep Biochem 22:87–103
Martinez-Hernandez A, Huffer WE (1974) Pseudoxanthoma elasticum: dermal polyanions and the mineralization of elastic fibers. Lab Invest 31:181–186
Maxwell WA, Spicer SS, Miller RL, Halushka PV, Westphal MC, Setser ME (1977) Histochemical and ultrastructural studies in fibrodysplasia ossificans progressiva (myositis ossificans progressiva). Am J Pathol 87:483–498
McClure J, Gardner DL (1976) The production of calcification in connective tissue and skeletal muscle using various chemical compounds. Calcif Tissue Res 22:129–135
McGregor DH, Mowry M, Cherian R, McAnaw M, Poole E (1995) Nonfamilial tumoral calcinosis associated with chronic renal failure and secondary hyperparathyroidism: report of two cases with clinicopathological, immunohistochemical, and electron microscopic findings. Hum Pathol 26:607–613
McKee MD, Nanci A, Khan SR (1995) Ultrastructural immunodetection of osteopontin and osteocalcin as major matrix components of renal calculi. J Bone Miner Res 10:1913–1929
Meyer WW, Stelzig H-H (1967) Verkalkungsformen der inneren elastischen Membran der Beinarterien und ihre Bedeutung für die Mediaverkalkung. Virchows Arch Pathol Anat 342:361–373
Michaud P-A (1978) Dèpôts cornéens et conjonctivaux lors des traitements d'insuffisance rénale par hémodialyse chronique (Etude de 47 patients). Klin Mbl Augenheilk 172:523–526
Miller VM, Rodgers G, Charlesworth JA, Kirkland B, Severson SR, Rasmussen TE, Yagubyan M, Rodgers JC, Cockerill FR III, Folk RL, Rzewuska-Lech E, Kumar V, Farell-Baril G, Lieske JC (2004) Evidence of nanobacterial-like structures in calcified human arteries and cardiac valves. Am J Physiol Heart Circ Physiol 287:H1115-H1124
Millis RR, Davis R, Stacey AJ (1976) The detection and significance of calcifications in the breast: a radiological and pathological study. Br J Radiol 49:12–26
Moe SM, Chen NX (2004) Pathophysiology of vascular calcification in chronic kidney disease. Circ Res 95:560–567
Moe SM, O'Neill KD, Duan D, Ahmed S, Chen NX, Leapman SB, Fineberg N, Kopecky K (2002) Medial artery calcification in ESRD patients is associated with deposition of bone matrix proteins. Kidney Int 61:638–647
Moe SM, Duan D, Doehle BP, O'Neill KD, Chen NX (2003) Uremia induces the osteoblast differentiation factor Cbfa1 in human blood vessels. Kidney Int 63:1003–1011
Mohler ER3, Adam LP, McClelland P, Hathaway DR (1997) Detection of osteopontin in calcified human aortic valves. Arterioscler Thromb Vasc Biol 17:547–552
Nimni ME, Myers D, Ertl D, Han B (1997) Factors which affect the calcification of tissue-derived bioprostheses. J Biomed Mater Res 35:531–537
Nishio S, Abe Y, Wakatsuki A, Iwata H, Ochi K, Takeuchi M, Matsumoto A (1985) Matrix glycosaminoglycans in urinary stones. J Urol 134:503–505
Nitta K, Ishizuka T, Horita S, Hayashi T, Ajiro A, Uchida K, Honda K, Oba T, Kawashima A, Yumura W, Kabaya T, Akiba T, Nihei H (2001) Soluble osteopontin and vascular calcification in hemodialysis patients. Nephron 89:455–458
Obenberger J, Ocumpaugh DE, Cubberly MG (1969) Experimental corneal calcification in animals treated with dihydrotachysterol. Invest Ophthal 8:467–474

Obenberger J, Cejkovà J, Brettschneider I (1970) Experimental corneal calcification. A study of acid mucopolysaccharides. Ophthal Res 1:175–186
Ogbuji LU, Finlayson B (1981) Crystal morphologies in whewellite stones: electron microscopy. Invest Urol 19:182–186
Ogilvie-Harris DJ, Fornasier VL (1980) Pseudomalignant myositis ossificans: heterotopic new-bone formation without a history of trauma. J Bone Joint Surg 62-A:1274–1283
Ortolani F, Petrelli L, Bonetti A, Contin M, Marchini M (2003a) Selective silver precipitation and malachite green uptake reveal calcium-binding sites and lipid involvment on calcified aortic valve thin sections. Eur J Histochem 47:5
Ortolani F, Petrelli L, Nori SL, Gerosa G, Spina M, Marchini M (2003b) Malachite green and phthalocyanine-silver reactions reveal acidic phospholipid involvement in calcification of porcine aortic valves in rat subdermal model. Histol Histopathol 18:1131–1140
Otkjaer-Nielsen A, Johnson E, Hentzer B, Danielsen L, Carlsen F (1977) Apatite crystals in pseudoxanthoma elasticum: a combined study using electron microscopy and selected area diffraction analysis. J Invest Dermatol 69:376–378
Pakasa NM, Kalengayi RM (1997) Tumoral calcinosis: a clinicopathological study of 111 cases with emphasis on the earliest changes. Histopathology 31:18–24
Parfitt AM (1969) Soft-tissue calcification in uremia. Arch Int Med 124:544–556
Patriarca P, Carafoli E (1968) A study of the intracellular transport of calcium in rat heart. J Cell Physiol 72:29–38
Paule WJ, Bernick S, Strates B, Nimni ME (1992) Calcification of implanted vascular tissues associated with elastin in an experimental animal model. J Biomed Mater Res 26:1169–1177
Pintér J, Lénárt G, Rischák G (1980) Histologic, physical and chemical investigation of myositis ossificans traumatica. Acta Orthop Scand 51:899–902
Placková A (1975) The ultrastructure of topical cutaneous calcinosis. Cell Tissue Res 159:523–529
Porter R, Crombie AL (1973) Corneal and conjunctival calcification in chronic renal failure. Brit J Ophthal 57:339–343
Povysil C, Matejovsky Z (1979) Ultrastructural evidence of myofibroblasts in pseudomalignant myositis ossificans. Virchows Arch A Path Anat Histol 381:189–203
Price PA, June HH, Buckley JR, Williamson MK (2001) Osteoprotegerin inhibits artery calcification induced by warfarin and by vitamin D. Arterioscler Thromb Vasc Biol 21:1610–1616
Price PA, June HH, Buckley JR, Williamson MK (2002) SB 242784, a selective inhibitor of the osteoclastic V-H^+ ATPase, inhibits arterial calcification in the rat. Circ Res 91:547–552
Price PA, Williamson MK, Nguyen TM, Than TN (2004) Serum levels of the fetuin-mineral complex correlate with artery calcification in the rat. J Biol Chem 279:1594–1600
Proudfoot D, Shanahan CM (2001) Biology of calcification in vascular cells: intima versus media. Herz 26:245–251
Proudfoot D, Skepper JN, Shanahan CM, Weissberg PL (1998) Calcification of human vascular cells *in vitro* is correlated with high levels of matrix Gla protein and low levels of osteopontin expression. Arterioscler Thromb Vasc Biol 18:379–388
Proudfoot D, Skepper JN, Hegyi L, Farzaneh-Far A, Shanahan CM, Weissberg PL (2001) The role of apoptosis in the initiation of vascular calcification. Z Kardiol 90:43–46
Proudfoot D, Davies JD, Skepper JN, Weissberg PL, Shanahan CM (2002) Acetylated low-density lipoprotein stimulates human vascular smooth cell calcification by promoting osteoblastic differentiation and inhibiting phagocytosis. Circulation 106:3044–3050
Qiao JH, Mertens RB, Fishbein MC, Geller SA (2003) Cartilaginous metaplasia in calcified diabetic peripheral vascular disease: morphologic evidence of endochondral ossification. Hum Pathol 34:402–407

Rajamannan NM, Subramaniam M, Richard D, Stock SR, Donovan J, Springett M, Orszulak T, Fullerton DA, Tajik AJ, Bonow RO, Spelsberg T (2003) Human aortic valve calcification is associated with an osteoblast phenotype. Circulation 107:2181–2184

Reslerova M, Moe SM (2003) Vascular calcification in dialysis patients: pathogenesis and consequences. Am J Kidney Dis 41:S96–99

Reynolds ES (1965) Liver paranchymal cell inhury. III. The nature of calcium-associated electron-opaque masses in rat liver mitochondria following poisoning with carbon tetrachloride. J Cell Biol 25:53–75

Reynolds JL, Joannides AJ, Skepper JN, McNair R, Schurgers LJ, Proudfoot D, Jahnen-Dechent W, Weissberg PL, Shanahan CM (2004) Human vascular smooth cells undergo vesicle-mediated calcification in response to changes in extracellular calcium and phosphate concentrations: a potential mechanism for accelerated vascular calcification in ESRD. J Am Soc Nephrol 15:2857–2867

Ringpfeil F, Pulkkinen L, Uitto J (2001) Molecular genetics of pseudoxanthoma elasticum. Exp Dermatol 10:221–228

Robinson MJ, Strebel RF, Wagner BM (1968) Experimental tissue calcification IV. Ultrastructural observations in vagal calciphylaxis. Arch Pathol 85:503–515

Rogers KM, Stehbens WE (1986) The morphology of matrix vesicles produced in experimental arterial aneurysms of rabbits. Pathology 18:64–71

Ryall RL, Fleming DE, Grover PK, Chauvet M, Dean CJ, Marshall VR (2000) The hole truth: intracrystalline proteins and calcium oxalate kidney stones. Mol Urol 4:391–402

Ryall RL, Fleming DE, Doyle IR, Evans NA, Dean CJ, Marshall VR (2001) Intracrystalline proteins and the hidden ultrastructure of calcium oxalate urinary crystals: implications for kidney stone formation. J Struct Biol 134:5–14

Scarpelli DG (1965) Experimental nephrocalcinosis: a biochemical and morphologic study. Lab Invest 14:123–141

Schäfer C, Heiss A, Schwarz A, Westenfeld R, Ketteler M, Floege J, Müller-Esterl W, Schinke T, Jahnen-Dechent W (2003) The serum protein α_2-Heremans-Schmid glycoprotein/fetuin-A is a systematically acting inhibitor of ectopic calcification. J Clin Invest 112:357–366

Schoen FJ, Hobson CE (1985) Anatomic analysis of removed prosthetic heart valves: causes of failure of 33 mechanical valves and 58 bioprostheses, 1980 to 1983. Hum Pathol 16:549–559

Schoen FJ, Levy RJ (1992) Bioprosthetic heart valve calcification: membrane-mediated events and alkaline phosphatase. Bone Miner 17:129–133

Schoen FJ, Levy RJ (1994) Pathology of substitute heart valves: new concepts and developments. J Card Surg 9:222–227

Schoen FJ, Levy RJ (2005) Calcification of tissue heart valve substitutes: progress toward understanding and prevention. Ann Thorac Surg 79:1072–1080

Schoen FJ, Levy RJ, Nelson AC, Bernhard WF, Nashef A, Hawley M (1985) Onset and progression of experimental bioprosthetic heart valve calcification. Lab Invest 52:523–532

Schoen FJ, Tsao JW, Levy RJ (1986) Calcification of bovine pericardium used in cardiac valve bioprostheses. Implications for the mechanisms of bioprosthetic tissue mineralization. Am J Pathol 123:134–145

Schoen FJ, Hirsch D, Bianco RW, Levy RJ (1994) Onset and progression of calcification in porcine aortic bioprosthetic valves implanted as orthotopic mitral valve replacements in juvenile sheep. J Thorac Cardiovasc Surg 108:880–887

Schoppet M, Preissner KT, Hofbauer LC (2002) RANK ligand and osteoprotegerin: paracrine regulators of bone metabolism and vascular function. Arterioscler Thromb Vasc Biol 22:549–553

Seemayer TA, Thelmo WL, Morin J (1973) Cartilaginous transformation of the aortic valve. Am J Clin Path 60:616–620
Seifert G (1970) Morphologic and biochemical aspects of experimental extraosseous tissue calcification. Clin Orthop Relat Res 69:146–158
Seifert G, Dreesbach HA (1966) Die calciphylaktische Artheriopatie. Frankf Z Path 75:342–361
Seifert G, v.Hentig O (1967) Histochemische und elektronenoptische Befunde zur Pathogenese der experimentellen Hautcalcinose. Beitr Path Anat 135:75–91
Selye H (1962) Calciphylaxis. The University of Chicago Press, Chicago
Selye H, Grasso S, Patmanabhan N (1960) Topical injury as a means of producing calcification at predetermined points with dihydrotachysterol. Proc Zool Soc 13:1–5
Shanahan CM, Proudfoot D, Farzaneh-Far A, Weissberg PL (1998) The role of Gla proteins in vascular calcification. Crit Rev Eukaryot Gene Expr 8:357–375
Shanahan CM, Proudfoot D, Tyson KL, Cary NR, Edmonds M, Weissberg PL (2000) Expression of mineralisation-regulating proteins in association with human vascular calcification. Z Kardiol 89:63–68
Singla A, Lee CH (2002) Effect of elastin on the calcification rate of collagen-elastin matrix systems. J Biomed Mater Res 60:368–374
Slavin RE, Wen J, Kumar D, Evans EB (1993) Familial tumoral calcinosis. A clinical, histopathologic, and ultrastructural study with an analysis of its calcifying process and pathogenesis. Am J Surg Path 17:788–802
Spronk HM, Soute BA, Schurgers LJ, Cleutjens JP, Thijssen HH, De Mey JG, Vermeer C (2001) Matrix Gla protein accumulates at the border of regions of calcification and normal tissue in the media of the arterial vessel wall. Biochem Biophys Res Comm 289:485–490
Srivatsa SS, Harrity PJ, Maercklein PB, Kleppe L, Veinot J, Edwards WD, Johnson CM, Fitzpatrick LA (1997) Increased cellular expression of matrix proteins that regulate mineralization is associated with calcification of native human and porcine xenograft bioprosthetic heart valves. J Clin Invest 99:996–1009
Steitz SA, Speer MY, Curinga G, Yang HY, Haynes P, Aebersold R, Schinke T, Karsenty G, Giachelli CM (2001) Smooth muscle cell phenotypic transition associated with calcification: upregulation of Cbfa1 and downregulation of smooth muscle lineage markers. Circ Res 89:1147–1154
Sugimoto T, Funae Y, Rubben H, Nishio S, Hautmann R, Lutzeyer W (1985) Resolution of proteins in the kidney stone matrix using high-performance liquid chromatography. Eur Urol 11:334–340
Suzuki K, Mayne K, Doyle IR, Ryall RL (1994) Urinary glycosaminoglycans are selectively included into calcium oxalate crystals precipitated from whole human urine. Scann Microsc 8:523–530
Sweatt A, Sane DC, Hutson SM, Wallin R (2003) Matrix Gla protein (MGP) and bone morphogenetic protein-2 in aortic calcified lesions of aging rats. J Trob Haemost 1:178–185
Szabó É, Módis L (1980) Histochemische Untersuchung der Matrix kalziumhaltiger Harnsteine an-hand topo-optischer Reaktionen. Z Urol u Nephrol 73:879–885
Tanimura A, McGregor DH, Anderson HC (1983) Matrix vesicles in atherosclerotic calcification. Proc Soc Exp Biol Med 172:173–177
Tanimura A, McGregor DH, Anderson HC (1986a) Calcification in atherosclerosis. I. Human studies. J Exp Pathol 2:261–273
Tanimura A, McGregor DH, Anderson HC (1986b) Calcification in atherosclerosis. II. Animal studies. J Exp Pathol 2:275–297

Thomas RS, Greenawalt JW (1968) Microincineration, electron microscopy, and electron diffraction of calcium-phosphate-loaded mitochondria. J Cell Biol 39:55–76

Tyson KL, Reynolds JL, McNair R, Zhang Q, Weissberg PL, Shanahan CM (2003) Osteo/chondrocytic transcription factors and their target genes exhibit distinct patterns of expression in human arterial calcification. Arterioscler Thromb Vasc Biol 23:489–494

Valente M, Bortolotti U, Thiene G (1985) Ultrastructural substrates of dystrophic calcification in porcine bioprosthetic valve failure. Am J Pathol 119:12–21

Valouch P, Obenberger J, Vrabec F (1974) Experimental corneal calcification. An ultrastructural study. Ophthal Res 6:6–14

Vandeputte DF, Jacob WA, Van Grieken RE (1990) Phosphorus, calcium and lead distribution in collagen in lead induced soft tissue calcification. An ultrastructural and X-ray microanalytical study. Matrix 10:33–37

Varma VA, Kim KM (1985) Placental calcification: ultrastructural and X-ray microanalytic studies. Scanning Electron Microsc 4:1567–1572

Vattikuti R, Towler DA (2004) Osteogenic regulation of vascular calcification: an early perspective. Am J Physiol Endocrinol Metab 286:E686–696

Vyavahare NR, Hirsch D, Lerner E, Baskin JZ, Zand R, Schoen FJ, Levy RJ (1998) Prevention of calcification of glutaraldehyde-crosslinked porcine aortic cusps by ethanol preincubation: mechanistic studies of protein structure and water-biomaterial relationships. J Biomed Mater Res 15:577–585

Vyavahare NR, Jones PL, Hirsch D, Schoen FJ, Levy RJ (2000) Prevention of glutaraldehyde-fixed bioprosthetic heart valve calcification by alcohol pretreatment: further mechanistic studies. J Heart Valve Dis 9:561–566

Wada T, McKee MD, Steitz S, Giachelli CM (1999) Calcification of vascular smooth muscle cell cultures: inhibition by osteopontin. Circ Res 84:166–178

Walton JN (1969) Disorders of voluntary muscle. Churchill Ltd., London

Walton RC, Kavanagh JP, Heywood BR (2003) The density and protein content of calcium oxalate crystals precipitated from human urine: a tool to investigate ultrastructure and the fractional volume occupied by organic matrix. J Struct Biol 143:14–23

Walzer C, Boivin G, Schönbörner AA, Baud CA (1980) Ultrastructural and cytochemical aspects of the initial phases of an experimental cutaneous calcinosis (calcergy) in the rat. Cell Tissue Res 212:185–202

Wang N-S, Steele AA (1979) Pulmonary calcification. Scanning electron microscopic and X-ray energy-dispersive analysis. Arch Pathol Lab Med 103:252–257

Weinbach EC, von Brand T (1965) The isolation and composition of dense granules from Ca^{++}-loaded mitochondria. Biochem Biophys Res Comm 19:133–136

15 Calcifying Matrices: Acquired or Experimental Diseases; Heritable Disorders; Genetically Modified Animals

15.1 Introduction

The changes that may occur in one or many components of calcified matrices in acquired or experimentally induced diseases, in heritable disorders, or in genetically modified animals, may shed light on the role each matrix component plays in the calcification process. This is especially true in the case of congenital diseases, most of which are due to an absence or abnormal expression of one or more matrix components. This feature has long been recognized, but it is only recently that the development of genetic techniques has opened the way to important new developments and insights. This is a broad, complex topic, and readers can refer to specialist journals if they wish to go further into the etiology, pathogenesis and genetics of these pathological conditions.

15.2 Acquired or Experimentally Induced Diseases

The best known acquired diseases involving calcified tissues are those of the skeleton, although tooth caries is probably the commonest of all. Even if they sometimes induce radical alterations in bone and cartilage, which may be severe enough to be recognized by simple X-ray inspection, and in spite of the frequent involvement of cells, they only rarely cause changes in the individual matrix components that play a key role in calcification. As a result, the deposition of inorganic substance may occur almost normally, even though the process often takes a disordered form (e.g., Paget's bone disease, hyperparathyroidisms, several tumors and tumor-like conditions). In some acquired skeletal diseases, however, matrix calcification is severely hampered.

15.2.1 Rickets

As it is a disease of the developing skeleton, rickets may arise at any time from almost immediately after birth to adolescence (the equivalent disease in adults is osteomalacia). It can be recognized by defective calcification of bone and cartilage, with a consequent increase in the volume of uncalcified matrix; it may be due to vitamin D deficiency (vitamin D-dependent rickets) or to various defects in the phosphate metabolism (vitamin D-resistant rickets) (Haussler and McCain 1977; Drezner and Harrelson 1979). Defective calcification is responsible for the main pathological changes which characterize the disease: increased amounts of uncalcified matrix in compact and trabecular bone, with the formation of wide osteoid borders, the abnormal development of maturing and hypertrophic zones in epiphyseal cartilage, and a low degree of matrix calcification.

In spite of these abnormalities, which may induce skeletal deformations and fractures, the basic structure and composition of rachitic bone and cartilage matrices resemble those in controls. It has, however, been reported that lysine hydroxylation occurs in the long bone collagen of rachitic rats and chicks to a greater extent than in control bone collagen (Barnes et al. 1973), that the quantitative relationships between collagen cross links and aldehydic precursors may be altered (Mechanic et al. 1972), and that proteoglycans isolated from rachitic-chick growth cartilage have the same average composition as, but are smaller than, those in controls, probably because of different proteolytic cleavage (Dickson and Roughley 1978; Roughley and Dickson 1980).

The rachitic changes that, because of their implications on the mechanism(s) of calcification, appear to be of greatest interest, are those affecting the early stages of the process; they, therefore, involve matrix vesicles. These appear structurally normal, so that rachitic and normal matrix vesicles cannot be distinguished under the electron microscope; moreover, they have a comparable composition, degree of alkaline phosphatase activity, and capacity to accumulate calcium ions in vivo and in vitro, and both are able to induce calcification, as shown by the presence of crystal clusters within them (Anderson and Sajdera 1975, 1976; Morris et al. 1990; Stechschulte et al. 1991; Anderson et al. 1992). The calcification process, however, stops at the early, intra-vesicular stage (Fig. 15.1): crystal formation continues until the vesicles have been filled, but does not spread into the surrounding matrix (Takechi and Itakura 1995).

The reason for this functional abnormality, which has been found in epiphyseal cartilage of rat fed on a calcium-free diet (Mocetti et al. 1997) and is common to other pathological conditions (scurvy, hypophosphatasia), is uncertain. In particular, the way in which calcification normally spreads from matrix vesicles through the matrix is not exactly known – the theory that the matrix vesicle membrane is mechanically broken by the crystals growing inside them (Anderson 1989) appears to be hard to sustain. Matrix vesicles contain proteases (Einhorn and Majeska 1991; Schmitz et al. 1996), so it may be speculated that the rupture of their membrane could be produced by enzymatic degradation brought about by Ca-activated neutral proteases. The

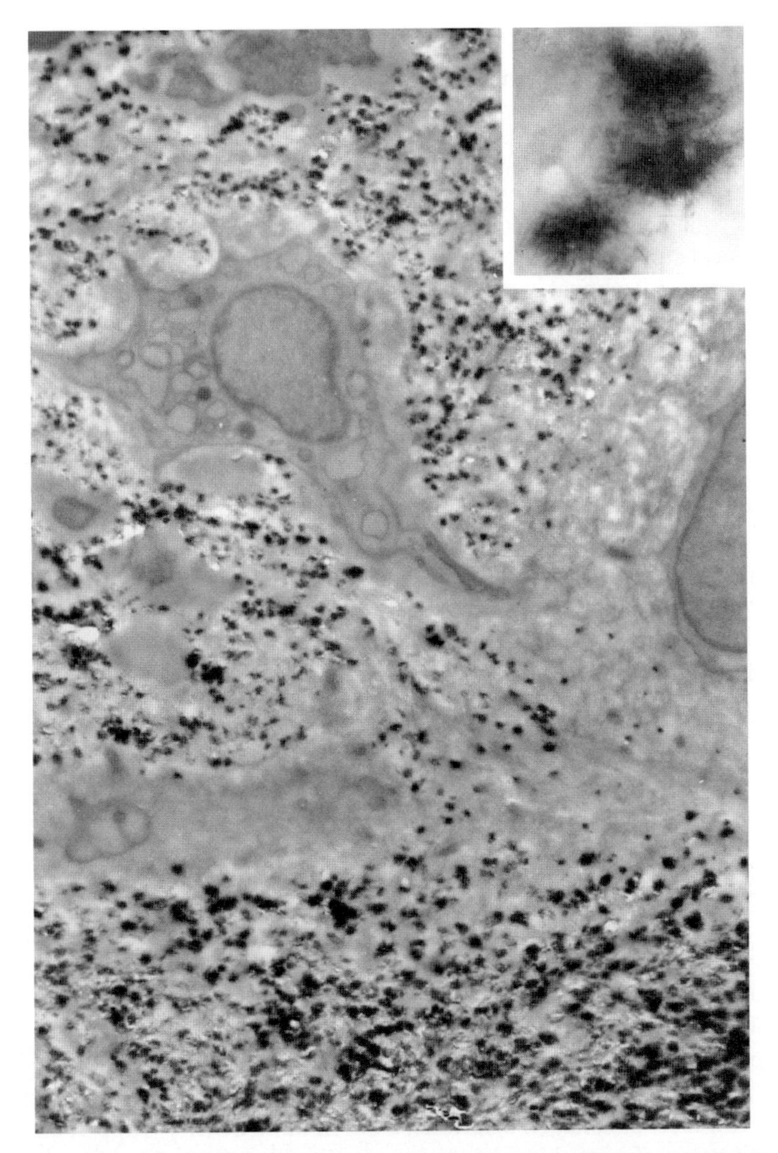

Fig. 15.1. Rat hypophosphatemic rickets: plenty of calcification nodules are scattered through the otherwise uncalcified metaphyseal bone matrix; an osteocyte and parts of other osteocytes are visible. *Inset*: detail of calcification nodules. Untreated, ×5850 and ×65,000

occurrence of the same abnormality in unrelated pathological conditions such as scurvy and hypophosphatasia does, however, suggest that the spread of calcification from matrix vesicles into the intercellular matrix depends on the normal structure and composition of the latter.

15.2.2
Fibrogenesis Imperfecta Ossium

Fibrogenesis imperfecta ossium is a very rare, suddenly acquired, potentially lethal bone disease of unknown etiology, which is clinically characterized by intense bone pain and osseous fragility. It is marked by the defective formation of collagen fibers (Baker et al. 1966), which are structurally normal in soft tissues and the periosteum, but are almost completely absent in the bone matrix (Fig. 15.2, above), where they are substituted by thin filaments or amorphous material (Baker 1956; Pinto et al. 1981; Sissons 2000). The calcification process is not completely inhibited but is severely hampered, and patients develop bone changes (such as increased amounts of uncalcified matrix) which mimic those of osteomalacia (Frame et al. 1971) to such an extent that a correct histological diagnosis requires bone sections to be checked under the polarizing microscope. This shows that the only detectable birefringence is that of the pre-existing axial trabeculae, while the wide borders of uncalcified osteoid tissue which cover them are isotropic (Thomas and Moore 1968; Swan et al. 1976), so confirming that they lack collagen fibrils. According to Henneman et al. (1972), total soluble collagen (as OH-proline) of bone rose sharply (from the normal value of 0.42 to 5.76 μg/100 mg of dried, defatted, decalcified matrix) in fibrogenesis imperfecta, whereas total collagen fell (from 99.3 to 65.0 μg/100 mg).

15.2.3
Scurvy

Scurvy, a disease induced by a prolonged lack of vitamin C, has practically disappeared in developed countries where fresh vegetables and citrus fruits are a readily available source of the vitamin. The interest in this disease, however, remains high, because the lack of ascorbic acid causes an abnormal development of collagen fibrils (Gould 1963), which itself offers insights into the role of fibrils in calcification.

Experimental scurvy can easily be produced in guinea pigs by feeding them on an appropriate, vitamin C-deprived diet: after about 30 days, they experience a low level of collagen synthesis, a fall in bone density (Kipp et al. 1996), fragmentation of metaphyseal trabeculae and loss of the seriated arrangement of chondrocytes in epiphyseal cartilage (Bonucci 1970). Degenerative changes can be detected in chondrocytes, the intercellular matrix reveals a reduction in numbers of collagen fibrils and the size of proteoglycan particles, whereas there are many matrix vesicles (Thyberg et al. 1971; Bonucci 1978). Because the synthesis of new collagenous matrix is impaired or absent, calcification occurs in the cartilage matrix already present at the beginning of the experiment and gradually expands through the normally uncalcified maturation zone to eventually reach the proliferation zone (Bonucci 1978). The scanty newly synthesized matrix is abnormal and consists of collagen fibrils that are thinner

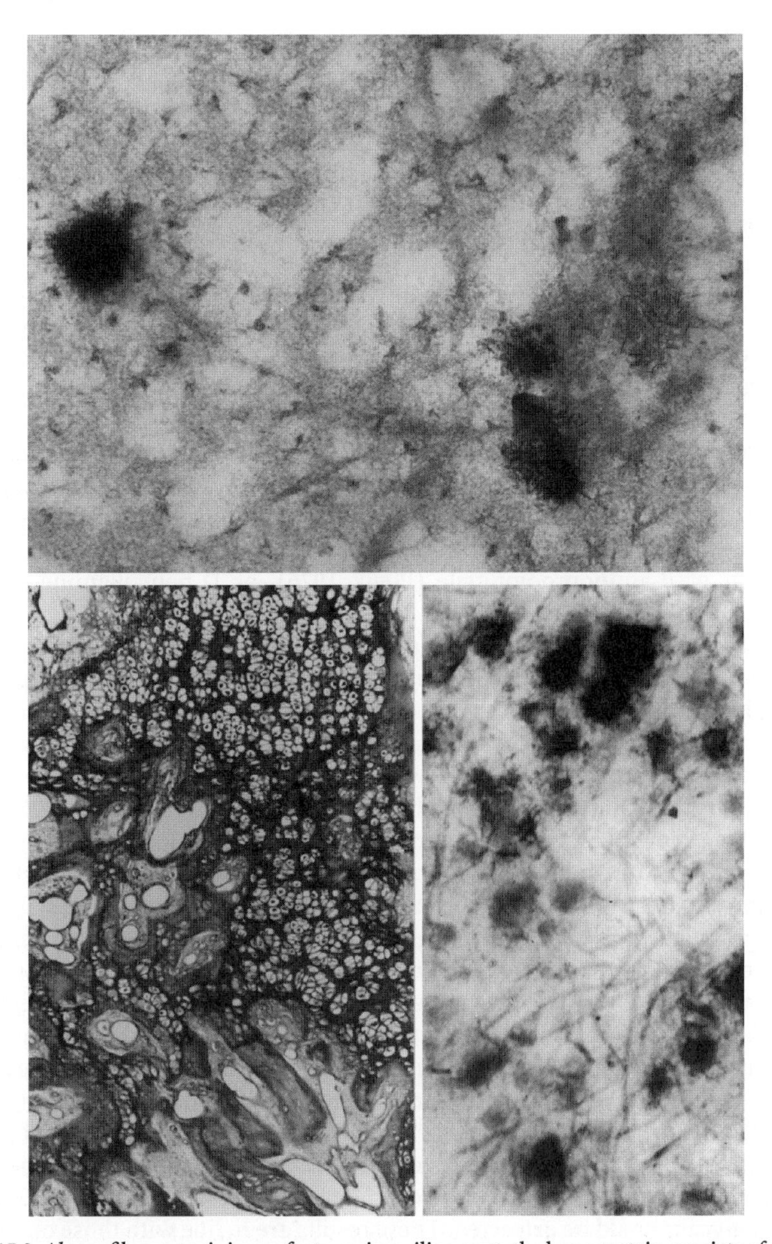

Fig. 15.2. *Above*: fibrogenesis imperfecta ossium, iliac crest: the bone matrix consists of a few abnormally thin collagen fibrils, a few proteoglycan granules, and plenty of granular and filamentous material which seems to have been calcifying, as suggested by the different degrees of electron density in the calcification nodules. Uranyl acetate and lead citrate, ×60,000. *Lower left*: rickets-like changes in metaphyseal bone from a newborn affected by hypophosphatasia: note excessive numbers of chondrocytes and distortion of cell columns and trabeculae. PAS, ×120. *Lower right*: same type of bone; calcification of cartilage matrix was limited to matrix vesicles. Uranyl acetate and lead citrate, ×65,000

than normal, filaments without any periodic banding, and/or amorphous substance. Calcification is confined to matrix vesicles and, as already described in rickets, is unable to spread into the surrounding matrix (Bonucci 1970).

15.2.4 Lathyrism

Lathyrism is a severe pathological condition that can be induced by the administration of lathyrogenic compounds to experimental animals (reviewed by Tanzer 1965). Interest in it arises from its resemblance to the human Marfan syndrome. It is characterized by pathological changes in several connective tissues, the severest of which include aortic aneurysms, hernias, and skeletal deformities. It is caused in chicken by the prolonged administration of peas (*Lathyrus odoratus*), which contain β-aminopropionitrile, an inhibitor of lysine oxidase and, therefore, of the formation of cross-links in collagen fibrils. Lathyrism can be induced experimentally by the administration of β-aminopropionitrile (Plenk 1976), or D-penicillamine (Riede 1970).

Within the skeleton, lathyritic changes include macroscopic deformities in long bones, subperiosteal bone formation and kyphoskoliosis, and microscopically irregular thickening of the hypertrophic zone of the epiphyseal cartilage and focal perilacunar calcification, the presence in the matrix of large collagen fibers (Matukas et al. 1967) with a periodicity of 64 nm (Riede 1970) and a cartilage picture that resembles that of rickets (Hartmann 1963). In bone, histomorphometric measurements have shown a significant reduction in areas of bone formation, total forming surface, numbers of osteoblasts and alkaline phosphatase activity (Kwong-Hing et al. 1991). A fall in the synthesis of tropocollagen by chondrocytes grown in vitro in the presence of penicyllamine has been detected by Merker et al. (1972). Lathyritic lesions, however, are mainly due to the defective maturation of newly synthesized collagen and elastin secondary to lathyrogen's interference with the formation of covalent interchain cross-links (Page and Benditt 1972). Impaired synthesis of proteoglycans has also been suggested (Hartmann 1963) and Glimcher et al. (1969) have shown that both collagen fibrils and proteoglycans may be abnormal: in lathyritic chicks, collagen and proteoglycans in cartilage were both almost completely soluble, whereas in controls collagen was insoluble, and only 37% of proteoglycans could be extracted. These results are in line with those obtained in the skin of rats fed with β-aminopropionitrile fumarate, in which a marked increase in soluble collagen and an increase in the extraction of hexosamine (both glucosamine and galactosamine) were observed (Orlowski and Orbison 1970). Histochemical studies on proteoglycans in lathyritic tissues have given uncertain results; in any case, the size of matrix proteoglycan granules seems to be much lower in the zone where epiphyseal cartilage proliferates and matures (Matukas et al. 1967).

Few studies have been carried out on the calcification of lathyritic bone. No marked changes have been reported in the calcification process and Sandhu and Jande (1982) have stated that the administration of β-aminopropionitrile does not interfere with calcification. Lee et al. (1990) added that the calcification process can apparently continue in the newly laid down bone matrix even when osteolathyritic conditions are severe.

15.3 Heritable Disorders

Heritable disorders of the skeleton may provide tools useful to morphological and biochemical studies on calcifying matrices. Actually, the main aim of this type of research is less that of verifying how the calcifying matrices are structured in a specific genetic condition, than of studying the effects that the lack of a protein, or its alteration, may have on the calcification process, a goal that is most easily attained by using genetically modified animals (see Sect. 15.4). There are a great many heritable disorders (see McKusik 1960 for a detailed description) and only those that are of major interest for the purposes of this book will be considered.

15.3.1 Osteogenesis Imperfecta

Osteogenesis imperfecta, a congenital disease characterized by brittle bones and frequent fractures, is one of the hereditary osteopathies most often studied (for reviews see Tsipouras 1993; Roughley et al. 2003; Plotkin 2004). It is of great clinical interest, because the frequent fractures, which in some cases can be induced even by very minor traumas, such as those exerted on a diseased newborn by uterine contractions, can severely impair individual life. It is also of great biological interest, because of the inferences that can be drawn from a knowledge of the relationship between abnormalities in bone matrix and its synthesis on one hand and the calcification process on the other.

Actually, osteogenesis imperfecta is a rather elusive disease which involves not only bone, but also uncalcified connective tissues (chiefly sclera, skin and ligaments). In his review on brittle bone syndromes, Plotkin (2004) has emphasized that no clear definitions of osteogenesis imperfecta have been formulated, that it can be subdivided into four different forms, that as many as 12 forms could be counted (seven according to Roughley et al. 2003), and that some of these may be congenital forms of brittle bones resembling osteogenesis imperfecta. He has proposed defining the disease as a syndrome of congenital brittle bone secondary to mutations in the genes that codify procollagen genes (COL1A1 and COL1A2). In reality, osteogenesis imperfecta is genetically heterogeneous (Sillence et al. 1979), and the alterations in collagen fibrils produced by those mutations are often hard to recognize. Missense base substitutions involving glycine codons in the exons that encode the central triple-helix forming domain are prevalent among type I collagen gene mutations that may occur in osteogenesis imperfecta (Roughley et al. 2003). According to Kuznetsova et al. (2001), α1-homotrimers could be generated by the formation of non-functional α2-chains, whose absence might prevent the binding of non-collagenous components, possibly glycosaminoglycans, which play a key role in bone matrix formation and calcification. The substitution of aspartic acid for glycine 700 in the α2(I) chain of collagen has been described in a lethal case of type II osteogenesis imperfecta by Cohen-Solal et al. (1994), and substitutions of aspartic acid for glycine-220 and of arginine for glycine-664 in the triple helix of the pro-α1(I) chain of type I procollagen have been reported by Culbert et al. (1995). The presence of type I procollagen with cysteine in the triple-helical domain of pro-α1(I) chains has been detected by de Vries and de Wet (1986) in an autosomal dominant form of osteogenesis imperfecta, and the same type of substitution has been found by Steinmann et al. (1986) in lethal and mild forms. According to Dominguez et al. (2005), genetic mutations distort triple helix conformation and slow the folding rate, so favoring overglycosylation. According to Fujii and Tanzer (1977), the formation of cross-links in collagen fibrils is delayed.

Uncertainties about the classification of the disease reflect not only the differences to be found in its pathogenesis, genetic traits, symptoms, age of onset, and accompanying bodily deformities, but also the variety of pathological changes to be found in bone which often overlap, and in some forms are hard to distinguish or apparently absent, so that osteogenesis imperfecta cannot be characterized on the basis of a single histological examination (Cassella et al. 1996). Without considering its different forms in detail, as Plotkin's review (Plotkin 2004) can be consulted on this, it is important for the purposes of this book to stress that the structure and constitution of the organic matrix in bone show the same degree of variability as the clinical characteristics. So, even if the bone matrix in osteogenesis imperfecta tarda shows no appreciable structural alterations (Doty and Mathews 1971), in most of the severe forms the spectrum of pathological changes (which have often been examined in single cases) is quite wide. The most evident of these comprise: low numbers of otherwise normal type I collagen fibrils; collagen fibrils with a thickness greater (Cassella et al. 1994), and a hydroxylysine content double (Trelstad et al. 1977), than in controls; collagen fibrils of unusually low thickness (mean thickness of 10–20 nm according to Cassella et al. 1996, of 24 nm according to Bonucci and De Matteis 1968, of 40 nm according to Haebara et al. 1969, and of 40–60 nm according to Jones et al. 1984); collagen fibrils without any periodic banding (Bonucci and De Matteis 1968; Haebara et al. 1969; Vitellaro-Zuccarello et al. 1984) – probably the expression of a defect in fibril aggregation (Teitelbaum et al. 1974). Likewise, the degree of calcification has been reported to range from that of normal calcified matrix to that of matrix partly covered by unusually small crystallites (Bonucci and De Matteis 1968; Vetter et al. 1991a), whose mean thickness in humans has been reported to be 1.2 nm (Bonucci and De Matteis 1968) and whose length in two bovine models of osteogenesis imperfecta ranged between 37.2 and 42.1 nm vs a normal value of 59.0 nm (Fisher et al. 1987a). In addition, calcification nodules are scattered through the osteoid matrix; they appear to be unrelated to the few thin collagen fibrils that are present in the matrix, and do not seem to coalesce (Bonucci and De Matteis 1968; Cohen-Solal et al. 1994). Granular inorganic aggregates and needle-like crystals have been reported in osteoblast mitochondria by Cassella et al. (1996). In some patients the Ca/P ratio has been found to be lower than that of hydroxyapatite (Cassella and Ali 1992; Cassella et al. 1995; Sarathchandra et al. 1999).

Besides collagen fibrils, other bone matrix components have been reported to be abnormal in osteogenesis imperfecta. A threefold increase in glycosaminoglycans, and the presence of chondroitin disulphate, which is normally found only in traces in cartilage, have been described by Engfeldt and Hjerpe (1976); an increase in chondroitin sulfate has been reported by Haebara et al. (1969). On the other hand, Nogami and Oohira (1988) found a 95.2% fall in the number of proteoglycan granules periodically associated with the cross-banding of collagen fibrils, as well as a strong accumulation of hyluronate in tubular bones (Oohira and Nogami 1989). According to Vetter et al. (1991b), unusually low levels of ON are found in bone in all cases of osteogenesis imperfecta, with the lowest levels in type III, whereas BSP is prominent in all patients, with the highest levels in type IV, OC and AHSG concentrations are high in all patients, and decorin levels are not significantly altered. On the other hand, the ON content in human bone specimens can vary from normal to severely depressed (Fisher et al. 1987b), and, in clinically identical but genetically unrelated types of bovine osteogenesis imperfecta, the bone ON content was normal in one type and severely depressed in the other (Fisher et al. 1986).

15.3.2 Dentinogenesis Imperfecta

Dentinogenesis imperfecta, an autosomal dominant inherited dental disease, is characterized by abnormal dentin production and calcification. It has many points in common with osteogenesis imperfecta, and both diseases may develop in the same patient; as with osteogenesis imperfecta, several different forms of dentinogenesis imperfecta are known.

An electron microscope study of Herold (1972) on the deciduous teeth of an eight-year-old male patient with clinical signs of dentinogenesis imperfecta showed coarse collagen fibrils with a diameter of 80–120 nm, a low degree of calcification, and apatite crystals mostly located in interfibrillary spaces. Abnormally thick collagen fibrils (hyperfibers) have been detected in dentin together with matrix vesicles, which in controls are only found in mantle dentin (Waltimo 1994). In a study of dentinogenesis imperfecta type II, carried out using high-resolution synchrotron radiation computed tomography (SRCT) and small-angle X-ray scattering (SAXS), Kinney et al. (2001) found that the inorganic component was 33% lower on average than in normal dentin, that during maturation the early needle-shaped crystals did not acquire a platelet-like shape as in controls, and that whereas SAXS diffraction peaks consistent with the nucleation and growth of the apatite phase inside gaps in collagen fibrils (intrafibrillar calcification) were detectable in normal dentin, they were absent in dentinogenesis imperfecta.

The gene for dentin sialoprotein is continuous with that of dentin phosphoprotein; thus, DSP and DPP must be secreted as a single protein (DSPP), which is then proteolytically processed to form the individual components found in dentin matrix (see Sect. 8.5.4); significantly, the DSP/DPP gene is localized in human chromosome 4q21 at a site involved in dentinogenesis imperfecta (Butler 1998). MacDougall et al. (2002) have reported that an overlapping segment of these disease loci contains a dentin/bone gene cluster which includes OPN, BSP, DMP1, DSP and MEPE. Their RT-PCR results show that MEPE is expressed during odontogenesis and that it might be a candidate gene for dentin structural disease mapping to human chromosome 4q21. The DMP-1 gene may be another candidate for dentinogenesis imperfecta type II (MacDougall et al. 1996).

15.3.3
Amelogenesis Imperfecta

Amelogenesis imperfecta is an often inherited, X-linked, autosomal dominant or recessive condition affecting the development and calcification of tooth enamel. As with osteogenesis imperfecta, it does not represent a unique, specific disease; a number of enamel abnormalities are, in fact, grouped under the name amelogenesis imperfecta. This explains why many classifications based on genetic traits and a molecular basis have been proposed, including at least 14 different forms (reviewed by Witkop 1988; Hart et al. 2002a; Aldred et al. 2003), some of them exceedingly rare. A clinical classification based on the predominant phenotype comprises three main groups, known as hypoplastic (when the enamel is thinner than normal), hypomaturation (when enamel has maturation defects), and hypocalcified (when enamel has calcification defects) (Witkop and Rao 1971). Although this subdivision is oversimplified, it is justified by the currently poor knowledge about matrix changes, which have only been studied in a small number of cases; it appears, in any case, to be sufficiently precise for the purposes of this book.

Hypoplastic enamel has been examined by Batina et al. (2004) using atomic force microscopy both in the contact and tapping modes. They found that the hypoplastic enamel surface, which normally shows closely packed, large, compact grains, had a porous structure and consisted of loosely packed, minute grains showing no tendency to associate, and were compatible with a defective synthesis of amelogenins. These hypoplastic changes were detected over small localized areas where the RMS[Rq] value (root-mean-square roughness, expressing surface roughness, whose normal value is 116.50 nm) was only 18.3 nm, or were found in larger enamel surfaces which were uniform, with a high number of small pores and grains (RMS[Rq] = 10 nm). Wright et al. (1993a) used light microscopy, scanning electron microscopy and microradiography to study the enamel from two related persons with X-linked amelogenesis imperfecta: in the heterozygous female, they found that the hypoplastic enamel showed a rough surface with marked depressions, poor organization and little prismatic structure; in the affected male, it was very thin (about 40 μm) and showed no organized structure. Abnormal prism structure and the presence of amorphous, presumably organic material (Wright et al. 1991), as well as short crystal segments (Robinson et al. 2003), have been identified as distinctive features of hypoplastic amelogenesis imperfecta. Small, closely packed crystallites (resembling those of a type of enameloid in teleost fishes) have been described in hypoplastic globular enamel (Kerebel and Kerebel 1981).

In hypomaturation amelogenesis imperfecta, enamel crystallites have been described as varying greatly in size and morphology and as coated by an organic, possibly partly calcified material that was not visible in normal enamel (Wright et al. 1992a). Abnormally large crystals alongside spherical subunits 50 nm in diameter have been reported by Robinson et al. (2003).

Scanning electron microscope studies on human hypocalcified amelogenesis imperfecta have shown that the enamel contains porosities (Wright et al. 1993b) and disoriented prisms, with the hypocalcification of prism sheaths (Sauk et al. 1972). Immunohistochemical studies on the enamel of patients suffering from hypocalcified amelogenesis imperfecta have detected considerable amounts of amelogenin peptides which, in one case examined by Western-blot transfer and immunobinding analysis, was found to consist of a 26-kDa molecule (Takagi et al. 1998). All mutations that alter the amelogenin C-terminus after the 157th amino acid result in an enamel hypoplastic phenotype, whereas other mutations cause what are predominantly calcification defects (Hart et al. 2002b). This subdivision, however, fails to account for all the patients who suffer from hypoplastic amelogenesis imperfecta (Snead 2003). Paine et al. (2002) have reported results supporting the conclusion that the amelogenin N-terminal self-assembly domain is essential for the formation of an organic matrix capable of directing the calcification process. Mutations in the human enamelin gene cause varying degrees of enamel hypoplasia (Hu and Yamakoshi 2003; Kim et al. 2005) and may be responsible for the autosomal-dominant hypoplastic form of amelogenesis imperfecta (Rajpar et al. 2001; Kida et al. 2002; Masuya et al. 2005).

In a comparison of the protein contents of enamel in different types of amelogenesis imperfecta, Wright et al. (1992b) have reported that while the protein content of normal mature enamel is only 0.3%, in hypoplastic enamel it rises to about 2%, keeping a practically normal amino acid profile, while in hypomaturation enamel it reaches about 5%, acquiring an amelogenin-like amino acid profile.

15.3.4 Hypophosphatasia

Hypophosphatasia (Fig. 15.2, below), a hereditary deficiency of the gene for human tissue non-specific alkaline phosphatase (TNSALP), is characterized by rickets-like changes in cartilage and bone (McCance et al. 1956; Whyte 1989) and high extracellular concentrations of inorganic pyrophosphate (PPi) which suppress the formation and growth of hydroxyapatite crystals (Hessle et al. 2002). As in other hereditary skeletal diseases, the clinical and pathological expression of the disease vary greatly, and a number of gene mutations are possible: Zurutuza et al. (1999) have proposed a classification of 32 gene mutations found in 23 European patients, 17 of whom had lethal hypophosphatasia; Brun-Heath et al. (2005) have added 11 new mutations and Taillandier et al. (2000) have reported 20 TNSALP gene mutations, 15 of them new, found in 12 families affected by severe or mild hypophosphatasia.

Matrix vesicles in the hypophosphatemic epiphyseal cartilage are reported to have a low alkaline phosphatase activity and, in two cases, to not contain this enzyme at all; despite this, they initiate the calcification process (Ornoy et al. 1985). This process, however, as already reported for rickets and scurvy, fails to spread into the surrounding organic matrix (Anderson et al. 1997). This may explain why pathological changes are more severe in cartilage, which has plenty of matrix vesicles, than in bone, which has few (Ornoy et al. 1985).

According to Genge et al. (1988), 65–70% of alkaline phosphatase activity is normally lost as Ca builds up in matrix vesicles; this loss is not due to protease activity, but to alkaline phosphatase interaction with the developing mineral phase, loss of metal ions (Zn and Mg) from active sites of the enzyme, and its concomitant irreversible denaturation. On this basis, it may be hypothesized that the calcification process can continue as long as it is mediated by a low level of vesicular enzymatic activity, whereas it subsides, and is unable to spread beyond the vesicle membrane, when that enzymatic activity is exhausted. It has, however, been reported that the lack of matrix calcification is due to its inhibition by an excess of pyrophosphate, which is no longer metabolized by alkaline phosphatase (Hessle et al. 2002), a topic further discussed in Sect. 15.4.2.

15.4 Animals with Genetic Defects

Genetic modifications which occur naturally in experimental animals, or can be induced in them through genetic engineering, may reproduce spontaneous human diseases or may make available new pathological pictures. Studies on these animals have been widely pursued in recent years and await further development; they allow verification of the effects induced by the abnormality or absence of one or more proteins on the structure and function of a tissue (Aszodi et al. 2000). Important information has been obtained in this way on the role of matrix components in promoting, inhibiting and regulating the calcification process. It should be noted, however, that the alteration or absence of single proteins may induce alterations of very different degrees of severity, which may develop over different time scales, and which may range from hardly apparent defects to lesions severe enough to cause an animal's death. Moreover, genetic defects may primarily induce the abnormal development or function of cells and the disruption of tissue organization, which may not necessarily affect calcification, or may only do so secondarily. This chapter deals with genetic alterations in the proteins that directly take part in, or may be supposed to take part in, the local mechanisms of the calcification process.

15.4.1 Oim/oim Mice

A transgenic animal which may serve as a human osteogenesis imperfecta model (oim) is the homozygous mouse known as oim/oim, which produces only $\alpha 1$(I) collagen homotrimers instead of the normal heterotrimers consisting of two $\alpha 1$(I) and one $\alpha 2$(I) chains. The lack of $\alpha 2$(I) chains in collagen fibrils results in a significant reduction in their mechanical strength and in their stabilizing intermolecular cross-links (reduced by 27%; Sims et al. 2003), and leads to a fall in bone mineral content and bone mineral density (Phillips et al. 2000). A SAXS study has shown that the compact bone of oim/oim mice contains needle-shaped crystals smaller, and less well aligned, than those in the compact bone of heterozygous (oim−/+) and wild (+/+) mice, a finding similar to that reported for human osteogenesis imperfecta (Fratzl et al. 1996). Fourier transform infrared microspectroscopy (FTIRM) and infrared imaging (FTIRI) analyses of the metaphyseal bone of oim/oim mouse have documented an increase in the ratio between amounts of organic matrix and of inorganic substance and a fall in crystallinity (pointing to less mature crystals) in the primary versus the secondary spongiosa, probably due to the presence of calcified cartilage (Camacho et al. 2003).

Oim/oim mice do not show coarse tooth abnormalities and their dentin tubule numbers, structure and orientation resemble those of oim+/+ control mice; their dentin calcification is, however, lower and the carbonate in crystals is more labile (Baechtold et al. 2000).

15.4.2 TNSALP −/− Mice

Mouse strains have been generated that are null for tissue non-specific alkaline phosphatase (TNSALP); their interest derives from the observation that they recapitulate the metabolic and skeletal defects of congenital infantile hypophosphatasia (Narisawa et al. 1997; Fedde et al. 1999). As in the human disease (Sect. 15.3.4), so in TNSALP −/− mice, rickets-like changes develop in the skeleton. This reveals osteopenia and fractures, the reduction or disappearance of the hypertrophic zone in the epiphyseal cartilage, a progressive increase in the volume of osteoid tissue (Fedde et al. 1999) and greatly delayed epiphyseal calcification (Tesch et al. 2003). Osteoblasts, which appear to be normally differentiated, synthesize osteopontin, osteocalcin, type I collagen and Cbfa 1, and produce cellular nodules in vitro, but are unable to induce matrix calcification (Wennberg et al. 2000). Moreover, the distribution of osteopontin in cement and reversal lines has been found to differ from that in controls, with osteopontin randomly dispersed throughout the uncalcified bone matrix, apart from a few focal densities (Tesch et al. 2003).

It is of special interest that matrix vesicles are able to initiate calcification in TNSALP −/− mice and that, as observed in human hypophosphatasia, rickets and scurvy, crystals accumulate within them without spreading to the surrounding matrix. It has, in fact, been shown that, even if ATP-dependent ^{45}Ca uptake turns out to be lower in the calvaria osteoblast matrix vesicles in TNSALP −/− mice (Johnson et al. 2000), these vesicles are able to induce the intravesicle formation of crystals as in controls, but the calcification process is unable to get beyond their membrane (Anderson et al. 2004). This abnormal evolution may, at least partly, depend on the inhibitory activity of inorganic pyrophosphate (PP_i). This, in fact, accumulates in the matrix for two reasons: on one hand, it is no longer metabolized by TNSALP (Harmey et al. 2004); on the other, it is produced in excessive amounts by the nucleoside triphosphate pyrophosphohydrolase activity of the plasma cell membrane glycoprotein-1 (PC-1, or NPP1), which occurs in matrix vesicles (Hessle et al. 2002) and is no longer antagonized by TNSALP (Johnson et al. 2000). The inhibitory activity of PP_i on calcification is clearly shown by the hypercalcified tissues seen in mice with spontaneous mutations into the NPP-1 gene (Hessle et al. 2002); on the other hand, the view that the conflicting activities of TNSALP and NNP-1 balance each other is supported by the observation that mice deficient in both TNSALP and NNP-1 genes show normal bone in calvaria and vertebrae (Hessle et al. 2002); the persistence of reduced calcification in femurs and tibiae has, however, been found not only in mice lacking either of these two genes, but also in those lacking both of them (Anderson et al. 2005).

15.4.3
Mice Defective in Non-collagenous Proteins

Osteopontin-deficient (OPN −/−) mice do not display appreciable alterations of compact and trabecular bone. FTIRM and FTIRI studies on femur and tibia compact and spongy bones have shown, however, that the size and perfection of crystals, as well as the relative amount of inorganic substance in the central zones of cortical bone, are significantly increased in OPN-knockout mice, probably as a result of the lack of OPN-inhibitory activity (Boskey et al. 2002).

Using FTIRM analysis, reduced crystal size and perfection have been detected in the bone of osteocalcin knockout mice, although they show no alterations in matrix calcification that are detectable histologically (von Kossa method) or by tetracycline uptake (Boskey et al. 1998).

Alterations in osteoblast formation, maturation and survival have been described in ON-null mice (Delany et al. 2003): there are fewer osteoblasts, the bone formation rate is lower than in controls, there are only a few bone marrow osteoblast precursors and, compared with those in wild animals, they generate a smaller number of calcified nodules, express less osteocalcin mRNA, and display an enhanced expression of adipocytic markers.

Studies of MGP −/− mice have given results of major importance: they have shown that the arteries of these animals calcify spontaneously (Luo et al. 1997), so that rupture of the aorta and death occur in a few months; in addition, cartilage develops in artery walls (El-Maadawy et al. 2003). Arterial calcification is associated with loss of smooth muscle markers and an increase in OPN and Cbfa 1 expression (Steitz et al. 2001; El-Maadawy et al. 2003). Increased susceptibility to arterial calcification, comparable with that found in MGP −/− mice, has been reported in osteoprotegerin-null mice, which also develop osteoporosis (Bucay et al. 1998), and in carbonic anhydrase II-deficient mice.

Compared with wild-type mice, biglycan-deficient mice had delayed osteogenesis (Chen et al. 2003) possibly caused by a lower capacity to produce bone cell precursors in bone marrow, and by a weakened cell response to TGF-β (Young et al. 2002). Thinner collagen fibrils in proximal predentin, thicker collagen fibrils in central and distal predentin, broader metadentin, and abnormally calcified dentin have been described by Goldberg et al. (2003) in the teeth of biglycan knockout mice. Mice deficient in the gene Bgn, which encodes for biglycan, develop osteoporosis which becomes more obvious with aging (Xu et al. 1998). Biglycan deficient mice develop collagen fibril abnormalities which mimic Ehlers-Danlos changes but do not show calcification defects (Corsi et al. 2002). Biglycan and decorin deficient newborn mice show hypomineralized dentin; the former also show a dramatic increase in enamel formation, DSP overexpression, upregulation of DMP-1, BSP and OPN, and unchanged enamelin expression, whereas the latter show delayed enamel formation, reduction in expression of enamelin and DSP, downregulation of DMP1, BSP and OPN (Goldberg et al. 2005). No differences were observed between the two genotypes and wild controls in adult mice.

Dentin matrix protein 1-deficient mice display no apparent gross abnormal phenotype (Feng et al. 2003). Ye et al. (2004) have, however, reported that DMP1 deletion leads to dentin hypocalcification, a partial failure in the maturation of predentin into dentin, and expanded cavities in the pulp and the root canal. The results of in vitro studies of Tartaix et al. (2004) support the view that DMP1 inhibits calcification in its native form, whereas it initiates calcification when cleaved or dephosphorylated. These results appear to be in line with the reported promoting role on calcification of DMP1 immobilized on type I collagen fibrils (He and George 2004), and with the observation that DMP1-deficient mice develop cartilage changes which point to a chondrodysplasia-like phenotype (Ye et al. 2005).

Dentin sialophosphoprotein null mice develop tooth changes similar to those of human dentinogenesis imperfecta, consisting of enlarged pulp chambers, increased width of predentin, reduced width and hypocalcification of dentin, an irregular calcification front deprived of the coalescence of its calcification nodules, and increased levels of decorin and biglycan in predentin and in uncalcified matrix around and between calcification nodules (Sreenath et al. 2003). The expression of DSPP by odontoblasts appears to be mediated by the transforming growth factor $\beta 1$ (TGF-$\beta 1$), because transgenic mise over-expressing TGF-$\beta 1$ show a significant down-regulation of DSPP (Thyagarajan et al. 2001).

The enamel of transgenic mice that express amelogenin proteins lacking self-assembly properties displayed a fall in surface indentation resistance (Fong et al. 2003). Induced amelogenin alterations, which prevent the self-assembly of its molecules, cause defects in enamel rod organization (Paine et al. 2003a). Amelogenin-null mice display an amelogenesis imperfecta phenotype (Gibson et al. 2001).

Transgenic mice that overexpress tuftelin show changes in the shape of the enamel crystals, which appear more plate-like than, and contrast with the ribbon-like appearance of, control crystals (Luo et al. 2004).

Transgenic mice that overexpress ameloblastin show a porous enamel whose crystal shape and dimensions range from normal, in areas with a rod and interrod organization, to completely abnormal in other sites, the abnormality consisting of crystals whose diameter is approximately twice that of control crystals (Paine et al. 2003b).

The function of the matrix metalloprotease-20 (MMP-20, or enamelysin) has been checked in MMP-20 null mice during enamel maturation (Bartlett et al. 2004). It was found that while, in relative terms, the percentage weight of the inorganic component in mature enamel was 7–16% less than in controls, overall it fell by about 50%. The most conspicuous difference with respect to controls was found in nearly mature enamel, where MMP-20 is no longer expressed, suggesting that this may directly remove, or indirectly facilitate the removal of, enamel proteins. Caterina et al. (2002) have found that MMP-20 knockout mice do not process amelogenin properly and display an amelogenesis imperfecta phenotype, characterized by hypoplastic enamel that delaminates from dentin and by an altered enamel matrix and rod pattern.

15.5 Concluding Remarks

Studies on acquired or congenital alterations of calcified tissues have provided plenty of information on the calcifying organic matrices and their role in the process. Many of the new data, however, have not yet found a precise role in the complex machinery that leads to matrix calcification. They make it clear the fact that the absence or abnormality of just one matrix component may cause a breakdown of the whole structure, but do not define its exact position or function in the general picture, and several conclusions await confirmation. The most interesting findings seem to be the following:

- Changes in the structure of collagen fibrils lead to severe disarray in the organization of bone and dentin, and a sharp fall in their strength and resistance to fractures, but do not prevent their calcification.
- Matrix vesicles can induce calcification and crystal formation even when these are completely inhibited in the matrix around them.
- In cartilage and bone, matrix vesicle calcification appears to be a distinct process with respect to matrix calcification; the former may occur normally even if the latter is totally inhibited.
- A scarcity or abnormality of non-collagenous proteins may induce disarray in crystal formation and/or maturation.
- Amelogenesis imperfecta is one example of this situation: structural abnormalities in amelogenin, or its retention, lead to the abnormal maturation of enamel crystals.
- Some of the non-collagenous proteins inhibit calcification, as shown by the widespread soft tissue calcification found in mice without matrix Gla protein or osteoprotegerin.
- It can be stated that transgenic and knockout animals provide efficient tools in studying the role of matrix organic components in calcification, even if their potentialities have not yet been completely explored.

References

Aldred MJ, Savarirayan R, Crawford PJM (2003) Amelogenesis imperfecta: a classification and catalogue for the 21st century. Oral Dis 9:19–23

Anderson HC (1989) Mechanism of mineral formation in bone. Lab Invest 60:320–330

Anderson HC, Sajdera SW (1975) Calcification of rachitic rat cartilage *in vitro* by extracellular matrix vesicles. Am J Pathol 79:237–254

Anderson HC, Sajdera SW (1976) Calcification of rachitic cartilage to study matrix vesicle function. Fed Proc 35:148–153

Anderson HC, Stechschulte DJ Jr, Hsu HHT, Morris DC (1992) Comparison of normal and rachitic rat matrix vesicles. Bone Miner 17:119–122

Anderson HC, Hsu HH, Morris DC, Fedde KN, Whyte MP (1997) Matrix vesicles in osteomalacic hypophosphatasia bone contain apatite-like mineral crystals. Am J Pathol 151:1555–1561

Anderson HC, Sipe JB, Hessle L, Dhamyamraju R, Atti E, Camacho NP, Millàn JL (2004) Impaired calcification around matrix vesicles of growth plate and bone in alkaline phosphatase-deficient mice. Am J Pathol 164:841–847

Anderson HC, Harmey D, Camacho NP, Garimella R, Sipe JB, Tague S, Bi X, Johnson K, Terkeltaub R, Millàn JL (2005) Sustained osteomalacia of long bones despite major improvement in other hypophosphatasia-related mineral deficits in tissue nonspecific alkaline phosphatase/nucleotide pyrophosphatase phosphodiesterase 1 double-deficient mice. Am J Pathol 166:1711–1720

Aszodi A, Bateman JF, Gustafsson E, Boot-Handford R, Fassler R (2000) Mammalian skeletogenesis and extracellular matrix: what can we learn from knockout mice? Cell Struct Funct 25:73–84

Baechtold AP, Wright JT, Yamauchi M, Spevak L, Camacho NP (2000) Dentin composition and structure in the *oim* mouse. In: Goldberg M, Boskey A, Robinson C (eds) Chemistry and biology of mineralized tissues. American Academy of Orthopaedic Surgeons, Rosemont, IL, pp 57–61

Baker SL (1956) Fibrogenesis imperfecta ossium. A generalised disease of bone characterized by defective formation of the collagen fibres of the bone matrix. J Bone Joint Surg 38-B:378–416

Baker SL, Dent CE, Friedman M, Watson L (1966) Fibrogenesis imperfecta ossium. J Bone Joint Surg 48-B:804–825

Barnes MJ, Constable BJ, Morton LF, Kodicek E (1973) Bone collagen metabolism in vitamin D deficiency. Biochem J 132:113–115

Bartlett JD, Beniash E, Lee DH, Smith CE (2004) Decreased mineral content in MMP-20 null mouse enamel is prominent during the maturation stage. J Dent Res 83:909–913

Batina N, Renugopalakrishnan V, Casillas Lavín PN, Guerrero JCH, Morales M, Garduño-Juárez R, Lakka SL (2004) Ultrastructure of dental enamel afflicted with hypoplasia: an atomic force microscopic study. Calcif Tissue Int 74:294–301

Bonucci E (1970) Fine structure of epiphyseal cartilage in experimental scurvy. J Pathol 102:219–227

Bonucci E (1978) Matrix vesicle formation in cartilage of scorbutic guinea pigs: electron microscope study of serial sections. Metab Bone Dis Rel Res 1:205–212

Bonucci E, De Matteis A (1968) Aspetti ultrastrutturali dell'ossificazione periostale nell'osteogenesi imperfetta congenita. Ortop Traumatol Appar Motore 36:309–318

Boskey AL, Gadaleta S, Gundberg C, Doty SB, Ducy P, Karsenty G (1998) Fourier transform infrared microspectroscopic analysis of bones of osteocalcin-deficient mice provides insight into the function of osteocalcin. Bone 23:187–196

Boskey AL, Spevak L, Paschalis E, Doty SB, McKee MD (2002) Osteopontin deficiency increases mineral content and mineral crystallinity in mouse bone. Calcif Tissue Int 71:145–154

Brun-Heath I, Taillandier A, Serre JL, Mornet E (2005) Characterization of 11 novel mutations in the tissue non-specific alkaline phosphatase gene responsible for hypophosphatasia and genotype-phenotype correlations. Mol Genet Metab 84:273–277

Bucay N, Sarosi I, Dunstan CR, Morony S, Tarpley J, Capparelli C, Scully S, Tan HL, Xu W, Lacey DL, Boyle WJ, Simonet WS (1998) *osteoprotegerin*-deficient mice develop early onset osteoporosis and arterial calcification. Genes Dev 12:1260–1268

Butler WT (1998) Dentin matrix protein. Eur J Oral Sci 106:204–210

Camacho NP, Carroll P, Raggio CL (2003) Fourier transform infrared imaging spectroscopy (FT-IRIS) of mineralization in bisphosphonate-trated *oim/oim* mice. Calcif Tissue Int 72:604–609

Cassella JP, Ali SY (1992) Abnormal collagen and mineral formation in osteogenesis imperfecta. Bone Miner 17:123–128

Cassella JP, Barber P, Catterall AC, Ali SY (1994) A morphometric analysis of osteoid collagen fibril diameter in osteogenesis imperfecta. Bone 15:329–334

Cassella JP, Garrington N, Stamp TCB, Ali SY (1995) An electron probe X-ray microanalytical study of bone mineral in osteogenesis imperfecta. Calcif Tissue Int 56:118–122

Cassella JP, Stamp TCB, Ali SY (1996) A morphological and ultrastructural study of bone in osteogenesis imperfecta. Calcif Tissue Int 58:155–165

Caterina JJ, Skobe Z, Shi J, Ding Y, Simmer JP, Birkedal-Hansen H, Bartlett JD (2002) Enamelysin (matrix metalloproteinase 20)-deficient mice display an amelogenesis imperfecta phenotype. J Biol Chem 277:49598–49604

Chen X-D, Allen MR, Bloomfield S, Xu T, Young M (2003) Biglycan-deficient mice have delayed osteogenesis after marrow ablation. Calcif Tissue Int 72:577–582

Cohen-Solal L, Zylberberg L, Sangalli A, Gomez Lira M, Mottes M (1994) Substitution of an aspartic acid for glycine 700 in the α2(I)chain of type I collagen in a recurrent lethal type II osteogenesis imperfecta dramatically affects the mineralization of bone. J Biol Chem 269:14751–14758

Corsi A, Xu T, Chen X-D, Boyde A, Liang J, Mankani M, Sommer B, Iozzo RV, Eichstetter I, Robey PG, Bianco P, Young MF (2002) Phenotypic effects of biglycan deficiency are linked to collagen fibril abnormalities, are synergized by decorin deficiency, and mimic Ehlers-Danlos-like changes in bone and other connective tissues. J Bone Miner Res 17:1180–1189

Culbert AA, Lowe MP, Atkinson M, Byers PH, Wallis GA, Kadler KE (1995) Substitutions of aspartic acid for glycine-220 and of arginine for glycine-664 in the proα1(I) chain of type I procollagen produce lethal osteogenesis imperfecta and disrupt the ability of collagen fibrils to incorporate crystalline hydroxyapatite. Biochem J 311:815–820

de Vries WN, de Wet WJ (1986) The molecular defect in an autosomal dominant form of osteogenesis imperfecta. Synthesis of type I procollagen containing cysteine in the triple-helical domain of pro-α1(I) chains. J Biol Chem 261:9056–9064

Delany AM, Kalajzic I, Bradshaw AD, Sage EH, Canalis E (2003) Osteonectin-null mutation compromises osteoblast formation, maturation, and survival. Endocrinology 144:2588–2596

Dickson IR, Roughley PJ (1978) A comparative study of the proteoglycan of growth cartilage of normal and rachitic chicks. Biochem J 171:675–682

Dominguez LJ, Barbagallo M, Moro L (2005) Collagen overglycosylation: a biochemical feature that may contribute to bone quality. Biochem Biophys Res Comm 330:1–4

Doty SB, Mathews RS (1971) Electron microscopic and histochemical investigation of osteogenesis imperfecta tarda. Clin Orthop Relat Res 80:191–201

Drezner MK, Harrelson JM (1979) Newer knowledge of vitamin D and its metabolites in health and disease. Clin Orthop Relat Res 139:206–231

Einhorn TA, Majeska RJ (1991) Neutral proteases in regenerating bone. Clin Orthop Relat Res 262:286–297

El-Maadawy S, Kaartinen MT, Schinke T, Murshed M, Karsenty G, McKee MD (2003) Cartilage formation and calcification in arteries of mice lacking matrix Gla protein. Connect Tissue Res 44:272–278

Engfeldt B, Hjerpe A (1976) Glycosaminoglycans of cartilage and bone tissue in two cases of osteogenesis imperfecta congenita. Acta Path Microbiol Scand Sect A 84:488–494

Fedde KN, Blair L, Silverstein J, Coburn SP, Ryan LM, Weinstein RS, Waymire K, Narisawa S, Millan JL, MacGregor GR, Whyte MP (1999) Alkaline phosphatase knock-out mice recapitulate the metabolic and skeletal defects of infantile hypophosphatasia. J Bone Miner Res 14:2015–2026

Feng JQ, Huang H, Lu Y, Ye L, Xie Y, Tsutsui TW, Kunieda T, Castranio T, Scott G, Bonewald LB, Mishina Y (2003) The Dentin matrix protein 1 (Dmp1) is specifically expressed in mineralized, but not soft, tissues during development. J Dent Res 82:776–780

Fisher LW, Denholm LJ, Conn KM, Termine JD (1986) Mineralized tissue protein profiles in the Australian form of bovine osteogenesis imperfecta. Calcif Tissue Int 38:16–20

Fisher LW, Eanes ED, Denholm LJ, Heywood BR, Termine JD (1987a) Two bovine models of osteogenesis imperfecta exhibit decreased apatite crystal size. Calcif Tissue Int 40:282–285

Fisher LW, Drum MA, Gehron Robey P, Conn KM, Termine JD (1987b) Osteonectin content in human osteogenesis imperfecta bone shows a range similar to that of two bovine models of OI. Calcif Tissue Int 40:260–264

Fong H, White SN, Paine ML, Luo W, Snead ML, Sarikaya M (2003) Enamel structure properties controlled by engineered proteins in transgenic mice. J Bone Miner Res 18:2052–2059

Frame B, Frost HM, Pak CYC, Reynolds W, Argen RJ (1971) Fibrogenesis imperfecta ossium. A collagen defect causing osteomalacia. New Engl J Med 285:769–772

Fratzl P, Paris O, Klaushofer K, Landis WJ (1996) Bone mineralization in an osteogenesis imperfecta mouse model studied by small-angle X-ray scattering. J Clin Invest 97:396–402

Fujii K, Tanzer ML (1977) Osteogenesis imperfecta: biochemical studies of bone collagen. Clin Orthop Relat Res 124:271–277

Genge BR, Sauer GR, Wu LNY, McLean FM, Wuthier RE (1988) Correlation between loss of alkaline phosphatase activity and accumulation of calcium during matrix vesicle-mediated mineralization. J Biol Chem 263:18513–18519

Gibson CW, Yuan ZA, Hall B, Longenecker G, Chen E, Thyagarajan T, Sreenath T, Wright JT, Decker S, Piddington R, Harrison G, Kulkarni AB (2001) Amelogenin-deficient mice display an amelogenesis imperfecta phenotype. J Biol Chem 276:31871–31875

Glimcher MJ, Seyer J, Brickley DM (1969) The solubilization of collagen and protein-polysaccharides from the developing cartilage of lathyritic chicks. Biochem J 115:923–926

Goldberg M, Rapoport O, Septier D, Palmier K, Hall R, Embery G, Young M, Ameye L (2003) Proteoglycans in predentin: the last 15 micrometers before mineralization. Connect Tissue Res 44:184–188

Goldberg M, Septier D, Rapoport O, Iozzo RV, Young MF, Ameye LG (2005) Targeted disruption of two small leucine-rich proteoglycans, biglycan and decorin, excerpts divergent effects on enamel and dentin formation. Calcif Tissue Int 77:297–310

Gould BS (1963) Collagen formation and fibrogenesis with special reference to the role of ascorbic acid. Int Rev Cytol 15:301–361

Haebara H, Yamasaki Y, Kyogoku M (1969) An autopsy case of osteogenesis imperfecta congenita - histochemical and electron microscopical studies. Acta Path Jap 19:377–394

Harmey D, Hessle L, Narisawa S, Johnson KA, Terkeltaub R, Millàn JL (2004) Concerted regulation of inorganic pyrophosphate and osteopontin by akp2, enpp1, and ank: an integrated model of the pathogenesis of mineralization disorders. Am J Pathol 164:1199–1209

Hart PS, Hart TC, Simmer JP, Wright JT (2002a) A nomenclature for X-linked amelogenesis imperfecta. Arch Oral Biol 47:255–260

Hart PS, Aldred MJ, Crawford PJ, Wright NJ, Hart TC, Wright JT (2002b) Amelogenesis imperfecta phenotype-genotype correlations with two amelogenin gene mutations. Arch Oral Biol 47:261–265

Hartmann F (1963) Die enchondrale Mineralisationsstörung beim experimentellen Lathyrismus. Verh Dtsch Ges Path 47:161–165

Haussler MR, McCain TA (1977) Basic and clinical concepts related to vitamin D metabolism and action. New England J Med 297:974–983

He G, George A (2004) Dentin matrix protein 1 immobilized on type I collagen fibrils facilitates apatite deposition *in vitro* . J Biol Chem 279:11649–11656

Henneman DH, Pak CYC, Bartter FC (1972) Collagen composition, solubility and biosynthesis in fibrogenesis imperfecta ossium. In: Frame B, Parfitt AM, Duncan H (eds) Clinical aspects of metabolic bone disease. Excerpta Medica, Amsterdam, pp 469–472

Herold RC (1972) Fine structure of tooth dentine in human dentinogenesis imperfecta. Archs Oral Biol 17:1009–1013

Hessle L, Johnson KA, Anderson HC, Narisawa S, Sali A, Goding JW, Terkeltaub R, Millàn JL (2002) Tissue-nonspecific alkaline phosphatase and plasma cell membrane glycoprotein-1 are central antagonistic regulators of bone mineralization. Proc Natl Acad Sci U S

A Hu JC, Yamakoshi Y (2003) Enamelin and autosomal-dominant amelogenesis imperfecta. Crit Rev Oral Biol Med 14:387–398

Johnson KA, Hessle L, Vaingankar S, Wennberg C, Mauro S, Narisawa S, Goding JW, Sano K, Millan JL, Terkeltaub R (2000) Osteoblast tissue-nonspecific alkaline phosphatase antagonizes and regulates PC-1. Am J Physiol Regul Integr Comp Physiol 279:R1365-R1377

Jones CJP, Cummings C, Ball J, Beighton P (1984) Collagen defect of bone in osteogenesis imperfecta (type I). An electron microscopic study. Clin Orthop Relat Res 183:208–214

Kerebel B, Kerebel L-M (1981) Enamel in odontodysplasia. Oral Surg Oral Med Oral Pathol 52:404–410

Kida M, Ariga T, Shirakawa T, Oguchi H, Sakiyama Y (2002) Autosomal-dominant hypoplastic form of amelogenesis imperfecta caused by an enamelin gene mutation at the exon-intron boundary. J Dent Res 81:738–742

Kim JW, Seymen F, Lin BP, Kiziltan B, Gencay K, Simmer JP, Hu JC (2005) ENAM mutations in autosomal-dominant amelogenesis imperfecta. J Dent Res 84:278–282

Kinney JH, Pople JA, Driessen CH, Breunig TM, Marshall GW, Marshall SJ (2001) Intrafibrillar mineral may be absent in dentinogenesis imperfecta type II (DI-II). J Dent Res 80:1555–1559

Kipp DE, McElvain M, Kimmel DB, Akhter MP, Robinson RG, Lukert BP (1996) Scurvy results in decreased collagen synthesis and bone density in the guinea pig animal model. Bone 18:281–288

Kuznetsova N, McBride DJ Jr, Leikin S (2001) Osteogenesis imperfecta murine: interaction between type I collagen homotrimers. J Mol Biol 309:807–815

Kwong-Hing A, Teasdale R, Sandhu HS (1991) Local effects of impaired mechanical properties of collagen on bone formation and resorption. Anat Embryol 184:93–97

Lee S, Eyre DR, Barnard SM (1990) BAPN dose dependence of mature crosslinking in bone matrix collagen of rabbit compact bone: corresponding variation of sonic velocity and equatorial diffraction spacing. Connect Tissue Res 24:95–105

Luo G, Ducy P, McKee MD, Pinero GJ, Loyer E, Behringer RR, Karsenty G (1997) Spontaneous calcification of arteries and cartilage in mice lacking GLA protein. Nature 386:78–81

Luo W, Wen X, Wang HJ, MacDougall M, Snead ML, Paine ML (2004) *In vivo* overexpression of tuftelin in the enamel organic matrix. Cells Tissues Organs 177:212–220

MacDougall M, Gu TT, Simmons D (1996) Dentin matrix protein-1, a candidate gene for dentinogenesis imperfecta. Connect Tissue Res 35:267–272

MacDougall M, Simmons D, Gu TT, Dong J (2002) MEPE/OF45, a new dentin/bone matrix protein and candidate gene for dentin diseases mapping to chromosome 4q21. Connect Tissue Res 43:320–330

Masuya H, Shimizu K, Sezutsu H, Sakuraba Y, Nagano J, Shimizu A, Fujimoto N, Kawai A, Miura I, Kaneda H, Kobayashi K, Ishijima J, Maeda T, Gondo Y, Noda T, Wakana S, Shiroishi,T (2005) Enamelin (Enam) is essential for amelogenesis: ENU-induced mouse mutants as models for different clinical subtypes of human amelogenesis imperfecta (AI). Human Mol Genet 14:575–583

Matukas VJ, Panner BJ, Orbison JL (1967) Ultrastructural evidence for an abnormality in proteinpolysaccharide metabolism in lathyritic cartilage. Lab Invest 16:726–735

McCance RA, Fairweather DV, Barrett AM, Morrison AB (1956) Genetic, clinical, biochemical, and pathological features of hypophosphatasia. Q J Med 25:523–538

McKusick VA (1960) Eritable disorders of connective tissues. The CV Mosby Comp, St Louis

Mechanic GL, Toverud SU, Ramp WK (1972) Quantitative changes of bone collagen crosslinks and precursors in vitamin D deficiency. Biochem Biophys Res Comm 47:760–765

Merker HJ, Zimmermann B, Günther T (1972) Elektronenmikroskopische Untersuchungen über die D-Penicillaminwirkung am Knorpel embryonaler Ratten (Tag 16) in vitro. Virchows Arch abt B Zellpath 12:51–60

Mocetti P, Silvestrini G, Lozupone E, Ballanti P, Bonucci E, Nanci A (1997) Effects of low calcium diet on rat epiphyseal cartilage calcification and matrix vesicle production. In: Bonucci E (ed) Biology and pathology of cell-matrix interactions. CLEUP Editrice, Padua, pp 199–208

Morris DC, Randall JC, Stechschulte DJ Jr, Zeiger S, Mansur DB, Anderson HC (1990) Enzyme cytochemical localization of alkaline phosphatase in cultures of chondrocytes derived from normal and rachitic rats. Bone 11:345–352

Narisawa S, Fröhlander N, Millàn JL (1997) Inactivation of two mouse alkaline phosphatase genes and establishment of a model of infantile hypophosphatasia. Dev Dyn 208:432–446

Nogami H, Oohira A (1988) Defective association between collagen fibrils and proteoglycans in fragile bone of osteogenesis imperfecta. Clin Orthop Relat Res 232:284–291

Oohira A, Nogami H (1989) Elevated accumulation of hyaluronate in the tubular bones of osteogenesis imperfecta. Bone 10:409–413

Orlowski WA, Orbison LJ (1970) Analysis of hexosamine and acid mucopolysaccharides in skins of normal and lathyritic rats. Lab Invest 23:246–252

Ornoy A, Adomian GE, Rimoin DL (1985) Histologic and ultrastructural studies on the mineralization process in hypophosphatasia. Am J Med Genet 22:743–758

Page RC, Benditt EP (1972) Diseases of connective and vascular tissues IV. The molecular basis for lathyrism. Lab Invest 26:22–26

Paine ML, Lei YP, Dickerson K, Snead ML (2002) Altered amelogenin self-assembly based on mutations observed in human X-linked amelogenesis imperfecta (AIH1). J Biol Chem 277:17112–17116

Paine ML, Luo W, Zhu DH, Bringas P Jr, Snead ML (2003a) Functional domains for amelogenin revealed by compound genetic defects. J Bone Miner Res 18:466–472

Paine ML, Wang HJ, Luo W, Krebsbach PH, Snead ML (2003b) A transgenic animal model resembling amelogenesis imperfecta related to ameloblastin overexpression. J Biol Chem 278:19447–19452

Phillips CL, Bradley DA, Schlotzhauer CL, Bergfeld M, Libreros-Minotta C, Gawenis LR, Morris JS, Clarke LL, Hillman LS (2000) Oim mice exhibit altered femur and incisor mineral composition and decreased bone mineral density. Bone 27:219–226

Pinto F, Bonucci E, Mezzelani P, Cetta G, De Sandre G (1981) Fibrogenesis imperfecta ossium (Clinical, biochemical and ultrastructural investigations). It J Orthop Traumatol 7:371–385

Plenk H Jr (1976) Osteolathyrismus. Quantitativ-morphologische Untersuchungen der experimentellen Skeletterkrankung bei der Ratte. In: Wolf-Heidegger G (ed) Bibliotheca Anatomica. Karger, Basel, pp 1–102

Plotkin H (2004) Syndromes with congenital brittle bones. BMC Pediatr 4:16

Rajpar MH, Harley K, Laing C, Davies RM, Dixon MJ (2001) Mutation of the gene encoding the enamel-specific protein, enamelin, causes autosomal-dominant amelogenesis imperfecta. Hum Mol Genet 10:1673–1677

Riede UN (1970) Penicillamine-induced changes in growing rats I. Epiphyseal plate. Am J Pathol 61:249–254

Robinson C, Shore RC, Wood SR, Brookes SJ, Smith DA, Wright JT, Connell S, Kirkham J (2003) Subunit structures in hydroxyapatite crystal development in enamel: implications for amelogenesis imperfecta. Connect Tissue Res 44:65–71

Roughley P, Dickson I (1980) Factors influencing proteoglycan size in rachitic-chick growth cartilage. Biochem J 185:33–39

Roughley PJ, Rauch F, Glorieux FH (2003) Osteogenesis imperfecta - clinical and molecular diversity. Eur Cell Mater 5:41–47

Sandhu HS, Jande SS (1982) Effects of β-aminoproprionitrile on formation and mineralization of rat bone matrix. Calcif Tissue Int 34:80–85

Sarathchandra P, Kayser MV, Ali SY (1999) Abnormal mineral composition of osteogenesis imperfecta bone as determined by electron probe X-ray microanalysis on conventional and cryosections. Calcif Tissue Int 65:11–15

Sauk JJ Jr, Cotton WR, Lyon HW, Witkop CJ Jr (1972) Electron-optic analyses of hypomineralized amelogenesis imperfecta in man. Archs Oral Biol 17:771–779

Schmitz JP, Dean DD, Schwartz Z, Cochran DL, Grant GM, Klebe RJ, Nakaya H, Boyan BD (1996) Chondrocyte cultures express matrix metalloproteinase mRNA and immunoreactive protein; stromelysin-1 and 72 kDa gelatinase are localized in extracellular matrix vesicles. J Cell Biochem 61:375–391

Sillence DO, Senn AS, Danks DM (1979) Genetic heterogeneity in osteogenesis imperfecta. J Med Genet 16:101–116

Sims TJ, Miles CA, Bailey AJ, Camacho NP (2003) Properties of collagen in OIM mouse tissues. Connect Tissue Res 44:202–205

Sissons HA (2000) Fibrogenesis imperfecta ossium (Baker's disease): a case studied at autopsy. Bone 27:865–873

Snead ML (2003) Amelogenin protein exhibits a modular design: implications for form and function. Connect Tissue Res 44:47–51

Sreenath T, Thyagarajan T, Hall B, Longenecker G, D'Souza R, Hong S, Wright JT, MacDougall M, Sauk J, Kulkarni AB (2003) Dentin sialophosphoprotein knockout mouse teeth display widened predentin zone and develop defective dentin mineralization similar to human dentinogenesis imperfecta type III. J Biol Chem 278:24874–24880

Stechschulte DJ Jr, Morris DC, Moylan PE, Davis LS, Anderson HC (1991) Increased matrix vesicle protein in rachitic rat epiphyseal growth plates. Bone Miner 14:121–129

Steinmann B, Nicholls A, Pope FM (1986) Clinical variability of osteogenesis imperfecta reflecting molecular heterogeneity: cysteine substitutions in the α1(I) collagen chain producing lethal and mild forms. J Biol Chem 261:8958–8964

Steitz SA, Speer MY, Curinga G, Yang HY, Haynes P, Aebersold R, Schinke T, Karsenty G, Giachelli CM (2001) Smooth muscle cell phenotypic transition associated with calcification: upregulation of Cbfa1 and downregulation of smooth muscle lineage markers. Circ Res 89:1147–1154

Swan CHJ, Shah K, Brewer DB, Cooke WT (1976) Fibrogenesis imperfecta ossium. Q J Med 45:233–253

Taillandier A, Cozien E, Muller F, Merrien Y, Bonnin E, Fribourg C, Simon-Bouy B, Serre JL, Bieth E, Brenner R, Cordier MP, De Bie S, Fellmann F, Freisinger P, Hesse V, Hennekam RC, Josifova D, Kerzin-Storrar L, Leporrier N, Zabot MT, Mornet E (2000) Fifteen new mutations (-195C>T, L-12X, 298–2A>G, T117N, A159T, R229S, 997+2T>A, E274X, A331T, H364R, D389G,1256delC, R433H, N461I, C472S) in the tissue-nonspecific alkaline phosphatase (TNSALP) gene in patients with hypophosphatasia. Hum Mutat 15:293

Takagi Y, Fujita H, Katano H, Shimokawa H, Kuroda T (1998) Immunochemical and biochemical characteristics of enamel proteins in hypocalcified amelogenesis imperfecta. Oral Surg Oral Med Oral Pathol Oral Radiol Endod 85:424–430

Takechi M, Itakura C (1995) Ultrastructural studies of the epiphyseal plate of chicks fed a vitamin D-deficient and low-calcium diet. J Comp Path 113:101–111

Tanzer ML (1965) Experimental lathyrism. Int Rev Connect Tiss Res 3:91–112

Tartaix PH, Doulaverakis M, George A, Fisher LW, Butler WT, Qin C, Salih E, Tan M, Fujimoto Y, Spevak L, Boskey AL (2004) *In vitro* effects of dentin matrix protein-1 on hydroxyapatite formation provide insights into *in vivo* functions. J Biol Chem 279:18115–18120

Teitelbaum SL, Kraft WJ, Lang R, Avioli LV (1974) Bone collagen aggregation abnormalities in osteogenesis imperfecta. Calcif Tissue Res 17:75–79

Tesch W, Vandenbos T, Roschger P, Fratzl-Zelman N, Klaushofer K, Beertsen W, Fratzl P (2003) Orientation of mineral crystallites and mineral density during skeletal development in mice deficient in tissue nonspecific alkaline phosphatase. J Bone Miner Res 18:117–125

Thomas WC Jr, Moore TH (1968) Fibrogenesis imperfecta ossium. Trans Am Clin Climat Ass 80:54–62

Thyagarajan T, Sreenath T, Cho A, Wright JT, Kulkarni AB (2001) Reduced expression of dentin sialophosphoprotein is associated with dysplastic dentin in mice overexpressing transforning growth factor-β in teeth. J Biol Chem 276:11016–11020

Thyberg J, Lohmander S, Friberg U (1971) Ultrastructure of the epiphyseal plate of the guina pig in experimental scurvy. Virchows Arch abt B Zellpath 9:45–57

Trelstad RL, Rubin D, Gross J (1977) Osteogenesis imperfecta congenita. Evidence for a generalized molecular disorder of collagen. Lab Invest 36:501–508

Tsipouras P (1993) Osteogenesis imperfecta. In: Beighton P (ed) McKusick's heritable disorders of connective tissue. Mosby, St. Louis, pp 281–314

Vetter U, Eanes ED, Kopp JB, Termine JD, Gehron Robey P (1991a) Changes in apatite crystal size in bones of patients with osteogenesis imperfecta. Calcif Tissue Int 49:248–250

Vetter U, Fisher LW, Mintz KP, Kopp JB, Tuross N, Termine JD, Gehron Robey P (1991b) Osteogenesis imperfecta: changes in noncollagenous proteins in bone. J Bone Miner Res 6:501–505

Vitellaro-Zuccarello L, Garino Canina G, De Biasi S, Cocchini F, Vitellaro-Zuccarello P (1984) Ultrastructural study on the stapes and skin in a case of type I osteogenesis imperfecta. J Submicrosc Cytol 16:779–786

Waltimo J (1994) Hyperfibers and vesicles in dentin matrix in dentinogenesis imperfecta (DI) associated with osteogenesis imperfecta (OI). J Oral Pathol Med 23:389–393

Wennberg C, Hessle L, Lundberg P, Mauro S, Narisawa S, Lerner UH, Millàn JL (2000) Functional characterization of osteoblasts and osteoclasts from alkaline phosphatase knockout mice. J Bone Miner Res 15:1879–1888

Whyte MP (1989) Alkaline phosphatase: physiological role explored in hypophosphatasia. In: Peck WA (ed) Bone and mineral research/6. Elsevier Science Publishers BV, Amsterdam, pp 175–218

Witkop CJ Jr (1988) Amelogenesis imperfecta, dentinogenesis imperfecta and dentin dysplasia revisited: problems in classification. J Oral Pathol 17:547–553

Witkop CJ, Rao S (1971) Inherited defects in tooth structure. Birth Defects 7:153–184

Wright JT, Robinson C, Shore R (1991) Characterization of the enamel ultrastructure and mineral content in hypoplastic amelogenesis imperfecta. Oral Surg Oral Med Oral Pathol 72:594–601

Wright JT, Lord V, Robinson C, Shore R (1992a) Enamel ultrastructure in pigmented hypomaturation amelogenesis imperfecta. J Oral Pathol Med 21:390–394

Wright JT, Robinson C, Kirkham J (1992b) Enamel protein in smooth hypoplastic amelogenesis imperfecta. Pediatr Dent 14:331–337

Wright JT, Alfred MJ, Crawford PJ, Kirkham J, Robinson C (1993a) Enamel ultrastructure and protein content in X-linked amelogenesis imperfecta. Oral Surg Oral Med Oral Pathol 76:192–199

Wright JT, Duggal MS, Robinson C, Kirkham J, Shore R (1993b) The mineral composition and enamel ultrastructure of hypocalcified amelogenesis imperfecta. J Craniofac Genet Dev Biol 13:117–126

Xu T, Bianco P, Fisher LW, Longenecker G, Smith E, Goldstein S, Bonadio J, Boskey A, Heegaard A-M, Sommer B, Satomura K, Dominguez P, Zhao C, Kulkarni AB, Gehron Robey P, Young MF (1998) Targeted disruption of the biglycan gene leads to an osteoporosis-like phenotype in mice. Nat Genet 20:78–82

Ye L, MacDougall M, Zhang S, Xie Y, Zhang J, Li Z, Lu Y, Mishina Y, Feng JQ (2004) Deletion of dentin matrix protein-1 leads to a partial failure of maturation of predentin into dentin, hypomineralization, and expanded cavities of pulp and root canal during postnatal tooth development. J Biol Chem 279:19141–19148

Ye L, Mishina Y, Chen D, Huang H, Dallas SL, Dallas MR, Sivakumar P, Kunieda T, Tsutsui TW, Boskey A, Bonewald LF, Feng JQ (2005) Dmp1-deficient mice display severe defects in cartilage formation responsible for a chondrodysplasia-like phenotype. J Biol Chem 280:6197–6203

Young MF, Bi Y, Ameye L, Chen XD (2002) Biglycan knockout mice: new models for musculoskeletal diseases. Glycoconj J 19:257–262

Zurutuza L, Muller F, Gibrat JF, Taillandier A, Simon-Bouy B, Serre JL, Mornet E (1999) Correlation of genotype and phenotype in hypophosphatasia. Hum Mol Genet 8:1039–1046

16 The Organic-inorganic Relationships in Calcifying Matrices

16.1 Introduction

There is now a consensus that biological calcifications occur through the deposition of inorganic substance in an organic matrix, one or many of whose components regulate the process by promoting or inhibiting crystals formation. In this view, crystals are very closely related to organic molecules that determine their shape, size, evolution and, in some cases, composition. As a result, studies on the relationships that link the organic and inorganic components of a calcified matrix – hypothesizing not merely a mixture of the two types of component, but steric configurations that make possible molecular connection and interaction between them – allow a deep understanding of how calcification begins and is regulated. In this connection, a major consideration is that these relationships evolve as the calcification process goes forward, so that those operative at its outset are found to have changed as it nears completion. The outstanding examples of this are the fall in protein content that takes place during enamel maturation (Blumen and Merzel 1972; Glimcher et al. 1977; Seyer and Glimcher 1977; Fukae and Tanabe 1998; Moradian-Oldak et al. 2002), and the similar loss of organic material that occurs in other hard tissues or can at least be hypothesized on the basis of protease concentrations in them (Pugliarello et al. 1970; Vaes et al. 1976; Mikuni-Takagaki and Cheng 1987; Martin-De Las Heras et al. 2000; Caron et al. 2001; Satoyoshi et al. 2001; Wu et al. 2002; Butler et al. 2003; Qin et al. 2004).

16.2 Organic-inorganic Relationships in Bone and Related Collagenous Tissues

Studies on the organic-inorganic relationships in calcifying tissues have preferentially been carried out on the calcified matrix of bones, probably because it is the major components of the vertebrate skeleton and is also easily available for laboratory investigations. These studies have led to a good knowledge of the relationships between inorganic substance and collagen fibrils, which are the predominant constituents of the organic matrix of bone and related collagenous tissues (calcified tendons, dentin, cementum). The best demonstration that the organic and inorganic components of the calcified matrix are intimately connected is probably given by the fact that the former can only be completely extracted if the latter have been solubilized by decalcification (Linde et al. 1980; Termine et al. 1980, 1981; Butler 1984a; Waddington et al. 1988); additional elements have been drawn from X-ray diffraction and electron microscopy. These are based on the natural crystallinity and intrinsic electron density of hydroxyapatite, which offer the advantage that no special treatments are needed to obtain diffractograms or to view the crystals under the electron microscope.

Early studies on this topic go back at least as far as 1935, when Caglioti (1935) reported that in the ox femur the X-ray diffractograms of the organic fraction coincided with those of the inorganic component, and suggested that they become so closely connected that they give rise to a new 'combined reticulum'. He later reported that the organic and inorganic substances of bone are linked by a sulfuric group of the organic matrix, a phosphate group of the inorganic substance, and a hydrogen bond (Caglioti et al. 1954). His view that the calcium salts are bound to organic structures of the matrix has also been supported by measurements of the intrinsic birefringence of bone (Ascenzi and Bonucci 1964).

The intimate relationship between bone organic and inorganic components was documented by Robinson (1952), Robinson and Watson (1952), and Robinson and Cameron (1956), using electron microscopy, and by Finean and Engström (1953) and Carlström and Finean (1954), using low angle X-ray diffraction. They observed that bone apatite crystals are lined up along collagen fibers and are arranged in such a way as to emphasize fibril periodicity. Under the electron microscope, the incompletely calcified areas of compact bone (for instance, incompletely calcified osteons, Ascenzi et al. 1965) show electron-dense bands occurring regularly at about 60–70 nm intervals and running perpendicular to the axis of the adjacent collagen fibrils which are in lateral register (see 'granular inorganic bands', Sect. 5.2.1). According to Glimcher and Krane (1968), the inorganic substance acts as a 'negative stain' on the collagen fibrils, emphasizing their axial repeat. These observations have been repeatedly confirmed and have been extended to include other collagenous tissues such as dentin, cementum and calcifying tendons and ligaments; they have also been interpreted as a precise indication that reinforcement of the collagen period depends on the development of crystals within the holes of the fibril gap-zones (Scott and Pease 1956; Fernández-Morán and Engström 1957; Molnar 1959; Nylen et al. 1960; Ascenzi et al. 1965, 1978; Myers and Engström 1965; White et al. 1977; Weiner and Traub 1986; Arsenault 1988, 1989, 1991; Bigi et al. 1988, 1996; Lee and Glimcher 1989; Traub et al. 1989, 1992; Heywood et al. 1990; Landis et al. 1991, 1996; Landis and Song 1991). The broad interest that has been dedicated to these topics (for reviews see Höhling et al. 1990, 1995; Bonucci 1984, 1992; Veis 2003) probably depends on the fact that the well-known theory of Glimcher (1959) – that bone and tendon calcification is initiated by heterogeneous nucleation catalyzed by specific atomic groups in the fibrils – was chiefly based on the concept that the inorganic substance is closely related to collagen periodic banding. This topic has been reviewed several times (Glimcher and Krane 1968; Glimcher 1976; Butler 1984b) and is further discussed in Sect. 18.2.

Granular inorganic bands can be clearly recognized in incompletely calcified matrix of compact bone (Fig. 7.2), such as that of osteons during the earliest phase of calcification (Ascenzi et al. 1965). Their evidence decreases with the increase of calcification of compact bone and of numbers of filament-like crystals; on the other hand, it is hard, in some cases impossible, to detect them in fully calcified bone tissues with loose collagen fibrils and wide interfibrillary spaces, such as embryonic woven bone, membranous bone, and the medullary bone of birds (Figs. 5.1 and 5.3). In these tissues, crystals may show some degree of alignment with collagen fibrils, but mostly appear to be unrelated to their periodic banding (Bonucci and Gherardi 1975; Bonucci and Silvestrini 1996), and to develop and to be located in interfibrillary spaces (Fig. 16.1). As early as 1959, an electron microscope study of fetal membranous ossification had shown that the earliest mineral deposits were associated with the ground (i. e., interfibrillary) substance and that crystals only came to be aligned with fibrils as calcification progressed; no granular inorganic bands were detected (Ascenzi and Benedetti 1959). The same negative result was obtained in the medullary bone of pigeons and ducks (Ascenzi et al. 1963). In the bone of embryonic fowl, crystal deposits could sometimes be identified in areas where collagen fibrils were apparently absent (Hancox and Boothroyd 1965). Neither a preferential orientation nor a specific alignment of the crystals with the collagen fibrils was seen in embryonic chick bone studied under the electron microscope, nor did crystals seem to be associated either with the major period or the sub-bands of collagen fibrils (Decker JD 1966). Investigations carried out on calcified turkey tendons using low temperature substitution, fixation and staining, followed by low temperature embedding, or on frozen, air dried cryosections, whether examined under the electron microscope at −165 °C or at normal temperature, showed that crystals were not confined to collagen axial periodicities and gap zones (Arsenault 1991). Some of crystals were found to be located and nucleated on the fibril surface (Landis et al. 1991; Bigi et al. 1997), a result that had already been reported by Nylen et al. (1960) in tendons and confirmed by Cameron (1963) in bone. Electron microscopic studies on bone tissue prepared anhydrously in organic solvents (Landis et al. 1977), high-resolution investigations on woven bone (Su et al. 2003), and atomic force microscopy applied to calcified tendons (Lees et al. 1994), all showed that crystals are located not only inside collagen fibrils, but also between and on them.

In analyzing these results, it should be borne in mind that the calcification process in woven bone differs from that in secondary compact bone. The process starts at the calcification front in both tissues, but in the woven bone it

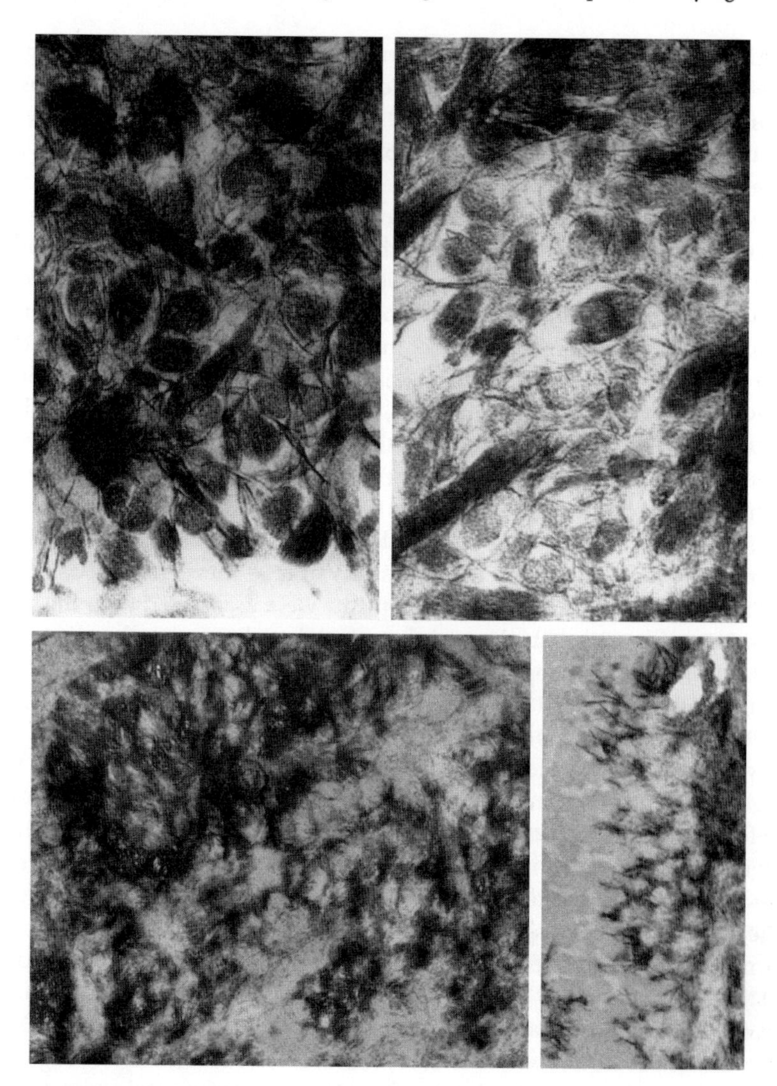

Fig. 16.1. Calcification front in normal bone of rat embryos (*above*) and in neoplastic bone (*below*): in both types of bone, crystals grow between and around uncalcified collagen fibrils, which appear in cross-section (*below*) as round, empty spaces. Unstained, ×90,000 (*above*) and ×60,000 (*below*)

reaches completion and attains its highest value very quickly, whereas in secondary compact bone it occurs in two stages: a rapid deposition of about 70% of inorganic substance is followed by slow progress toward completion (Amprino and Engström 1952). Thus the matrix of secondary compact bone remains incompletely calcified for some time, even at sites far from the calcification front, where woven bone is already fully calcified. A comparison between the organic-inorganic relationships in secondary and in woven bone during the early stage of calcification does, in fact, show that in the former the granular inorganic bands are easy to detect and needle- and filament-like crystals are relatively rare, even if they increase in numbers as the degree of calcification rises, whereas in the latter few granular inorganic bands and plenty of crystals can be recognized from the start. One suggestion has been that the granular inorganic bands may give rise to crystals, but the observations reported above support the view that crystals and the granular inorganic bands are formed by different, independent mechanisms at different sites (i. e., the granular inorganic bands are formed within collagen fibrils, the crystals between and on them). This view is supported not only by electron microscopy, but also by the finding that the shorter crystallites laid down inside the gap region of the collagen fibrils are richer in magnesium than the longer ones that fill the space between fibrils (Bigi et al. 1992). It is indirectly also supported by the finding that the expression of bone acidic glycoprotein-75 is restricted to woven bone (Gorski et al. 1990, 1998). Differences in the molecular composition of rapidly deposited and calcified woven bone with respect to the more slowly calcified lamellar bone imply, in fact, the existence of distinctive mechanisms of osteogenesis (Gorski 1998).

Comparative studies carried out on bone, dentin and calcified turkey leg tendons by Arsenault (1989) have shown that apatite crystals are located within both the gap and the overlap zones of collagen fibrils even at the early stage of calcification. If, however, the study of the organic-inorganic relationship is closely restricted to the narrow strip of bone matrix known as the calcification front, the granular inorganic bands are hardly detectable or not detectable at all, independently of bone type (Fig. 5.2). Both in woven and secondary lamellar bone, the calcification front is characterized by its content of small crystal aggregates which, as already discussed (Sect. 5.2.1), correspond to calcification nodules and calcification islands. Granular inorganic bands are hard to make out throughout the calcification front and, if any can be detected, are connected with calcification islands and not with calcification nodules. These nodules may be found together with calcification islands at the calcification front in the same bone, but the former are especially plentiful in woven bone, the latter in secondary, lamellar bone. Moreover, the smallest calcification nodules can be clearly made out in matrix vesicles, whereas calcification islands appear to be contained in interfibrillary spaces and to be related to interfibrillary substance. On the basis of these observations, calcification nodules and calcification islands appear to be different structures. According to Ørvig (1968), calcification nodules (spheritic mineralization) could be the precursors of the crystals that are oriented parallel to collagen fibrils (inotropic mineralization). In any case, the early stages of bone calcification may display three different prevalent pictures: calcification nodules and calcification islands, consisting of filament-like crystals, are prevalent at the calcification front, granular inorganic bands some distance away, and needle- and filament-like crystals in fully calcified matrix.

These findings imply differences in organic-inorganic relationships and, very probably, different modalities of formation. As discussed above, the granular inorganic bands are related to the periodic banding of the collagen, whereas the needle- and filament-like crystals are located between collagen fibrils and on their surface. The organic-inorganic relationships in calcification nodules and calcifications islands will now be discussed.

16.2.1 Organic-inorganic Relationships in Calcification Nodules of Bone

Bone calcification nodules, like those of calcifying cartilage (see Sect. 16.3), are the outcome of the progressive increase in numbers of crystals contained in matrix vesicles: these initially show single crystals, or small groups of crystals, scattered in their matrix, or laying on their membrane. As these increase in numbers, the whole space outlined by the vesicle membrane gradually becomes full of them. Any further increase leads to crystals breaking out of the membrane and becoming projected radially into the matrix, where they appear as the roundish interfibrillary aggregates known as calcification nodules (Figs. 5.2 and 5.3). The organic components of the matrix are masked by the natural electron density of the inorganic substance, so decalcification must be carried out before they can be studied morphologically.

The artifacts that may be caused by decalcification, and the methods needed to avoid them, have already been discussed (see Sect. 3.4.4). When decalcification is carried out on fresh, fixed, unembedded bone, a high percentage of organic matrix is solubilized and lost, along with the inorganic substance. Under the electron microscope, the calcification nodules will then appear as interfibrillary, almost empty areas, which contain only a few segments of collagen fibrils and are surrounded by a dense border, known as the 'lamina limitans' (Scherft 1978).

When decalcification is carried out using the PEDS method, the matrix of the calcification nodules is preserved and these appear under the electron microscope as roundish electron-dense areas (Figs. 3.5 and 16.2). At high enlargement, these can be resolved into aggregates of needle- or filament-like structures which have almost the same ultrastructural morphology as untreated crystals but lack their intrinsic electron density and only become visible after staining (Figs. 3.5, 16.2, 16.3, and 16.4). For all these reasons, they have been called crystal ghosts (Bonucci 1967). They have been found at the calcification front in bone (Bonucci 1971; Bonucci and Gherardi 1975; Bonucci and Reurink 1978; McKee and Nanci 1995; Aaron et al. 1999), dentin (Goldberg M et al. 1980; Hayashi 1988; Chardin et al. 1990a), cementum (Hayashi 1985, 1988; McKee and Nanci 1995), calcified tendons (Prostak and Lees 1996), in vitro cultures of cells from these tissues (Williams et al. 1980; Ecarot-Charrier et al. 1988; Nefussi et al. 1992; Hayashi et al. 1993), and in vitro calcified matrix vesicles (Hayashi and Nagasawa 1990). Even when they are not called crystal

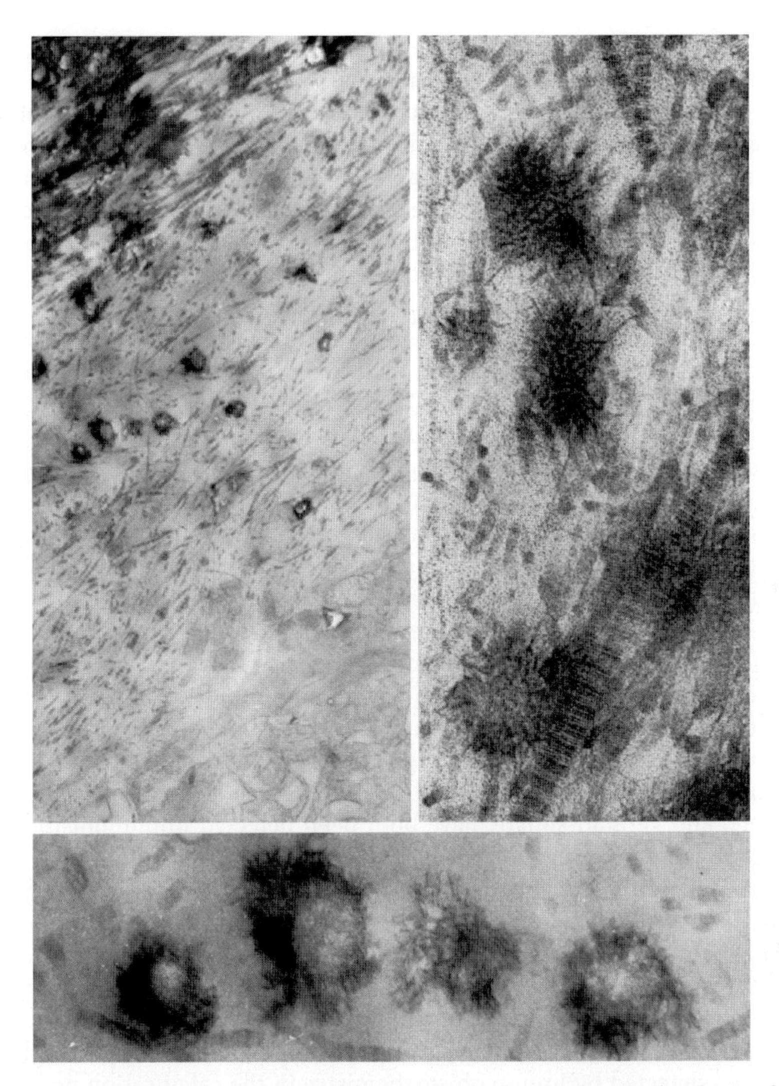

Fig. 16.2. *Upper left*: bone section decalcified according to the PEDS method and stained with acidic PTA: the calcification nodules (now nodules of crystal ghosts) appear as roundish, electron-dense areas scattered between the loose collagen fibrils of the osteoid border; calcified matrix in the *upper left corner*, part of an osteoblast in the *lower right corner*. PEDS, PTA, ×12,000. *Below*: detail of the previous figure, showing aggregates of crystal ghosts; ×60,000. *Upper right*: section decalcified according to the PEDS method and stained with uranyl acetate and lead citrate: the calcification nodules appear as crystal ghost aggregates; ×90,000

ghosts, they can be recognized in the figures reported in several other papers (Bernard and Pease 1969; Sauren et al. 1989).

When decalcification is carried out before embedding using the cationic-dye stabilization method, the decalcified nodules appear as roundish aggregates of organic filaments (see Sect. 16.3 and Fig. 16.5) which are thicker and less well outlined than, but have the same distribution and arrangement as, crystal ghosts (Scherft 1968; Shepard and Mitchell 1981; Bonucci et al. 1988; Shepard 1992; Bonucci 2002). This topic has been reviewed several times (Bonucci 1971, 1992, 2002) and is further discussed in Sect. 16.3.

Crystal ghosts are the main components, and make up the bulk, of the matrix of bone calcification nodules, whose histochemical properties can, therefore, reasonably be attributed to them. Calcification nodules have an acidic matrix, so they react with basic substances such as ruthenium red or colloidal thorium dioxide (Scherft 1968; Groot 1982a); all these reactions are reversed by digestion with hyaluronidase (Groot 1982b). This suggests that the crystal ghosts are acid proteoglycans, a view supported by the finding that they are stained by colloidal iron at pH 2.5 (Fig. 16.6) and that calcification nodules can be stained by alcian blue (Bonucci 2002). Sulfated glycoconjugates and vicinal glycol-containing glycoproteins have been demonstrated in osteoid interfibrillar focal circular profiles (which may reasonably be considered decalcified calcification nodules) using the high-iron diamine-thiocarbohydrazide-silver proteinate (HID-TCH-SP) method and Thiéry's periodate-TCH-SP method, respectively (Takagi et al. 1983). Their glycoprotein content has been confirmed by staining with acidic PTA (Fig. 16.2). Chondroitin 4-, but not chondroitin 6-sulfate, and keratan-sulfate, have both been detected immunohistochemically in calcification nodules (Hoshi et al. 2001a).

More precise results have been obtained when individual crystal ghosts, rather than bone calcification nodules, are examined by electron microscopy. This has confirmed not only that crystal ghosts contain acidic groups, but also that they are related to sulfate glycosaminoglycans. The electron density of crystal ghosts, at least the peripheral ones, is, in fact, increased by acidic colloidal iron (Bonucci 2002). Moreover, large, hyaluronate-binding proteoglycans have been detected by immunohistochemistry in the calcification nodules of developing mandibles in fetal rats, after their decalcification by the immunohistochemical procedure itself: specific immunogold particles marked peripheral, fine filamentous structures resembling crystal ghosts (Lee et al. 1998). The view that crystal ghosts are related to glycosaminoglycans is further supported by the observation that calcification nodules in bone react with cuprolinic blue (Nefussi et al. 1989; Hoshi et al. 2001b) and by the demonstration that, after this reaction, they contain filamentous structures shaped like crystal ghosts and digested by chondroitinase ABC and hyaluronidase (Sauren et al. 1989). Moreover, in areas of early calcification (apparently corresponding to calcification nodules) in mantle dentine, crystal ghosts react with soybean agglutinin-gold particles, which shows that they contain *N*-acetyl-D-galactosamine, a component of chondroitin sulfate (Hayashi 1988). The dentin calcification front is also labeled after applying the hyaluronidase-gold method for the ultrastructural demonstration of chondroitin sulfate (Chardin et al. 1990b); by contrast, calcospherites are not labeled, a finding that seems comparable to the lack of colloidal iron staining observed in the central zones of the widest calcification nodules (Bonucci 2002). In this connection, keratan sulfate has been demonstrated in the calcification nodules of the rat calvaria, where it is mainly localized in the peripheral region of the nodules as these enlarge and their calcification moves forward (Nakamura et al. 2001). The antibody MC22-33F, which displays a specific reaction with phosphatidylcholine, sphingomyelin and dimethylphosphatidylethanolamine, also stains the peripheral, but not the central, zone of the bone calcification nodules (Bonucci et al. 1997). Similar results were obtained in embryonic rat calvariae (Hoshi et al. 2001b), and in predentin and dentin (Chardin et al. 1990a): after fixation with a solution of glutaraldehyde and pyridinium chloride, which forms needle-like insoluble complexes with glycosaminoglycans, needle-like structures could be seen throughout small calcification nodules, whereas they could only be detected at the periphery of the largest ones. Treatment of sections with a chloroform-methanol mixture, which solubilizes lipidic material, completely removed the precipitate induced by cetylpyridinium chloride, which makes it likely that the glycosaminoglycans were combined with lipids (for a discussion on the possibility that crystal ghosts have a lipidic component see Nefussi et al. (1992).

Apart from these generic results, immunohistochemistry has provided clear evidence of the presence of specific molecules in bone calcification nodules (Fig. 7.5). Osteocalcin has been detected in them by Groot et al. (1986), and OPN and BSP have been found in them several times (McKee et al. 1992, 1993; Bianco et al. 1993; McKee and Nanci 1995, 1996; Riminucci et al. 1995; Hoshi et al. 2001a). These proteins have also been detected in the calcification nodules of cellular cementum (McKee and Nanci 1995, 1996; Bosshardt et al. 1998), calcifying cartilage, as well as in small calcification nodules and at the periphery of the largest ones in primary cultures of osteogenic cells derived from fetal rat cranial tissue (Nefussi et al. 1997; Irie et al. 1998).

Immunogold particles, which are specific to the non-collagenous proteins reported above, can be detected right over crystal ghosts. A direct relationship between them and the individual crystal ghosts cannot, however, be demonstrated because the resolution of the method is inadequate. According to Hoshi et al. (2001a), BSP and OPN are widely distributed in the calcification area, but are linearly arranged along electron-dense structures corresponding to crystal sheaths. In osteoblastic cultures to which organic phosphate had been added, BSP has been detected in extracellular matrix sites where hydroxyapatite crystals are nucleated at a later stage (Wang et al. 2000). In the same type of culture, as well as in membranous bone and the woven bone elicited by bone marrow ablation, condensed mesenchyme regions and spherical structures rich in alkaline phosphatase – both probably corresponding to calcification nodules –

contain BSP and also BAG-75 (Gorski et al. 2004; Midura et al. 2004). A major 66-kDa phosphoprotein synthesized by chicken osteoblasts during mineralization in vitro has been found to accumulate over electron-dense patches in vivo and in vitro (McKee et al. 1989, 1990; Gerstenfeld et al. 1990). The cytochemical immunoreactions for decorin, chondroitin sulfate and alkaline phosphatase proved to be positive in the calcification nodules of embryonic rat calvariae (Hoshi et al. 2001b).

It is noteworthy that the staining properties of bone calcification nodules, and of the crystal ghosts they contain, change with their development and degree of calcification. It has been reported above that, as calcification nodules increase in size and coalesce, crystal ghosts continue to be clearly visible at their periphery – where the lamina limitans can be detected in conventionally decalcified material, whereas they become only faintly visible, and lose their histochemical properties, in the central, fully calcified areas (see Sect. 16.3 and Fig. 16.4). The same pattern was reported in fetal porcine calvarial bone: organic crystal profiles – resembling crystal ghosts – were present in early calcification nodules, but were much less evident deeper in the mineralized matrix, where material of amorphous type was only detectable (Chen et al. 1994).

The weak evidence of crystal ghosts in central, fully calcified areas of the largest calcification nodules could depend either on their degradation and removal during calcification, on depletion of their histochemical reactivity, or on both. The first hypothesis is supported by the documented loss of organic material that occurs during bone calcification (Pugliarello et al. 1970; Nusgens et al. 1972), and indirectly by similar loss occurring in enamel (see Sect. 16.4). However, if crystal ghosts are, indeed, removed and lost as calcification increases, the visibility of the underlying collagen fibrils, which are distributed throughout the calcified areas (Takuma 1960), should progressively and proportionally increase. In reality, after the PEDS method, the central zone of the calcification nodules appears rather homogeneous and almost amorphous; not only crystal ghosts, but collagen fibrils too (Bonucci et al. 1988), along with their periodic banding (Prostak and Lees 1996), are poorly resolved there. Without excluding that a fraction of crystal ghosts, and/or non-collagenous interfibrillary substance, may be lost during calcification, this feature might be more plausibly accounted for by a fall in the chemical reactivity of crystal ghosts – and of collagen fibrils, too –, possibly depending on a more limited availability of free chemical groups.

Crystal ghosts in calcification nodules have a morphology so similar to that of untreated crystals, that they may be confused with one another and the former might be thought to be crystals left behind by incomplete decalcification (discussed in Sect. 18.4). The organic and inorganic components are so closely intermingled that the organic structures seem to be 'stained' by the inorganic substance. Moreover, crystals are so thin that it is not possible to determine what is inside them and what is outside, or what forms their core and what their surface. Crystal ghost-like organic substance has been described to be internalized and occluded in thick crystals, such as the calcite crystals of mollusk shells (see Sect. 16.6). In this case, too, it is hard to specify what is inside what: calcification apparently involves a stereochemical complementarity between specific crystal lattice planes and organic components (Goldberg HA et al. 2000), and it may well be that the inorganic substance is considered to be inside the organic substance simply because the organic substance is formed earlier and is then permeated by the inorganic. In any case, after decalcification and staining, organic substances located 'inside' inorganic crystals behave like crystal ghosts.

16.2.2 Organic-inorganic Relationships in Calcification Islands of Bone

At the bone calcification front, calcification islands are easy to confuse with calcification nodules, because both appear as small aggregates of crystals lying between collagen fibrils. Calcification nodules, however, are roundish, and their crystals appear as if radiating from a center, unlike calcification islands, which are elongated, with crystals running almost parallel to collagen fibrils (Fig. 5.2). As calcification proceeds, the former widen and coalesce; the latter widen, too, but, besides that, granular inorganic bands develop in the matrix.

After the PEDS method, calcification islands, like calcification nodules, show crystal ghosts and the two are hard to distinguish. On the other hand, calcification islands can be clearly distinguished in decalcified, uranyl acetate-lead citrate-stained sections from specimens of embryonic avian bone treated en bloc with $KMnO_4$ (Bonucci and Silvestrini 1996). They are probably identical with the structures that have been called 'grey patches', that is, islands of amorphous, electron-dense organic material located among collagen fibrils (McKee et al. 1990) that contain OPN and BSP (McKee and Nanci 1995). In chick tibiae (McKee et al. 1992; Chen et al. 1994; McKee and Nanci 1996; Nanci 1999), and at the bone-titanium interface of rat tibiae (Ayukawa et al. 1998), these patches have been immunolabeled using antibodies against OPN and OC.

16.3 Organic-inorganic Relationships in Cartilage

The relationship between the inorganic substance and the periodic banding of collagen fibrils that can be detected in incompletely calcified bone matrix (Fig. 7.2), indicating a relationship between crystals and collagen holes, cannot be detected in calcified cartilage. In this tissue, crystals are rather irregularly scattered through the territorial and interterritorial matrix, without any apparent relationship with collagen fibrils.

Resemblance with bone becomes evident at the calcification front. Here, exactly as in bone, the earliest crystals are found in matrix vesicles which are, however, much more plentiful than in bone. Single crystals, or small groups of them, may be located very close to, or even touching, the peripheral membrane of a matrix vesicle; most of them, however, lie within vesicles and show no special relationship with their membrane. As numbers of crystals increase, matrix vesicles become completely full of them; any further increase will then give rise to the formation of calcification nodules (Fig. 9.4). These are located between collagen fibrils and their crystals show no direct relationship with them; in this case, too, the organic components are masked by the inorganic substance and can only be studied after decalcification.

The decalcification of epiphyseal cartilage using the PEDS method clearly shows aggregates of crystal ghosts where calcification nodules had been (Figs. 3.5, 16.2, 16.3 and 16.4). Crystal ghosts were initially given this name in the calcification nodules of the epiphyseal cartilage (Bonucci 1967). They resemble the crystal ghosts found in the calcification nodules of bone (Sect. 16.2.1), from which they are indistinguishable from a morphological viewpoint. As reported several times (Appleton and Blackwood 1969; Bonucci 1969, 1971, 1995; Appleton 1971; Bonucci and Reurink 1978; Shepard and Mitchell 1981; Davis et al. 1982b; Scherft and Moskalewski 1984; Bonucci et al. 1988; Shepard 1992; Bonucci and Silvestrini 1994; Takagi et al. 2000), cartilage crystal ghosts, like those in bone, consist of filament-like structures with the same morphology and location as untreated crystals, but without their intrinsic electron density; they are only recognizable after PEDS-decalcified sections are treated with uranium and lead, acidic phosphotungstic acid, or other electron-dense 'stains'. Their similarity with bone calcification nodules persists if cartilage is decalcified before embedding in the presence of cationic dyes such as acridine orange, ruthenium red, or terbium chloride (Bonucci et al. 1988; Bonucci 1995, 2002): each nodule appears as an aggregate of filament-like structures that bear a close resemblance to crystal ghosts (Fig. 16.5). Intracrystalline, crystal ghost-like structures, as well as pericrystalline organic material, have been detected in cartilage calcification nodules that have been decalcified and stained using an aqueous solution of basic chromium(III) sulfate (Sundström and Takuma 1971).

The histochemical properties of cartilage crystal ghosts resemble those of bone crystal ghosts. Decalcified calcification nodules in the epiphyseal cartilage are, in fact, basophilic under the light microscope, and are stained by histochemical methods for glycoproteins (PAS method) and acid glycosaminoglycans (alcian blue, colloidal iron, ruthenium red). The acidity of crystal ghosts is confirmed under the electron microscope, which shows that they react with ruthenium red, ruthenium hexamine trichloride, therbium chloride, colloidal iron, or bismuth nitrate (Smith 1970; Davis et al. 1982b; Bonucci et al. 1988, 1989; Bonucci 2002). These reactions occurs at very low pH ($\sim$1.8), indicating that the acidic groups are very strong and that they probably correspond to sulfate groups of acid proteoglycans. This is in full agreement with the findings that calcification nodules contain sulfur (Davis et al. 1982b), that crystal ghosts

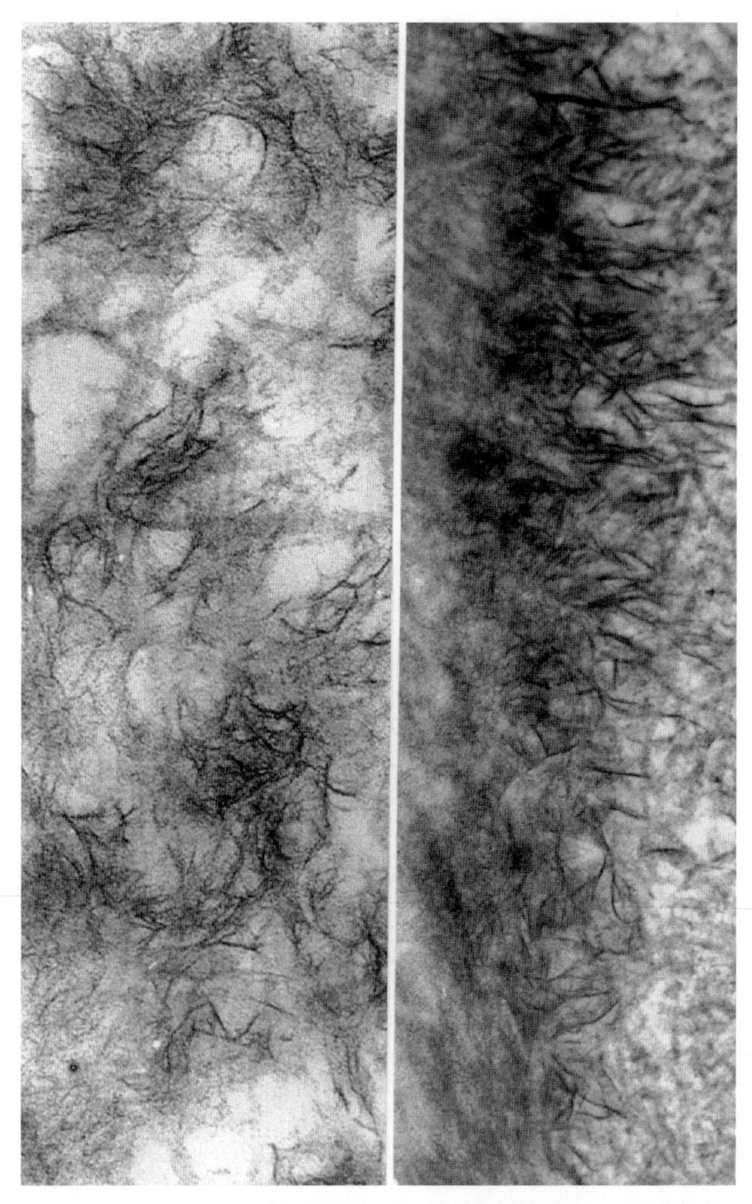

Fig. 16.3. *Left*: at the cartilage calcification front, crystal ghosts replace calcification nodules; note that they are filament-like and are sometimes irregularly branched. *Right*: in areas of advanced calcification, crystal ghosts protrude from the calcified (*left*) into the uncalcified (*right*) cartilage matrix; they again appear as filaments. PEDS, uranyl acetate and lead citrate, ×30,000

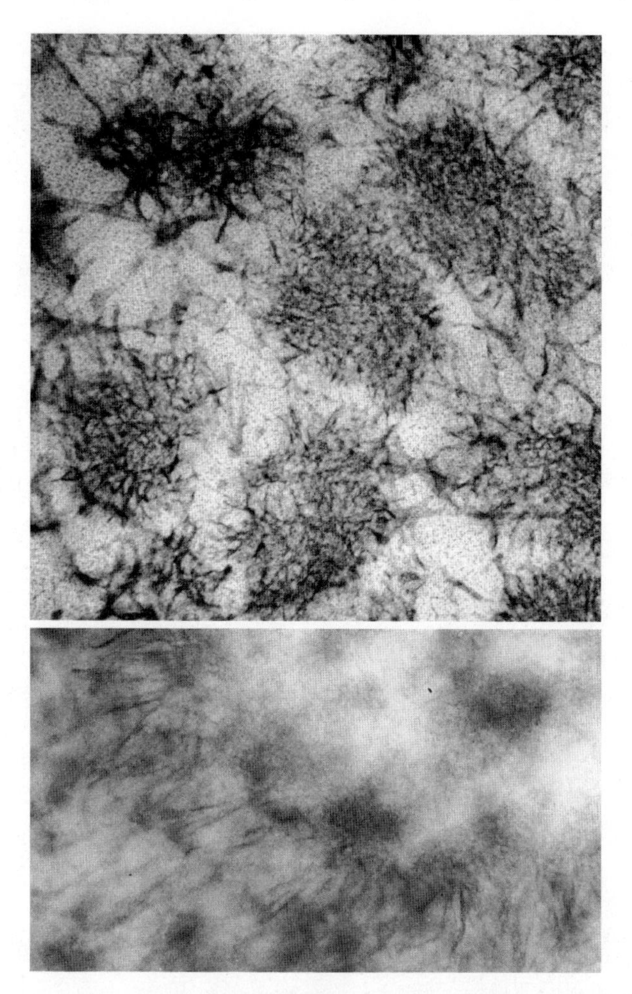

Fig. 16.4. Growth cartilage fixed in formaldehyde and glutaraldehyde, embedded in glycol-methacrylate, decalcified by floating sections on 1% formic acid, and stained with uranyl acetate and lead citrate. *Above*: aggregates of crystal ghosts simulate calcification nodules. *Below*: the central zone (*upper right*) in a developed calcification nodule is almost unstained and crystal ghosts only protrude from the border. PEDS, uranyl acetate and lead citrate, ×60,000 and ×78,000

react with CS-56 – an antibody specific for the glycosaminoglycan portion of chondroitin sulfate (Bonucci and Silvestrini 1992) – and that their reactions with basic dyes are inhibited by methylation, which blocks both sulfate and carboxy groups, and are not restored by saponification, which only reestablishes the carboxy groups (Bonucci et al. 1988). This behaviour is like that found in bone calcification nodules (Groot 1982a). The core protein of biglycan has been immunodetected in the calcification nodules of cartilage, where crystal ghosts were demonstrable too (Takagi et al. 2000).

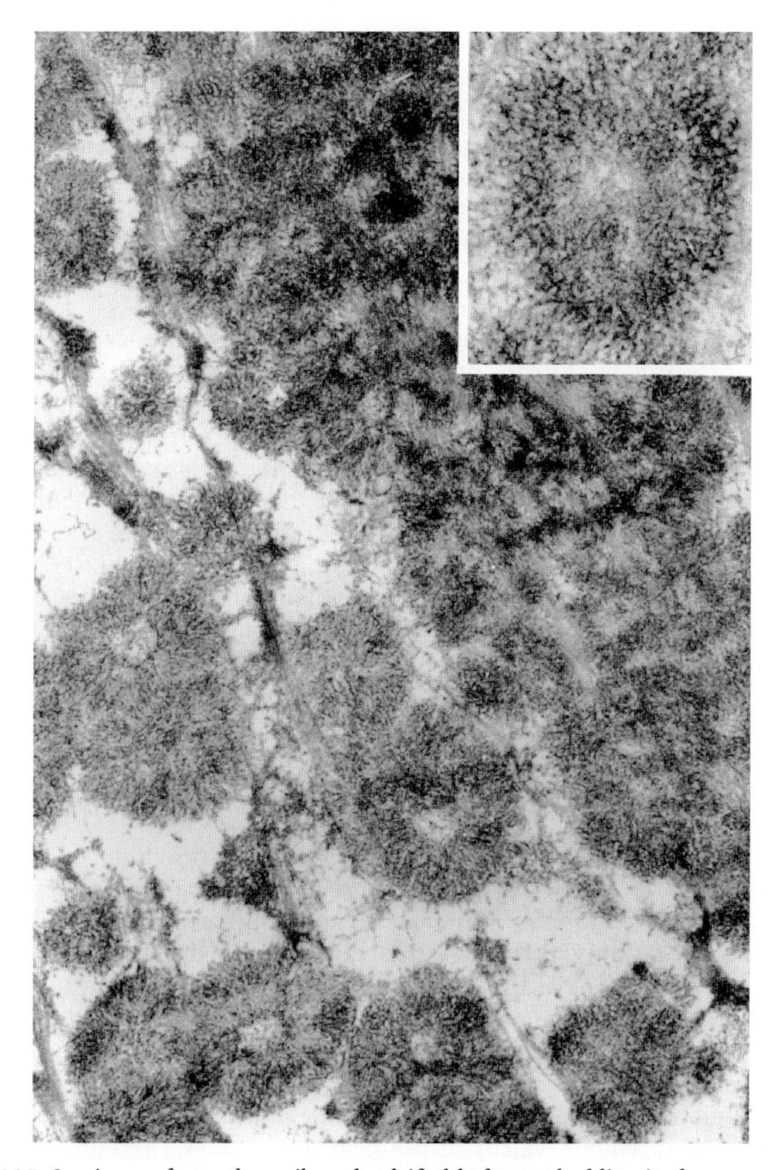

Fig. 16.5. Specimen of growth cartilage decalcified before embedding in the presence of acridine orange: the calcification nodules have been replaced by aggregates of filaments, which are similar to the crystal ghost aggregates shown by the PEDS method (compare with Fig. 16.4). Uranyl acetate and lead citrate, ×30,000 and, *inset*, ×75,000

It is worth remembering that, as in bone, cartilage crystal ghosts are hard to make out in the central zone within the largest calcification nodules (Fig. 16.4, below) – a zone that reveals no stainability with electron-dense 'stains' such as ruthenium red (Davis et al. 1982b), whereas they are easy to recognize at the periphery of the calcified areas. So too, the reaction with CS-56 in testing for chondroitin sulfate (Bonucci and Silvestrini 1992), and the immunoreaction for OPN (McKee and Nanci 1995), turn out to be stronger at the periphery than in the central zone. These findings suggest that sulfate proteoglycan molecules and non-collagenous proteins are either degraded during calcification, or segregated into macromolecular complexes and made unreactive, or both of these (Bonucci and Silvestrini 1992), so that the central, probably fully calcified areas, of the calcification nodules gradually acquire a molecular configuration different from that of the peripheral zones (Davis et al. 1982b).

As discussed in Sect. 9.3.1, the procedures used to process calcified specimens for electron microscopy profoundly change the morphology of proteoglycans. These have an extended, bottlebrush-like shape in their native state, whereas, after fixation and dehydration, they appear as small matrix granules (Campo and Phillips 1973; Thyberg et al. 1973; Newbrey and Banks 1975; Thyberg 1977; Hascall 1980; Takagi et al. 1982, 1983), a feature attributable to the collapse and shrinkage of their molecules resulting from the introduction of cross-links and the subtraction of water (Smith 1970; Serafini-Fracassini and Smith 1974; Hascall 1980). When these molecular changes are avoided by using cryogenic methods or anhydrous techniques, or by the addition of cationic dyes to the fixative solutions, or by using *N*-*N*-dimethylformamide after aldehyde fixation and before Spurr or Lowicryl embedding, proteoglycan molecules retain their extended, almost native state and appear as thin filaments that are interconnected in a dense network (Hunziker and Schenk 1984; Arsenault et al. 1988; Arsenault 1990; Kagami et al. 1990; Hagiwara 1992; Leng et al. 1998).

Are these findings compatible with the view that crystal ghosts are acid proteoglycans? To answer this question, the hypothesis can be put forward that calcium ions have the same stabilizing effects on proteoglycan molecules as cationic dyes. Crystal ghosts and matrix granules would then be different morphological expressions of the same molecules: as calcification goes forward, crystal ghosts would become the form taken by those molecules once immobilized and stiffened by calcium in their native, extended state – changes that would leave them able to resist the collapsing effects of specimen processing – whereas matrix granules would be those same molecules as modified by the preparation procedures, which collapse them into granules. These lines of interpretation, which are exemplified in the upper part of Fig. 16.6, are strengthened by the finding that both these structures (granules and filaments) are stained by very acidic colloidal iron (lower part of Fig. 16.6) – so both correspond to acid glycosaminoglycans.

These views would also explain some specific ultrastructural features shown by crystal ghosts. As noted by Bonucci (2002), some of them, especially those at the periphery of the calcification nodules, where calcification is in progress, appear as thick, apparently swollen, deeply stained, often ramified structures, so showing the same ultrastructure as that of acid proteoglycans treated with

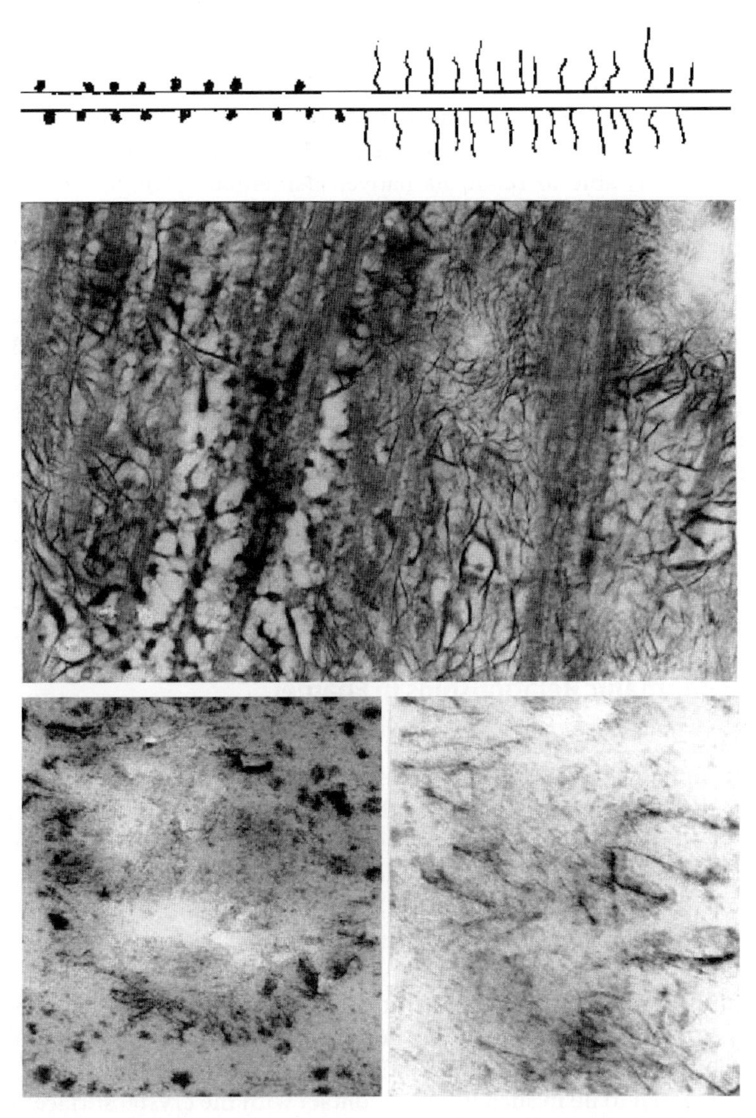

Fig. 16.6. *Above*: calcifying cartilage matrix processed according to the PEDS method: the ultrastructure of the granular aggregates is in agreement with Smith's suggestion (Smith 1970; see diagram) that the extended molecules of acid proteoglycans can collapse into granules as a result of fixation and dehydration (*left*), whereas in calcified cartilage they can maintain an extended configuration, and appear as crystal ghosts (*right*), because their link with calcium salts stabilizes their molecule, so preventing their collapse. PEDS, uranyl acetate and lead citrate, ×75,000. *Lower left*: the colloidal iron reaction stains the crystal ghosts at the periphery, but not in the central zone, of calcification nodules; proteoglycan granules in the uncalcified matrix are also stained; ×60,000. *Lower right*: detail of crystal ghosts at the periphery of a calcification nodule; note that some of them (*upper left corner* and *middle right*) have a swollen extremity. Colloidal iron, ×90,000

basic dyes (see, for instance, Fig. 16 in Hunziker and Schenk 1987). In addition, some crystal ghosts have a tadpole-like configuration because their outer, protruding extremity is thick and expanded (Fig. 16.6, lower right). This feature could be accounted for assuming that one part alone (the tadpole tail) of the proteoglycan molecule has become calcified and, having been stabilized by calcium, it is able to retain its native, filamentous configuration during the preparation procedures, whereas the other part (the tadpole head) is left uncalcified and therefore collapses into a granule. This view is reinforced by the finding that both filamentous and expanded parts react with acidic colloidal iron (Fig. 16.6, lower left), and that this sometimes stains a material surrounding crystal ghosts that appears to show periodicity (Bonucci 2002).

The calcification nodules of cartilage develop between type II collagen fibrils. The C-propeptide of type II procollagen (chondrocalcin; van der Rest et al. 1986) has been detected immunohistochemically in them (Poole et al. 1984; Poole 1991; Lee et al. 1996). It is of interest that this molecule is found during the early phases of cartilage calcification, whereas it is absent from completely calcified matrix, perhaps because it is removed by neutral degradation processes (Poole and Rosenberg 1986).

16.4 Organic-inorganic Relationships in Enamel

As discussed in Sect. 10.2 and shown in Fig. 10.1, the electron microscope study of immature enamel sections decalcified and stained according to the PEDS method unequivocally shows that inorganic crystals are replaced by crystal-like organic structures, or crystal ghosts. Even if named differently, structures morphologically identifiable as crystal ghosts have been reported in enamel since 1962. Rönnholm (1962), using a PEDS-like method (the EDTA decalcification of sections followed by staining with a saturated uranyl acetate solution at pH 7.0), described crystal-like, long, fairly broad septa that were assumed to represent the enamel stroma on whose surfaces crystallites were formed, so that the septa eventually became embedded in the crystallites. In that same year, Nylen and Omnell (1962) reported that part of the enamel matrix appeared to be in such intimate contact with the crystal surfaces that it could not be resolved microscopically in undecalcified sections and that, after decalcification, it appeared as thin borders surrounding empty spaces which had the same shape as that of crystals. Results of the same kind were reported by Travis and Glimcher (1964): they found that in decalcified sections the organic matrix of embryonic bovine enamel had the same overall organization as that of the undecalcified tissue and consisted of thin filaments encompassing tubular compartments looking like 'blueprints' of inorganic crystals. Tubular compartments in place of decalcified crystals were later described by Travis (1968), Warshawsky (1971), Decker JD (1973), Glimcher (1979), and Nylen (1979). Kemp and Park (1974) described them in the developing enamel of the shark *Carcharhinus menisorrah*, and Stern et al. (1992) in that of the opossum. Fibrillar, filamentous, lamellar or helical organic structures, with the same shape as untreated enamel crystals and, therefore, with the characteristics of crystal ghosts, have since been reported on various occasions (Kemp and Park 1974; Smales 1975; Warshawsky 1987; Frank 1979; Bishop and Warshawsky 1982; Nanci et al. 1983, 2000; Bai and Warshawsky 1985; Bonucci et al. 1986, 1994; Hayashi et al. 1986; Kallenbach 1986; Hayashi 1989; Bonucci 1995; McKee and Nanci 1995; Goldberg M et al. 1998).

Obviously, the distinction between tubules and filaments is of primary importance, because the presence of one or another of these structures, or of both, implies different mechanisms of crystal formation, either in a close compartment or in an open framework (discussed in Sect. 18.4). According to Kallenbach (1986), two matrix components are closely associated with enamel crystals: tubule-like coats which surround the crystals and may be due to proteins adsorbed on the crystal surface, and crystal ghosts themselves, which act as organic frameworks coextensive with the crystals; he found that, after brief decalcification and staining, apatite crystals and matrix components could be seen in the same field, and that some of the crystal ghosts were continuous with crystal fragments, while the crystal coats appeared as dense lines next to crystals and ghosts.

A very close relationship between inorganic and organic matrix components has been shown in decalcified enamel not only by electron microscopy, but also by X-ray and electron diffraction studies (Traub et al. 1985; Jodaikin et al. 1986). These have shown a β pleated-sheet conformation, associated with the acidic enamelin fractions, and a crystalline pattern the same size as the *hhl* plane of hydroxyapatite and with two regularly repeating calcium-binding sites. Moreover, electron-dense organic coats, between 3 and 10 nm in diameter, have been described around initial enamel crystals, measuring between 1 and 3 nm in thickness (Diekwisch et al. 1995).

Histochemical techniques have provided significant results about the composition of the enamel matrix and the relationship between organic and inorganic components. Using lectin-gold cytochemistry, *N*-acetyl-D-galactosamine was shown to be a component of the crystal ghosts in developing rat enamel (Hayashi 1989). Applying the same method, gold particles complexed with *Helix pomatia* agglutinin (HPA), specific to glycoconjugates containing *N*-acetyl-D-galactosamine, or complexed with wheat germ agglutinin (WGA), specific to *N*-acetyl-glucosamine/*N*-acetyl-neuraminic acid, were detected over organic matrix profiles, similar to crystal ghosts, left behind after decalcification (Nanci et al. 1989). The HPA-reactive proteins were lost before those reactive with WGA, and more than one component was probably present, but the results were consistent with the presence of glycosylated amelogenins throughout amelogenesis. Immunoreaction for amelogenin (Sect. 10.3.3) has, in fact, been reported to be positive in enamel matrix and areas showing crystal ghosts (Figs. 10.2 and 10.3), but these could not be definitively associated with the immunogold particles because of method's poor degree of resolution (McKee and Nanci 1995). Amelogenins could be demonstrated by immunohistochemistry in both decalcified and undecalcified immature enamel, whereas enamelins were reactive only after decalcification (Fig. 10.4); this suggests that amelogenins lie in the intercrystalline spaces, unlike enamelins, which appear to be linked to, and masked by, the inorganic substance (Bonucci et al. 1986; Hayashi et al. 1986). In line with this view, immunolabeling of amelogenins was found in enamel intercrystallline spaces, whereas labeling for enamelins was mainly attached to enamel crystals (Inage et al. 1989) and crystal ghosts (Hayashi et al. 1986). These findings are in accordance with the observation that amelogenins can be dissociated from enamel without physical disruption of the tissue, whereas enamelins can only be extracted after crystals have been dissolved (Termine et al. 1980; Bai and Warshawsky 1985; Bronckers et al. 1988).

Few of the studies just discussed have taken into account the fact that, in enamel, amounts of organic material fall during calcification and that the organic-inorganic relationships must change accordingly. The greatest difference between early, poorly calcified enamel, and mature, fully calcified enamel is, in fact, that crystal ghosts are almost totally lacking within the rods of the latter (Bonucci et al. 1994).

16.5 Organic-inorganic Relationships in Lower Vertebrates

Organic-inorganic relationships in the endoskeleton of lower vertebrates repeat those described in high vertebrates, with some differences. The most important of these involve the earliest phases of the calcification process.

On the basis of the results of a study carried out in pickerel and herring fish bone using high-voltage electron microscopy, stereomicroscopy, electron probe microanalysis, electron diffraction and three-dimensional computer reconstruction, it has been reported that, progressing from the earliest stages of the calcification process to its completion, crystals are never associated, or located within, membrane-bound structures – that is, matrix vesicles; instead, they are located within, and totally occupy, collagen fibrils, whose periodic banding is therefore totally obliterated (Lee and Glimcher 1989). Glimcher (1959) and Glimcher and Krane (1968) reported that collagen fibrils are widely separated in fish bone, so the localization in them of most crystals was easy to ascertain.

The electron microscope pictures produced in support of these studies can hardly be disputed, but they appear to conflict with those obtained from electron microscope studies of fish teeth or scales. These show that matrix vesicles are present in the growing or regenerating scales of several species of fish (Maekawa and Yamada 1970; Yamada J 1971; Kobayashi et al. 1972;

Schönbörner et al. 1979; Zylberberg and Nicolas 1982; Sire and Géraudie 1984), as well as in the tooth germ predentin of the killifish, *Oryzias latipes* (Yamada M and Ozawa 1978). In scales, as well as in some types of bone (Fig. 16.1), crystals do not develop within the collagen fibrils, as suggested; the earliest crystals are, on the contrary, located in the interfibrillary space, surrounding collagen fibrils like rings (Herold 1971; Lanzing and Wright 1976; Onozato and Watabe 1979; Schönbörner et al. 1979; Yamada J and Watabe 1979; Olson and Watabe 1980; Zylberberg and Nicolas 1982; Zylberberg et al. 1992), and they do not penetrate deeply into fibrils, which in cross section therefore appear as electron-lucent spaces (Zylberberg et al. 1992). Calcification of fibers may be completed through further crystal growth (Olson and Watabe 1980); in other cases, it is never attained, as is true of isopedine (Zylberberg et al. 1992). No definitive relation has been found in the bony fin rays of the trout *Oncorhynchus mykiss* between either the *c* or the *a*, *b* axis images of the crystals and collagen fibril periodicity (Landis and Géraudie 1990).

Crystals developing in interfibrillar spaces may aggregate into roundish globules which resemble bone calcification nodules; if decalcified in the presence of ruthenium red, their matrix appears to consist of filamentous structures (Zylberberg et al. 1992) bearing some resemblance to crystal ghosts. In sections of osteoderms fixed with a fixative containing ruthenium red and then decalcified, the organic matrix appears to be composed of thin microfibrils arranged in the same way as crystals (Zylberberg et al. 1992), so resembling crystal ghosts.

The organic-inorganic relationships in enameloid will be taken into consideration separately. They are not clearly understood, reflecting the doubts that still persist on the structure of this tissue and the shape of its crystals (Sects. 4.3.1 and 5.3.1), which may vary as a function of their fluoride content. According to Garant (1970), two layers can be detected in the developing teeth of the dogfish *Squalus acanthias*, an inner layer consisting of dentin, and an outer one corresponding to enameloid, whose hexagonal crystals are enveloped by a 6.0–7.0 nm thick membrane. Crystals, therefore, appear to grow within compartments and their membrane seems to be shed only after they are fully formed. Organic sheaths around individual enameloid crystals have also been described in sharks by Kemp and Park (1974) and Kemp (1985), in elasmobranch fish by Sasagawa (2002), and in teleost fish by Sasagawa and Ferguson (1990). It is worth noting that, in the maturation stage, the cell membrane of the inner dental epithelium shows plenty of complex infoldings (Sasagawa and Ishiyama 1988), whose cross-sections might easily be viewed as compartments. According to Miake et al. (1991), the way crystals grow in selachian enameloid – that is, by the formation of projections on the *c*-plane which then consolidate into a single crystal – casts doubt on the possibility that crystals grow within compartments. On Inage's interpretation (Inage 1975), the enameloid matrix consists of collagen fibrils and the deposition of small crystals, whether needle- or tube-like in shape, occurs in the circumference of fiber bundles. Crystals formed parallel to the long axis of the collagen fibrils have been reported by Prostak et al. (1993).

16.6 Organic-inorganic Relationships in Invertebrates

The possibility that organic material combines with, and stabilizes, amorphous calcium carbonate in sea urchin spicules and in calcified structures of other invertebrates has already been discussed in Sect. 4.4. Moreover, biophysical and morphological studies on invertebrate calcified structures have repeatedly shown that, besides an organic material (conchiolin) occurring as a "mortar" between polygonal, brick-like crystals of aragonite or calcite (Grégoire 1957; Watabe 1963; Towe and Hamilton 1968; Williams 1970; Marsh et al. 1978; Marsh and Sass 1981), other organic molecules can interact with certain crystal faces and can be occluded within growing crystals (Addadi and Weiner 1985).

The calcified integuments of most invertebrates are structured in such a way that they behave as single crystals when examined under the polarizing microscope or by X-ray diffraction (discussed in Sect. 5.4). Because only a minor amount of organic material is associated with them, they initially appeared to be completely inorganic structures and no evidence could be obtained for the existence of a coherent intracrystalline organic material. With the growing refinement of the electron microscopy techniques, Watabe (1963), who examined the same section before and after decalcification, was able to reach the important conclusions that in the shell of the fresh water mussel *Elliptio complanatus*: 1) the tabular crystals are not single crystals in the usual sense, but are made up of component blocks; 2) besides the intercrystalline, or interlamellar, organic matrix, there is also an intracrystalline organic matrix.

Watabe's conclusion was supported by the results of electron microscope studies on the prismatic region of *Mytilus edulis*: using ultrathin sections that had been decalcified on grids, Travis (1968), and Travis and Gonzales (1969), found interprismatic sheaths (mean thickness 29.2 nm; range 16.7–49.9 nm) and an intraprismatic matrix consisting of very thin filaments (mean thickness 2.2 nm; range 1.7–2.5 nm) which were structurally organized in such a way as to outline closely packed, sheet-like compartments whose shape was referred to as a "striking fingerprint" of intraprismatic crystals. Saleuddin (1971) confirmed that intracrystalline organic material is contained in the normal and regenerating shell of *Helix*. Intracrystalline soluble organic matrix was also reported by Watabe (1965), Mutvei (1970), Meenakshi et al. (1971), Crenshaw (1972), Wong and Saleuddin (1972), Albeck et al. (1996), Shen et al. (1997), Miyashita et al. (2000), Dauphin (2002), Tong et al. (2002). Crenshaw and Ristedt (1976) detected sulfate proteoglycans in the central zone of aragonite crystals in the nacre of *Nautilus pompilius*. Pokroy et al. (2006) have clearly shown that anisotropic lattice distortions, probably due to organic molecules intercalating into the aragonite lattice, can be detected in a number of mollusk shells. These results have been supported by X-ray and electron diffraction studies on the mollusk shell matrix showing that, as in enamel (see Sect. 16.4), proteins adopt an antiparallel β-sheet configuration which has two regularly repeating Ca-binding sites and whose crystallographic axes are aligned with those of the associated crystals (Weiner and Traub 1980, 1984; Weiner et al. 1983; Choi and Kim 2000).

The established occlusion of organic material within crystals of mollusk shells has been validated by researches carried out on the calcified structures of the sea urchin embryo. Chemical analyses (summarized by Wilt 2005) have shown that sea urchin spicules treated with NaOH to remove all the surface organic material still release soluble proteins when decalcified, showing that, as in bone and other calcified tissues (see above), proteins are linked to the inorganic substance and can only be released after decalcification. A small amount (about 0.1%) of integral, occluded proteins are released (Wilt 1999); these consist of glycoproteic material rich in Asx, Glx, Gly, Ser and Ala, so resembling the calcified matrix of mollusks, sponges, and arthropods (Benson SC et al. 1986). Moreover, the decalcification of cleaned spicules – that is, spicules whose surface organic material had been removed by treatment with urea and sodium dodecyl sulfate – releases an organic matrix with overall outlines similar to the shape of the intact spicule (Benson S et al. 1983). Intracrystalline incorporation of organic material has been experimentally demonstrated by growing calcite crystals in the presence of acidic glycoproteins extracted from the exoskeleton of sea urchins (Berman et al. 1988).

Electron microscope studies also showed the occlusion of organic material in crystals of sea urchin spicules. Studies by Travis (1970), reported above, on the skeletal plates and spicules of the post-larval sea urchin *Strongylocentrotus droebachensis* had previously detected a fibrous organic matrix associated with a PAS-positive, metachromatic amorphous material which, under the electron microscope, appeared to be, as in *Mytilus edulis*, a striking fingerprint of the polycrystalline structure of the calcified matrix. Lamellae of an irregular fibrillar nature which probably consisted of interconnected concentric sleeves, whose overall morphology was comparable with that of the intact spicule, have been detected in isolated, decalcified spicules (Benson S et al. 1983). Results pointing to the occlusion of organic material in the calcified matrix of sea urchin spicules have been reported several times (Berman et al. 1990; Benson SC and Wilt 1992; Cho et al. 1996; Ameye et al. 1998). A residual matrix which could be stained with ruthenium red or concanavalin A-gold was found in the space left by the spicule after its dissolution by decalcification (Decker GL et al. 1987; Decker G and Lennarz 1988).

More accurate results were provided by electron microscope immunohistochemistry. The presence of glycoprotein antigens in sea urchin spicules had already been detected using immunogold (Benson NC et al. 1989). Immunolocalization of the PM27 protein in the whole embryo of *Strongylocentrotus purpuratus* showed that the antigen is only produced by primary mesenchyme cells, and that during skeletogenesis its production is restricted to a sub-population of these cells associated with the growing tip of spicules, where it accumulates, while disappearing from the more mature mid-shaft regions (Harkey et al. 1995). Localization of the SM30 protein in the matrix occluded in the spicule has been clearly shown by Cho et al. (1996), Ingersoll et al. (2003) and Seto et al. (2004). These Authors based their study on the presence in the SM30 protein of a carbohydrate chain (the 1223 epitope) that has been implicated in the process of Ca^{2+} deposition as $CaCO_3$. A monoclonal antibody to this chain was prepared by them; antibodies to SM30, prepared by Killian and Wilt (1996), were also used. The treated spicules were first examined under a scanning electron microscope in the secondary electron imaging mode to observe their surface, and then in the backscattered electron imaging mode to detect the gold particles conjugated with the secondary antibody. These were not found on the surface of spicules that had been isolated from embryo homogenates and treated with alkaline hypochlorite to remove associated organic material. They were, by contrast, detected when spicules treated in this way were further etched with dilute acetic acid to expose their internal parts, so showing that SM30, and, to a lesser extent, SM50, proteins were occluded within them (Fig. 16.7).

The distribution of SM50 and SM30 proteins, as well as of another spicule matrix protein (29 kDa), has been studied by immunofluorescence and a 'spicule blot' procedure (Kitajima and Urakami 2000). This method, which is made possible by the small size of spicules, consists in separating isolated spicules from attached organic material and spreading them over a polyvinylidene fluoride (PVDF) membrane. This is overlaid with 0.8% agarose gel pre-soaked in phosphate-buffered saline (PBS) and placed in a vacuum blotter; the blotting buffer is PBS containing 10 mM EDTA. The membrane is then blocked with 3% BSA and allowed to react with the antibody (or antibodies) and with biotinylated antimouse IgG. After visualization of the reaction product, the stained pieces of membrane are immersed in 80% glycerol to make them translucent and, therefore, visible under the microscope. The SM30 protein was only faintly recognizable along the surface of the spicules when these were examined by immunofluorescence, whereas SM50 and 29-kDa proteins were enriched both outside and inside the triradiate spicules; SM30 did, however, become clearly visible inside the spicule when the 'spicule blot' procedure was used (see also Ingersoll et al. 2003). The SM50 and 29-kDa proteins were concentrated in radially growing portions of the spicules (perpendicular to the calcite *c*-axis), while SM30 protein was located in the longitudinally growing portions (parallel to the *c*-axis).

Further immunohistochemical studies on sea urchin embryo spicules and micromere cultures have shown that SM50, SM30 and PM27 proteins can be detected in the matrix surrounding the spicules, that the immunoreaction is completely eliminated if this matrix is removed by treatment with NaOH, and that the NaOH-treated spicules again show a positive immunoreaction

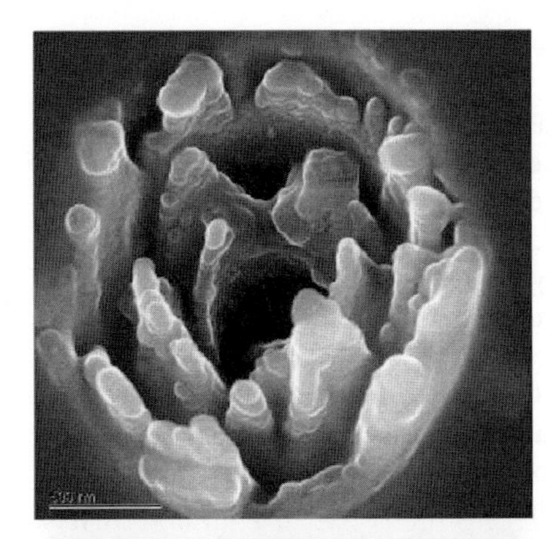

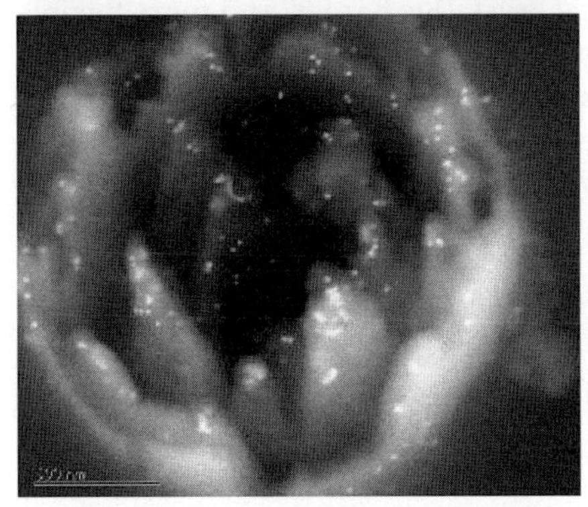

Fig. 16.7. Etched surface of fractured sea urchin spicule. *Above*: conventional SEM. *Below*: back scattered SEM image of the same surface after immunoreaction using antibody against the SM30 matrix protein; note the many immunogold particles (from Seto et al. (2004) J Struct Biol 148:123-130, courtesy of Fred Wilt, University of California, Berkeley, CA; with permission of Elsevier)

if etched by treatment with methanolic acetic acid to expose their internal parts (Urry et al. 2000). In this case, the most intense reaction was provided by SM30 antibodies. In discussing the results of their study on the expression of spicule matrix proteins in sea urchin, Urry et al. (2000) concluded that SM50, PM27 and, to a lesser extent, SM30 proteins are located in the extracellular space between the mesenchyme cell syncytium and the spicule surface, and that all three proteins are occluded in the inorganic phase within the spicules.

In sea urchin teeth, as in spicules, organic macromolecules are associated with the inorganic phase (Weiner 1985). They are acidic, contain phosphoserine and are rich in aspartic acid (Veis et al. 1986); some protein fractions show a relationship with vertebrate matrix proteins, as shown by immunoblots using antibodies to vertebrate tooth proteins (Veis et al. 2002). An immunohistochemical study carried out on the skeletal plates and teeth of the adult sea urchin showed that SM50 and SM30 antigens were confined to the Golgi apparatus of the preodontoblasts and the organic matrix of the calcification sites of both teeth and test plates (Ameye et al. 1999). SM30 mRNA had not been previously detected in the dermal spines of the adult skeleton (Kitajima et al. 1996), but this discrepancy may depend on differences in the methods of investigation used.

Intracrystalline organic material has also been reported in sponges and ascidians (Aizenberg et al. 1996; Lambert and Lambert 1996; Lambert 1998), corals (Allemand and Bénazet-Tambutté 1996; Rahman et al. 2005), gorgonians (Kingsley and Watabe 1982), and coccoliths (Westbroek et al. 1973; Marsh et al. 1992). Sulfur-containing proteoglycan material has been shown to be an intrinsic component of calcospherites in the mantle of the fresh-water mussel, *Amblema* (Davis et al. 1982a).

The occlusion of organic material in the crystalline structure has important functional implications (Aizenberg et al. 1995). Intracrystalline proteins can, in fact, regulate crystal growth and shape (Albeck et al. 1993). Moreover, internalized organic material is the factor that makes it clear why the biologically assembled calcite (or any other biological crystal) is harder, and more flexible and resistant, than pure calcite, and why its fractured surface has a conchoidal cleavage, whereas pure calcite fractures according to a linear cleavage (Emlet 1982; Berman et al. 1988). It may also explain how various inorganic materials come to be constructed biologically, even if they could only be produced experimentally under extreme conditions of pressure and temperature.

16.7 Organic-inorganic Relationships in Calcified Non-skeletal Tissues

The organic-inorganic relationships in calcified non-skeletal tissues have received less attention than those in skeletal tissues, and have not been studied at all in some of them, although the calcified areas in these tissues are known to contain organic material (see Chap. 13). This is mainly true of otoliths and otoconia, and of the pineal gland. This chapter will therefore only discuss the organic-inorganic relationships in the avian eggshell and unicellular organisms.

16.7.1 Organic-inorganic Relationships in the Eggshell

Although the organic matrix of the eggshell is well known, and the nature of its proteins and proteoglycans has been largely determined (see Sect. 13.4), the intimate relationship between organic and inorganic components is less well understood. The eggshell mammillae are the sites where the initiation of crystal formation takes place, as shown by the fact that the crystals appear to be nucleated in, and to grow radially from, their organic matrix cores (reviewed by Arias et al. 1993), but knowledge of the relationships between the organic molecules and the inorganic substance does not go beyond the identification of regions where organic matrix and calcite crystals are mixed, or beyond the immunohistochemical demonstration that specific proteins are components of the shell matrix (Arias et al. 1992; Carrino et al. 1996).

Using a PEDS-like procedure (specimen embedded in LR White and grid-mounted sections processed for immunohistochemistry), ovocleidin-116 has been localized in the organic matrix and vesicles of the palisade layer and in the calcium reserve assembly of the mammillary layer (Hincke et al. 1999), and ovocalyxin-32 in the outer palisade layer, the vertical crystal layer, and the cuticle (Gautron et al. 2001). Studies on frozen sections of partially or totally calcified eggshell specimens have shown that type X collagen is secreted during the first phase of shell membrane formation, keratan sulfate is detectable as soon as the first mammillae are formed, and dermatan sulfate is produced later in concomitance with shell matrix deposition (Fernandez et al. 1997). The same results were obtained by decalcifying the specimens after they had been embedded in a gel of 2% agar in water (Fernandez et al. 2001), a method which protects specimens from shear forces during processing, but is unable to completely avoid the solubilization of organic structures during decalcification. Using the same method, OPN was localized immunohistochemically in the uncalcified core of the shell membrane fibers, at the base of the mammillae, and in the outermost zone of the palisade (Fernandez et al. 2003).

The mantle morphology of the fibrils in the outer shell membrane changes with calcification and with the growth of $CaCO_3$ crystals. It shows stippled gaps, which are interpreted as being due to the growth of crystals that have become embedded in the mantle; after decalcification, this has a serrated appearance without uncalcified areas and is thought to be due to the appearance of gaps in sites formerly occupied by crystals (Dennis et al. 1996). The serrated shape might, in any case, be attributable to the presence of crystal ghost-like structures.

Findings in the decalcified cuticle layer are similar. After decalcification, this layer appears to consist of spherical vesicles with a characteristic ultrastructure: they are surrounded by an electron-dense border from which needle-, crystal ghost-like structures project inward yielding a picture which is the reverse image of the roundish aggregates of needle-shaped, hydroxyapatite crystals that are found in this layer (compare Figs. 10A and 11 of Dennis et al. 1996).

16.7.2 Organic-inorganic Relationships in Unicellular Organisms

Few data are available on the organic-inorganic relationship in unicellular organisms, but it is known that the organic and inorganic substance are closely associated during the development and growth of calcified structures. Important information has, however, been acquired from studies on magnetosomes. As discussed in Sect. 13.5.1, these iron-containing cellular particles are surrounded by electron-dense organic material which reacts strongly with cationic stains (Meldrum et al. 1993). Under the electron microscope, this material appears as a semicrystalline structure with lattice fringes aligned with the lattice planes of magnetosomes, across which a variable contrast can be detected: lattice fringes can be clearly seen where an inorganic, unidentified material, known as 'pre-magnetite' – possibly an iron hydroxide – is present, but as magnetite develops they show a continuum of lower contrasts ending in complete extinction (Taylor and Barry 2004). One hypothesis that has been put forward is that the matrix is a semicrystalline gel, possibly a polysaccharide, acting as a template for the spatially controlled formation of the unidentified material, which over time is transformed into magnetite (Taylor and Barry 2004).

Magnetosomes share some features with aurosomes, that is, gold-containing lysosomes which are generated in the cells of the synovial membrane (Lalonde and Ghadially 1981) and in the chondrocytes of articular cartilage (Oryschak and Ghadially 1974) by intra-articular injection of sodium chloroaurate or aurothiomalate: aurosomes contain sheet- and rod-like organic structures studded with gold particles, which on the whole seem to replicate the organic-inorganic relationships found in calcification nodules. Some similarities also exist with siderosomes, which contain hemosiderin iron and organic material consisting of proteins, lipids, phospholipids and other molecules (Ghadially et al. 1976), and with lead-containing intranuclear bodies which develop in kidney cells and other tissues (Bonucci and Silvestrini 1988) in cases of saturnism (Beaver 1961; Angevine et al. 1962; Hsu et al. 1973; van Mullem and Stadhouders 1974; Bonucci et al. 1983). These inclusions consist of acidic proteins and lead (Richter et al. 1968; Goyer et al. 1970; Moore et al. 1973; Choie et al. 1975; Shelton and Egle 1982); under the electron microscope they show a dense core from which thin filaments project outwards (Beaver 1961; Richter et al. 1968; Goyer et al. 1970). As in aurosomes, these filaments, which consist of a lead-protein complex (Richter et al. 1968; Moore et al. 1973), seem to replicate the organic-inorganic relationships in calcification nodules of bone and cartilage.

Coccolith crystals are enveloped by a narrow organic coat (reviewed by Young et al. 1999) which contains soluble acid polysaccharides (van der Wal et al. 1983). The distribution of a polysaccharide isolated from the coccoliths of *Emiliana huxleyi* has been studied immunohistochemically on coccolith sections that had been etched by flotation on distilled water, pH 5.5: immunogold particles were found on decalcified areas that had the same shape as coccoliths – and were therefore called coccolith ghosts (Van Emburg et al. 1986). An intra-crystalline, soluble, protein-free polysaccharide containing about 26% uronic acid, some methylpentose and no amino-sugars has been isolated from the coccoliths of *Coccolithus huxleyi* (Westbroek et al. 1973). The template production of long, curved, pseudo-single crystals of akaganeite (β-FeOOH) by microbial polysaccharides has been reported by Chan et al. (2004).

16.8 Organic-inorganic Relationships in Pathologically Calcified Tissues

Pathological calcification may involve such a wide range of organic structures that their relationship with inorganic substance can hardly be reviewed in detail. Despite these complications, mitochondria should probably be considered the site where ectopic calcification begins in most cases and the site where the organic-inorganic relationships can best be studied.

As reported in Sects. 14.3.1 and 14.3.2, both intrinsically electron-dense, inorganic granules and needle-shaped crystals can be detected in calcified mitochondria; both these types of structure are closely related to the organic substance. This has clearly been shown in the calcified liver mitochondria of CCl_4-intoxicated rats, using the PEDS method and electron microscopy: intramitochondrial inorganic granules are dissolved by decalcification and are no longer recognizable in unstained sections, but an amorphous substrate, which has the same shape as these granules, is found to have replaced them after staining with uranium and lead (Fig. 14.1); similarly, needle- and filament-like structures are recognizable where the now decalcified crystals had been located (Bonucci et al. 1973). Both these substrates are therefore identifiable as crystal ghosts; both can be stained using acidic phosphotungstic acid or periodic acid-silver nitrate-methenamine, which stain glycoproteins (Bonucci et al. 1973). These results are in accordance with the findings that intramitochondrial granules isolated from Ca^{2+}-loaded mitochondria contain significant amounts of organic material (Weinbach and von Brand 1965), and with the observation that the microincineration melting point of intact granules (500–600 °C) is lower than that (600–700 °C) of their mineral residues (Thomas and Greenawalt 1968).

Crystal ghosts have been demonstrated in various pathological calcifications. Sodium fluoride administered to rats induced in their ameloblasts the formation of intracellular calcifications characterized by the formation of crystal aggregates within vacuoles: after treatment based on the PEDS method, they were replaced by aggregates of organic filamentous material, looking like crystal ghosts (Kuhar and Eisenmann 1978). Crystal ghosts have also been found within calcified Bruch's membrane of the eye, after it has been decalcified with formic acid and stained with acidic phosphotungstic acid (Davis et al. 1981).

Proteins are selectively included in calcium oxalate crystals (Doyle et al. 1991), and the organic matrix of oxalate stones accounts for 2–3% of the total dry weight (reviewed by Khan 1992). As already discussed (see Sect. 14.2), it is partly occluded within the crystals, which makes it comparable with crystal ghosts (Doyle et al. 1991; Ryall et al. 2000, 2001; Khan et al. 2002; Walton et al. 2003). These have been detected in microliths experimentally induced in rat proximal and distal kidney tubules (Boevé et al. 1993) and in human calcium oxalate urinary stones (Khan and Hackett 1984, 1993), where they may appear either singly, or as agglomerates of elongated structures approximately 0.5–

1 μm in length and 20–50 nm in thickness (Khan and Hackett 1987). Crystal ghosts consisting of amorphous, malachite green-positive material have been found in calcium oxalate and calcium phosphate urinary crystals and stones (Khan et al. 1996). Osteopontin and osteocalcin have been reported to be components of crystal ghosts found in the nuclei, lamellae, and striations of the organic matrix of ethylene glycol-induced calcium oxalate stones (McKee et al. 1995).

16.9 Concluding Remarks

- Although the components of calcified tissues change with the nature of tissues and calcified matrices, and with types of crystals in them, a close organic-inorganic relationship is recognizable in all of them, especially during the early stage of calcification. This is clearly shown by X-ray diffraction and, above all, electron microscopy.
- There is a direct relationship between the periodic banding of the collagen fibrils and the granular inorganic bands in incompletely calcified bone, dentin, cementum, and calcified tendons.
- The granular inorganic bands are related to, and form within, the hole zones of the collagen fibrils; they are mainly found in bone with compact collagen fibrils (lamellar bone).
- Filament-like crystals form outside collagen fibrils; they are found between and on the surface of the fibrils. These crystals predominate in bone with loose and irregular collagen fibrils (embryonic bone, woven bone).
- Filament-like crystals form independently of the inorganic substance in bands. With increasing degree of matrix calcification, they increase in numbers and mask the inorganic substance in bands.
- Calcification nodules and calcification islands are distinctive features of the calcification front in bone.
- Calcification nodules develop from matrix vesicles; they are roundish crystal aggregates consisting of needle- and filament-like crystals which radiate from a center into the adjacent matrix. They are the chief feature of the calcification front in woven bone.
- Calcification islands develop between collagen fibrils; they consist of needle- and filament-like crystals, but, as calcification goes forward, granular inorganic bands become visible in them. They are the chief feature of the calcification front in woven bone.
- Taking osteon calcification (which occurs in two phases; see Sects. 2.2, page 10, and 5.2.1, page 110) as paradigmatic of the calcification of compact bone, it seems correct to state that, during its first, rapid phase, calcifi-

cation starts in interfibrillary spaces (with the formation of calcification nodules and islands) and only slightly later goes forward in intrafibrillar spaces (i. e., the collagen holes, with the formation of granular substance in bands). During the second, slow, completion phase, filament-like crystals grow in number in interfibrillar spaces until they almost completely mask the granular substance.

- If calcification nodules are decalcified using the PEDS method, or in the presence of cationic substances, organic structures, that had previously been masked by the inorganic material, become visible. These have the same ultrastructure and location as the now solubilized crystals and for this reason have been called crystal ghosts. After the PEDS method, calcification islands appear as electron-dense patches consisting of filamentous structures resembling crystal ghosts.
- In epiphyseal cartilage, early crystals develop in matrix vesicles and form calcification nodules like those of bone. Their coalescence leads to the calcification of the whole matrix. Crystals show no direct relationship with collagen fibrils; after decalcification and staining (PEDS), they are replaced by crystal ghosts.
- The early demonstrations of crystal ghost-like organic structures were made in enamel. The name crystal ghost could well be extended to all the organic structures which, after decalcification and staining, show the same morphology as, and replace, the untreated calcified structures, even if these have no crystal morphology (e. g., calcified mitochondrial granules).
- In bone, as well as in cartilage, crystals ghosts contain acidic groups belonging to sulfate glycosaminoglycans.
- The presence of bone sialoprotein, osteopontin and other specific proteins in calcification nodules and calcification islands provides evidence that crystal ghosts are in some way related to non-collagenous proteins.
- Several ultrastructural and immunohistochemical findings suggest that in enamel crystal ghosts correspond to enamelin.
- Although crystal ghosts have a solid structure, it has been suggested that they may outline compartments.
- In fish teeth and scales, early crystals develop between and on the fibril surface of collagen fibrils.
- In invertebrate calcified structures, an organic matrix acts as a “mortar” between polygonal crystals or crystal lamellae. In addition, organic matrix is occluded within the crystals and may change their biophysical properties.
- In mollusk shells, the intraprismatic organic material consists of nanofilaments, which are striking fingerprints of the intraprismatic crystals and may therefore be considered crystal ghosts.

- As in mollusk shells, the organic matrix occluded in the calcified spines of sea urchin embryos appears under the electron microscope as a fingerprint of the polycrystalline structure.
- Examples of organic structures closely connected with, and internalized within, the inorganic substance – in this respect comparable to crystal ghosts – can be found in magnetosomes, siderosomes, aurosomes, the inclusion bodies due to lead intoxication, coccoliths, the intramitochondrial calcified granules, several pathologically calcified areas in cells and tissues, and the matrix of calcium oxalate stones.
- Intracrystalline organic material can be found in all calcifying tissues during the early stages of calcification; it becomes less recognizable as the degree of calcification increases.

References

Aaron JE, Oliver B, Clarke N, Carter DH (1999) Calcified microspheres as biological entities and their isolation from bone. Histochem J 31:455–470

Addadi L, Weiner S (1985) Interactions between acidic proteins and crystals: stereochemical requirements in biomineralization. Proc Natl Acad Sci USA 82:4110–4114

Aizenberg J, Hanson J, Ilan M, Leiserowitz L, Koetzle TF, Addadi L, Weiner S (1995) Morphogenesis of calcitic sponge spicules: a role for specialized proteins interacting with growing crystals. FASEB J 9:262–268

Aizenberg J, Lambert G, Addadi L, Weiner S (1996) Stabilization of amorphous calcium carbonate by specialized macromolecules in biological and synthetic precipitates. Adv Mater 8:222–226

Albeck S, Aizenberg J, Addadi L, Weiner S (1993) Interactions of various skeletal intracrystalline components with calcite crystals. J Am Chem Soc 115:11691–11697

Albeck S, Addadi L, Weiner S (1996) Regulation of calcite crystal morphology by intracrystalline acidic proteins and glycoproteins. Connect Tissue Res 35:365–370

Allemand D, Bénazet-Tambutté S (1996) Dynamics of calcification in the Mediterranean red coral, *Corallium rubrum* (Linnaeus) (Cnidaria, Octocorallia). J Exp Zool 276:270–278

Ameye L, Compère P, Dille J, Dubois P (1998) Ultrastructure and cytochemistry of the early calcification site and of its mineralization organic matrix in *Paracentrotus lividus* (Echinodermata: Echinoidea). Histochem Cell Biol 110:285–294

Ameye L, Hermann R, Killian C, Wilt F, Dubois P (1999) Ultrastructural localization of proteins involved in sea urchin biomineralization. J Histochem Cytochem 47:1189–1200

Amprino R, Engström A (1952) Studies on X ray absorption and diffraction of bone tissue. Acta Anat 15:1–22

Angevine JM, Kappas A, DeGowin RL, Spargo BH (1962) Renal tubular nuclear inclusions of lead poisoning. Arch Path 73:486–494

Appleton J (1971) Ultrastructural observations on the inorganic/organic relationships in early cartilage calcification. Calcif Tissue Res 7:307–317

Appleton J, Blackwood HJJ (1969) Ultrastructural observations on early mineralization in cartilage. J Bone Joint Surg 51-B:385

Arias JL, Carrino DA, Fernandez MS, Rodriguez JP, Dennis JE, Caplan AI (1992) Partial biochemical and immunochemical characterization of avian eggshell extracellular matrices. Arch Biochem Biophys 298:293–302

Arias JL, Fink DJ, Xiao S-Q, Heuer AH, Caplan AI (1993) Biomineralization and eggshells: cell-mediated acellular compartments of mineralized extracellular matrix. Int Rev Cytol 145:217–250
Arsenault AL (1988) Crystal-collagen relationships in calcified turkey leg tendons visualized by selected-area dark field electron microscopy. Calcif Tissue Int 43:202–212
Arsenault AL (1989) A comparative electron microscopic study of apatite crystals in collagen fibrils of rat bone, dentin and calcified turkey leg tendons. Bone Miner 6:165–177
Arsenault AL (1990) The ultrastructure of calcified tissues: methods and technical problems. In: Bonucci E, Motta PM (eds) Ultrastructure of skeletal tissues. Kluwer Academic Publishers, Boston, pp 1–18
Arsenault AL (1991) Image analysis of collagen-associated mineral distribution in cryogenically prepared turkey leg tendons. Calcif Tissue Int 48:56–62
Arsenault AL, Ottensmeyer FP, Heath IB (1988) An electron microscopic and spectroscopic study of murine epiphyseal cartilage: analysis of fine structure and matrix vesicles preserved by slam freezing and freeze substitution. J Ultrastruct Mol Struct Res 98:32–47
Ascenzi A, Benedetti EL (1959) An electron microscopic study of the foetal membranous ossification. Acta Anat 37:370–385
Ascenzi A, Bonucci E (1964) A quantitative investigation of the birefringence of the osteon. Acta Anat 44:236–262
Ascenzi A, François C, Steve Bocciarelli D (1963) On the bone induced by estrogen in birds. J Ultrastruct Res 8:491–505
Ascenzi A, Bonucci E, Steve Bocciarelli D (1965) An electron microscope study of osteon calcification. J Ultrastruct Res 12:287–303
Ascenzi A, Bonucci E, Ripamonti A. Roveri N (1978) X-ray diffraction and electron microscope study of osteons during calcification. Calcif Tissue Res 25:133–143
Ayukawa Y, Takeshita F, Inoue T, Yoshinari M, Shimono M, Suetsugu T, Tanaka T (1998) An immunoelectron microscopic localization of noncollagenous bone proteins (osteocalcin and osteopontin) at the bone-titanium interface of rat tibiae. J Biomed Mater Res 41:111–119
Bai P, Warshawsky H (1985) Morphological studies on the distribution of enamel matrix proteins using routine electron microscopy and freeze-fracture replicas in the rat incisor. Anat Rec 212:1–16
Beaver DL (1961) The ultrastructure of the kidney in lead intoxication with particular reference to intranuclear inclusions. Am J Pathol 39:195–208
Benson NC, Benson SC, Wilt F (1989) Immunogold detection of glycoprotein antigens in sea urchin embryos. Am J Anat 185:177–182
Benson SC, Wilt FH (1992) Calcification of spicules in the sea urchin embryo. In: Bonucci E (ed) Calcification in biological systems. CRC Press, Boca Raton, pp 157–178
Benson S, Jones EM, Crise-Benson N, Wilt F (1983) Morphology of the organic matrix of the spicule of the sea urchin larva. Exp Cell Res 148:249–253
Benson SC, Benson NC, Wilt F (1986) The organic matrix of the skeletal spicule of sea urchin embryos. J Cell Biol 102:1878–1886
Berman A, Addadi L, Weiner S (1988) Interactions of sea-urchin skeleton macromolecules with growing calcite crystals - a study of intracrystalline proteins. Nature 331:546–548
Berman A, Addadi L, Kvick Å, Leiserowitz L, Nelson M, Weiner S (1990) Intercalation of sea urchin proteins in calcite: study of a crystalline composite material. Science 250:664–667
Bernard GW, Pease DC (1969) An electron microscopic study of initial intramembranous osteogenesis. Am J Anat 125:271–290

Bianco P, Riminucci M, Silvestrini G, Bonucci E, Termine JD, Fisher LW, Gehron-Robey P (1993) Localization of bone sialoprotein (BSP) to Golgi and post-Golgi secretory structures in osteoblasts and to discrete sites in early bone matrix. J Histochem Cytochem 41:193–203

Bigi A, Ripamonti A, Koch MHJ, Roveri N (1988) Calcified turkey leg tendon as structural model for bone mineralization. Int J Biol Macromol 10:282–286

Bigi A, Foresti E, Gregorini R, Ripamonti A, Roveri N, Shah JS (1992) The role of magnesium on the structure of biological apatites. Calcif Tissue Int 50:439–444

Bigi A, Gandolfi M, Koch MHJ, Roveri N (1996) X-ray diffraction study of *in vitro* calcification of tendon collagen. Biomaterials 17:1195–1201

Bigi A, Gandolfi M, Roveri N, Valdré G (1997) *In vitro* calcified tendon collagen: an atomic force and scanning electron microscopy investigation. Biomaterials 18:657–665

Bishop MA, Warshawsky H (1982) Electron microscopic studies on the potential loss of crystallites from routinely processed sections of young enamel in the rat incisor. Anat Record 202:177–186

Blumen G, Merzel J (1972) The decrease in the concentration of organic material in the course of formation of the enamel matrix. Experientia 28:545–548

Boevé ER, Ketelaars GA, Vermeij M, Cao LC, Schröder FH, de Bruijn WC (1993) An ultrastructural study of experimentally induced microliths in rat proximal and distal tubules. J Urol 149:893–899

Bonucci E (1967) Fine structure of early cartilage calcification. J Ultrastruct Res 20:33–50

Bonucci E (1969) Further investigation on the organic/inorganic relationships in calcifying cartilage. Calcif Tissue Res 3:38–54

Bonucci E (1971) The locus of initial calcification in cartilage and bone. Clin Orthop Relat Res 78:108–139

Bonucci E (1984) The structural basis of calcification. In: Ruggeri A, Motta PM (eds.) Ultrastructure of the connective tissue matrix. Martinus Nijhoff Publishing, Boston, pp 165–191

Bonucci E (1992) Role of collagen fibrils in calcification. In: Bonucci E (ed) Calcification in biological systems. CRC Press, Boca Raton, pp 19–39

Bonucci E (1995) Ultrastructural organic-inorganic relationships in calcified tissues: cartilage and bone vs. enamel. Connect Tissue Res 33:157–162

Bonucci E (2002) Crystal ghosts and biological mineralization: fancy spectres in an old castle, or neglected structures worthy of belief? J Bone Miner Metab 20:249–265

Bonucci E, Gherardi G (1975) Histochemical and electron microscope investigations on medullary bone. Cell Tissue Res 163:81–97

Bonucci E, Reurink J (1978) The fine structure of decalcified cartilage and bone: a comparison between decalcification procedures performed before and after embedding. Calcif Tissue Res 25:179–190

Bonucci E, Silvestrini G (1988) Ultrastructural studies in experimental lead intoxication. Contrib Nephrol 64:93–101

Bonucci E, Silvestrini G (1992) Immunohistochemical investigation on the presence of chondroitin sulfate in calcification nodules of epiphyseal cartilage. Eur J Histochem 36:407–422

Bonucci E, Silvestrini G (1994) Morphological investigation of epiphyseal cartilage after glutaraldehyde-malachite green fixation. Bone 15:153–160

Bonucci E, Silvestrini G (1996) Ultrastructure of the organic matrix of embryonic avian bone after en bloc reaction with various electron-dense 'stains'. Acta Anat 156:22–33

Bonucci E, Derenzini M, Marinozzi V (1973) The organic-inorganic relationship in calcified mitochondria. J Cell Biol 59:185–211

Bonucci E, Barckhaus RH, Silvestrini G, Ballanti P, Di Lorenzo G (1983) Osteoclast changes induced by lead poisoning (saturnism). Appl Pathol 1:241–250

Bonucci E, Bianco P, Hayashi Y, Termine JD (1986) Ultrastructural immunohistochemical localization of non-collagenous proteins in bone, cartilage and developing enamel. In: Ali SY (ed) Cell mediated calcification and matrix vesicles. Excerpta Medica, Amsterdam, pp 33–38

Bonucci E, Silvestrini G, Di Grezia R (1988) The ultrastructure of the organic phase associated with the inorganic substance in calcified tissues. Clin Orthop Relat Res 233:243–261

Bonucci E, Silvestrini G, Di Grezia R (1989) Histochemical properties of the “crystal ghosts” of calcifying epiphyseal cartilage. Connect Tissue Res 22:43–50

Bonucci E, Lozupone E, Silvestrini G, Favia A, Mocetti P (1994) Morphological studies of hypomineralized enamel of rat pups on calcium-deficient diet, and of its changes after return to normal diet. Anat Rec 239:379–395

Bonucci E, Silvestrini G, Mocetti P (1997) MC22–33F monoclonal antibody shows unmasked polar head groups of choline-containing phospholipids in cartilage and bone. Eur J Histochem 41:177–190

Bosshardt DD, Zalzal S, McKee MD, Nanci A (1998) Developmental appearance and distribution of bone sialoprotein and osteopontin in human and rat cementum. Anat Rec 250:1–21

Bronckers ALJJ, Lyaruu DM, Bervoets TJM, Wöltgens JHM (1988) Autoradiographic, ultrastructural and biosynthetic study of the effect of colchicine on enamel matrix secretion and enamel mineralization in hamster tooth germs *in vitro*. Archs Oral Biol 33:7–16

Butler WT (1984b) Dentin collagen: chemical structure and role in mineralization. In: Linde A (ed) Dentin and dentinogenesis, 2nd vol. CRC Press, Boca Raton, pp 37–54

Butler WT (1984a) Matrix macromolecules of bone and dentin. Collagen Rel Res 4:297–307

Butler WT, Brunn JC, Qin C (2003) Dentin extracellular matrix (ECM) proteins: comparison to bone ECM and contribution to dynamics of dentinogenesis. Connect Tissue Res 44:171–178

Caglioti V (1935) Sulla struttura delle ossa. Atti V Congr. Naz. Chimica Pura Applicata, Rome, Associazione Italiana di Chimica, pp 320–331

Caglioti V, Ascenzi A, Scrocco M (1954) Infrared spectrometric research on the relation between ossein and inorganic bone fraction. Experientia 10:371

Cameron DA (1963) The fine structure of bone and calcified cartilage. A critical review of the contribution of electron microscopy to the understanding of osteogenesis. Int Rev Cytol 11:283–306

Campo RD, Phillips J (1973) Electron microscopic visualization of proteoglycans and collagen in bovine costal cartilage. Calcif Tissue Res 13:83–92

Carlström D, Finean JB (1954) X-ray diffraction studies on the ultrastructure of bone. Biochim Biophys Acta 13:183–191

Caron C, Xue J, Sun X, Simmer JP, Bartlett JD (2001) Gelatinase A (MMP-2) in devoping tooth tissues and amelogenin hydrolysis. J Dent Res 80:1660–1664

Carrino DA, Dennis JE, Wu T-M, Arias JL, Fernandez MS, Rodriguez JP, Fink DJ, Heuer AH, Caplan AI (1996) The avian eggshell extracellular matrix as a model for biomineralization. Connect Tissue Res 35:325–329

Chan CS, De Stasio G, Welch SA, Girasole M, Frazer BH, Nesterova MV, Fakra S, Banfield JF (2004) Microbial polysaccharides template assembly of nanocrystal fibers. Science 303:1656–1658

Chardin H, Septier D, Goldberg M (1990a) Visualization of glycosaminoglycans in rat incisor predentin and dentin with cetylpyridinium chloride-glutaraldehyde as fixative. J Histochem Cytochem 38:885–894

Chardin H, Londono I, Goldberg M (1990b) Visualization of glycosaminoglycans in rat incisor extracellular matrix using a hyaluronidase-gold complex. Histochem J 22:588–594

Chen J, McKee MD, Nanci A, Sodek J (1994) Bone sialoprotein mRNA expression and ultrastructural localization in fetal porcine calvarial bone: comparisons with osteopontin. Histochem J 26:67–78

Cho JW, Partin JS, Lennarz WJ (1996) A technique for detecting matrix proteins in the crystalline spicule of the sea urchin embryo. Proc Natl Acad Sci USA 93:1282–1286

Choi CS, Kim YW (2000) A study of the correlation between organic matrices and nanocomposite materials in oyster shell formation. Biomaterials 21:213–222

Choie DD, Richter GW, Young LB (1975) Biogenesis of intranuclear lead-protein inclusions in mouse kidney. Beitr Path 155:197–203

Crenshaw MA (1972) The soluble matrix from *Mercenaria mercenaria* shell. Biomineralization 6:6–11

Crenshaw MA, Ristedt H (1976) The histochemical localization of reactive groups in septal nacre from *Nautilus pompilius* L. In: Watabe N, Wilbur KM (eds) The mechanisms of mineralization in the invertebrates and plants. University of South Carolina Press, Columbia, SC, pp 355–367

Dauphin Y (2002) Comparison of the soluble matrices of the calcitic prismatic layer of *Pinna nobilis* (Mollusca, Bivalvia, Pteriomorpha). Comp Biochem Physiol A Mol Integr Physiol 132:577–590

Davis WL, Jones RG, Hagler HK (1981) An electron microscopic histochemical and analytical X-ray microprobe study of calcification in Bruch's membrane from human eyes. J Histochem Cytochem 29:601–608

Davis WL, Jone RG, Knight JP, Hagler HK (1982a) An electron microscopic histochemical and X-ray microprobe study of spherites in a mussel. Tissue Cell 14:61–67

Davis WL, Jones RG, Knight JP, Hagler HK (1982b) Cartilage calcification: an ultrastructural, histochemical, and analytical X-ray microprobe study of the zone of calcification in normal avian epiphyseal growth plate. J Histochem Cytochem 30:221–234

Decker G, Lennarz WJ (1988) Skeletogenesis in the sea urchin embryo. Development 103:231–247

Decker GL, Morrill JB, Lennarz WJ (1987) Characterization of sea urchin primary mesenchyme cells and spicules during biomineralization in vitro. Development 101:297–312

Decker JD (1966) An electron microscopic investigation of osteogenesis in the embryonic chick. Am J Anat 118:591–614

Decker JD (1973) Fixation effects on the fine structure of enamel crystal-matrix relationships. J Ultrastruct Res 44:58–74

Dennis JE, Xiao S-Q, Agarwal M, Fink DJ, Heuer AH, Caplan AI (1996) Microstructure of matrix and mineral components of eggshells from white leghorn chickens (*Gallus gallus*). J Morphol 228:287–306

Diekwisch TGH, Berman BJ, Gentner S, Slavkin HC (1995) Initial enamel crystals are not spatially associated with mineralized dentine. Cell Tissue Res 279:149–167

Doyle IR, Ryall RL, Marshall VR (1991) Inclusion of proteins into calcium oxalate crystals precipitated from human urine: a highly selective phenomenon. Clin Chem 37:1589–1594

Ecarot-Charrier B, Shepard N, Charette G, Grynpas M, Glorieux FH (1988) Mineralization in osteoblast cultures: a light and electron microscopic study. Bone 9:147–154

Emlet RB (1982) Echinoderm calcite: a mechanical analysis from larval spicules. Biol Bull 163:264–275

Fernandez MS, Araya M, Arias JL (1997) Eggshells are shaped by a precise spatio-temporal arrangement of sequentially deposited macromolecules. Matrix Biol 16:13–20

Fernandez MS, Moya A, Lopez L, Arias JL (2001) Secretion pattern, ultrastructural localization and function of extracellular matrix molecules involved in eggshell formation. Matrix Biol 19:793–803

Fernandez MS, Escobar C, Lavelin I, Pines M, Arias JL (2003) Localization of osteopontin in oviduct tissue and eggshell during different stages of the avian egg laying cycle. J Struct Biol 143:171–180

Fernández-Morán H, Engström A (1957) Electron microscopy and X-ray diffraction of bone. Biochim Biophys Acta 23:260–264

Finean JB, Engström A (1953) The low-angle scatter of X-rays from bone tissue. Biochim Biophys Acta 11:178–189

Frank RM (1979) Tooth enamel: current state of the art. J Dent Res 58:684–693

Fukae M, Tanabe T (1998) Degradation of enamel matrix proteins in porcine secretory enamel. Connect Tissue Res 39:123–129

Garant PR (1970) An electron microscopic study of the crystal-matrix relationship in the teeth of the dogfish *Squalus acanthias* L. J Ultrastruct Res 30:441–449

Gautron J, Hincke MT, Mann K, Panhéleux M, Bain M, McKee MD, Solomon SE, Nys Y (2001) Ovocalyxin-32, a novel chicken eggshell matrix protein. Isolation, amino acid sequencing, cloning, and immunocytochemical localization. J Biol Chem 276:39243–39252

Gerstenfeld LC, Gotoh Y, McKee MD, Nanci A, Landis WJ, Glimcher MJ (1990) Expression and ultrastructural immunolocalization of a major 66 kDa phosphoprotein synthesized by chicken osteoblasts during mineralization in vitro. Anat Rec 228:93–103

Ghadially FN, Lalonde J-MA, Oryschak AF (1976) Electron probe X-ray analysis of siderosomes in the rabbit haemarthrotic synovial membrane. Virchows Arch B Cell Path 22:135–142

Glimcher MJ (1959) Molecular biology of mineralized tissues with particular reference to bone. Rev Modern Phys 31:359–393

Glimcher MJ (1976) Composition, structure, and organization of bone and other mineralized tissues and the mechanism of calcification. In: Greep RO, Astwood EB (eds) Handbook of Physiology: Endocrinology. American Physiological Society, Washington, pp 25–116

Glimcher MJ (1979) Phosphopeptides of enamel matrix. J Dent Res 58:790–806

Glimcher MJ, Krane SM (1968) The organization and structure of bone, and the mechanism of calcification. In: Gould BS (ed) Biology of collagen. Academic Press, London, pp 67–251

Glimcher MJ, Brickley-Parsons D, Levine PT (1977) Studies of enamel proteins during maturation. Calcif Tissue Res 24:259–270

Goldberg HA, Hunter GK, Mundy MA, Warner KJ, McKee MD (2000) Nature of hydroxyapatite crystals formed in the presence of bone sialoprotein, osteopontin and synthetic homopolymer analogues. In: Goldberg M, Boskey A, Robinson C (eds) Chemistry and biology of mineralized tissues. American Academy of Orthopaedic Surgeons, Rosemont, IL, pp 225–228

Goldberg M, Noblot MM, Septier D (1980) Effets de deux méthodes de démineralisation sur la préservation des glycoprotéines et des protéoglycanes dans les dentines intercanaliculaires et péricanaliculaires chez le cheval. J Biol Buccale 8:315–330

Goldberg M, Vermelin L, Mostermans P, Lécolle S, Septier D, Godeau G, LeGeros RZ (1998) Fragmentation of the distal portion of Tomes' processes of secretory ameloblasts in the forming enamel of rat incisors. Connect Tissue Res 38:159–169

Gorski JP (1998) Is all bone the same? Distinctive distributions and properties of non-collagenous matrix proteins in lamellar vs. woven bone imply the existence of different underlying osteogenic mechanisms. Crit Rev Oral Biol Med 9:201–223

Gorski JP, Griffin D, Dudley G, Stanford C, Thomas R, Huang C, Lai EL, Karr B, Solursh M (1990) Bone acidic glycoprotein-75 is a major synthetic product of osteoblastic cells and localized as 75- and/or 50-kDa forms in mineralized phases of bone and growth plate and in serum. J Biol Chem 265:14956–14963

Gorski JP, Wang A, Lovitch D, Law D, Powell K, Midura RJ (2004) Extracellular bone acidic glycoprotein-75 defines condensed mesenchyme regions to be mineralized and localizes with bone sialoprotein during intramembranous bone formation. J Biol Chem 279:25455–25463

Goyer RA, May P, Cates MM, Krigman MR (1970) Lead and protein content of isolated intranuclear inclusion bodies from kidneys of lead-poisoned rats. Lab Invest 22:245–251

Grégoire C (1957) Topography of the organic components in mother-of-pearl. J Biophys Biochem Cytol 3:797–806

Groot CG (1982a) An electron microscopic examination for the presence of acid groups in the organic matrix of mineralization nodules in foetal bone. Metab Bone Dis Rel Res 4:77–84

Groot CG (1982b) Acid groups in the organic matrix of foetal bone. An electron microscopical study. Thesis, Rijksuniversiteit te Leiden, pp 1–95

Groot CG, Danes JK, Blok J, Hoogendijk A, Hauschka PV (1986) Light and electron microscopic demonstration of osteocalcin antigenicity in embryonic and adult rat bone. Bone 7:379–385

Hagiwara H (1992) Immunoelectron microscopic study of proteoglycans in rat epiphyseal growth plate cartilage after fixation with ruthenium hexamine trichloride (RHT). Histochemistry 98:305–309

Hancox NM, Boothroyd B (1965) Electron microscopy of the early stages of osteogenesis. Clin Orthop Relat Res 40:153–161

Harkey MA, Klueg K, Sheppard P, Raff RA (1995) Structure, expression, and extracellular targeting of PM27, a skeletal protein associated specifically with growth of the sea urchin larval spicule. Dev Biol 168:549–566

Hascall GK (1980) Cartilage proteoglycans: comparison of sectioned and spread whole molecoles. J Ultrastruct Res 70:369–375

Hayashi Y (1985) Ultrastructural characterization of extracellular matrix vesicles in the mineralizing fronts of apical cementum in cats. Arch Oral Biol 30:445–449

Hayashi Y (1988) Ultrastructural demonstration of the carbohydrate in developing mantle dentine with soybean agglutinin-gold complexes. J Electron Microsc 37:150–154

Hayashi Y (1989) Ultrastructural demonstration of the carbohydrate in developing rat enamel using soybean agglutinin-gold complexes. Arch Oral Biol 34:517–522

Hayashi Y, Bianco P, Shimokawa H, Termine JD, Bonucci E (1986) Organic-inorganic relationships, and immunohistochemical localization of amelogenins and enamelins in developing enamel. Basic Appl Histochem 30:291–299

Hayashi Y, Nagasawa H (1990) Matrix vesicles isolated from apical pulp of rat incisors: crystal formation in low Ca x Pi ion-product medium containing β-glycerophosphate. Calcif Tissue Int 47:365–372

Hayashi Y, Imai M, Goto Y, Murakami N (1993) Pathological mineralization in a serially passaged cell line from rat pulp. J Oral Pathol Med 22:175–179

Herold RC (1971) Osteodentinogenesis. An ultrastructural study of tooth formation in the pike, *Esox lucius*. Z Zellforsch 112:1–14

Heywood BR, Sparks NH, Shellis RP, Weiner S, Mann S (1990) Ultrastructure, morphology and crystal growth of biogenic and synthetic apatites. Connect Tissue Res 25:103–119

Hincke MT, Gautron J, Tsang CPW, McKee MD, Nys Y (1999) Molecular cloning and ultrastructural localization of the core protein of an eggshell matrix proteoglycan, ovocleidin-116. J Biol Chem 274:32915–32923

Höhling HJ, Arnold S, Barckhaus RH, Plate U, Wiesmann HP (1995) Structural relationship between the primary crystal formation and the matrix macromolecules in different hard tissues. Discussion of a general principle. Connect Tissue Int 33:171–178

Höhling HJ, Barckhaus RH, Krefting E-R, Althoff J, Quint P (1990) Collagen mineralization: aspects of the structural relationship between collagen and the apatitic crystallites. In: Bonucci E, Motta PM (eds) Ultrastructure of skeletal tissues. Kluwer Academic Publishers, Boston, pp 41–62

Hoshi K, Ejiri S, Ozawa H (2001a) Organic components of crystal sheaths in bones. J Electron Microsc (Tokyo) 50:33–40

Hoshi K, Ejiri S, Ozawa H (2001b) Localizational alterations of calcium, phosphorus, and calcification-related organics such as proteoglycans and alkaline phosphatase during bone calcification. J Bone Miner Res 16:289–298

Hsu FS, Krook L, Shively JN, Duncan JR, Pond WG (1973) Lead inclusion bodies in osteoclasts. Science 181:447–448

Hunziker EB, Schenk RK (1984) Cartilage ultrastructure after high pressure freezing, freeze substitution, and low temperature embedding. II. Intercellular matrix ultrastructure – preservation of proteoglycans in their native state. J Cell Biol 98:277–282

Hunziker EB, Schenk RK (1987) Structural organization of proteoglycans in cartilage. In: Wight TN, Mecham P (eds) Biology of proteoglycans. Academic Press, Orlando (FL), pp 155–185

Inage T (1975) Electron microscopic study of early formation of the tooth enameloid of a fish (*Hoplognathus fasciatus*). I. Odontoblasts and matrix fibers. Arch Histol Jpn 38:209–227

Inage T, Shimokawa H, Teranishi Y, Iwase T, Toda Y, Moro I (1989) Immunocytochemical demonstration of amelogenins and enamelins secreted by ameloblasts during the secretory and maturation stages. Arch Histol Cytol 52:213–229

Ingersoll EP, McDonald KL, Wilt FH (2003) Ultrastructural localization of spicule matrix proteins in normal and metalloproteinase inhibitor-treated sea urchin primary mesenchymal cells. J Exp Zoolog Part A Comp Exp Biol 300:101–112

Irie K, Zalzal S, Ozawa H, McKee MD, Nanci A (1998) Morphological and immunocytochemical characterization of primary osteogenic cell cultures derived from fetal rat cranial tissue. Anat Rec 252:554–567

Jodaikin A, Traub W, Weiner S (1986) Protein conformation in rat tooth enamel. Arch Oral Biol 31:685–689

Kagami A, Takagi M, Hirama M, Sagami Y, Shimada T (1990) Enhanced ultrastructural preservation of cartilage proteoglycans in the extended state. J Histochem Cytochem 38:901–906

Kallenbach E (1986) Crystal-associated matrix components in rat incisor enamel. An electron-microscopic study. Cell Tissue Res 246:455–461

Kemp NE (1985) Ameloblastic secretion and calcification of the enamel layer in shark teeth. J Morphol 184:215–230

Kemp NE, Park JH (1974) Ultrastructure of the enamel layer in developing teeth of the shark *Carcharhinus menisorrah*. Arch Oral Biol 19:633–644

Khan SR (1992) Structure and development of calcific urinary stones. In: Bonucci E (ed) Calcification in biological systems. CRC Press, Boca Raton, pp 345–363

Khan SR, Hackett RL (1984) Microstructure of decalcified human calcium oxalate urinary stones. Scanning Electron Microsc Pt. 2:935–941

Khan SR, Hackett RL (1987) Crystal-matrix relationships in experimentally induced urinary calcium oxalate monohydrate crystals, an ultrastructural study. Calcif Tissue Int 41:157–163

Khan SR, Hackett RL (1993) Role of organic matrix in urinary stone formation: an ultrastructural study of crystal matrix interface of calcium oxalate monohydrate stones. J Urol 150:239–245

Khan SR, Atmani F, Glenton P, Hou Z-C, Talham DR, Khurshid M (1996) Lipids and membranes in the organic matrix of urinary calcific crystals and stones. Calcif Tissue Int 59:357–365

Khan SR, Glenton PA, Backov R, Talham DR (2002) Presence of lipids in urine, crystals and stones: implications for the formation of kidney stones. Kidney Int 62:2062–2072

Killian CE, Wilt FH (1996) Characterization of the proteins comprising the integral matrix of *Strongylocentrotus purpuratus* embryonic spicules. J Biol Chem 271:9150–9159

Kingsley RJ, Watabe N (1982) Ultrastructural investigation of spicule formation in the gorgonian *Leptogorgia virgulata* (Lamarck) (Coelenterata: Gorgonacea). Cell Tissue Res 223:325–334

Kitajima T, Urakami H (2000) Differential distribution of spicule matrix proteins in the sea urchin embryo skeleton. Dev Growth Differ 42:295–306

Kitajima T, Tomita M, Killian CE, Akasaka K, Wilt FH (1996) Expression of spicule matrix protein gene SM30 in embryonic and adult mineralized tissues of sea urchin *Hemicentrotus pulcherrimus*. Dev Growth Differ 38:687–695

Kobayashi S, Yamada J, Maekawa K, Ouchi K (1972) Calcification and nucleation in fishscales. Biomineralization 6:84–90

Kuhar KJ, Eisenmann DR (1978) Fluoride-induced mineralization within vacuoles in maturative ameloblasts of the rat. Anat Rec 191:91–102

Lalonde J-MA, Ghadially FN (1981) Aurosomes produced by sodium chloroaurate. Pathology 13:29–36

Lambert G (1998) Spicule formation in the solitary ascidian *Bathypera feminalba* (ascidiacea, pyuridae). Invertebrate Biol 117:341–349

Lambert G, Lambert CC (1996) Spicule formation in the New Zealand ascidian *Pyura pachydermatina* (chordata, ascidiacea). Connect Tissue Res 34:263–269

Landis WJ, Géraudie J (1990) Organization and development of the mineral phase during early ontogenesis of the bony fin rays of the trout *Oncorhynchus mykiss*. Anat Rec 228:383–391

Landis WJ, Song MJ (1991) Early mineral deposition in calcifying tendons characterized by high voltage electron microscopy and three-dimensional graphic imaging. J Struct Biol 107:116–127

Landis WJ, Paine MC, Glimcher MJ (1977) Electron microscopic observations of bone tissue prepared anhydrously in organic solvents. J Ultrastruct Res 59:1–30

Landis WJ, Moradian-Oldak J, Weiner S (1991) Topographic imaging of mineral and collagen in the calcifying turkey tendon. Connect Tissue Res 25:181–196

Landis WJ, Hodgens KJ, Arena J, Song MJ, McEwen BF (1996) Structural relations between collagen and mineral in bone as determined by high voltage electron microscopic tomography. Microsc Res Techn 33:192–202

Lanzing WJR, Wright RG (1976) The ultrastructure and calcification of the scales of *Tilapia mossambica* (Peters). Cell Tissue Res 167:37–47

Lee DD, Glimcher MJ (1989) The three-dimensional spatial relationship between the collagen fibrils and the inorganic calcium-phosphate crystals of pickerel and herring fish bone. Connect Tissue Res 21:247–257

Lee ER, Smith CE, Poole AR (1996) Ultrastructural localization of the C-propeptide released from type II procollagen in fetal bovine growth plate cartilage. J Histochem Cytochem 44:433–443

Lee I, Ono Y, Lee A, Omiya K, Moriya Y, Takagi M (1998) Immunocytochemical localization and biochemical characterization of large proteoglycans in developing rat bone. J Oral Sci 40:77–87

Lees S, Prostak KS, Ingle VK, Kjoller K (1994) The loci of mineral in turkey leg tendon as seen by atomic force microscope and electron microscopy. Calcif Tissue Int 55:180–189

Leng C-G, Yu Y, Ueda H, Terada N, Fujii Y, Ohno S (1998) The ultrastructure of anionic sites in rat articular cartilage as revealed by different preparation methods and polyethyleneimine staining. Histochem J 30:253–261

Linde A, Bhown M, Butler WT (1980) Noncollagenous proteins of dentin. A re-examination of proteins from rat incisor dentin utilizing techniques to avoid artifacts. J Biol Chem 255:5931–5942

Maekawa K, Yamada J (1970) Some histochemical and fine structural aspects of growing scales of the rainbow trout. Bull Fac Fish Hokkaido Univ 21:70–78

Marsh ME, Sass RL (1981) Matrix-mineral relationships in the scallop hinge ligament. J Ultrastruct Res 76:57–70

Marsh M, Hamilton G, Sass R (1978) The crystal sheaths from bivalve hinge ligaments. Calcif Tissue Res 25:45–51

Marsh ME, Chang D-K, King GC (1992) Isolation and characterization of a novel acidic polysaccharide containing tartrate and glyoxylate residues from the mineralized scales of a unicellular coccolithophorid alga *Pleurochrysis carterae*. J Biol Chem 267:20507–20512

Martin-De Las Heras S, Valenzuela A, Overall CM (2000) The matrix metalloproteinase gelatinase A in human dentine. Arch Oral Biol 45:757–765

McKee MD, Nanci A (1995) Postembedding colloidal-gold immunocytochemistry of noncollagenous extracellular matrix proteins in mineralized tissues. Microsc Res Techn 31:44–62

McKee MD, Nanci A (1996) Osteopontin at mineralized tissue interfaces in bone, teeth, and osseointegrated implants: ultrastructural distribution, and implications for mineralized tissue formation, turnover, and repair. Microsc Res Techn 33:141–164

McKee MD, Nanci A, Landis WJ, Gerstenfeld LC, Gotoh Y, Glimcher MJ (1989) Ultrastructural immunolocalization of a major phosphoprotein in embryonic chick bone. Connect Tissue Res 21:21–29

McKee MD, Nanci A, Landis WJ, Gotoh Y, Gerstenfeld LC, Glimcher MJ (1990) Developmental appearance and ultrastructural immunolocalization of a major 66kDa phosphoprotein in embryonic and post-natal chicken bone. Anat Rec 228:77–92

McKee MD, Glimcher MJ, Nanci A (1992) High-resolution immunolocalization of osteopontin and osteocalcin in bone and cartilage during endochondral ossification in the chicken tibia. Anat Rec 234:479–492

McKee MD, Farach-Carson MC, Butler WT, Hauschka PV, Nanci A (1993) Ultrastructural immunolocalization of noncollagenous (osteopontin and osteocalcin) and plasma (albumin and α_2HS-glycoprotein) proteins in rat bone. J Bone Miner Res 8:485–496

McKee MD, Nanci A, Khan SR (1995) Ultrastructural immunodetection of osteopontin and osteocalcin as major matrix components of renal calculi. J Bone Miner Res 10:1913–1929

Meenakshi VR, Hare PE, Wilbur KM (1971) Amino acids of the organic matrix of neogastropod shells. Comp Biochem Physiol 40B:1037–1043

Meldrum EC, Mann S, Heywood BR, Frankel RB, Bazylinski DA (1993) Electron microscopy study of magnetosomes in two cultured vibrioid magnetotactic bacteria. Proc R Soc London B 251:237–242

Miake Y, Aoba T, Moreno EC, Shimoda S, Prostak K, Suga S (1991) Ultrastructural studies on crystal growth of enameloid minerals in Elasmobranch and Teleost fish. Calcif Tissue Int 48:204–217

Midura RJ, Wang A, Lovitch D, Law D, Powell K, Gorski JP (2004) Bone acidic glycoprotein-75 delineates the extracellular sites of future bone sialoprotein accumulation and apatite nucleation in osteoblastic cultures. J Biol Chem 279:25464–25473

Mikuni-Takagaki Y, Cheng Y (1987) Metalloproteinases in endochondral bone formation: appearance of tissue inhibitor-resistant metalloproteinases. Arch Biochem Biophys 259:576–588

Miyashita T, Takagi R, Okushima M, Nakano S, Miyamoto H, Nishikawa E, Matsushiro A (2000) Complementary DNA cloning and characterization of Pearlin, a new class of matrix protein in the nacreous layer of oyster pearls. Mar Biotechnol 2:409–418

Molnar Z (1959) Development of the parietal bone of young mice 1. Crystals of bone mineral in frozen-dried preparations. J Ultrastruct Res 3:39–45

Moore JF, Goyer RA, Wilson M (1973) Lead-induced inclusion bodies. Solubility, amino acid content, and relationship to residual acidic nuclear proteins. Lab Invest 29:488–494

Moradian-Oldak J, Gharakhanian N, Jimenez J (2002) Limited proteolysis of amelogenin: toward understanding the proteolytic processes in enamel extracellular matrix. Connect Tissue Res 43:450–455

Mutvei H (1970) Ultrastructure of the mineral and organic components of molluscan nacreous layers. Biomineralization 2:48–72

Myers HM, Engström A (1965) A note on the organization of hydroxyapatite in calcified tendons. Exp Cell Res 40:182–185

Nakamura H, Hirata A, Tsuji T, Yamamoto T (2001) Immunolocalization of keratan sulfate proteoglycan in rat calvaria. Arch Histol Cytol 64:109–118

Nanci A (1999) Content and distribution of noncollagenous matrix proteins in bone and cementum: relationship to speed of formation and collagen packing density. J Struct Biol 126:256–269

Nanci A, Bai P, Warshawsky H (1983) The effect of osmium postfixation and uranyl and lead staining on the ultrastructure of young enamel in the rat incisor. Anat Rec 207:1–16

Nanci A, Ahluwalia JP, Pompura JR, Smith CE (1989) Biosynthesis and secretion of enamel proteins in the rat incisor. Anat Rec 224:277–291

Nanci A, Mocetti P, Sakamoto Y, Kunikata M, Lozupone E, Bonucci E (2000) Morphological and immunocytochemical analyses on the effects of diet-induced hypocalcemia on enamel maturation in the rat incisor. J Histochem Cytochem 48:1043–1057

Nefussi J-R, Septier D, Collin P, Goldberg M, Forest N (1989) A comparative ultrahistochemical study of glycosaminoglycans with cuprolinic blue in bone formed *in vivo* and *in vitro*. Calcif Tissue Int 44:11–19

Nefussi J-R, Septier D, Sautier J-M, Forest N, Goldberg M (1992) Localization of malachite green positive lipids in the matrix of bone nodule formed *in vitro*. Calcif Tissue Int 50:273–282

Nefussi JR, Brami G, Modrowski D, Oboeuf M, Forest N (1997) Sequential expression of bone matrix proteins during rat calvaria osteoblast differentiation and bone nodule formation in vitro. J Histochem Cytochem 45:493–503

Newbrey JW, Banks WJ (1975) Characterization of developing antler cartilage matrix II. An ultrastructural study. Calcif Tissue Res 17:289–302

Nusgens B, Chantraine A, Lapiere ChM (1972) The protein in the matrix of bone. Clin Orthop Relat Res 88:252–274

Nylen MU (1979) Matrix-mineral relationships – a morphologist's viewpoint. J Dent Res 58:922–926

Nylen MU, Omnell K-Å (1962) The relationship between the apatite crystals and the organic matrix of rat enamel. Fifth International Congress for Electron Microscopy, New York, Academic Press, pp QQ-4

Nylen MU, Scott DB, Mosley VM (1960) Mineralization of turkey leg tendon. II. Collagen-mineral relations revealed by electron and X-ray microscopy. In: Sognnaes RF (ed) Calcification in biological systems. American Association for the Advancement of Sciences, Washington, pp 129–142

Olson OP, Watabe N (1980) Studies on formation and resorption of fish scales. IV: Ultrastructure of developing scales in newly hatched fry of the sheepshead minnow, *Cyprinodon variegatus* (Atheriniformes: Cyprinodontidae). Cell Tissue Res 211:303–316

Onozato H, Watabe N (1979) Studies on fish scale formation and resorption III. Fine structure and calcification of the fibrillary plates of the scales in *Carassius auratus* (Cypriniformes: cyprinidae). Cell Tissue Res 201:409–422

Oryschak AF, Ghadially FN (1974) Aurosomes in rabbit articular cartilage. Virchows Arch B Cell Pathol 17:159–168

Ørvig T (1968) The dermal skeleton; general considerations. In: Ørvig T (ed) Current problems of lower vertebrate phylogeny, Nobel Symp, Vol 4. Wiley, New York, pp 373–397

Pokroy B, Fitch AN, Lee PL, Quintana JP, Caspi EN, Zolotoyabko E (2006) Anisotropic lattice distortions in the mollusk-made aragonite: a widespread phenomenon. J Struct Biol 153:145–150

Poole AR (1991) The growth plate: cellular physiology, cartilage assembly and mineralization. In: Hall B, Newman S (eds) Cartilage: molecular aspects. CRC Press, Boca Raton, pp 179–211

Poole AR, Rosenberg LC (1986) Chondrocalcin and the calcification of cartilage. A review. Clin Orthop Relat Res 208:114–118

Poole AR, Pidoux I, Reiner A, Choi H, Rosenberg LC (1984) Association of an extracellular protein (chondrocalcin) with the calcification of cartilage in endochondral bone formation. J Cell Biol 98:54–65

Prostak KS, Lees S (1996) Visualization of crystal-matrix structure. *In situ* demineralization of mineralized turkey leg tendon and bone. Calcif Tissue Int 59:474–479

Prostak KS, Seifert P, Skobe Z (1993) Enameloid formation in two tetraodontiform fish species with high and low fluoride contents in enameloid. Arch Oral Biol 38:1031–1044

Pugliarello MC, Vittur F, de Bernard B, Bonucci E, Ascenzi A (1970) Chemical modifications in osteones during calcification. Calcif Tissue Res 5:108–114

Qin C, Brunn JC, Cook RG, Orkiszewski RS, Malone JP, Veis A, Butler WT (2004) Evidence for the proteolytic processing of dentin matrix protein 1. Identification and characterization of processed fragments and cleavage sites. J Biol Chem 278:34700–34708

Rahman MA, Isa Y, Uehara T (2005) Proteins of calcified endoskeleton: II Partial amino acid sequences of endoskeletal proteins and the characterization of proteinaceous organic matrix of spicules from the alcyonarian, *Synularia polydactyla*. Proteomics 5:885–893

Richter GW, Kress Y, Cornwall CC (1968) Another look at lead inclusion bodies. Am J Pathol 53:189–217

Riminucci M, Silvestrini G, Bonucci E, Fisher LW, Gehron Robey P, Bianco P (1995) The anatomy of bone sialoprotein immunoreactive sites in bone as revealed by combined ultrastructural histochemistry and immunohistochemistry. Calcif Tissue Int 57:277–284

Robinson RA (1952) An electron-microscopic study of the crystalline inorganic component of bone and its relationship to the organic matrix. J Bone Joint Surg 34-A:389–434

Robinson RA, Cameron DA (1956) Electron microscopy of cartilage and bone matrix at the distal epiphyseal line of the femur in the newborn infant. J Biophys Biochem Cytol 2:253–263
Robinson RA, Watson ML (1952) Collagen-crystal relationships in bone as seen in the electron microscope. Anat Rec 114:383–409
Rönnholm E (1962) III. The structure of the organic stroma of human enamel during amelogenesis. J Ultrastruct Res 3:368–389
Ryall RL, Fleming DE, Grover PK, Chauvet M, Dean CJ, Marshall VR (2000) The hole truth: intracrystalline proteins and calcium oxalate kidney stones. Mol Urol 4:391–402
Ryall RL, Fleming DE, Doyle IR, Evans NA, Dean CJ, Marshall VR (2001) Intracrystalline proteins and the hidden ultrastructure of calcium oxalate urinary crystals: implications for kidney stone formation. J Struct Biol 134:5–14
Saleuddin ASM (1971) Fine structure of normal and regenerated shell of *Helix*. Can J Zool 49:37–41
Sasagawa I (2002) Mineralization patterns in elasmobranch fish. Microsc Res Tech 59:396–407
Sasagawa I, Ferguson MW (1990) Fine structure of the organic matrix remaining in the mature cap enameloid in *Halichoeres poecilopterus*, teleost. Arch Oral Biol 35:765–770
Sasagawa I, Ishiyama M (1988) The structure and development of the collar enameloid in two teleost fishes, *Halichoeres poecilopterus* and *Pagrus major*. Anat Embryol (Berl) 178:499–511
Satoyoshi M, Kawata A, Koizumi T, Inoue K, Itohara S, Teranaka T, Mikuni-Takagaki Y (2001) Matrix metalloproteinase-2 in dentin matrix mineralization. J Endod 27:462–466
Sauren YMHF, Mieremet RHP, Groot CG, Scherft JP (1989) An electron microscopical study on the presence of proteoglycans in the calcified bone matrix by use of cuprolinic blue. Bone 10:287–294
Scherft JP (1968) The ultrastructure of the organic matrix of calcified cartilage and bone in embryonic mouse radii. J Ultrastruct Res 23:333–343
Scherft JP (1978) The lamina limitans of the organic bone matrix: formation *in vitro*. J Ultrastruct Res 64:173–181
Scherft JP, Moskalewski S (1984) The amount of proteoglycans in cartilage matrix and the onset of mineralization. Metab Bone Dis Rel Res 5:195–203
Schönbörner AA, Boivin G, Baud CA (1979) The mineralization processes in teleost fish scales. Cell Tissue Res 202:203–212
Scott BL, Pease DC (1956) Electron microscopy of the epiphyseal apparatus. Anat Rec 126:465–495
Serafini-Fracassini A, Smith JW (1974) The structure and biochemistry of cartilage. Churchill Livingstone, Edinburgh
Seto J, Zhang Y, Hamilton P, Wilt F (2004) The localization of occluded matrix proteins in calcareous spicules of sea urchin larvae. J Struct Biol 148:123–130
Seyer JM, Glimcher MJ (1977) Evidence for the presence of numerous protein components in immature bovine dental enamel. Calcif Tissue Res 24:253–257
Shelton KR, Egle PM (1982) The proteins of lead-induced intranuclear inclusion bodies. J Biol Chem 257:11802–11807
Shen X, Belcher AM, Hansma PK, Stucky GD, Morse DE (1997) Molecular cloning and characterization of Lustrin A, a matrix protein from shell and pearl nacre of *Haliotis rufescens*. J Biol Chem 272:32472–32481
Shepard N (1992) Role of proteoglycans in calcification. In: Bonucci E (ed) Calcification in biological systems. CRC Press, Boca Raton, pp 41–58

Shepard N, Mitchell N (1981) Acridine orange stabilization of glycosaminoglycans in beginning endochondral ossification. A comparative light and electron microscopic study. Histochemistry 70:107–114

Sire J-Y, Géraudie J (1984) Fine structure of regenerating scales and their associated cells in the cichlid *Hemichromis bimaculatus* (Gill). Cell Tissue Res 237:537–547

Smales FC (1975) Structural subunit in prisms of immature rat enamel. Nature 258:772–774

Smith JW (1970) The disposition of protein-polysaccharide in the epiphyseal plate cartilage of the young rabbit. J Cell Sci 6:843–864

Stern DN, Song MJ, Landis WJ (1992) Tubule formation and elemental detection in developing opossum enamel. Anat Rec 234:34–48

Su X, Sun K, Cui FZ, Landis WJ (2003) Organization of apatite crystals in human woven bone. Bone 32:150–162

Sundström B, Takuma S (1971) A further contribution on the ultrastructure of calcifying cartilage. J Ultrastruct Res 36:419–424

Takagi M, Parmley RT, Toda Y, Austin RL (1982) Ultrastructural cytochemistry and immunocytochemistry of sulfated glycosamiglycans in epiphyseal cartilage. J Histochem Cytochem 30:1179–1185

Takagi M, Parmley RT, Toda Y, Denys FR (1983) Ultrastructural cytochemistry of complex carbohydrates in osteoblasts, osteoid, and bone matrix. Calcif Tissue Int 35:309–319

Takagi M, Kamiya N, Urushizaki T, Tada Y, Tanaka H (2000) Gene expression and immunohistochemical localization of biglycan in association with mineralization in the matrix of epiphyseal cartilage. Histochem J 32:175–186

Takuma S (1960) Electron microscopy of the developing cartilaginous epiphysis. Arch Oral Biol 2:111–119

Taylor AP, Barry JC (2004) Magnetosomal matrix: ultrafine structure may template biomineralization of magnetosomes. J Microsc 213:180–197

Termine JD, Belcourt AB, Christner PJ, Conn KM, Nylen MU (1980) Properties of dissociatively extracted fetal tooth matrix proteins I. Principal molecular species in developing bovine enamel. J Biol Chem 255:9760–9768

Termine JD, Belcourt AB, Conn KM, Kleinman HK (1981) Mineral and collagen-binding proteins of fetal calf bone. J Biol Chem 256:10403–10408

Thomas RS, Greenawalt JW (1968) Microincineration, electron microscopy, and electron diffraction of calcium-phosphate-loaded mitochondria. J Cell Biol 39:55–76

Thyberg J (1977) Electron microscopy of cartilage proteoglycans. Histochem J 9:259–266

Thyberg J, Lohmander S, Friberg U (1973) Electron microscopic demonstration of proteoglycans in guinea pig epiphyseal cartilage. J Ultrastruct Res 45:407–427

Tong H, Hu J, Ma W, Zhong G, Yao S, Cao N (2002) In situ analysis of the organic framework in the prismatic layer of mollusc shell. Biomaterials 23:2593–2598

Towe KM, Hamilton GH (1968) Ultrastructure and inferred calcification of the mature and developing nacre in bivalve mollusks. Calcif Tissue Res 1:306–318

Traub W, Jodaikin A, Weiner S (1985) Diffraction studies of enamel protein-mineral structural relations. In: Butler WT (ed) The chemistry and biology of mineralized tissues. Ebsco Media, Inc., Birmingham, pp 221–225

Traub W, Arad T, Weiner S (1989) Three-dimensional ordered distribution of crystals in turkey tendon collagen fibers. Proc Natl Acad Sci USA 86:9822–9826

Traub W, Arad T, Weiner S (1992) Origin of mineral crystal growth in collagen fibrils. Matrix 12:251–255

Travis DF (1968) Comparative ultrastructure and organization of inorganic crystals and organic matrices of mineralized tissues. Biology of the mouth. American Association for the Advancement of Sciences, Washington, pp 237–297

Travis DF (1970) The comparative ultrastructure and organization of five calcified tissues. In: Schraer H (ed) Biological calcification: cellular and molecular aspects. Appleton-Century-Crofts, New York, pp 203–311

Travis DF, Glimcher MJ (1964) The structure and organization of, and the relationship between the organic matrix and the inorganic crystals of embryonic bovine enamel. J Cell Biol 23:447–497

Travis DF, Gonsalves M (1969) Comparative ultrastructure and organization of the prismatic region of two bivalves and its possible relation to the chemical mechanism of boring. Am Zool 9:635–661

Urry LA, Hamilton PC, Killian CE, Wilt FH (2000) Expression of spicule matrix proteins in the sea urchin embryo during normal and experimentally altered spiculogenesis. Dev Biol 225:201–213

Vaes G, Eeckhout Y, Druetz JE (1976) A latent neutral protease released by bone in culture. Arch Int Physiol Biochim 84:666–668

van der Rest M, Rosenberg LC, Olsen BR, Poole AR (1986) Chondrocalcin is identical with the C-propeptide of type II procollagen. Biochem J 237:923–925

van der Wal P, de Jong EW, Westbroek P, de Bruijn WC, Mulder-Stapel AA (1983) Polysaccharide localization, coccolith formation, and Golgi dynamics in the coccolithophorid *Hymenomonas carterae.* J Ultrastruct Res 85:139–158

Van Emburg PR, de Jong EW, Daems WT (1986) Immunochemical localization of a polysaccharide from biomineral structures (coccoliths) of *Emiliania huxleyi.* J Ultrastruct Molec Struct Res 94:246–259

van Mullem PJ, Stadhouders AM (1974) Bone marking and lead intoxication. Early pathological changes in osteoclasts. Virchows Arch B Cell Path 15:345–350

Veis A (2003) Mineralization in organic matrix frameworks. Rev Mineral Geochem 54:249–289

Veis DJ, Albinger TM, Clohisy J, Rahima M, Sabsay B, Veis A (1986) Matrix proteins of the teeth of the sea urchin *Lytechinus variegatus.* J Exp Zool 240:35–46

Veis A, Barss J, Dahl T, Rahima M, Stock S (2002) Mineral-related proteins of sea urchin teeth: *Lytechinus variegatus.* Microsc Res Tech 59:342–351

Waddington RJ, Embery G, Last KS (1988) The glycosaminoglycan constituents of alveolar and basal bone of the rabbit. Connect Tissue Res 17:171–180

Walton RC, Kavanagh JP, Heywood BR (2003) The density and protein content of calcium oxalate crystals precipitated from human urine: a tool to investigate ultrastructure and the fractional volume occupied by organic matrix. J Struct Biol 143:14–23

Wang A, Martin JA, Lembke LA, Midura RJ (2000) Reversible suppression of in vitro biomineralization by activation of protein kinase A. J Biol Chem 275:11082–11091

Warshawsky H (1971) A light and electron microscopic study of the nearly mature enamel of rat incisors. Anat Rec 169:559–584

Warshawsky H (1987) External shape of enamel crystals. Scanning Microsc 1:1913–1923

Watabe N (1963) Decalcification of thin sections for electron microscope studies of crystal-matrix relationships in mollusc shells. J Cell Biol 18:701–703

Watabe N (1965) Studies on shell formation. XI. Crystal-matrix relationships in the inner layer of mollusk shells. J Ultrastruct Res 12:351–370

Weinbach EC, von Brand T (1965) The isolation and composition of dense granules from Ca^{++}-loaded mitochondria. Biochem Biophys Res Comm 19:133–136

Weiner S (1985) Organic matrixlike macromolecules associated with the mineral phase of sea urchin skeletal plates and teeth. J Exp Zool 234:7–15

Weiner S, Traub W (1980) X-ray diffraction study of the insoluble organic matrix of mollusk shells. FEBS Lett 111:311–316

Weiner S, Traub W (1984) Macromolecules in mollusc shells and their function in biomineralization. Phil Trans R Soc London 304B:425–434
Weiner S, Traub W (1986) Organization of hydroxyapatite crystals within collagen fibrils. Feder Eur Bioch Soc 206:262–266
Weiner S, Talmon Y, Traub W (1983) Electron diffraction of mollusk shell organic matrices and their relationship to the mineral phase. Int J Biol Macromol 5:325–328
Westbroek P, de Jong EW, Dam W, Bosch L (1973) Soluble intracrystalline polysaccharides from coccoliths of Coccolithus huxleyi (Lohmann) Kamptner (I). Calcif Tissue Res 12:227–238
White SW, Hulmes DJS, Miller A, Timmins PA (1977) Collagen-mineral axial relationship in calcified turkey leg tendon by X-ray and neutron diffraction. Nature 266:421–425
Williams A (1970) Spiral growth of the laminar shell of the brachiopod Crania. Calcif Tissue Res 6:11–19
Williams DC, Boder GB, Toomey RE, Paul DC, Hillman CC Jr, King KL, Van Frank RM, Johnston CC Jr (1980) Mineralization and metabolic response in serially passaged adult rat bone cells. Calcif Tissue Int 30:233–246
Wilt FH (1999) Matrix and mineral in the sea urchin larval skeleton. J Struct Biol 126:216–226
Wilt FH (2005) Developmental biology meets materials science: morphogenesis of biomineralized structures. Dev Biol 280:15–25
Wong V, Saleuddin ASM (1972) Fine structure of normal and regenerated shell of *Helisoma duryi duryi*. Can J Zool 50:1563–1568
Wu CW, Tchetina EV, Mwale F, Hasty K, Pidoux I, Reiner A, Chen J, Van W, Poole AR (2002) Proteolysis involving matrix metalloproteinase 13 (collagenase-3) is required for chondrocyte differentiation that is associated with matrix mineralization. J Bone Miner Res 17:639–651
Yamada J (1971) A fine structural aspect of the development of scales in the chum salmon fry. Bull Jpn Soc Sci Fish 37:18–29
Yamada J, Watabe N (1979) Studies on fish scale formation and resorption I. Fine structure and calcification of the scales in *Fundulus heteroclitus* (Atheriniformes: cyprinodontidae). J Morphol 159:49–66
Yamada M, Ozawa H (1978) Ultrastructural and cytochemical studies on the matrix vesicle calcification in the teeth of the killifish, *Oryzias latipes*. Arch Histol Jpn 41:309–323
Young JR, Davis SA, Bown PR, Mann S (1999) Coccolith ultrastructure and biomineralisation. J Struct Biol 126:195–215
Zylberberg L, Nicolas G (1982) Ultrastructure of scales in a teleost (*Carassius auratus* L.) after use of rapid freez-fixation and freeze-substitution. Cell Tissue Res 223:349–367
Zylberberg L, Géraudie J, Meunier F, Sire J-Y (1992) Biomineralization in the integumental skeleton of the living lower vertebrates. In: Hall BK (ed) Bone, volume 4: Bone metabolism and mineralization. CRC Press, Boca Raton, pp 171–224

17 Main Suggested Calcification Mechanisms: Cells

17.1 Introduction

The complexity of the mechanisms which lead to calcification in biological tissues is shown by the number of theories developed during the last century to explain how that process takes place, is regulated, and can be promoted or inhibited. There is still no unified overview on this topic, and, as pointed out by Posner (1987), the many hypotheses put forward can be grouped in three general categories on the basis of their central explanation: 1) spontaneous precipitation due to a rise in local inorganic ion concentration till supersaturation is attained; 2) the availability of factors able to promote calcification by creating nucleation sites; 3) the removal or neutralization of inhibitors. By now the value of some of these theories is little more than historical, even if they have made a major contribution to current knowledge about the mechanism of calcification. They have partly been anticipated in Sect. 2.2 and they are briefly reconsidered in this and next chapter, which are mainly devoted to current theories. These are divided into two main groups: 1) those focused on the function of specific cells and cell organelles, which are discussed in this chapter, and 2) those based on the role of extracellular matrix components, which are discussed in the next (Chap. 18). They are not mutually exclusive; it can be said that the principles of one theory can complement those of others.

The participation of cells in biological calcification is, clearly, fundamental, in the sense that cells are responsible for the production and transport of the components involved in the process. The claim that cells play a leading role in calcification seems, however, to derive mainly from the observation that the calcification of biological tissues cannot be accounted for simply by the precipitation of inorganic substance from a saturated solution, as the levels of calcium and phosphate stay well below the point that is critical for that in vertebrate interstitial fluids (Neuman and Neuman 1953). It soon became clear that some local factors must be operative to induce crystal formation. Alkaline phosphatase was probably the first to emerge as a candidate for this role.

17.2 Cells and Calcification: Tissue-nonspecific Alkaline Phosphatase (TNSALP)

As discussed in Sect. 2.2, Robison (1923) was the first to suggest that alkaline phosphatase could hydrolyze phosphate esters, so increasing local concentrations of inorganic phosphates and promoting calcium phosphate precipitation. The observation that concentrations of phosphate esters in calcifying tissues were insufficient to permit the process then led to the suggestion that their presence might be increased locally by glycogenolysis or that ATP could act as a phosphate donor. Several studies were, in fact, carried out to discover whether ATP acts in this way (Cartier and Picard 1955a,b,c; de Bernard et al. 1957). It was supposed that this substance, whose concentration reaches a maximum in the cartilage zones that are about to calcify, could play a role during the early phases of the calcification process by allowing tricalcium phosphate to be precipitated from orthophosphate or phosphoric esters (Cartier and Picard 1955c). Under the influence of an ATPase, ATP would then be reduced to AMP, with the fixation of the terminal pyrophosphate group in the cartilage (Cartier 1952; Cartier and Picard 1955b). Gutman and Yu (1950) did, however, report that rachitic cartilage fails to calcify in vitro when ATP alone is present in the medium. By contrast, Cartier and Picard (1955d) showed that, even if the ossifying cartilage of the ram embryo does not calcify in vitro if the medium contains either calcium or phosphate, it does become heavily calcified if ATP is present too. This finding was confirmed by Perkins and Walker (1958) in the cartilage of rachitic rats.

There are a number of conflicting opinions on the role of ATP in calcification, but it has been proved that the slow rate of cartilage calcification in vitro is greatly increased by the administration of ATP. Moreover, the findings reported above suggest that one or more components of the organic matrix must combine with phosphate before it/they can calcify (reviewed by Weidmann 1963).

Probably because of the uncertainty about its role in calcification, the tissue-non specific isoenzyme of alkaline phosphatase (TNSALP) has been considered to play a role in other processes, such as collagen and proteoglycan synthesis (Majno and Rouiller 1951; Cabrini 1961; Weidmann 1963) or matrix elaboration (Siffert 1951). In their review of the role of alkaline phosphatase in calcified tissues, Wuthier and Register (1985) listed six different possible functions: 1) increasing local concentrations of phosphates; 2) removing inhibitors of crystal formation and growth; 3) providing P_i-transport; 4) acting as a calcium-binding protein; 5) functioning as a Ca^{2+} pump; 6) regulating cell division or differentiation.

Whatever its role, TNSALP is indispensable to the initiation of the calcification process (Bellows et al. 1991). In this connection, many findings point to its role in removing inorganic pyrophosphate (PP_i), a calcification inhibitor

(Tenenbaum 1987). According to Felix and Fleisch (1976a,b), the function of alkaline phosphatase appears to be that of hydrolyzing the pyrophosphate, which inhibits calcification by preventing the precipitation of amorphous calcium phosphate and by slowing its transformation into hydroxyapatite (Fleisch et al. 1968; Wuthier et al. 1972). In reality, membrane-bound alkaline phosphatase from rat osseous plate does hydrolyze PP_i in the presence of magnesium ions (Rezende et al. 1998); on the other hand, PP_i inhibits the in vitro β-glycerophosphate-induced calcification of bone to a degree comparable with that induced by the inhibition of alkaline phosphatase by levamisole (Tenenbaum 1987).

This finding shifts the emphasis from TNSALP as an enzyme that promotes calcification by increasing local phosphate concentrations to TNSALP as an enzyme that permits calcification by disabling the inhibitory activity of PPi (discussed in Sect. 15.4.2 and by Murshed et al. 2005). It has, in fact, been hypothesized (Johnson et al. 2000) that TNSALP is able to promote calcification by neutralizing the inhibition of ATP-dependent calcification by PP_i generated by plasma cell membrane glycoprotein-1 (PC-1). Otherwise known as nucleotide pyrophosphatase phosphodiesterase 1 (NPP1), this enzyme generates PP_i by hydrolysis of the phosphodiester I of purine and pyrimidine nucleoside triphosphates (Goding et al. 1998). The upregulated activity of PC-1 promotes apoptosis and increases amounts of annexin V in matrix vesicles and the capacity of articular chondrocytes to precipitate calcium (Johnson et al. 2001). The conclusion is that TNSALP and PC-1 both appear to be key regulators of the calcification process (Hessle et al. 2002), but with antagonistic roles, even if their antagonism seems to be operative in calvaria and vertebrae but not in long bones (Anderson et al. 2005a). Moreover, in vitro studies on isolated chicken matrix vesicles have shown that TNSALP could contribute, acting as a phosphodiesterase, in the production of pyrophosphate from ATP, so playing the same function as PC-1 (Zhang et al. 2005).

Another TNSALP antagonist appears to be the so-called Ankylosis protein (Ank), which transports PP_i from cells to the extracellular matrix and controls PP_i levels in cultured cells. Mutation of the locus *progressive ankylosis* (*ank*) leads to progressive arthritis, ectopic calcification, bony outgrowths and severe arthrosis, probably induced by the derangement of PP_i regulation (Ho et al. 2000). The blocking of Ank activity or *ank* expression leads to an increase in intra- and extracellular PP_i concentrations and the inhibition of calcification – an effect similar to that induced by TNSALP inhibition by levamisole (Wang et al. 2005).

The demonstration that TNSALP is essential for calcification is given less compellingly by its constant presence in areas of calcification than by the fact that its absence prevents calcification occurring in vivo, as shown by the severe skeletal defects (rickets and osteomalacia) that develop in cases of congenital hypophosphatasia (see Sect. 15.3.4) or in TNSALP knockout mice (see Sect. 15.4.2). In these pathological conditions, contrary to all expectations, the numbers and distribution of matrix vesicles are approximately normal (Ornoy et al. 1985) and early calcification within them starts in an apparently normal way, but this process fails to spread to the surrounding matrix, which remains uncalcified (Anderson et al. 1997, 2004). The obvious inference is that TNSALP is not indispensable for crystal formation, at least in the closed environment of matrix vesicles, perhaps owing to the supplementary activity in them of other phosphatases (Anderson et al. 2004) or of the annexin system (see Sect. 9.4), whereas without it the propagation of the calcification process to the perivesicular matrix cannot proceed.

17.3 Cells and Calcification: Matrix Vesicles

The idea that matrix vesicles are early loci of calcification, and therefore possess the machinery needed to induce calcium phosphate formation, was expressed as early as the first morphological description of these structures as osmiophilic globules of cellular origin in epiphyseal cartilage (Bonucci 1967). Since then, these globules – now known as matrix vesicles – have been recognized to be organelles that promote early calcification in cartilage, bone, circumpulpal dentin and pathologically calcifying tissues (discussed in Sects. 7.3.5, 8.7 and 9.4, and reviewed by Wuthier 1982; Ali 1983; Bonucci 1984; Anderson 1989; Sela et al. 1992; Kirsch et al.2003; Anderson et al. 2005b).

It has, however, been maintained that no relationship can be detected between matrix vesicles and solid inorganic structures when bone sections are prepared using anhydrous techniques or ultracryomicrotomy (Landis et al. 1977a, 1980), and that, therefore, the presence of inorganic particles in conventionally prepared matrix vesicles may be an artifact (Landis and Glimcher 1982). For these reasons, their role in calcification has been questioned (Poole 1991). One view that has been expressed is that matrix vesicles can only be observed in the very early stages of embryonic bone development (Glimcher 1990), and that they could have a role in the calcification of this type of bone, but not in that of mature compact bone (Landis et al. 1977a). As previously discussed (see Sect. 9.4), these criticisms can easily be rejected. It is true that plenty of matrix vesicles can be found in calcifying cartilage and many of them can be detected in woven bone, whereas few are present in compact bone, so that their numbers are inversely correlated with the compactness of the calcifying matrix (Bonucci 1981). They are not, however, absent from diaphyseal lamellar bone, have been described in very compact calcified tissues, such as turkey tendons (Landis 1986; Landis and Arsenault 1989; Landis and Song 1991; Landis et al. 1992; Landis and Silver 2002), and can rapidly develop in fracture callus (Göthlin and Ericsson 1973) in such abundance that alkaline phosphatase can be purified from them (Kahn et al. 1978).

Obviously the calcification of matrix vesicles is not a universal process capable of explaining the deposition of inorganic substance in all hard tissues.

Although they have been described in a number of normal and pathological tissues, they have never been described in enamel, and have only been detected in circumpulpal dentin, not in intertubular dentin, to mention only hard tissues in vertebrates. The available results suggest that matrix vesicles are specially active in tissues where calcification is a very rapid process. Moreover, early crystals can develop independently of matrix vesicles, as shown in bone by the formation of granular inorganic bands in the holes of collagen fibrils and of filament-like crystals in interfibrillary spaces.

As already discussed (Sect. 9.4), and as shown in Fig. 9.5, alkaline phosphatase is always associated with matrix vesicles (except for those found in the uncalcified elastic cartilage of epiglottis (Nielsen 1978) and in the resting zone of the epiphysis (Ohashi-Takeuchi et al. 1990). Moreover, other phosphatases are detected in them, or in association with their membranes, which are able to yield orthophosphate by hydrolyzing organic phosphoesters such as ATP, ADP, AMP, 5′-AMP, pyrophosphate and phosphoethanolamine (reviewed by Anderson 1989; Ali 1992). Most of the theoretical constructs devised to explain matrix vesicle calcification are based on the activity of phosphatases. As reported above (Sect. 17.2), the local increase in phosphate by the hydrolysis of phosphate esters or the active transport and removal of pyrophosphates have been the mechanisms most often put forward (discussed by Anderson et al. 2004). Boskey et al. (1997) found that, when introduced into a gelatin gel within which calcium and phosphate ions could spread from the opposite ends, matrix vesicles significantly raised hydroxyapatite accumulation, even when proteoglycan aggregates or ATP – in vitro inhibitors of apatite formation – are included in the gel. They concluded that matrix vesicles promote calcification by providing enzymes that modify inhibitory factors in the extracellular matrix, or by creating a protected environment where mineral ions can accumulate.

Substantial amounts of lipids are contained in matrix vesicles, as already discussed (Sect. 9.4). Their role in matrix calcification, and the possible function of calcium-phospholipid-phosphate complexes and of membrane-associated proteolipids in nucleating hydroxyapatite (see Sect. 13.5.1), are considered below (Sect. 18.5). Intravesicle calcification can be enhanced by annexins because they exhibit proteolipid-like properties – a feature that might allow them to structure membrane phosphatidylserine molecules in a conformation conducive to mineral formation (Genge et al. 1991). According to Wu et al. (1993, 1997), matrix vesicles contain a nucleation core consisting of two main components: a membrane-associated complex of Ca, Pi, phosphatidylserine and the annexins, and a larger pool of Ca and Pi bound to proteins.

One possible mechanism of matrix vesicle calcification that has been put forward comprises the intravesicle accumulation of calcium, through its link to lipids located on the inner side of the vesicle membrane, and of phosphate, due to the activity of alkaline phosphatase. The intravesicular calcium-phosphate ionic product would be increased as a result, and hydroxyapatite crystals would be formed; these would gradually fill up the whole vesicle and eventually perforate the vesicle membrane, so exposing crystals to the extravesicular environment (Anderson 1989, 1995). It appears to be highly unlikely, however, that very thin crystals, even if behaving as structures with a tip, could overcome the plasticity of the membrane by perforating it mechanically.

Matrix vesicles are enclosed by a membrane which separates them from the surrounding matrix and at the same time gives them the configuration of closed, localized compartments (this term refers to local spaces outlined by a wall and not to generalized, compartmentalized systems such as skeletal compartments, vascular compartments, or even Ca^{2+} compartments). Closed, localized compartments could exert a strong influence on the calcification process, because calcium and phosphate ions can accumulate in them, so increasing their local concentration, whereas they would otherwise be dispersed in the interstitial fluids.

Besides matrix vesicles, other localized compartments have been described in calcifying tissues as sites where inorganic ions might accumulate, and early crystals might form. The following are examples of structures considered to be local compartments: the 'holes' that, together with the 'overlap' zones, are responsible for the periodic pattern of collagen fibrils (see Sects. 7.2 and 18.2); the small, tubule-like spaces which have been reported in the decalcified organic matrix of embryonic enamel (Travis and Glimcher 1964; Nylen 1979) and enameloid (Garant 1970; Kemp 1985; Sasagawa and Ferguson 1990); the membrane-delineated spaces in which spicules of the sea urchin embryo are generated (Beniash et al. 1999); the spaces outlined by pallial fluid-derived lamellae in mollusk shells (Bevelander and Nakahara 1969; Nakahara and Bevelander 1971); the spaces described in shell decalcified matrix (Travis 1968; Travis and Gonsalves 1969); the cytoplasmic vacuoles that house intracellular inorganic substance, such as those of coccoliths in coccolithophorids (Young et al. 1999), or those of magnetosomes in magnetic bacteria (Okamura et al. 2001; Komeili et al. 2004; Schüler 2004).

These findings all appear to indicate that calcification invariably occurs within local compartments and that these are indispensable for normal calcification. In his outstanding review of calcification in organic matrix frameworks, Veis (2003) expressed his viewpoint that biological calcifications take place within well-defined, restricted compartments or spaces, and that compartmentalization is crucial to the entire process, because it permits the mingling of organic and inorganic components needed for calcification in a single confined space.

Compartments appear to be indispensable for another reason: if calcification begins with the formation of nuclei (heterogeneous nucleation) from which crystals can then grow, it follows that a space is needed to house the definitive crystals ("Crystals can grow where they have space to grow", as Frank commented in discussing the papers presented at the IV Session of the meeting on 'Tooth morphogenesis and differentiation', Colloque INSERM, Strasbourg,

1984). This need explains why it has been postulated that pores, channels and grooves are located in collagen fibrils (see Sect. 18.2): crystals in bone and related tissues are nucleated in the small holes of the gap zones, from which they grow in length until their span exceeds one collagen period (Ascenzi et al. 1965); in so doing, they would necessarily sever intrafibrillary cross-links and modify their ultrastructure (Bonucci 1992), whereas preformed channels and grooves would allow them to grow freely.

In this connection, it must, however, be borne in mind that localized compartments are not empty spaces but contain molecules that fill up them completely. Depending on their structure and composition, these molecules may promote or inhibit calcification, and may leave no space for crystals to grow. As wittily argued by Wheeler (1992), the organisms do not prepare test tubes in which ion concentration can be increased until saturation and precipitation occur, and compartments are not empty balls which can expand to create free space for solutions or crystals. These are formed in compartments which are full of organic molecules, which on one hand may be responsible for their formation, but on the other may hinder their growth and maturation. The lack of space in these compartments may explain why organic material is partly occluded within growing crystals. This type of constraint may be compensated, as occurs in matrix vesicles, by the rupture of the enclosing membrane and by the development of crystals in extra-compartmental spaces. This seems to indicate that, although the accumulation of inorganic ions and other substances in a circumscribed space may make calcification easier, this can occur outside localized compartments. The long, filament-like crystals contained in bone interfibrillary spaces are not, in fact, confined to compartments; they are formed by the template activity of non-collagenous proteins. The topic is further discussed in Sect. 19.4.

17.4 Cells and Calcification: Cytoplasmic Vacuoles

Solid intracellular calcifications are a frequent event, occurring as scattered or circumscribed inorganic deposits. The former are characteristic of bacteria (Boyan et al. 1992), the latter of unicellular organisms (Pautard 1970). In this case, calcifications develop in cytoplasmic vacuoles which mostly derive from the Golgi apparatus, as exemplified by the early sites of skeleton formation in echinoderms (Ameye et al. 1998) and of spicules in gorgonians (Kingsley and Watabe 1982), coccoliths in coccolithophorids (Marsh 1999), laminar envelopes in mollusk shells (Nakahara and Bevelander 1971), calcospherites in the hepatopancreas of snails (Abolins-Krogis 1970), calcified particles in *Spirostomum ambiguum* (Pautard 1976), magnetosomes in magnetotactic bacteria (Komeili et al. 2004), and intracellular calcifications that develops in some pathological conditions (Kuhar and Eisenmann 1978).

It is not feasible, and would probably be dispersive, to consider in detail all the calcification mechanisms that may theoretically occur in these different settings. All of them can be viewed as processes taking place in compartments which resemble matrix vesicles not only because they are membrane-delimited spaces, but also because they have similar structural components, in particular alkaline phosphatase and several acidic proteins, phospholipids and proteolipids (see Sects. 12.2, 12.3 and 13.5).

17.5 Cells and Calcification: Mitochondria

The presence of mitochondrial inorganic granules in the cells of normal (Gonzales and Karnovsky 1961; Martin and Matthews 1969; Arsenis 1972; Holtrop 1972; Shapiro and Lee 1975; Burger and Matthews 1978) and pathologically altered (Bonucci and Sadun 1972, 1973; Brandt and Bässler 1972; Bonucci et al. 1973; Cavallero et al. 1974; Boivin 1975) calcified tissues (see Sects. 4.6.1 and 5.6.1) has prompted the proposal that they may play a direct role in calcification. On this topic, Arsenis (1972) reported that mitochondria from the hypertrophic zone of the cartilage can be divided into "light" and "heavy" groups, presumably on the basis of their content of inorganic calcium phosphate granules, so taking into account the finding that their "light" fraction is only detectable in resting, not in calcifying cartilage, unless isolated mitochondria are incubated in a calcium and phosphate medium, after which they too can be divided in two fractions, light and heavy. Potassium pyroantimonate investigations have shown that the calcium accumulated in cartilage mitochondria is gradually lost and transferred to extracellular structures as calcification proceeds (Brighton and Hunt 1974, 1976), a process that does not occur in rachitic animals (Brighton and Hunt 1978). Using microincineration procedures, Sayegh et al. (1974a,b) found that the cells at the tips of growing deer antlers showed a concentration gradient of mitochondrial granules as the cells proceeded from the primitive to the mature type, so indicating that they do take part in the calcification process: the mitochondria of antler chondrocytes built up inorganic granules in a zone lying a little above the matrix calcification zone and release them below, in the calcification zone itself. According to Plachot et al. (1986), the osteoblast mitochondria of the rat fetus could be extruded from cells, and mitochondrial granules could be the earliest mineral deposits to occur in osteoid.

These observations, especially the rising gradient of mitochondrial granules as calcification areas are approached (Matthews et al. 1970; Sayegh et al. 1974a; Sayegh and Abousy 1977), their depletion in mitochondria of hypertrophic chondrocytes with a concomitant increase in calcium concentration in matrix vesicles (Brighton and Hunt 1974, 1976, 1978), their sharp reduction in rachitic animals, and their restoration after vitamin D supplementation (Matthews et al. 1970), have all be taken to indicate that mitochondria may be involved in the calcification process (Sayegh et al. 1974a; Wuthier 1982), probably by producing a local rise in levels of mineral ions (Shapiro and Greenspan 1969).

It is universally accepted that mitochondria contribute to the transport of intracellular calcium, but serious doubts exist about their direct participation in the extracellular calcification process. In the mitochondria of normal tissues, granules appear to be especially plentiful when specimens are processed by freezing methods (Ali and Wisby 1975; Landis et al. 1977b; Ali et al. 1978; Landis and Glimcher 1982; Manston and Katchburian 1983; Takano et al. 1989), are prepared anhydrously (Landis et al. 1977a, 1980), or are treated with potassium pyroantimonate (Tandler and Kierszenbaum 1971; Schäfer 1973; Brighton and Hunt 1974, 1976, 1978; Oberc and Engel 1977; Burger and Matthews 1978; Morris and Appleton 1980; Lewinson and Silbermann 1982, 1990; Kogaya and Furahashi 1988). Regrettably, these methods of study, which apparently offer the best way of preserving calcium ions in cells, may cause changes in mitochondria that facilitate their calcium accumulation. It has, in fact, been shown that calcium phosphate granules are mainly found in the mitochondria of normal cells when these are damaged by excessive calcium concentration (Duffy et al. 1971), or mechanical trauma (Burger and de Bruijn 1979), or as a result of the freezing process (Appleton et al. 1985; Appleton 1987), and that divalent ions are massively accumulated in osmotically lysed rat liver mitochondria (Vasington and Greenawalt 1968). Canine fibroblasts in vitro only showed intramitochondrial inclusions if injured by freeze-thaw, anoxia or chemical modifications (Kim 1994).

These results show that mitochondria are certainly involved in the intracellular calcium metabolism and participate in the transport of calcium ions from inside cells to extracellular spaces and matrix vesicles (reviewed by Wuthier 1982). On the other hand, it has never been demonstrated that mitochondrial granules are directly involved in the generation of early extracellular crystal aggregates and the theories that favor this possibility seems to have no grounding in experimental findings.

17.6 Concluding Remarks

There are many theories on calcification in biological tissues but none of them are able to give a complete explanation of the mechanism of inorganic substance deposition in organic matrices. Those that deal with specific cell functions and organelles (alkaline phosphatase, matrix vesicles, cytoplasmic vacuoles, mitochondria) concern important aspects of the problem. In this context, the following remarks seem to be especially pertinent:

- The calcification of biological tissues is not due to the simple precipitation of inorganic substance from a saturated solution; some local factors must be operative to induce or prevent crystal formation.

- One active local factor is alkaline phosphatase, which is indispensable for the calcification process to spread through the matrix. Various possible functions have been assigned to this enzyme, the most important of which may be the hydrolysis of pyrophosphates, which inhibit calcification by forestalling the precipitation of amorphous calcium phosphate.
- Plenty of alkaline phosphatase is found in all areas of initial calcification and also in matrix vesicles. These are membrane-bound organelles of cellular origin that promote early calcification in several hard tissues (cartilage, bone, circumpulpal dentin, tendons, and several pathologically calcifying tissues), but not in all of them.
- Alkaline phosphatase might increase phosphate concentrations in matrix vesicles through the hydrolysis of phosphate esters or through active transport; a lack of alkaline phosphatase does not, however, completely prevent calcification within matrix vesicles, where a nucleation core containing annexin could be active.
- Matrix vesicles are extracellular compartments surrounded by a membrane; they, like other compartments detectable in calcified tissues, are larger than single crystals and may contain several of them.
- Mitochondria and cytoplasmic vacuoles behave as intracellular compartments. They contribute to intracellular calcium transport and, in some organisms (coccolithophorids, sea urchin embryos), provide the space needed for the development of skeletal structures (coccoliths, spines). They do not participate directly in the extracellular calcification process.
- Although cells and cell functions are indispensable for calcification, they do not take direct part in its very basic mechanism.

References

Abolins-Krogis A (1970) Electron microscope studies of the intracellular origin and formation of calcifying granules and calcium spherites in the hepatopancreas of the snail, *Helix pomatia* L. Z Zellforsch 108:501–515

Ali SY (1983) Calcification of cartilage. In: Hall BK (ed) Cartilage, vol. I. Structure, function and biochemistry. CRC Press, Boca Raton, pp 343–378

Ali SY (1992) Constitutive enzymes of matrix vesicles. Bone Miner 17:168–171

Ali SY, Wisby A (1975) Mitochondrial granules of chondrocytes in cryosections of growth cartilage. Am J Anat 144:243–248

Ali SY, Wisby A, Gray JC (1978) Electron probe analysis of cryosections of epiphyseal cartilage. Metab Bone Dis Rel Res 1:97–103

Ameye L, Compère P, Dille J, Dubois P (1998) Ultrastructure and cytochemistry of the early calcification site and of its mineralization organic matrix in *Paracentrotus lividus* (Echinodermata: Echinoidea). Histochem Cell Biol 110:285–294

Anderson HC (1989) Mechanism of mineral formation in bone. Lab Invest 60:320–330

Anderson HC (1995) Molecular biology of matrix vesicles. Clin Orthop Relat Res 314:266–280

Anderson HC, Hsu HH, Morris DC, Fedde KN, Whyte MP (1997) Matrix vesicles in osteomalacic hypophosphatasia bone contain apatite-like mineral crystals. Am J Pathol 151:1555–1561

Anderson HC, Sipe JB, Hessle L, Dhamyamraju R, Atti E, Camacho NP, Millàn JL (2004) Impaired calcification around matrix vesicles of growth plate and bone in alkaline phosphatase-deficient mice. Am J Pathol 164:841–847

Anderson HC, Harmey D, Camacho NP, Garimella R, Sipe JB, Tague S, Bi X, Johnson K, Terkeltaub R, Millàn JL (2005a) Sustained osteomalacia of long bones despite major improvement in other hypophosphatasia-related mineral deficits in tissue nonspecific alkaline phosphatase/nucleotide pyrophosphatase phosphodiesterase 1 double-deficient mice. Am J Pathol 166:1711–1720

Anderson HC, Garimella R, Tague SE (2005b) The role of matrix vesicles in growth plate development and biomineralization. Front Biosci 10:822–837

Appleton J (1987) X-ray microanalysis of growth cartilage after rapid freezing, low temperature freeze drying and embedding in resin. Scanning Microsc 1:1135–1144

Appleton J, Lyon R, Swindin KJ, Chesters J (1985) Ultrastructure and energy-dispersive X-ray microanalysis of cartilage after rapid freezing, low temperature freeze drying, and embedding in Spurr's resin. J Histochem Cytochem 33:1073–1079

Arsenis C (1972) Role of mitochondria in calcification. Mitochondrial activity distribution in the epiphyseal plate and accumulation of calcium and phosphate ions by chondrocyte mitochondria. Biochem Biophys Res Comm 46:1928–1935

Ascenzi A, Bonucci E, Steve Bocciarelli D (1965) An electron microscope study of osteon calcification. J Ultrastruct Res 12:287–303

Bellows CG, Aubin JE, Heersche JNM (1991) Initiation and progression of mineralization of bone nodules formed in vitro: the role of alkaline phosphatase and organic phosphate. Bone Miner 14:27–40

Beniash E, Addadi L, Weiner S (1999) Cellular control over spicule formation in sea urchin embryos: a structural approach. J Struct Biol 125:50–62

Bevelander G, Nakahara H (1969) An electron microscope study of the formation of the nacreous layer in the shell of certain bivalve molluscs. Calcif Tissue Res 3:84–92

Boivin G (1975) Étude chez le rat d'une calcinose cutanée induite par calciphylaxie locale I. – Aspects ultrastructuraux. Arch Anat Microsc Morphol Exp 64:183–205

Bonucci E (1967) Fine structure of early cartilage calcification. J Ultrastruct Res 20:33–50

Bonucci E (1981) The origin of matrix vesicles and their role in the calcification of cartilage and bone. In: Schweiger HG (ed) International Cell Biology 1980–81. Springer, Berlin Heidelberg New York, pp 993–1003

Bonucci E (1984) Matrix vesicles: their role in calcification. In: Linde A (ed) Dentin and dentinogenesis. CRC Press, Boca Raton, pp 135–154

Bonucci E (1992) Role of collagen fibrils in calcification. In: Bonucci E (ed) Calcification in biological systems. CRC Press, Boca Raton, pp 19–39

Bonucci E, Sadun R (1972) An electron microscope study on experimental calcification of skeletal muscle. Clin Orthop Relat Res 88:197–217

Bonucci E, Sadun R (1973) Experimental calcification of the myocardium. Ultrastructural and histochemical investigations. Am J Pathol 71:167–192

Bonucci E, Derenzini M, Marinozzi V (1973) The organic-inorganic relationship in calcified mitochondria. J Cell Biol 59:185–211

Boskey AL, Boyan BD, Schwartz Z (1997) Matrix vesicles promote mineralization in a gelatin gel. Calcif Tissue Int 60:309–315

Boyan BD, Swain LD, Everett MM, Schwartz Z (1992) Mechanisms of microbial mineralization. In: Bonucci E (ed) Calcification in biological systems. CRC Press, Boca Raton, pp 129–156

Brandt G, Bässler R (1972) Die Wirkung der experimentellen Hypercalcämie durch Dihydrotachysterin auf Drüsenfunktion und Verkalkungsmuster der Mamma. Licht-, elektronenmikroskopische und chemisch-analytische Untersuchungen. Virchows Arch Abt A Pathol Anat 356:155–172

Brighton CT, Hunt RM (1974) Mitochondrial calcium and its role in calcification. Histochemical localization of calcium in electron micrographs of the epiphyseal growth plate with K-pyroantimonate. Clin Orthop Relat Res 100:406–416

Brighton CT, Hunt RM (1976) Histochemical localization of calcium in growth plate mitochondria and matrix vesicles. Fed Proc 35:143–147

Brighton CT, Hunt RM (1978) The role of mitochondria in growth plate calcification as demonstrated in a rachitic model. J Bone Joint Surg 60-A:630–639

Burger EH, de Bruijn WC (1979) Mitochondrial calcium of intact and mechanically damaged bone and cartilage cells studied with K-pyroantimonate. Histochemistry 62:325–336

Burger EH, Matthews JL (1978) Cellular calcium distribution in fetal bones studied with K-pyroantimonate. Calcif Tissue Res 26:181–190

Cabrini RL (1961) Histochemistry of ossification. Int Rev Cytol 2:283–306

Cartier P (1952) Mécanisme enzymatique de l'ossification. Exp Ann Biochim Méd 14:73–86

Cartier P, Picard J (1955a) La minéralisation du cartilage ossifiable. II. - Le système ATPasique du cartilage. Bull Soc Chim Biol 37:661–675

Cartier P, Picard J (1955b) La minéralisation du cartilage ossifiable. III. - Le mècanisme de la réaction ATPasique du cartilage. Bull Soc Chim Biol 37:1159–1168

Cartier P, Picard J (1955c) La minéralisation du cartilage ossifiable: IV. - La signification de la réaction ATPasique. Bull Soc Chim Biol 37:1169–1176

Cartier P, Picard J (1955d) La minéralisation du cartilage ossifiable: I. - La minéralisation du cartilage *in vitro*. Bull Soc Chim Biol 37:485–494

Cavallero C, Spagnoli LG, Di Tondo U (1974) Early mitochondrial calcification in the rabbit aorta after adrenaline. Virchows Arch A Path Anat Histol 362:23–39

de Bernard B, Moretti A, Castellani AA, Zambotti V (1957) Studi sulla mineralizzazione *in vitro* della cartilagine preossea: azione mineralizzante dell'ATP e dell'UTP su fettine di cartilagine di accrescimento di giovani conigli. Boll Soc It Biol Sper 33:1772–1777

Duffy JL, Meadow E, Suzuki Y, Churg J (1971) Acute calcium nephropathy. Early proximal tubular changes in the rat kidney. Arch Pathol 91:340–350

Felix R, Fleisch H (1976a) Pyrophosphatase and ATPase of isolated cartilage matrix vesicles. Calcif Tissue Res 22:1–7

Felix R, Fleisch H (1976b) Role of matrix vesicles in calcification. Fed Proc 35:169–171

Fleisch H, Russell RGG, Bisaz S, Termine JD, Posner AS (1968) Influence of pyrophosphate on the transformation of amorphous to crystalline calcium phosphate. Calcif Tissue Res 2:49–59

Garant PR (1970) An electron microscopic study of the crystal-matrix relationship in the teeth of the dogfish *Squalus acanthias* L. J Ultrastruct Res 30:441–449

Genge BR, Wu LNY, Adkisson HDI, Wuthier RE (1991) Matrix vesicle annexins exhibit proteolipid-like properties. Selective partitioning into lipophilic solvents under acidic conditions. J Biol Chem 266:10678–10685

Glimcher MJ (1990) The nature of the mineral component of bone and the mechanism of calcification. In: Avioli LV, Krane SM (eds) Metabolic bone disease and clinically related disorders. W.B. Saunders, Philadelphia, pp 42–68

Goding J, Terkeltaub R, Maurice M, Deterre P, Sali A, Belli S (1998) Ecto-phosphodiesterase/pyrophosphatase of lymphocytes and nonlymphoid cells: structure and function of the PC-1 family. Immunol Rev 161:11–26

Göthlin G, Ericsson JLE (1973) Fine structural localization of alkaline phosphatase in the fracture callus of the rat. Histochemie 36:225–236

Gonzales F, Karnovsky MJ (1961) Electron microscopy of osteoclasts in healing fractures of rat bone. J Biophys Biochem Cytol 9:299–316

Gutman AB, Yu TF (1950) A concept of the role of enzymes in endochondral calcification. In: Reifenstein EC (ed) Metabolic interrelations. Josiah Macy Jr. Foundation, New York, pp 167–190

Hessle L, Johnson KA, Anderson HC, Narisawa S, Sali A, Goding JW, Terkeltaub R, Millàn JL (2002) Tissue-nonspecific alkaline phosphatase and plasma cell membrane glycoprotein-1 are central antagonistic regulators of bone mineralization. Proc Natl Acad Sci U S

A Ho AM, Johnson MD, Kingsley DM (2000) Role of the mouse *ank* gene in control of tissue calcification and arthritis. Science 289:265–270

Holtrop ME (1972) The ultrastructure of the epiphyseal plate. II. The hypertrophic chondrocyte. Calcif Tissue Res 9:140–151

Johnson KA, Hessle L, Vaingankar S, Wennberg C, Mauro S, Narisawa S, Goding JW, Sano K, Millan JL, Terkeltaub R (2000) Osteoblast tissue-nonspecific alkaline phosphatase antagonizes and regulates PC-1. Am J Physiol Regul Integr Comp Physiol 279:R1365-R1377

Johnson K, Pritzker K, Goding J, Terkeltaub R (2001) The nucleoside triphosphate pyrophosphohydrolase isozyme PC-1 directly promotes cartilage calcification through chondrocyte apoptosis and increased calcium precipitation by mineralizing vesicles. J Rheumatol 28:2681–2691

Kahn SE, Jafri AM, Lewis NJ, Arsenis C (1978) Purification of alkaline phosphatase from extracellular vesicles of fracture callus cartilage. Calcif Tissue Res 25:85–92

Katz EP, Li S-T (1973) Structure and function of bone collagen fibrils. J Mol Biol 80:1–15

Kemp NE (1985) Ameloblastic secretion and calcification of the enamel layer in shark teeth. J Morphol 184:215–230

Kim KM (1994) Cell death and calcification of canine fibroblasts *in vitro*. Cells and Materials 4:247–261

Kingsley RJ, Watabe N (1982) Ultrastructural investigation of spicule formation in the gorgonian *Leptogorgia virgulata* (Lamarck) (Coelenterata: Gorgonacea). Cell Tissue Res 223:325–334

Kirsch T, Wang W, Pfander D (2003) Functional differences between growth plate apoptotic bodies and matrix vesicles. J Bone Miner Res 18:1872–1881

Kogaya Y, Furahashi K (1988) Comparison of the calcium distribution pattern among several kinds of hard tissue forming cells of some living vertebrates. Scanning Microsc 2:2029–2043

Komeili A, Vali H, Beveridge TJ, Newman DK (2004) Magnetosome vesicles are present before magnetite formation, and MamA is required for their activation. Proc Natl Acad Sci U S A 101:3839–3844

Kuhar KJ, Eisenmann DR (1978) Fluoride-induced mineralization within vacuoles in maturative ameloblasts of the rat. Anat Rec 191:91–102

Landis WJ (1986) A study of calcification in the leg tendons from the domestic turkey. J Ultrastruct Mol Struct Res 94:217–238

Landis WJ, Arsenault AL (1989) Vesicle- and collagen-mediated calcification in the turkey leg tendon. Connect Tissue Res 22:35–42

Landis WJ, Glimcher MJ (1982) Electron optical and analytical observations of rat growth plate cartilage prepared by ultracryomicrotomy: the failure to detect a mineral phase in matrix vesicles and the identification of heterodispersed particles as the initial solid phase of calcium phosphate deposited in the extracellular matrix. J Ultrastruct Res 78:227–268

Landis WJ, Silver FH (2002) The structure and function of normally mineralizing avian tendons. Comp Biochem Physiol A Mol Integr Physiol 133:1135–1157

Landis WJ, Song MJ (1991) Early mineral deposition in calcifying tendons characterized by high voltage electron microscopy and three-dimensional graphic imaging. J Struct Biol 107:116–127

Landis WJ, Paine MC, Glimcher MJ (1977a) Electron microscopic observations of bone tissue prepared anhydrously in organic solvents. J Ultrastruct Res 59:1–30

Landis WJ, Hauschka BT, Rogerson CA, Glimcher MJ (1977b) Electron microscopic observations of bone tissue prepared by ultracryomicrotomy. J Ultrastruct Res 59:185–206

Landis WJ, Paine MC, Glimcher MJ (1980) Use of acrolein vapors for the anhydrous preparation of bone tissue for electron microscopy. J Ultrastruct Res 70:171–180

Landis WJ, Hodgens KJ, McKee MD, Nanci A, Song MJ, Kiyonaga S, Arena J, McEwen B (1992) Extracellular vesicles of calcifying turkey leg tendon characterized by immunocytochemistry and high voltage electron microscopic tomography and 3-D graphic image reconstruction. Bone Miner 17:237–241

Lewinson D, Silbermann M (1982) Landmarks in chondrocyte differentiation and maturation as envisaged by changes in the distribution of calcium complexes: an ultrastructural histochemical study. Metab Bone Dis Rel Res 4:143–150

Lewinson D, Silbermann M (1990) Ultrastructural localization of calcium in normal and pathologic cartilage. In: Bonucci E, Motta PM (eds) Ultrastructure of skeletal tissues. Kluwer Academic Publishers, Boston, pp 129–152

Majno G, Rouiller C (1951) Die alkalische Phosphatase in der Biologie des Knochengewebes. Histochemische Untersuchungen. Virchows Arch 321:1–61

Manston J, Katchburian E (1983) Demonstration of mitochondrial deposits in osteoblasts after anhydrous fixation and processing. J Microsc 134:177–182

Marsh ME (1999) Coccolith crystals of *Pleurochrysis carterae*: crystallographic faces, organization, and development. Protoplasma 207:54–66

Martin JH, Matthews JL (1969) Mitochondrial granules in chondrocytes. Calcif Tissue Res 3:184–193

Matthews JL, Martin JH, Sampson HW, Kunin AS, Roan JH (1970) Mitochondrial granules in the normal and rachitic rat epiphysis. Calcif Tissue Res 5:91–99

Morris DC, Appleton J (1980) Ultrastructural localization of calcium in the mandibular condylar growth cartilage of the rat. Calcif Tissue Int 30:27–34

Murshed M, Harmey D, Millàn JL, McKee MD, Karsenty G (2005) Unique coexpression in osteoblasts of broadly expressed genes accounts for the spatial restriction of ECM mineralization to bone. Genes and Developm 19:1093–1104

Nakahara H, Bevelander G (1971) The formation and growth of the prismatic layer of *Pinctada radiata*. Calcif Tissue Res 7:31–45

Neuman WF, Neuman MW (1953) The nature of the mineral phase of bone. Chem Rev 53:1–45

Nielsen EH (1978) Ultrahistochemistry of matrix vesicles in elastic cartilage. Acta Anat 100:268–272

Nylen MU (1979) Matrix-mineral relationships - A morphologist's viewpoint. J Dent Res 58:922–926

Oberc MA, Engel WK (1977) Ultrastructural localization of calcium in normal and abnormal skeletal muscle. Lab Invest 36:566–577

Ohashi-Takeuchi H, Yamada N, Hosokawa R, Noguchi T (1990) Vesicles with lactate dehydrogenase and without alkaline phosphatase present in the resting zone of epiphyseal cartilage. Biochem J 266:309–312

Okamura Y, Takeyama H, Matsunaga T (2001) A magnetosome-specific GTPase from the magnetic bacterium *Magnetospirillum magneticum* AMB-1. J Biol Chem 276:48183–48188
Ornoy A, Adomian GE, Rimoin DL (1985) Histologic and ultrastructural studies on the mineralization process in hypophosphatasia. Am J Med Genet 22:743–758
Pautard FGE (1970) Calcification in unicellular organisms. In: Schraer H (ed) Biological calcification: cellular and molecular aspects. Appleton-Century-Crofts, New York, pp 105–201
Pautard FGE (1976) Calcification in single cells: with an appraisal of the relationship between *Spirostomum ambiguum* and the osteocyte. In: Watabe N, Wilbur KM (eds) The mechanisms of mineralization in the invertebrates and plants. University of South Carolina Press, Columbia,SC, pp 33–53
Perkins HR, Walker PG (1958) The occurrence of pyrophosphate in bone. J Bone Joint Surg 40-B:333–339
Plachot JJ, Thil CL, Enault G, Halpern S, Cournot-Witmer G, Balsan S (1986) Mitochondrial calcium and bone mineralization in the rat fetus. Bone Miner 1:157–166
Poole AR (1991) The growth plate: cellular physiology, cartilage assembly and mineralization. In: Hall B, Newman S (eds) Cartilage: molecular aspects. CRC Press, Boca Raton, pp 179–211
Posner AS (1987) Bone mineral and the mineralization process. In: Peck WA (ed) Bone and mineral research/5. Elsevier Science Publisher, Amsterdam, pp 65–116
Rezende LA, Ciancaglini P, Pizauro JM, Leone FA (1998) Inorganic pyrophosphate-phosphohydrolytic activity associated with rat osseous plate alkaline phosphatase. Cell Mol Biol 44:293–302
Robison R (1923) The possible significance of hexosephosphoric esters in ossification. Biochem J 17:286–293
Sasagawa I, Ferguson MW (1990) Fine structure of the organic matrix remaining in the mature cap enameloid in *Halichoeres poecilopterus*, teleost. Arch Oral Biol 35:765–770
Sayegh FS, Abousy A (1977) Mitochondrial granule distribution in tooth germ cells. Anat Rec 189:451–466
Sayegh FS, Davis RW, Solomon GC (1974a) Mitochondrial role in cellular mineralization. J Dent Res 53:581–587
Sayegh FS, Solomon GC, Davis RW (1974b) Ultrastructure of intracellular mineralization in the deer's antler. Clin Orthop Relat Res 99:267–284
Schäfer H-J (1973) Ultrastructure and ion distribution of intestinal cell during experimental vitamin-D deficiency rickets in rats. Virchows Arch Abt A Path Anat 359:111–123
Schüler D (2004) Molecular analysis of a subcellular compartment: the magnetosome membrane in *Magnetospirillum gryphiswaldense*. Arch Microbiol 181:1–7
Sela J, Schwartz Z, Swain LD, Boyan BD (1992) The role of matrix vesicles in calcification. In: Bonucci E (ed) Calcification in biological systems. CRC Press, Boca Raton, pp 73–105
Shapiro IM, Greenspan JS (1969) Are mitochondria directly involved in biological mineralization? Calcif Tissue Res 3:100–102
Shapiro IM, Lee NH (1975) Calcium accumulation in chondrocyte mitochondria. Clin Orthop Relat Res 106:323–329
Shepard N (1992) Role of proteoglycans in calcification. In: Bonucci E (ed) Calcification in biological systems. CRC Press, Boca Raton, pp 41–58
Siffert RS (1951) The role of alkaline phosphatase in osteogenesis. J Exp Med 93:415–426
Takano Y, Matsuo S, Wakisaka S, Ichikawa H, Nishikawa S, Akai M (1989) Histochemical localization of calcium in the enamel organ of rat incisors in early-stage amelogenesis. Acta Anat 134:305–311

Tandler CJ, Kierszenbaum AL (1971) Inorganic cations in rat kidney. Localization with potassium pyroantimonate-perfusion fixation. J Cell Biol 50:830–839

Tenenbaum HC (1987) Levamisole and inorganic pyrophosphate inhibit β-glycerophosphate induced mineralization of bone formed in vitro. Bone Miner 3:13–26

Travis DF (1968) The structure and organization of, and the relationship between, the inorganic crystals and the organic matrix of the prismatic region of *Mytilus edulis*. J Ultrastruct Res 23:183–215

Travis DF, Glimcher MJ (1964) The structure and organization of, and the relationship between the organic matrix and the inorganic crystals of embryonic bovine enamel. J Cell Biol 23:447–497

Travis DF, Gonsalves M (1969) Comparative ultrastructure and organization of the prismatic region of two bivalves and its possible relation to the chemical mechanism of boring. Am Zool 9:635–661

Vasington FD, Greenawalt JW (1968) Osmotically lysed rat liver mitochondria. Biochemical and ultrastructural properties in relation to massive ion accumulation. J Cell Biol 39:661–675

Veis A (2003) Mineralization in organic matrix frameworks. Rev Miner Geochem 54:249–289

Wang W, Xu J, Du B, Kirsch T (2005) Role of the progressive ankylosis gene (*ank*) in cartilage mineralization. Mol Cell Biol 25:312–323

Weidmann SM (1963) Calcification of skeletal tissues. Int Rev Connect Tiss Res 1:339–377

Wu LNY, Yoshimori T, Genge BR, Sauer GR, Kiersch T, Ishikawa Y, Wuthier RE (1993) Characterization of the nucleational core complex responsible for mineral induction by growth plate cartilage matrix vesicles. J Biol Chem 268:25084–25094

Wu LNY, Genge BR, Dunkelberger DG, LeGeros RZ, Concannon B, Wuthier RE (1997) Physicochemical characterization of the nucleational core of matrix vesicles. J Biol Chem 272:4404–4411

Wuthier RE (1982) A review of the primary mechanism of endochondral calcification with special emphasis on the role of cells, mitochondria and matrix vesicles. Clin Orthop Relat Res 169:219–242

Wuthier RE, Register TC (1985) Role of alkaline phosphatase, a polyfunctional enzyme, in mineralizing tissues. In: Butler WT (ed) The chemistry and biology of mineralized tissues. Ebsco Media, Inc., Birmingham, pp 113–124

Wuthier RE, Bisaz S, Russell RGG, Fleisch H (1972) Relationship between pyrophosphate, amorphous calcium phosphate, and other factors in the sequence of calcification in vivo. Calcif Tissue Res 10:198–206

Young JR, Davis SA, Bown PR, Mann S (1999) Coccolith ultrastructure and biomineralisation. J Struct Biol 126:195–215

Zhang L, Balcerzak M, Radisson J, Thouverey C, Pikula S, Azzar G, Buchet R (2005) Phosphodiesterase activity of alkaline phosphatase in ATP-initiated Ca(2+) and phosphate deposition in isolated chicken matrix vesicles. J Biol Chem 280:37289–37296

18 Main Suggested Calcification Mechanisms: Extracellular Matrix

18.1 Introduction

The theories presented in Chap. 17, which are based on a direct cellular involvement in calcification, assign no function, or only a secondary one, to the extracellular organic matrix. It has long been known, however, that the matrix may well play a role in calcification. As early as 1923, Freudenberg and György had noted that the cartilage matrix calcifies in vitro if it is treated first with calcium chloride and then with a phosphate mixture, but not the reverse. This led them to put forward the view that in the early part of the process calcium becomes bound to a tissue "colloid", after which phosphate binding can take place, with the probable formation of calcium-phosphoprotein complexes in extracellular spaces.

Early investigations on the role of the extracellular organic matrix in calcification had to face various obstacles, firstly because of its being occluded in the inorganic substance, and, secondly, because of substantial variations in its composition in different types of hard tissue, as revealed by a comparison between bone, cartilage and enamel. Collagen fibrils are the one major component of bone matrix, constituting about 90% of the whole organic mass (reviewed by Bonucci 2000). They are also important constituents of the organic matrix in cartilage, even if they change from type I to type II collagen, but they only constitute about 10% of its wet weight and 40–50% of its dry weight, with a considerable amount of associated proteoglycans (reviewed by Mayne and von der Mark 1983). Collagen is not present at all in enamel matrix, which shows a distinctive composition featuring a high proline content, the virtual absence of cystine, hydroxyproline and hydroxylysine, and the predominance of histidine over lysine and arginine (reviewed by Eastoe 1968).

Probably to overcome the problems raised by these differences, investigations on the calcification process have often been carried out on bone, which is considered a paradigmatic calcified tissue, and on cartilage, which involves fewer technical difficulties than the study of calcification in bone or, more radically, than its study in highly calcified, extremely hard tissues like enamel or mollusk shells.

It was after studying hypertrophic cartilage in vitro that Robison and Rosenheim (1934), who recognized that the theory of the responsibility of alkaline phosphatase for the calcification process was untenable, because amounts of organic phosphates in calcifying matrices are too low for them to act as substrates for the enzyme (see Sect. 17.2), put forward the theory that calcification takes place as a result of the involvement of an enzymatic "second mechanism". More specifically, it was suggested that the onset of calcification can be attributed to a factor, called the "local factor", capable of promoting the linkage of calcium and/or phosphate ions even at low local concentrations (reviewed by Sobel 1955). This factor was thought to be either a specific component, or a complex of components, or a complex of components and/or functions, of the organic matrix.

In line with these concepts, DiStefano et al. (1953) proposed, as a working hypothesis, that calcification could be a process of catalyzed crystallization in which a specific surface or template acts as a center or "seed" inducing crystal formation. They thought that a template of this kind might consist of specific groups of collagen fibers to which calcium or phosphate ions become bound in a spatial conformation corresponding to the hydroxyapatite lattice, and that the association of collagen with chondroitin sulfate might allow calcium ions to form aggregates, so acting as a template, a hypothesis made improbable by the high degree of dissociation of the complex. In chick normal cartilage and in rat rachitic cartilage they identified a very stable phosphate ester of a hexosamine. This ester, tightly bound to material insoluble in trichloracetic acid, was considered to be part of the template, a view supported by the direct correlation between the presence of this phosphate ester and the calcification of cartilage, and by the fact that the complex disappears from fresh cartilage if this is left overnight in saline, a process that simultaneously prevents calcification of the tissue. In the same year, Neuman and Neuman (1953) specified that a "seeding" process may be initiated by the binding of calcium or phosphate ions to organic templates in accordance with the space relationships of the apatite lattice, so that the further aggregation of ions to the "seeds" would result in the growth and formation of definitive crystals, whose orientation would correspond to that of the template. These concepts shifted attention from cells and cellular activities to matrix components and, above all, to fibrils of type I collagen as the structures most probably responsible for triggering the calcification process.

18.2 Extracellular Matrix and Calcification: Collagen Fibrils

One of Glimcher's merits was that of working on a theory on bone calcification based on a process of heterogeneous nucleation catalysed by collagen fibrils (Glimcher 1959). Nucleation is the physical process by which a number of atoms or molecules in metastable equilibrium in a particular phase undergo spontaneous – or homologous – transformation into a new phase by aggregating into 'nuclei', or 'embryos', which can then grow into crystals. The nucleating process is called 'heterogeneous' when the phase transformation is catalysed by a substance – or by substances – added to a system in metastable equilibrium and possessing an atomic organization and molecular arrangement capable of promoting the aggregation of other atoms or molecules in exact alignment with the planes and lattice of the substance itself (for details, see Glimcher and Krane 1968; Eanes 1992). The addition of atoms or molecules to the nuclei, which necessarily occurs in alignment with the lattice of the nuclei, leads to the growth of crystals.

Glimcher's theory on bone calcification was based on three concepts: 1) bone matrix consists mostly of collagen fibrils; 2) the onset of calcification occurs in the holes of the gap zones within collagen fibrils; 3) the generation of the earliest aggregates of inorganic material (nuclei) takes place through a process of heterogeneous nucleation (Glimcher 1959). In Glimcher's view, nucleation centers are generated by special macromolecular aggregates and specific steric relationships between reactive amino acid side-chain groups from adjacent macromolecules within collagen fibrils (reviewed by Glimcher and Krane 1968; Glimcher 1976, 1990, 1992). These distinctive molecular and atomic relationships were thought to be set up in the holes of the gap zones within collagen fibrils (see Sect. 7.2). On this theory, calcification occurs in two phases: in the first, early nuclei are formed in the holes within collagen fibrils; in the second, these nuclei grow into crystals and gradually but completely fill the space constituting the fibril holes, so that, at the end of the process, their shape and size turn out to be identical with those of the holes. According to Glimcher and Krane (1968), approximately 50% of the inorganic substance may be contained in the holes; during a later phase of calcification, crystals may develop in regions outside those holes and eventually be found in interfibrillary spaces.

Evidence in support of this theory has been supplied by X-ray diffraction and electron microscope studies on the organic-inorganic relationships in bone, dentin, cementum and calcified tendons and ligaments (see Sect. 16.2). These studies have confirmed that a close relationship does exist between the inorganic substance and collagen periodic banding; more exactly, they have unequivocally shown that inorganic, electron-dense, granular substance is contained in the hole zone of the fibrils, where it forms structures arranged in bands, corresponding to the collagen fibril period or, more exactly, to holes of adjacent collagen fibrils arranged in lateral register (Glimcher 1959, 1976, 1989; Ascenzi et al. 1965, 1978, 1979; Glimcher and Krane 1968; Bigi et al. 1988, 1991, 1996; Christoffersen and Landis 1991).

In spite of these repeated confirmations, a succession of criticisms and observations has gradually built up, coming to limit the validity of some of the concepts implicit in the theory. Early electron microscope investigations of bone calcification had already shown, in fact, that the relationship between developing crystals and the periodic banding of collagen fibrils cannot always be detected (Ascenzi and Benedetti 1959; Decker 1966). Moreover, in his review of endochondral calcification, Wuthier (1982) had stressed that the native 64 nm-repeat of reconstituted collagen is a poor nucleator of apatite and requires high ion-products and a rather long incubation time to induce the deposition of inorganic structures, whose development is, in any case, independent of collagen periodic banding. His conclusion was that collagen fibrils act as receptors and propagators, rather than initiators, of the calcification process. Developing is own line of arguments, Bachra (1967) held the view that apatite crystals could hardly grow along the backbone of tropocollagen molecules, because the many side-chains of amino acid residues pointing away laterally at a distance of about 0.286 nm along the molecular axis would impede the development of the growing crystals.

Even if we take it for granted, as most electron microscope findings tend to imply, that hydroxyapatite is nucleated within the holes of the gap zones within collagen fibrils, other aspects of the crystal-collagen relationships, besides those discussed above, are still unsettled, especially those involving crystal growth within fibril holes and the relationship between these intrafibrillar crystals and those that are unequivocally located in extrafibrillar spaces.

It has been reported several times that, after their nucleation, crystals grow until they are long enough to span two or more major collagen periods (Fig. 18.1) – using the term 'period' to refer to the segments of collagen fibrils that are formed by a gap plus an overlap zone (Ascenzi et al. 1965, 1978, 1979; Höhling et al. 1990). This has been shown clearly in calcified turkey leg tendons visualized by selected-area dark field electron microscopy: at an early stage, calcification is characterized by rod-shaped crystals within the holes of the gap zones of collagen fibrils, whereas at a later stage crystals of larger size occupy not only the gap zones but the overlap zones, too (Arsenault 1988). As the gap zone in the direction of the collagen fibril axis is approximately 40 nm in length, the *c*-axis of the crystals tends to be aligned with crystal length

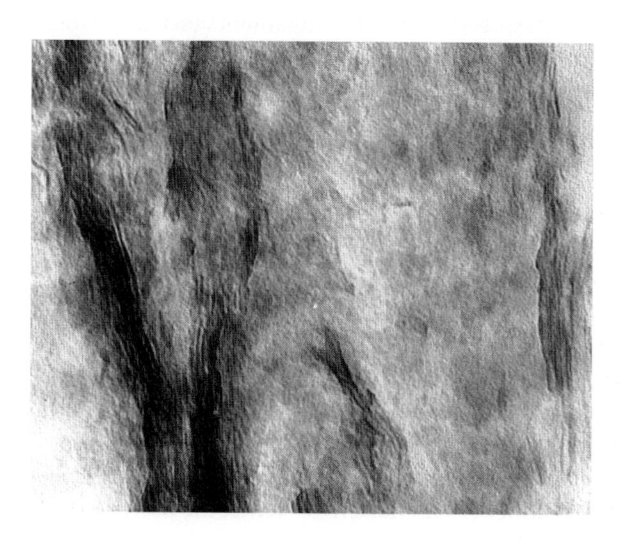

Fig. 18.1. Compact osteonic bone at low degree of calcification: note filament-like crystals longer than granular electron-dense bands. Unstained, ×240,000

(Moradian-Oldak et al. 1991), and this in its turn is parallel to the collagen fibril axis, it has been calculated that many crystals are so long that they show continuity between, and cover, both the gap and the overlap zones (Ziv and Weiner 1994). A comparative electron microscope study of crystals in collagen fibrils of rat bone and dentin, and in calcified turkey leg tendons, has confirmed that filament-like crystals cover both the gap and overlap zones of the collagen fibrils, and that about 80% of crystals are housed in the collagen holes at the beginning of tendon calcification, whereas, as calcification advances, 55% of them are located in the gap and 45% in the overlap zones, percentages also observed elsewhere, in cortical bone and in dentin (Arsenault 1989). Studies carried out by low temperature substitution, fixation and staining, followed by low temperature embedding, or by using frozen hydrated and air dried cryosections, all examined under the electron microscope at a temperature of −165 °C or at normal temperature, also showed that crystals are not confined to axial periodicities such as those of the gap zones but lie outside them, too (Arsenault 1991).

These results are far from being easy to interpret: if crystals are nucleated and grow in the holes of the gap zones within collagen fibrils, as Glimcher indicated, once they are fully developed they should have the same size as these holes. Any further growth should, in fact, imply an enlargement of the available space, which could only be produced by breaking the intermolecular cross links that maintain the structural arrangement and integrity of collagen fibrils. That cleavage would raise the solubility of bone collagen, which, conversely, falls with calcification (discussed by Bonucci 1992). On the other hand, a low-angle X-ray diffraction analysis of the inorganic phase of osteons did not show any change of the collagen periodic banding with increasing degree of matrix calcification (Ascenzi et al. 1985).

Actually, the length and width of crystals in bone and related tissues exceed those of the holes in the fibril gap zones (see Sect. 6.2.1). If crystals are nucleated within these holes, and if collagen molecules are packed as suggested by the model proposed by Hodge and Petruska, then the intermolecular cross links would have to be broken by growing crystals. It has been suggested that these crystals might fit into the pores, channels and grooves resulting from the confluence of contiguous holes (Christoffersen and Landis 1991; Landis and Song 1991; Weiner and Traub 1986; Landis et al. 1996a), but these spaces, if any, appear to be quite insufficient to contain all the long, filament-like crystals that can be detected in fully calcified bone, especially in that of woven type. On the other hand, the theory that collagen fibrils have a periodic structure in the axial direction, but behave like a two-dimensional fluid in the lateral plane, so explaining how crystals can be deposited inside the overlapping zone (Fratzl et al. 1993), still awaits confirmation.

The studies of Lees and Prostak (1988), besides showing that the dimensions of hydroxyapatite crystals exceed the intermolecular spacing in collagen fibrils, have also made it clear that most crystals are located in interfibrillar spaces.

These authors calculated that the ratio between the weight of the inorganic material that fills the collagen hole zones and the total inorganic content is limited to 20% in poorly calcified bones and falls to less than 5% in highly calcified ones. Weiner and Price (1986) measured the dimensions of isolated, platelet-like crystals from disaggregated bone, and found that few of the crystals are shorter than about 45.0 nm and that their width averages about 25.0 nm; in other words, they are quite often much larger than the holes in the gap regions of collagen fibrils (according to Bonar et al. (1985), the average equatorial spacing between hydrated collagen molecules fluctuates around 1.5 nm; according to Katz and Li (1973), the intermolecular spaces in fibrils have lateral dimensions of 2.5–7.5 nm and 0.6–1.2 nm); they concluded that crystals could not be located in the collagen holes, although they might fit into grooves created by contiguous adjacent gaps (Weiner and Traub 1986). According to Fratzl et al. (1991), however, calcium phosphate is nucleated within the hole zones of collagen fibrils and then grows in thickness to about 3.0 nm, which is exactly the maximum space available in the holes. Pidaparti et al. (1996) believe that about 75% of crystals reside outside collagen fibrils, although their *c*-axis runs parallel to the fibril long axis; according to Lees (1987), very little inorganic material is located within collagen fibrils. On the basis of the observation that collagen fibrils in calcified turkey leg tendons are surrounded by highly calcified material, Lees et al. (1997) came to the conclusion that the extrafibrillar material has a different composition from the intrafibrillar. On the basis of personal experience, the conclusion can be drawn that in bone only the granular inorganic substance arranged in bands lies within collagen fibrils, whereas the filament-like crystals are located in the interfibrillary space; these are crystals whose amounts vary with the type of bone, the highest being found in woven bone, the lowest in compact, lamellar bone (see Sect. 16.2).

Analysis of the organic-inorganic relationships in bone of the eso- and endo-skeleton of lower vertebrates has given partly conflicting results, but they seem to point to a common conclusions. On the basis of studies on pickerel and herring fish bone, already discussed in Sect. 16.5, it has been maintained that at the earliest stage of calcification, and continuing throughout the calcification process, the only crystals to be found are those located within collagen fibrils, which are completely occupied by them, except for a few that are located on, or close to, the fibril surface (Glimcher 1959; Lee and Glimcher 1989, 1991). On the basis of these results, it has been stated that crystals are not formed in matrix vesicles and that they are located within the collagen fibrils. Alongside these results, however, other ultrastructural studies have shown that early crystals in fish scales and dentin, as well as in some human neoplastic conditions, are exclusively located between and around collagen fibrils (Sect. 16.5), which therefore appear to be surrounded by, and enclosed in, rings of crystals (Fig. 16.1) (Herold 1971; Lanzing and Wright 1976; Onozato and Watabe 1979; Schönbörner et al. 1979; Yamada and Watabe 1979; Olson and Watabe 1980; Zylberberg and Nicolas 1982; Lees and Prostak

1988; Zylberberg et al. 1992). These rings fail to show any relationship with the periodic structure of collagen (Landis and Géraudie 1990), and crystals do not penetrate deeply into fibrils, which in cross section appear as electron-lucent spaces (Zylberberg et al. 1992; Su et al. 2003). This excludes the hypothesis that crystals could have been nucleated in the holes of collagen fibrils. Similar findings have been reported in calcified turkey leg tendons (Höhling et al. 1976; Arsenault et al. 1991) and in some pathological conditions (Bonucci and De Santis 1980; Boivin et al. 1987). In the osteoid areas of calcifying embryonic rat calvaria, calcification occurred both inside and outside collagen fibrils, but the interior areas displayed a lower density (Hoshi et al. 1999). The idea that crystals can have an independent extrafibrillar location is also strengthened by the finding that their degree of orientation is reduced by cyclical loads, while the orientation of collagen fibrils remains unchanged (Ascenzi et al. 1998), which suggests that these crystals are mechanically distinct from fibrils.

The convergence of so many results leaves little doubt that the long, filament-like crystals are formed outside the fibrils, are located on the fibril surface and in interfibrillary spaces, and are, in fact, independent of, and physically separate from, the granular inorganic substance in bands contained in the fibril holes. It follows that the formation of these filament-like crystals must be dependent on the catalytic activity of organic structures, which cannot include collagen fibrils and must be looked for among the non-collagenous components associated with, or located between, collagen fibrils. These components may have a specific relationship with the fibrils; they could act as nucleating sites within the hole regions of native collagen, as suggested for phosphoproteins (Veis and Perry 1967), and as shown for dentin phosphophoryn, which has been immunolocalized at the boundaries between the gap and overlap zones of the fibrils nearest the calcification front of dentin (Beniash et al. 2000). Alternatively, as emphasized by Traub et al. (1992a), the initiation of crystal formation might, at least partly, derive from the binding of a nucleating protein to specific sites on the collagen fibril surface; as an example, they selected the fact that phosphophoryn, which is considered a mediator of calcification in dentin (Linde and Goldberg 1993), links to the surface of collagen fibrils, mostly near their *e* band (Traub et al. 1992b).

To explain the formation of crystals outside collagen fibrils in interfibrillar spaces, a process of secondary nucleation has been proposed: on that view, the early crystals formed in the holes of the collagen fibrils act as secondary nucleation centers for the formation of crystals outside fibrils (Glimcher and Krane 1968). This is a theoretical possibility, but it has never been demonstrated experimentally.

In this connection, secondary nucleation has been considered a possible mechanism for the seeding of enamel crystals: at the dentino-enamel junction there is intimate contact between dentin and enamel crystals (Fearnhead 1979), and selected-area dark-field electron microscopy (Arsenault and Robinson 1989) and high resolution electron microscopy (Hayashi 1992; Tohda et al.

1997) have both shown that they may be connected, with continuity between their lattice fringes. On this basis, it has been proposed that enamel crystals are the outcome of secondarily nucleation by dentin crystals (Fearnhead 1979; Arsenault and Robinson 1989; Tohda et al. 1997), or even that the former could grow as extensions of the latter (passive seeding; discussed by Nanci and Smith, 1992). Such processes would explain why the enamel matrix can calcify as soon as it has been produced by ameloblasts, but, as reported by Kallenbach (1971), only after a thin layer of mantle dentin has been laid down. It has, however, been shown that the goniometric tilting of sections allows resolution of the apparently fused crystals with lattice fringe continuity into separate crystals (Dong and Warshawsky 1996) and that early enamel crystals can be spatially distinct from those of dentin (Diekwisch et al. 1995). Dentin sialophosphoprotein has been considered a candidate initiator for both dentin and enamel crystal formation (Ritchie et al. 1997; MacDougall et al. 1998).

As already discussed (Sect. 16.2), and on the basis of results given by transmission electron microscope (Ascenzi et al. 1965, 1967) and atomic force microscope (Lees et al. 1994) studies, it can reasonably be admitted that at least two types of inorganic structures exist in bone and related collagenous tissues, the first consisting of granular substance in bands connected to the collagen period and located within the holes of the gap zones, the second of filament-like crystals which are oriented parallel to collagen fibrils but are unrelated to their period; these are located in interfibrillary spaces and on the fibril surface. The relationships between these types of structures has not been fully defined, but the suggested derivation of the latter from the former through a process of secondary nucleation has not support in experimental findings.

Filament-like crystals can, in any case, be nucleated and develop on the surface of collagen fibrils (Cameron 1963; Decker 1966; Landis et al. 1996b; Bigi et al. 1997) without appreciably affecting their molecular packing or organization (Ascenzi et al. 1985). In this connection, neutron diffraction (Bonar et al. 1985), as well as electron microscope studies (Bonucci and Reurink 1978), have indicated that collagen fibrils in fully calcified compact bone are considerably more closely packed than had previously been assumed, so that their surfaces would hardly permit a nucleation process. Filament-like crystals are, however, clearly detectable between and on collagen fibrils in even the most compact bone (Bonar et al. 1985), although the granular inorganic substance arranged in band may be prevalent. Further support for the view that these crystals are interfibrillar in origin comes from the electron microscope study of the calcification front. At this site, crystals are aggregated in two different ways – as calcification nodules and calcification islands (see Sect. 16.2). The former are generated within matrix vesicles, the latter in connection with interfibrillar, non-collagenous matrix; both these components grow and coalesce, and crystals spread to fill the whole interfibrillar space. The granular inorganic substance in bands develops soon after within collagen fibrils.

If this view is correct, the filament-like crystals and the granular substance in bands are separate, autonomous structures. The former develop in matrix vesicles and in connection with the non-collagenous components of the matrix, and grow between collagen fibrils; the latter develop within the holes of the fibril gap zones, grow till they fill the hole spaces, and remain confined in them. In spite of its granular appearance, and probably as a result of its links with organic molecules, this substance has some degree of coherence, without necessarily being crystalline. This is suggested by the observation that it is possible to dissociate bone matrix, so isolating platelet-, crystal-like inorganic structures – probably the same structures that look like granular substance in bands – which have almost the same shape and size as the holes in the gap zones (see Sect. 5.2.1). These structures are not true crystals. It has, in fact, been shown that the platelet-like inorganic structures isolated from fully calcified bone have such a poor stochiometry and such a low degree of crystallinity that they cannot rightly be called crystals (the term 'mineralites' has been proposed for them; see Sect. 5.2.1 and Kim et al. 1995; Eppell et al. 2001; Tong et al. 2003).

A great deal remains to be discovered about the true nature and process of formation of the granular inorganic substance arranged in bands. It is mostly formed in very dense collagenous tissues, such as compact bone, dentin and tendons, where there is little interfibrillary space for the growth of filament-like crystals, which are predominant in woven bone; it is masked by filament-like crystals as these develop in interfibrillary spaces; it is restricted to the holes of the gap zones and does not reach the overlap zone, so that the intrafibrillar cross-links are not broken; most importantly, its crystallinity is poorly developed, so that collagen molecules are not spaced out by growing crystals. The formation of the granular inorganic substance arranged in bands is, therefore, mechanically favourable, because, contrary to the formation of filament-like crystals, it does not imply the rupture of the intermolecular cross-links.

The theory of heterogeneous nucleation and crystal growth in the holes of collagen fibrils was so simple, elegant and convincing, and, as commented by Wuthier (1982), was prompted so insistently for many years, that it became widely applauded and accepted. Ennever and Creamer (1967) did, however, report Pautard's view that it was unfortunate that the concentration of research on collagen had set back the study of other organic structures as possible promoters of calcification in biological tissues, and that the resulting, almost universal interest in type I collagen had dimmed the role that other proteins and polysaccharides probably play in inducing calcification.

This is true of type II collagen, whose role in calcification has been neglected, although collagen II mutants have severe or lethal chondrodysplasia characterized by a disorganized growth plate, an impaired collagen network, a defective endochondral bone formation, and an abnormal development of intervertebral disks (Gustafsson et al. 2003). In an exhaustive review on collagens in cartilage, including type II collagen, its relationship with calcification was not even considered (Mayne and von der Mark 1983); it must, in any case, be admitted that no direct relationship between type II collagen and crystals has ever been demonstrated so far. Rather than being a promoter or regulator of the calcification process, type II collagen appears to provide a supporting framework which the molecules of other matrix components become attached to (Lash and Vasan 1983). Even so, Mwale et al. (2002) have reported that a selective, progressive removal of types II and IX collagen, with the retention of proteoglycans, occurs during the calcification of fetal bovine epiphyseal cartilage.

Type X collagen has attracted attention because it is typically expressed during the differentiation of growth cartilage chondrocytes. As discussed in Sect. 9.2, it is a specific product of hypertrophic chondrocytes (although hypertrophy does not seem to be a necessary precondition for its expression, as reported by Ekanayake and Hall, 1994), is associated with collagen fibrils, and precedes matrix calcification. This is why it has been profiled as a possible mediator of the calcification process; according to Poole et al. (1989), it continues to be associated with type II collagen fibrils during cartilage calcification and induces the formation of inorganic substance in focal interfibrillary sites. According to Wu et al. (1991), collagen types II and X both bind to matrix vesicles and cosediment with them, so promoting interactions between matrix vesicles and the extracellular matrix which may play an important role in matrix calcification. Even if type X collagen binds calcium in a specific, dose dependent manner (Kirsch and Von der Mark 1991), results supporting its direct role in calcification are inconclusive. Type X collagen seems to have little impact on apatite deposition, apart from its probable function in providing a structurally organized tissue (Kwan et al. 1997; Boskey 1998). Its attachment to matrix vesicles (Kirsch and Claassen 2000) may, however, be necessary for their calcification (Kirsch et al. 1994), as suggested by the observation that isolated matrix vesicles, when treated with pure bacterial collagenase and 1 M NaCl in synthetic cartilage lymph to selectively and completely remove associated collagen types II and X , showed a marked fall in the uptake of $^{45}Ca^{2+}$, although the structure of the vesicle membrane and the nucleational core inside the vesicle lumen were still intact; the replenishment of either native type, II or X, collagen in these vesicles stimulated their Ca^{2+} uptake to levels close to those of untreated vesicles (Kirsch and Wuthier 1994). The stimulation of annexin V channel activity by type X (or II) collagen might explain the rapid uptake of Ca^{2+} by matrix vesicles (von der Mark and Mollenhauer 1997; Kirsch et al. 2000).

Type X collagen transgenic animals, which express a truncated collagen X protein, undergo the disruption of their hypertrophic chondrocyte pericellular matrix and abnormal proteoglycan-glycosaminoglycan distribution (Jacenko et al. 2001) and, when examined with FTIR spectroscopy, show poor crystal 'quality' (a term used to refer to the relationship between degree of crystallinity and crystal maturation on one hand, and acid phosphate content on the other); they also show the development of vertebral abnormalities in advanced age

(Paschalis et al. 1996a,b). Type X collagen-null mice do, however, show that long bones grow and develop normally (Rosati et al. 1994).

Type X collagen, along with acid proteoglycans and other proteins, is also a constituent of eggshell matrix (Carrino et al. 1996; Fernandez et al. 1997; see Sect. 13.4).

18.3 Extracellular Matrix and Calcification: Acid Proteoglycans

The failure of alkaline phosphatase as a local factor able to promote calcification, and the need to find one or more factors to take its place, very soon led to a consideration of the role in calcification of 'mucopolysaccharides', as the proteoglycans were then called. As early as 1971, Kobayashi was able to report four different theories on this topic, in all of which proteoglycans were viewed as to be promoters (Sobel 1955) or inhibitors (Glimcher 1959) of the process.

Interest in acid proteoglycans and glycosaminoglycans as possible representatives of the 'local factor' mainly derived from the findings that they strongly bind calcium ions, so maintaining calcium phosphate in a stable colloidal state (Campo 1970), that they are components of a number of calcifying tissues (without excluding non-calcifying ones), and of cartilage matrix and its calcification nodules, in particular (Campo 1974; Frasier et al. 1975), that a large proportion of cartilage calcium is bound to a fraction of the chondroitin sulfate associated with collagen (Bowness 1962), and that, during the early stage of calcification, most calcifying matrices are PAS-positive (Leblond et al. 1959), that is, contain glycoproteins, and are basophilic and metachromatic (Cabrini 1961), that is, contain acid glycosaminoglycans. The hypothesis has been put forward that proteoglycans can bind high concentrations of calcium and that this can be released locally by enzymatic modification or destruction of their molecules (Woodward and Davidson 1968), or simply by the ion-exchange effect of phosphates (Hunter 1987), so creating an environment permitting the precipitation of hydroxyapatite (see reviews by Poole 1991; Hunter 1992). According to Yang et al. (2003), polysaccharides not only direct the nucleation and growth of $CaCO_3$ crystals, but these in their turn have a certain impact on polysaccharide molecules.

Even if proteoglycans are a component of a wide variety of calcified matrices (ranging from bone to cartilage, from dentin to enamel, from sea urchin spicules to mollusk shells, and from eggshell to otoliths and otochonia, to mention only the commonest), investigations on their role in calcification have mainly been focused on those found in epiphyseal cartilage. This tissue does, in fact, contain high concentrations of proteoglycans, is easily available, is not too resistant to permit morphological studies, and its calcification process can be considered a paradigm for the calcification of other tissues. This is probably the underlying reason why chondroitin sulfate, the most abundant glycosaminoglycan in cartilage aggrecan (see Sect. 9.3.1), was one of the first molecules to be thought responsible for calcium binding and calcification. Results drawn from the in vitro reversible inactivation of calcification indicated that a complex of chondroitin sulfate and collagen in a critical conformation might be responsible for inducing cartilage calcification (Sobel 1955), a view supported by the in vitro findings that this is reversibly inhibited by substances (toluidine blue or protamine) which form links with chondroitin sulfate (Sobel and Burger 1954), and that bone cells in the rat, and pulp cells from human teeth, if grown on type I collagen sponge, synthesize a calcifiable matrix only when chondroitin sulfate is contained in the sponge (Bouvier et al. 1990a,b). According to Fernandez et al. (1997), keratan sulfate, due to its precise localization, temporal appearance and Ca-binding affinity, may be related to the maintenance of a Ca reserve, whereas dermatan sulfate may control crystal growth and orientation.

A number of studies (reviewed by Kobayashi 1971) have assessed the role of acid proteoglycans in calcification, but have failed to reach unambiguous conclusions. The role of chondroitin sulfate as a local factor, or as one component of a local factor, promoting calcification has been questioned mainly because it is present in non-calcifying tissues, and it is the main component of cartilage proteoglycans, which are regarded by many as inhibitors of the calcification process. That an inhibitory substance might be present in cartilage matrix, and that it might be removed before calcification, was suggested by Pita et al. (1970) on the basis of the effects on apatite precipitation of aspirates of epiphyseal cartilage interstitial fluid. It was later shown that proteoglycans, dissociatively extracted from bovine fetal or nasal cartilage and reaggregated in vitro, delay the growth of hydroxyapatite seeds into crystals (Chen and Boskey 1986). If the synthesis of proteoglycans is inhibited by retinoic acid, or if they are removed by digestion with chondroitinase ABC or hyaluronidase, the degree of in vitro cartilage matrix calcification shows a substantial increase (Boskey et al. 1997). It was also shown that proteoglycan monomers exert a stronger inhibitory effect than proteoglycan aggregates; the latter prevent in vitro calcium and phosphate precipitation at a concentration of 2 mg/ml (Dziewiatkowski and Majznerski 1985).

One possibility is that proteoglycans inhibit calcification by binding calcium ions (discussed by Hunter 1992); the pre-equilibration of chondroitin sulfate or proteoglycans with Ca^{2+} before the addition of phosphate does lead to levels of hydroxyapatite precipitation that are higher than in controls (Hunter and Szigety 1992). The in vitro effect of a variety of proteoglycans and glycosaminoglycans on crystal growth resulted in 93% inhibition with the addition of purified glycosaminoglycan chains, 55 and 37% inhibition with the addition of core proteins and intact proteoglycans, respectively, 6% inhibition with the addition of chondroitin-4-sulfate and no inhibition with the addition of chondroitin-6-sulfate (Rees et al. 2002). At a calcium concentration similar to that of serum, chondroitin-4-sulfate is reported to bind five times more calcium than does dermatan sulfate (Embery et al. 1998). According to

Blumenthal et al. (1979), proteoglycan aggregates are potent inhibitors of hydroxyapatite formation and their enzymatic cleavage into subunits may not be enough to allow calcification to occur; the subunit fractions themselves do, in fact, prove to be inhibitors of apatite formation and their further degradation or removal might be a precondition for calcification to take place.

At this point it should be recalled that the functional activity of polyanions, like that of acid proteoglycans, changes according to whether they are free in solution or are immobilized on a fixed substrate (Sect. 8.5.1). In the first case they are able to inhibit crystal induction and growth – a process that could, incidentally, be important in allowing the inhibition of stones in urines, whereas in the second they are, even in tiny quantities, capable of inducing calcification at physiological concentrations of calcium and phosphorus (Linde et al. 1989). It has also been reported that proteoglycans inhibit calcification in vitro, where Ca^{2+} availability is limited and Ca^{2+} ions are easily exhausted, but not in vivo, where Ca^{2+} ions are quickly replaced (Hunter 1991).

These results suggest that the role of acid proteoglycans in calcification depends both on their status (whether aggregate or monomer) and on the hydrodynamic size of their molecules (Chen et al. 1984), so that in vivo it may well change with every change in these parameters. The enzymatic breakdown of aggrecans may turn out to be a control mechanism during the calcification of growth plate cartilage (Eanes and Hailer 1994); changes in proteoglycans before or during calcification have, in fact, been reported in epiphyseal cartilage and other calcifying tissues (reviewed in Takagi 1990). With reference to the calcification of coccoliths and the function of polyanions as calcium ion carriers, Marsh (1994) has pointed out that the dissociation of calcium from acid proteoglycans and their transfer to crystals at the calcification front cannot occur by acidification, because the inorganic substance would be solubilized, or by chelation, because chelated calcium is not available for crystal formation; by contrast, the enzymatic degradation of polyanions raises calcium concentrations and allows calcium carbonate to be precipitated.

Hirschman and Dziewiatkowski (1966) were probably the first to suggest that protein-polysaccharides, or their protein component, are lost or drastically altered during or just before cartilage calcification. Using energy dispersive X-ray elemental analysis, Boyde and Shapiro (1980) found that matrix sulfur levels fall with cartilage calcification, indicating a loss of sulfated proteoglycans. By combining biochemical with X-ray microprobe analyses, Althoff et al. (1982) were able to confirm that the extracellular S content (per dry mass) is relatively constant at about 3.5% in the various zones of the epiphyseal cartilage, but falls to about 0.3% in the fully mineralized regions, indicating a loss of sulfur-containing substances with calcification. Lohmander and Hjerpe (1975) found that, with the onset of calcification, cartilage loses about half its content in proteoglycans, that those remaining in the calcified matrix differ in composition and size from those of non-calcified matrix, and that, with increasing calcification, gel chromatograms indicate falling proportions of very high molecular weight proteoglycans in the calcified matrix. A marked decrease in the size of granules, which are the form taken by cartilage protein-polysaccharides under the electron microscope (see Sect. 9.3.1), was reported to occur as one outcome of matrix calcification by Matukas and Krikos (1968) and by Takagi et al. (1983), a result confirmed in shark cartilage (Takagi et al. 1984). Silbermann and Frommer (1973) found that, at the calcification front, acid proteoglycan complexes can be enzymatically degraded to yield a highly anionic fraction which persists in the matrix and can be visualized by cationic dyes. Dean et al. (1992, 1994) reported that matrix vesicles have a high content of proteoglycan-degrading enzymes, Kuettner et al. (1974) proposed that proteoglycan degradation is due to lysozyme, and Armstrong et al. (2002) suggested that the same result can be produced by stromelysin, a metalloprotease they were able to demonstrate in bovine epiphyseal cartilage. Loss of glycosaminoglycans before or during cartilage calcification was confirmed by de Bernard et al. (1977), Vittur et al. (1979) and Mitchell et al. (1982), and basic ultrastructural differences between uncalcified and calcified cartilage matrix were described by Akisaka et al. (1998). Buckwalter et al. (1987) found that, before matrix calcification in the lower proliferative zone and right at the onset of calcification in the hypertrophic zone, the proteoglycan aggregate size and the proportion of aggregated monomers both fall, and the proportion of non aggregated monomers rises. A fall in decorin concentration, and a moderate fall in that of fibromodulin and biglycan, were detected in the maturation and hypertrophy zones of bovine growth plate by Alini and Roughley (2001).

Changes in proteoglycans with increasing calcification have also been found in bone (see Sect. 7.3.1) and dentin. Studies by Prince et al. (1983, 1984), based on the in vivo incorporation of ^{35}S into glycosaminoglycans within calvaria, femoral metaphyses, femoral diaphyses, and incisors, showed that 25–50% of the newly synthesized glycosaminoglycans are lost during matrix calcification. The existence of two distinct groups of ^{35}S-labeled proteoglycans, one exclusively related to predentin and disappearing with time, and the second located in dentin, was reported by Lormée et al. (1996). Active proteoglycanases have been found in predentin, where they were unable to induce proteoglycan degradation because of the low availability of calcium ions, whereas they did degrade proteoglycans at the calcification front, where calcium ions are plentiful (Fukae et al. 1994). A fall in the total amounts and molecular size of glycosaminoglycans was likewise reported by Engfeldt and Hjerpe (1976) in fractionated bone particles, and a loss of proteoglycans in the compact bone of isolated osteons was detected by Pugliarello et al. (1970). Using the HID-TCH-SP method for sulfated glycoconjugates, and Thiéry's method for vicinal glycol-containing glycoconjugates, Takagi et al. (1981) showed that both molecular types are secreted in the extracellular space by odontoblasts, and that vicinal-glycol containing substances concentrate at the calcification front, whereas sulfated ones accumulate in predentin and are removed or at least masked in dentin. Using cuprolinic blue to detect proteoglycans under the electron microscope, and immunohistochemistry for other substances, Hoshi et al. (2001a) found that, in embryonic rat calvaria, many cuprolinic blue particles were distributed around collagen fibrils in the osteoid tissue, that the size of these particles fell as calcification proceeded, and that small particles accumulated in early calcification nodules and at the periphery of the larger ones, where, according to Nakamura et al. (2001), keratan sulfate tends to accumulate too, and where electron microscopy shows protruding crystal ghosts (Bonucci 2002). Hoshi et al. (2001b) also found that osteoid tissue contains plenty of mesh-like proteoglycan fibers which include decorin, chondroitin-4-sulfate and hyluronan, that calcium tends to localize at proteoglycan sites, that Ca/P co-localize in and around calcified nodules, where ALP and small sized proteoglycans can be made out, and that during early calcification the proteoglycans that initially surround collagen fibrils disappear, so permitting lateral fibril fusion. A fall in decorin concentration, and side by side fusion of collagen fibrils, were found in the calcifying osteoid tissue of embryonic rat calvariae (Hoshi et al. 1999).

Subtle structural modifications in small leucine-rich proteoglycans have been reported to occur going from predentin to the predentin-dentin interface and from this interface to dentin (Waddington et al. 2003): decorin and biglycan were identified in all three sites, but an analysis of glycosaminoglycan constituents in predentin showed a high proportion (51.5%) of dermatan sulfate and minor concentrations of chondroitin sulfate (17.8%) and hyaluronan (30.7%); at the predentin-dentin interface the percentage of chondroitin sulfate rose sharply (62.5%), whereas there was a fall in concentrations of dermatan sulfate (21.4%) and hyaluronan (16.1%); in dentin matrix only chondroitin sulfate was identifiable. A sharp rise in decorin concentrations was reported in distal predentin, near the predentin/dentin transition zone, in rat teeth (Goldberg et al. 2003). In human aortic and coronary atherosclerotic lesions, calcified areas contain decorin, which appears to promote intimal calcification (Fischer et al. 2004).

It is worth stressing that the effect of biglycan, if any, on bone calcification appears to be limited, because disruption of the biglycan gene in mice causes osteoporosis (Xu et al. 1998) and leads to structural abnormalities, synergized by decorin deficiency, in collagen fibrils in bone, dermis, and tendon (Corsi et al. 2002), but does not induce calcification defects. Biglycan and decorin deficient newborn mice show hypomineralized dentin, which in the former is associated with an increase, and in the latter with a delay, in enamel formation, but no alterations are detectable in adult mice (Goldberg et al. 2005). On the other hand, studies by Parisuthiman et al. (2005) have shown that osteoblast differentiation and matrix calcification are severely impaired in MC3T3-E1 cell-derived clones expressing low levels of biglycan. According to Sugars et al. (2003), decorin and biglycan strongly associate with hydroxyapatite and inhibit crystal growth, but the degree of inhibition is curtailed when these proteoglycans are complexed with type I collagen fibrils.

On the basis of the results reported above, structural changes and a loss of proteoglycans seem to constitute a generalized process co-occurring with calcification in a number of tissues. Other findings, however, support the view that no proteoglycan loss occurs. Electron microscope microprobe analysis, carried out on sections of frozen and freeze-dried specimens by Barckhaus et al. (1981), have shown that sulfur content does not significantly decrease with calcification. Hargest et al. (1985) reported that sulfur levels greatly exceeded Ca levels in the upper hypertrophic zone, and Alini et al. (1992) found that the concentration per unit matrix volume of C-propeptide of type II collagen, aggrecan, and hyaluronic acid all progressively increased from maturing to hypertrophic zones. Using high resolution electron spectroscopic imaging, Arsenault and Ottensmeyer (1983) found that in calcifying and calcified zones there is a spatial overlap of calcium and sulfur, reaching a maximum value at the onset of calcification. On the basis of the number of matrix granules and the affinity of the matrix for colloidal thorium dioxide, Scherft and Moskalewsky (1984) concluded that the degradation of proteoglycans is not a necessary step in cartilage calcification. The results of a morphometric analysis of aggregating proteoglycans, carried out in the proliferative, maturing, hypertrophic, and calcifying zones of the bovine fetal epiphysis, led Matsui et al. (1991) to conclude that large aggregating proteoglycans reach their highest concentrations in the extracellular matrix of the hypertrophic zone, just when calcification starts, so that there is no need for their degradation or removal for calcification to begin, even if proteoglycans are progressively lost as matrix volume falls. According to Byers et al. (1997), chondrocytes in the proliferative zone synthesize proteoglycan monomers which are larger than those in articular or resting cartilage and increase further in size in maturing and, even more so, in hypertrophic cartilage, where they show a rising proportion of unsulfated residues. On the other hand, Campo and Dziewiatkowski (1963) concluded that, although proteoglycans are somehow removed before cartilage calcification, chondroitin sulfate is partly retained in the metaphyseal spicules of bone.

In agreement with the conclusions of Poole et al. (1982), these results suggest that there is no net loss of proteoglycans during cartilage calcification; they are, rather, depolymerized and modified, and are then entombed in the calcified cartilage matrix. In this connection, Campo and Romano (1986), after the gel chromatographic detection of small proteoglycan subunits in the calcified zones within cartilage, reached the conclusion that proteoglycan aggregates undergo partial degradation and disaggregation, but remain components of the calcified matrix, as shown by evidence that it incorporates ^{35}S-sodium sulfate. This is in accordance with the finding that the trypsin digestion of decalcified bone matrix releases a proteoglycan fragment which can be precipitated with $CaCl_2$, is closely associated both with the inorganic fraction and with the collagen matrix, and cannot be extracted by the trypsin digestion of undecalcified bone matrix (Hashimoto et al. 1995).

These results concur with those on the function of crystal ghosts, which, at least in cartilage, bone, and dentin, correspond to acid proteoglycan molecules, and with their changes in the central, fully calcified zones of the matrix (Sect. 16.2.1).

18.4 Extracellular Matrix and Calcification: Crystal Ghosts

Crystal ghosts – definable as the organic structures which after decalcification and staining, when viewed under the electron microscope, show the same ultrastructural morphology as the untreated calcified structures they replace, even when these do not appear as crystals (see Sect. 16.9) – have been found in all areas of early calcification, ranging from those in vertebrates to those in unicellular organisms, where they have been correctly investigated (Bonucci 1979), especially when the PEDS method has been used (reviewed by Bonucci 2002). Their presence is so constant and reproducible, and their morphological resemblance to untreated crystals is so striking, that their probable role in calcification could hardly be overlooked. They have, in fact, been considered organic structures involved in, and promoting, the earliest stages of the calcification process (Bonucci 1967, 1969, 1971, 1975, 1984, 1987, 1992b, 1995; Bonucci and Gherardi 1975; Bonucci and Reurink 1978; Bonucci et al. 1988, 1989; Bonucci and Silvestrini 1996). More exactly, it has been suggested that: 1) crystal ghosts are preformed organic structures whose reacting groups have a steric configuration that matches the hydroxyapatite crystal lattice; 2) when calcium ions are bound to these groups, apatite-like nanoparticles are formed which necessarily come to take up the same location as that of the reacting groups; 3) they are tridimensionally located along and around the template molecule if this has a filamentous conformation, and on its surface if it has a sheet-like structure; 4) as the nanoparticles grow and fuse, the template organic molecules are entrapped by them, so that they become internalized in crystal-like, organic-inorganic, poorly crystalline structures which acquire the same shape as that of the template crystal ghosts; 5) the later loss of organic components allows the growth, maturation and perfection of crystals (reviewed by Bonucci 2002).

At this point, the results of electron microscope studies carried out by Ascenzi et al. (1965) and Ascenzi and Bonucci (1966) on isolated osteons, by Ascenzi et al. (1963) and Bonucci and Gherardi (1975) on medullary bone, by Bonucci (1975) on epiphyseal cartilage, by Schönbörner et al. (1979) on fish scales, by Boivin (1975) on calcified skin, and above all on a variety of calcified tissues by Höhling and his group (for reviews see Höhling et al. 1981, 1990, 1995, 1997; Höhling 1984, 1989; Plate et al. 1998) should be taken into consideration, because of their pertinence to the topic under discussion in this chapter and their intrinsic scientific interest. These results support the concept that the primary inorganic aggregates formed in bone, dentine and other hard tissues consist of nano-sized apatite particles (referred to as 'granules', 'small spots' or 'islands'), which are nucleated along organic molecules and appear as beaded strands or ropes. They have an apatite structure, but are characterized by pronounced crystal lattice distortions, representing an intermediate state between amorphous and fully crystalline (Arnold et al. 2001). These nanoparticles rapidly coalesce into needle-like crystals, which then fuse, in Höhling's view, into platelet-like crystals. These granular strands are crystallographically oriented along the bipolar *c*-axis, so the center to center distances between the islands are approximately equal to the distances between nucleating sites. With reference to collagen fibrils and needle-like crystals, Höhling et al. (1971) calculated that the center-to-center distance between islands is in the 4.0–8.5 nm range (with a prevalent range of 5.5–7.5 nm), similar to that between collagen cross-bands, whose range is 4.0–8.0 nm, and that from six to eight islands may be located in each hole zone, and from three to five in each overlapping zone. On this view, crystals do not grow from single nuclei (i.e., with one nucleus becoming one crystal), as the theory of heterogeneous nucleation suggests, but are formed from several coalescing nuclei, corresponding to several active nucleating sites lying along collagen molecules. The nucleating activity begins at the collagen surface, so other molecules, including bound phosphoproteins and proteoglycans, could play an active part in this process (Höhling et al. 1990); this would explain the dot-like appearance of the earliest crystals in non-collagenous hard tissues, such as enamel (Höhling et al. 1982), salivary stones (Höhling and Schöpfer 1968; Höhling et al. 1970), or in matrix vesicles within the growth cartilage (Plate et al. 1996) and mantle dentin (Stratmann et al. 1996).

There appears to be almost total agreement between these results and those relevant to crystal ghosts – which, by the way, in cartilage and bone have the same granular appearance (Fig. 18.2) as that of the nanoparticles ropes, especially with reference to the concept that nanoparticles are formed along extended molecules and go on to fuse into crystals. Two main differences can be detected: first, according to Höhling, needle-like crystals fuse into platelet-like ones, whereas, in crystal ghosts theory, crystals keep the filament- or plate-like shape of the templates; second, in Höhling's view, the coalescence of nanoparticles can only be considered as a two-dimensional phenomenon, which allows crystals to be distributed in the spaces between adjacent collagen molecules, a modality in conflict with a three-dimensional development. In the latter case, nanoparticles should develop all around the organic molecules, which should therefore be enclosed in them after they coalesce, as proposed for crystal ghosts.

The theory that crystals are formed along, around or on crystal ghosts implies that they are organic-inorganic structures and that they initially possess only a low degree of crystallinity, which turns into a high degree at a later stage. This occurs through a maturation process which involves the almost complete loss of crystal ghosts. The best example of this evolution is found in

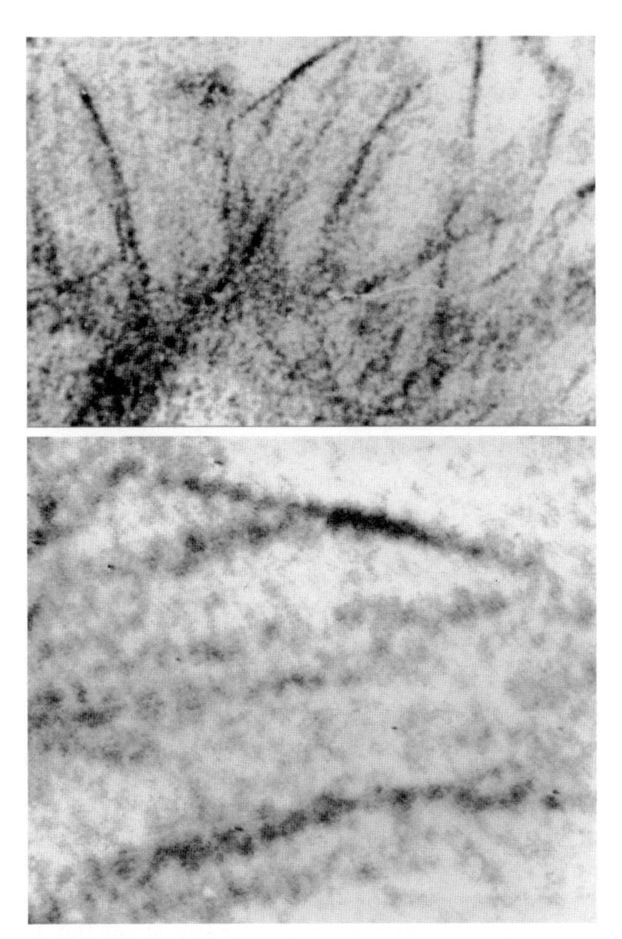

Fig. 18.2. High enlargement of crystal ghosts: note their granular, sometimes helical, appearance. PEDS, uranyl acetate and lead citrate; above ×335,000, below ×1,500,000

enamel calcification, where crystals mature, and reach their definitive structure and composition, through the loss of organic material (see Sect. 16.4): as this process reaches completion, crystal ghosts, which are easily detectable in immature enamel, are poorly recognizable (Bonucci et al. 1994). In fully calcified bone and dentin, as well as in the central zone of developed calcification nodules, crystal ghosts are likewise poorly recognizable, whereas they can be easily detected in early areas of calcification.

When this theory was first put forward, it was in such sharp contradiction with the current concepts on biological calcification, based on apatite nucleation by collagen fibrils, that the existence of crystal ghosts, except as artifacts, was considered to be untenable (Glimcher 1976). Several explanations were advanced: it was said that crystal ghosts may correspond to crystals left undecalcified in the decalcified areas, or to voids left by decalcification in the resin in place of crystals, or that they are illusory images due to the geometrical shape of crystals; they have also been thought to be produced by the adsorption of organic material on the crystal surface, or by the linking of heavy metals to EDTA. This, in fact, may persist in decalcified tissues despite exhaustive conventional dialysis (Wheeler et al. 1987). The findings that exclude crystal ghosts by classifying them as artifactual structures produced by the technical procedures used have been thoroughly discussed by Bonucci (2002), and are only summarized here.

Crystal ghosts cannot be residual undecalcified crystals because: 1) the complete decalcification of ultrathin sections is very easy to achieve, so much so that they are decalcified even by a short flotation on distilled water (Boothroyd 1964; Bishop and Warshawsky 1982); 2) the decalcified areas appear empty, i.e., under the electron microscope no crystals are visible (Bonucci 1967, 1969; Bishop and Warshawsky 1982; Nanci et al. 1983) and these areas are von Kossa negative under the light microscope (Bonucci and Reurink 1978); 3) the decalcified areas yield amorphous electron diffractograms (Bonucci 1967, 1969; Bishop and Warshawsky 1982; Nanci et al. 1983; Bonucci et al. 1988), which display no energy dispersive X-ray peaks for calcium (Nanci et al. 1983) or atomic absorption spectrophotometric signals for calcium and phosphate ions (Smales 1975).

Crystal ghosts are not voids left by the dissolution of crystals in the embedding resin (Khan 1992), which, by generating high capillary forces, might be tough to induce the penetration into them of staining solutions (Dong and Warshawsky 1995). Crystal ghosts can, in fact, be detected in embedded specimens decalcified until crystals are completely removed, and then re-embedded to fill up the clefts that may have been left by crystal solubilization (Bonucci et al. 1988). Moreover, structures similar to crystal ghosts are found in specimens decalcified before embedding (Sect. 16.3), provided that the organic components have been stabilized by a cationic substance (Shepard and Mitchell 1981; Groot 1982; Bonucci et al. 1988; Shepard 1992; Bonucci and Silvestrini 1996). On the other hand, if crystal ghosts were imprints of crystals in the resin, a direct correlation should exist between the density of crystals and that of crystal ghosts, whereas the former are plentiful in fully calcified areas, where the latter are hard to find. It should also be pointed out that the histochemical properties of crystal ghosts (Bonucci et al. 1989) and their digestion by proteases (Bonucci 1969; Appleton 1971) confirm that they are, indeed, organic structures.

Crystal ghosts are not illusory images of organic material located on the surface of rectangular crystals, whose shadow is projected under the electron microscope in such a way as to appear located within crystals (Bai and Warshawsky 1985; Warshawsky 1987, 1989). This reasoning might, in fact, be applied to enamel crystals – whose cross section may, in any case, be hexagonal (Kallenbach 1989, 1990) – but not to the thin, filament-like crystals which form bone and cartilage calcification nodules; those crystals, in fact, have the same shape and thickness as crystal ghosts.

Crystal ghosts are not organic material adsorbed on the surface of crystals. This hypothesis has an immediate plausibility, given the easy adsorption of proteins on hydroxyapatite (Hay and Moreno 1979) – for instance, in chromatographic apatite columns – and the in vitro adherence of acidic proteins to crystal surfaces (Fujisawa and Kuboki 1991), where they can give rise to globular structures (Wierzbicki et al. 1994; Kirkham et al. 2000; Wallwork et al. 2002) like those described on oyster shell folia (Sikes et al. 1998). This topic has been discussed by Bonucci and Reurink (1978), who support the concept that crystal ghosts are not proteins adsorbed on the crystal surfaces, because they are not desorbed by treatment with 1.2 M phosphate buffer – it must be noted, though, that 200 mM phosphate buffer fails to detach phosphophoryn from enamel crystal surface (Wallwork et al. 2002) – and because they resist denaturation due to fixation and dehydration. On the other hand, a material adsorbed on crystal surface during specimen processing could appear as a thin crystal coat in the case of isolated crystals, but not in that of crystals in a whole tissue; in this case, the other matrix components collapse, too, during fixation and dehydration, so coming to form, and appearing like, a homogeneous intercrystalline material continuous with, and undistinguishable from, that adsorbed on crystal surface. In any case, a material coated on needle- or filament-like crystals should behave like a tubule after decalcification; it should therefore appear as a ring in cross section. Tubular structures have never been found in bone, dentin or cartilage. They have sometimes been described in decalcified enamel (see Sect. 16.4) and enameloid (see Sect. 11.3.1), where they have been considered preformed compartments outlining spaces for crystal growth. According to Miake et al. (1991), the growth of crystals in selachian enameloid casts doubt on this possibility, and the tubular structures more probably derive from protein adsorption. According to Hoshi et al. (2001c), crystals in bones are enveloped by organic sheaths, 5–10 nm thick, which contain osteopontin and bone sialoprotein. Leach (1979) reports the results of Appleton, who found that synthetic hydroxyapatite soaked in albumin, washed in water, and treated with chromium sulfate, appears to be coated by a stained sheath some 2.5–5.0 nm thick. The existence of these coats appears to conflict with the electron microscope finding that the thickness of the untreated crystals – about 3.5 nm – does not change when sections are treated with heavy metals (Bonucci 1984), whereas these should stain organic coats and proportionally increase the apparent thickness of crystals. The fact that in epiphyseal cartilage the thickness of crystal ghosts is not significantly greater than that of untreated crystals (Bonucci 1969) points to the same conclusion. The findings of Kallenbach (1986), already discussed in Sect. 16.4, support the possibility that tubule-like coats and crystal ghosts can both be detected in the same field.

Crystal ghosts are not due to EDTA residues contaminating decalcified areas (Wheeler et al. 1987). They have, in fact, been detected after decalcification with formic acid, which is easily and completely washed away by flotation on distilled water and is neutralized by buffer solutions.

A further objection to the theory of crystal ghosts, seen as structures that promote calcification and as templates for crystal formation, was that the proposed mixing of organic and inorganic material, and the location of organic material within crystals, are crystallographically impossible (Warshawsky 1989). This opinion seems to derive not so much from a critical analysis of the reported findings, as from the conviction that the inorganic structures found in areas of calcification are true crystals and must, therefore, strictly obey the laws of mineralogy, totally disregarding the fact that the earliest calcium phosphate nanoparticles show a very low degree of crystallinity (see Sect. 4.2.1) and that only with maturation do they take on true crystalline features.

Actually, the existence of intracrystalline organic material is not easy to demonstrate in the very small, thin crystals that are found in bone, cartilage and enamel, but has repeatedly been substantiated in the large crystals that distinguish calcified structures in invertebrates. In these structures, organic molecules can interact with crystal faces and be occluded within growing crystals (Addadi and Weiner 1985). The wide-ranging literature on this topic has been reviewed in Sect. 16.6. The location of organic material within the crystals of calcium oxalate stones, and within intracellular inorganic concretions such as magnetosomes, siderosomes, aurosomes, the inclusion bodies attributable to lead intoxication, and intramitochondrial calcified granules, has been discussed in the same chapter. In particular, it should be noted here that in magnetotactic bacteria there is a precise alignment between the lattice fringes of the precursor semicrystalline material and those of the magnetosomes (see Sect. 16.7.2).

Crystal ghosts in bone, dentin and cartilage have the same filament-like shape as untreated crystals. They can have other shapes, as occurs, for instance, in the granular material that remains in place of the mitochondrial granules after these have been decalcified and stained by the PEDS method (see Sect. 16.8). Crystal ghosts in mollusk shells, if any, should have a laminar structure; Travis (1968), and Travis and Gonsalves (1969), have described an intraprismatic matrix consisting of very thin filaments (1.7–2.5 nm thick) whose electron microscope image was referred to as a "striking fingerprint" of intraprismatic crystals. The reconstructed model of chitin/silk fibroin templates described in the organic matrix of shells (see Sect. 12.2) might well be identified as a crystal ghost.

If the theory presented above is true, then crystal ghosts must pre-exist in the matrix before they calcify, or, at least, must be formed as calcification proceeds. While the second possibility is supported by the finding that glycoproteins adsorbed on specific planes of calcite crystals can be progressively occluded inside the solid phase (Berman et al. 1988, 1990), structures resembling crystal ghosts have never been found in not calcified matrices. It has already been noted, however, that the uncalcified organic components of the matrix – first of all acid proteoglycans – can be deeply modified by fixation, dehydration and embedding, and lose their extended conformation (see Sect. 16.3), so they may become unrecognizable (discussed by Bonucci 2002). The resemblance, or, possibly, identity, between cartilage crystal ghosts and acid proteoglycans make it more probable that, through fixation and dehydration, crystal ghosts lying in the still uncalcified matrix undergo the same collapse into granular structures as acid proteoglycans. By contrast, after calcification, they are transformed into stiff structures that retain their original extended morphology (Bonucci et al. 1988). Support for this perspective comes from the finding that calcium ions bound to protein and proteoglycan chains fix them into rigid structures whose geometry is epitaxial for apatite crystals (Schubert and Pras 1968), and by electron microscope results showing that the incubation of cartilage into lanthanum chloride allows the detection in uncalcified matrix of 'focal filament aggregates' (Gomez et al. 1996) – not visible in non-incubated cartilage – whose filamentous constituents correspond to acid proteoglycans (Fig. 18.3). It has been suggested that these filaments may have retained their extended conformation because the reaction with lanthanum chloride has transformed them into stiff structures, so making them closely resemble crystal ghosts, so much so that this process might be the same as that which occurs when crystal ghosts react with calcium ions (Bonucci 2002).

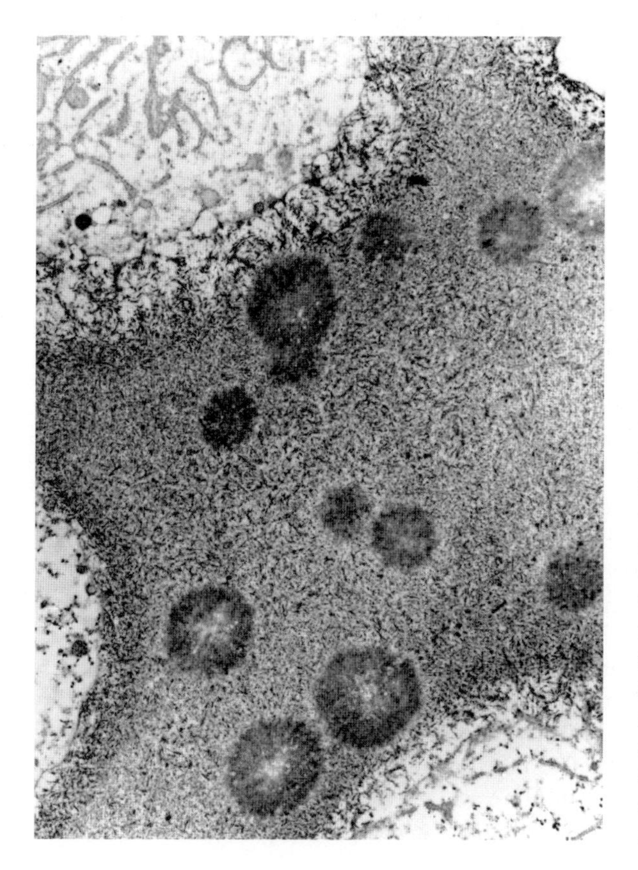

Fig. 18.3. Upper hypertrophic zone of not calcifying epiphyseal cartilage incubated in lanthanum chloride: spherical aggregates of very fine filaments (focal filament aggregates) are scattered through the matrix; they have a close resemblance to crystal ghost aggregates in the calcifying cartilage. Chondrocyte lacunae are partly recognizable in the corners of the figure. Uranyl acetate and lead citrate, ×12,000

18.5
Extracellular Matrix and Calcification: Lipids

As previously discussed (see Sects. 7.3.4, 8.6, 9.3.3 and 10.3.2), lipids are structural components of the calcified matrix from which they can only be extracted after decalcification. They are, of course, constituents of cell membranes and may contribute to cell signaling (Dziak 1992). Moreover, they are matrix vesicle components (Sect. 9.4) and are found in pathologically calcified tissues (see Chap. 14), especially the calcified walls of arteries (see Sect. 14.3.5). A unique, small (5.5-kDa) proteolipid, which precipitates calcium ions and contributes to the formation of dental calculus, has been extracted from *Corynebacterium matruchotii* and sequenced (Van Dijk et al. 1998). For all these reasons, lipid molecules have been considered to be active in calcification, although the way in which they induce apatite deposition is not exactly known and several different functions seem to be feasible. Their role in calcification is confirmed by the findings that membranes isolated from *C. matruchotii* induce calcification (Vogel and Smith 1976), that calcium-binding lipids extracted from these microorganisms are capable of promoting hydroxyapatite nucleation (Ennever et al. 1968; Takazoe et al. 1970; Vogel and Ennever 1971), that proteolipids isolated from calcifying bacteria and dental calculus induce calcification in vitro (Ennever et al. 1976), and that calcium-acidic phospholipid-phosphate (Ca-PL-P) complexes, isolated from bone (Boskey et al. 1982) and other calcifying tissues (Anghileri and Dermietzel 1973; Dmitrovsky and Boskey 1985; Boskey et al. 1988) in normal and pathological conditions, have the same effect.

Lipid molecules, especially acidic phospholipids, certainly play a pivotal role in calcification (Vogel and Boyan-Salyers 1976). They have a strong calcium-binding capability, which, when phosphate ions are present, may lead to the formation of Ca-PL-P complexes. As reviewed by Boyan et al. (1989), phospholipids may give rise to these complexes via the interaction of calcium and phosphate ions with phosphatidylserine, involving the participation of proteolipids, which are hydrophobic proteins in membranes characterized by solubility in organic solvents (Boyan 1985). The Ca-PL-P complexes so formed enhance calcification in vitro, are present in calcifying tissues but are absent from those that do not calcify, and are so closely associated with the inorganic substance that can only be extracted after decalcification. They do, in fact, play a role in initiating hydroxyapatite deposition in vitro (Boskey and Posner 1977; Boskey et al. 1982) and in vivo (Boskey and Reddi 1983; Raggio et al. 1986), the sequence involved in this process being Ca-PL-P complexes > proteolipid-acidic phospholipids > proteolipids > crude phospholipids > total lipids > whole cells, at least according to evaluations carried out on Ca-PL-P complexes extracted from *Bacterionema matruchotii* (Boyan-Salyers and Boskey 1980). The possible function of these complexes in inducing calcification is supported by their location in bone and cartilage. They do, in fact, seem to constitute a significantly higher proportion of lipids in young than in mature bone (Boskey and Posner 1976). In epiphyseal cartilage, their concentration rises from the reserve zone to the proliferative zone, reaches a maximum in the hypertrophic zone, where calcification is initiated, and falls in primary spongiosa and diaphyseal bone (Boskey et al. 1980).

As discussed in Sect. 9.4, and as reported by Genge et al. (1991), matrix vesicles contain annexins, which are membrane-associated, channel-forming proteins that allow the influx of Ca ions and, in acidic conditions, may undergo conformational changes which lead to a rise in their hydrophobicity, so that they acquire proteolipid-like properties and, like proteolipids, may structure membrane phosphatidylserine into a conformation conducive to apatite formation. Studies by Boskey and Posner (1977) strengthen the possibility that Ca-PL-P complexes are responsible for the onset of calcification in matrix vesicles. Kirsch et al. (2000), on the other hand, showed that annexin II, V and VI mediate calcium influx into lyposomes enriched with phosphatidylserine, liposomes containing lipids extracted from matrix vesicles, and intact matrix vesicles; they also showed that collagen types II and X only become bound to liposomes in the presence of annexin V, and that in so doing they stimulate the calcium channel activity, while increasing the calcium influx into liposomes. It is, however, remarkable that annexin V knock out mice do not reveal significant skeletal alterations, nor are the in vitro calcification properties of isolated chondrocytes altered by a lack of annexin V, which suggests that other annexins may exert a compensatory activity (Brachvogel et al. 2003).

18.6 Extracellular Matrix and Calcification: Non-collagenous Proteins

The finding that in bone filament-like crystals are located between the collagen fibrils of the matrix, the realization that matrix vesicles initiate calcification in bone, cartilage, mantle dentin, tendon and other calcifying tissues, but not in all of them, the finding that a protein fraction can only be extracted from calcified tissues after decalcification, the search for a localized basic factor that might regulate the mechanism of calcification in all calcifying tissues, and persistent uncertainties about the relationships between the organic and inorganic substances within them, all led to a broad development of studies on the role of non-collagenous proteins in calcification (reviewed by Butler 1984, 2000; Linde 1984; Termine 1985; Boskey 1989, 1998; Bianco 1990; Goldberg et al. 1995; Robey 1996; Nanci 1999).

The name 'non-collagenous proteins' was coined to indicate the noncollagenous organic components of bone and dentin matrix, with the aim of sharply differentiating them from collagen fibrils in a period during which these were considered pivotal in inducing bone and dentin calcification and were the almost exclusive object of investigation. This section of Chap. 18, which is dedicated to non-collagenous proteins, does, in fact, comprise other non-fibrillar proteins that can be found in calcified matrices, even if their inclusion, which is motivated by the need to simplify, may seem a little arbitrary.

The difficulties associated with these proteins mainly derive from their low amounts in the calcified matrix, their relatively high numbers, frequent new additions to the list of known proteins of this type, their possible processing, or the possible alternative splicing of the primary gene transcript during the calcification process, and the changes in their effects (as promoters or inhibitors of calcification) depending on their concentrations and the environmental conditions, and according to whether they are in solution or fixed to a substratum. In addition, as discussed by Boskey in her review of biomineralization (Boskey 1998), their function and role in calcification cannot simply be deduced from the fact that they are components of the calcifying matrix, that they have calcium-binding activities, or that they are expressed just prior to, or at the same time as, the earliest phases of calcification. Borelli et al. (2003), in referring to the soluble organic matrix of various calcified tissues, have stressed that biological calcifications are dynamic processes, and that an understanding of them cannot be attained simply by analytically studying the calcified matrices.

It is, however, true that, within the limitations set out above, some indications can be drawn from an evaluation of the sequential expression and localization of the matrix components involved in calcification (see Sects. 7.3.2 and 7.3.3). In this connection, Kasugai et al. (1991) and Cowles et al. (1998) showed that, in calcifying cultures, type I collagen, alkaline phosphatase, ON and OPN appear before, BSP concurrently with, and OC after calcification. Personal results (see Sect. 9.3.2) also point to the concurrent appearance of BSP and matrix vesicle calcification. Satoyoshi et al. (1995) found that, in odontoblast cultures, DPP precedes calcification. A study by Sasano et al. (2000) on the mandible development of Wistar rat embryos showed that calcification had not begun on day 14; type I collagen was first detected on day 15 in the uncalcified mandible; OC, ON, OPN and BSP appeared almost simultaneously in the calcified matrix of the 16-day mandible, and showed the results of continuous accumulation in the 18-day mandible. In the same type of material, Ishigaki et al. (2002) found the first expression of ON, OC and BSP mRNAs in newly differentiated osteoblasts of the developing mandible at embryonic day 15 (E15); by then, OC and BSP immunostaining had become visible in the same osteoblasts and uncalcified matrix and, concomitantly with the initiation of matrix calcification at E16, it became visible in the calcified bone matrix, too. Again in the developing mandible of Wistar rat embryos, Zhu et al. (2001) found that the cuboid osteoblasts that surround the uncalcified matrix in 15-day early mesenchymal condensations expressed pro-α1(I) collagen, ON and BSP intensely and OPN very weakly, whereas the flat osteoblasts surrounding the calcified bone expressed OPN very strongly. According to Sommer et al. (1996), ON, OPN and BSP mRNAs can be detected in osteoblasts and hypertrophic chondrocytes at the onset of calcification. According to Bosshardt and Nanci

(2000), BSP and OPN essentially show the same distribution in rat teeth: they are initially deposited before the cementum collagen fibrils, whereas, with advancing root length, a shift occurs toward the earlier deposition of collagen fibrils; in human teeth, the deposition of cementum collagen fibrils always precedes that of BSP and OPN.

It should be borne in mind that the detection of non-collagenous proteins in areas of early calcification is of little help in determining their functional role and does not necessarily testify that they are involved in the calcification process (Boskey 1998). In any case, immunohistochemical studies on the localization of various non-collagenous proteins appear to be of great interest: they have shown that OC is the only non-collagenous protein that is restricted to bone cells and calcified bone matrix, and that BSP is present both in bone and calcified cartilage, whereas other proteins (ON, OPN, MGP, decorin and biglycan) are distributed in soft tissues as well as in bone (Carlson et al. 1993). The same studies showed that staining for OC and BSP was widespread throughout the bone matrix, that staining for other proteins was concentrated along bone surfaces, and that osteoid proved to be negative for OC and BSP. During rat calvaria osteoblast differentiation and bone nodule formation in vitro, BSP was found to be associated with early clusters of needle-like crystals, but not with collagen fibrils, OC was limited to differentiated osteoblasts and was concentrated at the calcification front, and ON was evenly distributed throughout the osteoid and calcified matrix (Nefussi et al. 1997). According to Wang et al. (2000), BSP and apatite co-localize in the extracellular matrix of osteoblast-like cell cultures, with BSP expression preceding apatite deposition.

As discussed above, difficulties in studying non-collagenous proteins are aggravated by the fact that some of their functions may depend on post-translational modifications, such as phosphorylation, glycosylation, and proteolytic processing. According to Suzuki et al. (1996), non-collagenous proteins in bone and dentin may be post-translationally processed by limited proteolysis, and then be extracellularly processed by a casein kinase II into more deeply phosphorylated species that facilitate calcification. The proteolytic changes undergone by amelogenins during enamel maturation are well known (see Sect. 10.3.3). Qin et al. (2004) studied the components of the SIBLING group, finding that OPN and BSP occur in the extracellular matrix of bone and dentin in full-length forms, whereas amino acid sequencing shows that DMP1 and DSPP exist as proteolytically processed fragments resulting from the scission of X-Asp bonds. According to Razzouk et al. (2002), some form of OPN post-translational modification, possibly phosphorylation, is needed for in vitro osteoclastic bone resorption. On the other hand, Salih et al. (2002) reported that both OPN and BSP can undergo natural variations in the degree of phosphorylation as a function of the time needed for calcified tissue formation.

These results show not only that the function of these proteins can change as a result of post-translational changes, but also that each non-collagenous protein may have more than one function, each of which may be enhanced or reduced by their interactions. Studies of transgenic or knock out animals may be helpful in identifying the role of a single protein in calcification, but the integrated activity of different molecules is not deducible from the activity of just one of them. It must be recognized that at present we know something about what a protein does, but almost nothing of its interplay with other matrix components, despite our knowledge that several of them may occur together in a single calcifying areas and may be expressed concomitantly with calcification (Cowles et al. 1998). The available data are almost exclusively related to the function that single non-collagenous proteins play in the calcification process, independently of their interplay with other proteins.

Most of the non-collagenous proteins identified so far are acidic phosphoproteins, whose concentrations in calcified areas reach their highest levels simultaneously with the initial deposition of inorganic substance in the matrix (Gerstenfeld et al. 1990). They may form complexes with other matrix components, such as collagen fibrils in bone, or may be relatively free in the matrix. Their role in calcification has been documented by Saito et al. (1997), who showed that reconstituted collagen fibrils alone do not induce mineral formation, that the adsorption of phosphoproteins on collagen fibrils does not promote calcification, because phosphoproteins desorb in the calcifying medium, and that, by contrast, phosphoproteins cross-linked to collagen fibrils do induce crystal formation on the fibril surface. Veis and Perry (1967) suggested that phosphoproteins could function as sites within the "hole" regions of native dentin collagen for the nucleation of the dentin calcification process, a hypothesis supported by the observation that dentin phosphophoryn, in the presence of calcium ions, interacts with native turkey tendon collagen fibers, so forming globular particles predominantly at the *e* band in the collagen gap region (Traub et al. 1992b; see Sect. 18.2). Biotin-labeled bone sialoprotein (visualized with avidin conjugated with alkaline phosphatase) is also preferentially bound to the collagen $\alpha2$ chains and to peptides derived from $\alpha2$ chains and, in the case of reconstituted fibrils, is mainly located in the hole zones of fibrils (Fujisawa et al. 1995).

Phosphoprotein-induced calcification is thought to depend mostly on phosphate groups. In his review on bone mineral and the mineralization process, Posner (1987) clearly stated that in phosphoproteins the phosphate groups have a steric arrangement which appears to be an optimum for calcium binding and the subsequent formation of apatite crystals. He reported a diagram from one of his own publications, showing how phosphate groups can bind calcium ions, and these, in their turn, can bind other phosphate ions, so giving rise to a repetitive process ending in the formation of apatite crystals. According to Glimcher (1989, 1992), the protein-bound phosphomonoester groups could be components of both organic and inorganic phases and so act as a chemical and physical bridge linking the organic structural molecules of the matrix to the crystals: in that case, the phosphate groups of phosphoproteins would be integral components of the inorganic crystals, with some of the phosphate groups of the Ser(P) and Thr(P) located on the crystal surfaces. This hypothesis is supported by the finding that uncalcified samples of turkey leg tendons contain little or no Ser(P), Thr(P) or γ-carboxyglutamic acid, whereas all three of these Ca^{2+}-binding amino acids can be detected at the onset of tendon calcification and their concentrations increase as calcification proceeds (Glimcher et al. 1979). The importance of phosphate groups in calcification is shown by the finding that the dephosphorylation of insoluble collagen from bovine dentin induces a progressive fall in the degree of in vitro hydroxyapatite precipitation, which stops after 90% of ester phosphate has been removed (Saito et al. 1998). The concept that phosphoproteins play a significant role in initiating bone calcification is strengthened by the finding that, although the major phosphate components identified in the earliest stages of calcification are protein phosphoryl groups which are not complexed with calcium, small amounts of calcium-complexed protein phosphoryl groups as well as even smaller, trace amounts of apatite crystals can be detected in the earliest phases of calcification (Wu et al. 2003).

The interaction of acidic phosphoproteins with collagen, their strong calcium-binding activity, their possible nucleation of hydroxyapatite, their involvement in the regulation of morphology, growth rate, and stability of the inorganic crystals, have all been reviewed by Veis (1993, 2003).

Proteins with characteristics and properties similar to those of the non-collagenous proteins of bone, dentin and other calcifying collagenous tissues have been thought to be involved in the calcification of matrices containing no collagen fibrils, such as those of enamel, invertebrate integuments, and several non-skeletal tissues. There is little intrinsic justification for discussing them together with the non-collagenous proteins, but they have been included in this chapter for reasons of structural and functional analogy.

Plenty of studies have been devoted to determining the role that may be played by the organic components of the enamel matrix in inducing and regulating the formation and growth of enamel crystals (reviewed by Nanci and Smith 1992; Robinson et al. 1998; Fincham et al. 1999; Moradian-Oldak 2001; Zeichner-David 2001). After stressing the heterogeneity of the enamel matrix and the fact that each of its components is liable to undergo post-secretory sequential degradation, Robinson et al. (1998) assigned them the following possible functions: mineral nucleation/initiation (dentine sialoprotein, tuftelin), mineral ion binding as crystal precursors (amelogenin, enamelin), control over crystal growth (amelogenin, enamelin, ameloblastin), support for growing crystals (amelogenin, enamelin), determination of prismatic structure (ameloblastin), cell signaling (tuftelin, ameloblastin), control over secretion (breakdown products), and protection of the mineral phase (amelogenin, enamelin).

The role of matrix proteins in enamel calcification has still not been fully elucidated; by now, however, some functions can be considered reasonably well defined (Moradian-Oldak 2001). Amelogenins can initially self-associate into nanospheres and form parallel-oriented, 30–35 nm wide bundles (Goldberg et al. 1995), which are correlated with the finding of long parallel crystals in several species (Diekwisch et al. 2002) and might in a sense be considered the equivalents of the stromal components of the enamel matrix (Du et al. 2005). Crystal-like structures 10 nm wide are formed between these structures (Goldberg et al. 1995) and are thus forced to grow in parallel, along their *c*-axis, whereas they are prevented from fusing or growing in thickness. When amelogenins are processed into small fragments by proteolytic enzymes – in this differing from true stromal structures – the crystals can grow till they reach their definitive size. Non-amelogenin proteins, or at least some of them, induce crystal formation and appear to behave as crystal templates.

The many proteins found in calcified structures within invertebrates are heterogeneous, although many of them are acidic glycoproteins rich in Asp, Glu, Gly, Ser and Ala. Variable amounts of protein fractions can be extracted from mollusk shells, sea urchin embryo spicules, crustacean carapaces, corals, and other calcified structures; some of them are located around and between crystals (like mortar between bricks), and some are integral components of crystals, are occluded within them, and can modify their physical properties (see Sect. 16.6). For these reasons, these crystal proteins have been considered the best candidates for a leading role in inducing and regulating, i. e., promoting or inhibiting, the calcification process which, as in other hard tissues, seems to occur in two phases: the initial formation of crystal-like inorganic nanostructures, which could consist of stabilized amorphous calcium carbonate, and the later growth of these structures into crystals.

Actually, the first inorganic aggregates in the calcified structures of invertebrates appear to consist of amorphous calcium carbonate (Weiss et al. 2002; Gotliv et al. 2003; Politi et al. 2004) stabilized by organic material (Aizenberg et al. 1996, 2002), a finding analogous with that of amorphous calcium phosphate in vertebrate hard tissues (see Sect. 4.2.1). Calcium carbonate could be a transient precursor phase during the formation of the crystalline phase (Weiner et al. 2003) and, as reviewed by Weiner et al. (2005), in that form it could be active in a wide range of organisms. One striking hypothesis that has been put forward is that the transient, amorphous form taken by calcium carbonate is a disordered phase, comparable to a melt phase not requiring high temperatures; in this phase, large crystals would begin to take shape from small seeds or nucleation substrates (Weiner et al. 2005).

The development of the early inorganic aggregates into crystals appears to occur through a process which implies a structural correspondence (epitaxy) between their crystalline lattice spacings and the periodic structure – i. e., the distance that separates the ion-binding sites – of immobilized organic molecules. Electron diffraction studies on incompletely decalcified mollusk shells have shown that this kind of relationship occurs between the *a* and *b* axes of aragonite proteins in an antiparallel β-pleated sheet conformation and the aragonite crystal lattice (Weiner et al. 1983; Weiner and Traub 1984). The protein-inorganic substance interaction is highly specific: when immobilized on a substrate of β-chitin and silk fibroin, macromolecules from the mollusk shell aragonite layer induce aragonite formation, whereas, under the same in vitro conditions, macromolecules from the calcite shell layer induce calcite formation (Falini et al. 1996). In vitro studies by Blank et al. (2003) have shown that when the water-soluble proteins extracted from the nacre of the *Haliotis laevigata* snail are dialysed against a saturated calcium carbonate solution with 100 mM NaCl, the calcium carbonate crystals that are precipitated selectively contain occluded perlucin, whereas the other proteins (at least 10 different types) are left in the supernatant. Belcher et al. (1996) have observed that the abrupt transition from a prime layer of oriented calcite crystals into crystals of aragonite, which occurs during the initial stages of abalone shell calcification, is accompanied by the synthesis of specific polyanionic molecules.

The formation of the earliest inorganic aggregates can be mediated by immobilized polyanionic molecules (discussed in Sect. 8.5.1). It is not known exactly how this process is regulated; in his review on the mechanism of mollusk shell formation, Wheeler (1992) reported studies by E.M. Greenfield et al. suggesting that anionic groups can attract calcium to the immobilized substrate or accumulate calcium in it, so inducing the formation of the earliest inorganic nuclei. Wheeler himself (Wheeler et al. 1987) formulated a second hypothesis, that the immobilized proteins can bind inorganic nuclei preformed in the medium rather than calcium ions.

The idea that acidic groups (such as sulfate groups of glucosaminoglycans and proteoglycans, phosphate groups of phosphoproteins, and carboxylic groups of amino acids) can play a role in inducing calcification has been discussed above and in Sect. 18.2. That viewpoint is supported by the observation that acidic macromolecules appear to be ubiquitous in calcified tissues and have been detected in a number of them, including bone (Butler 1984), cartilage (Lash and Vasan 1983), dentin (Butler 1984), enamel (Deutsch et al. 1995), mollusk shells (Weiner 1979), spines of the sea urchin embryo (Benson and Wilt 1992), avian eggshell (Abatangelo et al. 1978) and other hard tissues. They can induce the formation of crystals and can regulate their growth (Moradian-Oldak et al. 1992; Gotliv et al. 2003). In reviewing this topic, Weiner and Addadi (1991) concluded that the deposition of crystals in biological tissues is a highly controlled process in which acidic proteins and glycoproteins located in the matrix play a crucial role by inducing oriented nucleation, interacting specifically with some crystal faces, and intercalating regularly into the crystal lattice. Acidic glycoproteins rich in aspartic acid are components of the matrix of mollusk shells (Weiner 1979) and other calcifying matrices. Specific amino acid sequences might induce crystal nucleation (Weiner and Traub 1984) and Weiner and Hood (1975) have suggested that recurrent sequences of aspartic acid, separated by either glycine or serine, might constitute the regularly spaced, negatively charged crystal templates. Actually, amino acid analyses of the shell proteins have shown about 30 mol % of aspartic acid in the aragonite layer and more than 50 mol % in the calcite layer (Albeck et al. 1993); specific, unusually acidic proteins can have a higher aspartic acid content: the Aspein gene, for instance, encodes a sequence of 413 amino acids including 60.4% Asp, 16.0% Gly, and 13.2% Ser, in which Asp sequences are punctuated with Ser-Gly dipeptides (Tsukamoto et al. 2004). Polyaspartate sequences could control the formation and growth of crystals by adopting a β-sheet conformation, a pattern described in the enamelin fraction of enamel proteins (Traub et al. 1985; Jodaikin et al. 1986) and, above all, in mollusk shell proteins (Weiner and Traub 1980; Weiner et al. 1983; Choi and Kim 2000).

In his reviews on molecular tectonics (Mann 1993) and on molecular recognition (Mann 1988) in biomineralization, after commenting that periodic secondary structures such as antiparallel β-pleated sheets can control the assembly of nuclei in specific crystallographic directions, and that crystal nucleation can occur if the lattice spacings of the crystals are in geometric register with the distance between ion-binding sites in the organic substrate, Mann has stressed that it is, precisely, aspartic acid β-pleated sheets that possess this type of favorable conformation, because the spacings between aspartic acid residues come to resemble the Ca–Ca distances in the (001) face of aragonite crystals of mollusk shells. Traub et al. (1985) had reported a similar arrangement with reference to crystal nucleation by enamelins. Boanini et al (2006) have recently synthesized hydroxyapatite-aspartic acid nanocrystals, in which a specific interaction of the amino acid carboxylic groups with the hydroxyapatite calcium ions is suggested by the reduction of the coherent length of the perfect crystalline domains along the crystal long dimension and cross section.

An attracting hypothesis on the nucleation of bone apatite crystals by aspartic acid has been put forward by Sarig (2004). Her suggestion is that aspartic acid can be enclosed inside a carbonated, calcium-deficient hydroxyapatite unit cell without modifying its external appearance; the insertion of four carbonate ions in sites previously occupied by four phosphate ions leads to the generation of four voids that can accommodate the four terminal oxygen of aspartic acid, so constituting a nucleating unit.

In the same way as immobilized proteins can promote crystal formation by attracting and accumulating inorganic ions in solution, crystal surfaces can adsorb organic material which, by interacting specifically with them, can affect their growth and morphology. The acidic, aspartic acid-rich mollusk shell matrix proteins can specifically interact with certain crystal faces during their growth, can induce oriented nucleation, or can intercalate in a regular manner into the crystal lattice itself, so that, as discussed in Sect. 16.6, they can be occluded within crystals and become integral components of them (Addadi and Weiner 1985; Addadi et al. 1989; Weiner and Addadi 1991). In this way they can prevent the growth of one face and promote that of others, so directing and modifying the overall crystal shape (Aizenberg et al. 1995). This mechanism has been confirmed by in vitro experiments by Albeck et al.

(1996): intracrystalline proteins were extracted from calcite crystals – treated with NaOH to digest the extracrystalline organic material – from shells of the *Atrina rigida* mollusk and from spines of the *Paracentrotus lividus* sea urchin. They obtained two families of macromolecules, the first consisting of highly acidic proteins, with an Asx+Glx content of over 80 mol %, the second of much less acidic glycoproteins. These two protein types were found to interact with in vitro growing calcite crystals in two distinct ways: the former interacted with calcite planes perpendicular to the *c* crystallographic axis and the latter with planes roughly parallel to the *c* axis. The Authors hypothesized that organisms alternate between the use of these two types of interaction to finely control crystal growth and shape. Fu et al. (2005) have shown that acidic, 8-kDa proteins isolated from the abalone shell nacre specifically modify the calcite rhombohedra into asymmetrically rounded crystals.

Avian eggshell has attracted special attention as a unique model for the study of calcification in biological tissues (see Sect. 13.4). Like other hard tissues, it has a composite structure comprising inorganic material – essentially calcium carbonate (calcite) – and organic molecules, which are considered to be regulators of calcification. As is true of other hard tissues, the interaction between inorganic and organic components has not yet been fully defined, and the mechanism of eggshell formation is still unknown. In their studies on the role of proteoglycans, based on the precise localization of these molecules in the oviduct, the timing of their secretion, and the effect of the inhibition of their sulfation by sodium chlorate, Fernandez et al. (2001, 2004) found indications that the earliest calcite crystals are nucleated by mammillan, that the growth and orientation of the crystals that form later are due to ovoglycan, and that carbonic anhydrase acts to increase carbonate availability. On the basis of ansocalcin's aggregating properties, Lakshminarayanan et al. (2002) believe that at low concentrations, single ansocalcin molecules are adsorbed on the growing crystal surface, whereas at high concentrations, protein aggregation, comparable to that of amelogenin, makes available a large surface area, so facilitating the crystal aggregation. During calcification, eggshell matrix proteins may be entrapped in the inorganic substance, as shown by Hincke et al. (1995) for ovocalyxin-17, which can only be extracted from eggshell after decalcification.

18.7 Concluding Remarks

It is by now clear that matrix components play a leading role in calcification. They are variable in structure, composition and constitution, but the available findings justify the chief conclusion that the calcification process occurs according to similar basic mechanisms in all of them. Moreover, other important conclusions can be put forward:

- The formation of early inorganic aggregates and their development into crystals appear to occur through a process which imposes a close structural correlation (epitaxy) between the crystal lattice spacings and the distance between ion-binding sites in the periodic structure of immobilized organic molecules (templates).
- The earliest inorganic aggregates bound to the reacting, epitaxial groups of acid proteoglycans or other polyanionic molecules appear as nanoparticles, which necessarily come to take up the same three-dimensional location as that of these groups, that is, they are distributed along and around filamentous templates and on the surface of laminar templates.
- This is why the growth and coalescence of nanoparticles give rise to the formation of crystal-like, organic-inorganic structures which repeat the shape of, and occlude, the templates. It is the subsequent loss of organic components that appears to allow early crystals to grow and mature.
- Under the electron microscope, the occluded molecules appear as crystal ghosts. These can be defined as preformed, polyanionic organic structures which function as templates for crystal formation.
- Crystals can be formed within (matrix vesicles) and outside preformed compartments, in connection with integral proteins of the matrix; these can self-aggregate into sheets – for example, amelogenin in enamel, conchiolin in mollusk shells – which may outline compartment-like spaces and help to mould the shape of crystals.
- Intra- and extra-cellular compartments are not empty spaces but contain molecules which fill them up completely and can promote or prevent crystal formation inside them; they must be modified or almost completely removed to allow crystals to grow.
- The early deposition of inorganic substance in collagenous matrices (bone, dentin, tendons) can occur, not only in matrix vesicles, but also in intrafibrillar compartments (holes in the gap zones of collagen fibrils) and outside fibrils (interfibrillar spaces). In the first case, inorganic substance fills up the compartments (granular inorganic substance in bands), giving rise to poorly crystalline structures which develop in the holes without altering the intermolecular cross-links and possibly mature into platelets; in the second, it is connected with components of the interfibrillar matrix, giving rise to filament-like crystals.
- These intra- and extra-fibrillar events seem to occur independently and with different modalities, but in both cases in accordance with the same basic mechanism of epitaxial formation along and around template molecules; neither appreciably affects the intrinsic molecular organization of collagen fibrils.

- Calcification can be induced and regulated by the proteic molecules of the matrix, mostly acidic phosphoproteins; they may be associated with other matrix components, such as collagen fibrils in bone, or may be relatively free in the matrix.
- Many organic components of invertebrate calcified structures are acidic glycoproteins rich in aspartic acid. Polyanions with repeating sequences of aspartic acid in a β-pleated sheet conformation might constitute regularly spaced, negatively charged crystal templates.
- The adsorption of proteins on crystal surfaces can promote or inhibit their growth, so regulating crystal shape.
- The organic components of the matrix can inhibit calcification when they are free in solution, or can promote it when they are immobilized on a fixed substratum.
- The role of acid proteoglycans and other polyanions in calcification depends on the status (aggregates, monomers) and the hydrodynamic size of their molecules. These can bind high concentrations of calcium ions which can be released locally once the molecules have been modified or depolymerized.
- Acidic phospholipids and calcium-acidic phospholipids-phosphate complexes are structural components of calcified matrices; they play a poorly understood role in the initiation of hydroxyapatite deposition. They, too, may act as templates for crystal formation.

References

Abatangelo G, Daga-Gordini D, Castellani I, Cortivo R (1978) Some observations on the calcium ion binding of the eggshell matrix. Calcif Tissue Int 26:247–252

Addadi L, Weiner S (1985) Interactions between acidic proteins and crystals: stereochemical requirements in biomineralization. Proc Natl Acad Sci USA 82:4110–4114

Addadi L, Berman A, Moradian Oldak J, Weiner S (1989) Structural and stereochemical relations between acidic macromolecules of organic matrices and crystals. Connect Tissue Res 21:127–135

Aizenberg J, Hanson J, Ilan M, Leiserowitz L, Koetzle TF, Addadi L, Weiner S (1995) Morphogenesis of calcitic sponge spicules: a role for specialized proteins interacting with growing crystals. FASEB J 9:262–268

Aizenberg J, Lambert G, Addadi L, Weiner S (1996) Stabilization of amorphous calcium carbonate by specialized macromolecules in biological and synthetic precipitates. Adv Mater 8:222–226

Aizenberg J, Lambert G, Weiner S, Addadi L (2002) Factors involved in the formation of amorphous and crystalline calcium carbonate: a study of an ascidian skeleton. J Am Chem Soc 124:32–39

Akisaka T, Nakayama M, Yoshida H, Inoue M (1998) Ultrastructural modifications of the extracellular matrix upon calcification of growth plate cartilage as revealed by quick-freeze deep etching technique. Calcif Tissue Int 63:47–56

Albeck S, Aizenberg J, Addadi L, Weiner S (1993) Interactions of various skeletal intracrystalline components with calcite crystals. J Am Chem Soc 115 :11691–11697
Albeck S, Addadi L, Weiner S (1996) Regulation of calcite crystal morphology by intracrystalline acidic proteins and glycoproteins. Connect Tissue Res 35:365–370
Alini M, Roughley PJ (2001) Changes in leucine-rich repeat proteoglycans during maturation of the bovine growth plate. Matrix Biol 19:805–813
Alini M, Matsui Y, Dodge GR, Poole AR (1992) The extracellular matrix of cartilage in the growth plate before and during calcification: changes in composition and degradation of type II collagen. Calcif Tissue Int 50:327–335
Althoff J, Quint P, Krefting E-R, Höhling HJ (1982) Morphological studies on the epiphyseal growth plate combined with biochemical and X-ray microprobe analyses. Histochemistry 74:541–552
Anghileri LJ, Dermietzel R (1973) Calcium-phosphate-phospholipid complexes in experimental tumors: their possible relationship with tumor calcification. Z Krebsforsch Klin Onkol 79:148–156
Appleton J (1971) Ultrastructural observations on the inorganic/organic relationships in early cartilage calcification. Calcif Tissue Res 7:307–317
Armstrong AL, Barrach HJ, Ehrlich MG (2002) Identification of the metalloproteinase stromelysin in the physis. J Orthop Res 20:289–294
Arnold S, Plate U, Wiesmann H-P, Straatmann U, Kohl H, Höhling H-J (2001) Quantitative analyses of the biomineralization of different hard tissues. J Microsc 202:488–494
Arsenault AL (1988) Crystal-collagen relationships in calcified turkey leg tendons visualized by selected-area dark field electron microscopy. Calcif Tissue Int 43:202–212
Arsenault AL (1989) A comparative electron microscopic study of apatite crystals in collagen fibrils of rat bone, dentin and calcified turkey leg tendons. Bone and Mineral 6:165–177
Arsenault AL (1991) Image analysis of collagen-associated mineral distribution in cryogenically prepared turkey leg tendons. Calcif Tissue Int 48:56–62
Arsenault AL, Ottensmeyer FP (1983) Quantitative spatial distributions of calcium, phosphorus, and sulfur in calcifying epiphysis by high resolution electron spectroscopic imaging. Proc Natl Acad Sci U S A 80:1322–1326
Arsenault AL, Robinson BW (1989) The dentino-enamel junction: a structural and microanalytical study of early mineralization. Calcif Tissue Int 45:111–121
Arsenault AL, Frankland BW, Ottensmeyer FP (1991) Vectorial sequence of mineralization in the turkey leg tendon determined by electron microscopic imaging. Calcif Tissue Int 48:46–55
Ascenzi A, Benedetti EL (1959) An electron microscopic study of the foetal membranous ossification. Acta Anat 37:370–385
Ascenzi A, Bonucci E (1966) The osteon calcification as revealed by the electron microscope. In: Fleisch H, Blackwood HJJ, Owen M (eds) Calcified tissues 1965. Springer, Berlin Heidelberg New York, pp 142–146
Ascenzi A, François C, Steve Bocciarelli D (1963) On the bone induced by estrogen in birds. J Ultrastruct Res 8:491–505
Ascenzi A, Bonucci E, Steve Bocciarelli D (1965) An electron microscope study of osteon calcification. J Ultrastruct Res 12:287–303
Ascenzi A, Bonucci E, Steve Bocciarelli D (1967) An electron microscope study on primary periosteal bone. J Ultrastruct Res 18:605–618
Ascenzi A, Bonucci E, Ripamonti A, Roveri N (1978) X-ray diffraction and electron microscope study of osteons during calcification. Calcif Tissue Res 25:133–143
Ascenzi A, Bonucci E, Generali P, Ripamonti A, Roveri N (1979) Orientation of apatite in single osteon samples as studied by pole figures. Calcif Tissue Int 29:101–105

Ascenzi A, Bigi A, Koch MH, Ripamonti A, Roveri N (1985) A low-angle X-ray diffraction analysis of osteonic inorganic phase using synchrotron radiation. Calcif Tissue Int 37:659–664

Ascenzi A, Benvenuti A, Bigi A, Foresti E, Koch MHJ, Mango F, Ripamonti A, Roveri N (1998) X-ray diffraction on cyclically loaded osteons. Calcif Tissue Int 62:266–273

Bachra BN (1967) Some molecular aspects of tissue calcification. Clin Orthop Relat Res 51:199–222

Baechtold AP, Wright JT, Yamauchi M, Spevak L, Camacho NP (2000) Dentin composition and structure in the *oim* mouse. In: Goldberg M, Boskey A, Robinson C (eds) Chemistry and biology of mineralized tissues. American Academy of Orthopaedic Surgeons, Rosemont, IL, pp 57–61

Bai P, Warshawsky H (1985) Morphological studies on the distribution of enamel matrix proteins using routine electron microscopy and freeze-fracture replicas in the rat incisor. Anat Rec 212:1–16

Barckhaus RH, Krefting E-R, Althoff J, Quint P, Höhling HJ (1981) Electron-microscopic microprobe analysis on the initial stages of mineral formation in the epiphyseal growth plate. Cell Tissue Res 217:661–666

Belcher AM, Wu XH, Christensen RJ, Hansma PK, Stucky GD, Morse DE (1996) Control of crystal phase switching and orientation by soluble mollusc-shell proteins. Nature 381:56–58

Beniash E, Traub W, Veis A, Weiner S (2000) A transmission electron microscope study using vitrified ice sections of predentin: structural changes in the dentin collagenous matrix prior to mineralization. J Struct Biol 132:212–225

Benson SC, Wilt FH (1992) Calcification of spicules in the sea urchin embryo. In: Bonucci E (ed) Calcification in biological systems. CRC Press, Boca Raton, pp 157–178

Berman A, Addadi L, Weiner S (1988) Interactions of sea-urchin skeleton macromolecules with growing calcite crystals – a study of intracrystalline proteins. Nature 331:546–548

Berman A, Addadi L, Kvick Å, Leiserowitz L, Nelson M, Weiner S (1990) Intercalation of sea urchin proteins in calcite: study of a crystalline composite material. Science 250:664–667

Bianco P (1990) Ultrastructural immunohistochemistry of noncollagenous proteins in calcified tissues. In: Bonucci E, Motta PM (eds) Ultrastructure of skeletal tissues. Kluwer Academic Publishers, Boston, pp 63–78

Bigi A, Ripamonti A, Koch MHJ, Roveri N (1988) Calcified turkey leg tendon as structural model for bone mineralization. Int J Biol Macromol 10:282–286

Bigi A, Dovigo L, Koch MHJ, Morocutti M, Ripamonti A, Roveri N (1991) Collagen structural organization in uncalcified and calcified human anterior longitudinal ligament. Connect Tissue Res 25:171–179

Bigi A, Gandolfi M, Koch MHJ, Roveri N (1996) X-ray diffraction study of *in vitro* calcification of tendon collagen. Biomaterials 17:1195–1201

Bigi A, Gandolfi M, Roveri N, Valdré G (1997) *In vitro* calcified tendon collagen: an atomic force and scanning electron microscopy investigation. Biomaterials 18:657–665

Bishop MA, Warshawsky H (1982) Electron microscopic studies on the potential loss of crystallites from routinely processed sections of young enamel in the rat incisor. Anat Rec 202:177–186

Blank S, Arnoldi M, Khoshnavaz S, Treccani L, Kuntz M, Mann K, Grathwohl G, Fritz M (2003) The nacre protein perlucin nucleates growth of calcium carbonate crystals. J Microsc 212:280–291

Blumenthal NC, Posner AS, Silverman LD, Rosenberg LC (1979) Effect of proteoglycans on in vitro hydroxyapatite formation. Calcif Tissue Int 27:75–82

Boanini E, Torricelli P, Gazzano M, Giardino R, Bigi A (2006) Nanocomposites of hydroxyapatite with aspartic acid and glutamic acid and their interaction with osteoblast-like cells. Biomaterials 27:4428–4433

Boivin G (1975) Étude chez le rat d'une calcinose cutanée induite par calciphylaxie locale I. – Aspects ultrastructuraux. Arch Anat Microsc Morphol Exp 64:183–205

Boivin G, Walzer C, Baud CA (1987) Ultrastructural study of the long-term development of two experimental cutaneous calcinoses (topical calciphylaxis and topical calcergy) in the rat. Cell Tissue Res 247:525–532

Bonar LC, Lees S, Mook HA (1985) Neutron diffraction studies of collagen in fully mineralized bone. J Mol Biol 181:265–270

Bonucci E (1967) Fine structure of early cartilage calcification. J Ultrastruct Res 20:33–50

Bonucci E (1969) Further investigation on the organic/inorganic relationships in calcifying cartilage. Calcif Tissue Res 3:38–54

Bonucci E (1971) The locus of initial calcification in cartilage and bone. Clin Orthop Relat Res 78:108–139

Bonucci E (1975) The organic-inorganic relationships in calcified organic matrices. Physicochimie et cristallographie des apatites d'intérêt biologique. Centre National de la Recherche Scientifique, Paris, pp 231–246

Bonucci E (1979) Presence of "crystal ghosts" in bone nodules. Calcif Tissue Int 29:181–182

Bonucci E (1984) The structural basis of calcification. In: Ruggeri A, Motta PM (eds) Ultrastructure of the connective tissue matrix. Martinus Nijhoff Publishing, Boston, pp 165–191

Bonucci E (1987) Is there a calcification factor common to all calcifying matrices? Scanning Electron Microsc 1:1089–1102

Bonucci E (1992a) Role of collagen fibrils in calcification. In: Bonucci E (ed) Calcification in biological systems. CRC Press, Boca Raton, pp 19–39

Bonucci E (1992b) Comments on the ultrastructural morphology of the calcification process: an attempt to reconcile matrix vesicles, collagen fibrils, and crystal ghosts. Bone Miner 17:219–222

Bonucci E (1995) Ultrastructural organic-inorganic relationships in calcified tissues: cartilage and bone vs. enamel. Connect Tissue Res 33:157–162

Bonucci E (2000) Basic composition and structure of bone. In: An YH, Draughn RA (eds) Mechanical testing of bone and the bone-implant interface. CRC Press, Boca Raton, pp 3–21

Bonucci E (2002) Crystal ghosts and biological mineralization: fancy spectres in an old castle, or neglected structures worthy of belief? J Bone Miner Metab 20:249–265

Bonucci E, De Santis E (1980) Ultrastructure of osteoblastoma, with particular reference to calcification and matrix vesicles. In: Donath A, Courvoisier B (eds) Bone and tumors. Medécine et Hygiène, Genève, pp 232–236

Bonucci E, Gherardi G (1975) Histochemical and electron microscope investigations on medullary bone. Cell Tissue Res 163:81–97

Bonucci E, Reurink J (1978) The fine structure of decalcified cartilage and bone: a comparison between decalcification procedures performed before and after embedding. Calcif Tissue Res 25:179–190

Bonucci E, Silvestrini G (1996) Ultrastructure of the organic matrix of embryonic avian bone after en bloc reaction with various electron-dense 'stains'. Acta Anat 156:22–33

Bonucci E, Derenzini M, Marinozzi V (1973) The organic-inorganic relationship in calcified mitochondria. J Cell Biol 59:185–211

Bonucci E, Silvestrini G, Di Grezia R (1988) The ultrastructure of the organic phase associated with the inorganic substance in calcified tissues. Clin Orthop Relat Res 233:243–261

Bonucci E, Silvestrini G, Di Grezia R (1989) Histochemical properties of the "crystal ghosts" of calcifying epiphyseal cartilage. Connect Tissue Res 22:43–50

Bonucci E, Lozupone E, Silvestrini G, Favia A, Mocetti P (1994) Morphological studies of hypomineralized enamel of rat pups on calcium-deficient diet, and of its changes after return to normal diet. Anat Rec 239:379–395

Boothroyd B (1964) The problem of demineralisation in thin sections of fully calcified bone. J Cell Biol 20:165–173

Borelli G, Mayer-Gostan N, Merle PL, De Pontual H, Boeuf G, Allemand D, Payan P (2003) Composition of biomineral organic matrices with special emphasis on turbot (*Psetta maxima*) otolith and endolimph. Calcif Tissue Int 72:717–725

Boskey AL (1989) Noncollagenous matrix proteins and their role in mineralization. Bone Miner 6:111–123

Boskey AL (1998) Biomineralization: conflicts, challenges, and opportunities. J Cell Biochem 30/31:83–91

Boskey AL, Posner AS (1976) Extraction of calcium-phospholipid-phosphate complex from bone. Calcif Tissue Res 19:273–283

Boskey AL, Posner AS (1977) The role of synthetic and bone extracted Ca-phospholipid-PO_4 complexes in hydroxyapatite formation. Calcif Tissue Res 23:251–258

Boskey AL, Reddi AH (1983) Changes in lipids during matrix-induced endochondral bone formation. Calcif Tissue Int 35:549–554

Boskey AL, Posner AS, Lane JM, Goldberg MR, Cordella DM (1980) Distribution of lipids associated with mineralization in the bovine epiphyseal growth plate. Arch Biochem Biophys 199:305–311

Boskey AL, Bullough PG, Posner AS (1982) Calcium-acidic phospholipid-phosphate complexes in diseased and normal human bone. Metab Bone Dis Rel Res 4:151–156

Boskey AL, Bullough PG, Vigorita V, Di Carlo E (1988) Calcium-acidic phospholipid-phosphate complexes in human hydroxyapatite-containing pathologic deposits. Am J Pathol 133:22–29

Boskey AL, Stiner D, Binderman I, Doty SB (1997) Effects of proteoglycan modification on mineral formation in a differentiating chick limb-bud mesenchymal cell culture system. J Cell Biochem 64:632–643

Bosshardt DD, Nanci A (2000) The pattern of expression of collagen determines the concentration and distribution of noncollagenous proteins along the forming root. In: Goldberg M, Boskey A, Robinson C (eds) Chemistry and biology of mineralized tissues. American Academy of Orthopaedic Surgeons, Rosemont, IL, pp 129–136

Bouvier M, Couble M-L, Hartmann DJ, Gauthier JP, Magloire H (1990a) Ultrastructural and immunocytochemical study of bone-derived cells cultured in three-dimensional matrices: influence of chondroitin-4 sulphate on mineralization. Differentiation 45:128–137

Bouvier M, Joffre A, Magloire H (1990b) *In vitro* mineralization of a three-dimensional collagen matrix by human dental pulp cells in the presence of chondroitin sulphate. Arch Oral Biol 35:301–309

Bowness JM (1962) Calcium binding by chondroitin sulfate associated with collagen. Biochim Biophys Acta 58:134–136

Boyan BD (1985) Proteolipid-dependent calcification. In: Butler WT (ed) The chemistry and biology of mineralized tissues. Ebsco Media, Inc., Birmingham, AL, pp 125–131

Boyan BD, Schwartz Z, Swain LD, Khare A (1989) Role of lipids in calcification of cartilage. Anat Rec 224:211–219

Boyan BD, Swain LD, Everett MM, Schwartz Z (1992) Mechanisms of microbial mineralization. In: Bonucci E (ed) Calcification in biological systems. CRC Press, Boca Raton, pp 129–156

Dean DD, Schwartz Z, Muniz OE, Gomez R, Swain LD, Howell DS, Boyan BD (1992) Matrix vesicles are enriched in metalloproteinases that degrade proteoglycans. Calcif Tissue Int 50:342–349

Dean DD, Schwartz Z, Bonewald L, Muniz OE, Morales S, Gomez R, Brooks BP, Qiao M, Howell DS, Boyan BD (1994) Matrix vesicles produced by osteoblast-like cells in culture become significantly enriched in proteoglycan-degrading metalloproteinases after addition of β-glycerophosphate and ascorbic acid. Calcif Tissue Int 54:399–408

Decker JD (1966) An electron microscopic investigation of osteogenesis in the embryonic chick. Am J Anat 118:591–614

Deutsch D, Catalano-Sherman J, Dafni L, David S, Palmon A (1995) Enamel matrix proteins and ameloblast biology. Connect Tissue Res 32:97–107

Diekwisch TGH, Berman BJ, Gentner S, Slavkin HC (1995) Initial enamel crystals are not spatially associated with mineralized dentine. Cell Tissue Res 279:149–167

Diekwisch TG, Berman BJ, Anderton X, Gurinsky B, Ortega AJ, Satchell PG, Williams M, Arumughan C, Luan X, McIntosh JE, Yamane A, Carlson DS, Sire J-Y, Shuler CF (2002) Membranes, minerals, and proteins of developing vertebrate enamel. Microsc Res Tech 59:373–395

DiStefano V, Neuman WF, Rouser G (1953) The isolation of a phosphate ester from calcifiable cartilage. Arch Biochem Biophys 47:218–220

Dmitrovsky E, Boskey AL (1985) Calcium-acidic phospholipid-phosphate complexes in human atherosclerotic aortas. Calcif Tissue Int 37:121–125

Dong W, Warshawsky H (1995) Failure to demonstrate a protein coat on enamel crystallites by morphological means. Archs Oral Biol 40:321–330

Dong W, Warshawsky H (1996) Lattice fringe continuity in the absence of crystal continuity in enamel. Adv Dent Res 10:232–237

Du C, Falini G, Fermani S, Abbott C, Moradian-Oldak J (2005) Supramolecular assembly of amelogenin nanospheres into birefringent microribbons. Science 307:1450–1454

Dziak R (1992) Role of lipids in osteogenesis: cell signaling and matrix calcification. In: Bonucci E (ed) Calcification in biological systems. CRC Press, Boca Raton, pp 59–71

Dziewiatkowski DD, Majznerski LL (1985) Role of proteoglycans in endochondral ossification: inhibition of calcification. Calcif Tissue Int 37:560–564

Eanes ED (1992) Dynamics of calcium phosphate precipitation. In: Bonucci E (ed) Calcification in biological systems. CRC Press, Boca Raton, pp 1–17

Eanes ED, Hailer AW (1994) Effect of ultrafilterable fragments from chondroitinase and protease-treated aggrecan on calcium phosphate precipitation in liposomal suspension. Calcif Tissue Int 55:176–179

Eastoe JE (1968) Chemical aspects of the matrix concept in calcified tissue organisation. Calcif Tissue Res 2:1–19

Ekanayake S, Hall BK (1994) Hypertrophy is not a prerequisite for type X collagen expression or mineralization of chondrocytes derived from cultured chick mandibular ectomesenchyme. Int J Dev Biol 38 :683–694

Embery G, Rees S, Hall R, Rose K, Waddington R, Shellis P (1998) Calcium- and hydroxyapatite-binding properties of glucuronic acid-rich and iduronic acid-rich glycosaminoglycans and proteoglycans. Eur J Oral Sci 106:267–273

Engfeldt B, Hjerpe A (1976) Glycosaminoglycans and proteoglycans of human bone tissue at different stages of mineralization. Acta Path Microbiol Scand Sect A 84:95–106

Ennever J, Creamer H (1967) Microbiological calcification: bone mineral and bacteria. Calcif Tissue Res 1:87–93

Ennever J, Vogel J, Takazoe I (1968) Calcium binding by a lipid extract of *Bacterionema matruchotii*. Calcif Tissue Res 2:296–298

Ennever J, Vogel JJ, Rider LJ, Boyan-Salyers B (1976) Nucleation of microbiologic calcification by proteolipid. Proc Soc Exper Biol Med 152:147–150
Eppell SJ, Tong W, Katz JL, Kuhn L, Glimcher MJ (2001) Shape and size of isolated bone mineralites measured using atomic force microscopy. J Orthop Res 19:1027–1034
Falini G, Albeck S, Weiner S, Addadi L (1996) Control of aragonite or calcite polymorphism by mollusk shell macromolecules. Science 271:67–69
Fearnhead RW (1979) Matrix-mineral relationships in enamel tissues. J Dent Res 58:909–916
Fernandez MS, Araya M, Arias JL (1997) Eggshells are shaped by a precise spatio-temporal arrangement of sequentially deposited macromolecules. Matrix Biol 16:13–20
Fernandez MS, Moya A, Lopez L, Arias JL (2001) Secretion pattern, ultrastructural localization and function of extracellular matrix molecules involved in eggshell formation. Matrix Biol 19:793–803
Fernandez MS, Passalacqua K, Arias JI, Arias JL (2004) Partial biomimetic reconstruction of avian eggshell formation. J Struct Biol 148:1–10
Fincham AG, Moradian-Oldak J, Simmer JP (1999) The structural biology of the developing dental enamel matrix. J Struct Biol 126:270–299
Fischer JW, Steiz S, Johnson P, Burke A, Kolodgie F, Virmani R, Giachelli C, Wight TN (2004) Decorin promotes aortic smooth muscle cell calcification and colocalizes to calcified regions in human atherosclerotic lesions. Arterioscler Thromb Vasc Biol 24:2391–2396
Frasier MB, Banks WJ, Newbrey JW (1975) Characterization of developing antler cartilage matrix I. Selected histochemical and enzymatic assessment. Calcif Tissue Res 17:273–288
Fratzl P, Fratzl-Zelman N, Klaushofer K, Vogl G, Koller K (1991) Nucleation and growth of mineral crystals in bone studied by small-angle X-ray scattering. Calcif Tissue Int 48:407–413
Fratzl P, Fratzl-Zelman N, Klaushofer K (1993) Collagen packing and mineralization. An X-ray scattering investigation of turkey leg tendon. Biophys J 54:260–266
Freudenberg E, György P (1923) III. Der Verkalkungsvorgang bei der Entwicklung des Knochens. Ergebn inn Med 24:17–28
Fu G, Valiyaveettil S, Wopenka B, Morse DE (2005) $CaCO_3$ biomineralization: acidic 9-kDa proteins isolated from aragonitic abalone shell nacre can specifically modify calcite crystal morphology. Biomacromolecules 6:1289–1298
Fujisawa R, Kuboki Y (1991) Preferential adsorption of dentin and bone acidic proteins on the (100) face of hydroxyapatite crystals. Biochim Biophys Acta 1075:56–60
Fujisawa R, Nodasaka Y, Kuboki Y (1995) Further characterization of interaction between bone sialoprotein (BSP) and collagen. Calcif Tissue Int 56:140–144
Fukae M, Tanabe T, Yamada M (1994) Action of metalloproteinases on porcine dentin mineralization. Calcif Tissue Int 55:426–435
Genge BR, Wu LNY, Adkisson HDI, Wuthier RE (1991) Matrix vesicle annexins exhibit proteolipid-like properties. Selective partitioning into lipophilic solvents under acidic conditions. J Biol Chem 266:10678–10685
Gerstenfeld LC, Gotoh Y, McKee MD, Nanci A, Landis WJ, Glimcher MJ (1990) Expression and ultrastructural immunolocalization of a major 66 kDa phosphoprotein synthesized by chicken osteoblasts during mineralization in vitro. Anat Rec 228:93–103
Glimcher MJ (1959) Molecular biology of mineralized tissues with particular reference to bone. Rev Modern Phys 31:359–393
Glimcher MJ (1976) Composition, structure, and organization of bone and other mineralized tissues and the mechanism of calcification. In: Greep RO, Astwood EB (eds) Handbook of physiology: endocrinology. American Physiological Society, Washington, pp 25–116
Glimcher MJ (1989) Mechanism of calcification: role of collagen fibrils and collagen-phosphoprotein complexes in vitro and in vivo. Anat Rec 224:139–153

Glimcher MJ (1990) The nature of the mineral component of bone and the mechanism of calcification. In: Avioli LV, Krane SM (eds) Metabolic bone disease and clinically related disorders. W.B. Saunders Company, Philadelphia, pp 42–68

Glimcher MJ (1992) The nature of the mineral component of bone and the mechanism of calcification. In: Coe FL, Favus MJ (eds) Disorders of bone and mineral metabolism. Raven Press, New York, pp 265–286

Glimcher MJ, Krane SM (1968) The organization and structure of bone, and the mechanism of calcification. In: Gould BS (ed) Biology of collagen. Academic Press, London, pp 67–251

Glimcher MJ, Brickley-Parsons D, Kossiva D (1979) Phosphopeptides and γ-carboxyglutamic acid-containing peptides in calcified turkey tendon: their absence in uncalcified tendon. Calcif Tissue Int 27:281–284

Goldberg M, Septier D, Lécolle S, Chardin H, Quintana MA, Acevedo AC, Gafni G, Dillouya D, Vermelin L, Thonemann B, Schmalz G, Bissila-Mapahou P, Carreau JP (1995) Dental mineralization. Int J Dev Biol 39:93–110

Goldberg M, Rapoport O, Septier D, Palmier K, Hall R, Embery G, Young M, Ameye L (2003) Proteoglycans in predentin: the last 15 micrometers before mineralization. Connect Tissue Res 44:184–188

Gomez S, Lopez-Cepero JM, Silvestrini G, Mocetti P, Bonucci E (1996) Matrix vesicles and focal proteoglycan aggregates are the nucleation sites revealed by the lanthanum incubation method: a correlated study on the hypertrophic zone of the rat epiphyseal cartilage. Calcif Tissue Int 58:273–282

Gotliv BA, Addadi L, Weiner S (2003) Mollusk shell acidic proteins: in search of individual functions. ChemBioChem 4:522–529

Groot CG (1982) Acid groups in the organic matrix of foetal bone. An electron microscopical study. Thesis, Rijksuniversiteit te Leiden, pp 1–95

Gustafsson E, Aszodi A, Ortega N, Hunziker EB, Denker HW, Werb Z, Fassler R (2003) Role of collagen type II and perlecan in skeletal development. Ann N Y Acad Sci 995:140–150

Hargest TE, Gay CV, Schraer H, Wasserman AJ (1985) Vertical distribution of elements in cells and matrix of epiphyseal growth plate cartilage determined by quantitative electron probe analysis. J Histochem Cytochem 33:275–286

Hashimoto Y, Lester GE, Caterson B, Yamauchi M (1995) EDTA-insoluble, calcium-binding proteoglycan in bovine bone. Calcif Tissue Int 56:398–402

Hay DI, Moreno EC (1979) Differential adsorption and chemical affinities of proteins for apatitic surfaces. J Dent Res 58:930–940

Hayashi Y (1992) High resolution electron microscopy in the dentino-enamel junction. J Electron Microsc 41:387–391

Herold RC (1971) Osteodentinogenesis. An ultrastructural study of tooth formation in the pike, *Esox lucius*. Z Zellforsch 112:1–14

Hincke MT, Tsang CPW, Courtney M, Hill V, Narbaitz R (1995) Purification and immunohistochemistry of a soluble matrix protein of the chicken eggshell (Ovocleidin 17). Calcif Tissue Int 56:578–583

Hirschman A, Dziewiatkowski DD (1966) Protein-polysaccharide loss during endochondral ossification: immunochemical evidence. Science 154:393–395

Hoshi K, Kemmotsu S, Takeuchi Y, Amizuka N, Ozawa H (1999) The primary calcification in bones follows removal of decorin and fusion of collagen fibrils. J Bone Miner Res 14:273–280

Hoshi K, Ejiri S, Ozawa H (2001a) Localizational alterations of calcium, phosphorus, and calcification-related organics such as proteoglycans and alkaline phosphatase during bone calcification. J Bone Miner Res 16:289–298

Hoshi K, Ejiri S, Ozawa H (2001b) Ultrastructural analysis of bone calcification by using energy-filtering transmission electron microscopy. It J Anat Embryol 106:141–150
Hoshi K, Ejiri S, Ozawa H (2001c) Organic components of crystal sheaths in bones. J Electron Microsc (Tokyo) 50:33–40
Höhling HJ (1984) Can the fine structure of the early crystal formations and of staining nuclei give information on the spatial distribution of nucleation sites at matrix molecules? INSERM 125:479–494
Höhling HJ (1989) Special aspects of biomineralization of dental tissues. In: Oksche A, Vollrath L (eds) Handbook of microscopic anatomy. Springer, Berlin Heidelberg New York, pp 475–524
Höhling HJ, Schöpfer H (1968) Morphological investigations of apatitic nucleation in hard tissue and salivary stone formation. Naturwissensch 55:545
Höhling HJ, Schöpfer H, Neubauer G (1970) Elektronenmikroskopie und Laserbeugungs-Untersuchungen zur Charakterisierung der organischen Matrix im Speichelstein und Hartgewebe. Z Zellforsch 108:415–430
Höhling HJ, Kreilos R, Neubauer G, Boyde A (1971) Electron microscopy and electron microscopical measurements of collagen mineralization in hard tissues. Z Zellforsch 122:36–52
Höhling HJ, Barckhaus RH, Krefting ER, Schreiber J (1976) Electron microscopic microprobe analysis of mineralized collagen fibrils and extracollagenous regions in turkey leg tendon. Cell Tissue Res 175:345–350
Höhling HJ, Barckhaus RH, Krefting ER, Althoff J, Quint P, Niestadtkötter R (1981) Relationship between the Ca-phosphate crystallites and the collagen structure in turkey tibia tendon. In: Veis A (ed) The chemistry and biology of mineralized connective tissues. Elsevier North Holland, Amsterdam, pp 113–117
Höhling HJ, Krefting ER, Barckhaus R (1982) Does correlation exist between mineralization in collagen-rich hard tissues and that in enamel? J Dent Res 61:1496–1503
Höhling HJ, Barckhaus RH, Krefting E-R, Althoff J, Quint P (1990) Collagen mineralization: aspects of the structural relationship between collagen and the apatitic crystallites. In: Bonucci E, Motta PM (eds) Ultrastructure of skeletal tissues. Kluwer Academic Publishers, Boston, pp 41–62
Höhling HJ, Arnold S, Barckhaus RH, Plate U, Wiesmann HP (1995) Structural relationship between the primary crystal formation and the matrix macromolecules in different hard tissues. Discussion of a general principle. Connect Tissue Int 33:171–178
Höhling HJ, Arnold S, Plate U, Stratmann U, Wiesmann HP (1997) Analysis of general principle of crystal nucleation, formation in the different hard tissues. Adv Dent Res 11:462–466
Hunter GK (1987) An ion-exchange mechanism of cartilage calcification. Connect Tissue Res 16:111–120
Hunter GK (1991) Role of proteoglycan in the provisional calcification of cartilage. A review and reinterpretation. Clin Orthop Relat Res 262:256–280
Hunter GK (1992) *In vitro* studies on matrix-mediated mineralization. In: Hall BK (ed) Bone, vol 4: Bone metabolism and mineralization. CRC Press, Boca Raton, pp 225–247
Hunter GK, Szigety S (1992) Effects of proteoglycan on hydroxyapatite formation under non-steady-state and pseudo-steady-state conditions. Matrix 12:362–368
Ishigaki R, Takagi M, Igarashi M, Ito K (2002) Gene expression and immunohistochemical localization of osteonectin in association with early bone formation in the developing mandible. Histochem J 34:57–66
Jacenko O, Chan D, Franklin A, Ito S, Underhill CB, Bateman JF, Campbell MR (2001) A dominant interference collagen X mutation disrupts hypertrophic chondrocyte peri-

cellular matrix and glycosaminoglycan and proteoglycan distribution in transgenic mice. J Pathol 159:2257–2269
Jodaikin A, Traub W, Weiner S (1986) Protein conformation in rat tooth enamel. Arch Oral Biol 31:685–689
Kallenbach E (1971) Electron microscopy of the differentiating rat incisor ameloblast. J Ultrastruct Res 35:508–531
Kallenbach E (1986) Crystal-associated matrix components in rat incisor enamel. An electron-microscopic study. Cell Tissue Res 246:455–461
Kallenbach E (1989) Critical comments on the article entitled "Organization of crystals in enamel" by H. Warshawsky. Anat Rec 224:263
Kallenbach E (1990) Evidence that apatite crystals of rat incisor enamel have hexagonal cross sections. Anat Rec 228:132–136
Kasugai S, Todescan R Jr, Nagata T, Yao K-L, Butler WT, Sodek J (1991) Expression of bone matrix proteins associated with mineralized tissue formation by adult rat bone marrow cells in vitro: inductive effects of dexamethasone on the osteoblastic phenotype. J Cell Physiol 147:111–120
Khan SR (1992) Structure and development of calcific urinary stones. In: Bonucci E (ed) Calcification in biological systems. CRC Press, Boca Raton, pp 345–363
Kim H-M, Rey C, Glimcher MJ (1995) Isolation of calcium-phosphate crystals of bone by non-aqueous methods at low temperature. J Bone Miner Res 10:1589–1601
Kirkham J, Zhang J, Brookes SJ, Shore RC, Wood SR, Smith DA, Wallwork ML, Ryu OH, Robinson C (2000) Evidence for charge domains on developing enamel crystal surfaces. J Dent Res 79:1943–1947
Kirsch T, Claassen H (2000) Matrix vesicles mediate mineralization of human thyroid cartilage. Calcif Tissue Int 66:292–297
Kirsch T, Von der Mark H (1991) Ca^{2+} binding properties of type X collagen. Fed Eur Bioch Soc 294:149–152
Kirsch T, Wuthier RE (1994) Stimulation of calcification of growth plate cartilage matrix vesicles by binding to type II and X collagens. J Biol Chem 269:11462–11469
Kirsch T, Ishikawa Y, Mwale F, Wuthier RE (1994) Roles of the nucleational core complex and collagens (types II and X) in calcification of growth plate cartilage matrix vesicles. J Biol Chem 269:20103–20109
Kirsch T, Harrison G, Golub EE, Nah HD (2000) The roles of annexins and types II and X collagen in matrix vesicle-mediated mineralization of growth plate cartilage. J Biol Chem 275:35577–35583
Kobayashi S (1971) Acid mucopolysaccharides in calcified tissues. Int Rev Cytol 30:257–371
Kuettner KE, Sorgente N, Croxen RL, Howell DS, Pita JC (1974) Lysozyme in preosseous cartilage VII. Evidence for physiological role of lysozyme in normal endochondral calcification. Biochim Biophys Acta 372:335–344
Kwan KM, Pang MK, Zhou S, Cowan SK, Kong RY, Pfordte T, Olsen BR, Sillence DO, Tam PP, Cheah KS (1997) Abnormal compartmentalization of cartilage matrix components in mice lacking collagen X: implications for function. J Cell Biol 136:459–471
Lakshminarayanan R, Kini RM, Valiyaveettil S (2002) Investigation of the role of ansocalcin in the biomineralization in goose eggshell matrix. Proc Natl Acad Sci U S A 99:5155–5159
Landis WJ, Géraudie J (1990) Organization and development of the mineral phase during early ontogenesis of the bony fin rays of the trout *Oncorhynchus mykiss*. Anat Rec 228:383–391
Landis WJ, Song MJ, Leith A, McEwen L, McEwen BF (1993) Mineral and organic matrix interaction in normally calcifying tendon visualized in three dimensions by high-voltage electron microscopic tomography and graphic image reconstruction. J Struct Biol 110:39–54

Landis WJ, Hodgens KJ, Arena J, Song MJ, McEwen BF (1996a) Structural relations between collagen and mineral in bone as determined by high voltage electron microscopic tomography. Microsc Res Techn 33:192–202

Landis WJ, Hodgens KJ, Song MJ, Arena J, Kiyonaga S, Marko M, Owen C, McEwen BF (1996b) Mineralization of collagen may occur on fibril surfaces: evidence from conventional and high-voltage electron microscopy and three-dimensional imaging. J Struct Biol 117:24–35

Lanzing WJR, Wright RG (1976) The ultrastructure and calcification of the scales of *Tilapia mossambica* (Peters). Cell Tissue Res 167:37–47

Lash JW, Vasan NS (1983) Glycosaminoglycans of cartilage. In: Hall BK (ed) Cartilage. Structure, function and biochemistry. Academic Press, New York, pp 215–251

Leach SA (1979) Enamel matrix and crystals. J Dent Res 58:943–947

Leblond CP, Lacroix P, Ponlot R, Dhem A (1959) Les stades initiaux de l'ostéogenèse. Nouvelles données histochimique et autoradiographiques. Bull Acad Roy Med Belgique 24:421–443

Lee DD, Glimcher MJ (1989) The three-dimensional spatial relationship between the collagen fibrils and the inorganic calcium-phosphate crystals of pickerel and herring fish bone. Connect Tissue Res 21:247–257

Lee DD, Glimcher MJ (1991) Three-dimensional spatial relationship between the collagen fibrils and the inorganic calcium phosphate crystals of pickerel (*Americanus americanus*) and herring (*Clupea harengus*) bone. J Mol Biol 217:487–501

Lees S (1987) Considerations regarding the structure of the mammalian mineralized osteoid from viewpoint of the generalized packing model. Connect Tissue Res 16:281–303

Lees S, Prostak K (1988) The locus of mineral crystallites in bone. Connect Tissue Res 18:41–54

Lees S, Prostak KS, Ingle VK, Kjoller K (1994) The loci of mineral in turkey leg tendon as seen by atomic force microscope and electron microscopy. Calcif Tissue Int 55:180–189

Lees S, Capel M, Hukins DWL, Mook HA (1997) Effect of sodium chloride solutions on mineralized and unmineralized turkey leg tendon. Calcif Tissue Int 61:74–76

Linde A (1984) Non-collagenous proteins and proteoglycans in dentinogenesis. In: Linde A (ed) Dentin and dentinogenesis, 2nd vol. CRC Press, Boca Raton, pp 55–92

Linde A, Goldberg M (1993) Dentinogenesis. Crit Rev Oral Biol Med 4:679–728

Linde A, Lussi A, Crenshaw MA (1989) Mineral induction by immobilized polyanionic proteins. Calcif Tissue Int 44:286–295

Lohmander S, Hjerpe A (1975) Proteoglycans of mineralizing rib and epiphyseal cartilage. Biochim Biophys Acta 404:93–109

Lormée P, Septier D, Lécolle S, Baudoin C, Goldberg M (1996) Dual incorporation of (^{35}S)sulfate into dentin proteoglycans acting as mineralization promotors in rat molars and predentin proteoglycans. Calcif Tissue Int 58:368–375

MacDougall M, Nydegger J, Gu TT, Simmons D, Luan X, Cavender A, D'Souza RN (1998) Developmental regulation of dentin sialophosphoprotein during ameloblast differentiation: a potential enamel matrix nucleator. Connect Tissue Res 39:329–341

Mann S (1988) Molecular recognition in biomineralization. Nature 332:119–124

Mann S (1993) Molecular tectonics in biomineralization and biomimetic materials chemistry. Nature 365:499–505

Marsh ME (1994) Polyanion-mediated mineralization – assembly and reorganization of acidic polysaccharides in the Golgi system of a coccolithophorid alga during mineral deposition. Protoplasma 177:108–122

Matsui Y, Alini M, Webber C, Poole AR (1991) Characterization of aggregating proteoglycans from the proliferative, maturing, hypertrophic, and calcifying zones of the cartilaginous physis. J Bone Joint Surg 73-A:1064–1074

Matukas VJ, Krikos GA (1968) Evidence for changes in protein polysaccharide associated with the onset of calcification in cartilage. J Cell Biol 39:43–48
Mayne R, von der Mark K (1983) Collagens of cartilage. In: Hall BK (ed) Cartilage. Structure, function and biochemistry. Academic Press, New York, pp 181–214
Miake Y, Aoba T, Moreno EC, Shimoda S, Prostak K, Suga S (1991) Ultrastructural studies on crystal growth of enameloid minerals in Elasmobranch and Teleost fish. Calcif Tissue Int 48:204–217
Mitchell N, Shepard N, Harrod J (1982) The measurement of proteoglycan in the mineralizing region of the rat growth plate. An electron microscopic and X-ray microanalytical study. J Bone Joint Surg 64-A:32–38
Moradian-Oldak J (2001) Amelogenins: assembly, processing and control of crystal morphology. Matrix Biol 20:293–305
Moradian-Oldak J, Weiner S, Addadi L, Landis WJ, Traub W (1991) Electron imaging and diffraction study of individual crystals of bone, mineralized tendon and synthetic carbonate apatite. Connect Tissue Res 25:219–228
Moradian-Oldak J, Frolow F, Addadi L, Weiner S (1992) Interactions between acidic matrix macromolecules and calcium phosphate ester crystals: relevance to carbonate apatite formation in biomineralization. Proc R Soc London B Biol Sci 247:47–55
Mwale F, Tchetina E, Wu CW, Poole AR (2002) The assembly and remodeling of the extracellular matrix in the growth plate in relationship to mineral deposition and cellular hypertrophy: an in situ study of collagens II and IX and proteoglycan. J Bone Miner Res 17:275–283
Nakamura H, Hirata A, Tsuji T, Yamamoto T (2001) Immunolocalization of keratan sulfate proteoglycan in rat calvaria. Arch Histol Cytol 64:109–118
Nanci A (1999) Content and distribution of noncollagenous matrix proteins in bone and cementum: relationship to speed of formation and collagen packing density. J Struct Biol 126:256–269
Nanci A, Smith CE (1992) Development and calcification of enamel. In: Bonucci E (ed) Calcification in biological systems. CRC Press, Boca Raton, pp 313–343
Nanci A, Bai P, Warshawsky H (1983) The effect of osmium postfixation and uranyl and lead staining on the ultrastructure of young enamel in the rat incisor. Anat Rec 207:1–16
Nefussi JR, Brami G, Modrowski D, Oboeuf M, Forest N (1997) Sequential expression of bone matrix proteins during rat calvaria osteoblast differentiation and bone nodule formation in vitro. J Histochem Cytochem 45:493–503
Olson OP, Watabe N (1980) Studies on formation and resorption of fish scales. IV: Ultrastructure of developing scales in newly hatched fry of the sheepshead minnow, *Cyprinodon variegatus* (Atheriniformes: Cyprinodontidae). Cell Tissue Res 211:303–316
Onozato H, Watabe N (1979) Studies on fish scale formation and resorption III. Fine structure and calcification of the fibrillary plates of the scales in *Carassius auratus* (Cypriniformes: cyprinidae). Cell Tissue Res 201:409–422
Parisuthiman D, Mochida Y, Duarte WR, Yamauchi M (2005) Biglycan modulates osteoblast differentiation and matrix mineralization. J Bone Miner Res 20:1878–1886
Paschalis EP, Jacenko O, Olsen B, Mendelsohn R, Boskey AL (1996a) FT-IR microscopic analysis identified alterations in mineral properties in bones from mice transgenic for type X collagen. Bone 18:151–156
Paschalis EP, Jacenko O, Olsen B, de Crombrugghe B, Boskey AL (1996b) The role of type X collagen in endochondral ossification as deduced by Fourier transform infrared microscopy analysis. Connect Tissue Res 35:371–377
Pidaparti RMV, Chandran A, Takano Y, Turner CH (1996) Bone mineral lies mainly outside collagen fibrils: prediction of a composite model for osteonal bone. J Biomechanics 29:909–916

Pita JC, Cuervo LA, Madruga JE, Muller FJ, Howell DS (1970) Evidence for a role of protein-polysaccharides in regulation of mineral phase separation in calcifying cartilage. J Clin Invest 49:2188–2197

Plate U, Tkotz T, Wiesmann HP, Stratmann U, Joos U, Höhling HJ (1996) Early mineralization of matrix vesicles in the epiphyseal growth plate. J Microsc 183:102–107

Plate U, Arnold S, Stratmann U, Wiesmann H-P, Höhling HJ (1998) General principle of ordered apatitic crystal formation in enamel and collagen rich hard tissues. Connect Tissue Res 38:149–157

Politi Y, Arad T, Klein E, Weiner S, Addadi L (2004) Sea urchin spine calcite forms via a transient amorphous calcium carbonate phase. Science 306:1161–1164

Poole AR (1991) The growth plate: cellular physiology, cartilage assembly and mineralization. In: Hall B, Newman S (eds) Cartilage: molecular aspects. CRC Press, Boca Raton, pp 179–211

Poole AR, Pidoux I, Rosenberg L (1982) Role of proteoglycans in endochondral ossification: immunofluorescent localization of link protein and proteoglycan monomer in bovine fetal epiphyseal growth plate. J Cell Biol 92:249–260

Poole AR, Matsui Y, Hinek A, Lee ER (1989) Cartilage macromolecules and the calcification of cartilage matrix. Anat Rec 224:167–179

Posner AS (1987) Bone mineral and the mineralization process. In: Peck WA (ed) Bone and mineral research/5. Elsevier Science Publisher, Amsterdam, pp 65–116

Prince CW, Rahemtulla F, Butler WT (1983) Metabolism of rat bone proteoglycans *in vivo*. Biochem J 216:589–596

Prince CW, Rahemtulla F, Butler WT (1984) Incorporation of [^{35}S]sulphate into glycosaminoglycans by mineralized tissues in vivo. Biochem J 224:941–945

Pugliarello MC, Vittur F, de Bernard B, Bonucci E, Ascenzi A (1970) Chemical modifications in osteones during calcification. Calcif Tissue Res 5:108–114

Qin C, Baba O, Butler WT (2004) Post-translational modifications of SIBLING proteins and their roles in osteogenesis and dentinogenesis. Crit Rev Oral Biol Med 15:126–136

Raggio CL, Boyan BD, Boskey AL (1986) *In vivo* hydroxyapatite formation induced by lipids. J Bone Miner Res 1:409–415

Razzouk S, Brunn JC, Qin C, Tye CE, Goldberg HA, Butler WT (2002) Osteopontin post-translational modifications, possibly phosphorylation, are required for in vitro bone resorption but not osteoclast adhesion. Bone 30:40–47

Rees SG, Shellis RP, Embery G (2002) Inhibition of hydroxyapatite crystal growth by bone proteoglycans and proteoglycan components. Biochem Biophys Res Commun 292:727–733

Ritchie HH, Berry JE, Somerman MJ, Hanks CT, Bronckers ALJJ, Hotton D, Papagerakis P, Berdal A, Butler WT (1997) Dentin sialoprotein (DSP) transcripts: developmentally-sustained expression in odontoblasts and transient expression in pre-ameloblasts. Eur J Oral Sci 105:405–413

Robey PG (1996) Vertebrate mineralized matrix proteins: structure and function. Connect Tissue Res 35:131–136

Robinson C, Brookes SJ, Shore RC, Kirkham J (1998) The developing enamel matrix: nature and function. Eur J Oral Sci 106:282–291

Robison R, Rosenheim AH (1934) Calcification of hypertrophic cartilage *in vitro*. Biochem J 28:684–698

Rosati R, Horan GS, Pinero GJ, Garofalo S, Keene DR, Horton WA, Vuorio E, de Crombrugghe B, Behringer RR (1994) Normal long bone growth and development in type X collagen-null mice. Nature Genet 8:129–135

Saito T, Arsenault AL, Yamauchi M, Kuboki Y, Crenshaw MA (1997) Mineral induction by immobilized phosphoproteins. Bone 21:305–311

Saito T, Yamauchi M, Crenshaw MA (1998) Apatite induction by insoluble dentin collagen. J Bone Miner Res 13:265–270

Salih E, Wang J, Mah J, Fluckiger R (2002) Natural variation in the extent of phosphorylation of bone phosphoproteins as a function of in vivo new bone formation induced by demineralized bone matrix in soft tissue and bony environments. Biochem J 364:465–474

Sarig S (2004) Aspartic acid nucleates the apatite crystallites of bone: a hypothesis. Bone 35:108–113

Sasano Y, Zhu JX, Kamakura S, Kusunoki S, Mizoguchi I, Kagayama M (2000) Expression of major bone extracellular matrix proteins during embryonic osteogenesis in rat mandibles. Anat Embryol (Berl) 202:31–37

Satoyoshi M, Koizumi T, Teranaka T, Iwamoto T, Takita H, Kuboki Y, Saito S, Mikuni-Takagaki Y (1995) Extracellular processing of dentin matrix protein in the mineralizing odontoblast culture. Calcif Tissue Int 57:237–241

Scherft JP, Moskalewski S (1984) The amount of proteoglycans in cartilage matrix and the onset of mineralization. Metab Bone Dis Rel Res 5:195–203

Schönbörner AA, Boivin G, Baud CA (1979) The mineralization processes in teleost fish scales. Cell Tissue Res 202:203–212

Schubert M, Pras M (1968) Ground substance proteinpolysaccharides and the precipitation of calcium phosphate. Clin Orthop Relat Res 60:235–255

Shepard N (1992) Role of proteoglycans in calcification. In: Bonucci E (ed) Calcification in biological systems. CRC Press , Boca Raton, pp 41–58

Shepard N, Mitchell N (1981) Acridine orange stabilization of glycosaminoglycans in beginning endochondral ossification. A comparative light and electron microscopic study. Histochemistry 70:107–114

Sikes CS, Wheeler AP, Wierzbicki A, Dillaman RM, De Luca L (1998) Oyster shell protein and atomic force microscopy of oyster shell folia. Biol Bull 194:304–316

Silbermann M, Frommer J (1973) Dynamic changes in acid mucopolysaccharides during mineralization of the mandibular condylar cartilage. Histochemie 36:185–192

Smales FC (1975) Structural subunit in prisms of immature rat enamel. Nature 258:772–774

Sobel AE (1955) Local factors in the mechanism of calcification. Ann N Y Acad Sci 60:713–731

Sobel AE, Burger M (1954) Calcification XIV. Investigation of the role of chondroitin sulfate in the calcifying mechanism. Proc Soc Exper Biol Med 87:7–13

Sommer B, Bickel M, Hofstetter W, Wetterwald A (1996) Expression of matrix proteins during the development of mineralized tissues. Bone 19:371–380

Stratmann U, Schaarschmidt K, Wiesmann HP, Plate U, Höhling HJ (1996) Mineralization during matrix-vesicle-mediated mantle dentine formation in molars of albino rats: a microanalytical and ultrastructural study. Cell Tissue Res 284:223–230

Su X, Sun K, Cui FZ, Landis WJ (2003) Organization of apatite crystals in human woven bone. Bone 32:150–162

Sugars RV, Milan AM, Brown JO, Waddington RJ, Hall RC, Embery G (2003) Molecular interaction of recombinant decorin and biglycan with type I collagen influences crystal growth. Connect Tissue Res 44:189–195

Suzuki Y, Kubota T, Koizumi T, Satoyoshi M, Teranaka T, Kawase T, Ikeda T, Yamaguchi A, Saito S, Mikuni-Takagaki Y (1996) Extracellular processing of bone and dentin proteins in matrix mineralization. Connect Tissue Res 35:223–229

Takagi M (1990) Ultrastructural cytochemistry of cartilage proteoglycans and their relation to the calcification process. In: Bonucci E, Motta PM (eds) Ultrastructure of skeletal tissues. Kluwer Academic Publishers, Boston, pp 111–127

Takagi M, Parmley RT, Denys FR (1981) Ultrastructural localization of complex carbohydrates in odontoblasts, predentin, and dentin. J Histochem Cytochem 29:747–758

Takagi M, Parmley RT, Denys FR (1983) Ultrastructural cytochemistry and immunocytochemistry of proteoglycans associated with epiphyseal cartilage calcification. J Histochem Cytochem 31:1089–1100

Takagi M, Parmley RT, Denys FR, Yagasaki H, Toda Y (1984) Ultrastructural cytochemistry of proteoglycans associated with calcification of shark cartilage. Anat Rec 208:149–158

Takazoe I, Vogel J, Ennever J (1970) Calcium hydroxyapatite nucleation by lipid extract of *Bacterionema matruchotii*. J Dent Res 49:395–398

Termine JD (1985) The tissue specific proteins of the bone matrix. In: Butler WT (ed) The chemistry and biology of mineralized tissues. EBSCO Media, Birmingham,AL, pp 94–97

Tohda H, Yamada M, Yamaguchi Y, Yanagisawa T (1997) High-resolution electron microscopical observations of initial enamel crystals. J Electron Microsc 46 :97–101

Tong W, Glimcher MJ, Katz JL, Kuhn L, Eppell SJ (2003) Size and shape of mineralites in young bovine bone measured by atomic force microscopy. Calcif Tissue Int 72:592–598

Traub W, Jodaikin A, Weiner S (1985) Diffraction studies of enamel protein-mineral structural relations. In: Butler WT (ed) The chemistry and biology of mineralized tissues. Ebsco Media, Inc., Birmingham, pp 221–225

Traub W, Arad T, Weiner S (1992a) Origin of mineral crystal growth in collagen fibrils. Matrix 12:251–255

Traub W, Jodaikin A, Arad T, Veis A, Sabsay B (1992b) Dentin phosphophoryn binding to collagen fibrils. Matrix 12:197–201

Travis DF (1968) Comparative ultrastructure and organization of inorganic crystals and organic matrices of mineralized tissues. Biology of the mouth. American Association for the Advancement of Sciences, Washington, pp 237–297

Travis DF, Gonsalves M (1969) Comparative ultrastructure and organization of the prismatic region of two bivalves and its possible relation to the chemical mechanism of boring. Am Zoolog 9:635–661

Tsukamoto D, Sarashina I, Endo K (2004) Structure and expression of an unusually acidic matrix protein of pearl oyster shells. Biochem Biophys Res Commun 320:1175–1180

Van Dijk S, Dean DD, Liu Y, Zhao Y, Chirgwin JM, Schwartz Z, Boyan BD (1998) Purification, amino acid sequence, and cDNA sequence of a novel calcium-precipitating proteolipid involved in calcification of *Corynebacterium matruchotii*. Calcif Tissue Int 62:350–358

Veis A (1993) Mineral-matrix interactions in bone and dentin. J Bone Miner Res 8:S493-S497

Veis A (2003) Mineralization in organic matrix frameworks. Rev Mineral Geochem 54:249–289

Veis A, Perry A (1967) The phosphoprotein of the dentin matrix, Biochemistry 6:2409–2416

Vittur F, Zanetti M, Stagni N, de Bernard B (1979) Further evidence for the participation of glycoproteins to the process of calcification. Perspect Inherit Metab Dis 2:13–30

Vogel JJ, Ennever J (1971) The role of a lipoprotein in the intracellular hydroxyapatite formation in *Bacterionema matruchotii*. Clin Orthop Relat Res 78:218–222

Vogel JJ, Smith WN (1976) Calcification of membranes isolated from *Bacterionema matruchotii*. J Dent Res 55:1080–1083

Vogel JJ, Boyan-Salyers BD (1976) Acidic lipids associated with the local mechanism of calcification. A review. Clin Orthop Relat Res 118:230–241

von der Mark K, Mollenhauer J (1997) Annexin V interactions with collagen. Cell Mol Life Sci 53:539–545

Waddington RJ, Hall RC, Embery G, Lloyd DM (2003) Changing profiles of proteoglycans in the transition of predentine to dentine. Matrix Biol 22:153–161

Wallwork ML, Kirkham J, Chen H, Chang SX, Robinson C, Smith DA, Clarkson BH (2002) Binding of dentin noncollagenous matrix proteins to biological mineral crystals: an atomic force microscopy study. Calcif Tissue Int 71:249–255

Wang A, Martin JA, Lembke LA, Midura RJ (2000) Reversible suppression of in vitro biomineralization by activation of protein kinase A. J Biol Chem 275:11082–11091
Warshawsky H (1987) External shape of enamel crystals. Scanning Microsc 1:1913–1923
Warshawsky H (1989) Organization of crystals in enamel. Anat Rec 224:242–262
Weiner S (1979) Aspartic acid-rich proteins: major components of the soluble organic matrix of mollusk shells. Calcif Tissue Int 29:163–167
Weiner S, Addadi L (1991) Acidic macromolecules of mineralized tissues: the controllers of crystal formation. Trends Biochem Sci 16:252–256
Weiner S, Hood L (1975) Soluble protein of the organic matrix of mollusk shells: a potential template for shell formation. Science 190:987–989
Weiner S, Price PA (1986) Disaggregation of bone into crystals. Calcif Tissue Int 39:365–375
Weiner S, Traub W (1980) X-ray diffraction study of the insoluble organic matrix of mollusk shells. FEBS Lett 111:311–316
Weiner S, Traub W (1984) Macromolecules in mollusc shells and their function in biomineralization. Phil Trans R Soc London 304 B:425–434
Weiner S, Traub W (1986) Organization of hydroxyapatite crystals within collagen fibrils . Feder Eur Bioch Soc 206:262–266
Weiner S, Talmon Y, Traub W (1983) Electron diffraction of mollusk shell organic matrices and their relationship to the mineral phase. Int J Biol Macromol 5:325–328
Weiner S, Levi-Kalisman Y, Raz S, Addadi L (2003) Biologically formed amorphous calcium carbonate. Connect Tissue Res 44:214–218
Weiner S, Sagi I, Addadi L (2005) Choosing the crystallization path less traveled. Science 309:1027–1028
Weiss IM, Tuross N, Addadi L, Weiner S (2002) Mollusc larval shell formation: amorphous calcium carbonate is a precursor phase for aragonite. J Exp Zool 293:478–491
Wheeler AP (1992) Mechanisms of molluscan shell formation. In: Bonucci E (ed) Calcification in biological systems. CRC Press, Boca Raton, pp 179–216
Wheeler AP, Rusenko KW, George JW, Sikes CS (1987) Evaluation of calcium binding by molluscan shell organic matrix and its relevance to biomineralization. Comp Biochem Physiol 87B:953–960
Wierzbicki A, Sikes CS, Madura JD, Drake B (1994) Atomic force microscopy and molecular modeling of protein and peptide binding to calcite. Calcif Tissue Int 54:133–141
Woodward C, Davidson EA (1968) Structure-function relationships of protein polysaccharide complexes: specific ion-binding properties. Proc Natl Acad Sci 60:201–205
Wu LNY, Genge BR, Lloyd GC, Wuthier RE (1991) Collagen-binding proteins in collagenase-released matrix vesicles from cartilage. J Biol Chem 266:1195–1203
Wu Y, Ackerman JL, Strawich ES, Kim KM, Glimcher MJ (2003) Phosphate ions in bone: identification of a calcium-organic phosphate complex by 31P solid-state NMR spectroscopy at early stage of mineralization. Calcif Tissue Int 72:610–626
Xu T, Bianco P, Fisher LW, Longenecker G, Smith E, Goldstein S, Bonadio J, Boskey A, Heegaard A-M, Sommer B, Satomura K, Dominguez P, Zhao C, Kulkarni AB, Gehron Robey P, Young MF (1998) Targeted disruption of the biglycan gene leads to an osteoporosis-like phenotype in mice. Nature Genet 20:78–82
Yamada J, Watabe N (1979) Studies on fish scale formation and resorption I. Fine structure and calcification of the scales in *Fundulus heteroclitus* (Atheriniformes: cyprinodontidae). J Morphol 159:49–66
Yang L, Zhang X, Liao Z, Guo Y, Hu Z, Cao Y (2003) Interfacial molecular recognition between polysaccharides and calcium carbonate during crystallization. J Inorg Biochem 97:377–383
Zeichner-David M (2001) Is there more to enamel matrix proteins than biomineralization? Matrix Biol 20:307–316

Zhu J-X, Sasano Y, Takahashi I, Mizoguchi I, Kagayama M (2001) Temporal and spatial gene expression of major bone extracellular matrix molecules during embryonic mandibular osteogenesis in rats. Histochem J 33:25–35

Ziv V, Weiner S (1994) Bone crystal sizes: a comparison of transmission electron microscopic and X-ray diffraction line width broadening techniques. Connect Tissue Res 30:165–175

Zylberberg L, Géraudie J, Meunier F, Sire J-Y (1992) Biomineralization in the integumental skeleton of the living lower vertebrates. In: Hall BK (ed) Bone, volume 4: Bone metabolism and mineralization. CRC Press, Boca Raton, pp 171–224

Zylberberg L, Nicolas G (1982) Ultrastructure of scales in a teleost (*Carassius auratus* L.) after use of rapid freez-fixation and freeze-substitution. Cell Tissue Res 223:349–367

19 Conclusions

19.1 Introduction

Calcified tissues in living organisms are so widespread and diversified, and their structure, assembly and function are so divergent in different species, that a comprehensive study of how they are generated and reach maturity is a demanding task. For this reason, any attempt to formulate a unified theory able to explain all the aspects of the calcification process, or, less ambitiously, only those aspects that are strictly connected with its earliest phases, may appear to be utopian and pretentious. A careful analysis of the data made available by the literature does, however, allow some definitive progress to be recorded; a number of general principles can cautiously be stated, and some coherent conclusions and provocative inferences can be set out, even if further verification is, in any case, needed and several gaps must still be filled.

19.2 Organic Components Play a Primary Role in Calcification

The organic components of the calcifying matrices – whether contained in compartments or located outside them – play a primary role in initiating and directing calcification in biological tissues. This conclusion is justified not only by the observation that the calcification process is not a matter of the simple precipitation of inorganic substance from a saturated or metastable solution, that the inorganic particles have recurrent structure and nano-size in different species (Chaps. 4, 5 and 6), and that a very close organic-inorganic relationship persists throughout the early phases of the process (Sect. 16.9), but also, as perceptively noted by Pautard (1970), by the finding that, over the centuries, inorganic deposits have maintained such a highly consistent chemical composition, repeated morphology, and meticulously executed generation in each species, that any degree of randomness in the process of inorganic substance deposition appears to be excluded. Moreover, several biological apatites have shapes, sizes and composition that, if there were no active synergy with organic molecules, could only originate under conditions comprising extremely high temperatures and pressures (Eanes and Posner 1970).

The role of the matrix components in calcification has not yet been precisely mapped out, but it definitely appears to be an active one. Several findings suggest that these components, in contradiction with what has been supposed over a long period, do not constitute a simple permeable frame, and do not outline micro- or nanospaces, that are destined to be passively filled with inorganic crystals; by contrast, there are reasons to conclude that their organic molecules play a direct, active role in inducing or inhibiting the early phases of the calcification process and in regulating the development of inorganic structures in an oriented way, in this way setting up direct links with the inorganic substance and giving rise to organic-inorganic, crystal-like structures with an apatite-like conformation.

It should be noted that not all matrix components are actively engaged in the local process of calcification; some of them simply act as 'stromal' or supporting structures. The assessment of Wuthier (1982) was that calcifiable collagens act less as initiators than as receptors and propagators of calcification, and Veis (1985) rightly considered that the organic matrix contains two distinct types of component: components of the first type define and support the architecture of the whole system and the ultimate orientation of crystal axes, whereas those of the second type consist of proteins that perform the dual role of interacting specifically with the components of the first type and – as a direct outcome – inducing crystal formation. Evidence of two organic component types – the first acting as a supporting framework and the second as triggering elements able to promote calcification, can in fact be identified in several systems: in bone, dentin, cartilage and tendons, the first type is represented by collagen fibrils, the second by acid proteoglycans and glycoproteins; in mollusk shells, the first type is recognized in chitin, the second in silk fibroin and acidic glycoproteins. In other systems, this distinction is less clear-cut. This is true of enamel, in which authentic 'stromal' components cannot be detected, but the amelogenin lamellae could be considered the equivalent of a structural scaffolding. In a very broad sense, the walls of compartments, such as the membranes of matrix vesicles and cytoplasmic vacuoles, may be considered stromal structures which enclose calcification-promoting proteins.

Several findings indicate that the more prominent the stromal components are in the matrix, the lower its degree of calcification. So compact bone, whose collagen fibrils adhere closely to each other, leaving very few interfibrillary spaces (as happens in young osteons, for instance), has a degree of calcification that is lower than that of woven bone, whose fibrils are irregularly and loosely arranged and leave wider spaces, in which non-collagenous components can be accommodated. Woven bone, in its turn, shows a lower degree of calcification than that of tissues containing no collagen fibrils at all, such as enamel or invertebrate hard tissues. The degree of calcification higher in the peritubular dentin, which displays very few collagen fibrils, than in intertubular dentin, which is comparable to woven bone, provides another example of this difference (Sect. 8.2).

19.3 The Earliest Stage of Calcification is an Epitaxial Process

Each of the processes listed above implies integration with the others. So, during the earliest phases of calcification, adequate amounts of free inorganic ions must be locally available – but without reaching concentrations high enough to induce spontaneous precipitation. This localized rise in ion concentration might be attained through the activity of alkaline phosphatase (Sect. 17.2) and/or through that of calcium-chelating molecules (those of aggrecan, for instance, in their native form). These may bind calcium ions, but then release them once they have been degraded by proteolytic enzymes, a process that is in line with the acid proteoglycan changes that occur at the calcification front (Sect. 18.3). At the same time, all molecules that are capable of inhibiting links between ions and reacting groups must be neutralized. In this connection, the importance of the role of alkaline phosphatase in removing pyrophosphates has already been stressed (Sect. 17.2).

The very first phase of calcification probably involves the formation of amorphous calcium phosphate or carbonate, both stabilized by organic molecules (Sects. 4.2.1 and 4.4). They can be detected under the electron microscope as an amorphous or finely granular, intrinsically electron-dense substance located in circumscribed areas along the calcification front.

As already discussed (Chap. 18), there seems to be a general consensus that the formation of the early solid inorganic aggregates occurs through a process of epitaxy, i.e., the oriented overgrowth of apatite-like particles on atomic groups in organic molecules in the matrix. Many of these atomic groups are found in each single organic molecule, and, as one nanoparticle grows on each group, an identical number of inorganic nanoparticles are formed; these take on the same crystal lattice spacing as that of the reacting groups in the molecules. Electron microscope studies on several hard tissues, especially those carried out by Höhling and coworkers (Sect. 18.4), have shown that the earliest crystals appear as rows of discrete nanoparticles. These have an apatite-like structure, but are marked out by conspicuous crystal lattice distortions, reflecting a state intermediate between amorphous and fully crystalline. It is arguable that each of these particles is assembled in connection with reacting groups within the organic matrix and that the definitive crystals are formed when these particles coalesce and fuse. It should be noted that the small differences in enthalpy and free energy between metastable nanoscale phases open up controlled thermodynamic and mechanistic pathways which make nanoparticle assembly a valid means of crystal formation (discussed by Navrotsky 2004).

It has been hypothesized that the assembly sites for the earliest inorganic aggregates may prove to be the acidic groups within the organic molecules of the matrix – a view strengthened by the frequent finding in calcifying areas of polyanions with repetitive sequences of aspartic acid (discussed in Sect. 18.6), and by the observation that almost all calcified tissues contain acidic – in some cases strongly acidic (Gotliv et al. 2005) – proteins. This is true of acidic phosphoproteins in bone (Sect. 7.3.3) and dentin (Sect. 8.5), and of the acid proteoglycans in cartilage (Sect. 9.3.1), but also applies to acidic proteins in mollusk shells (Sect. 12.2), sea urchin spines (Sect. 12.3), and other invertebrates or unicellular organisms. These findings lead one to conclude that the calcification process is initiated by a variety of organic molecules that occur in a wide spectrum of different calcifying tissues. This differentiation does not rule out the hypothesis that the basic calcification mechanism may be the same in all of them.

The assembly of discrete nanoparticles in connection with specific domains of organic matrix components, in line with the modalities depicted above, may also be responsible for the formation of the 'granular inorganic bands' which are typically found in connection with the period of the collagen fibrils in the calcifying matrix in compact bone, dentin and tendons (Sect. 5.2.1). These bands, which form within the holes of collagen fibrils – i. e., in enclosed compartments – have, in fact, been described as consisting of very small, dot-like inorganic granules, which proceed to fuse into poorly crystalline structures (see Sects. 4.2.1, 5.2, 16.2 and 18.2). Even if these findings do not differ from those reported for other sites of early calcification, they merit separate discussion because they have long provided the foundation for accredited theories of calcification, and, most of all, because they bring into the problem the question, already discussed in Sect. 17.3, of whether preformed compartments are a precondition for the onset of the process – a question considered so vital that any deviation from the idea that they are indispensable has been viewed as a severe limitation on any proposed theory of calcification, which would, in fact, have been seen as largely invalidated if it failed to center on them (for instance, Wheeler 1992).

19.4 Compartments are not an Absolute Prerequisite for Calcification

As reported in Sect. 17.3, the topic has recently been discussed by Veis (2003), who believes that the construction of confined spaces is a priority for the regulation of the flux of organic components and inorganic ions into them, but it had already been examined by Glimcher and Krane (1968) whose assessment was that compartments can be thought of as fundamental structural features in calcifying tissues, because calcification can occur within them in a highly organized manner. Besides this, compartments have often been thought to be indispensable on the grounds that they are the spaces where crystals can grow.

There is no doubt that early crystals are often formed in spaces surrounded by a membrane, such as matrix vesicles (see Sect. 9.4), Golgi vesicles (Sect. 13.5.3), cytoplasmic vacuoles (Sect. 17.4), mitochondria (Sect. 17.5), and specific organelles such as magnetosomes (Sect. 13.5.1). It is also clear that crystals can be built up in compartments which are not surrounded by a membrane, but are outlined by protein walls (Sects. 10.3 and 12.2). Early inorganic nanoparticles are, however, assembled in connection with organic molecules which, even in one and the same tissue, can be located inside compartments, but also outside them. In epiphyseal cartilage, for instance, early rows of particles (Plate et al. 1996), as well as early crystals (Bonucci 1971), first develop within matrix vesicles, but are then formed outside them, in connection with the acid proteoglycans of the intercellular matrix (Bonucci 2002); likewise, early inorganic particles in bone are formed in matrix vesicles and in the 'holes' within collagen fibrils, but long, filament-like interfibrillary crystals develop outside these compartments. This leads to the conclusion that preformed compartments are not absolute prerequisites to calcification, and that, in any case, their most distinctive contribution might not be that of regulating the flux of substances and allowing inorganic ion accumulation till a metastable equilibrium is reached, or even of providing the space needed for crystals to grow, but that of enclosing the organic molecules needed for the formation of early nanoparticles. A concept like this had earlier been formulated by Towe and Cifelli (1967), who assumed that calcite crystal formation occurs in a matrix comprising two distinct organic components, which they called 'passive' and 'active': the former, a polysaccharide-like material, was thought to give form to a compartment in which the latter, probably a template protein, allows active calcification to occur epitaxially. To this, Veis (2003) added the view that the compartments contain a finely filamentous organic matrix whose molecules induce calcification and are then occluded in the inorganic phase as that develops and fills the available spaces. It should be born in mind, too, that the segregation of proteins in a restricted space might help to immobilize them while activating their calcifying properties.

The fact that compartments are not an absolute precondition for the formation of early inorganic nanoparticles suggests that preformed spaces are not needed for the growth of crystals. The spaces within which crystals develop and come to be located may, in fact, be the end-result of the enzymatic digestion of matrix components, as already noted in the case of amelogenin and growing enamel crystals (Sect. 10.3.3).

19.5 Calcification is not a Process of Heterogeneous Nucleation

The considerations just set out prompt the conclusion that compartments do, certainly, play an active role in areas of early calcification but are not indispensable to the occurrence of the process, while raising the question whether crystal formation really requires inorganic ion concentration in tissue fluids to reach a metastable equilibrium, as predicted by the theory of heterogeneous nucleation. As just noted, inorganic nanoparticles definitely can be formed outside compartments, where inorganic ions can easily spread; under these circumstances they could hardly reach a metastable concentration. In addition, if the earliest inorganic particles are formed through ongoing linkage between inorganic ions and acidic or other active groups of organic molecules, then the whole process seems to depend on the availability of these groups – made reactive, for instance, by the removal of inhibiting molecules – rather than on the sheer quantity of inorganic ions; these would be linked to organic groups even where local inorganic ion concentrations are low. Rather than conventional nucleation and crystal growth, assembly of nanoparticles may provide alternative mechanisms for crystal formation (Navrotsky 2004). The decisive question appears to be whether the earliest nanoparticles are formed as predicted by the theory of heterogeneous nucleation and are inorganic nuclei destined to grow into crystals through a phase transformation – i. e., through processes which need compartmentalized space and metastable concentrations of ions – or if they are nanoaggregates of inorganic material which are able to assemble thanks to their relationships with chemical groups in organic substrates – a process which only calls for these groups to be free and reactive. The possibility that the earliest inorganic aggregates are formed through a chemical reaction has been discarded in the past on the basis of the belief that biological tissue calcification occurs when a solution phase becomes a solid one, that is, undergoes a phase transformation which implies nucleation and growth (e. g., Glimcher 1992). The findings just discussed, as well as those on the occurrence of crystal ghosts in areas of early calcification in a number of hard tissues (Sect. 18.4), argue strongly for the view that chemical reactions provide the foundation for the calcification process (see Chap. 16).

19.6 Crystal Development is a Template-guided Process

The epitaxial formation of nanoparticles in connection with reacting groups of organic molecules stands as the first phase of the calcification process, which is only completed when these structures evolve into definitive apatite crystals. Electron microscope studies of areas of early calcification in bone, cartilage and dentin show dotted, crystal-like ropes, which seem to mark an intermediate stage between rows of single nanoparticles and definitive crystals and whose X-ray and electron diffractograms begin to show a poorly defined apatite-like pattern. These ropes appear to form when nanoparticles coalesce in a configuration determined by that of the organic molecules (templates) to which they are linked, and whose reacting groups have a periodic arrangement conducive to the definitive development of apatite-like crystals. In this connection, it appears to be significant that divalent cations can induce a conformational transition from a 'random coil' state to a predominantly β-sheet structure, and can therefore generate molecular complementarity to apatite surfaces in proteins, such as phosphophoryn, which are involved in calcification (Evans and Chan 1994).

The coalescence and fusion of nanoparticles give rise to organic-inorganic, crystal-like structures whose shape and dimensions mirror those of the templates. Generally speaking, these crystalline structures should have a filament-like shape (good examples are those found in the calcification nodules of bone, cartilage and dentin) when their templates are filament-like, or a polygon-like shape (good examples are those found in mollusk shells) when templates are flat and sheet-like. A number of studies have been made to find out which specific organic molecules may trigger the first phase of the calcification process. Long lists of proteins that have been extracted from the calcified matrix and are supposedly involved in calcification testify to the efforts made in this direction (reported in Chaps. 7–15). It must be recalled, however, that areas of early calcification contain many different molecules, whose structure and composition vary in different species, and that their effects in promoting or inhibiting the deposition of inorganic substance may be the resultant of the activities of some or all of them. Transgenic and knock-out animals and pathologically calcified tissues could greatly favor an understanding of the role of single proteins in normal and pathological calcifications, but their use has been disappointing so far, even if they have yielded invaluable information in isolated cases. Thus, the role of some proteins in calcification seems to be well on the way to being defined: osteopontin, matrix Gla protein and fetuin-A, for instance, appear to be major inhibitors, though the proofs of their function are still indirect and incomplete. At present, it seems preferable to indicate general domains that favor links with calcium ions – for instance, polymeric anionic molecules – rather than attributing an exact function to specific proteins.

During the coalescence and fusion of nanoparticles, the template molecules with which they are connected necessarily become entrapped, internalized and occluded in the inorganic substance. They can be detected under the electron microscope as crystal ghosts and can be extracted from the calcified matrix as crystal-bound proteins (discussed in Sects. 16.2.1, 16.6, 18.4 and reviewed by Bonucci 2002).

These findings point to the conclusion that the shape and size of early crystals depend on, and are copies of, the shape and size of crystal ghosts which, in turn, are genetically determined; if this is true, the replication feature would explain why crystals sometimes have strange shapes, as in the case of coccoliths, or unusual sizes, as in the case of enamel crystals, and why these characteristics are maintained unchanged in the course of centuries. These modalities, which are operative in the building of organic-inorganic, apatite- and crystal-like structures, might open the way to their synthetic reproduction and to the construction of nanostructures whose shape, size and texture could all be predetermined (Mann 1993; Champ et al. 2000; Chen et al. 2005; Cheng and Gower 2006).

19.7 Crystal Growth and Maturation Imply the Degradation of Organic Components

Crystal shape and size are not regulated by crystal ghosts alone, but are under the control of proteins that can help to mould the growth of crystal surfaces. Mollusk shell acidic proteins, for instance, can interact specifically with certain facets of various dicarboxylate crystals, so promoting or inhibiting their growth (Addadi and Weiner 1985; Addadi et al. 1989). The growth of enamel crystals points to the same possibility: as discussed in Sects. 10.3.3 and 18.6, amelogenin nanospheres can aggregate into sheets which are interposed between adjacent crystals, so preventing their fusion through undue growth in width and thickness, and allowing, but regulating, their growth in length.

The way the organic-inorganic, crystal-like structures described above grow and mature into definitive inorganic crystals – or, at least, into crystals that contain only minimum amounts of organic material – makes it highly likely that this material is almost completely eliminated by the activity of proteases. This is, again, exemplified by enamel (Sect. 10.5), but the process is probably active in all calcifying tissues, although, to the Author's knowledge, it does not seem to have been studied in invertebrates, apart from the report that in *Lingula unguis* – whose crystals, however, consist of calcium phosphate – a rising degree of calcification is associated with a fall in the percentage of proteins (Kelly et al. 1965). The enzymatic breakdown of organic components of the matrix produces the space required for the ultimate crystal growth; the digestion of crystal ghosts, on the other hand, allows the crystal lattice to be perfected and crystals to complete their maturation.

This sequence of events could be challenged on the grounds that, once crystal ghosts are occluded in the inorganic substance, they could no longer be reached by proteolytic enzymes. It is, however, possible that, due to the extreme thinness of the early crystal-like structures, their organic components, even if occluded in them, remain partly exposed at their surface. In this regard, Ryall et al. (2000, 2001) have reported that the occluded proteins facilitate the enzymatic deconstruction of phagocytosed crystals. It is, in any case, certain that crystal ghosts are modified, and become unrecognizable under the electron microscope, as calcification proceeds (Sect. 16.2.1).

19.8 A Unified Theory of Calcification

A unified theory on the basic processes that regulate calcification, extendable to all biological tissues in which the process can occur physiologically or pathologically, appears to be unattainable at this state of the art, not only because more information is needed, but mainly because of the immense scope and variety of the calcifying systems in the animal kingdom. Having said that, the currently available findings still allow the following general conclusion to be put forward.

The calcifying matrices comprise two sets of components; the first has stromal properties, and plays a passive role in the calcification process, whereas the second, which is connected with the first, plays an active role. This active set, whose components differ in different matrices, may include proteoglycans, phosphoproteins, phospholipids and acidic proteins of various types, that is, it mostly consists of polymeric, polyanionic molecules. The acidic groups, or other reacting groups, in these molecules, once they are activated by the removal of inhibitors, bind inorganic ions and are epitaxial for the building of apatite-like nanoparticles. These grow and coalesce according to the configuration of the polymeric molecules, which behave as templates of organic-inorganic, crystal-like structures possessing an apatite-like configuration. The organic components of these structures are recognizable as crystal ghosts under the electron microscope and can be sequentially extracted as crystal-bound proteins using chemical methods. The further addition of inorganic ions to these organic-inorganic structures leads to their growth, which can occur according to one preferential direction. This may be determined not only by the orientation provided by the template molecules, but also by the fact that either the crystals are forced to assume the shape and dimensions of the compartments – if any – within which they are formed and constrained, or the growth of one or other of their surfaces is inhibited by adsorbed proteins. The proteolytic removal of these barriers may be necessary for crystals to reach their ultimate size; the maturation of the organic-inorganic structures into definitive crystals also implies that their organic ghosts will be partly removed or modified.

This outline of the basic processes which lead to calcification in biological tissues may need to be corrected and improved, but it currently seems to be supported by the vast majority of the available evidence. One general concept that should be born in mind is that the structures which are called 'crystals' could not be so classified from a pure mineralogical standpoint (see Sect. 4.2.1). Their formation is not a simple phase transformation and presupposes a complex organic-inorganic interaction; their growth and maturation are not due simply to additions of inorganic ions, but imply the participation of enzymes and the removal of organic components.

References

Addadi L, Weiner S (1985) Interactions between acidic proteins and crystals: stereochemical requirements in biomineralization. Proc Natl Acad Sci USA 82:4110–4114

Addadi L, Berman A, Moradian Oldak J, Weiner S (1989) Structural and stereochemical relations between acidic macromolecules of organic matrices and crystals. Connect Tissue Res 21:127–135

Bonucci E (1971) The locus of initial calcification in cartilage and bone. Clin Orthop Relat Res 78:108–139

Bonucci E (2002) Crystal ghosts and biological mineralization: fancy spectres in an old castle, or neglected structures worthy of belief? J Bone Miner Metab 20:249–265

Champ S, Dickinson JA, Fallon PS, Heywood BR, Mascal M (2000) Hydrogen-bonded molecular ribbons as templates for the synthesis of modified mineral phases. Angew Chem Int Ed Engl 39:2716–2719

Chen H, Clarkson BH, Sun K, Mansfield JF (2005) Self-assembly of synthetic hydroxyapatite nanorods into an enamel prism-like strucrure. J Coll Interface Sci 288:97–103

Cheng X, Gower LB (2006) Molding mineral within microporous hydrogels by a polymer-induced liquid-precursor (PILP) process. Biotechnol Prog 22:141–149

Eanes ED, Posner AS (1970) Structure and chemistry of bone mineral. In: Schraer H (ed) Biological calcification. Appleton-Century-Crofts, New York, pp 1–26

Evans JS, Chan SI (1994) Phosphophoryn, an "acidic" biomineralization regulatory protein: conformational folding in the presence of Cd(II). Biopolymers 34:1359–1375

Glimcher MJ (1992) The nature of the mineral component of bone and the mechanism of calcification. In: Coe FL, Favus MJ (eds) Disorders of bone and mineral metabolism. Raven Press, New York, pp 265–286

Glimcher MJ, Krane SM (1968) The organization and structure of bone, and the mechanism of calcification. In: Gould BS (ed) Biology of collagen. Academic Press, London, pp 67–251

Gotliv BA, Kessler N, Sumerel JL, Morse DE, Tuross N, Addadi L, Weiner S (2005) Asprich: a novel aspartic acid-rich protein family from the prismatic shell matrix of the bivalve *Atrina rigida*. ChemBioChem 6:304–314

Kelly PG, Oliver PTP, Pautard FGE (1965) The shell of *Lingula unguis*. In: Richelle LJ, Dallemagne MJ (eds) Calcified tissues. Université de Liège, Liège, pp 337–345

Mann S (1993) Molecular tectonics in biomineralization and biomimetic materials chemistry. Nature 365:499–505

Navrotsky A (2004) Energetic clues to pathways to biomineralization: precursors, clusters, and nanoparticles. Proc Natl Acad Sci U S A 101:12096–12101

Pautard FGE (1970) Calcification in unicellular organisms. In: Schraer H (ed) Biological calcification: cellular and molecular aspects. Appleton-Century-Crofts, New York, pp 105–201

Plate U, Tkotz T, Wiesmann HP, Stratmann U, Joos U, Höhling HJ (1996) Early mineralization of matrix vesicles in the epiphyseal growth plate. J Microsc 183:102–107

Ryall RL, Fleming DE, Grover PK, Chauvet M, Dean CJ, Marshall VR (2000) The hole truth: intracrystalline proteins and calcium oxalate kidney stones. Mol Urol 4:391–402

Ryall RL, Fleming DE, Doyle IR, Evans NA, Dean CJ, Marshall VR (2001) Intracrystalline proteins and the hidden ultrastructure of calcium oxalate urinary crystals: implications for kidney stone formation. J Struct Biol 134:5–14

Towe KM, Cifelli R (1967) Wall ultrastructure in the calcareous foraminifera: crystallographic aspects and a model for calcification. J Paleontol 41:742–762

Veis A (1985) Phosphoproteins of dentin and bone. Do they have a role in matrix mineralization. In: Butler WT (ed) The chemistry and biology of mineralized tissues. EBSCO Media, Birmingham,AL, pp 170–176

Veis A (2003) Mineralization in organic matrix frameworks. Rev Mineral Geochem 54:249–289

Wheeler AP (1992) Mechanisms of molluscan shell formation. In: Bonucci E (ed) Calcification in biological systems. CRC Press, Boca Raton, pp 179–216

Wuthier RE (1982) A review of the primary mechanism of endochondral calcification with special emphasis on the role of cells, mitochondria and matrix vesicles. Clin Orthop Relat Res 169:219–242

Subject Index

Printing: Krips bv, Meppel
Binding: Stürtz, Würzburg